Karst Geomorphology and Hydrology

D. C. Ford
Dept of Geography, McMaster University

P. W. Williams
Department of Geography, University of Auckland

CHAPMAN & HALL
London · Glasgow · Weinheim · New York · Tokyo · Melbourne · Madras

Published by Chapman & Hall, 2-6 Boundary Row, London SE1 8HN, UK

Chapman & Hall, 2-6 Boundary Row, London SE1 8HN, UK

Blackie Academic & Professional, Wester Cleddens Road, Bishopbriggs, Glasgow G64 2NZ, UK

Chapman & Hall GmbH, Pappelallee 3, 69469 Weinheim, Germany

Chapman & Hall Inc., One Penn Plaza, 41st Floor, New York, NY10119, USA

Chapman & Hall Japan, Thomson Publishing Japan, Hirakawacho Nemoto Building, 6F, 1-7-11 Hirakawa-cho, Chiyoda-ku, Tokyo 102, Japan

Chapman & Hall Australia, Thomas Nelson Australia, 102 Dodds Street, South Melbourne, Victoria 3205, Australia

Chapman & Hall India, R. Seshadri, 32 Second Main Road, CIT East, Madras 600 035, India

First edition 1991
First published in Paperback 1993
Reprinted 1994, 1996

Typeset in 10/12 pt Times by Columns, Reading
Printed in Great Britain by the University Press, Cambridge

ISBN 0 412 44590 5

A Catalogue record for this book is available from the British Library

Library of Congress Cataloging-in-Publication Data available

To Gwyn' and Margaret

Acknowledgements

We are grateful to the following individuals and organizations who have kindly given permission for the reproduction of copyright material (figure numbers in parentheses):

American Association of Petroleum Geologists (2.1, 2.5, 2.6, 2.11); K.-H. Pfeffer (2.10); W. B. White (2.10, 3.15, 3.16, 3.17, 3.18, 7.42B); U. Holbye (2.12); Figures 3.1, 5.8, 5.11 and Table 5.2 reproduced by permission from Freeze and Cherry, *Groundwater*, © 1979 Prentice-Hall, Inc; J. Cherry (3.1); A. N. Palmer (3.5, 7.25, 7.27, 7.33, 7.34, 7.36, 7.38, 7.41, 8.6, 8.12, 8.14, 8.15, 9.2, 10.22); R. G. Picknett (Table 3.6, Figure 3.11); Institution of Civil Engineers (3.19); the Editor, *Proceedings of the Geological Association* (4.2); Table 4.2 reproduced by permission from Gunn, *Earth Surface Processes and Landforms* **6**, © 1981 John Wiley & Sons; Tables 4.3 and 4.4 reproduced by permission from Spate et al., *Earth Surface Processes and Landforms* **10**, © 1985 John Wiley & Sons; Academic Press (4.4, 5.3, 5.5, 5.12); J. Gunn (4.5); Centre National de la Récherche Scientifique, Aix-en-Provence (4.6, 9.40, 9.41); Figures 4.7 and 4.11 reproduced by permission from Williams and Dowling, *Earth Surface Processes and Landforms* **4**, © 1979 John Wiley & Sons; Gebrüder Borntraeger (4.7, 4.10, 5.16, 9.15, 9.16, 9.22, 9.23, 9.28, Table 9.5); Canadian Association of Geographers (4.8); *New Zealand Geographer* (4.13, 9.27); Figure 5.1 reproduced by permission from Dunne and Leopold, *Water in Environmental Planning*, © 1978 W. H. Freeman & Co; Centre for Cave and Karst Studies, Western Kentucky University (Table 5.5); Figures 5.6, 6.5, 6.6, 6.7, 6.9, 6.11 and Table 6.2 reproduced by permission from *Guide to the Hydrology of Carbonate Rocks*, © Unesco 1984; the Editor, *Ground Water* (Table 5.8); the Editors, *The Journal of Geology*, by permission of the University of Chicago Press (5.9); McGraw-Hill Book Company (UK) (5.10); British Cave Research Association (5.14, 6.29); Elsevier Science Publishers (5.19, 6.12, 6.13, 6.16, 6.18, 6.19, 6.20, 6.21, 6.27B, 9.14, 9.27B); University of Bristol Speleological Society (5.23); Blackwell Scientific Publications (6.2); Water Resources Publications (6.4, 6.8, 6.15); the Editor, *Annales des Mines* (Table 6.5); North Illinois University Press (Table 6.6); the Editor, *Die Erde* (6.13, 9.42); Regents of the University of Colorado (6.13); International Atomic Energy Commission (6.13); Editions du BRGM (6.13, 11.2); Académie des Sciences de Paris (6.23); Societé Géologique de France (6.24); American Geophysical Union (6.30); the Editor, *Canadian Journal of Earth Sciences* (6.32); A. Eavis (7.2); S.-E. Lauritzen (7.10, 10.4); C. A. Hill (7.28, 8.9, 8.11); D. Craig and J. E. Mylroie (7.30); R. L. Curl (7.39C); the Editor, *Geografiska Annaler* (8.2);

the Editor, *Transactions of the Institute of Chemical Engineers* (8.3); D. Gillieson (8.4); A. A. Balkema (Table 9.3); Edward Arnold (Table 9.6); I. Gams (9.11B); Methuen & Co (9.12); I. Lewis (9.19B); the Editor, *Carsologica Sinica* (9.21); Longman Paul (9.25); the Editor, *Geological Society of America Bulletin* (9.26); N. Stephens (9.31); Figures 9.34, 9.36 and 9.37 reproduced by permission from Williams, *Earth Surface Processes and Landforms* **12**, © 1987 John Wiley & Sons; Basil Blackwell (9.42, 9.43, 9.49); A. Goede (10.1A); J. Schroeder (10.3); E. Christiansen (10.11); D. Dunkerley (10.17, 11.15); Figure 10.21 reproduced by permission from *Hydrology of Fractured Rocks*, © Unesco 1967; K. Hofius (11.4, 11.5); R. Rader (11.7); B. Beck (11.8, 11.10).

We wish to thank two very special friends and colleagues who have guided our work over the years: Marjorie Sweeting and Joe Jennings. Marjorie, who supervised both our doctoral theses, nurtured and directed our early enthusiasm for karst research and introduced us to the world of karst literature and ideas. Joe, regrettably now deceased, infected us – as he did many others – with his boisterous enthusiasm and provided an inspiring example of scientific integrity and scholarly activity. It has been an enormous privilege to have been guided and encouraged by these two most distinguished karst scholars of their generation. We see our book as building on the firm foundations that they laid for us.

Tim Atkinson, Michel Bakalowicz, Alain Mangin and Will White are old friends and scientific colleagues from whom we have received hospitality, visited many karst areas, and spent many hours in discussion. We thank them sincerely for considerably extending our insight into our subject. Tim's careful and constructively critical review of the manuscript of this book was of especial value to us.

We also wish to thank our graduate students and technicians for their refreshing approaches, good ideas, timely knocks and general support. Special thanks to Don Branch and others for draughting the numerous figures and to Peter Crossley for much help in the field.

Contents

Acknowledgements *page* vii

List of Tables xiii

1 Introduction to karst 1

1.1 Definitions 1
1.2 The global distribution of karst 4
1.3 The growth of ideas 6

2 The karst rocks 10

2.1 Carbonate rocks and minerals 10
2.2 Limestone compositions and depositional facies 14
2.3 Diagenesis and metamorphism of limestones; formation of dolomite 20
2.4 The evaporite rocks 27
2.5 Quartzites and siliceous sandstones 29
2.6 Effects of lithologic properties upon karst development 30
2.7 Interbedded clastic rocks 34
2.8 Bedding planes, joints, faults and fracture traces 35
2.9 Fold topography 41

3 Dissolution chemical and kinetic behaviour of the karst rocks 42

3.1 Introduction 42
3.2 Aqueous solutions and chemical equilibria 46
3.3 The dissolution of anhydrite, gypsum and salt 51
3.4 Bicarbonate equilibria and the solution of carbonate rocks 53
3.5 Measurements in the field and lab; computer programs 60
3.6 Chemical complications in carbonate solution 62
3.7 Two examples of the chemical evolution of simple calcium carbonate solutions 79
3.8 Dissolution and precipitation kinetics of the karst rocks 81

4 Distribution and rate of karst denudation 96

4.1 Global variations in the solution of carbonate terrains 96
4.2 Measurement and calculation of solution rates 104
4.3 Solution rates in non-carbonate rocks 114
4.4 Interpretation of measurements 115

5 Karst hydrology 127

5.1 Basic hydrological concepts, terms and definitions 127
5.2 Applicability of Darcy's law to karst 142
5.3 Controls on the development of karst aquifers 148
5.4 Energy supply for karst aquifer development 162
5.5 The rate of development of flow paths 165
5.6 Classification and characteristics of karst aquifers 166

6 Analysis of karst drainage systems 171

6.1 The 'grey box' nature of karst 171
6.2 Exploration and survey techniques 173
6.3 Aquifer zonation and thickness 177
6.4 Borehole analysis 178
6.5 Spring hydrograph analysis 193
6.6 Spring chemograph interpretation 204
6.7 Interpretation of the degree of organization of a karst aquifer 210
6.8 Polje hydrograph analysis 214
6.9 Water balance estimation 218
6.10 Water tracing techniques 219

7 Cave systems 242

7.1 Classifying cave systems 242
7.2 Formation of plan patterns of common caves 249
7.3 The common cave systems in depth 261
7.4 System modifications occurring within a single phase 271
7.5 Multi-phase cave systems 274
7.6 Meteoric water caves developed where there is confined circulation or basal injection of water 278
7.7 Hypogene caves A. Hydrothermal caves associated with CO_2 282

7.8 Hypogene caves B. Caves formed by waters containing H_2S 287
7.9 Sea coast mixing zone cavities 290
7.10 Massive sulphide deposits in karst cavities 291
7.11 Passage cross-sections and smaller features of erosional morphology 294
7.12 Breakdown in caves 309

8 Cave interior deposits 316

8.1 Introduction 316
8.2 Clastic sediments 318
8.3 Calcite, aragonite and other carbonate precipitates 330
8.4 Other cave minerals 347
8.5 Ice in caves 351
8.6 Dating and paleo-environmental analysis of calcite speleothems and other interior deposits 355
8.7 Mass flux through a cave system; the example of Friars Hole, W. Virginia 372

9 Karst landform development in humid regions 374

9.1 Coupled hydrological and geochemical systems 374
9.2 Small scale solution sculpture 375
9.3 Dolines – the 'diagnostic' karst landform? 396
9.4 The origin and development of solution dolines 399
9.5 The origin of collapse and subsidence depressions 405
9.6 Morphometric analysis of dolines 413
9.7 Landforms associated with allogenic inputs 423
9.8 Karst poljes 428
9.9 Corrosional plains and shifts in baselevel 432
9.10 Residual hills on karst plains 440
9.11 Depositional and constructional karst features 448
9.12 Sequences of carbonate karst evolution in humid terrains 451
9.13 Special features of evaporite terrains 458

10 The influence of climate, climatic change and other environmental factors on karst development 466

10.1 The precepts of climatic geomorphology 466

10.2 The hot arid extreme 467
10.3 The cold extreme: 1 karst development in glaciated terrains 472
10.4 The cold extreme: 2 karst development in permafrozen terrains 489
10.5 Sea level changes, tectonic movement and implications for the development of coastal karst 496
10.6 Polycyclic and polygenetic karsts 506
10.7 Relict karsts and paleokarsts 507

11 Karst resources, their exploitation and management 513

11.1 Karst hydrogeological mapping and water resources assessment 513
11.2 Pollution of karst aquifers 518
11.3 Problems of construction on and in karst rocks – expect the unexpected! 521
11.4 Urban hydrology of karst 537
11.5 Industrial exploitation of karst rocks and minerals 538
11.6 Recreational and scientific values of karstlands 543

References 547

Index 583

List of Tables

2.1 Properties of the principal karst rock minerals *page* 12
2.2 Crystal cationic radii (Å) 13
2.3 Some representative limestone and dolomite bulk chemical compositions 14
2.4 The principal components of limestone 16
2.5 Compressive strength of some common rocks 34
2.6 Terminology for bed thickness and joint spacing 36
2.7 Geomorphic rock mass strength classification and ratings (r = rating of parameter) 39
3.1 Dissociation reactions and solubilities of some representative minerals that dissolve congruently in water, at 25°C and 1 bar (10^5 Pa) pressure 42
3.2 Common chemical classifications of waters 44
3.3 Parameters for the Debye–Huckel equation at one atmosphere 49
3.4 The solubility of CO_2 54
3.5 Equilibrium constants for the carbonate solution system, gypsum and halite at 1 atmosphere pressure 56
3.6 A. Solution of calcite in a sequential system at 10°C. Initial P_{CO_2} = 4.3% B. Solution and deposition of calcite in a coincident system where CO_2 is being lost to air 80
4.1 Sources of solute load (Ca+Mg) in the Riwaka Basin, New Zealand 99
4.2 Data used in computing rates of solutional erosion in the Cymru Basin, New Zealand 110
4.3 Corrected limestone erosion rates for Cooleman Plain and Yarrangobilly Caves, Australia 112
4.4 Summary of reported micro-erosion meter erosion rates 112
4.5 Vertical distribution of solutional denudation 116
4.6 Approximate rates of solutional lowering of limestone surfaces, from limestone pedestals and micro-erosion meter measurements 117
4.7 Magnitude and frequency parameters for dissolved solids transport 125
5.1 Karst hydrographic zones 128
5.2 Range of values of hydraulic conductivity and permeability 131
5.3 Values for fresh water of fluid density, dynamic viscosity and kinematic viscosity for different water temperatures at one atmosphere pressure 133

5.4 Specific storage and transmissivity values for some karstic aquifers 137
5.5 Effect of hydrogeologic setting on carbonate aquifers 150
5.6 Discharges of some of the world's largest karst springs 155
5.7 Flow velocities through karst conduits for straight line plan distances of more than 10 km 160
5.8 Hydrologic classification of carbonate aquifers 167
6.1 Assumptions and decisions made about the nature of a karst aquifer and the most appropriate method for its analysis 172
6.2 Borehole recharge and pumping methods and their applicability to field problems 180
6.3 Classification of rocks according to their specific permeability 185
6.4 Flood dilution mixing model calculations for four karst springs in Mexico 209
6.5 Classification of karst aquifers 213
6.6 Interpretation of tritium data for the Waikoropupu Springs, New Zealand 223
6.7 Fluorescent dyes used in water tracing with an indication of their properties 232
6.8 Comparison of *Lycopodium* and fluorescent dye tracer techniques 240
7.1 Some classifications of solution caves 243
7.2 The longest and deepest caves as at April 1988 244
7.3 Classification of karst solution caves 248
8.1 Cave interior deposits 317
8.2 Sediment characteristics in three West Virginia karst basins 328
8.3 The principal minerals deposited in caves 343
8.4 Conditions of calcite speleothem growth, destruction or decay 344
8.5 Representative mean extension rates for sample stalagmites (candlestick and tapered types) that have been established by two or more radiometric dates per sample 345
8.6 Estimated dimensions and mass fluxes through the Friars Hole Cave System, West Virginia 373
9.1 Classification of karren forms 376
9.2 Some morphometric measures applied to karst depressions 416
9.3 Average depth and density of karst depressions by region 416
9.4 Depression density and nearest neighbour statistics for various polygonal karsts 419

9.5 Selection of the most distinctive poljes of the Dinaric karst 430
9.6 A tentative typology of biokarst forms 451
10.1 Categories of karst landforms in glaciated terrains 477
10.2 Effects of glacier action upon karst systems 483
11.1 Examples of dam grouting costs in the Spanish-speaking countries 532

1 Introduction to karst

1.1 Definitions

Karst is terrain with distinctive hydrology and landforms arising from a combination of high rock solubility and well developed secondary porosity. Considerable rock solubility alone is insufficient to produce karst. Rock structure is also important. The key to karst is the development of its unusual subsurface hydrology. The 'engine' that powers its natural processes is the hydrological cycle. Soluble rocks with extremely high primary porosity usually have poorly developed karst. Yet soluble rocks with negligible primary porosity that have later evolved a very large secondary porosity support excellent karst. The distinctive landforms above and below ground that are a hallmark of karst result from solution along pathways provided by the structure.

Hydrological and chemical processes associated with karst are best understood from a systems perspective. This has determined the approach adopted in this book. Karst can be viewed as an open system composed of two clearly integrated hydrological and geochemical subsystems operating upon the karstic rocks. Karst landforms are the products of the interplay of processes in these linked subsystems.

Some karsts are hydrologically decoupled from the contemporary system. We refer to these as *paleokarsts*. They have usually experienced tectonic subsidence and lie unconformably beneath clastic cover rocks. Occasionally they are exhumed and re-integrated into the active system, thus resuming a development that was interrupted for perhaps millions of years. Contrasting with these are *relict karsts*, which exist within the contemporary system but are removed from the situation in which they are developed, just as river terraces – representing floodplains of the past – are remote from the river that formed them. Relict karsts have often experienced a major change in baselevel. A high level corrosion surface with residual hills now located far above the modern water table is one example; drowned karst on the coast another. Drained phreatic cave passages are further instances, although their emptying need not necessarily involve a change in baselevel.

Karst-like landforms produced by processes other than solution or corrosion-induced subsidence and collapse are known as *pseudokarst*. Caves in glaciers are pseudokarst, because their development in ice involves a change in phase, not dissolution. *Thermokarst* is a related term applied to topographic depressions resulting from thawing of ground ice. *Vulcanokarst* comprises tubular caves within lava flows plus mechanical collapses of the roof into them. *Piping* is the mechanical washout of caves in gravels, soils, loess, etc. plus associated collapse. On the other hand, solution forms such as karren on outcrops of granite and basalt are karst features, despite their occurrence on lithologies that are of low solubility when compared to typical karst rocks.

The word *karst* can be traced back to pre-Indoeuropean origins (Gams 1973a). It stems from *karra* meaning stone, and its derivatives are found in many languages of Europe and the Middle East. In northern Yugoslavia the word evolved via *kars* to *kras*, which in addition to meaning stony, barren ground is also a regional name for a district on the Yugoslavian/Italian border in the vicinity of Trieste. This district is sometimes referred to as the 'classical karst', being the type site where its natural characteristics first received intensive scientific investigation. In the Roman period the regional name was Carsus and Carso. When it became part of the Austro–Hungarian empire it was germanicized as the Karst. The geographical and geological schools of Vienna exercised a decisive influence on the word as an international scientific term, its technical use being established by the mid-19th century. The unusual natural features of the Kras (or Karst) region became known as 'karst phenomena' and so too, by extension, did similar features found elsewhere in the world. Thus we now consider karst to comprise terrain typically characterized by sinking streams, caves, enclosed depressions, fluted rock outcrops and large springs.

The main features of the karst system are illustrated in Figure 1.1. The primary division is into erosional and depositional zones. In the erosional

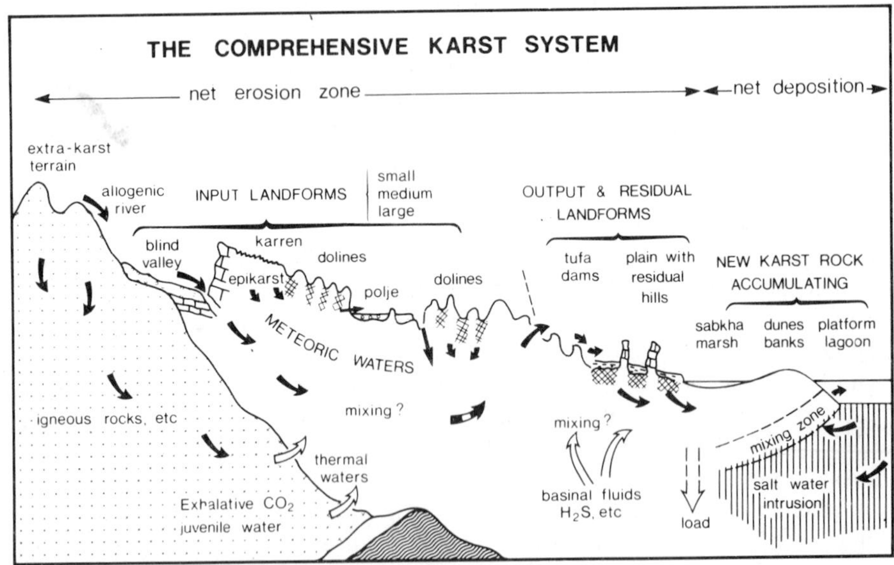

Figure 1.1 The comprehensive karst system: a composite diagram illustrating the major phenomena encountered in active karst terrains.

zone there is net removal of the karst rocks, by dissolution alone or by dissolution serving as the trigger mechanism for other processes; some re-deposition of the eroded rock occurs in the zone, mostly in the form of precipitates, but this is transient. In the net deposition zone, which is chiefly offshore or on marginal (inter- and supra-tidal) flats, new karst rocks are created. Many of these rocks display evidence of transient episodes of dissolution within them. This book is concerned primarily with the net erosion zone, the deposition zone being the field of sedimentologists.

Within the net erosion zone dissolution along groundwater flow paths is the diagnostic characteristic of karst. Most groundwater in most karst systems is of meteoric origin, circulating at comparatively shallow depths and with short residence times underground. Deep circulating, heated waters or waters originating in igneous rocks or subsiding sedimentary basins mix with the meteoric waters in many regions, and dominate the karstic dissolution system in a small proportion of them. At the coast, mixing between sea water and fresh water can be an important agent of accelerated dissolution.

In the erosion zone most dissolution occurs at or proximate to the bedrock surface where it is manifested as surface karst landforms. Specific forms may be termed *small scale* where their characteristic dimensions (such as diameter) are commonly less than ~ 10 m, *intermediate scale* in the range, 10 to 1000 m, and *large scale* where dimensions are greater than 1000 m. In a general systems framework most surface karst forms can be assigned to *input*, *throughput* or *output* roles. Input landforms are predominant; they discharge water into the underground and their form differs distinctly from that of e.g. fluvial- or glacial-created landforms because of this function. Some distinctive valleys and flat-floored depressions termed *poljes* convey water across a belt of karst rocks at the surface and so serve in a throughput role. Varieties of erosional gorges and of precipitated or constructional landforms such as travertine dams may be created where karst groundwater is discharged as springs, i.e. they are output landforms. Residual karstic hills, sometimes of considerable height and abruptness, may survive on the alluvial plains below receding spring lines. Our cover shows examples from southern China, the world's most spectacular karst terrain.

Gypsum, anhydrite and salt are so soluble that they have comparatively little exposure at the Earth's surface in net erosion zones. Instead, they are protected by less soluble or insoluble cover strata such as shales. Despite this protection, circulating waters are able to attack them and selectively remove them over large areas, even where they are buried as deeply as 1000 m. The phenomenon is termed *interstratal or intrastratal karsification* and may be manifested by collapse or subsidence structures in the overlying rocks or at the surface. Interstratal karstification occurs in carbonate rocks also, but is of less significance there.

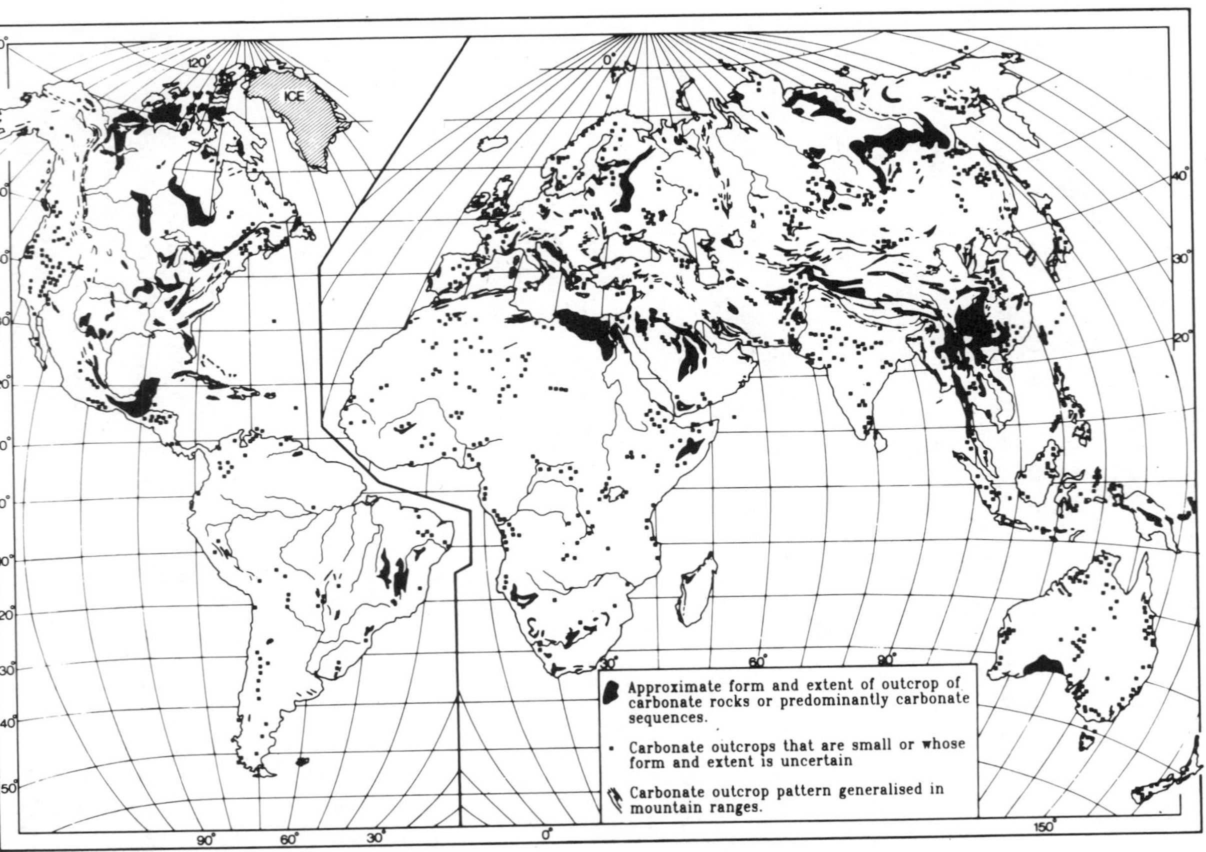

Figure 1.2 Major outcrops of the carbonate rocks.

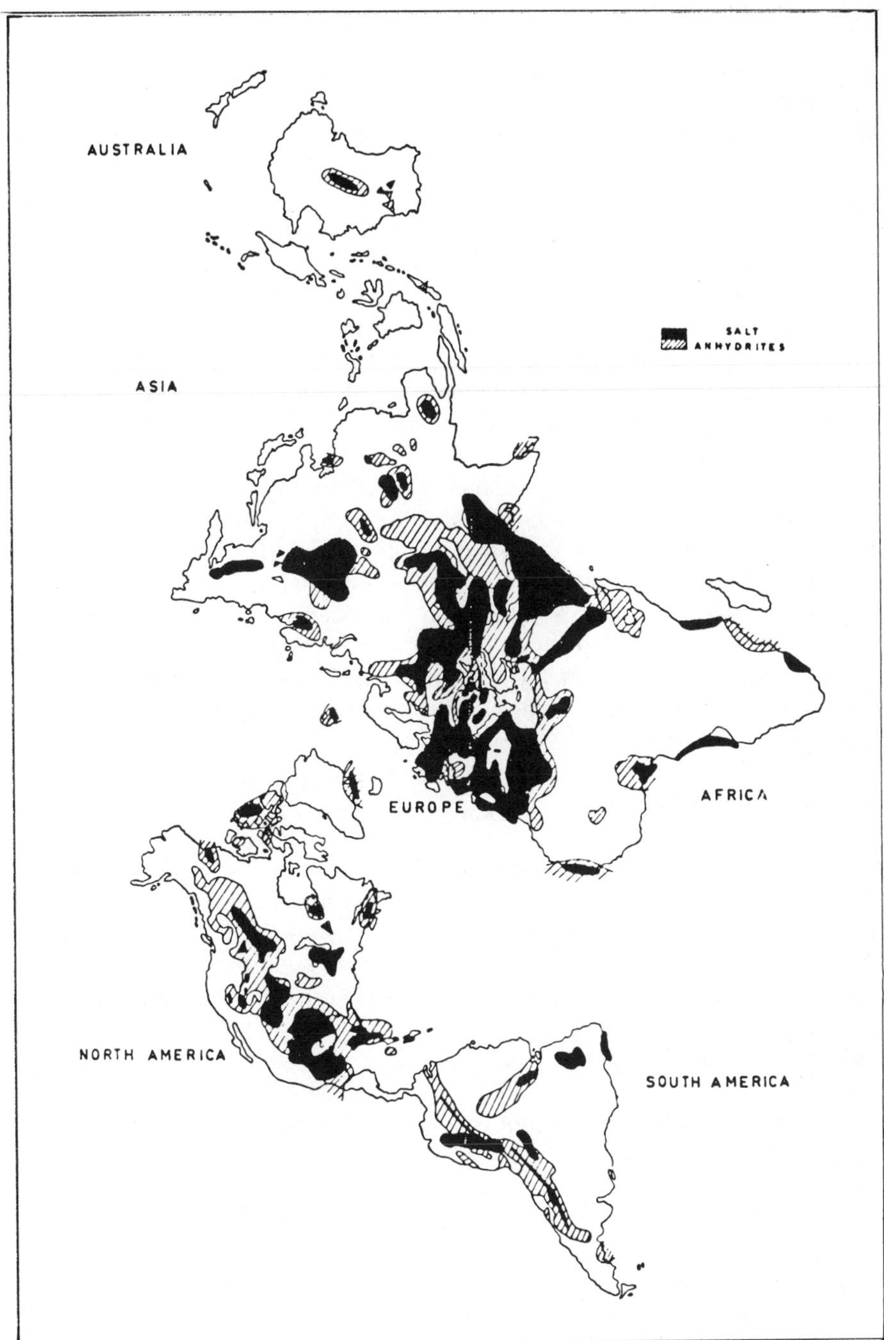

Figure 1.3 The global distribution of evaporite rocks. From Kozary *et al.* (1968) with permission. Note that more than 90% of the anhydrite/gypsum and more than 99% of the salt displayed here does not outcrop. It is covered by later rocks. A small part of the covered subcrop is subject to modern interstratal dissolution.

1.2 The global distribution of karst

Figure 1.2 displays the distribution in outcrop on the continents of the carbonate rocks, which are the principal karst rocks. The mapping is most approximate and generalized; many very small outcrops are omitted, and possibly some large ones. In the aggregate they occupy ~ 12% of the planet's dry, ice-free land. The extent of carbonate terrains displaying distinctive karst landforms and/or significant karst groundwater circulation is less than this; we estimate it to fall between 7–10% of the area. Carbonates are more abundant in the Northern Hemisphere; the old Gondwana continents expose comparatively small outcrops except around their margins where there are some large spreads of Cretaceous or later age (post-breakup of the super-continent). Large numbers of people live on the carbonate rocks. It is estimated that 25% of the global population is supplied largely or entirely by karst waters.

Figure 1.3 shows the maximum aggregate extent of gypsum, anhydrite and salt known to have accumulated over geologic time. Most of it is now buried beneath later carbonate or clastic (detrital) rocks. Also, many occurrences have been partially removed by dissolution or much reduced in geographic extent by folding and thrusting e.g. in the Andes. Nevertheless, there is gypsum and/or salt beneath ~ 25% of the continental surfaces. Gypsum and salt karst that is exposed at the surface is much smaller in extent than the carbonate karst, but interstratal karst is of the same order of magnitude.

1.3 The growth of ideas

The Mediterranean basin is the cradle of karstic studies. Ancient Greek and Roman philosophers made the first known contributions to our scientific ideas on karst, as well as contributing to a mythology that, like the River Styx, lives on in the place names given by cavers and others. Pfeiffer (1963) identified five epochs in the development of ideas about karst groundwaters, in the interval 600–400 BC until the early 20th century. Thales (624–548? BC), Aristotle (385–322 BC) and Lucretius (96–45 BC) formulated concepts on the nature of water circulation. Flavius (1st century AD) described the first known attempt at karst water tracing in the River Jordan basin (Milanovic 1981). Pausanias (2nd century AD) also reported experiments that were interpreted as proving the connection between a stream-sink beside Lake Stymphalia and Erasinos spring (Burdon & Papakis 1963). The conceptual understanding of hydrology established by Greek and Roman scholars remained the basis of the subject until the 17th century, when Perrault (1608–70), Mariotte (1620–84) and Halley (1656–1742) commenced its transformation into a quantitative science, showing the relationships between evaporation, infiltration and streamflow (Biswas

1970). Also in the 17th century, the understanding of karst caves was being advanced by scholars such as Xu Xiake in China (Yuan 1981) and Valvasor in Yugoslavia (Herak 1976, Milanovic 1981).

By the end of the 18th century, the role of carbonic acid in the dissolution of limestones was understood (Hutton 1795). Experimental work on carbonate solution in water followed a few decades later. The concept of chemical denudation was advanced in 1854 by Bischof's calculation of the dissolved calcium carbonate load of the River Rhine and in 1890 by Goodchild's estimation of the rate of surface weathering of limestone in northern England from his observations of gravestone corrosion. By 1883, the first modern style study of solution denudation had been completed by Spring & Prost in the Meuse river basin in Belgium.

The mid- to late-19th century was a very significant period for the advancement of our understanding of limestone landforms. In Britain, Prestwich (1854) and others investigated the origin of swallow-holes, while on the Continent impressive progress was made in the study of karren. But truly outstanding among the many excellent contributions of that time was the work of Cvijic. His 1893 exposition, *Das Karstphaenomen*, laid the foundation of modern ideas in karst geomorphology, ranging over landforms of every scale from karren to poljes. His contribution to our understanding of dolines is rightly considered of 'benchmark' significance by Sweeting (1981). Cvijic's thorough investigation provides the first instance of morphometry in geomorphology, and his conclusion that most dolines have a solutional origin has withstood the test of time.

The mid-19th century was also a time of notable advance in our understanding of groundwater flow. Although the experiments of Hagen (1839), Poiseuille (1846) and Darcy (1856) were not specifically related to karst, they nevertheless provided the theoretical foundation for later quantitative explanation of karst groundwater movement. And in 1874 the first attempt was made to analyse the hydrogeology of a large karst area. This was an investigation by Beyer, Tietze and Pilar of the 'lack of water in the karst of the military zone of Croatia'. Herak & Stringfield (1972) consider their ideas to be the forerunners of those that emerged more clearly in the early 20th century. In particular they foreshadowed the heated and long-lasting debate that erupted on the relative importance of isolated conduit flow as opposed to integrated regional flow.

In 1903 Grund proposed that groundwater in karst terrain is regionally interconnected and ultimately controlled by sea level (Fig. 1.4). He envisaged a saturated zone within karst, the upper level of which coincides with sea level at the coast, but rises beneath the hills inland (today we call this surface the water table). In the saturated zone, only water above sea level was considered to move, and that was termed *Karstwasser*. The water body below sea level was assumed stagnant and was called *Grundwasser*. It was conceived to continue downwards until impervious rocks were

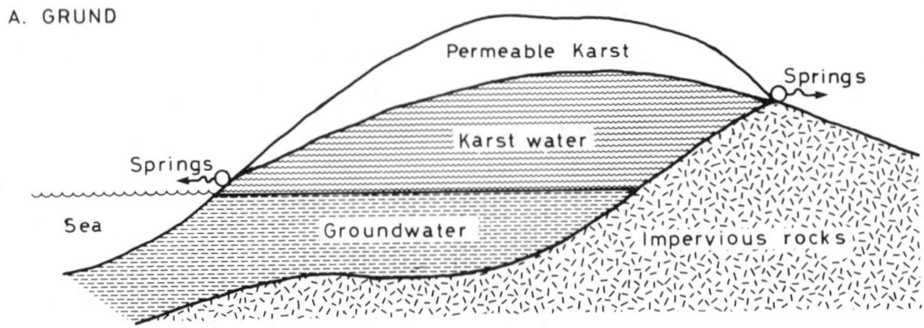

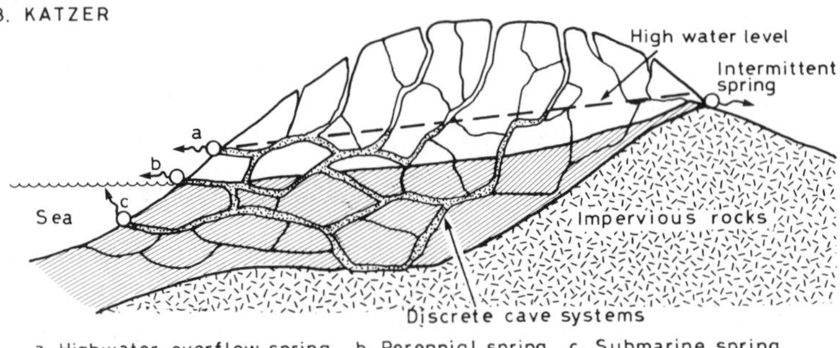

Figure 1.4 A. Essential features of the karst groundwater system according to Grund (1903). He envisaged a fully integrated circulation, although with stagnant water below sea level. B. The karst water system according to Katzer (1909), who stressed the operation of essentially independent subterranean river networks.

ultimately encountered. Grund had a dynamic view of the karst water zone and imagined that its upper surface would rise following recharge by precipitation. Should recharge be particularly great, the saturated zone would in places rise to the surface and cause the inundation of low-lying areas. In this way he explained the flooding of poljes.

However, field evidence showing the lack of synchronous inundation in neighbouring poljes of about the same altitude was used by Grund's critics to argue against the mechanism he proposed. Katzer (1909) observed, for example, that when springs are at different heights, it is not always the upper one that dries up first. He also noted that the responses to rainfall of springs with intermittent flow are unpredictable; some react, others appear not to. On the phenomenon of polje inundation, he stated that during their submergence phase water may sometimes still be seen flowing into adjacent stream-sinks (ponors) even when the flooding of the polje floor is becoming deeper; thus a general explanation of polje inundation through rising

Karstwasser cannot hold. Katzer did not accept the division of Karstwasser and Grundwasser. Instead he interpreted karst as consisting of shallow and deep types. In the former, karstification extends down to underlying impermeable rocks, while in the latter it is contained entirely within extensive carbonate formations. Katzer was apparently influenced by results of the impressive subterranean explorations of the French speleologist, Martel (1894), and particularly by his ideas on cave rivers, for deep within karst Katzer imagined water circulation to occur in essentially independent river networks (Fig. 1.4) with different water levels and with separate hydrological responses to recharge. His work therefore represents an important integration of the emerging ideas of groundwater hydrology and speleology.

One may imagine that Cvijic was stimulated by this controversy, as well as by the extra dimension added by Grund's (1914) publication on the cycle of erosion in karst. Thus we see the appearance in 1918 of another of Cvijic's now famous papers that drew together his maturing ideas on the nature of subterranean hydrology and its relation to surface morphology. He too rejected the division of underground water into Karstwasser and Grundwasser, although he implicitly accepted the occurrence of groundwater as we now understand it. He believed in a discontinuous water table, the level of which was controlled by lithology and structure, and proposed the notion of three hydrographic zones in karst: a dry zone, a transitional zone, and a saturated zone with permanently circulating water. He maintained that the characteristics of these zones would change over time, the upper zones moving successively downwards and replacing those beneath as the karst develops. The idea of a dynamically evolving karst hydrology was thus born. The very circulation of water enhances permeability and thereby progressively and continuously modifies the groundwater hydrologic system. This characteristic is now recognized as an important and unique feature of karst. Thus over a few decades around the turn of this century Cvijic laid the theoretical foundation of many of our current ideas. Without doubt he must be regarded as the father of modern karstic research.

2 The karst rocks

2.1 Carbonate rocks and minerals

Carbonate rocks outcrop over approximately 12% of the continental areas. They can accumulate to thicknesses of several kilometres and volumes of thousands of cubic kilometres. They are forming today in temperate and tropical seas and are known from strata as old as 3.3×10^9 years (Blatt *et al.* 1980). Approximately 25% of the world's population obtain their domestic water from these rocks. They host 50% of the known petroleum and natural gas reserves, plus bauxite, silver, lead, zinc and other economic minerals. They supply agricultural lime, Portland cement, fine building stones, and aggregate. As a consequence they are studied from many different perspectives. There are innumerable descriptive terms and many different classifications.

Figure 2.1 gives a basic classification. Carbonate rocks contain > 50% carbonate minerals by weight. There are two common, pure mineral end members, limestone (composed of calcite or aragonite) and dolostone (dolomite). Most authors neglect 'dolostone' and describe both the mineral and the rock as 'dolomite'; we shall follow this practice.

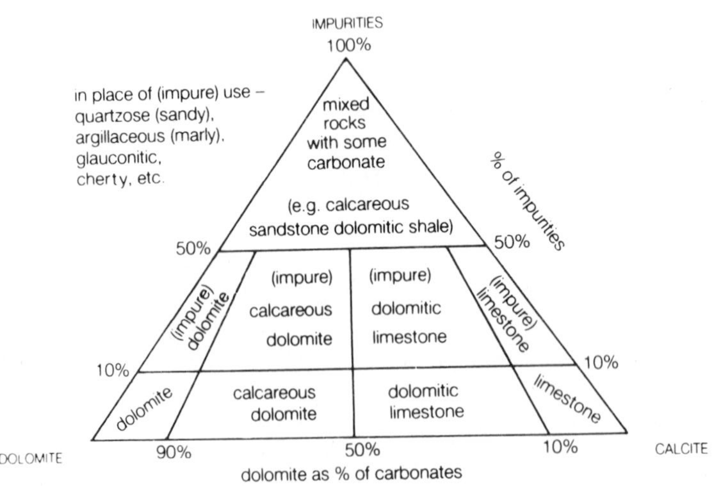

Figure 2.1 A bulk compositional classification of carbonate rocks by Leighton & Pendexter (1962), reprinted by permission of American Association of Petroleum Geologists.

Carbonate rocks are distinctive because their accumulation is highly dependent upon organic activity and they are more prone to post-depositional alteration than other sediments. Ancient carbonate rocks preserved on the continents are mainly deposits of shallow marine platforms (the 'knee deep environment'), but 95% of modern carbonates accumulate on deeper oceanic slopes and floors. The ancient equivalents of the latter have been subducted.

Carbonate deposits that are forming today consist of approximately equal proportions of calcite and aragonite. There appears to be little primary precipitation of dolomite, though this is controversial. Lithified carbonates as young as mid-Tertiary age have little aragonite remaining; it has inverted to calcite. In bulk terms, calcite : dolomite ratios are about 80 : 1 amongst Cretaceous strata, 3 : 1 in rocks of Lower Palaezoic age and ~ 1 : 3 in the Proterozoic. However, the ratio is about 1 : 1 in Archean rocks, negating any simple time trend.

Calcite, aragonite and dolomite mineralogy

Some important characteristics of the principal karst minerals are given in Table 2.1. In Nature there is a continuous range between pure calcite or aragonite and dolomite. These are the abundant and important carbonate minerals. Blends between dolomite and magnesite do not occur. Dolomite may become enriched in iron to form ankerite ($Ca_2FeMg(CO_3)_4$) but this is rare. Also rare are pure magnesite and iron carbonate (siderite) and their blends.

In carbonate structures the CO_3 anions can be considered three overlapping oxygen atoms with a small carbon atom tightly bound in their centre. In pure calcite, the anions are in layers that alternate with layers of calcium cations (Fig. 2.2). Each Ca ion has six CO_3 anions in octahedral coordination, building hexagonal crystals. Divalent cations smaller than Ca^{2+} may substitute randomly in the cation layers; larger cations such as Sr^{2+} can only be accepted with difficulty (Table 2.2). The common calcite crystal forms are rhombohedrons and scalenohedrons, but more than 300 variations are known. Calcite may also be massive.

In aragonite the Ca and O atoms form unit cells in cubic coordination, building orthorhombic crystal structures. These will not accept cations smaller than Ca. Sr is the most common substitute atom. Many aragonites are strontium-rich. Aragonite is only metastable. In the presence of water it may dissolve and re-precipitate as calcite, expelling most Sr ions. Aragonite is 8% less in volume than calcite. Inversion to calcite therefore normally involves a reduction of porosity. Aragonite crystals display acicular (needlelike), prismatic or tabular habits; there is frequent twinning.

In ideal dolomite layers of Ca^{2+} and Mg^{2+} ions alternate regularly between the CO_3 planes. The reality is more complex. Some Ca atoms

Table 2.1 Properties of the principal karst rock minerals.

Mineral	Chemical composition	Specific gravity	Hardness
	Carbonates		
A.			
Calcite	$CaCO_3$	2.71	3
Trigonal subsystem; rhombohedral. Habit – massive, scalenohedral, rhombohedral. Twinning is rare. > 300 m varieties of crystals reported. Colourless or wide range of colours. Effervesces vigorously in cold, dilute acids.			
Aragonite	$CaCO_3$	2.95	3.5–4
Orthorhombic system; dipyramidal. Habit – acicular, prismatic or tabular. Frequent twinning. Metastable polymorph of calcite. Colourless, white or yellow. Effervesces in dilute acids.			
Dolomite	$CaMg(CO_3)_2$	2.85	3.5–4
Hexagonal system; rhombohedral. Habit – rhombohedrons, massive, sugary (saccharoidal) or powdery. Colourless, white or brown, usually pink tinted. Effervesces slightly in dilute acid.			
Magnesite	$MgCO_3$	3.0–3.2	3.5–5
Hexagonal system; rhombohedral. Habit – usually massive. White, yellowish or grey. Effervesces in warm HCl.			
B.	Sulphates		
Anhydrite	$CaSO_4$	2.9–3.0	3–3.5
Orthorhombic system; crystals rare, usually massive; tabular cleavage blocks. White; may be pink, brown or bluish. Slightly soluble in water.			
Gypsum	$CaSO_4.2H_2O$	2.32	2
Monoclinic system; habit – needles, or fibrous or granular or massive. Colourless, or white and silky, grey. Slightly soluble in water.			
Polyhalite	$K_2Ca_2Mg(SO_4)_4.2H_2O$	2.78	3–3.5
Triclinic system; granular, fibrous or foliated. White, grey, pink or red; bitter taste. Slightly soluble in water.			
C.	Halides		
Halite	NaCl	2.16	2.5
Cubic euhedral crystals; usually massive or coarsely granular. Colourless or white. Very soluble in water.			
Sylvite	KCl	1.99	2
Cubic system; cubic and octahedral crystals or massive. Colourless or white. Very soluble in water.			
Carnallite	$KCl.MgCl_2.6H_2O$	1.6	1
Orthorhombic system; massive or granular. White; bitter taste. Very soluble in water.			

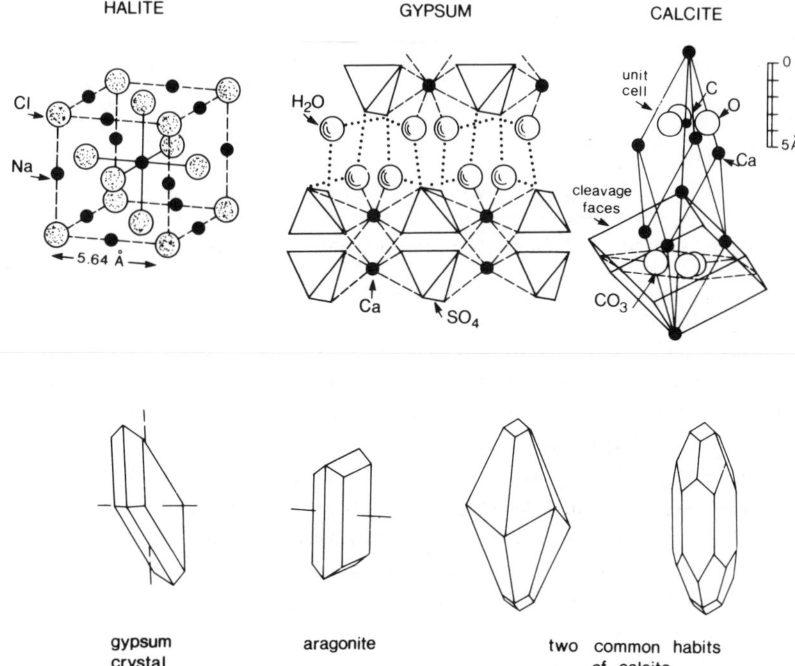

Figure 2.2 The unit cell configurations of halite, gypsum and calcite, and their characteristic crystal habits.

Table 2.2 Crystal cationic radii (Å)

Ba^{2+}	1.34	Fit into orthorhombic structures
K^{+}	1.33	
Pb^{2+}	1.20	
Sr^{2+}	1.12	
Ca^{2+}	0.99	
Na^{+}	0.97	
Mn^{2+}	0.80	Fit into rhombohedral structures
Fe^{2+}	0.74	
Zn^{2+}	0.74	
Mg^{2+}	0.66	

Table 2.3 Some representative limestone and dolomite bulk chemical compositions.

Per cent oxides	1	2	3	4	5	6	7	8
CaO	56.0	55.2	40.6	42.6	37.2	30.4	29.7	34.0
MgO	–	0.2	4.5	7.9	8.6	21.9	20.3	19.0
Fe and Al oxides	–	–	2.5	0.5	1.6	–	0.2	0.2
SiO_2	–	0.2	14.0	5.2	8.1	–	1.5	–
CO_2	44.0	44.0	35.6	41.6	43.0	47.7	46.8	46.8

1. Ideally pure limestone. 2. Holocene Coral, Bermuda. 3. average of 500 building stones. 4. average of 345 samples (from Clarke 1924). 5. Hostler Limestone (Ordovician), Pennsylvania. 6. ideally pure dolomite. 7. Niagaran dolomite, Silurian. 8. Triassic dolomite, probably hydrothermal, Budapest.

substitute into the Mg layers and trace quantities of Zn, Fe, Mn, Na and Sr atoms may be present in Ca or Mg planes. Most dolomites are slightly Ca-rich so that the formula is properly written: $Ca_{(1+x)}Mg_{(1-x)}(CO_3)_2$. In addition, because Fe^{2+} is intermediate in size it fits readily into either Ca or Mg layers. As a result, dolomite typically contains more iron than does calcite. This is why it often weathers to a pinkish or buff colour, Fe^{2+} being oxidised to Fe^{3+}. In the past such disordered dolomites have been termed 'protodolomite' or 'pseudodolomite' but these names are now out of favour. As disorder increases so does solubility.

Dolomite may be massive or powdery or sugary crystalline ('sucrose' or 'saccharoidal') in texture. The crystals are rhombic and opaque. Staining with alizarin red is the standard means of distinguishing it from calcite (Adams *et al.* 1984).

The proportions of Ca and Mg ions within a given crystal can vary between the ideal calcite and dolomite extremes, with a few per cent of Sr and Fe being substituted in as well. It follows that at the larger scale there may be variations between adjoining crystals, or larger patches of rock. A much used classification has evolved for this latter scale of variation: calcite with zero weight per cent $MgCO_3$ is pure calcite; $>0< 4\%$ is a 'low-Mg calcite' and $> 4 < 25\%$ is 'high-Mg calcite'. Some representative bulk compositions are given in Table 2.3.

2.2 Limestone compositions and depositional facies

Limestones are the most significant karst rocks. Here we consider the nature and environmental controls of their deposition. These determine much of the purity, texture, bed thickness and other properties of the final rock.

Limestone may be formed by chemical precipitation from waters in almost every environment between high mountains and the deep sea. However, most that survives as consolidated older rocks was formed in

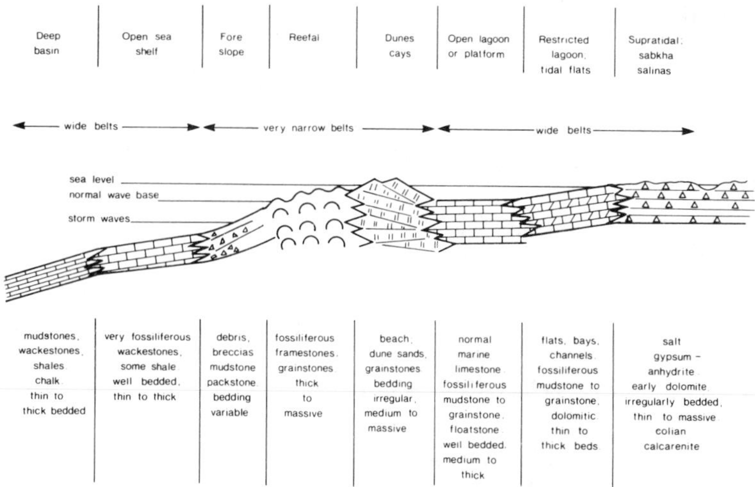

Figure 2.3 Composite facies model to illustrate deposition of limestone, early dolomite, and evaporite rocks. Modified from J. L. Wilson (1974). This is a generalized, simplified picture. Not all facies will be present in any given transect. Narrow belts range from a few metres to a few kilometres in width, wide belts from hundreds of metres to more than 100 km.

shallow tropical to warm temperate marine environments. Figure 2.3 is a composite model of these latter that extends from supratidal saltmarshes through lagoons, reefs and shelves out to the deep sea. Each differing environment within it may generate deposits with distinct characteristics termed 'facies'. Note that sulphate and halide rocks can also be formed in some of these model environments.

Most of the original limestone material is aragonite or calcite precipitated by marine animals for shell and skeleton building or expelled as faeces or precipitated in the tissues of algal plants. Some aragonite is formed inorganically at the sea surface by homogeneous precipitation (see section 3.8). Rate of production by these various means can aggregate 500–1000 g $m^{-2}y^{-1}$. Later vadose encrustations may be formed if the limestone is raised above sea level (see 'case hardening', section 9.10). Cementation and recrystallization during diagenesis add inorganic calcite spar. These principal components of limestones are summarized in Table 2.4.

Carbonate mud or micrite is the most important bulk constituent. It can compose entire beds or formations, or serve as matrix or infilling. Much originates as aragonite needles from algae, some is precipitated directly, the rest is the finest fragmentary matter produced by abrasion, faunal burrowing, excretion, etc.

Carbonate sand is formed mainly of faecal pellets, ooliths and fragments of skeletons and shells. It may accumulate in higher energy environments

Table 2.4 The principal components of limestone.

Textural type	Description	Origin
Micrite	0.5–5 μm diameter lime mud and silt particles. The largest component by volume in a majority of limestones	Clay- and silt-sized original marine grains. Ooze
Peloids	faecal pellets, micro-ooliths, 30–100 μm diam. By volume, the most important larger particles.	
Ooliths	sand-sized spherical accretions	Sand-sized or larger original marine grains, skeletons and growths
Lumps or grapestones	clumped peloids, ooliths	
Oncolites	algal accretionary grains up to 8 cm diameter	
Skeletal	corals, vertebrates, shell fauna, etc. Algal stems and other flora *in situ* or transported. Fragments of all genera	
Intraclasts	eroded fragments of partly lithified local carbonate sediment e.g. beach rock	
Lithoclasts	consolidated limestone and other fragments; often allogenic	
Frameworks	Constructed reefs, etc. mounds, bioherms, biostromes	
Vadose silt	carbonate weathering silt	Formed during vadose exposure
Pisoliths	large ooliths or concretions e.g. nodular caliche, cave pearls	
Stalactites, crusts	dripstones and layered accretions, caliche	
Sparite	medium to large calcite crystals as cementing infill; drusy, blocky, fibrous, or rim cements. > 20 μm diameter	Diagenetic cements
Microspar	5–20 μm grains replacing micrite	

(beaches, bars, deltas) and build to sand ripples or dunes. More frequently it is dispersed within carbonate mud. Intraclasts and lithoclasts are larger erosional fragments. The former are produced when storm waves break up the local sea bed; the latter have been transported greater distances by longshore, delta or turbidity currents and may include large particles such as pebbles, cobbles and boulders, or may consist of cliff or reef talus.

Reefs make only a small volumetric contribution to the world's limestones but they can be spectacular. They range from those having a complete framework tens to hundreds of metres in height built of successive generations of coral or algae (framestone) to carbonate sand or silt piles containing scattered, but unlinked, corals and algal mats. Modern coral grows between 30°N and 25°S in the photic zone (upper layer of the sea where photosynthesis occurs). It grows to the sea surface and may be exposed at low tide. Growth rates are commonly 1–7 mm a^{-1}. Algal mats (or stromatolites) have a somewhat greater environmental range and can flourish in the supratidal zone. Reefs may grow continuously along marine platform edges. On platforms and in lagoons they may occur as scattered, isolated mounds termed 'patch reefs' when small, 'mounds', 'bioherms' or 'pinnacle reefs' when large. 'Biostromes' are horizontal, tabular spreads of coral or algae.

When first laid down or built up these carbonate sediments contain large amounts of organic matter. This rapidly decomposes and is removed during lithification. Modern consolidated limestones contain ~ 1% of organic matter and ancient limestones average only 0.2%.

Most initial limestone deposits will contain some insoluble mineral impurities derived from near or distant eroding terrains. Type and proportion clearly will vary with differing environments. At the least there is volcanic ash and other dust settling into deep sea oozes where it may supply only 0.1% by volume. At the other extreme, river clay, sand and gravel deposited in deltas and estuaries can exceed the local carbonate production rate and so create calcareous shales, sandstones, etc. (Fig. 2.1).

Petrological classification of limestone

Sedimentological classifications of limestone have been based upon grain size, composition and perceived facies. A scheme by Grabau (1913) with

<table>
<tr><td colspan="6">Allochthonous limestones
original components not organically bound during deposition</td><td colspan="3">Autochthonous limestones
original components organically bound during deposition</td></tr>
<tr><td colspan="4">Less than 10% > 2 mm components</td><td colspan="2">Greater than 10% > 2 mm components</td><td rowspan="4">By organisms which act as baffles</td><td rowspan="4">By organisms which encrust and bind</td><td rowspan="4">By organisms which build a rigid framework</td></tr>
<tr><td colspan="3">Contains lime mud (< .03 mm)</td><td>No lime mud</td><td rowspan="3">Matrix supported</td><td rowspan="3">> 2 mm component supported</td></tr>
<tr><td colspan="2">Mud supported</td><td colspan="2" rowspan="2">Grain supported</td></tr>
<tr><td>Less than 10% grains (> .03 mm < 2 mm)</td><td>Greater than 10% grains</td></tr>
<tr><td>Mudstone</td><td>Wackestone</td><td>Packstone</td><td>Grainstone</td><td>Floatstone</td><td>Rudstone</td><td>Bafflestone</td><td>Bindstone</td><td>Framestone</td></tr>
</table>

Figure 2.4 The R. J. Dunham (1962) classification of carbonate rocks, as modified by Embry & Klovan (1971).

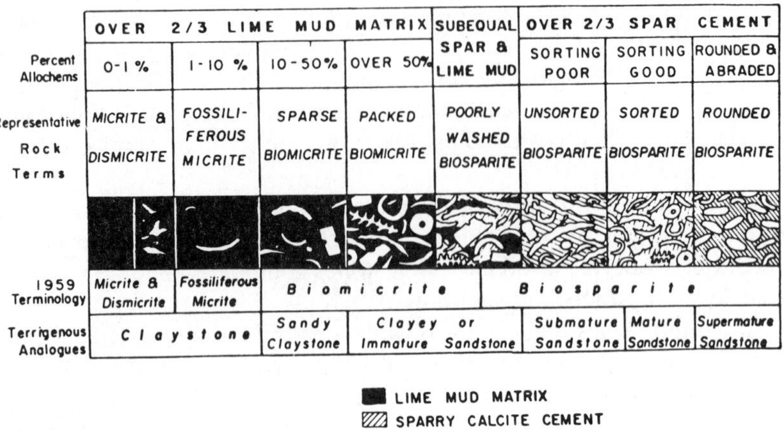

	OVER 2/3 LIME MUD MATRIX				SUBEQUAL SPAR & LIME MUD	OVER 2/3 SPAR CEMENT		
Percent Allochems	0-1 %	1-10 %	10-50%	OVER 50%		SORTING POOR	SORTING GOOD	ROUNDED & ABRADED
Representative Rock Terms	MICRITE & DISMICRITE	FOSSILIFEROUS MICRITE	SPARSE BIOMICRITE	PACKED BIOMICRITE	POORLY WASHED BIOSPARITE	UNSORTED BIOSPARITE	SORTED BIOSPARITE	ROUNDED BIOSPARITE
1959 Terminology	Micrite & Dismicrite	Fossiliferous Micrite	Biomicrite			Biosparite		
Terrigenous Analogues	Claystone		Sandy Claystone	Clayey or Immature Sandstone		Submature Sandstone	Mature Sandstone	Supermature Sandstone

■ LIME MUD MATRIX
▨ SPARRY CALCITE CEMENT

Figure 2.5 Carbonate textural spectrum from Folk (1962), reprinted by permission of American Association of Petroleum Geologists.

dominant grain size is still widely used, especially in Europe, and guides later schemes. It recognizes *calcilutites* (carbonate mudstones), *calcarenites* (sandstones) and *calcirudites* (conglomerates). Later, others added *calcisiltites*. Dunham (1962) refined this principle. He first divided carbonates into those preserving recognizable depositional texture and those where it has been destroyed. The latter are simply 'crystalline limestone' or 'crystalline dolomitic limestone', etc. The former are further subdivided into the organically bound and the loose sediment classes. A further amplification of this scheme (Fig. 2.4) is now widely used. It recognizes nine original texture types, plus crystalline limestone. Some authors also designate the dominant particle kinds e.g. 'pellet wackestone', 'intraclast rudstone', etc.

Another popular modern classification is that of Folk (1959, 1962) given in Figure 2.5. The two end members are *biolithites* (framed and bound coral and algal rocks) and *micrite* or *dismicrite* (pure carbonate mud or mud disturbed by burrows). In between, other types are defined by combining different proportions of mud and later spar cement with differing kinds of original grains (allochems) having differing degrees of sorting as functions of the wave or current energies locally available. Where bottom currents are weak, occasional allochems are dispersed amongst the mud matrix, which supports them. These are *biomicrites*, *oomicrites*, etc. Where currents are stronger the mud is partly or entirely winnowed from the skeletal fragments, piled pellets or intraclasts etc. to produce grain-supported layers where the voids may be filled with clear calcite spar during diagenesis. These are *biosparites*, *pelsparites* and *intrasparites*; 'packstone', 'grainstone' and 'rudstone' are approximately equivalent in the Dunham classification but do

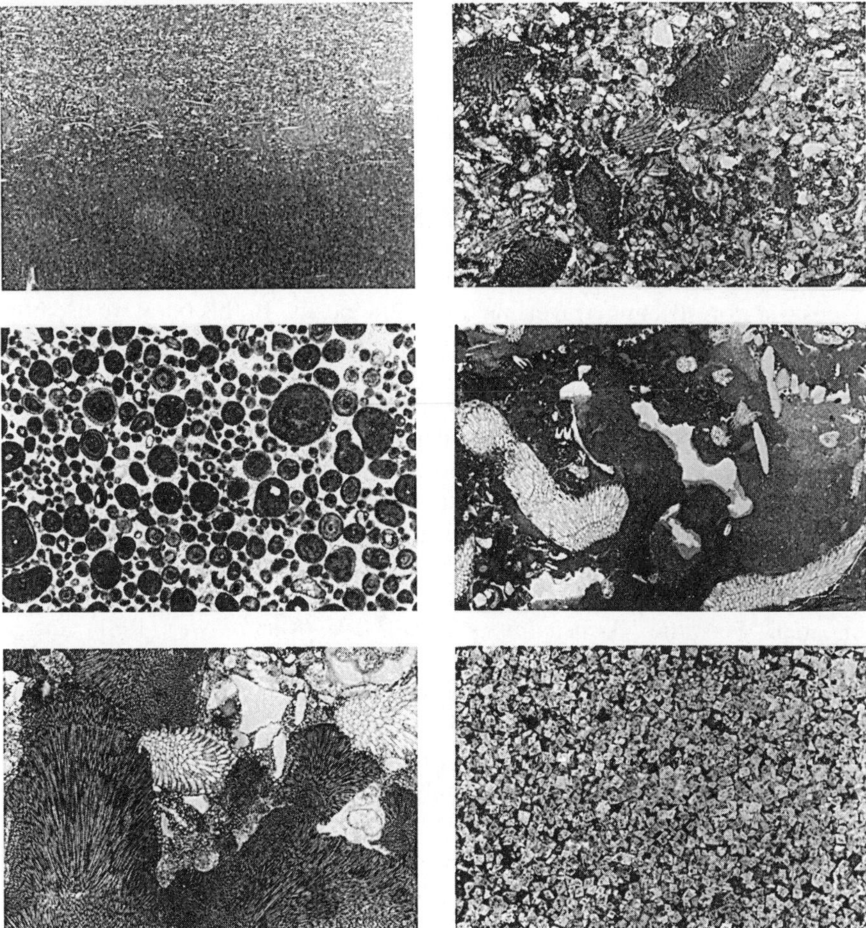

Figure 2.6 Thin sections of characteristic limestones and a dolomite. *Top left*; mudstone (Dunham) or micrite (Folk) coarsening upwards to wackestone (Dunham) or biomicrite (Folk). Ordovician, Quebec. Width of photo 1 cm. *Top right*: fossiliferous packstone (Dunham) or fossiliferous micrite (Folk) composed of foraminifers (large fossils), pelmatozoans and bryozoans. Lower Cretaceous, United Arab Emirates. Width of photo 1.5 cm. *Middle left*: oolitic grainstone (Dunham) or oosparite (Folk). Jurassic, Louisiana. Width of photo 0.8 cm. *Middle right*: bryozoan floatstone (Embry & Klovan); irregular cavities are borings filled with a second generation of geopetal mud and calcite spar. Middle Ordovician, Quebec. Width of photo 1.5 cm. *Lower left*; coral boundstone (Dunham), framestone (Embry & Klovan) or biolithite (Folk). Note that the large fossils are in stylolitic (pressure solution) contact. *Lower right*; dolomite composed of euhedral zoned crystals almost completely replacing limestone. Miocene, south Australia. Width of photo 5.2 mm. (From photographs and descriptions kindly supplied by Professor N. P. James, Queen's University, Canada).

not indicate the kind of grains that are most important in rock. Sample thin sections are given in Figure 2.6; see Adams, Mackenzie & Guilford (1984) for comprehensive illustration.

Terrestrial carbonates

As noted, some carbonate can be precipitated in almost every terrestrial environment. The most widespread types are tufa and travertine. There is confusion in terminology here, some authors using 'tufa' for all surficial deposits and restricting travertine' to cave deposits. In this book we use tufa for granular deposits accreting to algal filaments, plant stems and roots at springs, along river banks, lake edges, etc. Tufa is often a sort of framestone. It is typically dull and earthy in texture, and is highly porous once the vegetal frame rots out. In contrast, travertine is crystalline, quite dense calcite that is usually well layered, lustrous and lacks visible plant content. That formed underground or at hot springs is largely or entirely inorganic. Other surficial travertines may be bacterially precipitated (Chafetz & Folk 1984) and are usually mixed with tufas. These deposits may accumulate to tens of metres in thickness and cover a few square kilometres in area. See the *Association Française de Karstologie* (1981) for a comprehensive discussion, and Chapters 8 and 9 for further details.

Calcium carbonate is deposited in some fresh water lakes and in hypersaline water bodies such as the Dead Sea. Fresh water deposits are laminated micrites with a high silicate mud content, termed *marls*. They are common in temperate regions but rarely accumulate to great thicknesses. At Great Salt Lake, Lake Chad, and many smaller salt lakes, carbonate sands form locally around the shores and out to ~ 3 m depth where they are replaced by sulphates and halides. Algal mats are well developed on the carbonates. In the Dead Sea both gypsum and aragonite are precipitated at the water surface. The gypsum is immediately replaced by calcite which settles with the aragonite to accumulate as micrite.

2.3 Diagenesis and metamorphism of limestones; formation of dolomite

Limestone deposits begin as unconsolidated muddy sediments with a porosity of 40–80%, that are 'bathed in their embryonic fluids' (James & Choquette 1984). Diagenesis describes their alteration to consolidated rocks with a porosity rarely more than 15% and usually less than 5%. The processes are compaction, cementation, dissolution and replacement. The diagenetic environment may remain the shallow to deep submarine site of deposition, or it may become subaerial or phreatic with meteoric water as a consequence of uplift, etc. It can be deep subsurface because of burial by later sediments, or tectonic or thermal metamorphic or hydrothermal. In many examples several of these environments succeed one another. All introduce somewhat different rock fabrics or other features, so that there is a great deal of variation in consolidated limestone. Scholle *et al.* (1983) give a most comprehensive review.

Submarine diagenesis is slow and imperfect. There is compaction if shallow overburden is added and some aragonite and calcite spar is precipitated into voids as sea water is expelled. Chalk is an extreme example of a limestone from this environment, having undergone minimal diagenesis. The chalks of NW Europe retain a porosity often $\geqslant 40\%$ and a density of only 1.5–2.0. Cementation is weak. Fresh water marls are similar when drained.

Diagenesis in the subaerial and meteoric groundwater environment is rapid and extensive. More than half of all ancient limestones have experienced one or more such episodes. When sea level falls meteoric water invades the marine sediment and depresses the salt water interface in the proportion 40 : 1 (see section 5.1). The limestone deposit, which was stable in a sea water chemical environment, is now exposed to circulating fresh waters and all degrees of mixture, fresh : salt, as the salt solution is progressively expelled. There is now much dissolution, including the removal of most aragonite. It is replaced by calcite spar cements (Table 2.4). Slow dissolution and re-precipitation (crystal cell by cell) is fabric selective, preserving aragonite skeletal fossils, etc. as calcite 'ghosts'; this is 'replacement'. Rapid dissolution obliterates the fossil and substitutes a new, coarser crystalline spar. In the vadose zone new lenticular voids may be created at bedding planes and filled with vadose silt; they appear rather like stromatolite biolithites and are termed 'stromatactic'. If there is strong evaporation (e.g. a semi-arid coast) pisolites, travertine nodules and crusts (caliche or calcrete) replace the uppermost limestone. Esteban & Klappa (1983) provide a comprehensive review of them, and see Chapter 9.

Breccias

Shallow-water and supratidal limestones often contain interbeds of gypsum or salt. In vadose diagenesis these may be partly or wholly dissolved, causing collapse of semi-lithified overlying carbonates. Many limestone sequences contain local or regional breccias of pebble to boulder grain sizes, and ranging from a few centimetres to many metres in thickness (Fig. 2.7). Spar fills the voids in the finer-grained varieties. Coarse breccias display complex fillings.

Coarse breccias also accumulate as foreslope and reef foot talus or landslide deposits, with micritic fillings (see McIlwreath & James 1978).

Figure 2.7 Some representative limestones and dolomites in outcrop. *Top*. Thick bedded, platformal micritic limestone with regular tension joint systems. *Bottom left*. Argillaceous (clay-rich) basinal limestone; the recessive beds have a higher clay content. *Bottom centre*. Cyclic alternation of thick and thin beds, characteristic of shallow platform facies. Rocks in this photo are fully dolomitized, as suggested by the high frequency of vugs. *Bottom right*. A dolomite solution breccia created by the preferential dissolution of gypsum in a sabkha sequence of dolomite and gypsum beds. It is now a firmly re-cemented, resistant rock.

Stylolites

Stylolites are pressure solution seams. They are a striking feature of many well-bedded limestones and dolomites. They can develop in all diagenetic environments but are most common where there is deep burial with high overburden pressures. The carbonate dissolves at highest pressure points. These may be at the bottom and top, respectively, of adjoining crystals or larger masses (Fig. 2.6). The results are highly irregular dissolution seams with relief from millimetres to a metre or more. They stand out in exposures because they are darkened by residual insoluble minerals and organic matter concentrated along them. Micro-stylolites tend to develop in clayey limestones, creating nodular weathering habits (Wanless 1979). As much as 40% of the mass of some carbonates may have been destroyed at their stylolites. The dissolved and expelled material may be precipitated as cement elsewhere.

Chert

Some silica will accumulate in many lime sediments, from transported quartz or from siliceous sponges, radiolaria, and diatoms. If very alkaline conditions occur during diagenesis this is dissolved. It is re-precipitated as accretionary nodules or lenses of chert (flint). These usually accumulate along bedding planes where they may coalesce to form continuous sheets. Nodules as great as 1 m in diameter are known. Sheets are rarely as thick as 10 cm and are normally perforated or fractured so that fluid flow across them is not often prohibited. Chert tends to be more abundant in the older limestones.

The formation of dolomite

This has been a subject of great debate, but it remains poorly understood in many respects. Morrow (1982) and Zenger *et al.* (1980) provide recent comprehensive reviews. Dolomite forms by replacing earlier calcite and aragonite. There are few authenticated instances of primary dolomite precipitation. It is now believed that this is because the solute Mg^{2+} ion is strongly hydrated (Hanshaw & Back 1979). Its electrostatic bond with H_2O molecules is 20% stronger than that of Ca^{2+}. This bond is a barrier that must be broken to absorb the ion into the crystal lattice. Where there is a high Mg : Ca ratio and abundant CO_3^{2-} in the solution this can occur. Standard sea water approaches these conditions because high-Mg calcite can be precipitated, but it does not attain them. That occurs during diagenesis.

The principal modern models for dolomitization are shown in Figure 2.8.

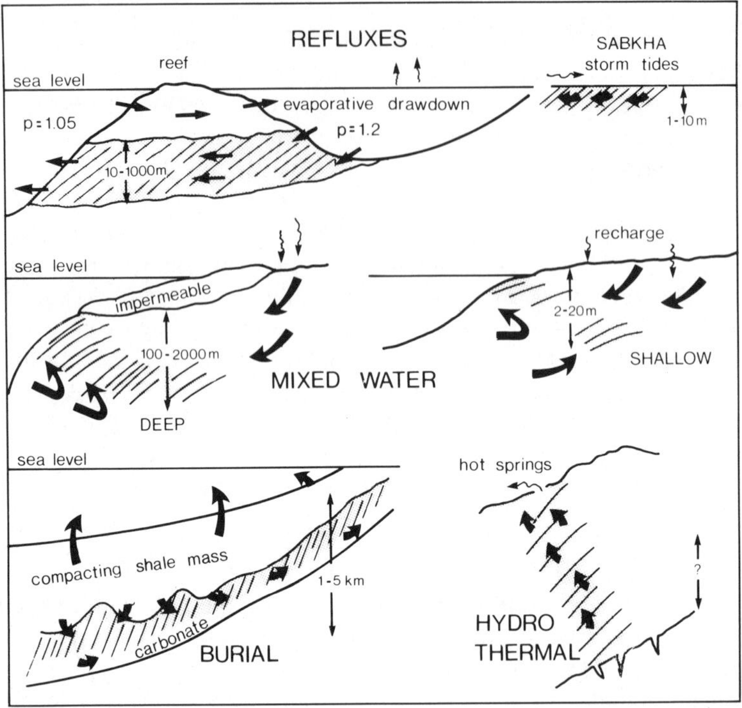

Figure 2.8 Dolomitization models, illustrating the depth to which the process may operate in different circumstances. p = density of salt waters.

In reflux models, sea water is first concentrated to hypersaline levels by evaporation in lagoons and then fluxes through lagoonal, reefal or supratidal lime sediments, exchanging ions with them. Dolomite is formed very early in diagenesis. These models explain the frequent association of dolomite with gypsum beds in sabkha facies.

Mixing models portray mixing of fresh and salt waters to produce conditions where calcite is soluble but dolomite is not and where Mg : Ca ratio requirements are met. Hanshaw & Back (1979) hypothesize that regional scale dolomitization is occurring today in the mixing zone that underlies much of Florida, USA.

It is disputed whether burial and compaction can achieve regional dolomitization via the expulsion of basinal fluids. It may explain the preferential dolomitization of individual, more permeable, beds. Some dolomite is of a hydrothermal origin; a Ca : Mg molar ratio as high as 10 : 1 will yield dolomite at 300°C. Hydrothermal dolomite tends to be localized along fractures, etc.

It is important to understand that as a consequence of these various modes of dolomitic replacement, the scale, extent and patterns of dolomitization within limestone masses can be highly variable. Because consolidated dolomite is normally less soluble than limestone, the effect upon karst morphogenesis and groundwater circulation can be profound.

Dolomite composition

Dolomicrite displays grains $< 10\ \mu m$ in diameter and is believed to be early replacement of aragonite. Much more abundant is medium crystalline or 'sucrosic' dolomite (10–20 to 100 μm in diameter). It is the standard dolomite. White, sparry crystalline-to-megacrystalline (centimetres in diameter) dolomite is found in association with lead-zinc deposits and other situations where it appears that several episodes of dolomitization are possible.

Where dolomitization is complete but proceeded slowly ghosts of fossils and other allochems or framework may be well preserved. Older English-language classifications (pre-1960) define 'primary dolomite' as that preserving some original depositional texture. In 'secondary dolomite' this is entirely destroyed (Fig. 2.6).

As dolomitization of limestone proceeds through the range 5–75% there is, in general, progressive reduction in porosity due to infilling. There after, porosity increases again because dolomite rhombs are smaller than the calcite crystals they have now largely or entirely replaced. There is an increase in large vug porosity. The high initial and early diagenetic porosity of reef rocks (attributable to their framework) makes them preferred sites of dolomitization – which again enhances their porosity. This is why buried reefs are prime targets in petroleum exploration.

Dedolomitization

Dolomitization can occasionally be thrown into reverse during diagenesis if calcium ion in solution should be greatly enriched. The solution may then be supersaturated with $CaCO_3$ but undersaturated with regard to $MgCO_3$. The process creates beds or patches of etched, crumbly rock. It requires maintenance of a particular and rather delicate hydrochemical balance (see discussion of 'incongruent solution' in section 3.4).

Marble

Marble is produced when limestone or dolomite are metamorphosed by heat. In the lowest grade of metamorphism the smaller grains plus the outer surfaces of larger ones become annealed. They recrystallize as spar upon cooling. Fossil form, lithologic texture, etc. are quite well preserved. In the next grade there is complete annealing and subsequent recrystallization.

This produces hard, dense, saccharoidal crystalline rock of rather even grain size. Grains are irregular in form with sinuous or zigzag surfaces. If originally limestone this is simply 'marble'; if from dolomite it is 'dolomitic marble' which has a finer grain size.

Marble has very low porosity and often there is negligible permeability so that it is difficult for karst waters to penetrate. But where penetrated (e.g. at a joint or a contact) it tends to give the sharpest, cleanest solutional morphology.

With greater heating there is partial expulsion of CO_2 and bound water, the volatile components. At 400°C dolomite is converted to periclase marble:

$$CaMg(CO_3)_2 \xrightarrow{\text{heat}} CaCO_3 + MgO + \overset{\uparrow}{CO_2} \qquad (2.1)$$

The MgO (periclase) occurs as insoluble octahedra embedded in the recrystalized calcite.

The dissociation temperature for calcite is much higher, ~ 900°C. In a limekiln this produces quick lime:

$$CaCO_3 = CaO + \overset{\uparrow}{CO_2} \qquad (2.2)$$

It is not found in natural metamorphism of pure limestone except in rare instances where a block falls onto burning lava, etc. Limestones with a high chert content may display reaction around the chert:

$$CaCO_3 + SiO_2 = \underset{\text{wollastonite}}{CaSiO_3} + \overset{\uparrow}{CO_2} \qquad (2.3)$$

Wollastonite appears as big, lustrous white, insoluble crystals in many cherty marbles. Talc is also formed in this fashion.

The highest grade of metamorphism involves invasion by magmatic fluids. Schistose or gneissic mixtures are formed. Their solubility is low and they rarely function as karstic rocks.

Carbonatites

Carbonatites are peralkaline igneous rocks. They are normally intrusive, associated with pyroxenites and amphibolites. They are rare (<< 1% of igneous rocks in outcrop) but examples are known in all continents. They are composed of 60% to > 90% carbonate minerals, principally calcite. They tend to be enriched in rare earths and base metals.

Karst features develop well on high-calcite carbonatites. Sandvik & Erdosh (1977) describe a carbonatite intrusion containing up to 17% apatite ($Ca_5(F, Cl, OH)3PO_4$) where the phosphate has been concentrated as a weathering residuum in large sinkholes in the calcitic rock.

2.4 The evaporite rocks

Evaporite rocks are present beneath ~ 25% of the continental surfaces (Fig. 1.3). They are much less common in outcrop than the carbonates but extensive gypsum karsts occur in parts of Canada, the USA and USSR, and there are smaller examples in many other countries. Salt karst landscapes are limited to small patches in deserts. Deep interstratal solution of these rocks (with or without some surface expression) is widespread and important.

The rocks are formed by homogeneous or heterogeneous precipitation in sea water that has been concentrated by partial evaporation, or as residues left by complete evaporation. The former is much more important. Restricted lagoons and sabkhas (Figs 2.3 & 2.9) are the chief depositional environments. Gypsum is the more common deposit because it precipitates first, lagoonal waters being generally renewed before there is much

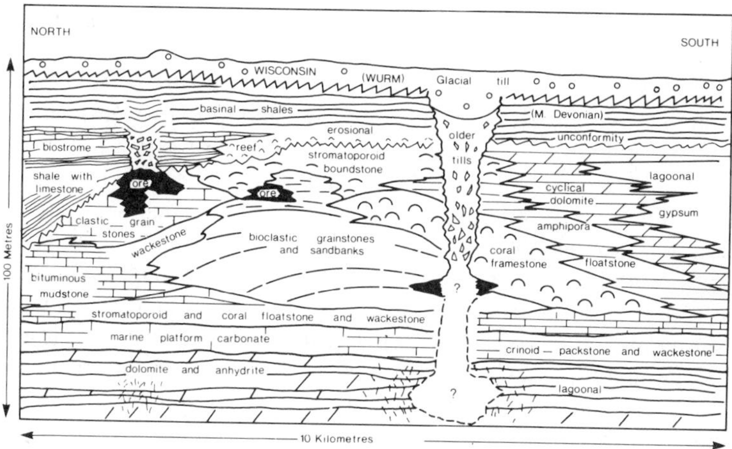

Figure 2.9 Schematic section through the Presqu'ile Reef at Pine Point, NWT, Canada, to illustrate lateral and vertical facies changes in limestone, dolomite, gypsum and anhydrite formations. The rocks are of Middle Devonian age. They display paleokarst cavities which are believed to be younger and which contain sulphide deposits of zinc/lead ore grade. During the Quaternary the sinkholes were rejuvenated; glacial deposits from the last ice advance (Wisconsinan) are intruded through the section. Devonian section is generalized from the work of Rhodes *et al.* (1984).

deposition of salt (Fig. 2.12). Sylvite, sylvinite ($KNa(Cl)_2$) and carnallite (Table 2.1) are rare, and so of little karst significance. In contrast to the carbonates, all of these rocks are wholly inorganic in origin and composition. Evaporite sequences may exceed 1000 m in thickness.

Gypsum and anhydrite

Gypsum is the mineral normally first precipitated. Primary anhydrite is rare. When buried beneath 200–300 m or more of overburden most gypsum is converted to anhydrite, although some is reported to survive at $\geqslant$ 3000 m. If the overburden is then stripped by erosion, re-hydration to gypsum normally takes place. Most gypsum that is exposed in karst has been through the cycle of dehydration and rehydration in differing temperature and pressure environments. As a consequence, there is a wide range of petrologic and lithologic forms.

Gypsum occurs as isolated crystals or clusters of crystals in some carbonate rocks. It appears as rare to frequent interbeds in sequences of medium- to thin-bedded series containing dolomite, clay or shale interbeds. Where it is interbedded it is common to see dikes, diapirs and other intrusions of gypsum penetrating the other rocks.

Gypsum also occurs as coarsely crystalline to granular or amorphous, massive beds of the pure mineral. Individual beds 10–40 m thick are known. Sequences of them may exceed 200 m. These are rarer than the thin, interbedded sequences but support most of the prominent surficial gypsum karsts.

Dehydration of gypsum to anhydrite occurs at pressures of $18\text{–}75\times10^{-5}$ Pa and results in a volume reduction of 38% in the laboratory (Priesnitz 1972). In the field, anhydrite contains no measurable porosity, and joints and bedding planes are annealed. Flow as diapirs may occur. Overlying strata (e.g. dolomite) are often brecciated as a result of this sulphate volume reduction and compaction.

Hydration of anhydrite is reported as deep as 600 to 2000 m in Texas, but most of it probably occurs within 100 m of the surface and much within the topmost few tens of metres. A. N. Pechorkin (1986) stresses that it advances along re-opening joints or new fractures so that the hydration front is highly irregular. It is common to find small patches of anhydrite surviving in gypsum cliffs where there is rapid exposure. Quinlan (1978) argues that hydration involves solution with immediate reprecipitation. It generates a pressure of $\sim$ 20 kg cm^{-2}. Below 150–200 m this is probably dispersed in fluid flow without expansion of volume (Gorbunova 1977b, Quinlan 1978). Above that limit mineral volume expansion of 30% to 67% may occur. This causes flow and intrusion at depth, brittle fracture towards the surface, and recreates some bulk porosity. A. N. Pechorkin (1986) questions whether there is much expansion in most instances, because ions will be removed (i.e. the system, anhydrite+water→gypsum is not a closed one); instead he suggests the loading of the newly created gypsum causes its rapid deformation and flow. At the surface tightly folded (corrugated) or small-scale blockfaulted topography may appear, or tent-like blisters (section 9.13).

Salt

The mode of occurrence of rock salt is similar to that of gypsum. It may be disseminated in carbonates, sulphates or shales, occur as thin interbeds, or as massive units up to 1000 m in thickness that contain only a few anhydrite or shale beds. Most disseminated or thin-bedded salt is dissolved during areal diagenesis so that its presence is signified only by voids, breccia or disconformities in surviving rocks such as dolomite. In massive salt all joints and other openings become annealed by lithostatic pressure; the rock is made quite impervious, prone to flow and intrude to form diapirs and smaller structures. The deep solution of salt can only occur where water approaches it via mechanically strong aquifer strata immediately above or below it and can sap it at the contact. No groundwater flow occurs within salt itself. This is also true of most anhydrite.

2.5 Quartzites and siliceous sandstones

Pure quartzites and silica sands cemented by silica can develop a wide range of solutional karst landforms at all scales (Mainguet 1972, Jennings 1985, Pouyllan & Seurin 1985). In the mineral habit of quartz, the solubility of silica is very low in meteoric waters; but quartzite strongly resists most other forms of weathering attack, as well. Amorphous silica (which occurs in many cements) is more soluble. Solubility of all forms of silica greatly increases in water above 50°C (section 3.1).

Sandstone and quartzite karst is comparatively rare but may be spectacular indeed, as in the great shafts of the Sarisariñama Plateau, Venezuela (Szczerban & Urbani 1974) or the karst corridor landscape of Arnhem Land, Australia (Jennings 1985). It appears that there are three principal requirements for its full development: (a) high mineral purity, so that initial solution channels do not become blocked by the insoluble aluminosilicates etc. that are present in a majority of sandstones; (b) thick to massive bedding with strong but widely spaced fracturing, and (c) absence of strong competing geomorphic processes such as frost shattering or wave attack. The absence of effective competition permits the comparatively slowly developing solution landforms to become dominant.

2.6 Effects of lithologic properties upon karst development

The remainder of this chapter considers factors of 'rock control' of karst morphogenesis. Karst geomorphologists are concerned with a much narrower range of rock types than most other geomorphologists, so it is somewhat ironic that they find it necessary to investigate local rock properties in more detail in a majority of field studies. The need is real; petrologic, lithologic and structural features greatly influence all aspects of karst genesis. Here we outline some main points. Others are illustrated throughout all later chapters of the book.

Significant properties at the scale of individual crystals were summarized previously. This section notes properties that are important at the scale of hand specimens. Later sections then consider local and regional scale factors.

Rock purity

Clay minerals and silica are the most common insoluble impurities in carbonate rocks. It is a widespread finding that limestones with more than 20–30% clay (argillaceous limestones) form little karst; probably the abundant clay particles clog proto-conduits. There is not such a clearcut relationship with respect to sand content. Large and diversified carbonate karst assemblages do not commonly develop where silica exceeds 20–30%, but shallow dolines and well formed, small caves are known in some calcareous sandstones.

The best karst rocks are > 70% pure. Studies of such limestone and dolomite specimens have been made in many countries. They have established that laboratory dissolution rates in carbonated water may vary by more than × 5. Fastest dissolution has been recorded where per cent insolubles were nil and where they were as great as 14%, although most investigations show a clear positive correlation between per cent CaO and dissolution rate (e.g. Fig. 2.10). Pure dolomites are normally slowest to dissolve. But there is always much variation about any trend that cannot be explained by simple bulk purity. For example, in an exhaustive study in Pennsylvania Rauch & White (1977) found that the greatest solubility in pure carbonates occurred where MgO = 1 to 3% and was present within silty streaks. It was suggested that the latter might increase the roughness (exposed area) of the dissolving surface. This is a textural effect. James & Choquette (1984) suggest that high-Mg calcite is normally the most soluble because of severe distortion of the calcite lattice, followed by aragonite, low-Mg calcite, calcite, and dolomite, in descending order of solubility.

In gypsum, anhydrite and salt there is normally a simpler, positive correlation between purity and solubility.

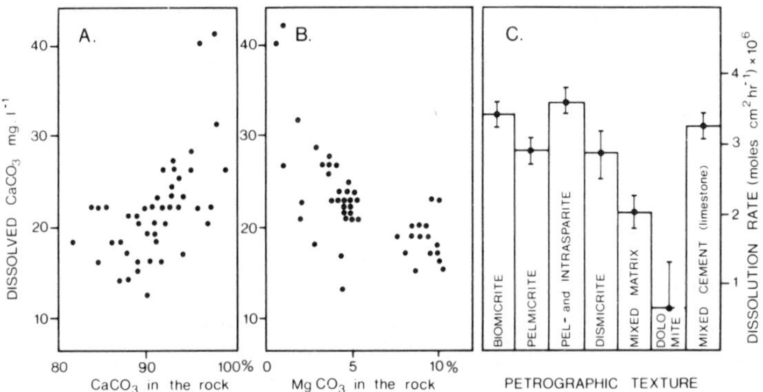

Figure 2.10 Experimental dissolution rates of sample carbonate rocks. A and B. 28-h solubilities of 46 different limestones in water equilibrated to atmospheric P_{CO_2}, plotted against $CaCO_3$ content and $MgCO_3$ content in the rock. From Gerstenhauer & Pfeffer (1966). C. Mean carbonate dissolution rates and error bars based on three replicate runs at 22% of calcite saturation, compared to the petrographic texture of some Pennsylvania carbonate rocks. From Rauch & White (1970).

Grain size and texture

The finer its grain size the more rapidly soluble a rock tends to be because the area of exposed grain surfaces is increased. Sweeting & Sweeting (1969) studying in Yorkshire, England, and Rauch & White (1977 and Fig. 2.10) both reported that biomicrites were most readily soluble and that rate of dissolution decreased substantially where sparite (coarse crystals) was greater than 40–50% by volume. In a discriminant analysis of the different purity, grain size, texture and porosity measures applied to cavernous limestones and dolomites in Missouri, Dreiss (1984) found grain size to be the most significant, the finer-grained rocks being more soluble. However, the finest-grained limestones are often *less* soluble if the grains are uniform in their size and packing because surfaces are then smooth, with exposed grain areas being reduced. Such rocks are termed 'porcellaneous' or 'aphantic' in texture. The Porcellaneous Band is a distinctive very fine-grained micrite that obstructs cave genesis and perches passages in Gaping Ghyll Cave, England. It is sandwiched between coarser biomicrites.

The greater the heterogeneity of grain size, the greater is the roughness of a dissolving surface. This increases solubility, up to a limit. Biomicrite is more soluble than pure micrite because the tiny fossil fragments protrude as roughnesses.

Karren, because they are small, are strongly affected by texture. As homogeneity increases so do the number of karren types that may be hosted and also the regularity of form displayed by any particular type. No channel karren develop well on very heterogeneous rocks such as conglomerates or reefs. Solution scallops (round forms) and rillenkarren (linear) require fine grain size and high homogeneity. Trittkarren, a mini-cirque form, are largely confined to aphanitic rocks and marble. The morphology of these features is discussed in section 9.2.

Porosity

Much of the variation in erosional behaviour of carbonate rocks is due to variation in the nature, scale and distribution of voids within them. This is termed porosity. It is a subject of the greatest importance. There are now many different terms and classifications, e.g. Reeckmann & Friedmann (1982) provide a thorough review from the perspective of petroleum exploration. Sedimentologists define primary porosity as that created during deposition of the rock (i.e. created first) and secondary porosity as that produced during diagenesis (Bebout *et al.* 1979). For many hydrogeologists, all types of bulk rock porosity are 'primary' and only fissure and conduit porosity are classified as secondary; we adopt this convention in later chapters – see section 5.1. A popular general classification given in Figure 2.11 avoids these problems of precedence by distinguishing porosity that is related to petrofabric from that which is not.

In general it is true to write that karst morphology and hydrology are

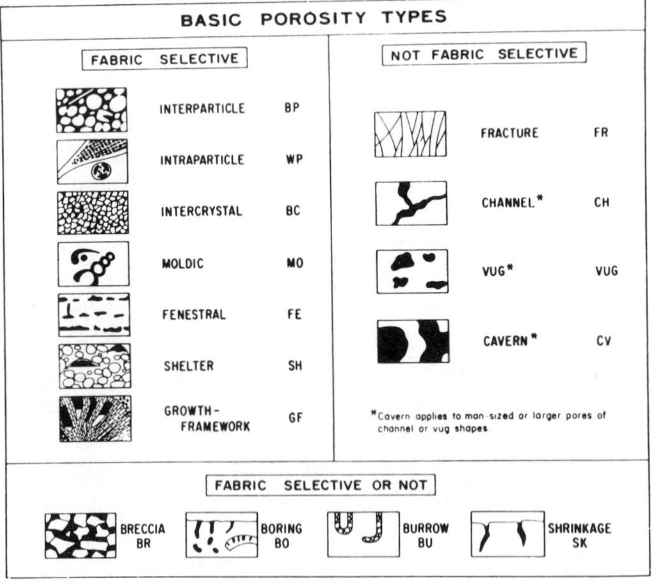

Figure 2.11 Classification of porosity in sedimentary carbonates (Choquette & Pray 1970). Reprinted by permission of American Association of Petroleum Geologists.

concerned largely or entirely with large-scale, interconnected, non-fabric selective porosity (fissures, channels and caverns of Fig. 2.11) in rocks where the fabric-selective porosity is low (< 15%). This is because the hydraulic head gradients experienced also tend to be low, and may be insufficient to drive fluids through tiny, poorly connected, pores within the rock fabric itself. The converse is true in much petroleum exploration, which is concerned with high pressure environments where fissures are closed but fabric porosity is preserved. Large-scale karstic porosity is further discussed in section 5.1 and later.

Fabric selective porosity is important in determining such features as the form, scale and distribution of solution pits and some other karren, the distribution of many stalactites, etc. In a limestone or dolomite it tends to be positively correlated with grain size and textural heterogeneity, although there is much variation. The porosity of micrite is generally less than 2%, that of sparite between 5 and 10%. Dolomitization increases porosity by 5–15% in most instances. The porosity of most marble is < 1%.

Anhydrite and salt anneal readily and so have negligible porosity. Where gypsum is formed by hydration, high intercrystal and breccia porosity may be created.

Mechanical strength

At small scale the strength of a rock is a function of its inter-particle bonding. Such strength can be measured in the laboratory by compression, shear or hammer tests. At larger scale in sedimentary rocks strength is more obviously a function of the density of fissures such as joints or bedding planes. This kind of strength is not amenable to machine testing.

Compressive strength is probably the most significant of the laboratory measures (Table 2.5). It conveys some idea of how a given rock bed will respond if it must bear extra load where there is no buttressing support, as at the base of a cliff or at a cave passage junction. Weaker rocks yield by platy fracture parallel to unsupported faces. This undermines higher parts of the cliff or cave wall, etc. and may induce more widespread block failure (Figure 7.42). A majority of carbonate rocks are quite strong and will support vertical cliffs and cave roofs for long periods unless they are thinly bedded and highly fissured. Many chalks and some other poorly cemented, particle-supported limestones (e.g. eolianites or oolites of Pleistocene age) are too weak to support big cliffs or caves of enterable dimensions.

The Schmidt hammer is a field tool designed to measure the hardness of concrete on a scale of 10–100, as a surrogate for compressive strength. Day & Goudie (1977) showed that its field values for natural rocks also correlate quite well with compressive strength. Day (1982) found that sample dolomitic limestones in the Caribbean area had mean Schmidt hardness, '*R*', of 40–41 whereas limestone micrites and sparites scored

Table 2.5 Compressive strength of some common rocks.

	Uniaxial compressive strength (bars)		Uniaxial compressive strength (bars)
Limestones (excluding chalk and breccias)	340–3450	Shales	300–2300
Dolomites	620–3600	Sandstones	120–2400
Marbles	460–2400	Basalts	800–3600
Anhydrite	220–800	Granites	1600–3000
		Quartzites	1500–6300

Suggested terms:		
	Very weak	< 350
	Weak	350–700
	Strong	700–1750
	Very strong	> 1750

Modified from Jennings (1985).

34–35. There was some positive correlation with topographic ruggedness. On older crystalline limestones in West Virginia, USA, and the Rocky Mountains of Canada we have obtained measures ranging from 35 to 70.

Gypsum, anhydrite and salt are weak. In most instances they will support cliffs and cave roofs, but with excessive rates of block and slab breakdown due to mechanical failure.

2.7 Interbedded clastic rocks

Here we consider the bulk solubility of rocks at the scale of geologic units, members and formations. Many carbonate, sulphate and salt formations are without significant clastic interbeds for thicknesses of tens to hundreds of metres, even thousands of metres in some carbonate groups. These strata generally yield the best karst development. However, the geologic record contains many more examples of formations with frequent beds of clay, shale, sandstone or coal between the soluble strata. These grade to shales, etc. with limestone interbeds, etc. and the geomorphic and hydrologic systems grade from wholly karstic to non-karstic.

It is difficult to offer valid generalizations concerning karst development in the intermediate conditions. As frequency of shale increases in a formation, so does the likelihood that intervening limestones will be argillaceous and non-karstic but this is not always true. Groundwater penetration to initiate karst is often easier at the contact between limestone and shale than it is at bedding planes, joints, etc. within limestone. As a consequence small, independent or poorly-connected, solution conduit systems may develop in adjoining, sandwiched limestones. The same is true of gypsum. Karren develop where the soluble rocks outcrop. There may be small dolines. Collapse features are important where gypsum is interbedded.

2.8 Bedding planes, joints, faults and fracture traces

Bedding planes, joints and faults are of the greatest importance because they host and guide almost all parts of the underground solution conduit networks that distinguish the karst system from all others. When all is said and done about properties of karst rocks it is these entities where rock is absent that determine much of the variety of form and behaviour that occurs in the system.

Bedding planes

Bedding or parting planes in sedimentary rocks are produced by some change in sedimentation or by an interruption of it. The change may be minor, e.g. from one size of carbonate grain to another a little bigger. Major changes are represented by big differences in grain size and, more often, by the introduction of clay by a storm or flood, etc. that leaves a paper-thin or thicker parting between the successive regular carbonate layers. Interruption is usually a brief marine emergence with some erosion and the start of meteoric diagenesis before there is submergence again. Such bedding planes are minor geological disconformities. Sub-parallel pseudo-bedding is created by current scour where carbonate shell banks are reworked by the tides. This gives rise to a platiness in weathered outcrops.

To a hydrogeologist bedding planes are only significant if they are sufficiently open to be penetrable by water under reasonable pressure gradients. Only a minority of sedimentologic planes will be penetrable in most cases. To the karst geomorphologist these planes are also the most important, but impenetrable planes that will rupture under mechanical stress are also significant.

Table 2.6 presents a standard classification of bed thickness. In karst work it defines the separation between successive bedding planes that are penetrable by water in the prevailing conditions.

The areal extent of penetrable bedding planes varies considerably. Where bedding is thin to very thin, they may cover only a few square metres. Where it is medium to thick the extent is normally 10^3–10^6 m^2 or much more. Truly major planes may be followed throughout a formation, sometimes for hundreds of kilometres. As a consequence major bedding planes can be considered to be *continuous* entities when solution caves are propagating through them, whereas joints and most faults are *discrete* (they terminate in comparatively short distances). This enhances the importance of bedding planes in cave genesis (Ford 1971a).

The penetrable plane itself can be considered to comprise two rock surfaces in planar to undulating contact, with some greater interlocked prominences and depressions. Voids in the bedding plane are tiny, thread-like, sinuous and partly interconnected, with openings of 10 μm diameter or greater.

Bedding planes most readily exploited by groundwater include those with substantial depositional disconformities, plus planes with shale laminae or thicker partings (often with disseminated pyrite) and planes with nodules or sheets of chert. Perhaps the most important are those that have served as surfaces of differential slippage during tectonic events. Even if the slippage is just a few centimetres there is some slickenside striation and brecciation that enhances openings. Most steeply tilted and all folded strata display some measure of differential slip.

It is widely recognized that the finest karst landforms require medium to massive bedding. The solutional attack is dispersed where beds are thin. They also lack the mechanical strength to sustain steep slopes and enterable caves in most instances.

The Mammoth–Flint Ridge–Roppel system (Kentucky, USA) and Holloch (Switzerland) are the two most extensive limestone caves currently known. In both the great majority of conduits are guided by bedding planes.

Joints and joint systems

Joints are simple fractures without significant vertical or lateral displacement of strata. They occur during diagenesis, later tectonism, erosional loading and unloading. They may be caused by tensional and shear forces.

Most joints are oriented normal to bedding planes, but they may be inclined. In plan view a majority are straight, but sinuous and curved joints are quite common. Parallel joints constitute a joint set. Two or more sets intersecting at regular angles compose a system. Rectangular and 60°/120° systems are most common, caused by simple tension and shear forces respectively. Major joints extend through several or many beds; they are often terms 'master joints'. They terminate at other master joints. Cross joints are confined to one or a few beds and terminate at master joints.

Table 2.6 gives the scale of joint spacing. This is broadly proportional to bed thickness but the correlation is not precise. An individual bed may contain just one set or system, or several systems imposed at different times.

Table 2.6 Terminology for bed thickness and joint spacing.

Bed thickness (cm)		Joint spacing (cm)	
100–1000	very thick bedded or massive	> 300	very wide
30–100	thick	100–300	wide
10–30	medium	30–100	medium
3–10	thin	5–30	close
1–3	very thin	< 5	very close
< 1.0	laminated		

Adjoining beds frequently display different patterns and densities.

Joint fracture openings may be latent or tiny and impenetrable to water, or larger but filled by secondary calcite or quartz that renders them effectively impermeable. Most master joints in a set will be penetrable, plus many cross-joints. Before any solutional modification, it appears that such openings are angular and jagged, with many irregular points and patches of rock contact. Under lithostatic pressure, joints are more readily closed to impenetrable dimensions than are bedding planes.

All consolidated carbonate rocks display regular patterns of jointing in outcrop, including reef and bank rocks that often lack any bedding planes. Master joints in thick to massive rocks may be as long as several hundred metres, exceptionally extending for several kilometres.

It is important to understand that new joints are created as a karst terrain is eroding. This is not true of bedding planes or of most faulting. Large tensional joints form parallel to steep faces, especially at the rims of plateaus. Small compressional joints form at the base of cliffs and cave walls.

As with bedding, the best development of karst features is found where joint spacing is wide to very wide. Many caves are rectangular mazes guided rigidly by joint patterns, including Optimists' Cave (Ukraine, USSR), a gypsum cave that is second in aggregate known length in the world (Fig. 7.24).

Faults and fracture traces

Faults are fractures with some displacement of rock up, down and/or laterally. Where this is less than about 1 cm they may be considered to grade into joints. At the greatest, vertical displacement extends several kilometres while lateral displacement may amount to 10^2–10^3 km.

Normal faults are produced by tension and, therefore, a wide opening is possible (as much as a few centimetres) though it may fill with breccia, secondary calcite, etc. Reverse faults and lateral or transcurrent faults are compressional features, and so may be impenetrably tight. However, formation of breccia or slickenside grooves can open them, while displacement may bring together recessed facets to create wide spaces. Thrust faults are low angle reversed faults. They are often particularly important because they are areally extensive (nearly vertical faults are not) and so emulate very penetrable bedding planes in their capacity to host interconnected solution conduits. In regions of mild tectonic activity it is common to see a thrust fault originate in one slipped bedding plane, pass through a few beds as a curvilinear surface and terminate in a second disturbed plane.

Large faults are rarely represented by a single fracture surface. There may be a few closely spaced, parallel fractures with lesser ones feathering from them.

Where large vertical faults are present in a karst area it is common to find sinkholes and larger landforms aligned along them or close to them. The situation is variable and less predictable with respect to solution caves. Big underground galleries and rooms often form along vertical faults (e.g. Kiraly *et al.* 1971) but it is comparatively rare to find an entire system of conduits that is directed primarily by faults or contained within them. Because they are initially mostly widely opened, faults attract migrating solutions during and after deep tectonism. Parts of them become sealed by secondary calcite so that, although substantial voids remain elsewhere, these cannot be connected to permit groundwater flow along, up or down the entire fault plane.

Fracture traces (or linears or lineaments) are narrow linear trends detectable on high altitude aerial photography. All karsts appear to display them. On the ground they are zones of closely spaced major joints, or faults of minor displacement plus their feathering fractures and joints. Research in USSR, in Pennsylvania and West Virginia, in Belize, Jamaica, and elsewhere has shown that, once again, major karst depressions are guided by them and may be centred where two traces intersect (e.g. Brook & Allison 1983, Pechorkin & Bolotov 1983, Wadge & Dixon 1984). In dolomite terrains drilling on fracture traces yields the greatest volumes of groundwater. In most instances little or no correlation with the position and orientation of cave systems in limestones has been found, but Barlow & Ogden (1982) report a close association of orientation in Arkansas.

Geomorphic Rock Mass Strength Classification

Taking into consideration factors of rock composition, texture and compressive strength, and the frequency of penetrable bedding planes, joints and faults, the Geomorphic Rock Mass Strength Classification and Rating as proposed by Selby (1980) is a useful guide to the strength of karstifiable rocks at the scale of the principal karst landforms i.e. cave systems, dolines, karren fields, residual hills and towers. The classification is intended to rate the strength of hillslope masses. However, it is developed from mining engineering applications (Beniawski 1976, Brady & Brown, 1985) thus it is also pertinent to the stability of cave roofs, the likelihood of catastrophic sinkhole collapse, etc.

The Selby classification is presented in a slightly modified form in Table 2.7. Most karst terrains with well-developed landforms seem likely to fall into the categories Strong–Weak. Moon (1985), applying the classification to hillslopes of quartzite or shale in South Africa, considered that it should contain a further parameter for the roughness of fissures of all kinds (bedding planes, joints, faults). This should be a particularly complex parameter in karst because the roughness (interlocking) on the fissure planes becomes progressively reduced by dissolution. Limestone and dolomite dip slopes are preferred sites for large landslides because of this factor (e.g. Cruden 1985).

Table 2.7 Geomorphic Rock Mass Strength Classification and Ratings [r = rating of parameter].

Parameter	1 Very strong	2 Strong	3 Moderate	4 Weak	5 Very weak
Intact rock strength (N-type Schmidt Hammer 'R')	100–60 r: 20	60–50 r: 18	50–40 4: 14	40–35 r: 10	35–10 r: 5
Weathering	unweathered r: 10	slightly weathered r: 9	moderately weathered r: 7	highly weathered r: 5	completely weathered r: 3
Spacing of fissures	> 3 m r: 30	3–1 m r: 28	1–0.3 m r: 21	300–50 mm r: 15	< 50 mm r: 8
Fissure orientations	Very favourable. Steep dips into slope, cross joints interlock r: 20	Favourable. Moderate dips into slope r: 18	Fair. Horizontal dips, or nearly vertical (hard rocks only) r: 14	Unfavourable. Moderate dips out of slope r: 9	Very unfavourable. Steep dips out of slope r: 2
Width of fissures	< 0.1 mm r: 7	0.1–1 mm r: 6	1–5 mm r: 5	5–20 mm r: 4	> 20 mm r: 2
Continuity of fissures	none continuous r: 7	few continuous r: 6	continuous, no infill r: 5	continuous, thin infill r: 4	continuous, thick infill r: 1
Outflow of groundwater	none r: 6	trace r: 5	slight $<25\ \mathrm{lmin}^{-1}\ 10\ \mathrm{m}^{-2}$ r: 4	moderate $25–125\ \mathrm{lmin}^{-1}\ 10\ \mathrm{m}^{-2}$ r: 3	great $>125\ \mathrm{lmin}^{-1}\ 10\ \mathrm{m}^{-2}$ r: 1
Total rating	100–91	90–71	70–51	50–26	<26

Adapted from Selby (1980); by permission.

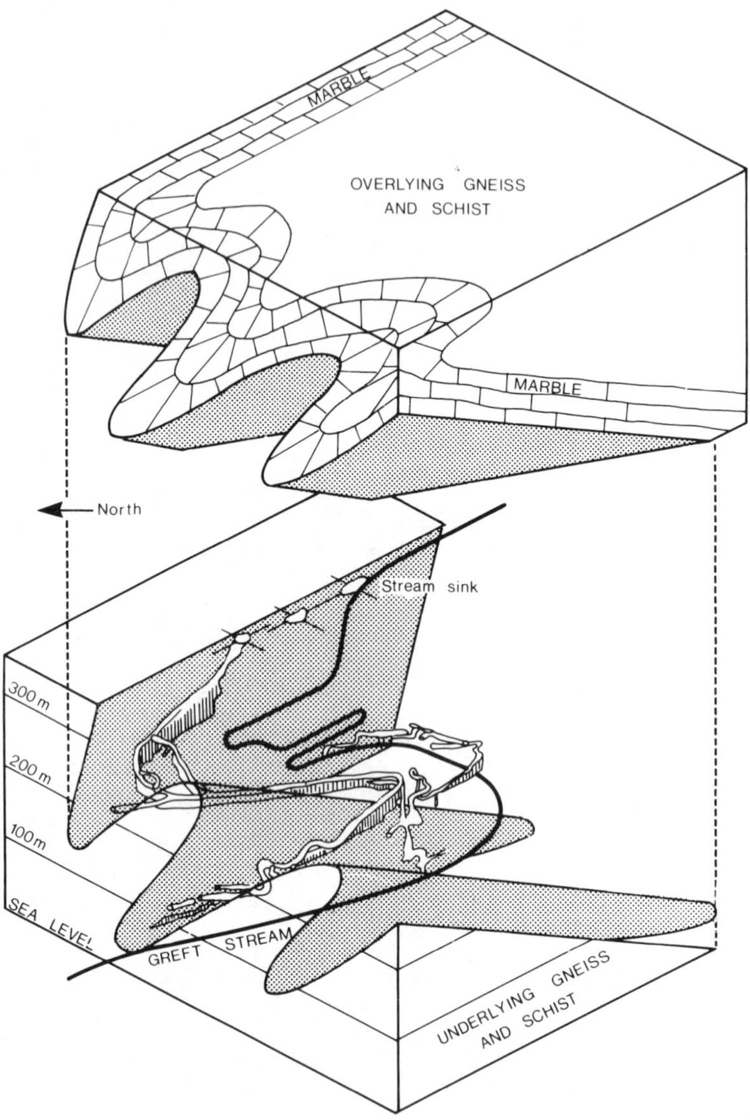

Figure 2.12 Development of Greft Stream cave in tightly folded marbles in the mountains of Nordland, Norway. This spectacular diagram shows the cave draining north and extending up the surface input channel in two or more successive levels, against tight folds plunging south. Adapted from Holbye (1983), by permission.

2.9 Fold topography

The world's karst terrains encompass every type of larger structure. These include plains and plateaus with horizontal or sub-horizontal strata, steep and gentle homoclines, simple and multiple fold topographies, nappe structures, diapiric domes, etc. These may create differing styles of karst at the surface and of geohydrological organization underground.

Rock folding requires plastic deformation. It is generally associated with diagenetically mature karst rocks of Cretaceous or greater age. It is rare to encounter significant folding in Tertiary and Quaternary limestones, although there are spectacular examples in New Guinea.

The amplitude of folds ranges from a few centimetres to several kilometres. High folds may extend for hundreds of kilometres along the strike. The kind of influence that they can exert on cave development is illustrated in Figure 2.12.

Tensional forces tend to create strike-aligned master joint sets at the crests of anticlines and in the troughs of synclines. Differential slipping of beds may be more important on the flanks. Where cave systems extend across one or several anticlines or synclines it is common to find trunk passages centred in the troughs. However, the converse does occur and there are instances in Britain, Kentucky and elsewhere of principal passages extending around the noses of plunging anticlines.

Where karstic beds or formations are mingled with siliclastic strata, tilting and folding often create conditions of artesian confinement. A proportion of the recharge water entering the karst rock may become trapped beneath an impermeable seal and circulate slowly to remote springs. The longest karst groundwater flow systems that are known are created in this manner. Water is believed to flow for more than 1000 km through carbonates beneath confining shales on the Canadian Prairies, with an underground residence time of more than 30 000 years.

3 Dissolution chemical and kinetic behaviour of the karst rocks

3.1 Introduction

When a rock dissolves its different minerals (or parts of them) disintegrate into individual ions or molecules which diffuse into the solution. Of necessity, study of dissolution focuses upon the specific minerals rather than the aggregate rock. Hence, this chapter is concerned with mineral solubility. Because the best karst rocks are nearly pure, monomineralic aggregates there is often little difference between discussion of e.g. calcite solubility, and limestone solubility.

Dissolution is said to be congruent when all components of a mineral dissolve together. Table 3.1 gives the congruent dissolution reactions for a range of minerals. Dissolution is incongruent where only a part of the

Table 3.1 Dissociation reactions and solubilities of some representative minerals that dissolve congruently in water, at 25°C and 1 bar (10^5 Pa) pressure. Modified from Freeze & Cherry (1979).

Mineral	Dissociation reaction	Solubility at pH7 (mg l^{-1})	Common range of abundance in meteoric waters (mg l^{-1})
Gibbsite	$Al_2O_3 . 2H_2O + H_2O = 2Al^{3+} + 6OH^-$	0.001	trace
Quartz	$SiO_2 + 2H_2O = Si(OH_4)$	12	1–12
Amorphous silica	$SiO_2 + 2H_2O = Si(OH)_4$	120	1–65
Calcite	$CaCO_3 = Ca^{2+} + CO_3^{2-}$	100*,500†	10–300
Dolomite	$CaMg(CO_3)_2 = Ca^{2+} + Mg^{2+} + 2CO_3^{2-}$	90*,480†	10–300 (as $CaCO_3$)
Gypsum	$CaSO_4 . 2H_2O = Ca^{2+} + SO_4^{2-} + 2H_2O$	2 400	0–1500
Sylvite	$KCl = K^+ + Cl^-$	264 000	0–10 000
Mirabilite	$Na_2SO_4 . 10H_2O = 2Na^+ + SO_4^{2-} + 10H_2O$	280 000	5–10 000
Halite	$NaCl = Na^+ + Cl^-$	360 000	5–10 000

* $P_{CO_2} = 10^{-3}$ bar
† $P_{CO_2} = 10^{-1}$ bar.

components dissolve. The alumino silicate minerals are the great example of the incongruent class, releasing Na^+, K^+, HCO_3^-, etc. ions in reaction with water but retaining most of their atoms in re-ordered solids such as kaolinite. The karst minerals are all congruent in normal conditions. Incongruent solution of dolomite and precipitation of calcite may occur in some exceptional conditions mentioned later.

The sample of congruent minerals in Table 3.1 contains all the common elements of crustal rocks except Fe, and furnishes a majority of the common dissolved inorganic species. The range of solubility is enormous. Gibbsite is an example that is insoluble to all intents and purposes; even in the most favourable circumstances encountered on the surface of this planet physical processes will disaggregate it and remove it as colloids or larger grains before there is significant solution damage. Rock salt (halite) is so soluble that it is rapidly destroyed in outcrop except in the driest places; it is principally important for its role in interstratal karstification. Sylvite and mirabilite are rarely encountered and never in great bulk. They occur as minor secondary cave minerals (see section 8.4). Gypsum and anhydrite are quite common in outcrop. Karst features develop upon them rapidly because of their comparatively high solubility.

Limestone and dolomite are common in outcrop. Their maximum solubility varies with environmental conditions but never approaches that of gypsum. Quartzite and siliceous sandstones are equally common in outcrop. In terms of solubility and of common solute abundance in water there is a big overlap with the range exhibited by the carbonate rocks. Yet siliceous rocks are not normally considered to be karstic. This raises the question of what is the lower limit of solubility for the development of karst? The answer is that a transitional situation exists in reality, though it is rarely considered by karst specialists. Karst landforms as defined in Chapter 1 develop at all scales on siliceous sandstones and at the small scale on many rocks of yet lower specific (mineral) solubility. However, the dissolution process is rarely predominant on these rocks and thus, at the global scale, these karst landforms must be considered rare and of minor importance. There are less than 40 mg l^{-1} of dissolved silica in most meteoric waters sampled on sandstones and more than 50 mg l^{-1} of calcite in most samples from carbonate terrains. Karst becomes abundant above the latter concentration. Silica solution will not be considered further in this chapter, except to note that its solubility increases rapidly with temperature. At 100°C, quartzite is more soluble than calcite.

Table 3.2 presents the specifications underlying some common chemical and environmental classifications of waters. On the continents solutions notably stronger than seawater are rare; most examples that are encountered are long resident basinal waters intercepted in deep drilling. In a majority of gypsum karsts, it is unusual for concentrations to exceed 2000 mg l^{-1} $CaSO_4$. In carbonate terrains concentrations higher than

Table 3.2 Common chemical classifications of waters.

	Total dissolved solids ($mg\ l^{-1}$)
'soft water'	< 60
'hard water'	> 120
Brackish water	1 000– 10 000
Saline water	10 000–100 000
(Seawater)	(35 000)
Brines	> 100 000
potable water – for humans	< 1000 or < 2000*
potable water – for livestock	< 5000

* Varies between jurisdictions; these are the two most frequent limits.

NB Total dissolved solids in potable waters are presumed to be only the bicarbonates, sulphates, chlorides and their associated species as discussed in this chapter.

350 $mg\ l^{-1}$ $CaCO_3$ will almost invariably prove to be enriched by sulphates or chlorides. The great majority of karst waters contain only a few tens or hundreds of $mg\ l^{-1}$ of dissolved solids. As a result, their chemistry is that of very dilute solutions.

Definition of concentration units

Mass concentrations of dissolved solids measured in water samples are commonly reported in milligrams per litre ($mg\ l^{-1}$). These are equivalent to; parts per million (ppm) and to grams per cubic metre ($g\ m^{-3}$). One 'German Degree of hardness' (GD) = 17 $mg\ l^{-1}$ $CaCO_3$.

The chemical reactions are studied in molar units. A solution of 1 *mole* of calcium (atomic weight = 40.08) = 40.08 grams of calcium per litre. This is a large quantity; thus, many concentrations are reported in millimoles per litre (mmol). To convert from $mg\ l^{-1}$

$$Molarity = \frac{mg\ l^{-1}}{1000\ .\ \text{formula weight}}$$

Molality is defined as moles of solute per kilogram of solvent. In these dilute solutions it can be considered identical to molarity. Because the formula weight of $CaCO_3$ adds up to 100, 1.0 mmol Ca^{2+} is the equivalent of 100 $mg\ l^{-1}$ $CaCO_3$ dissolved.

$$Equivalents\ per\ litre = \frac{\text{moles of solute . charge of ionic species}}{\text{litre of solution}}$$

Data are often reported in milliequivalents per litre (meq l^{-1}).

Readers should be careful when reading scales of calcium carbonate concentration in the literature. $CaCO_3$ may be reported as mg l^{-1} Ca^{2+} or as mg l^{-1} $CaCO_3$. Total hardness (calcium plus magnesium carbonates) is normally reported as mg l^{-1} $CaCO_3$ equivalent.

Example: a solution contains 250 mg l^{-1} $CaCO_3$. The solution contains
100 mg l^{-1} Ca^{2+} = 0.0025 mol l^{-1} or $10^{-2.68}$
150 mg l^{-1} CO_3^{2-} = 0.0025 mol l^{-1} or $10^{-2.68}$

$$\textit{Ionic strength}, I = \frac{1}{2} \sum m_i z_i^2$$

where m_i is the molarity of species i and z_i its valence or charge. In most karst waters there will be only six constituents in significant concentration:

$$I = \frac{1}{2} [(Na^+) + (K^+) + 4(Mg^{2+}) + 4(Ca^{2+}) + (HCO_3^- + (Cl^-) + 4(SO_4^{2-})]$$

In limestone and dolomite areas, Na^+, K^+, Cl^- and SO_4^{2-} can often be neglected, as well, but it is important to establish this by measurement. It should not be assumed.

As a rule of thumb ionic strength of brackish waters ~ 0.1; fresh waters ~ 0.01.

Use of negative logarithms

Because karst waters are very dilute solutions numbers involved in calculations may be inconveniently small. To reduce the likelihood of arithmetical errors arising from misplaced decimal points, it is conventional to do much of the calculation with negative logarithms.

The symbol for a negative logarithm is lower case p. In the example given above, 100 mg l^{-1} Ca^{2+} = 0.0025 mol l^{-1}. Log_{10} of the molarity is $10^{-2.68}$; thus pCa^{2+} = 2.68.

Source books

In this book we use the thermodynamic equilibrium approach and saturation indices to investigate problems of mineral dissolution. It is a comprehensive approach, giving information on the evolution of water from an initial state towards its state when sampled at a karst spring, etc. Accuracy of results is dependent on precision of pH measurements, which may be difficult to achieve in the field. Hence many European karst workers have preferred simpler approaches. These may yield less insight but are also less prone to error.

The 'classic' text is Garrels & Christ (1965) *Solutions, Minerals and Equilibria*. Most later works use the format and conventions adopted by these authors. A most comprehensive recent treatment is by Stumm & Morgan (1981) *Aquatic Chemistry*. Other useful works are by Drever (1982), Butler (1982) and Plummer & Busenberg (1982).

3.2 Aqueous solutions and chemical equilibria

Dissociation, hydration and the Law of Mass Action

Water is an effective conductor because it is a dipole. Cation–anion electric bonds are weakened in solids in contact with it. Their normal thermal agitation suffices to detach some ions, which diffuse away into the solution. For example, for halite

$$NaCl \overset{H_2O}{\rightleftharpoons} Na^+ + Cl^- \tag{3.1}$$

where $\overset{H_2O}{\rightleftharpoons}$ means 'in the presence of water'. This most simple process of solution is termed *dissociation*. It adequately describes the dissolution of rock salt, gypsum and quartz.

A more complex solution process involves the partial or complete neutralizing of either the cation or the anion charge. This unbalances the solution, requiring further dissociation (or equivalent back reaction by precipitation) to restore it.

Pure water itself dissociates to a small extent

$$H_2O \rightleftharpoons H^+ + OH^- \tag{3.2}$$

Comparatively little dissociation occurs with $CaCO_3$. But if a free proton, H^+, approaches the solid we may write the sequence of reactions

$$CaCO_3 \rightleftharpoons Ca^{2+} + CO_3{}^{2-} \tag{3.3}$$

$$Ca^{2+} + CO_3{}^{2-} + H^+ \rightleftharpoons Ca^{2+} + HCO_3{}^- \tag{3.4}$$

Unless an OH^- is within a few nanometres of the site of these reactions close to the solid/liquid interface, the solution is unbalanced there and a further $CO_3{}^{2-}$ ion can dissociate to restore it. This is the process of acid dissolution. It dominates solution of the carbonate minerals.

Systems of such reactions proceed in a forward direction with rates proportional to the concentration of reactants. Accumulation of reaction products increases the rate of back reaction until forward and backward rates are equal. The system then has reached a dynamic equilibrium for the given set of physical conditions imposed upon it, i.e. temperature, pressure. Variation of any of these conditions induces systematic change in the concentrations of each reacting species until equilibrium is again attained. This is the *Law of Mass Action*, which may be written

$$bB + cC \rightleftharpoons dD + eE \tag{3.5}$$

where $bB = b$ moles (or mmol) of reactant species B, and $dD = d$ moles (or mmol) of product D, etc. At dynamic equilibrium this relation becomes

$$K_{\text{eq}} = \frac{(D)^d(E)^e}{(B)^b(C)^c} \tag{3.6}$$

where K_{eq} is a coefficient termed the *thermodynamic equilibrium constant* (or solubility product or stability constant or dissociation constant by different authors). As an example

$$H_2O = \frac{(H^+)\,(OH^-)}{(H_2O)} = K_w \tag{3.7}$$

By convention the value assigned to water is unity, yielding

$$K_w = (H^+)(OH^-) \tag{3.8}$$

K_w, the dissociation constant of water, has the molar value 10^{-14} at 25°C and 1 bar (10^5Pa), and $10^{-14.9}$ at 0°C.

A system at equilibrium is in a state of minimum energy. Systems not at equilibrium therefore proceed towards it by releasing energy. In the aqueous systems being considered the energy measure is ‘Gibbs free energy’ (G);

$$G = H - TS \tag{3.9}$$

where H = heat content (enthalpy), S = entropy, and T = degrees Kelvin. ΔG° is the *standard free energy* of a reaction, the change in free energy where, e.g. b mmol of $B + c$ mmol of C are converted into d mmol of D and e mmol of E.

Activity

Water with ions diffusing through it is a weak electrolyte. Some ions of opposite charge will combine to form ion pairs with reduced charge or zero charge. These are less able or unable, respectively, to take part in further forward reactions. Hence, the number of potentially reactive ions of a given species (e.g. Ca^{2+}) that is present in an aqueous solution is always somewhat less than the molar sum of ions of that species in the solution. The proportion of potentially reactive (or free) ions is termed the 'activity' of the species. As ionic strength (I) increases from ~ 0 to 0.1, activity decreases. This is a reflection of the increasing opportunity for ion combination to occur. For many species it increases again between $I = 0.1$ and 1.0.

Determination of activity is fundamental to the correct computation of all equilibria for solute species. Activity itself is symbolized by '*a*' in most texts. In our text (and most others – but be careful) standard brackets () signify that it is the activity of the contained species that is being considered; square brackets [] signify molarity (or molality) of a species.

The *activity coefficient* γ is defined

$$\gamma_i = \frac{(a_i)}{[m_i]} \tag{3.10}$$

$$\gamma_i \rightarrow 1 \text{ as } \Sigma\, m_i \rightarrow 0$$

Approximate values of the activity coefficients for the dissolved species of interest in most karst work are given in Figure 3.1. Normally they are not read off graphically but are computed with variants of the Debye–Huckel equations that are contained as sub-routines within larger programs computing the equilibrium state of reported dissolved species. A standard extended form of the equation is satisfactory for most purposes in normal karst waters. Consult Plummer & Busenberg (1982), Butler (1982) or Stumm & Morgan (1981) for better precision. Where $I \leqslant 0.1$, the standard form is

$$\log \gamma_i = \frac{-Az_i^2\sqrt{I}}{1 + Br_O\sqrt{I}} \tag{3.11}$$

z = valence of the ion. A and B are constants depending upon temperature and pressure and r_O is the hydrated radius of the ith ion. Relevant values for A, B and r_O appear in Table 3.3. Where I is > 0.1 but < 0.5, Davies' variant (1962) is recommended by Stumm & Morgan

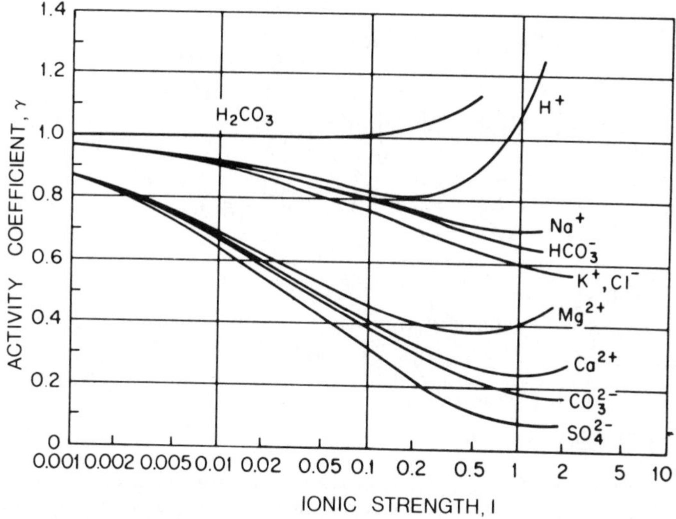

Figure 3.1 Activity coefficients and ionic strength of the common ionic constituents in karstic waters. From Freeze & Cherry (1979); by permission.

Table 3.3 Parameters for the Debye–Huckel equation at one atmosphere.

Temperature (°C)	*A*	*B*
0	0.488	0.324
5	0.492	0.325
10	0.496	0.326
15	0.500	0.326
20	0.504	0.327
25	0.508	0.328
30	0.513	0.329
40	0.522	0.330
50	0.532	0.332
60	0.542	0.334

r_O = 2.5(NH_4^+): 3.0(K^+, Cl^-, NO_3^-): 3.5(OH^-, HS^-): 4.0(SO_4^{2-}, PO_4^{3-}): 4.0–4.5(Na^+, HCO_3^-): 4.5(CO_3^{2-}): 5(Sr^{2+}, Ba^{2+}, S^{2-}): 6.0(Ca^{2+}, Fe^{2+}, Mn^{2+}): 8(Mg^{2+}): 9(H^+, Al^{3+}, Fe^{3+}).

$$\log \gamma_i = -Az_i^2\left(\frac{\sqrt{I}}{1 + \sqrt{I}} - 0.21\right) \tag{3.12}$$

Activity is related to Gibbs free energy

$$\frac{(D)^d\,(E)^e}{(B)^b\,(C)^c} = \exp\frac{-\Delta G^\circ}{RT} = K_{eq} \tag{3.13}$$

where R is the gas constant (1.987×10^{-3} kcal/°/mol) and T = degrees K. Values for G, enthalpy and entropy for minerals and solute species are given in Stumm & Morgan (1981), pp. 749–56.

Saturation indices

From the Law of Mass Action a solution containing a given mineral will be in one of three conditions with respect to a given mineral solid phase:

(a) forward reaction predominates. There is net dissolution of the mineral; the solution is said to be 'undersaturated' or 'aggressive' with respect to the mineral.
(b) there is a dynamic equilibrium; the solution is 'saturated' with the mineral.
(c) back reaction predominates and there may be net precipitation of the mineral. The solution is 'supersaturated'.

Few sampled waters are precisely at equilibrium. Saturation indices measure the extent of their deviation, i.e. their aggressivity or supersaturated condition. The measured product of ion activity in a sample is compared to the K_{eq} value. The standard form of the *saturation index* (*SI*) used by many karst workers is that of Langmuir (1971)

$$SI = \log\frac{K_{IAP}}{K_{eq}} \tag{3.14}$$

where K_{IAP} is the *ion activity product*. Here, a solution is at equilibrium at 0.0, aggressive waters have negative values, etc. as illustrated in Figure 3.2. An alternative index that is occasionally used is the *saturation ratio* (SR); this is simply the non-logarithmic version, where SR is 1.0 at equilibrium. Readers are urged to use the *SI* index in order that results can be more immediately compared.

Figure 3.2 illustrates a further point that is most important. For the mineral species of interest in karst research, the approach to dynamic equilibrium (SI = 0.0) is asymptotic where boundary conditions of

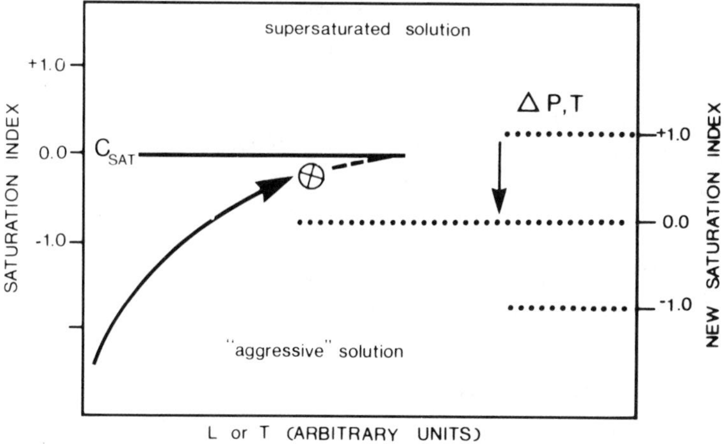

Figure 3.2 Evolutionary path of a water sample, X, approaching thermodynamic equilibrium with respect to a given mineral. C_{sat} = concentration at saturation = an *SI* value of 0.0. Change of boundary conditions will shift the *SI* scale, as illustrated at right.

temperature, etc. remain constant. Ideal equilibrium is difficult to attain; a comparatively long timespan or long flow path through the rock will be required to effect net addition of the last few ions. In karst, a supersaturated water almost invariably indicates that a significant change of boundary conditions has occurred. In Figure 3.2 the change is indicated by ΔP, T.

3.3 The dissolution of anhydrite, gypsum and salt

Anhydrite ($CaSO_4$) may dissociate directly in the presence of water. In field conditions it normally hydrates first, becoming gypsum which dissolves by dissociation

$$CaSO_4 \,.\, 2H_2O \overset{H_2O}{\rightleftharpoons} Ca^{2+} + SO_4^{\,2-} + 2H_2O \tag{3.15}$$

or

$$Kg = \frac{(Ca^{2+})(SO_4^{\,2-})}{(CaSO_4)_s} \tag{3.16}$$

where s = solid.

The equilibrium constant value, K_g, is $10^{-4.61}$ at 25°C, declining to $10^{-4.65}$ at 0°C. Similarly, the expression for halite is

$$K_h = \frac{(Na^+)(Cl^-)}{(NaCl)_s} \qquad (3.17)$$

The activity of a solid is assigned a value of unity in equilibrium expressions. In the case of halite as an example, it is normally written

$$K_h = (Na^+)(Cl^-) \qquad (3.18)$$

$K_h = 10^{-1.58}$ at 25°C, declining to $10^{-1.51}$ at 0°C.

The saturation index for gypsum is

$$SI_g = \log \frac{(Ca^{2+})(SO_4{}^{2-})}{K_g} \qquad (3.19)$$

or

$$SI_g = \log (Ca^{2+}) + \log (SO_4^{2-}) + pKg \qquad (3.20)$$

It takes the same form for salt; however, because the solubility is so great, even brines are strongly aggressive in most instances. The salt index is of little practical utility in karst studies.

The environmental controls of the rate and amount of gypsum or halite solution may be summarized very simply. As with most other dissociation reactions there are positive correlations with pressure and temperature. However, the minor effects of pressure changes can be ignored even where groundwaters circulate to depths of several kilometres. The same is broadly true of temperature. The natural environmental range of 0° to 30°C for meteoric waters has an effect that is inconsequential in the case of salt. For gypsum (Fig. 3.3), it produces an increase of approximately 20% in the

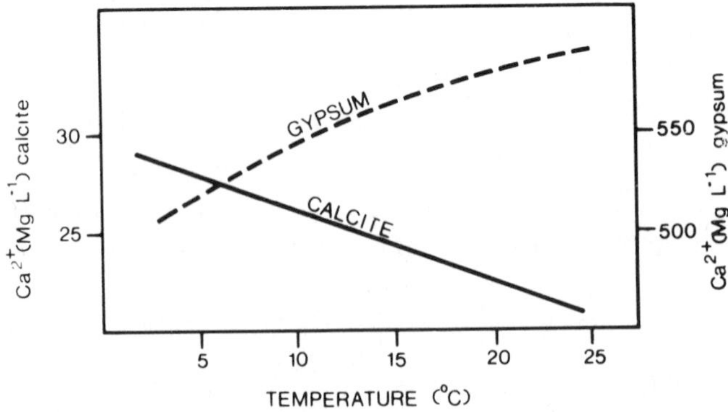

Figure 3.3 The solubility of calcite and gypsum in water and the standard atmosphere, between 2 and 25°C.

solubility product. It has not been shown that this is an important boost in terms of any effect on karst morphology or rate of development: see Cigna (1986) for discussion of thermal mixing effects.

Solution rates and concentrations are controlled primarily by the amount of water contacting these minerals and, to a lesser extent, by the mode of supply – as laminar or turbulent flows or as impacting raindrops and spray.

The prime significance of Figure 3.3 is that it is seen that, in similar conditions, the solubility of calcite is inversely related to temperature. The same is true of aragonite and dolomite. It is just one part of the much more complex story of carbonate solution, to which we now turn.

3.4 Bicarbonate equilibria and the solution of carbonate rocks

Bicarbonate waters

The solubility of calcite and dolomite by dissociation in pure, de-ionized water is very low. With the addition of H^+ from the dissociation of water (Eqn 3.2) it is increased to a total of only 14 mg l^{-1} (as $CaCO_3$) at 25°C. This is scarcely more than the solubility of quartz.

Investigations in many countries have long established that most of the enhanced solubility of carbonate minerals that occurs is due to the hydration of atmospheric CO_2 (Roques 1962, 1964). This produces carbonic acid which, in turn, dissociates to provide abundant H^+. Other acids may furnish additional H^+, and other complexing effects may further increase solubility. These are summarized in the next section. Here we consider the effect of CO_2, which is quite predominant in most carbonate terrains.

CO_2 is the most soluble of the standard atmospheric gases, e.g. 64 times more soluble than N_2. Its solubility is proportional to its partial pressure (Henry's Law) and inversely proportional to temperature. *Partial pressure* is that part of the total pressure exerted by a mixture of gases that is attributable to the gas of interest.

For the dissolution of CO_2 in water, Henry's Law may be written

$$CO_{2(\text{aqueous})} = C_{ab} \times P_{CO_2} \times 1.963 \tag{3.21}$$

CO_2 is in g l^{-1} and P_{CO_2} = partial pressure of CO_2. 1.963 is the weight in grams of one litre of CO_2 at one atmosphere and 20°C. C_{ab} is the temperature-dependent absorption coefficient (Table 3.4).

In the standard atmosphere, P_{CO_2} lies between 0.02 and 0.04% at sea level, with a mean value of 0.033% or 0.000338 atmosphere. This is equivalent to 0.5–0.6 mg CO_2 per litre of air. P_{CO_2} declines very slightly with increasing altitude. It may also be reduced a little in forests (by assimilation)

Table 3.4 The solubility of CO_2. Adapted from Bogli (1980).

A. Absorption coefficients of CO_2

Temperature of solution (°C)	0	10	20	30
Absorption coefficient C_{ab}	1.713	1.194	0.878	0.665

B. Equilibrium solubility of CO_2 (mg l^{-1})

	Temperature (°C)			
P_{CO_2}	0	10	20	30
0.0003	1.01	0.7	0.52	0.39
0.001	3.36	2.34	1.72	1.31
0.003	10.10	7.01	5.21	3.88
0.01	33.6	23.5	17.2	13.1
0.05	168	117	86	65.3
0.10	336	235	172	131
0.20	673	469	342	261

and over fresh snow. However, effects of these reductions appear to be trivial.

Of the greatest importance is the increase of P_{CO_2} that may occur in soil atmospheres as a consequence of organic release in the rooting zones. In principle, CO_2 can entirely replace O_2 there, i.e. increasing P_{CO_2} to 21%. Soil CO_2 is discussed at the beginning of the next section.

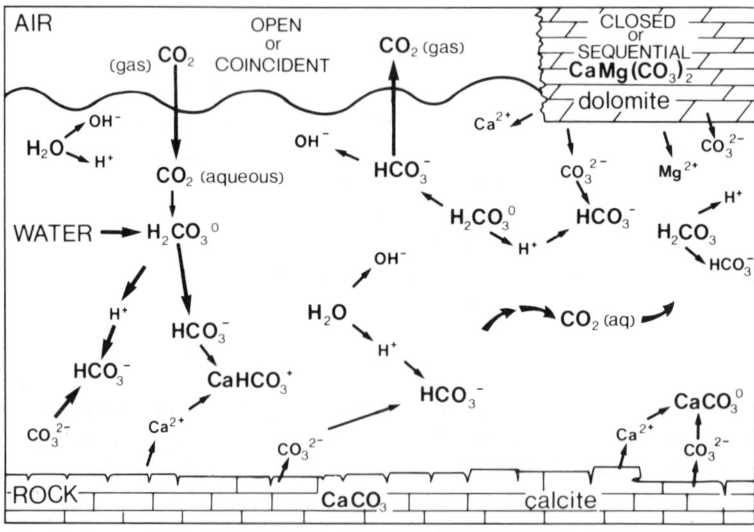

Figure 3.4 Cartoon depicting the dissolved species and reactions involved in the dissolution of calcite and dolomite under coincident and sequential conditions.

The role of CO_2 is illustrated in Figure 3.4. Its dissolution, hydration and consequent dissociation proceed

$$CO_2 \text{ (gas)} \rightarrow CO_{2(\text{aqueous})} \tag{3.22}$$

$$CO_{2(aq)} + H_2O \rightarrow \underset{\text{(carbonic acid)}}{H_2\,CO_3{}^{\circ}} \tag{3.23}$$

Carbonic acid dissociates rapidly; nevertheless, it is conventional to combine these reactions to obtain one equilibrium expression:

$$K_{CO_2} = \frac{(H_2CO_3{}^{\circ})}{P_{CO_2}} \tag{3.24}$$

The carbonic acid dissociates

$$H_2\,CO_3{}^{\circ} \rightarrow H^+ + HCO_3{}^- \tag{3.25}$$

or

$$K_1 = \frac{(HCO_3{}^-)\,(H^+)}{(H_2CO_3)} \tag{3.26}$$

Bicarbonate may then dissociate

$$HCO_3{}^- \rightleftharpoons H^+ + CO_3{}^{2-} \tag{3.27}$$

$$K_2 = \frac{(H^+)(CO_3{}^{2-})}{(HCO_3{}^-)} \tag{3.28}$$

All fresh waters exposed to the ordinary atmosphere will contain these different species of dissolved inorganic carbon, whether or not there are carbonate rocks in the watershed.

Dissolution of calcite and dolomite

The pH of water in limestone and dolomite terrains usually falls between 6.5 and 8.9. In this range, $HCO_3{}^-$ is the predominant species and CO_3 is negligible. It is more appropriate, therefore, to approach Equation 3.27 as a back reaction. This requires introducing the minerals

$$CaCO_{3_s} \rightleftharpoons Ca^{2+} + CO_3{}^{2-} \tag{3.29}$$

where $CaCO_{3_s}$ is solid calcite.

$$K_{\text{calcite or aragonite}} = (Ca^{2+})(CO_3^{2-}) \tag{3.30}$$

Then,

$$Ca^{2+} + CO_3^{2-} + H^+ \rightleftharpoons Ca^{2+} + HCO_3^- \tag{3.31}$$

From laboratory experiments Plummer *et al.* (1978) consider calcite solution to be the sum of three forward rate processes which are reaction 3.31 plus direct reaction with carbonic acid;

$$CaCO_3 + H_2CO_3^\circ \rightarrow Ca^{2+} + 2HCO_3^- \tag{3.32}$$

and dissolution in water (a double dissociation);

$$CaCO_3 + H_2O \rightarrow Ca^{2+} + HCO_3^- + OH^- \tag{3.33}$$

This full sequence of reactions is often summarized

$$CaCO_3 + CO_2 + H_2O \rightleftharpoons Ca^{2+} + 2HCO_3^- \tag{3.34}$$

For dolomite, the dissociation reaction is

$$CaMg\ (CO_3)_2 \rightarrow Ca^{2+} + Mg^{2+} + 2CO_3^{2-} \tag{3.35}$$

Table 3.5 Equilibrium constants for the carbonate solution system, gypsum and halite at 1 atmosphere pressure. From Garrels & Christ (1965), Langmuir (1971), and Plummer & Busenberg (1982).

Temperature °C	pK_{CO_2}	pK_1	pK_2	$pK_{calcite}$	$pK_{aragonite}$	$pK_{dolomite}$	pK_{gypsum}	pK_{halite}
0	1.12	6.58	10.63	8.38	8.22	16.56	4.65	1.52
5	1.19	6.52	10.56	8.39	8.24	16.63		
10	1.27	6.46	10.49	8.41	8.26*	16.71		
15	1.34	6.42	10.43	8.42	8.28	16.79		
20	1.41	6.38	10.38	8.45	8.31*	16.89		
25	1.47	6.35	10.33	8.49	8.34	17.0	4.61	1.58
30	1.52	6.33	10.29	8.52*	8.37*	17.9		
50	1.72	6.29	10.17	8.66	8.54*	–		
70	1.85*	6.32*	10.15	8.85*	8.73*	–		
90	1.92*	6.38*	10.14	9.36	9.02	–		
100	1.97	6.42	10.14	–	–			

log K $CaHCO_3^+$ = 1.11 at 25°C. log K $MgHCO_3^+$ = −0.95 at 25°C
log K $CaCO_3$ = 3.22 at 25°C
* = interpolation.

or

$$K_d = (Ca^{2+})(Mg^{2+})(CO_3{}^{2-})^2 \quad (3.36)$$

and the summary is

$$Ca\,Mg\,(CO_3)_2 + 2CO_2 + 2H_2O \rightleftharpoons Ca^{2+} + Mg^{2+} + 4HCO_3 \quad (3.37)$$

Equilibrium constants for these reactions at a range of temperatures are given in Table 3.5.

If the solution of calcite or aragonite alone is considered (and if the ion pairs, $CaHCO_3{}^+$ and $CaCO_3{}^\circ$, that appear in Figure 3.4 are ignored for the present), the water contains six dissolved species: Ca^{2+}, H^+, $H_2\,CO_3$, $CO_3{}^{2-}$, $HCO_3{}^-$, OH^-. These are defined by Equations 3.2, 3.21, 3.26, 3.28

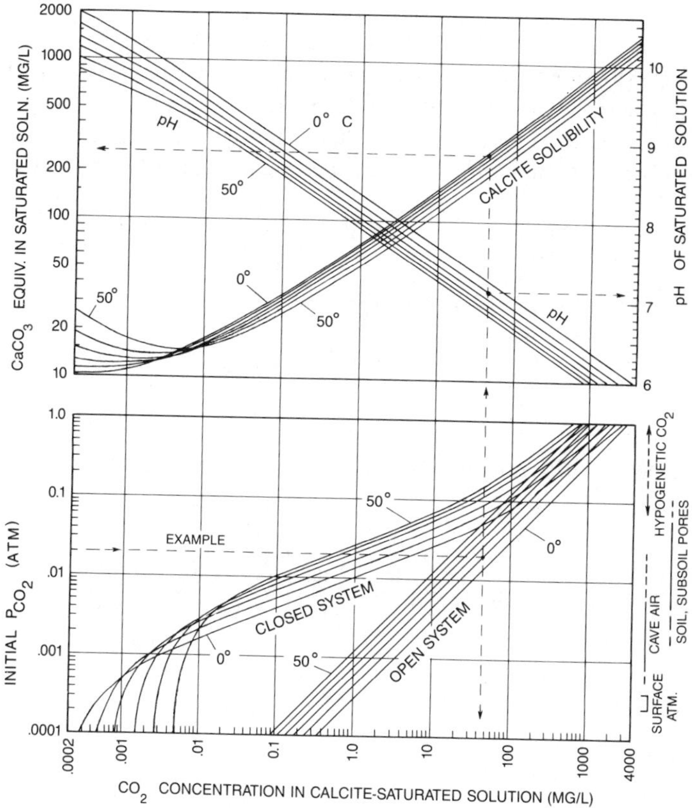

Figure 3.5 Saturation values of dissolved calcite in water at various values of CO_2 partial pressure, and for *coincident* (here termed 'open') and *sequential* (here termed 'closed') system conditions. From Palmer (1984) by permission.

and 3.30. The molar abundancies of these species at equilibrium in given conditions are calculated by adding a further equation and then solving the set simultaneously. The additional equation is for charge balance

$$m_i z_{i\ \mathrm{cations}} = m_i z_{i\ \mathrm{anions}} \tag{3.38}$$

For calcium carbonate solutions this equation is

$$2m_{Ca^{2+}} + m_{H^+} = 2m_{CO_3^{2-}} + m_{HCO^{3-}} + m_{OH^-} \tag{3.39}$$

A more comprehensive charge balance equation (one that will serve for almost any water encountered in karst terrains) is

$$2m_{Ca_{2+}} + 2m_{Mg^{q+}} + m_{Na^+} + m_{K^+} + m_{H^+} = 2m_{CO_3^{2-}} + 2m_{SO_4^{2-}} + m_{HCO^{3-}} + m_{Cl^-} + m_{\ OH^-} \tag{3.40}$$

Solutions are obtained by iterative approximations (Garrels & Christ 1965). Figure 3.5 shows the approximate solutions for a coincident or 'open' system and a sequential or 'closed' system, with just the six species noted; see section 3.6 for an explanation of coincident and sequential conditions.

The saturation indices

The saturation index for calcite or aragonite is

$$SI_{c(\mathrm{or\ a})} = \frac{\log\ (Ca^{2+})(CO_3^{2-})}{K_{c(\mathrm{or\ a})}} \tag{3.41}$$

However, CO_3^{2-} is not easily measured; using known or measurable parameters,

$$SI_c = \log \frac{(Ca^{2+})(HCO_3^-)\ K_2}{(H^+)(K_c)} \tag{3.42}$$

or

$$SI_c = \log\ (Ca^{2+}) + \log\ (HCO_3^-) + \mathrm{pH} - \mathrm{p}K_2 + \mathrm{p}K_c \tag{3.43}$$

For dolomite the saturation index is

$$SI_d = \log(Ca^{2+}) + \log(Mg^{2+}) + 2\log\ (HCO_3^-) + 2\mathrm{pH} - 2\mathrm{p}K_2 + \mathrm{p}K_d \tag{3.44}$$

A very significant parameter (for it reveals much of the provenance or history of a karst water) is the P_{CO_2} with which an analysed sample would be in equilibrium. This is given by

$$P_{CO_2} = \frac{(HCO_3^-)(H^+)}{K_1 \ K_{CO_2}} \qquad (3.45)$$

or

$$\log P_{CO_2} = \log (HCO_3^-) - pH + pK_{CO_2} + pK_1 \qquad (3.46)$$

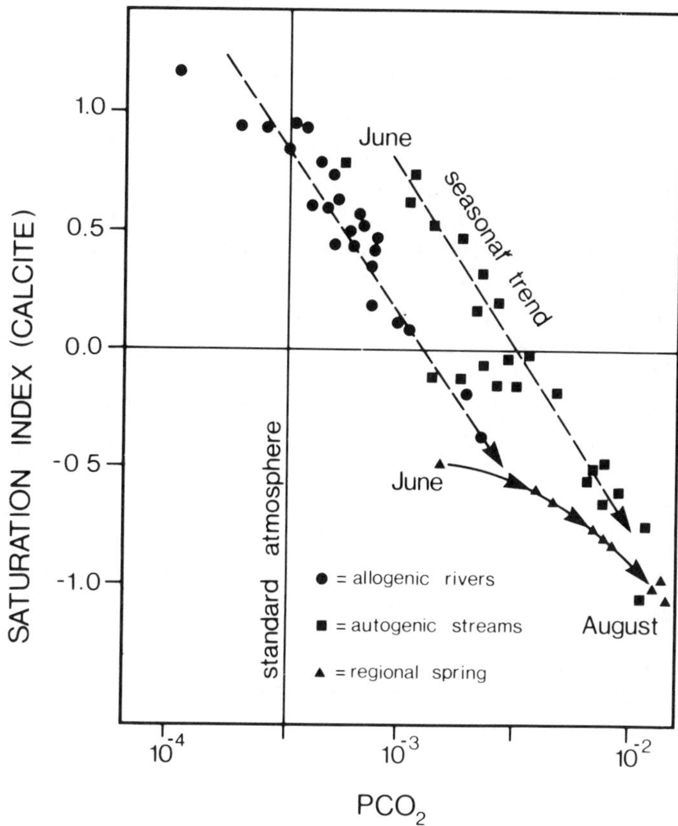

Figure 3.6 Illustrating the use of SI_c and P_{CO_2} parameters; analysis of some simple calcium bicarbonate water samples collected between June and August in a limestone basin on Anticosti Island, Quebec. Allogenic rivers drain lime-rich glacial soils and flow for several kilometres before sinking underground. Autogenic streams drain similar local soils and sink within a few hundred metres. These two types of water plus direct subsoil percolation combine to discharge at a major regional spring. As the summer season advances a progressive increase in the effect of soil CO_2 comes to dominate the evolution of these waters. (From Roberge 1979).

Figure 3.6 presents an example of the Saturation Index and the calculated P_{CO_2} being used to analyse field data.

3.5 Measurements in the field and lab; computer programs

Most nations measure bicarbonate, sulphate and chloride concentrations as part of their water quality monitoring programmes, and so publish handbooks of standard methods. Our purpose here is to give only the briefest summary, with some tips on practice for karst purposes, and to refer to some other useful sources.

Field Measurement of Temperature and pH

The accurate measurement of the pH of a water sample is the gravest analytical problem encountered in the carbonate equilibrium approach that we advocate. pH is a log scale quantity that must be known in order to solve Equations 3.26 and 3.28, and to determine the saturation indices and theoretical P_{CO_2} values. It is the problem that has led many to shun this approach.

pH should be measured in the field to two decimal places, using a solid state digital pH meter with a glass electrode that is shielded, and a reference electrode (Stumm & Morgan 1980, pp. 480–6). Electrodes must be at the ambient water temperature. The instrument should be calibrated with pH buffer solutions before each determination, the solutions being brought to the water temperature by immersion. With this care, reproducibility to ± 0.05 pH is obtainable. Where it is not practicable to take the instrument to the sampling site (e.g. deep in a cave), pH should be determined as soon as possible afterwards. Ek (1973) has shown that for some karst waters a very good linear correlation exists between field pH and lab pH so that the former can be omitted. In any study area this fortunate circumstance must be established. It cannot be assumed. Water temperature measurement to $\pm$ 0.5°C is sufficient for most applications. Most modern conductivity meters include a temperature probe.

Electrical conductivity

At low I the electrical conductivity of water is proportional to total ionic concentration. Modern portable meters can measure a wide range of conductivities on linear scales. Results must be corrected to a standard temperature (normally 25°C) as indicated by the manufacturer. Thus water temperature at the time of measurement must always be recorded. Under laboratory conditions Herman & White (1984) found

$$\text{calcite (as mg l}^{-1}\text{ CaCO}_3) = 0.78\ SpC - 21 \tag{3.47}$$

and

$$\text{dolomite (as mg l}^{-1}\text{ CaCO}_3) = 0.66\ SpC - 13 \tag{3.48}$$

In the field in central Pennyslvania they obtained

$$\text{mg l}^{-1}\text{ CaCO}_3 = 0.59\ SpC - 30 \tag{3.49}$$

For the same geographical area Langmuir (1971) determined the relation

$$I = 1.88 \times 10^{-5}\ (SpC). \tag{3.50}$$

Conductivity meters can be adapted for continuous recording. Where an accurate relationship between electrical conductivity and the dissolved species of interest has been established, therefore, continuous estimates of the latter can be obtained.

Specific determination of dissolved species

Ideally, all variables should be measured at the sampling site in order to avoid disturbances to the system that occur during transport and storage. In practice, it is quite feasible to assemble portable apparatus for a back pack or a small field lab and to obtain adequately accurate results with it. Ca^{2+} and Mg^{2+} are determined by complexometric titration with EDTA. With care, results are reproducible to $\pm$ 1.0 mg l^{-1}.

Carbonate alkalinity (HCO_3^- plus a little CO_3^{2-}) can also be determined by acid-base titration using a weak acid and commercial indicators such as 'BDH 4.5'. It is better done by potentiometry, using 0.01 or 0.02 N HCl and the field pH meter, with an endpoint at pH 4.5 or between pH 3.0 and 4.0 (see Butler 1982).

Sulphate and chloride may be determined by similar colorimetric titrations.

Laboratory methods

All of the above methods can be used in a fully equipped laboratory. Bench instruments will permit greater precision but may not give accurate results because of the disturbance of field equilibria. It is conventional to determine Ca^{2+} and Mg^{2+} by atomic absorption spectrophotometry.

Few, if any, of the available specific ion electrodes are suitable for this work, because they do not measure the low concentrations encountered with sufficient accuracy.

Analytical accuracy

The completeness and/or accuracy of the measurement of a water sample are checked by calculating the ion balance error

$$E = \frac{\Sigma\, m_i z_{i_{\text{cations}}} - \Sigma\, m_i z_{i_{\text{anions}}}}{\Sigma\, m_i z_{i_{\text{cations}}} + \Sigma\, m_i z_{i_{\text{anions}}}} \times 100 \tag{3.51}$$

This formula gives results in per cent. Given the problems of field science, we find any error up to 5% acceptable in very dilute solutions. Where error is greater, either a mistake has been made in the determinations or there are one or more major ionic species present that have not been determined.

Because of the field difficulties, saturation index values for calcite, aragonite and dolomite will normally have an error of $\pm$ 0.1 to 0.2. Greater accuracy can be achieved with gypsum if special care is taken in the determination of sulphate ion.

Handbooks and computer programs

Handbooks and other more detailed outlines include US Geological Survey (1970), US Environmental Protection Agency (1979), Picknett *et al.* (1976), Goudie (1981), Butler (1982).

In North America major programs from at least three different origins are currently in use. They have been modified through several generations. They are: WATEQ (water equilibria) available in Fortran IV as WATEQF and revised as WATEQ2 (Ball *et al.* 1979); WATSPEC (Wigley 1977); SOLMNEQ (solution–mineral equilibrium computations) by Kharaka & Barnes (1973). Nordstrom *et al.*. (1979) compare these and their variants.

WATSPEC was written primarily for carbonate equilibria but includes the equilibria of the other solute species likely to be encountered in natural waters. It is an economical program in FORTRAN. WATEQ2 is a larger general purpose program that contains comprehensive data on many minerals, solute species, their thermodynamic properties and equilibria.

3.6 Chemical complications in carbonate solution

This section could well be subtitled ‘boosters and depressants’. It is concerned with special conditions and effects that may significantly increase or decrease the solubility of the carbonate minerals. Some of these effects also work on gypsum but because its solubility is so great, they are of little importance. Most of the analysis has been concerned with calcite solubility but dolomite is similarly affected in most cases. Figure 3.7 summarizes the principal effects.

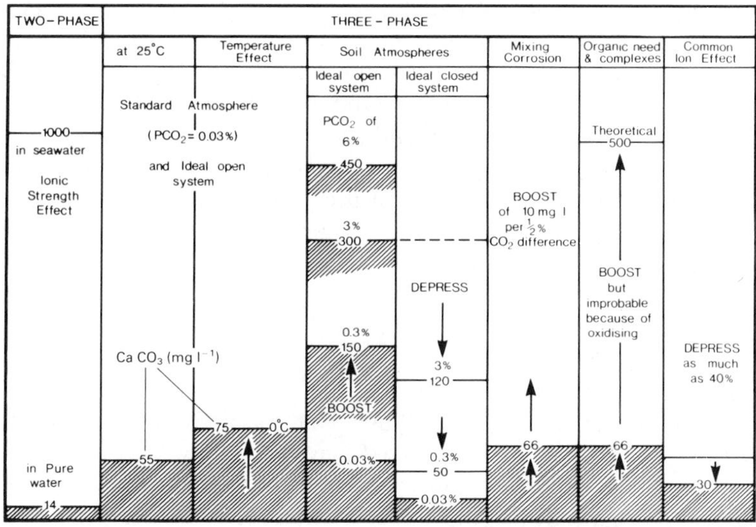

Figure 3.7 Summary of the principal complicating effects in carbonate dissolution chemistry. Values indicated refer to $CaCO_3$ mg l^{-1}. (Prepared by Ford & J. M. James).

The section begins with effects that occur within the carbonate solution system already described, then continues with effects when foreign acids, ions or molecules are introduced.

Simple temperature and pressure effects

From Figure 3.5 and Henry's Law, the solubility of calcite in water equilibrated to the standard atmosphere (P_{CO_2} = 0.03%) at 25°C is 55 mg l^{-1}. This increases to 75 mg l^{-1} at 0°C.

Water often cools as it passes underground. This enhances its solvent potential. For a water saturated at 240 mg l^{-1} $CaCO_3$ and cooling from 20° to 10°C, Bogli (1980) cites a boost of 17.7 mg l^{-1} $CaCO_3$.

Increase of hydrostatic pressure has negligible effects on dissolved species, including gases. However, if any CO_2 bubbles can be introduced into water under pressure, CO_2 solubility increases at a rate of approximately 6 mg l^{-1} per 100 m depth of water (at 25°C) until a depth of ~ 400 m. At greater depths solubility increases at ~ 0.3 mg l^{-1} 100 m^{-1}.

When a cave is flooding rapidly, much air may be trapped and dissolved at pressures up to 2–3 atmospheres. Bogli (1980) and others have suggested that this boost may explain the development of ceiling half-tubes in cave passages because bubbles will be dissolved against the ceiling. Much more significant, we believe, is the combination of pressure and cooling where crustal exhalative CO_2 in active volcanic or tectonic areas is added (initially as gas bubbles) to deeply circulating waters that emerge via karst hot springs. Added CO_2 may greatly boost the solvent capacity of deep, hot water, creating deep karst. As the water ascends and pressure falls, gas is released as bubbles. The solution may become supersaturated with calcite, or the cooling effect may predominate so that the gas becomes re-dissolved to create a second zone of boosted solvent capacity. The complex association of corrosional cavities with precipitated $CaCO_3$ linings in many thermal water caves is explained by changing permutations of this cooling and/or de-gassing relationship.

Soil CO_2

The porosity of normal soils is greater than 40% but part is occupied by bound water. The maximum volume available for air storage and circulation ranges from ~ 17% in clay soils to 31% in very sandy soils (Drake 1984). Gases produced in soils will tend to accumulate because rapid diffusion or drainage are retarded by the high friction of tortuous, intergranular pathways.

Into the pore space green plants respire approximately 40% of the CO_2 that they extract from the atmosphere above ground. Their roots are CO_2 pumps. Yet greater quantities of CO_2 are respired by soil fauna, microfauna and microflora, principally bacteria, actinomycetes and fungi. Their greatest densities occur in and above the rooting zone. From it the gas diffuses out to the surface (and O_2 diffuses in) and also to the soil base and below.

CO_2 productivity of roots and soil bacteria increases with temperature. Optimum temperatures for different species range from 20°C to as high as 65°C (Miotke 1974). Cold-adapted bacteria continue to respire down to −5°C; for example, Cowell & Ford (1980) recorded a sharp drop in soil P_{CO_2} after the first hard frost of winter in central Canada.

The field capacity of a soil is notionally defined as the amount of water retained after free drainage. Soil CO_2 production is greatest at 50–80% capacity but may continue as soil dries to as little as 5%. CO_2 production and retention tends to be greatest in fine-grained soils with swelling clays that retain water. For example, in a loess with a mature brown earth profile in western Germany, Miotke (1974) found that maximum P_{CO_2} was stable at 0.3% atm during dry spells in the growing season. After rains it rose quickly to 4% in the B horizon, presumably because water sealed pore outlets in the A horizon.

These observations indicate that patterns of soil CO_2 abundance will be most variable. They vary with soil type, texture and horizon, depth, drainage and exposure, types of vegetation cover, soil flora and fauna, with seasonal and shorter period warming and wetting. There is a considerable literature on the subject because it is also of great interest to botanists, zoologists, agronomists, etc. In tropical areas reported common ranges of soil P_{CO_2} are from 0.2 to 11.0% (Smith & Atkinson 1976) with an extreme of 17.5% that may be suspected of error. In temperate areas the usual range is from 0.1 to 3.5% but 10% is occasionally reported. In an arctic tundra Woo & Marsh (1977) report 0.2–1.0% over the brief thaw season. In alpine tundras in the Rocky Mountains of Canada, Miotke (1974) obtained a range of 0.04–0.5%.

As noted in section 3.4 the soil CO_2 effect can be studied also by back-calculating the equilibrium P_{CO_2} of groundwaters that have drained through the soil. By this means Drake & Wigley (1975) investigated limestone and dolomite spring waters in Canada and the USA that were just saturated, i.e. very close to equilibrium with respect to calcite. The sites ranged from the sub-arctic to Texas, a mean annual temperature range of 20°C. They obtained the linear relation

$$\log P_{CO_2} = -2 + 0.04T \tag{3.52}$$

where T represents mean annual temperature (°C). Because P_{CO_2} of the standard atmosphere is 5.10^{-3} this signifies a soil enrichment effect of ×5, onto which the positive temperature effect is added.

Although it might appear that CO_2 could accumulate in a soil until all O_2 was replaced (P_{CO_2} = 21% by volume), Drake and others suggest that this cannot occur in practice. Root respiration begins to be impaired at P_{CO_2} = 6% and at slightly higher concentrations aerobic bacteria begin to be killed; hence the process is self-inhibiting. Drake (1980) suggests

$$P_{CO_2} = \frac{(0.21 - P_{CO_2})}{0.21} P_{CO_2}{}^{*} \tag{3.53}$$

where P_{CO_2} represents the actual (inhibited) bulk soil air value and $P_{CO_2}{}^{*}$ is that which would be obtained if there were no inhibition.

Use of mean annual temperature will give an inaccurate estimate where there are strong seasonal variations in temperature–recharge–growing season relationships (Bakalowicz 1976). Brook *et al.* (1983) investigated this question, using soil CO_2 field data from arctic to tropical locales in place of P_{CO_2} of carbonate waters. Equation 3.52 was essentially confirmed. The predictive power of mean annual precipitation was poor

$$\log P_{CO_2} = 2.55 + 0.0004 \text{ Precipitation} \tag{3.54}$$

but there was a good correlation with the logarithm of actual evapotranspiration, which is related to the length of the growing season and thus to the period during which soil CO_2 is being produced

$$\log P_{CO_2} = -3.47 + 2.09\ (1-e^{-0.00172AET}) \qquad (3.55)$$

where *AET* is the calculated mean annual actual evapotranspiration of a site. Brook *et al.* (1983) concluded that ~ 50% of variation of soil P_{CO_2} was explained by temperature, 20% by precipitation, and the balance by seasonal water availability and by growth factors in Equation 3.53.

CO_2 *in caves and fissures of the vadose zone*

Soil CO_2 may drain down into any underlying fissures or caves because it is a heavy gas. Vegetal matter washed into caves, and bat droppings, etc. also decompose and release CO_2 directly into fissure or cave atmospheres. At gravitational trap sites such as pits with no outlets for air through the bottom CO_2 can accumulate to lethal levels. There have been many local studies.

In comprehensive accounts, Renault (1979) and Ek & Gewelt (1985) have reviewed some thousands of measurements of CO_2 in caves world-wide. Cave air is generally enriched 2–20 times with respect to the standard atmosphere but P_{CO_2} as high as 6% has been recorded. It is highest where air circulation is weakest (e.g. in the narrowest accessible fissures) or at sites closest to overlying soils. Northern Hemisphere maxima occur in July–September, when CO_2 concentrations may be 2–4 times as great as in winter. Bakalowicz *et al.* (1985) report precise flux rates for summer CO_2 in the large Grotte de Bedeilhac, Pyrenees. Rates approximated 4–16 kg m^{-2} d^{-1} CO_2 for the surface area of the cave. The gas was derived from gravitational drainage and the de-gassing of saturated infiltration waters.

Cave streams, especially floodwaters, will pick up this excess CO_2 and so boost their solvent capacity. In river caves of warm, humid regions where much vegetal debris is carried underground it may make a significant contribution to corrosional enlargement.

Coincident (or open) and sequential (or closed) systems

In the present context a system is *coincident* when all three phases, solid, liquid and gas, are able to react together. An ideal coincident system exists when such conditions are maintained until thermodynamic equilibrium is achieved. A system is *sequential* when only two phases can interact at a given site.

The terms, 'coincident' and 'sequential', were proposed by Drake (1983) to replace 'open' and 'closed' in previous usage. The former are preferred because these particular systems are always *physically* open. Smith (1965) used 'aerobic' and 'anaerobic' in the same sense but these terms also create confusion in many minds.

The application to carbonate solution is suggested in Figures 3.4 , 3.5 and 3.7. In the ideal coincident case, as H^+ and H_2CO_3 are converted to bicarbonate by reaction with $CaCO_3$, more CO_2 is able to dissolve from the air and so replenish the H_2CO_3 until equilibrium is reached. An open pool on limestone is such a system. P_{CO_2} is fixed at 0.03% and is interactive until equilibrium which, at 25°C, occurs when 55 mg l^{-1} $CaCO_3$ have been dissolved (Fig. 3.7).

In an ideal sequential system, air and water alone react until the solution is saturated with dissolved CO_2 plus derived H_2CO_3 and HCO_3^-. The water then flows away from the atmosphere and first contacts carbonate minerals where no air is present e.g. in a waterfilled capillary or joint. H^+ and $H_2CO_3^0$ that are withdrawn by association with $CaCO_3$, cannot be replenished. For 25°C and P_{CO_2} = 0.03%, the solution becomes saturated at only 25 mg l^{-1} $CaCO_3$, or 40% of that achieved in the coincident system.

In reality, we may expect that many karst waters will evolve under hybrid conditions i.e. where the system is part coincident, part sequential. The cartoon, Figure 3.4, is an example. Drake (1984) suggests that ideal coincident system conditions may not apply in soils with low air volume or in low temperatures, because the rate of dissolution of CO_2 into recharge waters exceeds the CO_2 production rate. Very high rates of recharge will have the same effect. However, where they have been adequately studied, it is found that karst waters at equilibrium tend toward one or the other of the ideal extremes.

A fundamental point to appreciate is that the initiation and early expansion of solutional conduits in the rock will occur under sequential conditions.

Figure 3.8 A and B presents conceptual models by Drake (1984) that relate P_{CO_2} and other relevant soil conditions to the operation of coincident or sequential systems. The models have great explanatory power. In particular they demonstrate why dissolved carbonate concentrations are often much higher in cold northern regions where low P_{CO_2} values are measured in the soil than they are in comparable waters in tropical regions of higher soil P_{CO_2} and temperature. In the cold regions the soil is glacial or thermoclastic detritus in origin (e.g. till) that contains many carbonate clasts.

Much dissolution is accomplished under ideal coincident system conditions in the soil itself. In the tropics, deep soils on limestone are usually dissolutional residua. These are clay-rich (favouring high P_{CO_2}) but without limestone fragments, so that dissolution commences only at their base, where the system becomes partly or fully sequential. Figure 3.9 is a global model unifying the P_{CO_2} and coincident–sequential system effects.

Gunn (1986) criticizes this model on two grounds. The first is that it computes solvent capability from theoretically derived soil CO_2 concentrations rather than from the wide range of field measurements now available. The second is that a significant ground air reservoir may be interposed between the soil atmosphere and the water table in many karsts, especially where the vadose zone is deep and cave systems are well developed

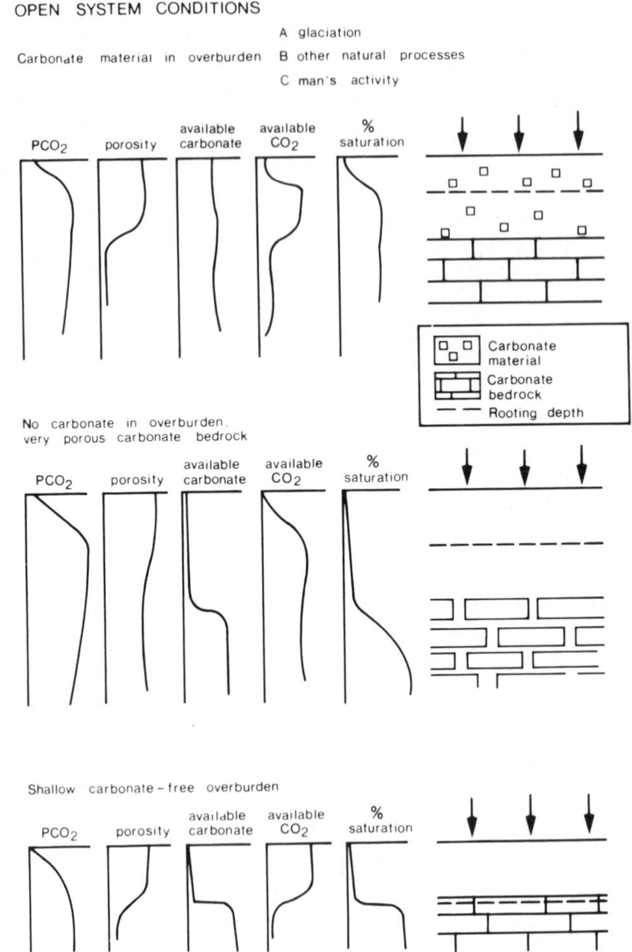

Figure 3.8 Conceptual models of Drake (1984) depicting relationships between carbonate available for dissolution in bedrock and overburden, and controlling variables such as P_{CO_2} and porosity under coincident (A) and sequential (B) system conditions.

(Atkinson 1977a). As a consequence the system is coincident and may equilibriate at the lower P_{CO_2} concentrations reported in cave and fissure air. Both of these grounds are correct *per se*, but they overestimate the accuracy that is claimed for Drake's model. Its functions rough out the limits for the simple system, $CaCO_3$–CO_2–H_2O, before it is complicated by effects considered below.

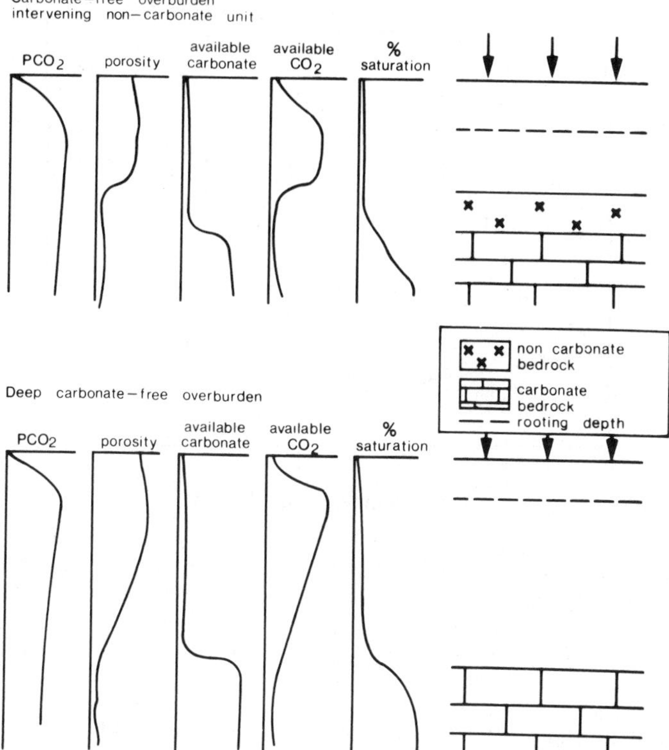

Mixing corrosion

The important concept of *mixing corrosion* was introduced into karst studies by Bogli in 1964 (see also Bogli 1980 pp. 35–7). As he defined it, this is an effect produced by the mechanical mixing of two karst waters that are from different sources, but both are saturated with calcite and therefore, acting alone, are incapable of further dissolution. In general, in such mixing cases the resulting mixture will be somewhat aggressive if one or more of the waters contains less than 250 mg l^{-1} CaCo$_3$.

The principle is illustrated with an example in Figure 3.10. In the dissolved $CaCO_3$ range of 0–350 mg l^{-1} that is found in most carbonate karsts the equilibrium relationship between P_{CO_2} and calcite is non-linear. Groundwater A has equilibrated with a rich soil CO_2 atmosphere and is saturated with 300 mg l^{-1} $CaCO_3$. Water B is also saturated, but for the open atmosphere (60 mg l^{-1} $CaCO_3$). They mix along the line AB and if the two waters are of equal volume (mixing ratio = 1 : 1) the mixture has concentration of 180 mg l^{-1} $CaCO_3$. This is undersaturated. It contains 55 mg l^{-1} $CO_{2(aq)}$ of which only 24 mg l^{-1} are required for thermodynamic

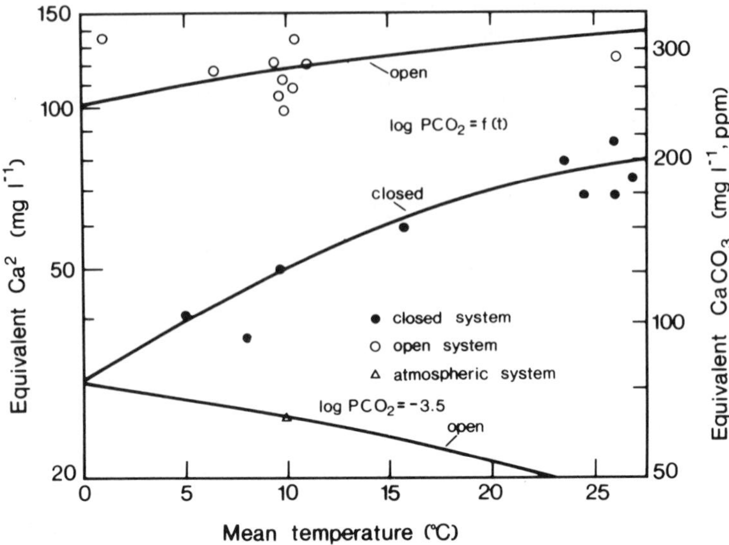

Figure 3.9 Global model for the dissolution of calcite under *ideal* coincidental (open) and sequential (closed) system conditions and for the open atmosphere (log P_{CO2} = −3.5). Coincidental and sequential functions are calculated from Equations 3.52 and 3.53. Data points are means of 20 different groundwater sample sets where waters are at or very close to equilibrium with respect to calcite. Sets range from alpine to tropical environments. (From Drake 1984).

equilibrium at the stated temperature. The balance will react $CO_{2(aq)} \rightarrow H_2CO_3 \rightarrow H^+ + HCO_3^-$ in proportion to the mixing ratio. In the example, the solution becomes saturated again at $CaCO_3$ = 210 mg l^{-1}, a solvent boost of 30 mg l^{-1} or 17%. Angle α is defined by the mixing ratio; A : B = 1 : 1 = 45°; A : B = 2 : 1 = 60°; etc.

Bogli's proposal has been confirmed by calculations by many others. Wigley & Plummer (1976) objected to the term 'mixing corrosion' because it does not include similar boosts that are obtained by mixing water containing different dissolved species (chemical mixing as in 'foreign ion effects' – see below). Others argue that few waters will be truly saturated in the phreatic environment in which they are considered to play their most important role; where undersaturated, any mechanical mixing boost will be less.

Bogli (1964) contended that the bicarbonate mixing corrosion mechanism described here is crucial in the initiation of caves in limestone and dolomite because waters are saturated before they have advanced more than a few metres along their host fissures. Dreybrodt (1981) supported this argument with an important analysis of the solution kinetics. But other kinetic studies imply that saturation is difficult to obtain, while we foresee grave dynamic difficulties in obtaining appropriate mixing ratios (see next section and section 5.3).

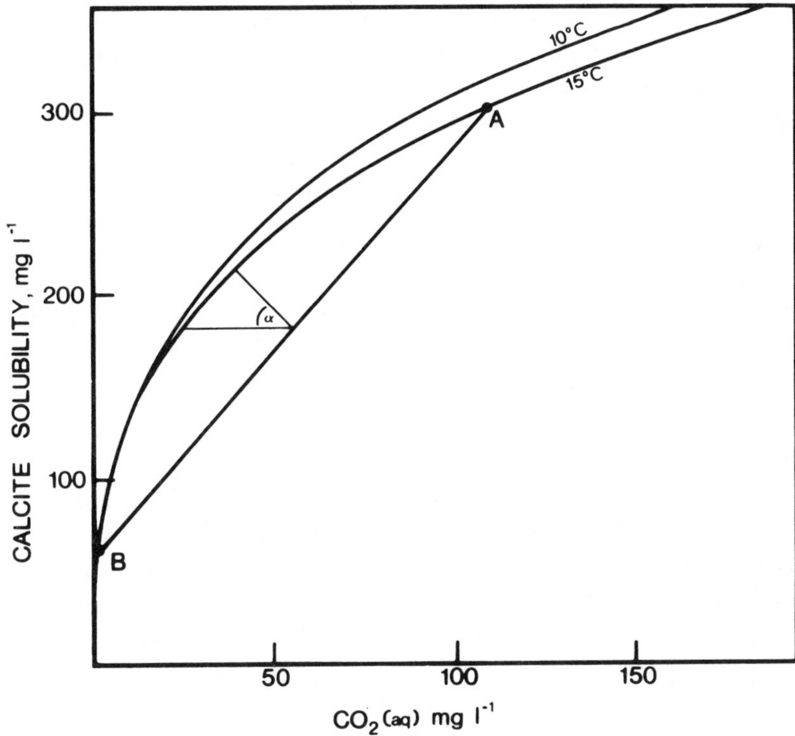

Figure 3.10 Graphical representation of the mixing corrosion effect in simple bicarbonate waters, showing the equilibrium relationships between dissolved CO_2 and calcite at 10 and 15°C. See the text for explanation.

However, the mechanism is very appealing. It, or some similar effect, offers the best explanation of wall and ceiling solution pockets that are common in phreatic caves where penetrable joints intersect major passages.

Ion pairs

Ion pairs normally found in simple bicarbonate waters are shown in Figure 3.4. They are weakly associated cation–anion pairs within a solution that also contains many free ions. As ionic strength increases, pairing will increase. This reduces the activity of ions of interest and so increases mineral solubility.

Chloride ions do not pair significantly. Ion pairs in karst and saline waters will be combinations of the cations Ca^{2+}, Mg^{2+}, K^+, Na^+ and H^+, with CO_3^{2-}, HCO_3^-, OH^- and SO_4^{2-}. Thus, $H_2CO_3^\circ$ (carbonic acid) is an ion pair. Complex pairings such as Ca $(HCO_3)_2^\circ$ are possible, but minor. In bicarbonate and sulphate waters the significant pairs appear to be: $CaHCO_3^+$, $CaCO_3^\circ$, $MgHCO_3^+$, $MgCO_3^\circ$, $CaSO_4^\circ$ and $MgSO_4^\circ$. As an example, Wigley (1971) studied spring water from a gypsum and carbonate basin in British Columbia. Total dissolved solids were 1700 mg l^{-1}. He found that 70.6% of Ca^{2+} ions were free, 26.7% paired with SO_4^{2-}, 1.7% with HCO_3^- and 1% with CO_3^{2-}. Mg^{2+} ion pairing was almost identical.

Although it may increase carbonate and gypsum solubility a little (generally < 10%) ion pairing is truly more important for its effect on calculated saturation indices. If pairing is not allowed for the index values are overestimated; solutions appear more saturated than they are. Standard programs mentioned above (WATSPEC, WATEQ2) compute all probable pairs. It is essential that this be done where total dissolved solids exceed ~150 mg l^{-1}.

Common ions

The principle of the common ion effect is that if one of the ions created by dissolution of a given mineral should be introduced from some other source, the amount of that mineral which can be dissolved at equilibrium is reduced. For calcite and dolomite, this normally implies alternative sources of Ca^{2+} ions because other carbonate and magnesium minerals are rare. Ca^{2+} is furnished by gypsum, calcic feldspars and rare compounds such as $CaCl_2$.

$$K_{\text{calcite}} = \gamma_{Ca^{2+}} \gamma_{CO_3^{2-}} [Ca^{2+}][CO_3^{2-}] \tag{3.56}$$

Addition of Ca^{2+} from gypsum decreases the activity (the product of $\gamma_{Ca^{2+}} \gamma_{CO_3^{2-}}$) but increases the molar product by a much greater amount. Less calcite can be dissolved before equilibrium is attained as a consequence. At 10°C addition of 100 mg l^{-1} Ca^{2+} from gypsum reduces a given calcite solubility of 100 mg l^{-1} $CaCO_3$ to 66 mg l^{-1}. Where total ionic strength is much less than 0.1, the solubility of calcite and dolomite is considerably reduced if waters have already had substantial contact with gypsum.

Ionic strength effect and seawater mixing

This effect is sometimes termed the foreign ion effect. Addition of large quantities of foreign ions such as Na^+, K^+, Cl^- to a bicarbonate water decreases the activity of Ca^{2+}, HCO_3^-, etc. and so increases calcite and dolomite solubility (Figs 3.11 & 3.12). Note also Picknett *et al.*'s (1976) conclusions that 0–10% Mg^{2+} reduces Ca^{2+} solubility, whereas > 10% Mg^{2+} increases Ca^{2+} solubility.

The ionic strength effect is primarily associated with addition of salt. Solubility of gypsum is tripled in a sea water strength solution (Fig. 3.11A).

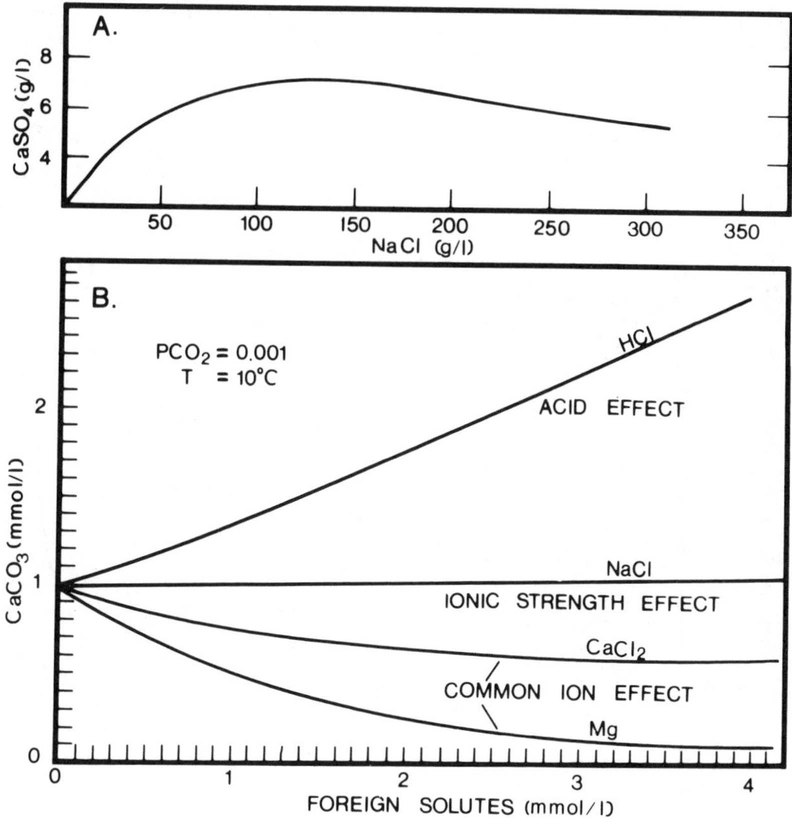

Figure 3.11 Illustrating common ion, foreign ion and ionic strength effects. A. Increase of gypsum solubility with addition of NaCl. B. Ionic strength and common ion effects upon calcite solubility at 10°C. After Picknett *et al.* (1976). By permission.

In the low concentrations of normal limestone karst water (Fig. 3.11B) the effect appears modest. Approximately 250 mg l^{-1} NaCl need to be added to boost $CaCO_3$ solubility by 10 mg l^{-1}. When thousands of mg l^{-1} NaCl are added in salt water mixing situations effects can be considerable. Figure 3.12 shows Plummer's analysis (1975) for sea water mixing at 25°C. With high P_{CO_2}, calcite solubility can be boosted to ~ 1000 mg l^{-1}.

On limestone coasts a mixing zone between fresh and marine groundwaters exists (see section 5.1). Back *et al.* (1984) report upon the behaviour of groundwaters in the Yucatan Peninsula of Mexico, a young, permeable limestone plain. Waters from the interior flow for as much as 100 km underground and become saturated with calcite at ~ 250 mg l^{-1}. In the final kilometre of their journey to the sea, a further 120 mg l^{-1} $CaCO_3$ is added from the ionic strength effects as mixing with sea water occurs.

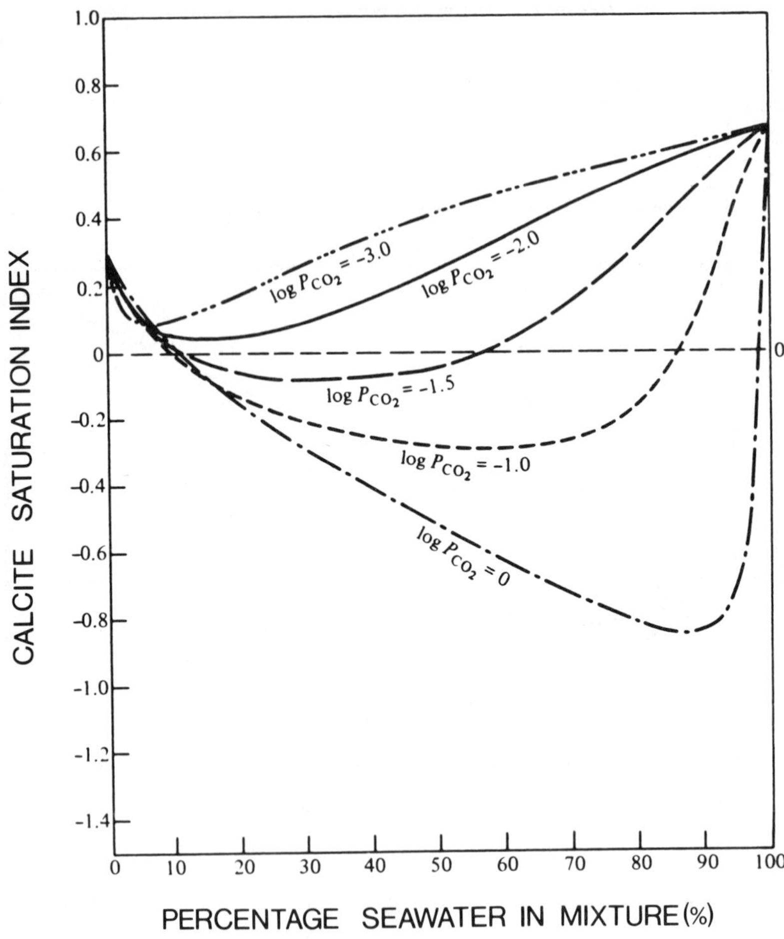

Figure 3.12 Effect of sea water–fresh water mixing upon calcite solubility, at 25°C and log P_{CO_2} as specified. From Plummer (1975) by permission.

Inorganic exotic acids

Here we refer to acids generated outside of the bicarbonate system by reaction with other minerals. This is illustrated in Figure 3.11B where the increase in solubility that is shown for HCl will be true for all acids introduced in lN solution.

Weak solutions of hydrochloric acid occur in nature and the reaction may be represented

$$CaCO_3 + 2HCl \rightleftharpoons Ca^{2+} + 2Cl^- + H_2O + CO_2 \qquad (3.57)$$

It is a first example of a *doubled solvency effect* common in dissolution by these acids; i.e. in the first step (Eqn 3.57) some calcite has been dissolved and some CO_2 produced that may dissolve more calcite in its turn.

Most exotic production involves oxidation, e.g. the case of manganese

$$2Mn^{2+} + O_2 + 2H_2O = 2MnO_2 + 4H^+ \qquad (3.58)$$

The oxide will appear as a precipitate veneer.

The most significant acid effects are those derived from oxidation and other reactions with iron compounds and with H_2S. Iron compounds are very common constituents of shale interbeds in limestone and dolomite. For pyrite in shale

$$FeS_2 + \frac{7}{2}O_2 + H_2O = Fe^{2+} + 2H_2SO_4 \qquad (3.59)$$

$$CaCO_3 + 2H_2SO_4 = Ca^{2+} + SO_4{}^{2-} + CO_2 + H_2O \qquad (3.60)$$

i.e. each mole of pyrite oxidized yields two moles of sulphuric acid and further CO_2 is made available for a potential double solvency.

In the case of siderite

$$2FeCO_3 + 2CO_2 + 2H_2O = 2Fe(HCO_3)_2 \qquad (3.61)$$

$$2Fe(HCO_3)_2 + \frac{1}{2}O_2 + H_2O = 2Fe\,(OH)_3 + 4CO_2 \qquad (3.62)$$

or

$$4\,FeCO_3 + O_2 + H_2O = \underset{\text{limonite}}{2Fe_2O_3.nH_2O} + 4CO_2 \qquad (3.63)$$

i.e. these reactions double or quadruple the CO_2, which is then available for boosted dissolution of calcite, etc.

Molecular oxygen has a low solubility. Therefore, these effects will be of limited importance in phreatic environments unless they are much disturbed (e.g. by mining). Physical effects are readily seen in vadose environments. Water seeping from a limestone bedding plane that contains a little pyritiferous shale may be conspicuously aggressive, etching a pattern of solutional micro-rills at its emergence (see Eraso (1975) for details).

In low pH environments these effects can be further enhanced. Take Fe^{2+} (from Eqn 3.59)

$$Fe^{2+} + \frac{1}{4} O_2 + H^+ = Fe^{3+} + \frac{1}{2} H_2O \qquad (3.64)$$

$$Fe^{3+} + 3H_2O = Fe(OH)_3 + 3H^+ \qquad (3.65)$$

i.e. there is net production of two hydrogen ions to dissolve further calcite. In very acid waters we may find

$$FeS_2 + 14Fe^{3+} + 8H_2O = 15\ Fe^{2+} + 2SO_4{}^{2-} + 16H^+ \qquad (3.66)$$

(from Stumm & Morgan 1980, p. 470). This is the kind of reaction occurring in acid drainage from coal mines, etc. It will be rare in natural conditions but, obviously, it is potent where it does occur!

H_2S is liberated in many volcanic regions and may enter groundwaters as a gas. It can also be created by reduction processes operating on fluids being expelled from sedimentary basins, e.g. around some oil fields, where natural gas has been found composed of as much as 55% CO_2 and 28% H_2S. Currently there is great interest in the latter possibility because it may explain evolution of certain large, exotic caves such as Carlsbad Caverns; see Hill (1987) for a thorough review that pertains to the Permian Basin of New Mexico, including Carlsbad.

Hill (1987) and Egemeier (1981) have proposed the following reactions

$$\underset{\text{(petroleum)}}{C_nH_m} + SO_4 \rightarrow H_2S + CO_2 + H_2O \qquad (3.67)$$

This can be a deep reaction occurring where petroleum fluids contact sulphate strata. It is driven by sulphur bacteria (Kemp & Thode 1968). H_2S ascending to a water table then may become oxidized

$$H_2S + 2O_2 = H_2SO_4 \qquad (3.68)$$

or it may dissociate

$$H_2S = H^+ + HS^- \qquad (3.69)$$

These reactions have generated CO_2, H_2SO_4 and free H^+ for carbonate solution. However, balances are delicate in such systems. If proportions of reactants are varied, replacement or net precipitation may occur, e.g.

$$CaCO_3 + H_2SO_4 + 2H_2O = CaSO_4 \,.\, 2H_2O + H^+ + HCO_3^- \qquad (3.70)$$

Here gypsum replaces calcite, whereas native sulphur is produced in the reaction

$$2H_2S + O_2 = 2S + 2H_2O \tag{3.71}$$

Gypsum in the form of wall crusts that are replacing limestone is quite common in some New Mexico and Soviet caves (see Fig. 7.29). The converse process also occurs (Dreybrodt 1983). For example, in the great gypsum caves of Podolia (section 7.6) water with dissolved CO_2 is condensed onto ceilings, dissolving the gypsum and precipitating a crust of calcite in its place; here, Equation 3.70 is operating in reverse.

Substantial deposits of native sulphur are reported in the Big Room (principal oxidizing zone) of Carlsbad Caverns and in a few other caves. It may react further with water and air to form H_2SO_4.

Evaluation of the quantitative potential of these processes is only just beginning. Hill (1987) calculates that the H_2S required to generate Carlsbad Big Room (Fig. 7.43) is less than 10% of one year's commercial production of natural gas from the adjoining New Mexico gas fields. This demonstrates that significant dissolution by naturally leaking H_2S is feasible.

H_2S may be created and discharged from sedimentary basins in association with fluids containing metal chlorides. Where these contact carbonates there is, once again, the potential for a powerful double solvency effect; e.g.

$$CaMg(CO_3)_2 + 2ZnCl + 2H_2S \rightarrow 2ZnS + Ca^{2+} + Mg^{2+} + 2Cl^- + 2H_2O + 2\,CO_2 \tag{3.72}$$

This reaction and similar ones involving iron and lead may be responsible for the emplacement of many massive sulphide ores in limestones and dolomites: see section 7.10.

'Acid rain'

The pH of normal rainwater is between 5.6 and 6.4. In industrialized regions and for hundreds of kilometres downwind of them it is now often below 5.0 and has been recorded below 3.5. There are two principal sources of the additional acidity:

(a) H_2SO_4 from atmospheric oxidation of SO_2 produced in the burning of fossil fuels and the smelting of sulphide ores, and
(b) HNO_3, produced when atmospheric nitrogen is oxidized in internal combustion engines and then vented as exhaust gas, or from inorganic fertilizers.

Acid rain has been attacking limestone buildings in Europe for more than 150 years. The H_2SO_4 reaction (Eq 3.70) is the more noticeable because surficial skins of stone are spalled off by the expansion involved; the process is often termed 'sulphation'. However, N is now raining out in amounts as great as 10 kg ha^{-1} a^{-1} in southern England (J.I. Pitman, personal communication, 1987). It is now recognized as a problem in many other parts of the world (Reddy 1988). Attention is focused on its deleterious effects on forests, rivers and lakes in regions where there are no carbonate rocks to buffer the acid, e.g. much of central Sweden or the eastern Canadian Shield. So far as we are aware, acid rain has not yet produced any notable new karst, but students of karst water equilibria and erosion rates must be alert to its potential effects in the water samples that they analyse. Acid rain at pH = 3.0 will dissolve 50% more $CaCO_3$ than 'normal' rainwater falling on to bare rock.

Organic complexes and acids

Bacteria oxidize organic matter to produce CO_2, e.g. for formaldehyde

$$CH_2O + O_2 = CO_2 + H_2O \tag{3.73}$$

This process is believed to be the principal source of the soil CO_2 discussed above. It also operates in most natural waters because they contain organic material in suspension. Bacteria may survive indefinitely in a dormant state and organic matter contained in sedimentary rocks degrades very slowly. These factors have led some karst researchers to suggest that deeply circulating groundwaters may be replenished continuously with local CO_2. The limiting factor, once again, would appear to be the supply of O_2, which is never large in amount. The initiating role of bacteria CO_2 in karst genesis has not been evaluated fully but it appears likely to be minor.

Humic and fulvic acids are long chain, high molecular weight (> 10 000) compounds produced by respiration and decay in soils and leaf litter. They are soluble in water where they may chelate (complex) cations directly (see Trudgill 1985, p. 34). They also react to produce CO_2

$$2R\text{–COOH} + CaCO_3 \overset{H_2O}{=} (R\text{–COO})_2\,Ca + H_2O + CO_2 \tag{3.74}$$

where R represents any organic radical (e.g. CH_3). This is another doubled solvency dissolution process.

The effectiveness of humic and fulvic acids draining through the soil base and into fissured limestone below has scarcely been evaluated. Recent research has established that they are co-precipitated in calcite speleothems (they furnish much of the colour in these deposits), so they are certainly passed into groundwater flows at all scales including the mere threads that feed small stalactites. At much larger scales, the drainage from acidic peat bogs is known to be highly corrosive; it is attributed to presence of these acids. J.M. James (personal communication 1977) has suggested that calcite solubility in the standard atmosphere can be boosted to as much as 500 mg l^{-} by them in principle but that, in practice, rapid oxidation may reduce their effect considerably.

Trace element effects

The minor element content of limestone or trace elements from other sources that are present in the water have been found to have substantial effects on calcite solubility in laboratory experiments. The principal study is by Terjesen *et al.* (1961), confirmed by some later work. The presence of tiny amounts of certain metals (one micromole or less) reduces calcite solubility. The inhibiting effect increases with increase of trace content and is believed to be due to absorption of the metal ions onto dislocations in the calcite crystal surface that otherwise are the sites of dissolution as explained in section 3.8. In decreasing order of effectiveness, important metals investigated were scandium, lead, copper, gold, zinc, manganese, nickel, barium and magnesium. As examples, 6 mg l^{-1} Cu^{2+} or 1 mg l^{-1} Pb^{2+} reduced calcite solubility by a factor of 2 at $P_{CO_2} = 1$ atmosphere. No research has been conducted at more appropriate P_{CO_2} levels. No allowance for these trace metal effects is made in saturation index calculations. The effects are probably insignificant in most waters.

Some magnesium is present in most calcites. Effects of differing proportion of Mg in the solid solution are discussed in section 3.8 and Chapter 2. Phosphate ion may also be a strong inhibitor.

3.7 Two examples of the chemical evolution of simple calcium carbonate solutions

Discussion of karst solution chemistry closes with the two calculated examples of the solution of pure calcite that are given in Table 3.6 (modified from Picknett *et al.* 1976). Temperature is fixed at 10°C and only water, carbon dioxide and calcite are present so that common and exotic ion effects, etc. may be neglected. We need consider only P_{CO_2}, the state of the system, major ions and ion pairs.

In Case A specified conditions are an ideal sequential (closed) system, in which the solution will contain 200 mg l^{-1} calcite when exactly saturated. This is a representative value commonly encountered in limestone karst waters. We see that for the ideal sequential case, initial P_{CO_2} must be as high as 0.043 atm. Most of the CO_2 is retained as $CO_{2(aq)}$ but there are 4.76 μmol of carbonic acid ion pair present and approximately five times that amount of the acid has dissociated to form HCO_3^- and H^+. The latter lowers the pH to 4.5.

Table 3.6 A. Solution of calcite in a sequential system at 10°C. Initial P_{CO_2} = 4.3%.

	In mmol l^{-1}			In μmol l^{-1}						
	Ca_{total}	Ca^{2+}	HCO_3^-	CO_2°	$H_2CO_3^{\circ}$	CO_3^{2-}	$CaHCO_3^+$	$CaCO_3^{\circ}$	pH	SI_c
1	0	0	0.0286	2.35	4.76	0	0	0	4.55	0
2	0.50	0.497	0.997	1.88	3.81	0.054	3.23	0.19	6.17	−0.60
3	1.50	1.47	2.95	0.89	1.81	1.06	25.5	8.98	6.95	−0.12
4	2.00	1.92	3.87	0.43	0.87	3.86	42.2	40.0	7.39	0.0

B. Solution and deposition of calcite in a coincident system where CO_2 is being lost to air.

	P_{CO_2}(%)									
1	10	1.95	3.95	5.26	10.7	0.33	43.7	3.43	6.30	−0.39
2	0.81	1.92	3.87	0.43	0.87	3.86	42.2	40.0	7.39	0.00
3	0.10	1.73	3.44	0.05	0.11	24.6	34.3	236	8.25	0.31
4	0.03	1.22	2.25	0.005	0.01	103	16.5	760	9.07	0.63

Row A_2 shows the system now closed and with 25% of its solvent potential exhausted (SI_c = −0.60). CO_2° and $H_2CO_3^{\circ}$ are being depleted. The predominant new species are Ca^{2+} and HCO_3^-, in a ratio which is close to 1 : 2 throughout the evolution. Row A_3 gives the system when 75% of the solvent potential is exhausted. Note that the $CaHCO_3^+$ and $CaCO_3^{\circ}$ ion pairs have increased substantially but that CO_3^{2-} remains very small. These two major pairs will clearly affect saturation index calculations. The solution becomes saturated (row A_4) at pH = 7.39 with about 20% of its initial CO_2 remaining.

Case B shows dissolution and degassing of CO_2 in a coincident system, proceeding until there is supersaturation and deposition of calcite. Imagine water flowing through a rich soil CO_2 atmosphere (P_{CO_2} = 0.1 atm) and then trickling into a vadose (airfilled) cave releasing its excess gas into the open air. In row B1 it is equilibrated with the 10% P_{CO_2} atmosphere and has exhausted about 40% of its calcite solvent potential. In row B2 it has emerged into the cave and degassed to thermodynamic equilibrium with respect to calcite; concentrations are as in row A_4, the solution contains 200 mg l^{-1} of calcite, equilibrium P_{CO_2} = 0.81% and note that 0.03 mmol Ca^{2+} (3 mg l^{-1} $CaCO_3$) have been lost. Rows B_3 and B_4 show the solution at two and four times the saturation values (supersaturated). Calcite precipitation is expected from all solutions where $SI_c > 0.3$ (Plummer & Busenberg 1982), so there will be vigorous deposition unless circumstances are very peculiar. Ca^{2+}, HCO_3^- and $CaHCO_3^+$ are depleted in fixed proportions, $CO_2{}^{\circ}{}_{(aq)}$ + $H_2CO_3^{\circ}$ are much depleted in fixed proportions, and $CaCO_3^{\circ}$ and CO_3^{2-} begin a rapid increase.

3.8 Dissolution and precipitation kinetics of karst rocks

Solution kinetics refer to the dynamics of dissolution. Processes are at their most vigorous when a solution is far from equilibrium. Forward processes to dissolve a mineral and back reactions to precipitate it are governed by the same rules and so may be considered together. The central problem in kinetics is to determine what controls the rate of reaction in specified conditions. Using solutions to the problem, karst specialists may then devise numerical models to estimate, for example, rates of growth of stalagmites or rates of extension of proto-caves. Most relevant kinetic studies have been limited to calcite so it is emphasized here. Dolomite, gypsum and salt are discussed more briefly.

Reactions are *homogeneous* when they take place in one phase e.g. $CO_{2(aq)} \rightarrow H_2CO_3^{\circ}$. They are *heterogeneous* when two phases are involved. All rock solution reactions are heterogeneous and so is all karstic precipitation.

Heterogeneous reactions are extremely sensitive to surface conditions. Their behaviour is investigated by controlled physico–chemical experiments and by theoretical modelling. One of the foremost experimentalists writes 'Actual dissolution rates cannot be predicted from laboratory experiments because experiments ordinarily fail to reproduce the composition and structure of natural mineral surfaces . . .' (Berner 1978). Bogli (1980) and Drever (1982) issue similar warnings. There are many conflicting experimental results. Plummer *et al.* (1979) are able to reconcile them to an extent, but their model rates often differ by some orders of magnitude from other results, as they point out. Karst genetic models built from these kinetic models simplify further, for example by assuming straight uniform flows in capillaries or cracks. Kinetics is a fundamental and most significant area of karst research but readers should be aware that, unlike the Laws of Mass Action that govern species equilibria, there is little that is wholly confirmed. It is an area of vigorous debate and ongoing research (cf. Buhmann & Dreybrodt 1985a, b; Dreybrodt 1988).

Homogeneous reactions, diffusion and surface reactions

In a static liquid, ions and molecules of dissolved species move from regions of higher concentration to lower concentration by the process of molecular diffusion. If the liquid is flowing or is disturbed by waves or currents, dissolved species are dispersed by eddy diffusion which typically is several orders of magnitude more rapid than molecular diffusion.

In most karst situations the water is in motion and therefore eddy diffusion dominates. However, a boundary layer is assumed at the liquid/solid interface where the water is static because of friction and where

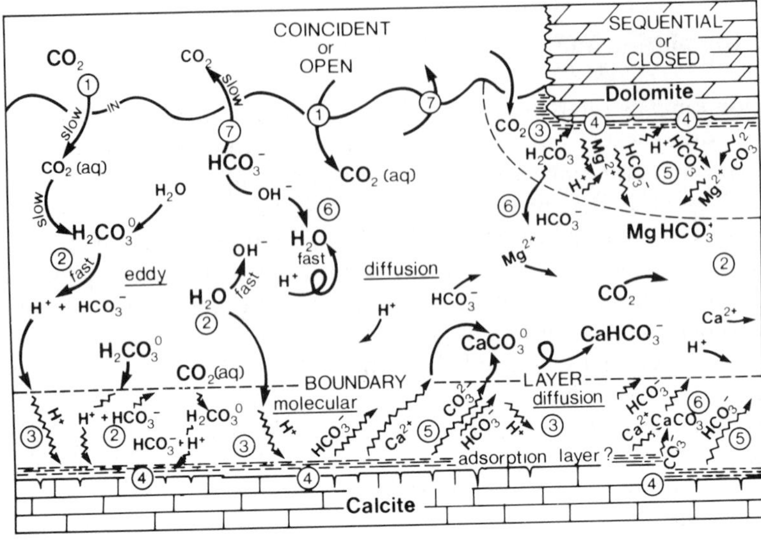

Figure 3.13 Species and reactions involved in the dissolution of calcite and dolomite, depicting the different species that are of significance in the boundary (or molecular diffusion) layer and comparative rates of reaction in the eddy diffusion layer.

molecular diffusion must operate. The boundary layer will be saturated with respect to the mineral, or nearly so. It is very thin, ranging from perhaps 1×10^{-6} m to 1×10^{-3} m. Its thickness is determined by surface roughness, fluid viscosity and velocity of flow in the bulk liquid. Plummer *et al.* (1979) proposed that there may be a further 'adsorption layer' of solute ions and molecules that is loosely bound to the solid surface. This sub-layer is only a few molecules deep.

Figure 3.13 displays these concepts. It is the equilibria cartoon (Fig. 3.4) re-drawn to emphasize kinetic features. The boundary layer can be conceived as a weak shield between the aggressive bulk liquid and the soluble solid.

In a coincident system, the complete bicarbonate dissolution–equilibration sequence may be divided into seven kinetic steps:

(1) CO_2 dissolves into the water.
(2) Dissolved CO_2 hydrates and dissociates. This may occur in the bulk flow or in the boundary layer; most will occur in the bulk flow because this is the region of rapid eddy diffusion.
(3) Potentially aggressive species (H^+, $H_2CO_3^\circ$, $CO_{2(aq)}$) cross the boundary layer and adsorption layer by molecular diffusion.
(4) Reaction occurs; a Ca^{2+} or CO_3^{2-} ion dissociates from the crystal surface or H^+ combines with a CO_3^{2-} ion to dislodge it.

(5) Solute species diffuse out of the boundary layer.
(6) Further equilibration (e.g. $Ca^{2+} + HCO_3^- \rightarrow CaHCO_3^+$) must occur in the bulk liquid.
(7) The saturated liquid degasses, returning CO_2 to the atmosphere.

The dissolution rate will be determined by whichever of these seven steps is the slowest. Where sequential system conditions only apply, steps 1 and 7 are eliminated.

Rates of dissolution of CO_2 (step 1) and of degassing (step 7) are not well known. There is rapid diffusion of the species in air and flowing water so that conditions at the interface are uniform at a scale of centimetres. Roques (1969) found that a drop of water forming at the tip of a soda straw stalactite lost 10% of its CO_2 in the first second, 30% in 90 s, 70% in 15 min. Dissolution and degassing of CO_2 do not appear to be rate-controlling in most circumstances.

Steps 2 and 6 are homogeneous reactions. Most are effectively instantaneous but the hydration of CO_2 takes approximately 30 s at 25°C. Great importance is now attached to this slow reaction by several authorities (see below) if it occurs within the boundary layer. Roques (1969) showed that all equilibration within the bulk liquid occurs within 5 min. This is faster than steps 1 and 7 and so is not considered to be rate-controlling.

Molecular diffusion of species in or out through a static boundary layer (steps 3 and 5) is described by Fick's first law:

$$F = -D \frac{dC}{dx} \tag{3.75}$$

where F is mass flux (M/L^2T); D = a diffusion coefficient (L^2/T); C is the concentration of the solute and dC/dx, its gradient. For the species of interest in karst work, values of D are $1 - 2 \times 10^{-5}$ cm^2 s^{-1} at 25°C, falling to about one half of these rates at 0°C. Different ionic strengths (I) have little effect. The diffusion coefficient of CO_2 in still air is ~ 1 cm^2 s^{-1} at 25°C, 0.14 cm^2 s^{-1} at 0°C. Eddy diffusivity rates in water range from 10^{-1} to 10^{-3} cm^2 s^{-1}. We see that, where diffusion processes are rate controlling, they must be those of the liquid boundary layer.

Where water is flowing through a porous medium D is reduced because there cannot be simple linear diffusion as bulk flow twists and turns around grains. Freeze & Cherry (1979) suggest the empirical relation

$$D^* = \omega D \tag{3.76}$$

where ω is less than 1.0. It is found to fall between 0.5 and 0.01 in laboratory studies of different sands and clays. Berner (1971) gives

$$D^* = \frac{D\varnothing}{\theta^2} \tag{3.77}$$

where $\varnothing$ = porosity and θ^2 = tortuosity. θ is difficult to measure.

Published studies modelling the earliest stages of solvent water flow through fissures have neglected tortuosity, the D^* correction. We suspect that this simplification introduces a significant error into calculations of quantities such as penetration distance where a diffusion constant is used (see below). The effect of correcting for tortuosity will be to increase values of time or distance that are obtained.

To understand reactions at the solid surface it is best to picture a 'step and kink' model such as Figure 3.14. Atoms and molecules in calcite, etc. are ordered in layers. Atoms at a step have higher free energy because they expose two 'sides', at a kink three 'sides'. These will be the preferred sites of solution or precipitation. An H^+ ion that has diffused to the crystal surface will move across it until encountering a $CO_3{}^{2-}$ molecule at such a site. The $HCO_3{}^-$ ion created then diffuses away, laying bare a Ca^{2+} atom which dissociates in its turn.

Most crystals have imperfections. These can be viewed as small fault planes cutting through the regularly stacked boxes of Figure 3.14 so that there is offsetting extending through all layers. These are 'screw dislocations'. In calcite dissolution, destruction passes from one atomic layer to the next by way of them, much like unravelling successive rows of knitting. In cave calcite precipitation, building of layers with screw dislocations may create the disordered speleothems termed 'helictites' or 'eccentrics'.

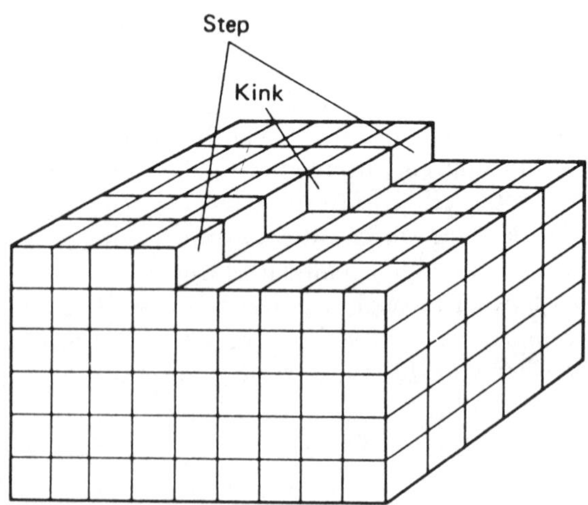

Figure 3.14 Step and kink model of atomic structure in a crystal. See text for explanation.

We can now appreciate the dissolution inhibiting effects of trace elements such as lead, copper, etc. mentioned above. In eroding back, a kink or screw dislocation arrives at one of these atoms which will not dissociate or combine with H^+. Erosional exploitation of that particular kink, etc. ceases.

Where the dissolution (or precipitation) rate is controlled entirely by reactions at the surface we might expect surfaces that are angular, etched and pitted at the microscale because effects of steps, kinks, dislocations and inhibitors dominate. Where there is control entirely by molecular diffusion more rounded forms are expected (corners are preferentially eroded because the boundary layer thins around them, Fig. 3.13). It is possible to have a combination of both diffusion and reaction processes controlling the rate.

Kinetics of calcite solution

The first studies to provide controlled data were those of Erga & Terjesen (1956). A major influence has been the work of Weyl (1958). He concluded that diffusion of Ca^{2+} (step 5) was the rate control in sequential systems and developed time and space models describing the penetration of groundwater into capillaries at the initiation of karstification. Curl (1965) provided a more elegant model and Roques (1969) supplied important new experimental data. But today the foremost contributions are seen to be comprehensive series of experiments of the 1970s, with seawater by Berner & Morse (e.g. 1974) and with carbonated distilled water by Plummer and his associates (1976, 1978, 1979). These cover the complete range of expected natural temperatures and a wide range of ionic strengths.

The rate of solution of calcite as a function of undersaturation is shown in Figure 3.15. Note that

$$\Delta pH = \tfrac{1}{2}\, SI_c.$$

The range of pH change in the experiments (four units) is much greater than will occur in normal karst waters. The solution rate changes through five orders of magnitudes.

Following White (1977a) the graph may be divided into three regions. In region 1 rate of solution is proportional to hydrogen ion concentration. Actual rates of solution are high and the pH will be very low. We would expect to encounter region 1 behaviour where acid mine drainage spills onto limestone, and perhaps where humic acids drain from peat bogs. Region 2 is appropriate to normal aggressive waters. Rate of solution is nearly independent of pH but is dependent upon P_{CO_2}. Region 3 encompasses the final 0.3 units of pH change (SI of -0.15 to 0.00); the rate of solution falls nearly one thousand fold; either it is controlled by molecular diffusion in a nearly saturated medium or forward reactions are close to being overwhelmed by back reactions. In the final 0.1 pH change Berner & Morse found that net reaction had ceased; it could only be induced by adding acid

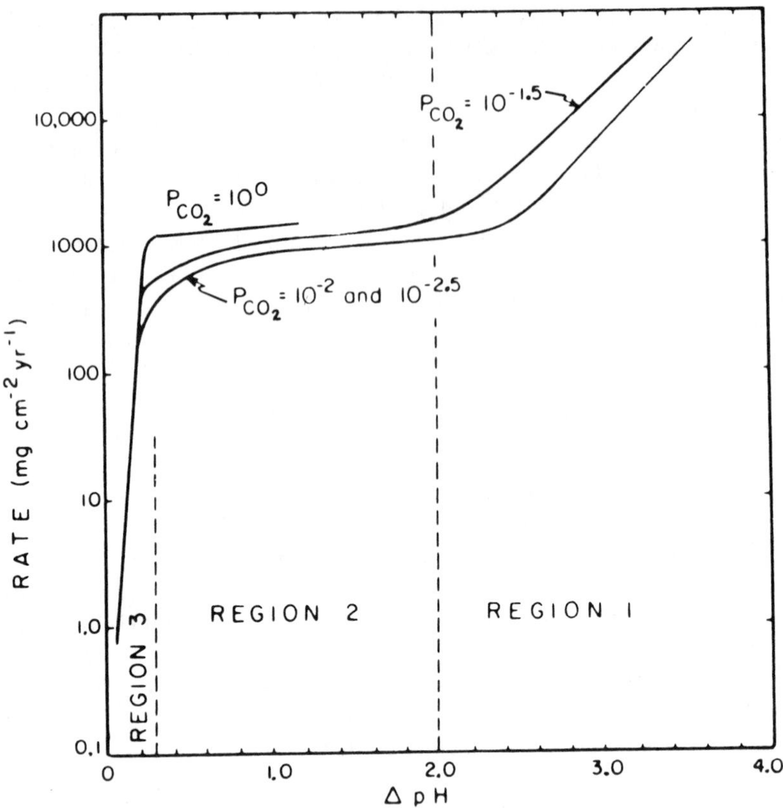

Figure 3.15 The rate of solution of calcite as a function of aggressivity, and defining three kinetic regions. $\Delta pH = \frac{1}{2}SI_c$. From White (1984); data of Berner & Morse (1974).

to perturb the system. Region 3 is roughly equivalent to the 90%+ saturation in the earlier work of Weyl.

Plummer & Wigley (1976) expressed their results as a standard second order rate equation

$$\frac{dC}{dt} = k_c \frac{A}{V}(C_s - C)^2 \qquad (3.78)$$

where k_c is the reaction rate constant for calcite, C_s is the concentration at saturation (i.e. at the solid surface) and C is the concentration in the bulk flow (steps 2 and 6); A is the surface area and V the volume of water upon it. This form is used by White (1977a), Palmer (1984) and others in speleogenetic models. Following further experiments, Plummer *et al.* (1978) proposed that there are three significant forward rate processes – reaction with H^+, with $H_2CO_3^\circ$, and by $CaCO_3$ dissociation alone. The comprehensive rate equation is (Plummer *et al.* 1978)

$$r = k_1\, a_{H^+} + k_2\, a_{H_2CO_3^\circ} + k_3\, a_{H_2O} - k_4\, a_{Ca^{2+}}\, a_{HCO_3^-} \qquad (3.79)$$

where a represents activity. The three forward rates were fitted as functions of temperature: $\log k_1 = 0.98 - 444/T$, $\log k_2 = 2.84 - 217/T$, and $\log k_3 = -5.86 - 317/T$, where T is in degrees Kelvin. The final term is the expression for back reaction. The derivation of k_4 is complicated but can be approximated by standard rate equations for calcite precipitation (Plummer *et al.* 1978).

In region 1 and the k_1 term of Equation 3.79, the rate control is believed to be diffusion of H^+ into the solid surface (step 3). This is dominant in strong acid solutions, as noted. Using a somewhat more realistic experimental simulation (single crystals rather than powdered calcite) Herman (1982) was unable to obtain region 1 rates with carbonated water, reinforcing the supposition that these rates are most appropriate for exotic acids (Fig. 3.16). Where carbonic acid solution predominates, White (1984) suggests that far from equilibrium the rate control is again hydration of CO_2 at or near the crystal surface. This was earlier proposed by Roques (1969). However, Herman & White (1985) found that in such conditions the rate increases quite abruptly at the transition from laminar to turbulent flow

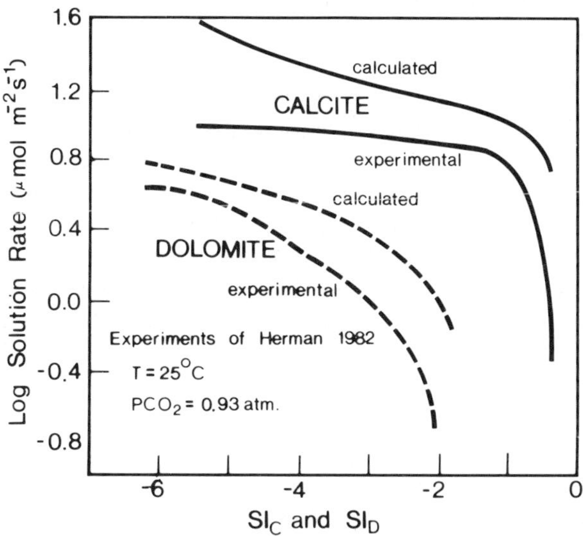

Figure 3.16 Rates of calcite and dolomite solution as functions of saturation index. Calculated curves are from Plummer *et al.* (1978) for calcite, from Busenberg & Plummer (1982) for dolomite. Adapted from White (1984) by permission.

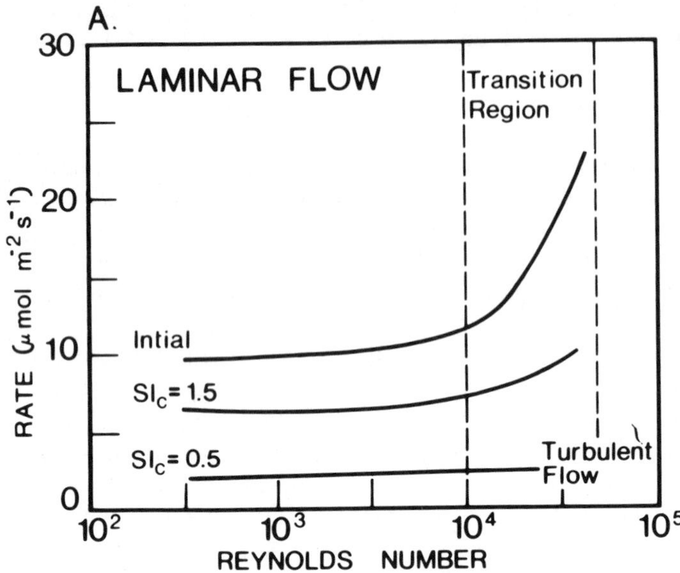

Figure 3.17 Solution rate as a function of Reynolds Number for various values of saturation. From White (1984) by permission.

(Fig. 3.17) which supports a diffusion control model.

For region 2 and the second term of Equation 3.79 it is now quite widely held that the hydration reaction within the boundary layer or at the surface itself is the principal control.

Region 3 may be diffusion controlled as Weyl (1958) supposed or there may be forward and backward reaction control competition to so drastically slow the approach to equilibrium. Equation 3.79 predicts a faster approach to saturation than Herman (1982) obtained with her crystal experiments (Fig. 3.16). Using the equation alone Dreybrodt (1981) was unable to generate caves within geologically reasonable timespans. He had to invoke the mixing corrosion effect because his calculated solutions became saturated too quickly. White (1984) showed that with Herman's rates for region 3, penetration distances and speleogenesis timespans became more realistic.

The solution kinetics of dolomite

Busenberg & Plummer (1982) studied dolomite dissolution over the P_{CO_2} range, 0–1 atmosphere, and between 1.5 and 65°C. They worked with cleavage fragments using fixed and free-drift pH experimental designs. They were unable to come closer to saturation than $SI_d = -3$. Herman & White (1985) used a spinning dolomite disc design at $P_{CO_2} = 0.93$ atm and 25°C.

Rate of solution became infeasibly slow above $SI_d = -2$.

Results of the two investigations are compared in Figure 3.16. Initial dolomite dissolution rates are between 5 and 6 mol μm^{-2} s^{-1} (at the unrealistically high P_{CO_2} value of 0.93 atm) and decline in an exponential fashion through one order of magnitude until $SI_d = -2$. There are no distinct 'regions 1 and 2' as in the case of calcite (Fig. 3.15). When varying P_{CO_2}, Busenberg & Plummer (1982) found significant differences of dissolution rate between different samples of dolomite. These differences were greatest at lowest temperatures. Clear crystalline dolomite of hydrothermal origin dissolved at a rate an order of magnitude lower than cloudy, low temperature dolomite which is expected to contain a greater density of lattice defects.

Busenberg & Plummer (1982) concluded that dolomite dissolution proceeds in two steps. The first is reaction of the $CaCO_3$ component with H^+, $H_2CO_3^{\circ}$ and H_2O as in Equation 3.79. The second sees the same reactions occurring much more slowly with the $MgCO_3$ component: this is rate controlling. As solute concentrations increase (but still at SI_d below −3 or 0.1%) there is significant back reaction of HCO_3^-. It is being adsorbed onto positively charged sites (i.e. protruberances of Ca^{2+} and Mg^{2+}) at the surface. This explains the exponential slowing of the dissolution rate. Their comprehensive kinetic equation is of the same form as that for calcite

$$\lambda = k_1 a^n{}_{H^+} + k_2 a^n{}_{H_2CO_3^{\circ}} + k_3 a^n{}_{H_2O} - k_4 a_{HCO_3^-} \qquad (3.80)$$

'n' = 0.5 at temperatures below 45°C, i.e. the square root of activity in a double carbonate.

We can now propose a mental picture of what happens at the solid surfaces when dolomites and Mg-calcites dissolve. Dolomite comprises $CaCO_3$ and $MgCO_3$ molecules in alternate layers. A layer of $CaCO_3$ is quickly peeled away by dissolution. $MgCO_3$ is more strongly bonded. Its exposed layer resists dissolution while any residual pinnacles of $CaCO_3$ plus $MgCO_3$ protruding from it attract HCO_3 ions in back reaction. As density of lattice defects increases so do the opportunities to 'unravel' the crystal by following screw dislocations that breach its $MgCO_3^-$ layers.

Mg-calcite is a solid solution. Molecules of $MgCO_3$ are scattered from place to place in the $CaCO_3$ lattice; there is no regular alternation of layers. In 'low Mg-calcite' the greater resistance of the few $MgCO_3$ clumps creates many steps and kinks in calcite about them. Low Mg-calcite is more soluble than pure calcite as a consequence. In 'high Mg-calcite' the density of clumps may become too great; the step and kink forward effects are overwhelmed by the back reaction of HCO_3^- adsorbing onto protruberances. Most workers have found that high Mg-calcite is less soluble than pure calcite, although it is still more soluble than true dolomite in most instances.

Penetration distances and relaxation times in limestone and dolomite

For calcite, net forward reactions slow drastically in region 3 when the bulk solution is about 90% saturated. For dolomite there is greater slowing at $SI_d = -2$ or less. Weyl (1958) introduced the most useful concept of L_9 penetration distance – the distance a given solution will flow through a capillary or crack before it is 90% saturated and effectively incapable of further enlarging that capillary at a reasonable rate. T_9 is the time required to extend L_9 a specific distance.

Before being modified by dissolution, fissures in limestone are highly irregular voids offering some tortuous but continuously open paths that water may flow along. Following Davis (1968) and Bocker (1969) if the continuous opening is less than ~ 10 μm in width the fissure is effectively impenetrable within a reasonable timespan. Karst groundwater networks begin with the dissolutional penetration enlargement of continuous openings probably 10–100 μm in width within fissures.

Minimum necessary penetration distances are about 1 m for many types of karren. Maximum distances are 10–100 km for regional karst drainage systems. Early penetration must occur under sequential system conditions and with laminar flow.

Weyl (1958) developed the expression

$$L_9 = \frac{0.572 - v\, r^2}{D} \tag{3.81}$$

where v = mean flow velocity, r = the radius of a capillary or half-width of a fissure, and D = diffusion coefficient. This presumes that the calcite rate control is diffusion of Ca^{2+}. White (1977a) took the reaction control result of Equation 3.79 to derive

$$L_9 = \frac{9}{16}\,\frac{r^3\, g \sin \phi}{\nu\, C_s\, k_c} \tag{3.82}$$

ν = kinematic viscosity; g = acceleration due to gravity; ϕ is slope (hydraulic gradient) in degrees, k_c is a rate constant equivalent to $k_1 + k_2 + k_3$ of Equation 3.79, and C_s is Ca^{2+} concentration at saturation.

Application of both Weyl's and White's equations are given in Figure 3.18. Fissure half-width begins at 10 μm. L_9 penetration distances of interest in most karst work range from 10^2 to 10^7 cm. There is not a radical difference in result produced by the two equations. The width of the fissure is the most important parameter: for White (1977a) and Palmer (1984) penetration distances are proportional to the cube of half-width; for Weyl (1958) and Dreybrodt (1981), to the square. Other parameters are proportional only to the first power.

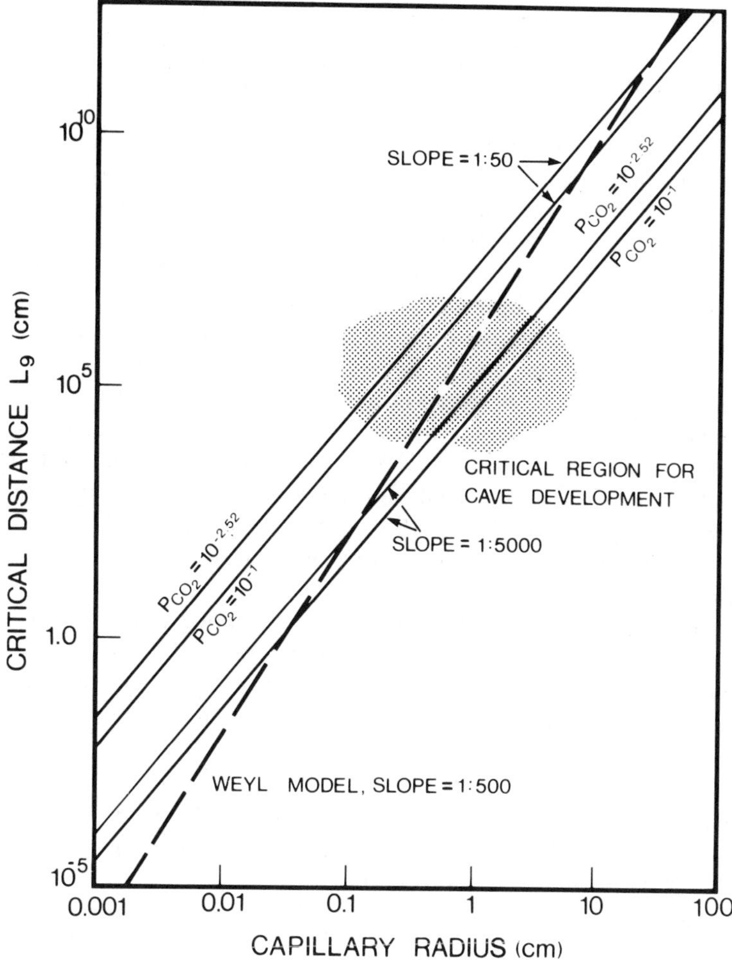

Figure 3.18 L_9 penetration distances with respect to calcite for conduits of specified radius, showing the model of Weyl (1958) and those of White (1977). Hydraulic gradients and equilibrium P_{CO_2} values are also specified. From White (1977a) by permission.

Reading Figure 3.18, water equilibrated to $P_{CO_2} = 10^{-2.5}$ (ten times the standard atmosphere) and entering a 10 μm crack at a slope of 1 : 50 will be 90% saturated after approximately 0.5 mm. Water entering an especially wide crack (100 μm) at the same P_{CO_2} and slope penetrates about 50 cm. It is clear that most karst groundwater nets must be initiated by very slow dissolution at *SI* between −0.15 and 0.00 unless (as above) mixing corrosion effects, etc. are invoked. If diffusion is truly rate-controlling (Eqn 3.81, Weyl) penetration distances are increased by adopting the D* convention (Eqns 3.76 & 3.77).

When laminar flow breaks down to turbulent flow the dissolution rate increases quite rapidly for SI_c less than −1.0 (Fig. 3.17). Depending on hydraulic gradient and roughness this will occur between approximately $r = 3$ mm and $r = 8$ mm. Solutional flowpaths enlarged to these radii potentially can expand rapidly if within the L_9 penetration distance. From Figure 3.18 that distance for $r = 3$, $P_{CO_2} = 10^{-2.5}$, slope = 1 : 50, is about 10 km, the scale appropriate to major caves and karst aquifers. To state it another way, if the flowpath between potential sink points and a spring is 10 km long, that path will be enlarged at the very slow rates of region 3 until it is some 3–8 mm in radius or half-width throughout its length. Then it escapes into the realm of vigorous reaction kinetics (region 2 or even region 1).

The time that elapses between initiation of dissolution and this escape has been investigated by several authors. For a flowpath that is 1.0 km long, and assuming reasonable ranges of slope, P_{CO_2}, etc., White (unpublished) has estimated 3000–5000 years from Herman's results (Fig. 3.16). Ford (1980) and Palmer (1984) estimate 10^4–10^5 years. Dreybrodt (1981) obtains 10^3–10^7 years by fitting different mixing corrosion ratios to his development of Equation 3.79.

Because the dissolution rate is very slow above $SI_d = -3$ or -2, L_9 penetration distances and T_9 times will be much longer in the case of dolomite. One comparative picture is as follows: for calcite, the kinetics produce a micro-conduit that is broad at the inlet and tapers sharply downstream. Its rate of broadening and the rate of advance of the tapered tip (the solution front) are quite rapid. For dolomite, the conduit is narrower at the inlet and has a very low taper. Rate of broadening and advance of the front are much slower. White (1984) suggests that this explains why fissured dolomites are often excellent aquifers; the long penetration distances permit a little dissolution everywhere within them. In limestone aquifers solvent capacity is exhausted at the inlet and in enlarging a few selected caves.

The solution kinetics of gypsum, anhydrite and salt

There minerals dissolve by molecular dissociation alone. For gypsum and salt the surface reaction rates are effectively instantaneous, so that the dissolution rate is controlled by simple diffusion i.e. is proportional to the thickness of the boundary layer. Dissolution rates are described by a first order equation

$$\frac{dC}{dt} = k\frac{A}{V}(C_s - C) \tag{3.83}$$

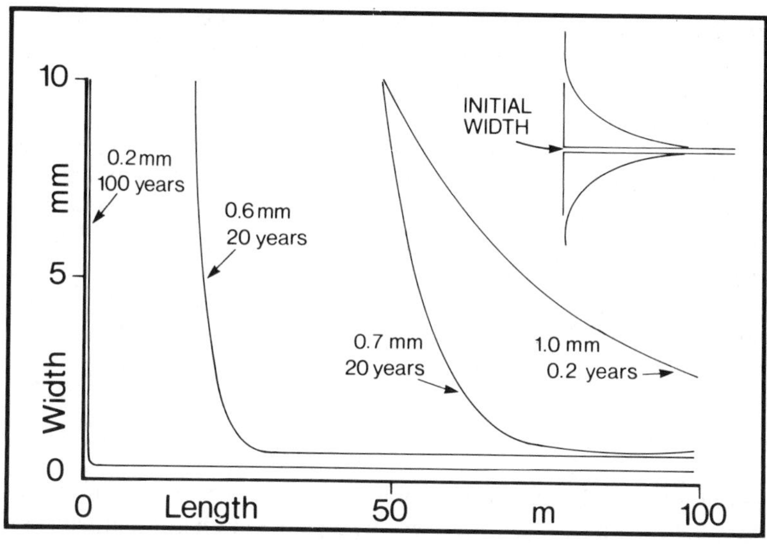

Figure 3.19 Penetration distances or progress of the dissolution front for ~ L_{99} in massive gypsum, calculated for initial fissure widths ranging 0.21–1.0 mm. Time elapsed since initiation is in years. The hydraulic gradient is 0.2 and water temperature is 10°C. *Inset*: the form of the dissolutional taper into the fissure that is obtained with theoretical calculations such as these. Adapted from James & Lupton (1978) with permission.

where k is a rate constant that varies with the mineral, its purity and surface roughness, and with velocity of flow (Pechorkin *et al.* 1982).

The dissolution rate of anhydrite is described by a second order equation

$$\frac{dC}{dt} = k\frac{A}{V}(C_s - C)^2 \tag{3.84}$$

This suggests that a surface reaction is also partly rate-controlling (James & Lupton 1978); this is probably hydration.

The potential significance of diffusion control and, in turn, its dependence upon velocity of flow is indicated dramatically in Figure 3.19. This is a set of penetration distances for fissures of differing initial width, calculated by James & Lupton (1978) from their experimental data for a massive gypsum.

Berner (1978) compiled data for 18 different minerals. He suggested that those examples with solubilities less than 10^{-4} mol l^{-1} in standard conditions are all surface reaction controlled. Calcite is included in this group. All with solubilities greater than ~ 2×10^{-3} mol l^{-1} are transport controlled. They include gypsum, salt and most of the exotic carbonates, sulphates and chlorides occasionally deposited in caves. Minerals falling between the two solubility ranges display mixed kinetic control.

We express reservations concerning the opinion that the rate control for calcite in solutions far from equilibrium is hydration of CO_2 at the crystal surface. These reservations arise from considering dissolutional scallops and rillenkarren, small-scale erosional forms common to limestone, gypsum, salt and other materials. They are discussed in sections 7.11 and 9.2 respectively. In both cases, highly specific forms are created because boundary layers are breached to expose the solid surface to an aggressive bulk fluid. It is the elimination of boundary layer diffusion control that permits scallops on ice, snow, salt and gypsum and rillenkarren on salt and gypsum. It appears that this effect must also apply to limestone and dolomite, which can host the same forms in the same hydrodynamic conditions.

The precipitation of calcite

By definition a solution at thermodynamic equilibrium is in the lowest energy state. An energy barrier (the nucleation potential) must be overcome, therefore, before precipitation can begin. As a consequence precipitation does not commence as soon as a solution is carried above $SI_c = 0.00$.

Homogeneous precipitation can occur in the bulk liquid, producing a cloud of crystallites. The required nucleation potential is high and SI_c values about +1.5 are demanded before there is any significant production. These are readily achieved in some lagoonal seawaters, etc. but rarely in fresh waters. Homogeneous precipitation is of little significance in karst studies.

The nucleation potential for precipitation onto existing solid surfaces (heterogeneous precipitation) is much less. It is least where there is the best lattice match between substrate and nucleus. As a consequence significant precipitation may begin at $SI_c = +0.30$ in optimum circumstances. Any solid may serve as a host, for example a speck of silicate dust supported on the meniscus of a saturated pool can be the nucleus of a veneer of 'calcite ice'. The fastest precipitation occurs onto earlier calcite because of the lattice match.

Homogeneous nucleation begins with the formation of a cluster of molecules in the liquid. This is enlarged by the regular or irregular accretion of further molecules to form a micro-crystal (or 'crystallite'). Crystallites 'ripen' by regular accretion to become crystals that settle out of the liquid.

In heterogeneous nucleation ions, molecules and ion pairs diffuse into the absorption layer where they are adsorbed directly at steps, kinks and other dislocations in the lattice. Slow adsorption builds the most regular crystals. If the degree of supersaturation is increased by processes such as evaporation, there may be a simultaneous formation of clusters, crystallites and even small crystals within the boundary layers. These then attach to the substrate to create a solid that is much less regularly ordered and is of higher porosity. It lacks the lustre of crystalline surfaces and may crumble in the fingers; it is frequently described as 'earthy'. This is the calcareous *tufa* found at springs or in cave mouths, sites where there may be vigorous evaporation, although biological processes are also involved (see section 9.11).

The calcite lattice is quite robust, able to absorb a wide variety of foreign ions and molecules without becoming totally disordered. For example, the very large uranyl ions are absorbed from the soluble complex, $UO_2(CO_3)_2^{4+}$. Recent research has established that humic and fulvic acids (long chain molecules with atomic weights up to 30 000) can be taken up in cave calcite; they furnish much of its variety of colours. However, as in the case of dissolution, other substances inhibit precipitation. They are often termed 'poisons' and include some organic matter, phosphates and a variety of trace metals. Most important is Mg^{2+} which strongly inhibits calcite when present in high molar ratio. It can not absorb onto the aragonite lattice to inhibit that polymorph. This is why much aragonite is precipitated from seawater, which is enriched in Mg^{2+}. Some freshwater aragonite precipitation occurs where first there is deposition of gypsum to deplete Ca^{2+} and so increase the molar proportion of Mg^{2+} (e.g. Harmon *et al.* 1983).

A comprehensive equation for calcite crystal growth is given by Nancollas & Reddy (1971)

$$\lambda = -k_G\left(m_{Ca^{2+}} + m_{CO_3^{2-}} - \frac{k_c}{\gamma^2}\right) \tag{3.85}$$

where k_G is a crystal growth rate constant and $\gamma^2 = \gamma_{Ca^{2+}} \cdot \gamma_{CO_3^{2-}}$. Assuming that the rate of precipitation is surface reaction controlled (dissociation of HCO_3^- at the interface), Plummer *et al.* (1979) obtained good agreement with results predicted by the equation at high P_{CO_2}. At low P_{CO_2} (as in most karst situations) the calculated rates were 10–20 times faster than observed experimental rates. For low P_{CO_2} Stumm & Morgan (1980) use a simpler diffusion equation

$$\frac{dC}{dt} = -k\frac{A}{V}(C - C_s)^n \tag{3.86}$$

A = surface area, and $n = 2$ for $CaCO_3$.

Theoretical precipitation rates calculated from such equations, using values of k or k_G from different experiments, range from 10^{-5} mmol cm^{-2} s^{-1} to 10^{-2} mmol cm^{-2} s^{-1} (cited by Plummer *et al.* 1979). Growth rates (thickening rates) for flowstone sheets of crystalline calcite deposited in caves range at least 0.1–1 m ka^{-1} and tufa growth rates can be much faster (see section 8.3).

4 Distribution and rate of karst denudation

4.1 Global variations in the solution of carbonate terrains

The power of rain water to dissolve karst rocks has been appreciated for almost 200 years, as is evident from James Hutton's (1795) comments on solution forms on limestones in the Alps. Estimates of natural solution rates have been made since at least 1854, when by some extra-ordinary computation Bischof asserted that the annual dissolved $CaCO_3$ load of the River Rhine was equivalent to '332 539 millions of oysters of the usual size'! Credit for being the pioneers of modern methods for estimating solution must go to Spring & Prost (1883) in Belgium and Ewing (1885) in the USA. After 366 days of sampling, Spring & Prost determined the annual solute load of the R. Meuse at Liege to be 1 081 844 tonnes, whereas Ewing calculated limestone denudation in a river basin in Pennsylvania to be equivalent to 1 ft in 9 000 years (34 mm ka^{-1}). These figures are of the same order as more recent estimates in the regions concerned.

About 30 years ago, the French geomorphologist Jean Corbel made a great impact on conventional thinking when he published results derived from analyses of thousands of field samples. He concluded (1) that cold high mountains provide the most favourable environment for limestone solution and (2) that there is a factor of ten in the difference of solution rates between cold and hot regions for a given annual rainfall, hot regions having the lowest karst solution rates (Corbel 1959). He inferred from this that the principal control of solution is temperature, probably operating through its inverse effect on the solubility of carbon dioxide (see section 3.4).

These conclusions really rocked the boat. They ran contrary to both morphological evidence and conventional wisdom – that weathering processes in general are most rapid in hot humid conditions. Corbel's findings were strongly disputed and so stimulated numerous other process studies on karst.

The published results of some 200 later investigations were synthesized by Priesnitz (1974), who found a significant, positive relationship (r=0.74) between the rate of limestone solution and the amount of runoff. This confirmed conclusions by Pulina (1971), who identified a linear relationship between chemical denudation and precipitation, and Gams (1972), who showed Yugoslav solution denudation rates to be dependent on runoff.

Smith & Atkinson (1976) used two data sets: 134 estimates of the rate of solutional denudation in different regions of the world and 231 reports on the mean hardness of spring and river waters. They confirmed the above conclusions and added more detail and critical interpretation. Runoff probably accounts for 50–77% of the variation in total solution rates, the remaining variation being mainly accounted for by solute concentration. Smith & Atkinson supported Corbel's hypothesis in that they found the greatest solutional denudation to occur in alpine and cold temperate regions, but they stressed that the climatic effect is much less marked than Corbel claimed. For a given runoff, it appears to involve an increase not of ten times between the tropics and the cool temperate alpine zones, but only 36%. The greatest limetone solution in the world occurs where it is wettest. Hence precipitation rather than temperature is the principal control. In five areas of New Britain in Papua New Guinea, where rainfalls lie in the range 5700–12 000 mm a^{-1}, Maire (1981b) calculates solutional denudation rates of 270–760 mm ka^{-1}.

Solution rates and denudation rates

The term 'solution rate' as widely used in the geomorphic literature is ambiguous. The solubilities of various minerals in water are listed in Table 3.1, in which the values refer to the maximum concentration that can be achieved under a particular set of conditions given unlimited time. They represent an equilibrium state but tell us nothing about the rate at which that equilibrium was reached. This is determined by the solution kinetics or reaction dynamics and can be visualized as the slope of a curve plotting solute concentration against time (Fig. 3.2). This is the meaning of 'solution rate' understood by chemists and is to be preferred to the looser usage by geomorphologists, who also take it to mean the annual rate of chemical denudation, i.e. the *solutional denudation rate*. This latter must also be distinguished from the *karst denudation rate*, which is the sum of both chemical and mechanical erosion processes. However, since in practice it is much easier to estimate chemical than mechanical erosion, only solutional denudation rates are usually considered. This ignoring of an often significant part of the picture must be conceded as a major deficiency in karst process research.

Although karst erosion occurs underground as well as on slopes, by convention and for ease of comparison denudation rates are quoted as an equivalent thickness of rock removed per unit time across a horizontal surface. Millimetres per thousand years (mm ka^{-1}) are most commonly used. 1 mm ka^{-1} is equivalent to 1 m^3 km^{-2} a^{-1}. However, to generalize a mass transfer rate calculated for one year to an equivalent rate per 1000 years is only justified if there has been no significant change in environmental conditions in that period. This is an increasingly untenable assumption given the intensity of human impact on the ecosystem, although under natural conditions the whole of postglacial time (10 000 years or more) can be considered metastable (Ford & Drake 1982).

Autogenic, allogenic and mixed denudation systems

In interpreting results of karst erosion studies, it is important to know where the water and its load has come from. An *autogenic* (or autochthonous) system is one composed entirely of karst rocks and derives its water only from meteoric precipitation falling onto them (Fig. 4.1A). By contrast, a

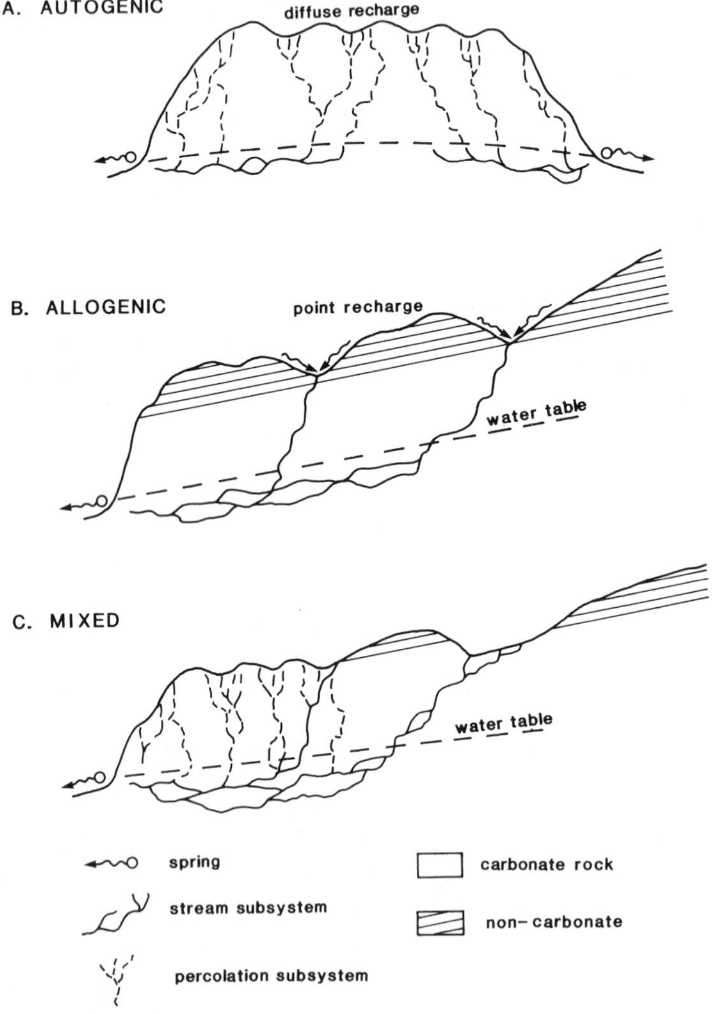

Figure 4.1 Three karst denudation systems: A autogenic and B allogenic are end members with C the mixed autogenic–allogenic intermediate case being the most common.

Table 4.1 Sources of solute load (Ca+Mg) in the Riwaka basin, New Zealand. From Williams & Dowling (1979).

Source of Ca+Mg load	Tonnes per annum	Percentage of Karst solution	Percentage of Total solute load
From solution of marble by autogenic waters	1709*	79.5*	68
From solution of marble by allogenic waters	440**	20.5**	17.5
NET KARST SOLUTION	2149†	100	85.5
From solution of non-karst rocks	250	–	9.9
Introduced by precipitation	116	–	4.6
TOTAL SOLUTE LOAD	2515	–	100

* Mainly surface lowering
** Mainly cave conduit development
† Equivalent to marble removal rate of 100 ± 24 m^3 km^{-2} a^{-1}.

purely *allogenic* (or allochthonous) system derives its water entirely from that running off a neighbouring non-karst catchment area. In practice, many if not most karst systems have a mixture of autogenic and allogenic components (Fig. 4.1C). Jakucs (1977) discusses this in detail.

Allogenic waters flowing into a karst area represent an import of energy capable of both chemical and mechanical work. Thus at the output boundary the autogenic and allogenic components of denudation must be separated if sense is to be made of the landform development and if valid comparisons of denudation rates are to be made with other areas. Clearly a

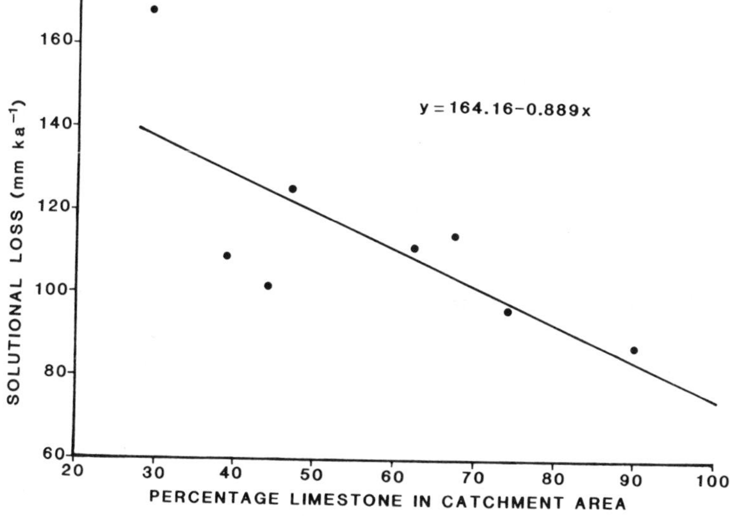

Figure 4.2 Relationship of solutional denudation rates to the percentage of limestone in a catchment. Data for the southern Pennine Hills, England. From Pitty (1968a).

small karst area with a massive throughput of allogenic water (such as Mulu in Borneo) will experience very much more erosion than it would if it operated only as an autogenic system. In the Riwaka basin, New Zealand, autogenic solution is about 79 mm ka^{-1} but karst rocks cover only 46.6% of the catchment. The large allogenic input increases net solution of this karst by over 20% (Table 4.1). Meaningful comparisons of solution denudation rates from different areas thus can only be made when the relative proportions of carbonate rock are taken into account. From an analysis of data for 18 basins where this proportion was known, Smith & Atkinson (1976) concluded that estimates of denudation from mean annual runoff information yielded values roughly twice as great (for the same runoff) as those where the proportion of limestone was not taken into account. Pitty's (1968a) results are particularly illuminating (Fig. 4.2).

Gross and net solution

Published estimates of karst denudation are not always consistent in what they represent. The reason for this is that solute load may be derived from various sources, some non-karstic, and some re-precipitation of previously dissolved materials may have occurred upstream of the sampling site.

The total solute discharge is the product of river discharge Q at the outflow of the basin and the corresponding solute concentration C. Each has a value that comprises the relative contributions made by its different components. Thus

$$Q = (P - E)_{\text{autogenic}} + (P - E)_{\text{allogenic}} \pm \Delta S$$

where P is precipitation, E is evapotranspiration and ΔS is change in storage. $(P - E)_{\text{autogenic}}$ is karst runoff and $(P - E)_{\text{allogenic}}$ is non-karst runoff. Salts introduced by precipitation onto autogenic and allogenic components contribute to solute load at the outflow (see Cryer 1986). Solutes derived from the corrosion of non-karst rocks by allogenic runoff also contribute. Therefore

> Solute Load = (autogenic + allogenic karst corrosion − karst deposition) + allogenic non karst corrosion + solute load from rain and snowfall.

Gross karst solution comprises autogenic plus allogenic solution; net karst solution is autogenic plus allogenic solution less karst deposition. To gain a realistic appraisal of the rate of transformation of relief, estimates are required of gross karst solution. Where re-precipitation is not important, as in most cold climates, net solution approximates the gross. But in tropical and warm temperate regimes speleothem and tufa deposition can be profuse; so net solution can significantly underestimate gross rates. Solutional denudation will be over estimated where the proportion of solutes introduced by rainfall and non-karst rocks is not subtracted. In arid regions the introduction of carbonate in dew and dust by airborne fallout may also complicate the picture (Yaalon & Ganor 1968, Gerson 1976).

Factors influencing global variations in solution

In the presence of water, there is no dynamic threshold of calcite or dolomite solution (Ford 1980). Solutional denudation rates have also been found to depend linearly on runoff. Hence it might be supposed that because runoff values form a world-wide continuum there can be no regional discontinuities in solutional denudation. This would be so if variations in solute concentrations are minimal. But this is not always the case, as illustrated for instance by the differing $CaCO_3$ values of groundwaters from various limestone lithologies in southern Britain (e.g. Paterson 1979). The underlying reasons for variations in solute concentrations need closer scrutiny.

In section 3.6 we considered factors that boost or depress the solution of carbonate minerals and stressed the importance of coincident and sequential (or 'open' and 'closed') systems. Important variables superimposed upon system conditions that influence the per cent saturation of percolating groundwaters were identified as occurrence of carbonate in the soil, rooting depth, porosity, CO_2 concentration and availability, and the residence time of the water. The models presented in Figures 3.10 and 3.11 were concluded to have great power to explain why dissolved carbonate concentrations are often much higher in some regions than others, e.g. coincident solution in carbonate-rich glacial tills in cold northern regions explains why very high carbonate values are recorded in groundwaters there despite relatively low P_{CO_2} values sometimes found in the soils. The models are believed to incorporate the significant factors in the $CaCO_3$–H_2O–CO_2 system that account for spatial variations in calcium or carbonate in groundwaters in karst areas, with the possible exception of ground–air CO_2 in the atmosphere of the unsaturated zone (see section 3.6).

The 'porosity' variable in the models subsumes many lithological, mineralogical and structural characteristics of the rock that affect the surface area exposed to corrosional attack, its solubility, and the flow through time of water. The term requires further refinement and probable redefinition before the influence of different lithologies on ground water carbonate values can be adequately explained. Theoretical chemical considerations bearing on the solubility of rocks with different mineral assemblages were discussed in sections 3.6 and 3.7, and geological factors in sections 2.6, 2.7 and 2.8.

The link between chemical and environmental factors in the solutional denudation of limestones is explored by White (1984), who develops a theoretical expression for their relation

$$D_{max} = \frac{100}{\rho^3 4} \frac{K_C K_1 K_{CO_2}{}^{1/3}}{K_2} P_{CO_2}{}^{1/3} (P-E) \tag{4.1}$$

where D_{max} is the solutional denudation rate (mm ka^{-1}) for the system at equilibrium, P is precipitation and E evapotranspiration (both in mm a^{-1}), ρ is rock density, and K refers to the equilibrium constants cited in Table 3.5. All terms in the equation can be calculated. It combines in a single statement the rock and equilibria factors and important climatic variables of precipitation and temperature. The equation expresses the linear variation of solutional denudation with runoff ($P-E$), and indicates solution to vary with the cube root of P_{CO_2}. Carbon dioxide volume in the atmosphere and soil varies from about 0.03% to 10% or more (section 3.6) but, as White points out, because of the cube root dependence, a factor of 100 P_{CO_2} only admits a factor of 5 in the denudation rate. The complex effects of temperature on solutional denudation rates are incorporated in the equilibrium constants. White concludes that denudation increases by about

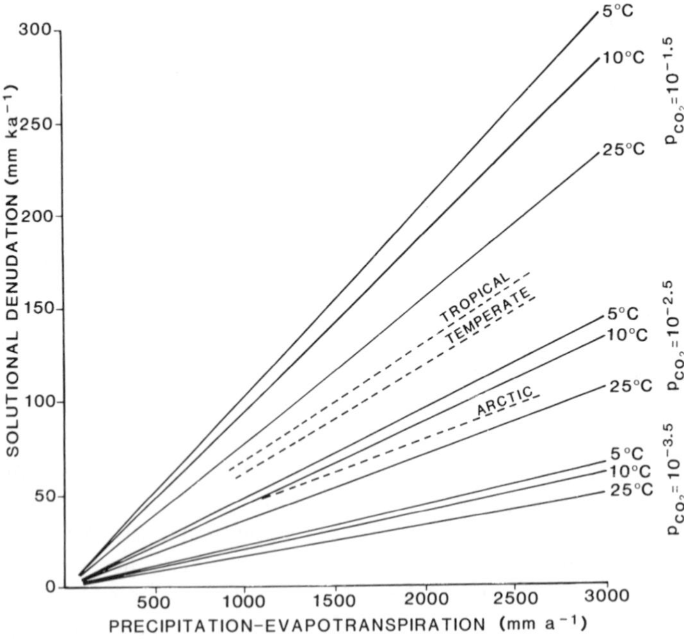

Figure 4.3 Theoretical relationships of solutional denudation of limestone to water surplus and CO_2 availability under coincident conditions from White (1984) with empirical relationships derived by Smith & Atkinson (1976) superimposed as dashed lines.

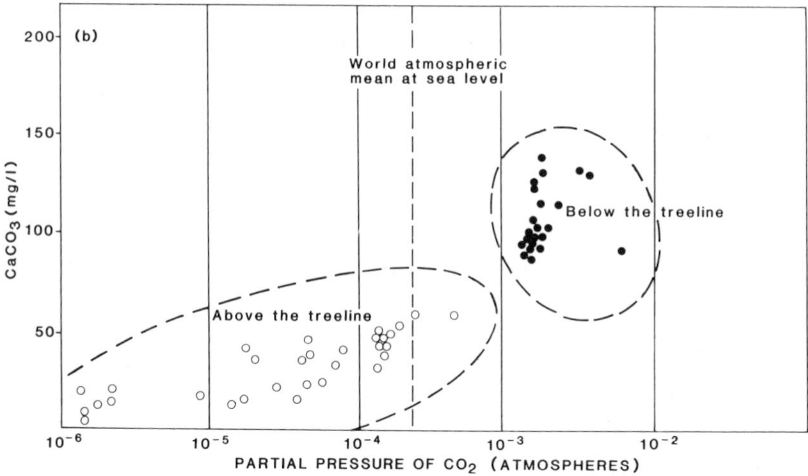

Figure 4.4 Hardness of waters from the Rocky Mts plotted against P_{CO_2} in the gas phase in which they were equilibriated. From Atkinson & Smith (1976) after Ford (1971a).

30% as mean temperatures decrease from 25 to 5°C, which broadly concurs with Smith & Atkinson's (1976) empirical findings. Temperature is thus the least important of the climatic variables in the equation and in practice is often more than offset by the influence of other factors. Figure 4.3 illustrates the theoretical relationships embodied in Equation 4.2, with empirical trends derived from field studies being superimposed on them.

The difference between the theoretical and observed trends of Figure 4.3 can be explained by several factors termed 'boosters and depressants' in section 3.6. Firstly, as White points out, the theoretical curves assume that the water and carbonate rocks are in equilibrium whereas in reality most karst waters are undersaturated. More important, the model assumes ideal coincident or open conditions, which often do not occur.

For a given water surplus and under coincident system and autogenic conditions, the effect of increasing latitude on solutional denudation in theory is to increase it, provided waters do not freeze. Increasing altitude will have a similar effect because P_{CO_2} only decreases marginally in the open atmosphere with height above sea level (Fig. 3.7). However, the tree-line provides an important threshold for soil CO_2 values that in turn affect calcium concentrations in groundwaters (Fig. 4.4). Thus for a given runoff, solutional denudation tends to be greater below the tree-line than above it.

4.2 Measurement and calculation of solutional denudation rates

Objectives

Erosion rates are measured: (1) to obtain a generalized value for the overall rate of denudation or transformation of relief; (2) to compare denudation rates in contrasting environments and by different processes; (3) to gain more understanding of the evolution of landforms; and (4) to understand the processes themselves.

Where the principal objective is to obtain a number to compare with other karsts, perhaps in other climatic zones, then an estimate of the autogenic solutional denudation rate is required (for reasons discussed in the previous section). Where erosion by karst solution is being compared to erosion by mechanical processes, autogenic rates are also needed unless the role of allogenic water is being explicitly assessed.

The study of denudation in karst is best undertaken in a systems context, the karst drainage basin being treated as an open system with a definable boundary and identifiable inputs, throughputs and output. Secondary objectives then emerge as being to define the system boundary (usually the watershed) and to measure the flux of solvent and solutes.

Hundreds of studies of solutional denudation have been completed since about 1960. A major shortcoming of much of the work is that autogenic rates have often not been distinguished from mixed autogenic–allogenic rates so that there is still no unequivocal answer to the question posed long ago by climatic geomorphologists: in which climatic zone does karst evolve most rapidly? Wet regions are certainly more important than dry ones, but the field data are too 'noisy' to confirm with confidence what theory suggests (c.f. Fig. 4.3). Thus an important objective of solution studies over the next few years will be to compare autogenic rates in different environments with a view to clarifying this issue. At the same time it is also necessary to recognize that high rates of solutional denudation do not necessarily imply rapid karstification.

The value of denudational data is much enhanced if information is available on where the erosion has taken place. The karst denudation system is three-dimensional, and if an understanding of the transformation of relief is to be acquired, the relative importance of sites contributing to the total denudation must be identified. Hence, another important objective is to describe quantitatively the spatial distribution of corrosion throughout the system. To achieve this requires a more sophisticated experimental design than is needed merely to estimate the total solutional denudation.

Experimental design and field installations

The hydrogeochemical system of karst can be conceptualized as consisting of autogenic and allogenic systems, each with percolation and stream subsystems (Fig. 4.1). Since the solute load is the product of discharge and concentration both must be carefully measured. The most important measurements to characterize the hydrological system are rainfall input and streamflow output.

Topographic complexity determines the number of rain gauges that are required to obtain an acceptable estimate of basin rainfall. Thiessen polygons or the isohyetal method will probably be used to calculate rainfall input (Dunne & Leopold 1978). The representativeness of the rainfall record may be assessed by comparison with long-term records from a nearby meteorological station. Under the very best conditions, the mean annual rainfall estimate is likely to have an error of at least 5%. Point measurements for individual storms can be in error up to 30%.

In many karst basins a convenient outfall point is a resurgence. The measurement of its discharge is usually made in a straight channel reach a short distance downstream from the spring. The hydrological literature (e.g. Ward 1975) provides considerable guidance on site selection criteria, appropriate measurement techniques and the associated errors. Ideally a solid rock channel is desirable to avoid cross-section change and to minimize leakage. Where a watertight weir is used (Fig. 4.5), discharge values of small streams can be accurate to 1%. Correlation of the outflow data with longer term records at nearby hydrological stations will permit assessment of the representativeness of the karst data set.

The need for secondary discharge stations depends on the objectives of the research and the nature of the catchment. Estimation of mean annual solutional denudation from an autogenic basin requires only a master discharge measurement site at the outfall. But in a mixed autogenic–

Figure 4.5 *Left* A V-notch weir measuring the flow of the autogenic Cymru stream in Mangapohue Cave, New Zealand. Photo by J. Gunn. *Right* Rectangular section weir at Puding experimental karst basin, Guizhou Province, China.

allogenic basin, secondary sites are required to gauge the allogenic inputs. Where the spatial distribution of corrosion is being assessed, secondary sites are also desirable, but not always practicable if the system is entirely subterranean.

At the principal flow measuring site discharge records should be continuous. Equipment for the continuous measurement of water quality at present is less reliable, but the technology is advancing very rapidly. Continuous recording of electrical conductivity is particularly valuable when it correlates highly with Ca^{2+} concentration (see section 3.5).

A minimum requirement at the outflow site is to establish the relationship between solute concentration and discharge. This is often achieved satisfactorily by constructing a rating curve, although the relationship of solute concentration to discharge frequently shows much scatter due in particular to pulses of hard water emerging at the beginning of storms (see section 6.6). Errors in the rating curve also arise from the precision of the chemical techniques used to determine Ca^{2+} etc. and from the representativeness of conditions when samples were taken, which must include a very wide range of flows in every season. The standard error of estimate on the regression assumes that the Ca^{2+} and discharge values are accurate, which may not be the case; thus, a large number of samples is required to reduce the problem. Sampling at regular intervals (e.g. weekly) seldom obtains the information needed, except over a very long period, because most flows encountered will be below mean discharge. The most efficient technique is event sampling. Having constructed a statistically acceptable rating curve, one must then be alert to the probability that it will not be equally valid for every season (Hellden 1973) and to the possibility that the general relationship may change from one year to the next, as Douglas (1968) showed for the Green River at Munford, Kentucky.

Calculation of solutional denudation

The best known formula for calculating solutional denudation is that by Corbel (1956, 1959)

$$X = \frac{4ET}{100} \tag{4.2}$$

where X is the value of limestone solution ($m^3\ km^{-2}\ a^{-1}$ or mm/1000a), E is runoff (dm) and T is the average $CaCO_3$ content of the water ($mg\ l^{-1}$). This formula is important because it forms the basis of many of the earlier published figures of solutional denudation and its use gives an insight into the methods used. The equation has been criticized for (1) assuming all carbonate rocks to have a density 2.5 (density can range from 1.5 to 2.9), (2) ignoring $MgCO_3$ (although it can be accommodated) and solute accession from rainfall, (3) ignoring the possibility that sulphate rocks contribute to Ca^{2+}, and (4) for generalizing carbonate hardness to the mean value, thus overlooking its possible variation with flow.

Drake & Ford (1973) point out that a strict formulation of the solute discharge rate D could be expressed as

$$D = \frac{\int CQ\,dt}{\int dt} \tag{4.3}$$

where $C = C(t)$ is the solute concentration, $Q = Q(t)$ is the instantaneous runoff and t is time. This may be approximated by

$$D = \frac{\sum^{m} C_i\, Q_i}{m} \tag{4.4}$$

where i refers to equal time intervals in which C and Q may be considered constant, and m is the number of such periods. Since the sum of deviations of concentration and discharge from the mean are usually negative, Corbel's formula generally overestimates solutional denudation. Drake & Ford (1973) demonstrated a 31% overestimate for the Athabasca and North Saskatchewan Rivers of Canada by comparing results from Equations 4.2 and 4.4. They also pointed out that assuming T to be attributable to carbonate solution overlooks the important contribution of sulphates in the Athabasca basin where about 40% of the mean solutional denudation is derived from sulphate rocks that do not outcrop. Schmidt (1979) demonstrated a 24% overestimate for a German karst area.

Very often karst rocks occupy only part of the basin being considered. Where the fraction is $\frac{1}{n}$, Corbel incorporated it into his equation thus

$$X = \frac{4ETn}{100} \tag{4.5}$$

However, because it is assumed that T is entirely derived from the carbonate rocks, another source of error will be introduced if some of the solute load has an allogenic origin. Corrosion by autogenic and allogenic waters cannot be separated.

Modifications to Corbel's method by Williams (1968), Gams (1967), Pulina (1971) and others resulted only in minor improvements. Considerable increases in accuracy are achieved by applying mass flux rating curves

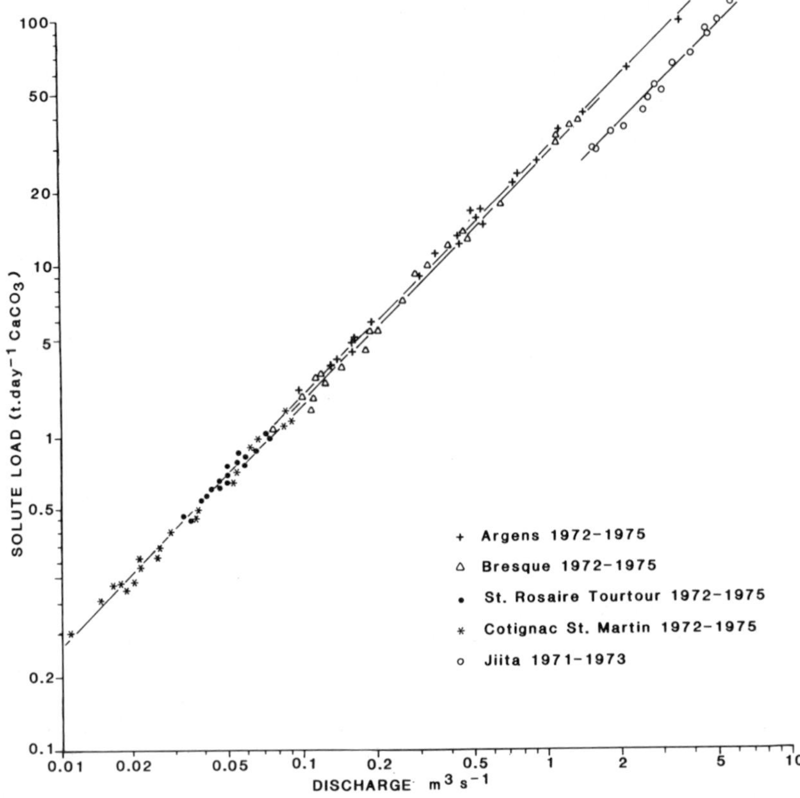

Figure 4.6 Mass flux rating curves showing the relationships of equivalent calcium carbonate loads to discharge in karst springs in lower Provence, France, and Lebanon. From Julian *et al.* (1978). The overlapping curves for the four French sites indicate the operation of a regional carbonate solution regime.

(Fig. 4.6) to the flow duration curve (Williams 1970, Smith & Newson 1974) or better still to the outflow hydrograph (Drake & Ford 1973, Gunn 1981a). As automatic equipment improves in reliability, it is becoming feasible to apply instantaneous determinations of solute concentration to the corresponding instantaneous discharge. Summation over the year then yields the solute discharge rate, from which deductions can be made for non-karst inputs to give the net solutional denudation rate. The hydrological representativeness of a given sample year can be assessed by correlating runoff with rainfall data (Fig. 4.7), the latter usually having by far the longer record.

Gunn (1981a) found that applying mass flux rating curves to hourly discharge data and summing over a year yielded values 4% lower than those estimated from an equation using mean solute concentration (the equivalent

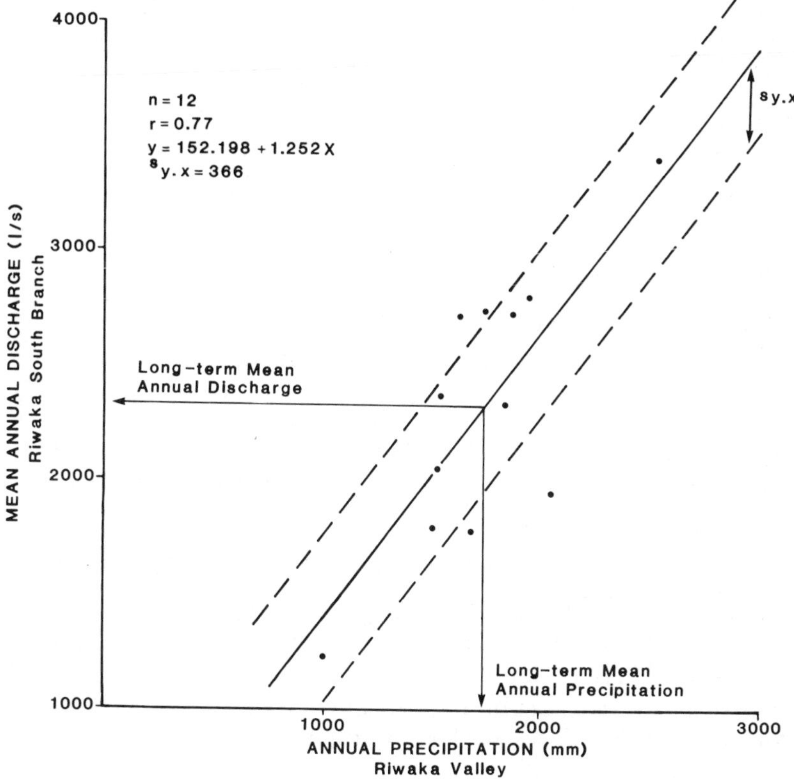

Figure 4.7 Assessment of medium-term mean annual discharge by correlation of short-term (12 years) streamflow records with rainfall data. Mean discharge is predicted from the average rainfall over the 25 year meteorological record. From Williams & Dowling (1979).

of T in Equation 4.2). This is because changes in solute concentration with discharge are not adequately taken into account in the more traditional methods which, incidentally, can also produce results differing by up to about 9% from the same data.

Errors in estimating solutional denudation can arise from many sources (Williams & Dowling 1979). Table 4.2 illustrates potential errors in a particularly carefully conducted study by Gunn (1981a). His best estimate of net autogenic limestone solution was 69 $m^3\ km^{-2}\ a^{-1}$, with qualitative assessment of potential errors indicating that the true value could lie between 61 and 88. The numerous sources of error that can occur in applying the Corbel technique suggest that published solutional denudation values derived by that method may be up to 100% or more in error. Given the highly variable quality of the data that were available to Smith &

Table 4.2 Data used in computing rates of solutional erosion in the Cymru basin, New Zealand. From Gunn (1981a).

Parameter	Measured value	Potential error (%)	Probable max. value	Probable min. value
Basin area (m^2)	95 350.0	+5, −2	100 117.5	93 443.0
Precipitation (mm) (water year total)	2366.0	±7.5	2543.5	2188.8
Discharge ($l \times 10^3$) (water year total)	155 454.7	±5	163 227.4	147 682.0
Mean Ca concentration (mg l^{-1})	48.13	±3	49.57	46.69
Mean Mg concentration (mg l^{-1})	1.26	±3	1.30	1.22
Mean Ca concentration in rain (mg l^{-1})	1.48		1.48	0.53
Mean Mg concentration in rain (mg l^{-1})	0.32		0.56	0.32
Limestone density (g cm^{-3})	2.66		2.66	2.50

Atkinson (1976) for their synthesis, it is hardly surprising that Gunn (1981b) finds their equations inadequate for predicting solutional denudation.

Limestone tablets and micro-erosion measurements

Alternative methods to assess limestone corrosion rates include (a) measurement of weight loss in standard rock tablets and (b) direct measurement of surface lowering by a micrometer gauge. Gams (1981) summarizes an international project using limestone tablets of standard size and lithology which was designed to determine:

(1) the effect of different climates by means of suspending the tablets in the open air;
(2) the rate of solution of bare limestone by laying the tablets on either rock or grass;
(3) the corrosion rate in soil by burying the tablets at various depths; and
(4) the variability of corrosion within a single karst area by placing tablets in a variety of sites.

More than 1500 were distributed around the world. Preliminary results from eight countries are reported by Gams (1981) and a careful analysis of about five years' data from Wisconsin is presented by Day (1984a). Data are recorded as weight loss after different periods of exposure at the various sites. Their conclusions indicate that:

(a) Weight loss in tablets placed in the soil is generally greater than that from tablets in the air or on the surface and seems directly dependent on water surplus ($P - E$) rather than on temperature.
(b) Solution rates show a distinct climatic control, with generally higher rates being recorded in the humid tropics.

(c) The effect of varying lithology is often greater than that of varying climate.
(d) Mechanical erosion may be significant even in areas with high solution rates.
(e) At higher elevations in Slovenia and the French Alps solution of tablets on the surface is less than at lower elevations.
(f) In arid climates weight loss in the air is often higher than on the ground surface, although the opposite appears to hold for humid climates.
(g) Weight loss at any site in Slovenia is considerably less than that expected from solution rates calculated from basin runoff and solute data.

This last point is reinforced by the findings of Crowther (1983), who concluded for a site in West Malaysia that rates derived from tablet weight loss are one to two orders of magnitude less than those calculated from water hardness and runoff data. Solution tablet data must therefore be interpreted cautiously. Results tend to confirm our understanding of processes rather than to add to it.

The micro-erosion meter (MEM) was developed by High & Hanna (1970) and improved by Coward (1975) and Trudgill *et al.* (1981). Its use and limitations have been assessed by Spate *et al.* (1985), using a traversing version which permits measurement of a large number of points within a triangular area of 12–200 cm^2. The instrument consists of a probe connected to a micrometer gauge and locks precisely into stainless steel studs set into the rock surface. Selected points in this surface can be repeatedly measured for erosional lowering. Results have been claimed accurate to the nearest 10^{-4} mm.

Spate *et al.* (1985) undertook experiments in temperature-controlled rooms to assess three possible sources of error in the use of the traversing MEM. They found (1) different instruments to have differing temperature correction factors; (2) differential expansion and contraction of the rock and rock/stud interface which produced an apparent lowering of the surface with increasing temperature; and (3) a considerable range in erosion of rock by the probe from point to point within a site (up to an order of magnitude). Instrument wear, particularly of the probe tip, is also a problem but no reliable data are available. Table 4.3 presents the corrected results from measurements on Palaeozoic limestone surfaces in New South Wales, Australia (annual rainfall about 950 mm). Since in most cases the error term is shown to be of the same order of magnitude as the natural solutional lowering rate, published results for other sites (Table 4.4) have to be treated with great caution, e.g. the figures for bare rock surfaces in the well known karsts of Clare, Ireland, and Yorkshire, England, are similar to those of New South Wales. However, Spate *et al.* conclude that the values reported

Table 4.3 Corrected limestone erosion rates for Cooleman plain and Yarrangobilly caves, Australia. From Spate *et al.* (1985).

Site	Period of observation (y)	No. of times read	Erosion (mm a^{-1})	Standard deviation	Estimated temperature error (mm)	Probe erosion Rate per reading	Rate (mm a^{-1})	Corrected erosion rate	Estimate of error
FLY	4.46	5	0.0083	0.0139	±0.0072	0.0004	0.0005	0.008	±0.016
OFFY	4.46	5	0.0027	0.0039	±0.0072	0.0009	0.0012	0.002	±0.008
ONY	3.43	4	0.0024	0.0130	±0.0072	0.0020	0.0027	0.000	±0.014
SPY	4.46	5	0.0033	0.0104	±0.0072	0.0004	0.0006	0.004	±0.013
RIPCP	4.42	5	0.0053	0.0078	±0.0072	0.0084	0.0012	0.004	±0.008
OFFCP	4.42	5	0.0077	0.0023	±0.0072	0.00055	0.0008	0.007	±0.008
ONCP	4.42	5	0.0139	0.0142	±0.0072	0.0008	0.0011	0.013	±0.016
LITY	4.46	5	0.0078	0.0100	±0.0072	0.00035	0.0005	0.007	±0.012
LITCP	4.42	5	0.0067	0.0108	±0.0072	0.0009	0.0012	0.006	±0.013
GSY	4.46	4	0.0112	0.0085	±0.0022	0.0010	0.0010	0.010	±0.009
GSCP	4.42	4	0.0208	0.0075	±0.0022	0.0011	0.0011	0.020	±0.008

Table 4.4 Summary of reported micro-erosion meter erosion rates. From Spate *et al.* (1985).

	No. of sites (t = traversing MEM)	No. of measuring points		Mean erosion rate (mm a^{-1})	Range, individual maximum and minimum (mm a^{-1})
Intertidal sites					
Aldabra Atoll, coral limestones	NAt	NA	Sandy sites	1.25	–
(Trudgill 1976)	NAt	NA	Non-sandy sites	1.01	–
Aldabra Atoll, coral limestones	4	–	Ramp top (11 years)	1.53	1.22 –1.73
(Viles & Trudgill 1984)		–			
	5		Ramp edge	1.27	0.09 –2.27

Table 4.4 continued

N. Yorks, UK, shales (Robinson 1977)	106	318		1.0	9.0 –0.00
S. Island, NZ, mudstone and limestone (Kirk 1977)	5	15	Limestone (2 sites)	1.35–0.38	4.68 –0.12
	26	78	Mudstone (5 sites)	2.50–0.64	8.05 –0.09
Co. Clare, Ireland, limestone (Trudgill *et al.* 1981)	1t	14		0.20	0.70 –0.07
Cayman Is, calcerenite (Spencer 1981)	3t	60		0.383	3.024–0.0138
Otways, Victoria, greywacke and siltstone (Gill & Lang 1983)	62	186		0.37	1.8 –0.02
Limestone, non-marine sites					
Co. Clare, Ireland (High 1970)	6	17	Stream passage	0.41	–
Co. Clare, Ireland (High & Hanna 1970)	1	3	Stream sink	0.5	–
	1	3	Spring	0.05	–
Co. Clare, Ireland (Trudgill *et al.* 1981)	NA	NA	Overall surface lowering	0.005	–
Pikhagan, north Norway (Lauritzen 1984a)	NA	NA	Bare marble	0.025	NA
Aldabra Atoll (Trudgill 1976)	1t	NA	Subsoil, deep organic	12.53	20.37 –5.55
	1t	NA	Subsoil, shallow organic	0.107	0.13 –0.09
	1t	NA	Subsoil, brown earth	0.41	0.70 –0.12
			Surface, calcarenite	0.39	
	10t	240	Surface coral fragments	0.09	–
			Surface algal limestone	0.10	–
Aldabra Atoll, coral limestones (Viles & Trudgill 1984)	3	–	Rock basins (12 years)	0.04	0.11 –0.75
	2	–	Surface calcarenite	0.27	0.06 –0.47
Yorks, UK (Trudgill *et al.* 1981)	2	6	Surface exposure (lichens)	0.013	0.41 –0.027
St Paul's Cathedral, London (Sharp *et al.* 1982)	6t	156	Portland Limestone	0.139	1.151–0.000

NA – detailed data not available

for coastal sites are more reliable because the rates there are far in excess of the inland rates and so the errors are not likely to be as significant.

4.3 Solution rates in gypsum and salt

There appear to have been few attempts to obtain basin or other regional solutional denudation rates for gypsum and salt terrains. The development of solution features on salt is so rapid, because of its great solubility, that authors stress the speed of appearance of individual features rather than mean regional rates. For example, multi-level caves of enterable dimensions and profusely decorated with speleothems have developed in two or three centuries where allogenic streams were diverted into salt domes during mining operations in Europe.

A priori, solutional denudation rates in gypsum karsts will be about one order of magnitude more rapid than in limestone karsts because the equilibrium solubility is 2100–2400 mg l^{-1} in the normal temperature range. However, in purely sulphate regions (i.e. no salt is present) it is comparatively rare to measure spring waters that are saturated; most are aggressive with respect to gypsum and anhydrite. This suggests that dissolution often has not attained the equilibrium maximum before the waters are discharged from the karst. For example, in the great interstratal gypsum karst of Podolia, USSR, Klimchouk & Andrejchouk (1986) report that both vadose and phreatic waters are undersaturated where the groundwaters flow with normal vigour. Where downfaulting has produced deeper artesian flow that is very sluggish, waters may be saturated.

Dissolution of gypsum tablets is being investigated in the Podolian karst. Tablets are exposed to (a) surface precipitation, (b) flowing cave streams, and (c) static or semi-static water in caves, at 30 stations. The highest solution rates are recorded in the flowing streams and range from -5 to -70 mg d^{-1} at the different stations. Rates of -1 to $+1$ mg d^{-1} are recorded in the static waters (A.B. Klimchouk, personal communication, 1986).

The Pinero–Odvinsky gypsum karst lies to the east of the White Sea in subarctic Russia (Gorbunova 1977b). The mean annual temperature is close to 0°C and mean annual precipitation ranges from 400 to 500 mm. The sulphate solutional denudation rate is ~ 220 m^3 km^2 a^{-1} (V.I. Tanasijtshuk, personal communication, 1986) of which approximately 50% must be attributed to throughflow of allogenic waters. This is similar to the 0.26 m ka^{-1} estimated for the dry Almeria region of southern Spain by Pulido-Bosch (1986). In the French Alps, Nicod (1976) estimated a rate of more than 1 m ka^{-1} in a region with an annual rainfall exceeding 1670 mm. The roughly 10 : 1 ratio in the solution of gypsum and limestone is thus broadly borne out by the field evidence. However, this is not always

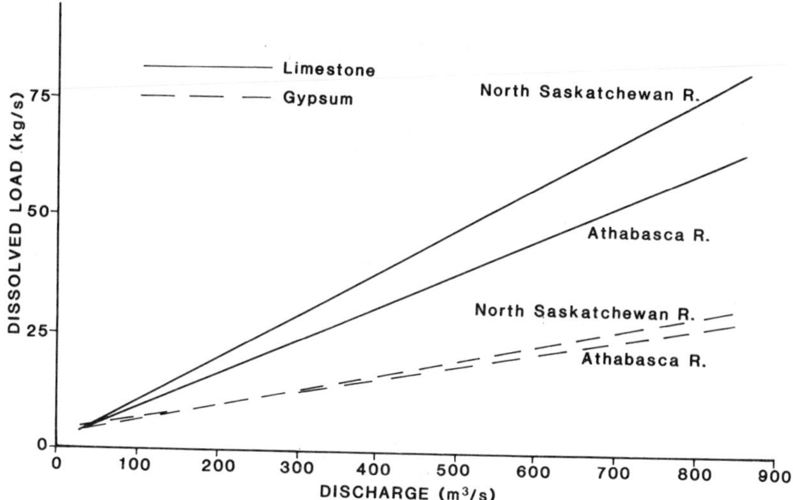

Figure 4.8 Relationship of dissolved limestone and gypsum transport to discharge in the Athabasca and North Saskatchewan rivers, Canada. From Drake & Ford (1976). The lower gypsum loads are due to their smaller outcrop areas.

immediately obvious from a comparison of mass flux rating curves. Figure 4.8 shows two cases where limestone mass flux rates are greater than those for gypsum. The explanation is that the outcrop areas of the sulphate-bearing rocks are much smaller than those of the carbonate units. The actual rate of solutional denudation per unit area of outcrop is considerably greater on the gypsum (Drake & Ford 1976).

4.4 Interpretation of measurements

The large accumulated error that is likely to arise in the calculation of a solutional denudation rate demands that caution be exercized in its interpretation and in its extrapolation over long periods of *time*. Furthermore, point estimates from micro-erosion meter measurements or from tablet weight loss are difficult to extrapolate over *space* and provide less reliable areal estimates of solution than do basin studies. 'The problem of the validity of karst erosion data is, therefore, the problem of transforming unique values into values representative of some defined space and time' (Ford & Drake 1982, p. 157).

The vertical distribution of solution

The contrasts in karst landform styles in humid tropical and temperate zones appear not to be attributable to radically different solutional

Table 4.5 Vertical distribution of solutional denudation.

Area	Overall rate ($m^3\ km^{-2}\ a^{-1}$)	Remarks	Source
Fergus R., Ireland	55	60% at surface, up to 80% in the top 8 m	Williams 1963, 1968
Derbyshire	83	Mostly at surface	Pitty 1968a
Northwest Yorkshire	83	50% at surface	Sweeting 1966
Jura Mountains	98	33% on bare rock, 58% under soil, 37% in percolation zone, 5% in conduits	Aubert 1967, 1969
Cooleman Plains, NSW, Australia	24	75% from surface and percolation zone, 20% from conduit and river channels, 5% from covered karst	Jennings 1972 a & b
Somerset Island, NWT Canada	2	100% above permafrost layer	Smith 1972
Riwaka South Branch, New Zealand	100	80% in to 10–30 m, 18% in conduits mainly by allogenic streams	Williams & Dowling 1979
Waitomo, New Zealand	69	37% in soil profile, most of remainder in top 5–10 m of bedrock	Gunn 1981a
Caves Branch, Belize	90	60% on surface and in percolation zone, 40% in conduits (this site has large allogenic rivers invading conduits)	Miller 1982

denudation rates (Smith & Atkinson 1976). Therefore it follows that contrasts in the spatial distribution of corrosion must be the explanatory factor – unless our comparative denudation estimates are seriously in error. For this reason it is essential to gain an insight into the vertical distribution of corrosion, but unfortunately data on this are scarce (Table 4.5). The general conclusion is that in vegetated soil-covered karst most autogenic solution will occur near the top of the profile; that is, in the soil, at the soil/bedrock interface, and in the uppermost bedrock. The data are insufficiently precise and too few to separate tropical and temperate corrosion regimes.

Direct solutional lowering of bare rock surfaces has been estimated from *Karrentische*, which are pedestals of limestone protected from corrosion by a non-carbonate boulder that functions as an umbrella (Fig. 4.9). They are found on glacially scoured rock surfaces, the boulders being glacial erratics. The height of the pedestal is a measure of the corrosion of the surface since

A

B

Figure 4.9 A. Limestone pedestal with a glacial erratic caprock protecting it from solution by rainfall, County Leitrim, Ireland. B. A percolation measurement station in Planina Cave, Slovenia.

the last glaciation. By this method Bogli (1961) estimated the rate of surface lowering in the Swiss Alps to be 1.51 cm ± 10% per 1000 years. Peterson (1982) measured pedestal heights at 4300 m in the glaciated tropical mountains of West Irian, where surface solution appears to have been twice as fast as in the Alps (Table 4.6). In glaciated terrains currently below the tree-line pedestals can be even taller, e.g. 49 cm in northwest Yorkshire, England, and an average of 51 cm in County Leitrim, Ireland. However, in the relatively low-lying Burren of County Clare, Ireland, pedestals are generally similar in height to those found in the Alps.

The heights of emergent quartz veins standing proud of a limestone

Table 4.6 Approximate rates of solutional lowering of limestone surfaces, from limestone pedestals and micro-erosion meter measurements.

Area	Average height of pedestals (cm)	Time since ice retreat (a)	Surface lowering (mm ka^{-1})	Extrapolated micro-erosion meter rate (mm ka^{-1})
Maren Mts Switzerland	15	10 000	15[1]	–
Clare-Galway, Ireland	15	12 000	12[2]	5[5]
Leitrim, Ireland	51	12 000	42[2]	–
Craven, England	50	12 000	42[3]	13[5]
Mt Jaya West Irian	30	9500	32[4]	–

Data from: (1) Bogli 1961, (2) Wiliams 1966a, (3) Sweeting 1966, (4) Peterson 1982, (5) Trudgill *et al.* 1981.

surface because of differential solution yield similar data to pedestals. By this means Akerman (1983) estimated surface lowering of dolomitic limestone in Spitzbergen (latitude 78°N) to have averaged 2.5 mm ka^{-1} since isostatic uplift raised the area above sea level following the last glaciation. His interesting data set also shows that 4000–9000 years ago the lowering rate was 3.5 mm ka^{-1} compared to 1.5 mm ka^{-1} during the last 2000–4000 years. The latter rate represents about 11% of the present total solutional denudation in the area, estimated by Hellden (1973) as 11–15 mm ka^{-1}.

Rates of surface lowering derived from pedestal heights are compared with extrapolated MEM figures in Table 4.6. Because of the uncertainties associated with MEM results not too much should be made of minor differences and it should be recognized that most MEM results are from bare rock sites, whereas the limestone around the foot of pedestals is usually soil covered. The MEM results for northwest Yorkshire are more than double those for County Clare, as also indicated by the pedestals, but the absolute MEM rates for both sites appear to underestimate surface lowering rates by a factor of at least two. Evidently the 20th century phenomenon of acid rain (in Yorkshire particularly) has not compensated in corrosional terms for the relative devegetation of the karst by human agricultural activity.

Rates of surface lowering can also be computed from rainfall volume and solute concentrations achieved in runoff over rock outcrops, e.g. Miotke (1968) in the mountains of Picos de Europa, Spain, and Dunkerley (1983) on fluted tower karst in Queensland. By comparing values thus derived with total basin denudation calculated from outflow discharge and solute concentration, a reasonable estimate can be gained of the proportion of net corrosion occurring on bare rock at the very surface.

Most rock surfaces are in fact not completely bare, but are covered with a thin layer of lichens, algae and other epilithic, chasmolithic and endolithic organisms. Viles (1984) provides details and also points out that the associated bioerosive mechanisms are poorly understood. Excellent studies of the role of cyanobacteria and lichens in the weathering of limestones in Israel have recently been published by Danin (1983) and Danin & Garty (1983). Cyanobacteria are shown to lead to rock weathering at a rate of 5 mm ka^{-1} (Fig. 4.10). Features produced by the action of flora and fauna on karst rocks are termed *biokarst* (Viles 1984) or *phytokarst* (Folk *et al.* 1973, Bull & Laverty 1982). Trudgill (1985) provides a particularly good discussion of biokarst and of the role of organic acids as weathering agents (see also section 3.6).

Further information on the vertical distribution of corrosion can be obtained by following the evolution of the chemical characteristics of water as it percolates through the soil and underlying bedrock (Fig. 4.9B). In this way Gams (1962) found the bulk of the corrosion of limestone in Slovenia

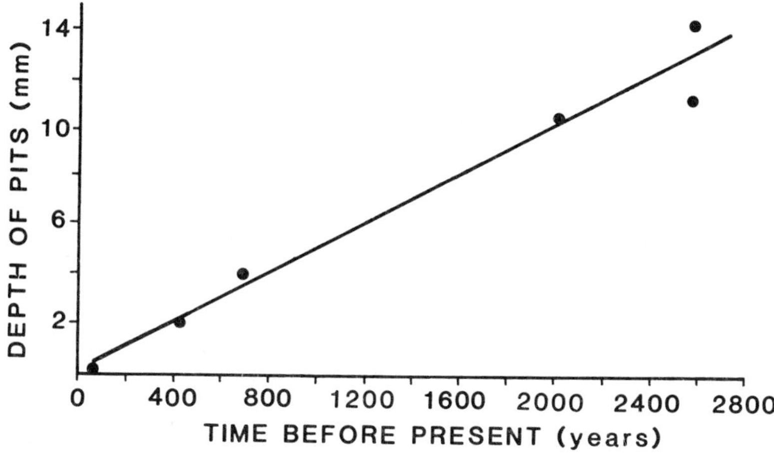

Figure 4.10 Depths of pits resulting from cyanobacteria-induced weathering in comparable limestone walls of varying age in Jerusalem. From Danin (1983).

to occur in the top 10 m of the percolation zone. This important conclusion has since been found to apply to most other places where the vertical distribution of corrosion has been studied (Table 4.5). The consensus is that about 70% of autogenic solution takes place in the uppermost 10 m of the percolation zone, although actual figures vary from about 50–90% depending on lithology and other factors. The implications are (a) that most solutional denudation leads to surface lowering and (b) that solutional activity in cave conduits is of relatively minor significance to total denudation although fundamental to the development of the karst landform system.

The vertical distribution of denudation depends upon two factors: (a) the distribution of water flow and (b) the distribution of solute concentrations. Efforts to date have focused on the latter, with data being used in a budget approach to deduce the relative contributions to total solutional denudation (Table 4.1). Illustrations of results are provided by Jennings' (1972a) research in New South Wales, Australia, Atkinson's & Smith's (1976) in the Mendip Hills, England, and Williams & Dowling's (1979) in New Zealand (Fig. 4.11).

In entirely autogenic systems, much field evidence suggests that corrosion in vadose cave passages of enterable dimensions is negligible except during occasional floods (when corrasion may also be important). In two systems at Waitomo, New Zealand, Gunn (1981a) found water contributing to cave streams to be saturated or supersaturated in most conditions, as did Miller (1982) in caves draining a tropical cockpit karst in Belize. Gunn concluded 37% of the solution to occur within the soil, at the soil–rock interface, or on

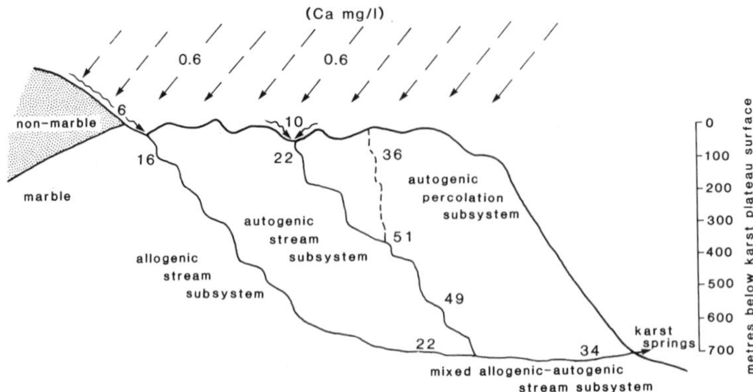

Figure 4.11 Dissolved calcium concentrations in the Pikikiruna karst, New Zealand. From Williams & Dowling (1979).

occasional limestone outcrops, with most of the remainder concentrated in 5 to 10 m of weathered bedrock beneath. Difficulties in estimating discharge through the subcutaneous zone precluded the calculation of an erosion rate for that zone. This is a major problem, but even more significant is the difficulty of obtaining information on the spatial variation of corrosion within the subcutaneous zone, for it is there that landforms like solution dolines are made. It is at this point that we must refer back to the first of the two factors mentioned previously but not examined: the distribution of water flow.

The solutional denudation rate is the product of water flow and solute concentration. In an autogenic percolation system a wide range of flows occur on a continuum from very slow seepages through trickles to showering cascades. Discharges of observed flows range over several orders of magnitude, whereas their associated solute concentrations are unlikely to vary by more than one order of magnitude. Although direct relationships have been reported between flow-through time and calcium hardness (e.g. Pitty 1968a & b), the small volumes of highly mineralized seepages do not account for much denudation, although they make an impressive contribution to speleothem deposition. Recent observations show that in well developed karsts the upper part of the percolation zone (termed the subcutaneous or epikarstic zone) has a significant water storage capacity and sufficient interconnection to diffuse tracer dye quite widely. Drainage from this zone is not uniform, but down preferred paths which act as foci for subcutaneous streamlines. Therefore it is probable that corrosion within the epikarst is greatest where flow paths converge above the more efficient percolation routes (Fig. 4.12). Because solute concentrations do not vary widely compared to percolation discharge rates, corrosion denudation in a given area (or volume) of convergent flow can be many times greater

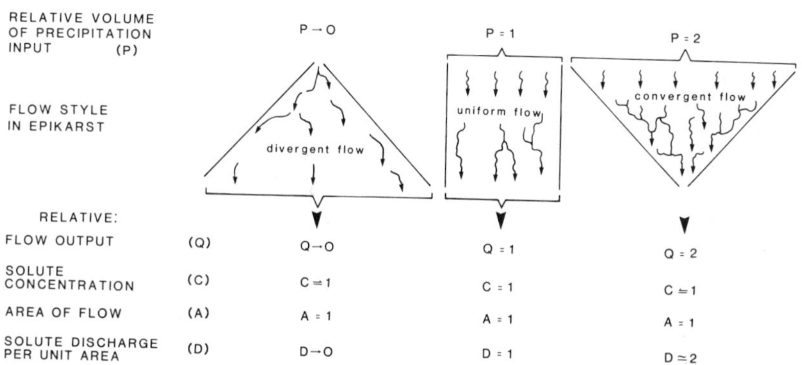

Figure 4.12 Relationship between percolation flow characteristics and the distribution of corrosion within the subcutaneous zone.

than in an equivalent area (or volume) of divergent flow. Karsts with uniform percolation through the subcutaneous zone will experience uniform surface lowering, whereas divergence also implies neighbouring convergence and the development of accentuated relief. Inequalities in subcutaneous corrosion become increasingly manifested topographically as the surface lowers over time. The implications are developed further in section 9.4.

Figure 4.12 is simplistic to the extent that it overlooks many other factors that may contribute to spatial variations in corrosion. These will include variations in vegetation cover, soil thickness and outcrop lithology. Their relative importance is not easily quantified.

Rates of denudation and downcutting

The rate of incision of cave streams has been estimated by Gascoyne *et al.* (1983) using the ages of speleothems in various positions above active streamways. Stalagmites can only accumulate when they are not subject to erosion by flood waters as explained by section 8.6. Hence the height of the base of a stalagmite above a bedrock stream channel divided by the stalagmite age yields a maximum downcutting rate for the stream. Rates of about 20–50 mm ka^{-1} over the last 350 000 years were determined in this manner for cave channels in northwest Yorkshire, England, with a mean maximum rate of valley entrenchment (aided by glaciation) in the area of 50 to < 200 mm ka^{-1}. This compares to valley deepening rates calculated by the same method by Ford *et al.* (1981) in the Canadian Rocky Mountains of 40–70 mm ka^{-1} as a minimum, with maximum values to 2 m ka^{-1}. Williams (1982b, 1987b) determined incision in Metro Cave, New Zealand, to have proceeded at a rate of 280 mm ka^{-1}. Since part of the erosional activity in the caves concerned was mechanical, as a consequence of large allogenic

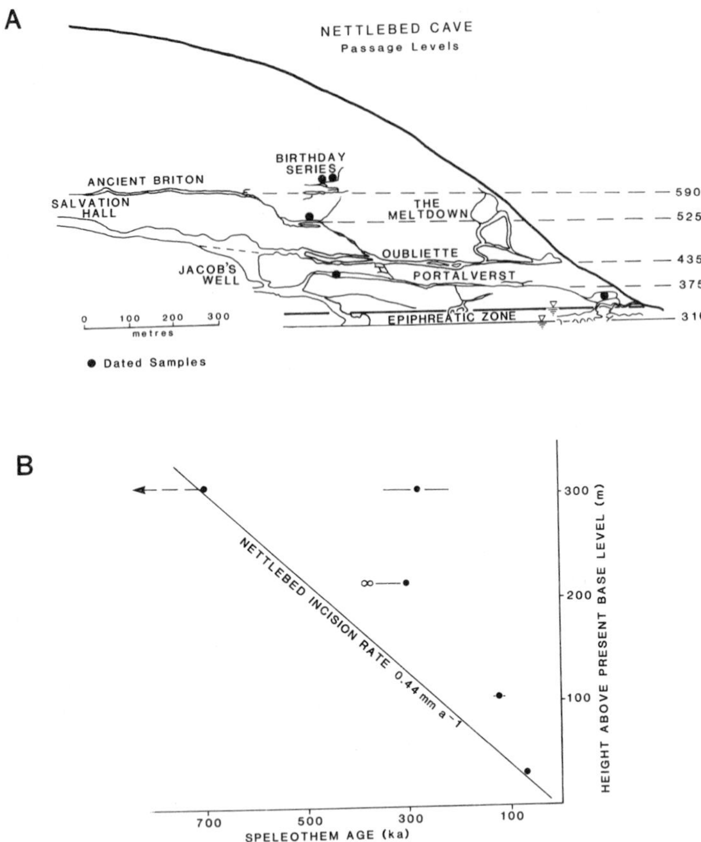

Figure 4.13 A Cave levels plotted against height above the outflow spring, and B speleothem ages plotted against height above the outflow spring, Nettlebed Cave, New Zealand. From Williams 1987b.

components, the downcutting rates cannot easily be compared to solutional denudation figures.

This problem is overcome where speleothems are found at various heights above a phreatic zone where significant mechanical action is precluded. In Nettlebed Cave, New Zealand, a series of abandoned phreatic levels occur above the present flooded zone, each level being related to a former position of the spring. Speleothem dates have been obtained up to 350 m above the present active system (Fig. 4.13), the uppermost being beyond the range of U–Th dating and paleomagnetically reversed (Lyons 1983). The collected information indicates a maximum incision rate of 500 mm ka^{-1} attributable entirely to corrosion over a period of at least 700 000 years.

The maximum downcutting rate at Nettlebed represents linear corrosion and is about six times greater than – though of the same order as – the average areal solutional denudation of the surface zone in the nearby Riwaka basin (Table 4.1), where the karst is in the same rock and the climate is comparable. This supports the general conclusion of Gascoyne *et al.* (1983), who observed in the Yorkshire Dales that the maximum cave channel entrenchment rates are of the same order of magnitude as limestone areal lowering rates calculated from solute budgets in temperate to subarctic localities, although Ford (1985) noted that some cave channel entrenchment rates can be surprisingly low, as in West Virginia and Belize, where cave streams are close to local baselevels.

The vertical distribution of carbonate deposition

Chemical processes in the solution and precipitation of calcite were discussed in Chapter 3; crystal growth of carbonate precipitates in caves is examined in Chapter 8 and case hardening and tufa deposition in rivers and lakes in Chapter 9. This section focuses on general aspects of the vertical distribution of carbonate deposition by meteoric waters.

Most corrosion occurs in the epikarst, the upper part of the percolation zone. It may even occur entirely in the soil should it be rich in carbonate fragments, such as a calcareous glacial till. Consequently, most autogenic percolation water is close to saturation with respect to calcite as it moves downwards through the remainder of the vadose zone. Ground–air CO_2 may sustain a slight degree of undersaturation, although when percolating water leaves the soil zone it is likely that most solution will occur under essentially closed (or sequential) system conditions.

Carbonate precipitation most commonly occurs when water percolating beneath soil enters a cavity that has some connection with the outside atmosphere. Cavities may be large caverns or small vugs, but gaseous exchange with the external air is essential because it ensures that the P_{CO_2} of the cave air is broadly of the same order as that measured in the normal atmosphere (section 3.6). The percolating water is equilibriated with a higher, soil P_{CO_2}, with the result that carbonate is precipitated if significant ($SI_c > +0.3$) supersaturation is attained. Should cave ventilation also be strong, relative humidity may be less than 100%, in which case evaporation will also occur. This accelerates deposition as well as encouraging the growth of a different suite of speleothem forms.

In arid and seasonally arid environments carbonate deposition can occur in the soil as a result of evaporation. In some humid environments carbonate precipitation may closely follow dissolution due to a combination of CO_2 degassing and evaporation, resulting in formation of calcrete case hardening (see section 9.10). However, most carbonate deposition takes place beneath the solum and commences in the first cavity where P_{CO_2} is less than in the soil atmosphere. There are countless instances of caves profusely decorated with actively growing speleothems within a few metres of the surface. In raised atolls carbonate deposits fill interconnecting vugs in the rock matrix like butter melting down into hot toast; under such circumstances deposition reduces primary (bulk) porosity although in the standard vadose cave it reduces secondary (fissure) porosity.

As the degree of supersaturation of the percolating groundwater reduces, so the tendency to deposit carbonate diminishes. Thus with depth through the vadose zone the amount of speleothem deposition may decrease (section 8.3). However, the secondary porosity of well karstified terrains is often no more than 1%; thus percolating water often does not encounter aerated vadose cavities and the first opportunity to precipitate carbonate may be in the saturated zone. In a partly flooded cavern communicating directly with the outside atmosphere, heterogeneous carbonate precipitation can take place onto solid surfaces such as cave walls or other crystals. It is comparatively rare because it only occurs where currents are slow moving. Deposits are commonly destroyed if the passage is liable to turbulent floods.

These points on the vertical distribution of deposition apply to autogenic karst. Under other geological situations differing hydrological and hydrochemical conditions dictate where deposition is possible. Impervious caprocks preclude percolation and hence limit deposition to streamways, provided that mechanical action is also not too strong. Streams in allogenic systems, however, are commonly undersaturated with respect to calcite so that carbonate deposition is not usually a feature. This contrasts strongly with vadose streams in autogenic systems, which are often very favoured sites for cascades of speleothems.

Carbonate deposition in the vadose zone has received remarkably little quantitative study. However, it is generally acknowledged that the pattern of deposition discussed is much better displayed in tropical and warm temperate zones than in cold regions. This is presumed to be a consequence of the greater contrast achieved between soil and atmospheric CO_2 in the tropics compared to sub-arctic and alpine environments.

Magnitude, duration and frequency of solution

Gunn (1982) has reviewed the magnitude and frequency properties of dissolved solids transport. He compares results from 24 basins of which ten are underlain by carbonate rocks (Table 4.7) and concludes:

(a) The greatest variation in solute transport work achieved is shown by high flows operational for only 5% of the time. In the carbonate basins these flows account for less than a quarter of the work done (except in one case where it is 44%) compared to 24–57% in the non-carbonate catchments.

(b) Flows less than mean discharge occur for 60–75% of the time and account for 20–55% of the dissolved load transport. Flows less than

Table 4.7 Magnitude and frequency parameters for dissolved solids transport. From Gunn (1982).

Drainage basin		Percentage of annual solute load transported by: (1) Flows equalled or exceeded 5% of time	(2) Flows less than the mean discharge	(3) Flows less than the median discharge	Percentage of time required to remove 50% of solute load
Mellte	(CO_3)	–	–	<33	–
Shannon	(CO_3)	–	32	28	–
Rickford	(CO_3)	5	34	23	24
Langford	(CO_3)	10	48	35	30
S. Rockies	(CO_3)	13	26	19	20
	(SO_4)	12	35	26	23
Riwaka	(CO_3)	44	33	20	10
Honne	(CO_3)	16	–	26	26
Cymru	(CO_3)	18	34	21	22
Glenfield	(CO_3)	15	33	24	25
Cooleman Plain	(CO_3)	21	55	29	29
S.E. Devon (1)	(TDS)	57	20	7	5
(2)	(TDS)	29	–	17	15
(3)	(TDS)	29	46	25	18
(4)	(TDS)	29	36	23	19
(5)	(TDS)	31	39	24	18
Slapton Ley	(TDS)	28	26	12	12
Ei Creek	(TDS)	>25	–	–	10
East Twin GP1	(TDS)	27	–	11	1.5
GP2	(TDS)	24	–	20	18
New England (1)	(TDS)	50	–	–	5
(2)	(TDS)	–	–	–	7
(3)	(TDS)	–	–	–	10
Avon	(TDS)	–	–	–	20
Creedy	(TDS)	25	–	15	12

median discharge usually account for less than a quarter of the solute load.

(c) High flows are less significant for dissolved load transport in carbonate basins than in non-karst basins. Nevertheless, Gunn refutes Wolman & Miller's (1960) suggestion that a very large part of solute load is transported by flows as low as the mean or even the median discharge. In the carbonate basins flows greater than the mean account for 45–74% of the dissolved load transport.

The evidence is not sufficiently comprehensive to distinguish with confidence between the magnitude and frequency behaviour of autogenic as compared to allogenic basins. However, theoretical considerations suggest that the greater the autogenic component, the less significant will be the relatively high magnitude but low frequency discharges in transporting solute load. This is because in autogenic basins the Ca versus discharge relation usually has a lower slope than in allogenic basins, i.e. the dilution effect is less marked; and in autogenic basins the discharges are in any case less variable, i.e. the flow duration curve shows a narrower range. In 'pure' autogenic karsts the solute magnitude and frequency relationships are essentially controlled by the regime of the outflowing stream.

Solutional denudation, karstification and inheritance

In discussing the relevance of solution rates to geomorphology, Priesnitz (1974) quite rightly observed that an average annual denudation rate does not characterize karstification. For example, there is a strong discrepancy between average corrosional lowering of the surface and the solutional modelling of the terrain. Regions with high solutional denudation rates do not necessarily display well developed karst. He suggests that factors important in modelling karst include the size and form of the solution front, its location, the intensity of solution at each point across the front, and the form and location of eventual redeposition. Priesnitz also proposes the use of a surface lowering: surface modelling ratio as an indicator of the morphological effectiveness of solution. Estimating the surface modelling effect from the volume of dolines and applying this idea to an area of gypsum and limestone karst near Bad Gandersheim in Germany, he concludes that for both rock types 98–100% of the considerable solution in the Holocene has produced only surface lowering. The relationship between karstification and solutional denudation clearly requires much more study.

Another factor barely recognized and certainly unresolved in the interpretation of denudation rates is that of inheritance. Imagine two areas of dense limestone near baselevel that are identical in every respect except that one is already karstified and the other is not. They are uplifted an equal amount and subjected to similar climatic regions. What would be the effect of the different geomorphic inheritance on the amount and distribution of solutional denudation that results in these two karsts? Although this question is unresolved, different geomorphic inheritances must be significant in steering erosion processes and resultant landform development. In geomorphology as a whole, time scales are so long that there is seldom a recognizable beginning for a landscape, only an inheritance. In karst terrains more than most we can sometimes identify a beginning; perhaps a time when an impervious caprock was first breached or when a coral reef was uplifted from the sea. But the commencement of karstification in some of the world's greatest karsts, as in southern China for instance, is so far back (see section 10.7) that landforms developing in response to modern processes must depend in part on preconditioning that provides ready made avenues for solute attack.

5 Karst hydrology

5.1 Basic hydrological concepts, terms and definitions

Aquifers

Rock formations that store, transmit and yield economically significant amounts of water are known as *aquifers*. Karst aquifers like those of other rocks may be *confined*, *unconfined* and *perched* (Fig. 5.1). A confined aquifer is contained like a sandwich between relatively impervious rocks that overlie and underlie it. By contrast to the formation containing the aquifer, an impermeable rock incapable of absorbing or transmitting significant amounts of water is known as an *aquifuge*. Other rocks such as clay and mudstone may absorb large amounts of water, but when saturated are unable to transmit it in significant amounts. These are termed *aquicludes*. A relatively less permeable bed in an otherwise highly permeable sequence is referred to as an *aquitard*; a calcareous sandstone in a karstified limestone sequence could provide such a case.

Aquifers can be differentiated into three end member types according to the nature of the voids in which the water is stored and through which it is transmitted, namely *porous* (or granular), *fissure* (or fracture) and *conduit*.

In an unconfined aquifer the upper boundary of the aquifer is the *water table*. This is an English language term for the surface defined by the level of free standing water in fissures and pores delimiting the top of the phreatic

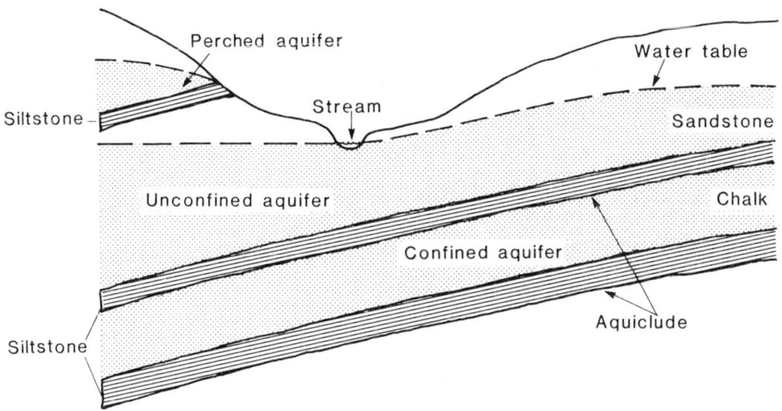

Figure 5.1 Confined, unconfined and perched aquifers. From Dunne & Leopold (1978).

zone. It is an equilibrium surface at which fluid pressure in the voids is equal to atmospheric pressure. An equivalent and acceptable alternative term frequently used in continental European literature is the *piezometric surface*.

By contrast, in a confined aquifer the water in a bore tapping the water-bearing formation usually rises up the bore hole to a level that is above the top of the aquifer. The theoretical surface fitted to the water levels in such bores is termed the *potentiometric surface*. Where water rises up a bore as just described, it is sometimes called an *artesian* well and the water is said to be confined under artesian conditions.

The lower boundary of an aquifer is commonly an underlying impervious formation. But should the karst rocks be very thick, the effective lower limit of the aquifer occurs where no significant porosity has developed. This may be because the rocks have only recently been exposed to karstification or because lithostatic pressure at depth is so great that there is no penetrable fissuring and groundwater is consequently precluded. In regions of extensive or continuous permafrost, the aquifer may be restricted to a surficial 'active' layer from about 0.5 to 1 m in depth. However, in karstic rocks the permafrost aquifuge beneath the active layer is often breached by groundwater drains termed *taliks* and there may be an underlying circulation (Figs 10.13 & 10.14).

In the unsaturated zone above the water table, voids in the rock are only partially occupied by water, except after heavy rain when some fill completely. Water percolates downwards in this zone by a multiphase process, air and water co-existing in the pores and fissures. Air bubbles may even impede percolation by blocking capillary channels. More significant impediments to downwards flow are sometimes provided, however, by localized impermeable layers, such as shale or chert bands in a limestone sequence. Ponding occurs above these layers, producing a localized saturated zone known as a perched aquifer, suspended above the main

Table 5.1 Karst hydrographic zones.

1 Unsaturated (vadose) zone
- 1a Soil
- 1b Subcutaneous (epikarstic) zone
- 1c Free draining percolation zone

2 Intermittently saturated (epiphreatic or floodwater) zone

3 Saturated (phreatic) zone
- 3a Shallow phreatic zone
- 3b Deep phreatic (bathyphreatic) zone
- 3c Stagnant phreatic zone

[Each of the above may be traversed by caves, permanently flooded in zone 3]

water table (Fig. 5.1). Subdivisions of the unsaturated and saturated zones of an unconfined aquifer are listed in Table 5.1, although not all categories may be present in any given karst.

Flow through pores and pipes

Conventional groundwater hydrology usually considers aquifers to be porous media; so it must be questioned whether the normal laws of groundwater hydrology are applicable to fractured rocks perforated by large solution pipes as in limestones and gypsum. The consideration raised by this is whether a better explanation of groundwater movement in karst will be achieved by understanding the flow of water through individual conduits or by treating the rock as an idealized continuum of saturated voids in a solid matrix. The work of Hagen (1839) and Poiseuille (1846) as compared to Darcy (1856) helps to illustrate the point.

The pressure at a given point below the water table is equal to the product of the depth and the unit weight of water, termed the *pressure head*, plus atmospheric pressure (Fig. 5.2). When there is no flow, pressures are equal in all directions and conditions are termed *hydrostatic*. The *hydraulic head* h is the sum of the elevation head and the pressure head. The product of hydraulic head and acceleration due to gravity g yields the *hydraulic potential* ϕ, an expression of the mechanical energy of water per unit mass. But since gravitational acceleration is practically constant near the earth's surface, hydraulic head and hydraulic potential are very closely correlated.

Hagen and Poiseuille studied the flow of water through small tubes and discovered its *specific discharge* u to be directly proportional to the hydraulic head loss by friction Δh between one end of the tube and the other

$$u = \frac{\pi r^4}{8\, l\, \mu} \cdot \Delta h \qquad (5.1)$$

where r is the radius of the tube, μ is the dynamic viscosity of the water, and l is the length of the tube. This equation is known as *Poiseuille's law*.

Darcy (1856) performed experiments passing water through a vessel filled with saturated sand. His results confirmed Hagen and Poiseuille's findings by showing that the quantity of water flowing through a porous medium is proportional to the difference in total hydraulic head, as represented by a *hydraulic gradient* $\frac{dh}{dl}$ between the inflow and outflow points. The relationships are expressed in what is termed *Darcy's law*

$$u = -K\frac{dh}{dl} = \frac{Q}{a} \qquad (5.2)$$

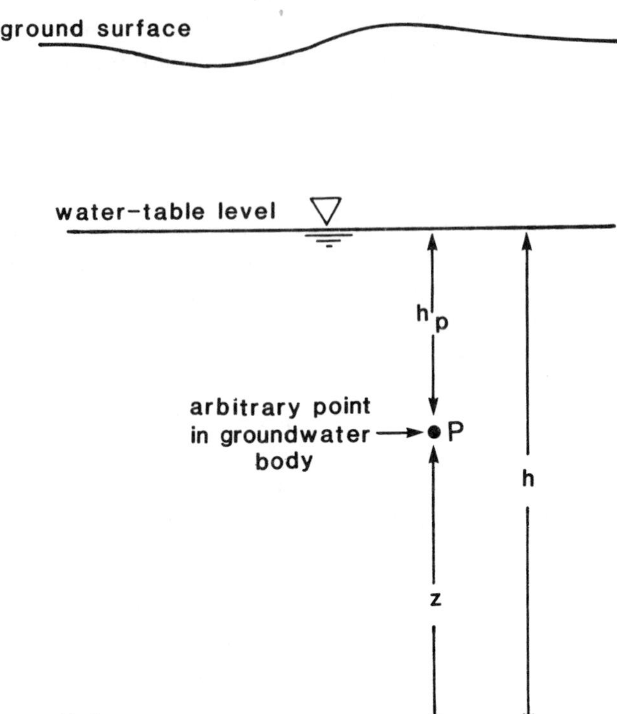

AT POINT P, THE HYDRAULIC HEAD $h = h_p + z$

WHERE h_p = PRESSURE HEAD
AND z = ELEVATION HEAD

Figure 5.2 Definition of hydraulic head, pressure head and elevation head for an unconfined aquifer.

where Q is the *discharge*, a is the cross-sectional *area* of porous medium through which it flows, and K is the *hydraulic conductivity* (or coefficient of permeability). Water in fact moves in response to changes in hydraulic potential; the negative sign on the right hand term of the equation indicating a loss in hydraulic potential in the direction of flow over the distance l travelled.

Hydraulic conductivity K is sometimes loosely (and incorrectly) referred to as the *permeability* k. The latter, also called the intrinsic permeability, is a measure of the ability of a material to transmit fluids. It depends on the physical properties of the material, especially pore sizes, shapes and distribution. By contrast, the hydraulic conductivity reflects the properties of *both* the medium and the fluid. The two terms are related as follows

Table 5.2 Range of values of hydraulic conductivity and permeability. From Freeze & Cherry (1979).

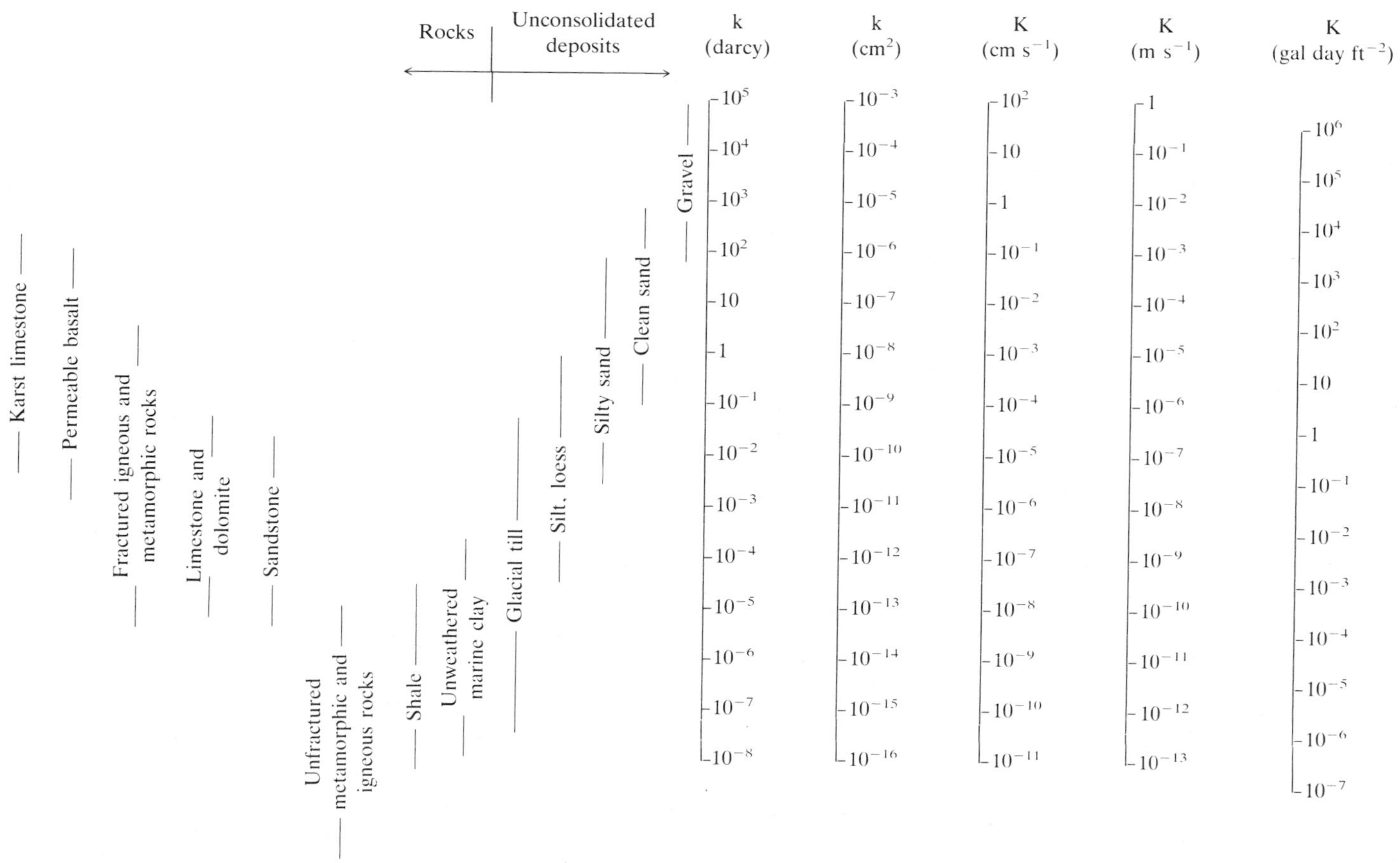

Rocks ← | → Unconsolidated deposits

Rocks: Karst limestone; Permeable basalt; Fractured igneous and metamorphic rocks; Limestone and dolomite; Sandstone; Unfractured metamorphic and igneous rocks; Shale; Unweathered marine clay

Unconsolidated deposits: Glacial till; Silt, loess; Silty sand; Clean sand; Gravel

k (darcy)	k (cm^2)	K ($cm\ s^{-1}$)	K ($m\ s^{-1}$)	K ($gal\ day\ ft^{-2}$)
10^5	10^{-3}	10^2	1	10^6
10^4	10^{-4}	10	10^{-1}	10^5
10^3	10^{-5}	1	10^{-2}	10^4
10^2	10^{-6}	10^{-1}	10^{-3}	10^3
10	10^{-7}	10^{-2}	10^{-4}	10^2
1	10^{-8}	10^{-3}	10^{-5}	10
10^{-1}	10^{-9}	10^{-4}	10^{-6}	1
10^{-2}	10^{-10}	10^{-5}	10^{-7}	10^{-1}
10^{-3}	10^{-11}	10^{-6}	10^{-8}	10^{-2}
10^{-4}	10^{-12}	10^{-7}	10^{-9}	10^{-3}
10^{-5}	10^{-13}	10^{-8}	10^{-10}	10^{-4}
10^{-6}	10^{-14}	10^{-9}	10^{-11}	10^{-5}
10^{-7}	10^{-15}	10^{-10}	10^{-12}	10^{-6}
10^{-8}	10^{-16}	10^{-11}	10^{-13}	10^{-7}

$$K = \frac{k \rho g}{\mu} \tag{5.3}$$

where ρ, mass density, and μ, dynamic viscosity, are functions of the fluid alone. Whereas K has the dimensions of a velocity $[L/T]$ and is commonly expressed in m s^{-1}, k has the dimensions $[L^2]$ and is sometimes expressed in *darcy* units, 1 darcy being approximately equal to 10^{-8} cm^2 (Freeze & Cherry 1979). Table 5.2 shows the range of values of hydraulic conductivity and permeability encountered in common earth materials.

Darcy's law assumes flow to be *laminar*. Under these conditions individual 'particles' of water move in parallel threads in the direction of flow, with no mixing or transverse component in their motion. This can most easily be visualized in a straight cylindrical tube of constant diameter (Fig. 5.3), but it also applies to granular media. At the tube wall flow velocity v is zero, because of adhesion, and it rises to a maximum at the centre. But as the radius of the tube and flow velocity increase, so fluctuating eddies develop and transverse mixing occurs. The flow is then termed *turbulent*. Such conditions frequently arise in pipes and fissures in karst, and can be considered predominant in cave systems.

Where laminar flow conditions exist in jointed rock, the hydraulic conductivity of a single fissure with plane parallel sides can be determined from the expression

$$K = \frac{g w^3}{12 \nu} \tag{5.4}$$

where w is the width of the fissure and ν is kinematic viscosity of the fluid. Brady & Brown (1985) observe that for water at 20°C, if w increases by a factor of 10 from 0.05 mm to 0.5 mm, then K increases by a factor of 1000 (from $K = 1.01 \times 10^{-7}$ to 1.01×10^{-4} m s^{-1}). Values of fluid density, dynamic viscosity and kinematic viscosity for different water temperatures are given in Table 5.3.

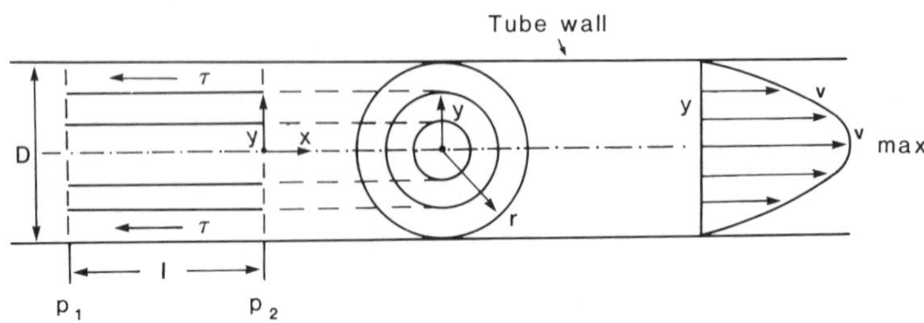

Figure 5.3 Laminar flow through a cylindrical tube. From Hillel (1982).

Table 5.3 Values for fresh water of fluid density, dynamic viscosity and kinematic viscosity for different water temperatures at one atmosphere pressure.

Temperature (°C)	Fluid density, ρ ($g\ cm^{-3}$)	Dynamic viscosity, μ ($g\ cm^{-1}\ s^{-1}$)	Kinematic viscosity ν ($cm^2\ s^{-1}$)
0	1.000	0.01787	0.01787
10	1.000	0.01307	0.01307
20	0.998	0.01002	0.01004
30	0.996	0.00797	0.00800

Karstic enhancement of porosity

A most important characteristic of karst that sets it apart from the hydrology of other rocks is that its permeability changes with time, a point that seems first to have been appreciated by Cvijic (1918). When a carbonate sediment is formed it acquires a fabric selective porosity, usually termed *primary porosity* of about 25–80% from interstitial spaces between its mix of materials, as discussed in section 2.3. These primary voids are diminished during compaction and/or by cementation during diagenesis. Indeed, the preservation of effective primary porosity under deep burial conditions is an exception rather than the rule (Moore 1979). Intercrystalline porosity that occurs between mineral crystals may account for up to 1% of the total porosity. However, later chemical diagenetic processes such as dolomitization and subsequent fracturing by tectonic movements result in the acquisition of *secondary porosity*, which is considerably enhanced by karst solution along penetrable fissures by circulating groundwaters. These voids are not fabric selective and may continue to enlarge while groundwater circulation persists. Their enlargement can impart a secondary porosity of up to about 3.5% in otherwise dense rocks with negligible primary porosity.

A distinction is made between the *porosity* n of a rock and its *effective porosity* n_e. Porosity is defined as the ratio of the aggregate volume of pores V_p to the total bulk volume V_b of a rock specimen. Thus

$$n = V_p/V_b \tag{5.5}$$

Effective porosity refers only to those voids which are hydrologically interconnected. For a fully saturated rock, it can be expressed as the ratio of the aggregate volume of gravitation water that will drain from the rock V_a to the total bulk volume of the rock V_b

$$n_e = V_a/V_b \tag{5.6}$$

In unconfined aquifers, it amounts to the volume of water that will drain freely under gravity from a unit volume of the aquifer. Castany (1984b) describes measurement techniques.

Effective porosity is influenced by pore size. Thus clays with a porosity of 30–60%, but with pores of only 1×10^{-2} to 10^{-3} mm in width yield almost no water when able to drain freely under gravity. This is because the force of molecular adhesion involved in the adsorption of water in clay is sufficient to overcome the force of gravity. But karstified limestone with a porosity of perhaps only 2%, but with interconnected voids measuring 1–10^3 mm or more in diameter, will yield almost all the water in storage if freely drained.

Homogeneous and isotropic aquifers

Well sorted sand and gravel aquifers have essentially constant values for porosity and permeability throughout their extent. Under such conditions hydraulic conductivity is independent of *position* within the formation. The aquifer is then considered *homogeneous*. But it is *heterogeneous* if K varies with location within the formation. If hydraulic conductivity is the same regardless of *direction* of measurement the aquifer is considered *isotropic*. But it is *anisotropic* if K varies with direction. Karst aquifers become both heterogeneous and anisotropic with time. The measurement and definition of karst aquifer heterogeneity is discussed by Yuan (1985b). Kiraly (1975) and Castany (1984a) point out that average total porosity is a function of the reference volume of the rock considered, and that three orders of magnitude can be defined for the heterogeneity: macroscopic (1st order, i.e. regional scale), mesoscopic (2nd order, i.e. pumping test scale), and microscopic (3rd order, i.e. petrographic scale) (Fig. 5.4). Darcy's law will not apply to an aquifer where it is anisotropic and heterogeneous, unless the length scale of the region over which it is applied is large enough – often it is not.

If a typically anisotropic karst aquifer is conceptualized as behaving

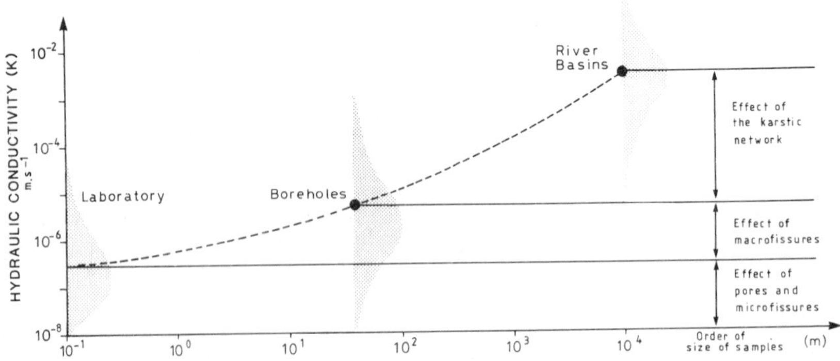

Figure 5.4 Schematic representation of the effect of scale on the hydraulic conductivity of karst. From Kiraly (1975).

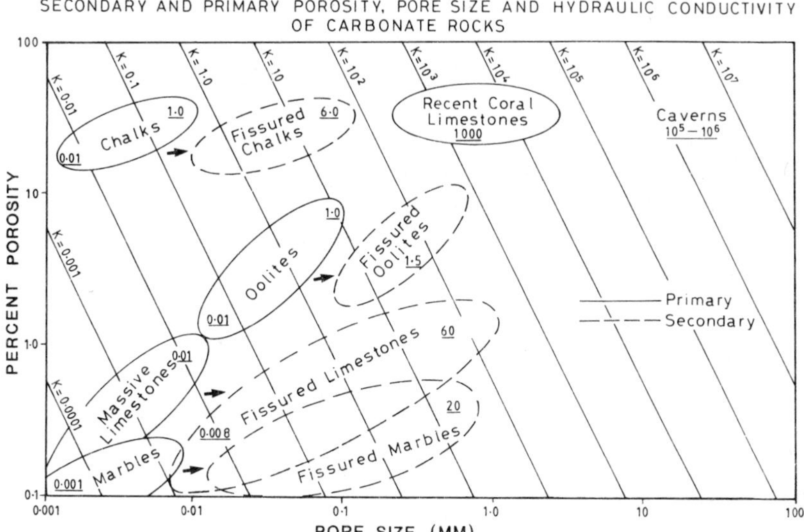

Figure 5.5 Relationship of primary and secondary porosity and pore size to the hydraulic conductivity of karst rocks. From Smith *et al.* (1976).

hydrologically like rock pierced by straight, parallel capillary tubes, it is possible to show the way in which tube diameter and per cent porosity affect hydraulic conductivity in the direction of the tubes, assuming laminar flow (Fig. 5.5). The effect of the secondary enhancement of permeability is also illustrated. In the case of a massive limestone with typically very low primary porosity, secondary enlargement of the fissure network can lead to as much as a 10^6 increase in hydraulic conductivity (Smith *et al.* 1976)

Transmissivity and storage

The ease with which water flows through a karst aquifer usually varies according to direction. This may be approximated by the parallel capillary tube model, where hydraulic conductivity is greatest in the direction of the tubes but very much less at right angles to them. Thus hydraulic conductivities K_x, K_y, and K_z may be envisaged for different directions, although only one value for K will apply in an isotropic aquifer. Horizontal values (K_x, K_y) are of interest in the analysis of flow in the saturated zone, and vertical values K_z of importance in characterizing recharge. However, it is simplistic to assume that the hydraulic conductivity in any particular direction will remain constant over a large distance. Vertical hydraulic conductivity, in particular, diminishes considerably with depth below the surface in most well karstified rock, because secondary permeability is usually greatest near the surface.

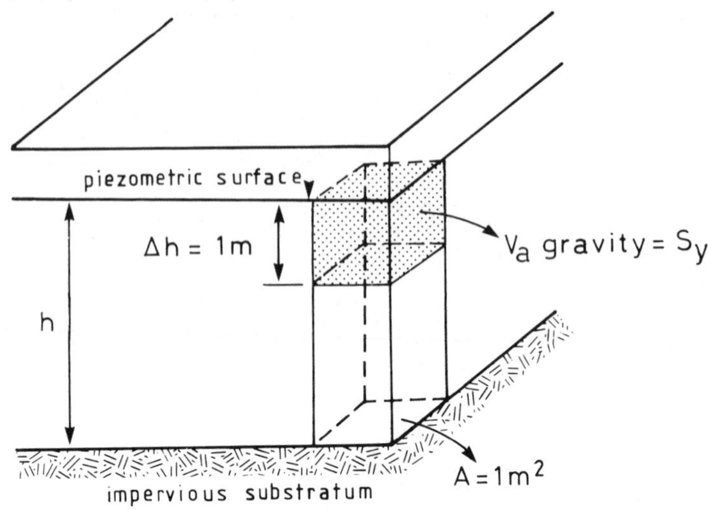

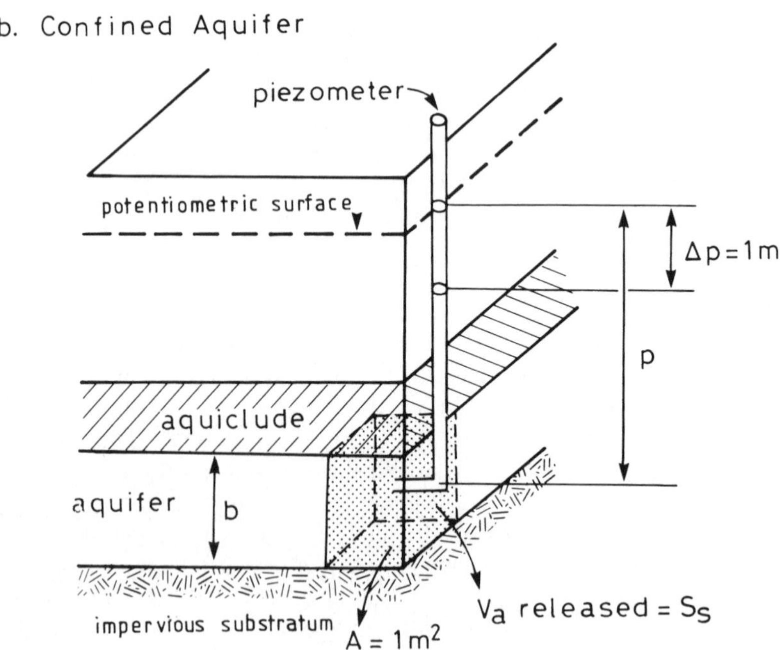

Figure 5.6 Specific storage defined for (a) unconfined and (b) confined aquifers. From Castany (1984a).

Table 5.4 Specific storage and transmissivity values for some karstic aquifers. From Castany (1984a).

Carbonate Rock Category	Age	Location	S (%)	T (m^2 s^{-1})
Fissured limestone	Turonian–Cenomanian	Israel	1	0.1×10^{-2} to 1.3×10^{-1}
	Upper Cretaceous	Tunisia	0.5 to 1	1
	Jurassic	Lebanon	0.1 to 2.4	0.1×10^{-2} to 6×10^{-2}
Karstic fissured limestone	Urgonian	Salon (France)	1 to 5	10^{-3}
	Jurassic	Parnassos (Greece)	5	1 to 2×10^{-3}
Fissured dolomite	Jurassic	Grandes Causses (France)		10^{-3}
	Lias	Morocco		10^{-2} to 10^{-4}
	Jurassic	Parnassos (Greece)		3×10^{-5}
Fractured marble		Almeria (Spain)	10 to 12	
Fissured dolomite		Murcie (Spain)	7	
Marly limestone	Jurassic	Grandes Causses (France)		10^{-3}

The ability of an aquifer to transmit water is defined by its *transmissivity* T, which is dependent on the thickness b of the aquifer and its hydraulic conductivity

$$T = Kb \tag{5.7}$$

Clearly, T varies with direction in an anisotropic aquifer.

The specific discharge from an aquifer was seen, according to Darcy's law, to be proportional to the hydraulic gradient, a decrease in hydraulic head implying a decrease in outflow. The volume of water that a unit volume of saturated rock releases following a unit decline in head is a measure of the storage capacity of the aquifer. In an unconfined aquifer it is termed *specific yield* S_y and in a confined aquifer *specific storage* S_s (Fig. 5.6). The *storativity* S of a confined aquifer is defined as the product of specific storage and aquifer thickness

$$S = S_s\, b \tag{5.8}$$

Storativity and transmissivity values for some carbonate aquifers are presented in Table 5.4. The specific yields of unconfined aquifers are considerably larger than the storativities of confined aquifers (Freeze & Cherry 1979).

Specific yield can be expected to vary according to vertical position in the karst rock, because effective porosity also varies vertically. It will be highest in the epikarst (Table 5.1) and usually diminishes with depth. Castany (1984a) cites an example from South Africa where effective porosity diminishes from 9% at 60 m below the surface to 5.5% at 75 m, 2.6% at 100 m, 2% at 125 m and 1.3% at 150 m. Eventually lithostatic pressure prevents the penetration of groundwater and precludes the development of secondary porosity.

Flow nets

A map of a water table or of a potentiometric surface provides a two-dimensional view of the aquifer. The general movement of groundwater in an isotropic aquifer can be deduced as being perpendicular to the contours on these surfaces, down the slope of the water table in the direction of the steepest hydraulic gradient. However, aquifers are three-dimensional

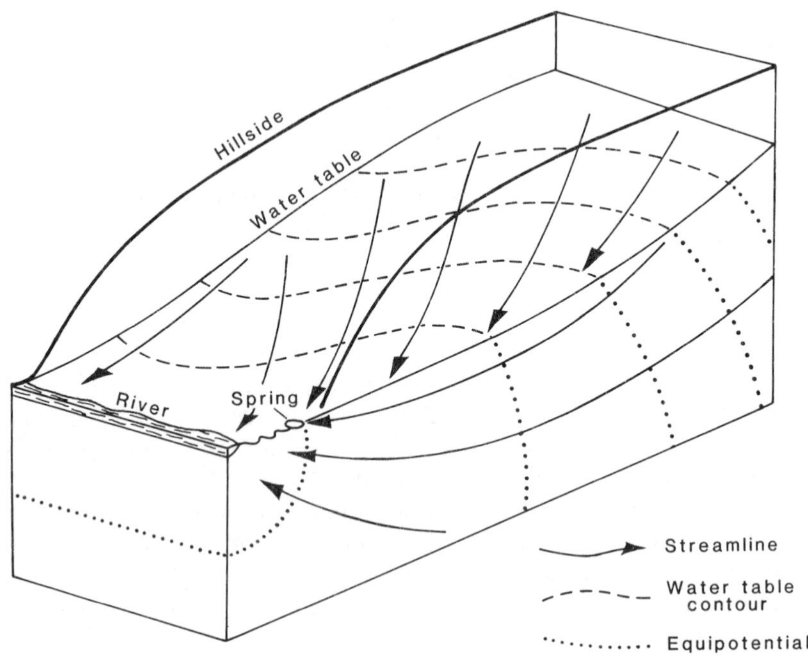

Figure 5.7 Water table contours, streamlines, equipotentials and flow net.

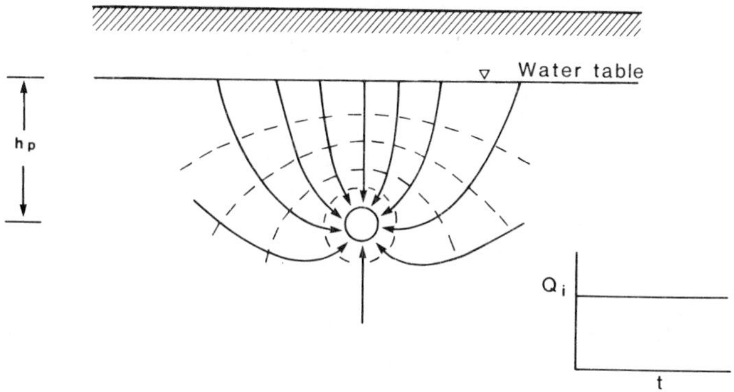

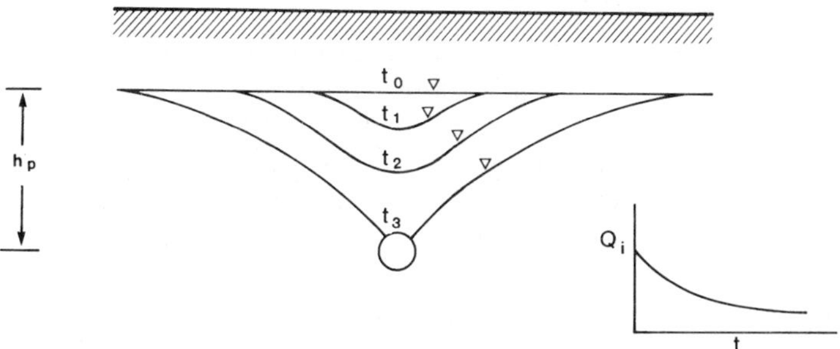

Figure 5.8 Tunnel or cave as (a) a steady-state and (b) a transient drain. From Freeze & Cherry (1979).

phenomena; thus a more complete view of groundwater movement will consider variations in *hydraulic potential* throughout the aquifer.

Points of equal hydraulic potential within an aquifer may be depicted by equipotential surfaces. In the horizontal plane, these can be mapped in two-dimensions as *equipotential lines*, and are represented in the case of an unconfined aquifer by contours of equal vertical interval on the water table (Fig. 5.7). In the vertical plane, and also in two-dimensions, the cross-sections of equipotential surfaces may likewise be mapped.

Water flow in an isotropic aquifer is always orthogonal to the equipotentials, the path followed by a particle of water being known as a *streamline*. A mesh formed by a series of equipotentials and their corresponding streamlines is known as a *flow net*.

Since water flows from zones with high potential to places where it is lower, if fluid potential increases with depth then groundwater flow will be towards the surface (Hubbert 1940). Flow nets in the vertical plane parallel to the hydraulic gradient of the water table usually show flow to converge near valley floors or along the coast. If a large conduit traverses the saturated zone, it may also be a zone of relatively low fluid potential and thus may cause flow to converge on it (Fig. 5.8). Goodman *et al.* (1965) show that the rate of groundwater inflow Q_i along a unit length of such a conduit can be estimated by

$$Q_i = \frac{2\ \pi K h_z}{2.3 \log (2\ h_z/r)} \tag{5.9}$$

where h_z is the depth of the conduit of radius r beneath the water table. The rock is assumed isotropic with hydraulic conductivity K. Freeze & Cherry (1979) provide a useful discussion of this approach.

The fresh water/salt water interface

Near the coast the water table declines towards sea level. Water quality analyses from samples taken at various depths from bores just inland show that fresh water overlies salt water, which penetrates the aquifer at depth. This interesting phenomenon was first investigated at the turn of this century by two European scientists, Ghyben and Herzberg, whose names are lent to the relationship they found (Reilly & Goodman 1985). The depth below sea level Z_s at which the fresh water/salt water interface occurs is related to the elevation of the water table above sea level h_f and to the density of the fresh ρ_f and salt ρ_s waters respectively. The *Ghyben–Herzberg principle* can be stated as

$$Z_s = \frac{\rho_f}{\rho_s - \rho_f} \cdot h_f \tag{5.10}$$

Thus if the density of the fresh water is 1.0 and that of the salt water is 1.025, then under hydrostatic equilibrium the depth to the salt water interface is 40 times the height of the water table above sea level. A practical consequence of this is that if the pumping of a bore in a coastal aquifer causes the water table to be drawn down by 1 m, then salt water will intrude upwards beneath the well by a distance of 40 m. Excessive pumping can therefore risk contamination by saline water.

The Ghyben–Herzberg principle simplifies the relationship usually found in nature, because the two fluids are treated as immiscible and groundwater conditions are normally dynamic rather than static. Hubbert (1940) showed that the interface is deeper under dynamic conditions than under static. He treated the interface as a boundary surface that couples two separate flow fields, with continuity of pressure being maintained across the interface

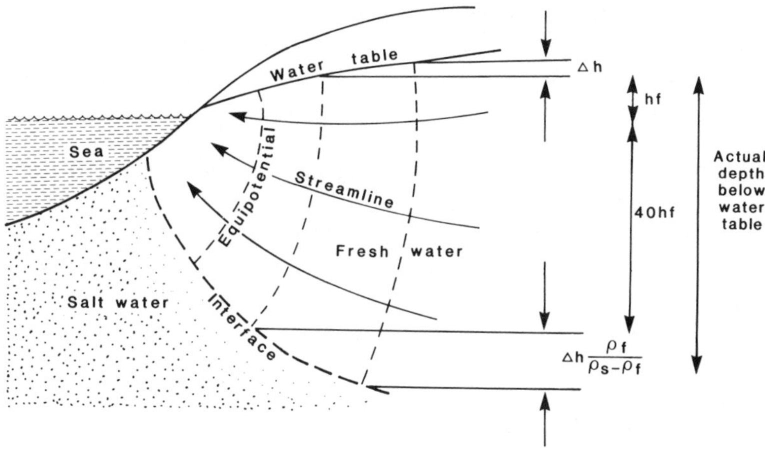

Figure 5.9 Illustration of the Ghyben–Herzberg principle under hydrodynamic conditions. From Hubbert (1940).

(Fig. 5.9). Mijatovic (1984a) discusses the application of this approach to the Dinaric karst. Recent general advances in the mathematical description of fresh water–salt water relationships are reviewed by Reilly & Goodman (1985).

The interface between fresh and salt water can be seen from Figure 5.10 to be a zone of transition rather than an abrupt discontinuity. It is also

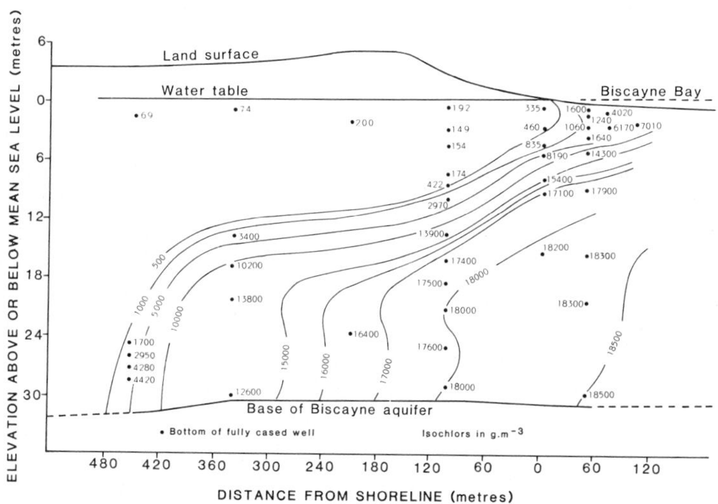

Figure 5.10 (A) isochlores and (B) equipotential lines in the Biscayne aquifer, Florida. From Ward (1975).

evident from the equipotential lines that much of the fresh water must escape through the sea bed in the near shore zone. The existence of submarine springs (vruljas) in karst terrains is a well known phenomenon recognized from at least the 1st century BC (Herak & Stringfield 1972b). Their occurrence implies confined pipe flow at depth and their location below present sea level may be partly related to the position of springs developed during low stands of the sea in glacial episodes (see section 10.3). High secondary porosity below present sea level in the zone of mixing is also a consequence of the geochemistry of brackish water (see section 3.6). The geochemistry and its karst geomorphological significance is discussed by Back *et al.* (1984) for a site on the Yucatan Peninsula.

5.2 Applicability of Darcy's law to karst

It is of fundamental importance to establish whether it is justified to treat a karst aquifer as a porous medium in the Darcian sense. In Darcy's experiment, discharge was measured from a given cross-sectional area a of the saturated medium. Hence in Equation 5.2 Q/a is an expression of the discharge per unit area. It therefore has the dimensions of a velocity and is sometimes simply denoted by u, the specific discharge (*filtration velocity* or *Darcy flux*). However, the flow does not issue from the entire cross-sectional area, but only from the voids between the solid grains. It follows then that the real microscopic velocities of flow through the interstitial spaces must be considerably larger than the average, macroscopic velocity denoted by u. Thus an implication of accepting the Darcian approach is that the rock is considered as a continuum of voids and solid matter for which certain generalized macroscopic parameters (such as K) can be defined, that represent and in some sense describe the true microscopic behaviour. In karst this means that the fractured rock penetrated by solution conduits would be replaced by a conceptual representative continuum for which it is assumed possible to determine hydrologically meaningful macroscopic parameters.

Reviewing earlier experimental evidence, Bocker (1973/4) concluded that if characteristic fracture size is 3 mm, then Darcy's law is not valid if hydraulic gradient is $\geqslant 0.01$. In a comprehensive series of experiments, Ewers (1972, 1982) showed that the law applies strictly when solutional proto-conduits up to 1 mm diameter are first extending through a fissure (Fig. 7.6). It ceases to apply once the extension is completed and the proto-conduit is connected to others. Mangin (1975) also suggested that in karst the range of conditions under which Darcy's law can be considered valid is very restricted.

An alternative to Darcian-based techniques of aquifer analysis is to accept the anisotropic and heterogeneous nature of a karst aquifer and to treat it as

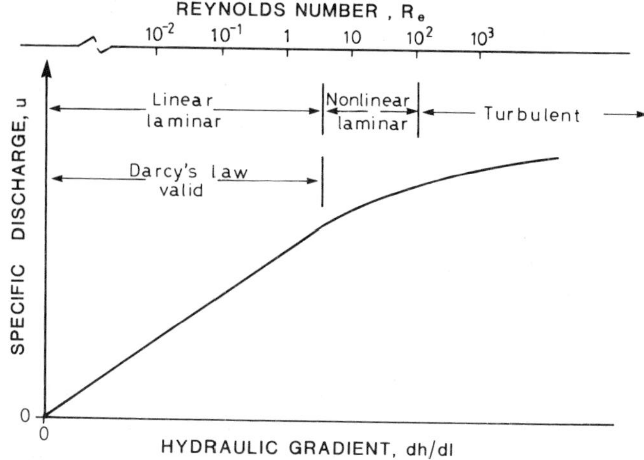

Figure 5.11 Range of validity of Darcy's law. From Freeze & Cherry (1979).

an interconnected conduit system in a more-or-less porous (or fissured) matrix. This is closer to the Hagen–Poiseuille approach in which the hydraulics of flow in individual fractures and pipes are considered. This method of aquifer analysis takes a dynamic input–output systems perspective and focuses particularly on the spring hydrograph (Bakalowicz & Mangin 1980). The assumption made is that the spring hydrograph provides an integrated representation of the network of stores and passages delivering water to the aquifer outflow point, the quality of the water as well as its quantity being of diagnostic importance (see section 6.5).

Because of the wide range of aquifer types encountered in karst, one must conclude that both approaches to aquifer analysis have their place. For the purposes of regional groundwater resources assessment and management, the continuum representation of the aquifer is often an acceptable generalization, particularly when hydraulic gradients are low and storage very large. It is largely a matter of fissure frequency and scale (Snow 1968 & 1969, Kovacs 1983), the lower the fissure density the larger the volume required for macroscopic generalization (Fig. 5.4). The considerable significance of fissure frequency for cave development is discussed in Chapter 7.

Another problem associated with the application of Darcy's law to karst is that of the flow regime, for the law is only valid when flow is laminar. Freeze & Cherry (1979) point out that Darcy's law is a linear law and that if it were universally valid a plot of the specific discharge against hydraulic gradient would reveal a straight line gradient for all hydraulic gradients. This is not the case (Fig. 5.11). At relatively high rates of flow Darcy's law breaks down.

The *Reynold's Number* R_e is usually used to help identify the critical velocity at which laminar flow gives way to turbulent flow (Vennard & Street 1976). It is thus valuable in helping to define the upper limit to the validity of Darcy's law. The Reynold's Number is expressed as:

$$R_e = \frac{\rho \, v \, d}{\mu} \tag{5.11}$$

where v is the mean velocity of a fluid flowing through a pipe of diameter d. In a porous or fissured medium, the macroscopic velocity u may be substituted for v and d becomes a representative length dimension characterizing interstitial pore space diameter or fissure width.

Bear (1972) concluded from experimental evidence that Darcy's law remains valid provided R_e does not exceed 1 to 10. Since fully turbulent flow does not occur until velocities are high and R_e is in the range 10^2 to 10^3, there is an interval between the turbulent and linear laminar regimes characterized by non-linear laminar flow. It should be noted that dynamic viscosity μ varies markedly with temperature, in the tropics being about half that encountered in the cool temperate zone (Table 5.3). Thus under some conditions, turbulent flow will occur in the tropics when it would be laminar in a cooler groundwater environment.

While it may be readily accepted that flow through a granular aquifer is usually laminar, less confidence may attach to this assumption if the rocks are fractured and penetrated by solution pipes, especially when considering the microscopic velocities that may be attained. Since the specific discharge ($u = Q/a$) defines the macroscopic velocity through the medium, the average microscopic velocity u^* can be determined by taking into account the actual cross-sectional area of voids through which the flow occurs. This depends upon the porosity n (Eqn 5.5) and hence

$$u^* = \frac{Q}{na} \tag{5.12}$$

If pore spaces through which water flows comprise 20% of the rock ($n = 0.2$), then u^* is about five times the Darcy flux. However, since water follows relatively long tortuous flow paths through the rock, the actual velocities must be greater still (cf. tortuosity term in Eqn 3.37).

Rock cores extracted from bore holes provide evidence of the size range of fissures and pipes encountered in a karst aquifer. Water tracing tests can be used to determine how quickly water moves. From such evidence the appropriateness of the laminar flow assumption can be judged. Conduit flow may remain in the laminar regime in pipes up to about 0.5 m diameter provided velocity does not exceed 1 mm s^{-1} (Fig. 5.12).

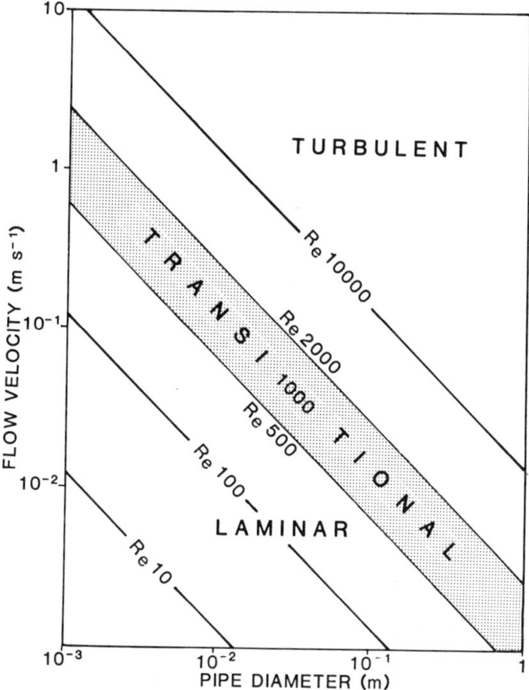

Figure 5.12 Values of Reynolds Number at various velocities and pipe diameters, with fields of different flow regimes. From Smith *et al.* (1976).

Karst aquifers often comprise an extensive, thick, fissured mass of rock traversed by the occasional large branching pipe, transmitting the flow of an underground river and its tributaries. Within the main and areally extensive body of rock the continuum approach using Darcy's law is usually justified to obtain spatially defined values of hydraulic characteristics, but this is totally inappropriate for describing flow in individual major conduits.

Under laminar flow conditions, discharge through a pipe can be evaluated by Poiseuille's law (Eqn 5.1) which shows specific discharge to vary directly with the hydraulic gradient and with the fourth power of the radius, but inversely with the length of the tube and the viscosity of the fluid; or with the Hagen–Poiseuille equation (Vennard & Street 1976)

$$Q = \frac{\pi\ d^4 \rho g}{128\ \mu} \cdot \frac{dh}{dl} \qquad (5.13)$$

where $\frac{dh}{dl}$ is the head loss over a unit length of the pipe. Since the discharge is proportional to the fourth power of the diameter, large capillary tubes are very much more conductive than small ones. A tube of 2 mm diameter will conduct the water passed by 10 000 capillaries of diameter 0.2 mm under the same hydraulic gradient. This factor dominates the processes of cave pattern construction as they are developed in Chapter 7.

Increasing velocity, sinuosity and roughness may eventually result in flow through the tube becoming turbulent. When this occurs, the specific discharge may be calculated using the Darcy–Weisbach equation (Thrailkill 1968)

$$u^2 = (2dg/f) \,.\, (dh/dl) = Q^2/a^2 \tag{5.14}$$

and hence

$$Q = (2dga^2/f)^{1/2} \,.\, (dh/dl)^{1/2} \tag{5.15}$$

where f is a *friction factor* or coefficient. Spring & Hutter (1981a, b) explain the determination of the friction factor and discuss the relationship of the Darcy–Weisbach and Manning–Gauckler–Strickler approaches. They point out that in turbulent pipe flow, wall friction gives rise to a quadratic dependence of shear stress τ on mean water velocity v, as expressed by the Darcy–Weisbach friction law:

$$\tau = \frac{f\,\rho_f}{8}\,v^2 \tag{5.16}$$

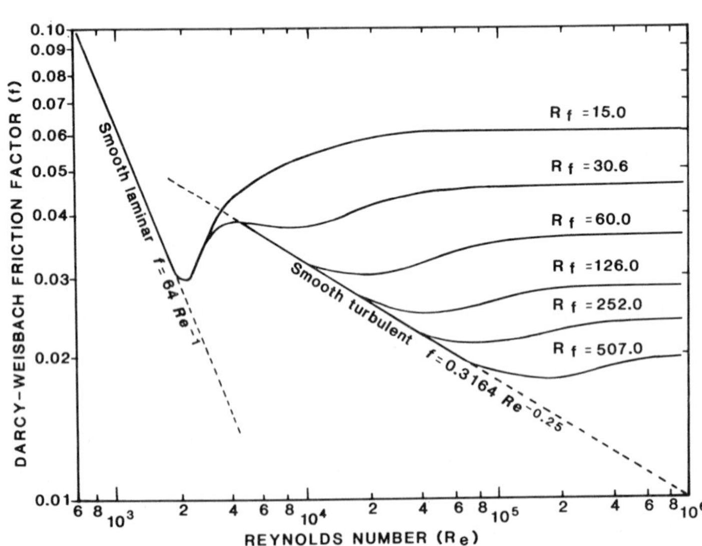

Figure 5.13 Experimental curves of the Darcy–Weisbach friction coefficient as a function of Reynolds Number and relative roughness ($R_f = r/e$) for flow through circular pipes. From Allen (1977).

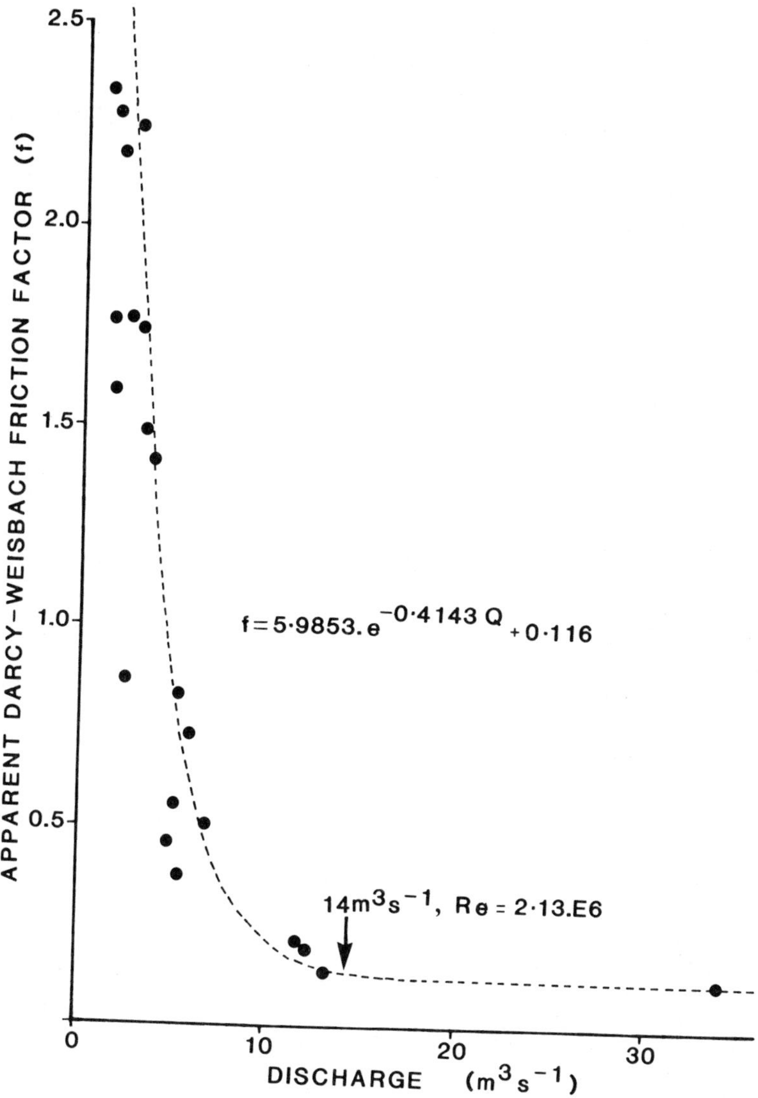

Figure 5.14 Relationship between f and Q in an active phreatic conduit in Norway. From Lauritzen *et al.* (1985).

where ρ_f is the density of fresh water. The friction factor can thus be estimated by rearranging the equation. Relative roughness R_f on a flow boundary (e.g. walls of a pipe) can be described by the ratio of pipe radius r to a characteristic length e representing the size of the features contributing to roughness, such as projections, cavities and loose grains. Its relation to f and the Reynolds Number is illustrated in Figure 5.13.

An excellent study by Lauritzen *et al.* (1985) reveals the relationship between f and Q in an active phreatic conduit in Norway (Fig. 5.14). Apparent friction shows a dramatic decrease with discharge, until attaining a constant value. Friction is affected by complexity of conduit geometry, tube dimensions, roughness of walls and occurrence of breakdown. In the comparatively simple conduit studied in Norway, Lauritzen *et al.* found f to be particularly related to the hydraulic radius of the passage and the characteristic scallop size on its walls. Darcy–Weisbach f values determined in karst investigations lie in the range 0.039 to 340, whereas 0.25 was found appropriate for a cave in ice (Spring & Hutter 1981b).

5.3 Controls in the development of karstic aquifers

The development of karstic aquifers depends principally upon geological, geomorphological, climatic and biological controls. The relationship of these controlling factors to the physical characteristics of an aquifer, such as porosity, hydraulic conductivity and storage capacity, is illustrated schematically in Figure 5.15. There is a considerable interaction amongst the principal controlling factors and between them and the chemical and mechanical processes whose rates of activity they determine. Geomorphological, climatic and biological factors also dictate the boundary conditions of the aquifer through their control of the sites and quantities of recharge and discharge.

The relative influence of the geomorphological, geological and process factors determines the distribution of voids in the karst rock and through that the physical characteristics of effective porosity, hydraulic conductivity and specific storage. For a given set of boundary conditions, hydraulic gradient and specific discharge can then be estimated. But since the very process of karst groundwater circulation modifies effective porosity, specific storage and hydraulic conductivity, and lowering of the outlet spring modifies hydraulic potential, then the karst circulation system must undergo a continuous process of self-adjustment.

The most abrupt changes to karstic aquifers are brought about by geomorphically rapid events culminating in major alterations to hydraulic gradient because of modifications to boundary conditions. For example, valley deepening by glaciation increases the hydraulic potential of the system, whereas submergence of coastal springs by glacio-eustatic sea level rise reduces it.

Geological control on the development of karst aquifers operates in several ways (Table 5.5) and is discussed in more detail in Chapter 2. Boundary conditions are influenced at the regional scale through the definition of outcrop pattern, thickness and properties of karst rocks and their relationship to other lithologies. Tectonism affects the balance between

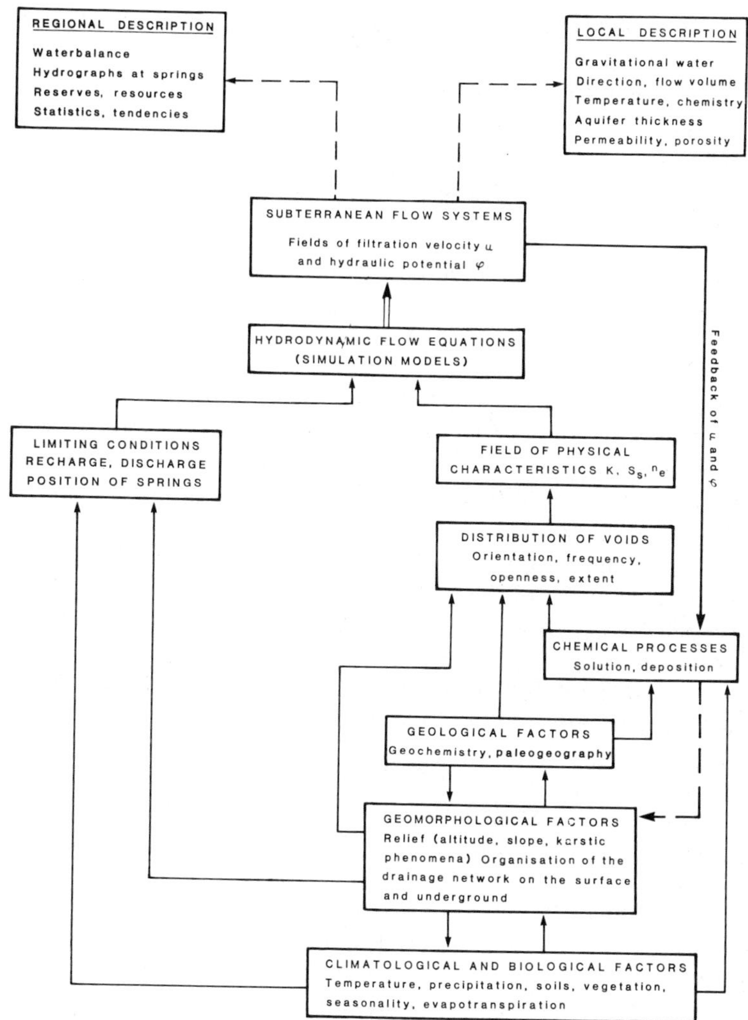

Figure 5.15 Schematic relationship between hydrologic factors, physical properties of the aquifer and geological characteristics of the karst. From Kiraly (1975).

rates of uplift and denudation and thus has a major influence on hydraulic potential. Regional structure is also important for its control of folding and faulting. Anticlinal and synclinal structures are associated with tension and compression respectively and thus with joint patterns (Fig. 2.16) that reflect these condition (Price 1966). Joints are most readily penetrated by percolating waters when under tension; thus anticlines (and domes) are potentially important sites for aquifer recharge. On the other hand, synclinal

Table 5.5 Effect of hydrogeologic setting on carbonate aquifers. From White (1977b).

Geologic element	Control
Macro-structure (folds, faults)	Placement of carbonate rock units relative to other rocks
Topographic setting	Placement of recharge and discharge regions
Stratigraphic sequence	Thickness and chemical character of aquifer
Mini-structure (joints, fractures)	Orientation and transmissibility of primary flow paths
Relief	Defines hydraulic gradients

troughs favour flow convergence and the accumulation of groundwater. The significance of fracture traces and lineaments in karst hydrology has been investigated by Parizek (1976) and Lattman & Parizek (1964).

Bedding-planes also play a role in linking joint dominated routes for downward percolation in the vadose zone, but are more important in the phreatic zone because of their great lateral continuity. As dip becomes steeper, so bedding-plane partings can increasingly provide recharge routes. Water confined in major bedding-planes between dense, thick sheets of rock may be led to great depths in what amounts to artesian conditions, before cross-joints permit lateral movement.

Faults often operate hydrologically like major joints. Their vertical and lateral continuity can render them particularly important features in the orientation of water flow in both the vadose and phreatic zones. However, many fault planes are highly compressed or filled with secondary calcite of low porosity; they are then barriers to groundwater flow. Further, they are also sometimes important in introducing blocks of other lithologies that may act as barriers to water movement. This may arise from normal or reverse faulting of adjacent non-karst rocks or may involve fault lane guided intrusions of igneous material. These impose an impervious curtain across an aquifer, considerably interrupting groundwater flow and aquifer development.

As a consequence of the effects of fissuring and differential solution, permeability may be greater in some directions than in others, as well as in certain preferred stratigraphic horizons. However, while geological factors dictate where storage is greatest, local relief normally exerts a still greater influence on the direction in which groundwater flows, because hydraulic gradient is strongly influenced by it. It is local relief that determines both the highest positions where recharge can occur and the lowest points at which groundwater outflow can take place. Where the horizontal distances between the point of input and output are minimized, the hydraulic gradient is steepest. Thus, the shortest flow path determines the direction of groundwater movement in an isotropic aquifer, because its innumerable pores and fissures provide pathways for water flow in any direction; i.e. their control on flow is merely of secondary importance, a rate control. However, where strongly preferred fissure patterns cause the aquifer to be markedly anisotropic, the orientation of maximum hydraulic gradient at an arbitrary point within the flow field will reflect a balance between the direction in which resistance to flow is least (i.e. where hydraulic conductivity maximized) and the direction in which the rate of energy loss is maximized (i.e. the shortest and steepest route).

Within a given karst drainage system, the trunk conduit acts as the local discharge focus for tributaries. Hence local flow directions may diverge from the general trend. Between adjacent karst systems in the same karst region, competitive advantage stemming from different geomorphic histories and allogenic inputs may also be significant in determining basin shape and drainage orientation. Thus if we critically assess, for example, the factors controlling the development and character of the karstic aquifer below the Sinkhole Plain of Kentucky (Fig. 6.31), we may appreciate that geology is just one element amongst the several factors involved. When relief energy is high and a karst has experienced a long period of evolution, the route that maximizes hydraulic gradient is usually also the shortest path. However, even when relief energy is very low, such as in the Gort Lowland of western Ireland (Fig 9.31 & 9.32), the regional hydraulic gradient in a well karstified rock may still be more strongly dominated by the shortest path than by intervening geological factors, even though anisotropy is marked and has a discernible local influence on flow directions.

Input control

A karst aquifer can be envisaged as an open system with a boundary defined by the catchment limits and with input, throughput and output flows, mechanisms and controls. In the simplest case, only karst rocks are found within the catchment and recharge is derived solely from precipitation falling directly on them – termed *autogenic recharge*. However, commonly more complex geological circumstances occur and runoff from neighbouring or overlying non-karst rocks drains into the karst aquifer – termed *allogenic recharge*. Whereas autogenic recharge is often quite diffuse, down many fissures across the karst outcrop, allogenic recharge normally occurs as concentrated point inputs of sinking streams. Both the water chemistry and the recharge volume per unit area are different in these two styles of recharge, with considerable consequences for the scale and distribution of the development of secondary permeability.

Emerged coral reefs provide natural examples of simple autogenic systems. Recharge is spatially uniform and distributed through innumerable pores and fissures across the outcrop. Where thick soils cover the bedrock recharge conditions are modified. If the soil is less permeable than the rock beneath, then it provides a recharge regulator, limiting recharge to the infiltration capacity of the soil. Permeable rock formations overlying the karst also act as percolation 'governors' in much the same way, their vertical saturated hydraulic conductivity being the principal control. Percolation input from a permeable soil is considered autogenic, whereas that from a permeable non-karstic caprock is considered diffuse allogenic in origin.

Relatively concentrated recharge occurs in autogenic systems only where solution dolines are well developed (Fig. 5.16). This is because solution dolines reflect underlying spatial inequalities in vertical hydraulic conductivity that result in the development of preferred percolation paths or zones. The funnelling of rainwater by enclosed depressions positively reinforces the significance of the underlying percolation path by an autocatylitic process (Williams 1983 & 1985). However, the volumes of point input recharge are small compared to those derived from allogenic basins because of the relatively small surface areas of individual dolines.

Concentrated inflows of water from allogenic sources sink underground at *swallow holes* (also known as swallets, stream-sinks or ponors). They are of two main types: vertical point inputs from perforated overlying beds and lateral point inputs from adjacent impervious rocks. A perforated impermeable caprock will funnel water into the karst in much the same way as solution dolines, except that the recharge point is likely to be more precisely defined and the peak inflow larger. Inputs of this kind favour the development of large shafts beneath. Lateral point inputs are usually much greater in volume, often being derived from large catchments, and are commonly associated with major river caves. The flow may come from: (1) a retreating overlying caprock; (2) the up-dip margin of a stratigraphically lower impermeable formation that is tilted; or (3) an impermeable rock across a fault boundary (Fig. 5.17).

The capacity of the input passages is the ultimate regulator of the volume of recharge; thus if inflow from surface streams is too great then ponding occurs, giving rise to overflow via surface channels or to surface flooding in blind valleys and poljes.

A more unusual form of input control is provided by head fluctuations caused by floods. This may result in discharge conduits reversing their function temporarily to become inflow passages. This occurs where outflow springs discharge at or beneath river level into a major river flowing through the karst. If a flood wave generated by heavy rain in an upstream part of its basin passes down the major river, the springs will be more deeply submerged and the hydraulic gradient in the karst will reverse, especially if the tributary karst catchment was unaffected by the storm. Inflow into the karst will then occur, like a form of bank storage. Water intruded by backflooding will later be withdrawn from storage as the floodwave in the main river passes and the hydraulic gradient reverts to normal. Reversing springs

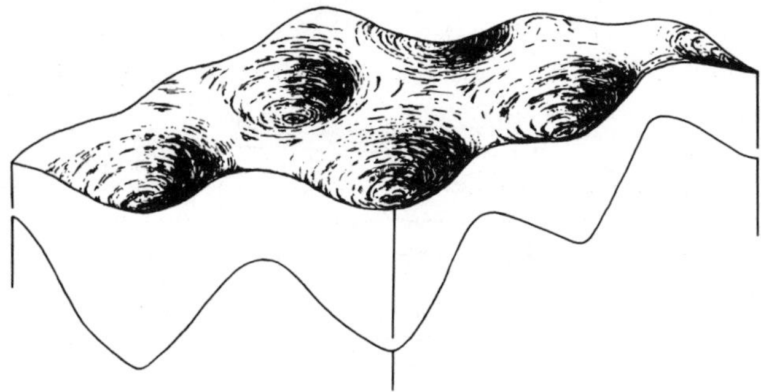

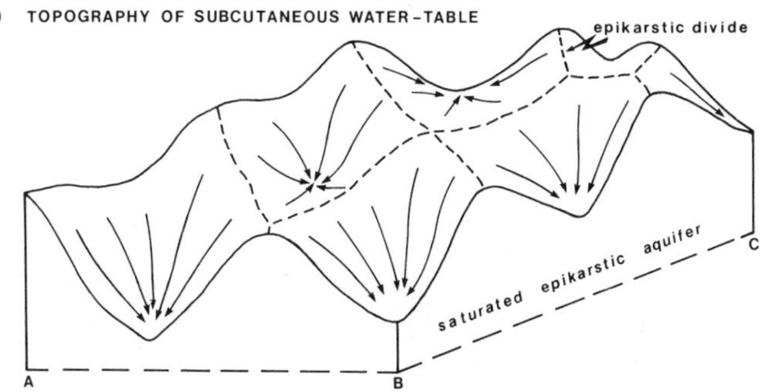

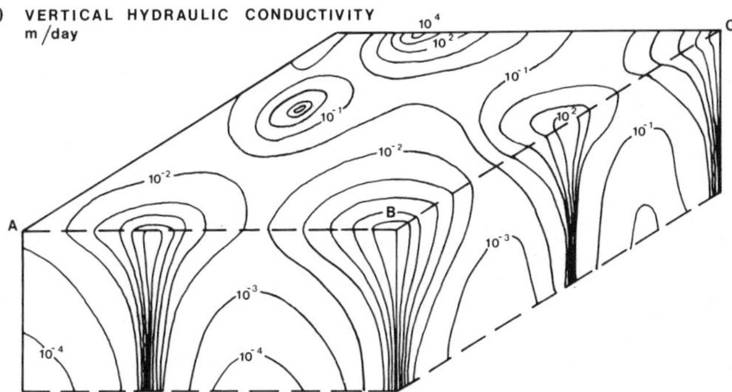

Figure 5.16 The relationship between (A) surface solution doline topography, (B) underlying relief on the subcutaneous water table, and (C) vertical hydraulic conductivity near the base of the subcutaneous zone. From Williams (1985).

1.

2.

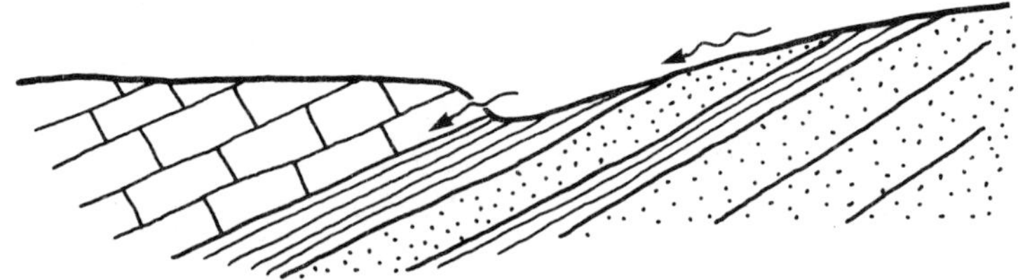

3.

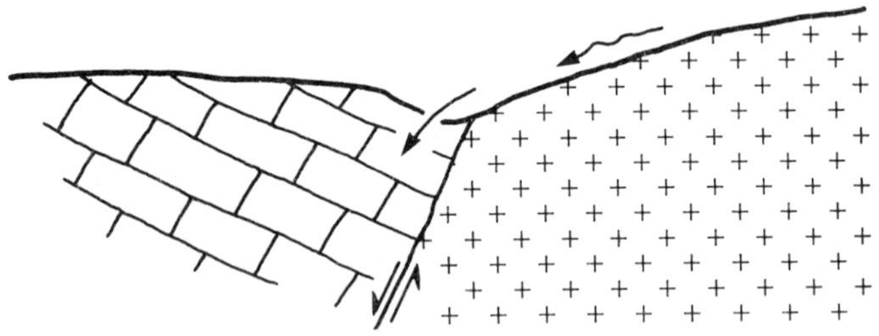

Figure 5.17 Recharge by allogenic streams flowing (1) from overlying beds, (2) from underlying beds exposed up-dip, and (3) across a faulted contact with impervious rocks.

of this sort that can temporarily function as sink points are known as *estavelles*. There are many examples along the Green River of Kentucky (Quinlan 1983). The converse is also common: some stream-sinks become temporary springs as the water table rises. These are also estavelles.

Output control

Most of the largest springs in the world are karst springs (Table 5.6). Only those from volcanic rocks rival their discharge output. They represent the

Table 5.6 Discharges of some of the world's largest karst springs.

Spring	Discharge ($m^3 s^{-1}$) mean	max	min	Basin area	Reference
Tobio, Papua New Guinea	85–115	–	–	–	Bourke 1981, Maire 1981c
Matali, Papua New Guinea	90	>240	20	350	Maire 1981b
Bussento, Italy	–	117	76	–	Bakalowicz 1973
Dumanli*, Turkey	50	–	25	2800	Karanjac & Gunay 1980
Trebišnjica, Yugoslavia	50	250	3	–	Maćejka 1976
Buna, Yugoslavia	40	440	2	–	Maćejka 1976
Galowe, New Britain	40	–	–	–	Maire 1981a
Ljubljanica, Yugoslavia	39	132	4.25	1100	Gospodarič & Habić 1976
Ras-el-Ain, Syria	39	–	–	–	Burdon & Safadi 1963
Stella, Italy	37	–	23	–	Burdon & Safadi 1963
Chingshui, China	33	390	4	1040	Yuan 1981
Yedi Miyear, Turkey	–	–	25	–	Bakalowicz 1973
Vaucluse, France	29	200	4.5	2100	Paloc 1970
Frió, Mexico	28	515	6	>1000?	Fish 1977
Coy, Mexico	24	200	13	>1000?	Fish 1977
Silver, USA	23.25	36.5	15.3	1900	Faulkner 1976
Lulangdong, China	–	74.6	8.9	1000	Yuan 1981
Ombla, Yugoslavia	–	165	4.1	>600	Milanović 1976
Velika Ruda, Yugoslavia	20	50	2	–	Maćejka 1976
Vrela Pliva, Yugoslavia	20	32	8	–	Maćejka 1976
Peschiera, Italy	18	–	15	–	Nicod 1972
Timava, Italy	17.4	138	9	980	Gams 1976
Kavakuna, Papua New Guinea	15	–	–	–	Bourke 1981
Waikoropupu, New Zealand	15	21	5.3	450	Williams 1977
Niangziguan	13.5	~17	~11	3600	Zhang 1980
Maligne, Canada	13.5	45	1	730	Kruse 1980

(i) *Dumanli spring is the largest of a group of springs that collectively yield a mean flow of 125–130 $m^3 s^{-1}$ at the surface of the Manavgat River.

(ii) Data for Papua New Guinea are not good because of field conditions, but other large springs are known with discharges of >15$m^3 s^{-1}$, e.g. Nare and Minye.

(iii) Details of other significant but smaller springs may be found in Bakalowicz (1973).

termination of underground river systems and mark the point at which surface fluvial processes become dominant. The vertical position of the spring controls the elevation of the water table at the output of the aquifer, whereas the hydraulic conductivity and throughput discharge determines the slope of the water table upstream and its variation under different discharge conditions.

The difference in elevation between the spring and the water table upstream determines the head in the system and thus the energy available to drive a deep circulatión. Hence springs exercise considerable control on the operation of karst groundwater bodies. Furthermore, that control can vary markedly, because springs are most susceptible to geomorphic events such as glacio–eustatic sea level fluctuations, valley aggradation, and valley deepening by glacial scour.

The influence which springs exert on the aquifer which they drain depends principally upon the topographic and structural context of the spring. Springs may be classified in several ways (Trombe 1952, Sweeting 1972, Bogli 1980), but when considering their hydrological control function the following perspective is important:

FREE DRAINING SPRINGS (Fig. 5.18a, b)
In these cases the karst rock slopes towards and lies above the adjacent valley, into which karst water drains freely under gravity. The karst system is entirely or dominantly vadose, and is sometimes termed shallow karst (Bogli 1980). Complications may arise where the underlying impermeable rock is folded or has an irregular surface, because then subterranean ponding can occur with the consequent development of isolated phreatic zones (Fig. 5.18b).

DAMMED SPRINGS (Fig. 5.18c–e)
These are the most common type of karst springs. They result from the location of a major barrier in the path of underground drainage. Impoundment may be by another lithology, either faulted or in conformable contact, or be caused by valley aggradation, such as by glacifluvial deposits. The denser salt water of the sea also forms a barrier to fresh water outflow. In each case, temporary overflow springs may form in response to high water tables. The type of cave upstream of the spring will determine whether its discharge spills from a flat passage developed close to the water table or wells up from a great depth within the phreatic zone. Thus a dammed karst outflow site typically consists of a main low water spring with one or more associated high water relief springs; Smart (1983a, b) has termed these overflow–underflow systems.

CONFINED SPRINGS (Fig. 5.18f, g)
Artesian conditions prevail where karst rocks are confined by an overlying

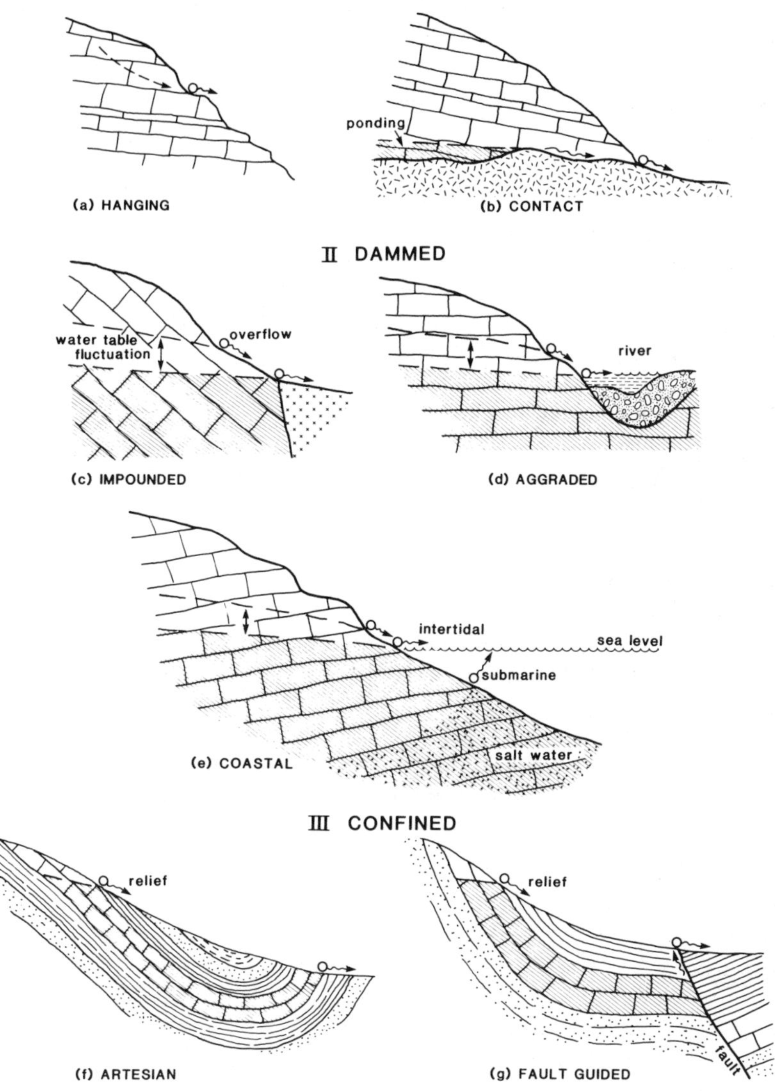

Figure 5.18 Types of springs encountered in karst.

impervious formation. Fault planes sometimes provide exit routes for the water; elsewhere it may escape where the caprock is breached by erosion. Since the emerging water is usually under hydrostatic pressure, an up-domed turbulent 'boil' is particularly characteristic of spring pools in this class, although dammed springs that are semi-confined by a particularly thick bed may also 'boil', especially during flood. Artesian springs are also sometimes termed *vauclusian* after the type example, La Fontaine de Vaucluse (Table 5.6), in the south of France (Durozoy & Paloc 1973).

The discharge capacity of the artesian spring determines the elevation of the potentiometric surface in the aquifer and hence the depth of the phreatic zone. Artesian springs may also have associated high water relief springs.

Other characteristics have also been used to classify karst springs, as noted by Bogli (1980); these include:

(a) according to the outflow
   - perennial
   - periodic
   - rhythmic (ebb and flow)
   - episodic

(b) according to the supposed origin of the water
   - emergence (no evidence of origin)
   - resurgence (re-emergence of a known swallet stream)
   - exsurgence (autogenic seepage water)

Periodic and rhythmic springs, sometimes referred to as ebbing and flowing wells, are particularly interesting natural phenomena. They are usually dammed springs with a siphoning reservoir system controlling their outflow. They are discussed by Trombe (1952), Mangin (1969a, b) and Gavrilovic (1970).

Throughput control

The factors which determine the density, size and distribution of voids are of fundamental importance in controlling the throughput and storage of water in a karst aquifer, because they dictate the potential flow paths. Effective porosity (Eqn 5.6) is strongly influenced by pore size and determines specific storage and storativity (Eqn 5.8). Void space (diameter or width) available for water movement ranges over seven or eight orders of magnitude up to tens of metres and since permeability is a function of void size it also varies within a wide range. As void size and continuity increase, permeability increases and resistance to flow diminishes; thus hydraulic conductivity (Eqn 5.3) is enhanced and, for a given aquifer thickness, transmissivity (Eqn 5.7) also grows. Babushkin *et al.* (1975) discuss these points further.

Voids of different size, distribution and density result from: (a) the properties of the aquifer rock, (b) the style of recharge and (c) time, which permits the evolution of input and output passages. Voids are partly inherited (primary porosity) and partly acquired (secondary porosity), as discussed earlier.

The intrinsic properties of some rocks, such as young, highly porous corals and calcareous aeolianites, immediately permit significant aquifer storage and throughput even before a significant amount of solution, whereas at the other extreme marble and evaporites may be almost impermeable before karstification takes place. Rocks with an initially low void density will tend eventually to develop high conduit permeability with minimal fissure storage and diffuse phreatic flow. Whereas rocks that are porous, thin bedded and highly jointed will still develop conduit permeability if point recharge occurs, but will always have a relatively high diffuse component in the saturated zone.

The development of secondary voids is very strongly influenced by characteristics of recharge; so much so that simply because of this throughput conditions in allogenic and autogenic systems are often radically different, even in the same rock type. Allogenic point recharge favours the growth of large stream passages, whereas spatially diffuse autogenic recharge enhances pore and fissure porosity, but has little impact on the development of conduits. Thus for a given lithology, in the first case rapid turbulent throughput can occur, while in the second case flow may be mainly laminar and diffuse.

As karstification proceeds and large secondary cavities (i.e. cave systems) develop in the phreatic zone, there is a progressive decoupling of flow between that passing relatively rapidly through the karst pipes and that in the surrounding porous and fissured matrix (White 1977b). Water movement may be rapid and turbulent in one, while slow and laminar in the other. This makes the analysis of the aquifer and its response to recharge particularly difficult. Thus, for example, the speed with which a spring responds to recharge is not a simple reflection of the velocity of water flow through the aquifer; there will be a range of velocities.

If significant secondary porosity develops in the shallow phreatic zone two things happen: (a) the increased horizontal hydraulic conductivity favours greater water movement there rather than at depth and (b) the increased storage lowers the water table gradient. Circulation in the deep phreatic zone thus becomes less vigorous, and as karstification continues still farther, the increasingly active shallow phreatic zone may lead to relatively stagnant conditions in the deepest phreatic voids. A stagnant phreas can also be produced as a consequence of positive baselevel changes. A rising sea level, valley aggradation or tectonic subsidence can submerge a previously active phreatic region beyond the zone of contemporary circulation. It then becomes a variety of paleokarst and may be simply a passive store or vessel

Table 5.7 Flow velocities through karst conduits for straight line plan distances of more than 10 km. From Aley (1975), Bakalowicz (1973), Kruse (1980) and Williams (1977).

Locality	Distance km	Descent m	Time h	Velocity $m\ h^{-1}$
Taurus, Turkey	134	1095	8860	15.2
	103	735	5424	19.4
	81	985	8740	9.2
	35	1030	816	42
	22	920	456	48
Vaucluse, France	46	545	600	76.6
	24	420	1990	12
	22	500	840	26.2
	40	360	2200	18
Jura	21	521	180	117
Herault	21	112	2950	7.2
	16.5	60	405	54
	15.4	85	672	23
	11.6	600	96	120
Aude	16.6	550	408	39.6
	12.75	531	336	50.4
	12.25	429	408	28.8
Lot	11	136	2400	4.5
Aach, Germany	12.5	170	56	221
Timava, Italy	41	322	210	195
Ombrie	11.5	326	15	766
	11	240	15	733
Maligne, Canada	16	380	11–80	200–1450
Missouri, USA	25	<200		13-213
Waikoropupu, NZ	20.2	50	$3–7 \times 10^4$	0.3–0.7*

* This compares to a pulse through velocity of 2020.

for the receipt of the finest suspended sediment. Such features are known to depths greater than 3 km in oil wells.

Flooding allogenic streams inject water and sediment into an aquifer. The flood passes as a *kinematic wave* down the vadose cave passages. But once the saturated zone is reached, the recharge wave causes a rise in the water table and a *pressure pulse* is forced through phreatic conduits, giving a hydrograph peak at the spring. Kinematic waves in open channels travel about 30% faster than the water itself, in the order of tens to thousands of metres per hour, whereas pressure pulses through flooded pipes are propagated almost instantaneously (at the speed of sound). The *flow through time* of the water responsible for the hydrograph rise at a spring is much longer than either and can be estimated from the travel velocity of tracers such as dyes injected into the system (Table 5.7). The pressure pulse mechanism is also known expressively as *piston flow*.

Diffuse autogenic recharge can also generate pulses in percolation throughput, although some of the water displaced may be many months old due to storage in the epikarstic aquifer. Pitty (1966, 1968a & b), Gunn (1981c, 1983), Friederich & Smart (1981), Bakalowicz & Jusserand (1986) and Even *et al.* (1986) provide field evidence that percolation into caves has variable flow through times from weeks to decades, depending on the site. Variable lags of weeks to months were shown by Friederich & Smart (1981) and Williams (1983) to apply to pulse through times. Thus with a range of flow through and pulse through rates in the unsaturated and saturated zones, the output response of a karst aquifer to recharge is complex.

The development of conduit permeability in aquifers where secondary or fissure porosity is dominant is explained in Chapter 7. Figures 7.6–7.14 illustrate the principal features. Conduits propagate in a competitive manner from input boundaries. Flow through them is very slow and laminar until they become connected to an output point such as a previously developed conduit or a spring, when turbulence may commence after enlargement to about 5–15 mm diameter.

When primary porosity (chiefly fabric selective) is predominant, as is the case in many Pleistocene and Holocene reefs, a very diffuse flow and slow development of poorly integrated micro-conduits may occur. Flow is laminar in general but becomes turbulent at springs around the coast, where water typically emerges through fractures. In general, these aquifers behave like the granular Darcian model. Yet significantly they are transformed into relatively decoupled karstic aquifers if allogenic point recharge occurs along an input boundary. The cavernous calcareous aeolianite karst of Western Australia (Jennings 1968) illustrates this perfectly. Allogenic runoff from volcanic inliers on coral islands has the same effect (Stoddart *et al.* 1985).

The spacing and volume of recharge is a question of scale; high density plus low volume is the 'diffuse' end of the spectrum. Diffuse recharge into a porous medium dissipates corrosive activity with the result that fabric voids dominate flow routes. Point recharge into the same medium focuses corrosion, overcomes a threshold inhibiting cave development, and conduit permeability results. The very much greater potential energy of the large point recharge volume is probably the key factor.

We may conclude from this discussion that, given sufficient hydraulic potential, both the style of recharge and rock fabric have an influence on the occurrence, density and size of conduit permeability (although not on the *process* of conduit development). End member conditions are:

(a) diffuse recharge onto a carbonate rock with high primary porosity, e.g. rain on uplifted coral, when few or no karst conduits form; and
(b) widely spaced, large volume, point recharge into a dense carbonate rock with well developed fissures, e.g. recharge windows in a breached caprock over massive limestones, when a few very large diameter conduits form commensurate in size with their throughput discharge.

> Competition is limited to corridors downstream of recharge points that are separated by unkarstified rock beneath the umbrella of still intact caprock.

Doline karst falls between these two extremes, there being a large number of point inputs of modest volume (Fig. 5.16).

Time is a third important factor in the development of conduit permeability, particularly influencing successive ranks (or generations) of point inputs, as explained in detail in Chapter 7, and the size of individual passages.

If the rate of recharge during heavy rain is greater than the maximum rate of vertical throughput in the vadose zone, then excess recharge is stored in the void space of the highly corroded *subcutaneous zone*, which lies immediately beneath the soil but above the main mass of largely unweathered rock (Table 5.1). Water stored in that zone constitutes an *epikarstic aquifer* perched above a leaky capillary barrier. Once available storage is full, any continuing recharge overflows across the surface. The direction of subcutaneous flow is down the hydraulic gradient of the epikarstic water table.

Thus throughput in the vadose zone is both vertical by leakage from epikarstic storage and horizontal by subcutaneous water movement through the epikarstic aquifer. The ratio of vertical to horizontal flows depends upon the contrast in hydraulic conductivity in the upper and lower parts of the vadose zone. In well bedded, near horizontal carbonates, this depends pre-eminently on the frequency and pattern of solutionally corroded joints and bedding planes. Illustrations of vadose throughput processes are provided by Gunn (1981c, 1983) and Williams (1983).

5.4 Energy supply for karst aquifer development

The development of flowpaths in karst aquifers depends upon the energy supply available. This derives principally from the throughput volume of water and the difference in elevation between the recharge and discharge areas. The principal forms of fluid energy are potential energy, kinetic energy and internal energy.

Most of the potential energy is realized as kinetic energy as the water descends through the vadose zone, where much mechanical work is done by fluvial processes. Flowing water can be regarded as a transporting machine, the stream power of which can be determined by Bagnold's (1966) equation

$$\Omega = \rho g Q \theta \qquad (5.17)$$

where Ω is the gross stream power, ρ the fluid mass density, g gravitational acceleration, Q discharge and θ is slope. From this, the energy available per unit area of channel bed – termed *specific stream power* – can be derived from

$$\omega = \frac{\Omega}{W} = \tau v \tag{5.18}$$

where ω is specific stream power, W stream width, and τ is mean shear stress at the bed (see also section 8.1). Most of this power is dissipated in overcoming fluid shear resistance to flow; so relatively little energy surplus is available for mechanical erosion and transport by the stream.

Once the water table is reached the energy of water per unit mass is determined by gravitational acceleration times the local head, i.e. by the height of the water table above the outflow spring. It is this energy coupled with the vertical hydraulic conductivity characteristics of the phreatic zone that determines the depth to which groundwater can circulate.

The velocity of water flow in karst varies considerably both within a given aquifer and between aquifers. Thus, for example, Friederich & Smart (1981) found from dye tracing in the Mendip Hills that the horizontal flow through rates of water through the epikarstic store is of the order of 4 m h^{-1} with vertical rates in the range of 2–100 m h^{-1}. By contrast Stanton & Smart (1981) found by repeated (43) dye tests in three largely phreatic swallet to resurgence systems on Mendip that average flow rates are about 50–100 m h^{-1}, although actual travel time is inversely proportional to mean resurgence discharge over the same period. This relationship is brought out still more strongly in the Maligne River system (Table 5.7) (Kruse 1980). Milanovic (1981) reports that from 281 dye tests over distances of 10–15 km or more in the Dinaric karst, 70% of cases have a flow velocity less than 180 m h^{-1} although they vary over a range of 7.2 to 1880 m h^{-1}. Velocities greater than 500 m h^{-1} are rare, and those less than 18 m h^{-1} involve long water retention underground. Highest velocities are in vadose cave streams. Milanovic concludes that although the energy gradient affects the average flow velocity, it is not the only factor determining it.

Internal energy, best expressed by the fluid's temperature, is a secondary factor that moderates mechanical processes through its influence on the dynamic (and kinematic) viscosity of water (Table 5.3), which is more than twice as viscous at 0°C as at 30°C. The lower viscosity at higher temperatures permits a greater discharge through capillary tubes (Poiseuille's law, Eqn 5.1), increases hydraulic conductivity (Eqn 5.3) in porous media, and increases the Reynolds Number for a given specific discharge (Eqn. 5.11). The influence of temperature on the physics of flow may therefore help to explain some of the differences in karst encountered in the cool temperature and tropical zones, by influencing penetration distances of capillary water and the consequent work that can be done by chemical processes.

Total chemical energy for solution can be viewed as the product of the aggressivity and volume of the solvent throughput, although the energy available for individual chemical reactions in a limestone karst is applied by chemical potential differences between water, CO_2 and calcite on the one hand and dissolved calcium bicarbonate on the other. Kinetic factors, which are largely temperature dependent, determine the rate at which solution occurs; while equilibrium factors determine the ultimate solute concentrations that can be achieved given sufficient time (Chapter 3). In a conduit dominated groundwater system, throughput is fast and there is likely to be insufficient time for most of the water to reach equilibrium solute concentrations; thus for a given discharge kinetic factors dictate both the location and the total amount of work done by chemical processes. In a preponderantly porous karst rock with a diffuse, largely laminar flow system, kinetic factors still determine where most chemical energy is expended (in the upper part of the vadose zone), but for a given water

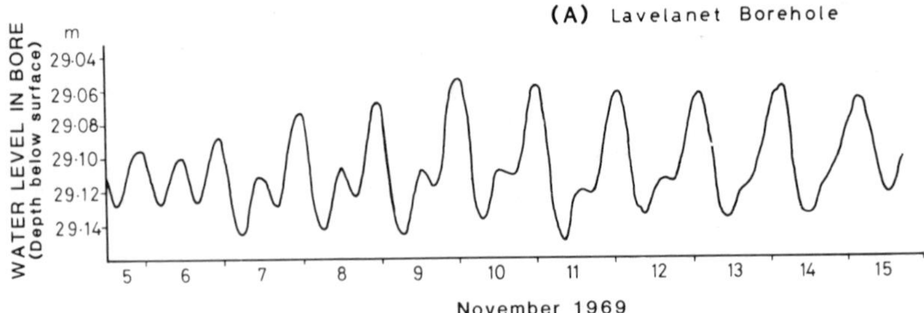

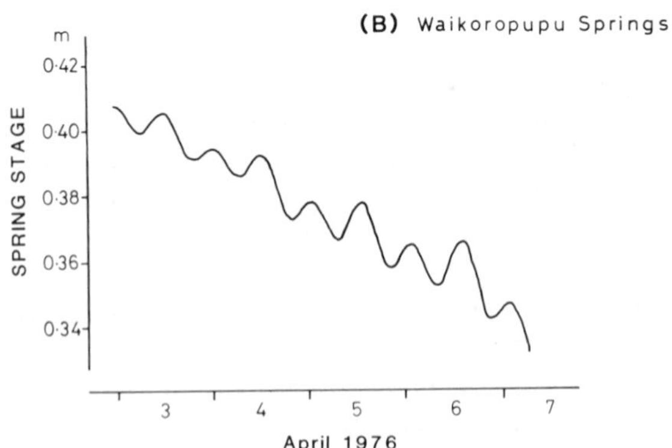

Figure 5.19 Earth tide effects as measured (A) in a bore and (B) at a spring in karst terrains. From Mangin (1975) and Williams (1977).

throughput it is equilibrium considerations that determine the total chemical load discharged from the system.

Where water is stored under confined artesian conditions, the influence of another form of energy can sometimes be observed. Water levels in artesian bores can fluctuate twice daily by several centimetres as a consequence of *earth tides* (Fig. 5.19). These are caused by the same mechanism that produces marine tides, the moving tidal bulge in the solid earth causing the reservoir rocks to compress and relax and thus the fissures to close and open. The pumping action of earth tides may be especially significant in the earliest stages of karstification, when it is difficult to initiate a secondary permeability because of the short penetration distance of groundwater before it becomes fully saturated with respect to calcium carbonate (Davis 1966).

5.5 Rate of development of flowpaths

There are two rates of development to be considered. First is the extension of the initial solutional conduit front through a fissure until it becomes connected to a spring or to another conduit that is already connected (Figs 7.6–7.8). And second is the rate of cross-sectional enlargement in an already connected fissure. The extension rate is the more significant; it defines what may be termed the 'gestation period' of the karst aquifer.

The extension rate is a function of (a) solvent capacity at the front (b) the width plus sinuosity and roughness of the initial fissure, and (c) the pressure head. For reasonable values of solvent capacity and head, White (1984) obtained an extension rate of 200–300 m per 1000 years (or 3000–5000 years per kilometre) (Fig. 5.20).

Rates of this order of magnitude are upheld by many examples of new, short conduit systems that developed in western Ireland and lowland Canada during the 8000 to 12 000 years that have elapsed since the end of the last glaciation. However, the penetrated fissures may have been unusually wide because of glacial rebound flexure, especially in the Canadian cases. Ford (1980) and Palmer (1984) suggest that extension times of 10 000 to 100 000 years per kilometre may have prevailed in a majority of established karsts.

These estimates permit us to specify conditions where karstic rocks fail to develop important karstic porosity, i.e. connected conduit porosity. Either the extension is incomplete because insufficient time has elapsed since initiation (the aquifer is not yet 'born'), or the rate of surficial erosion processes exceeds the extension rate and thus destroys the rock first. The latter case often applies on steep mountain slopes. If surficial erosion is negligible, most carbonate and sulphate rocks will become conduit aquifers given sufficient time. As an extreme example, sufficient time appears to be

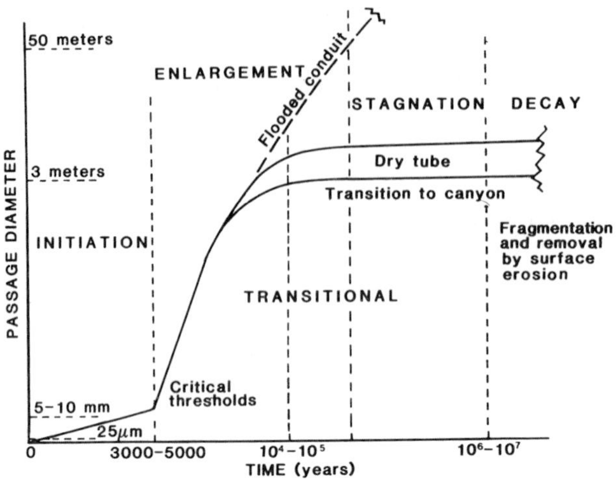

Figure 5.20 Thresholds in the development of cave passages. Proto-cave initiation is in the laminar regime when expansion is slow and linear. Three thresholds are then crossed at about the same proto-cave diameter, marking a boundary between fracture permeability and conduit permeability: a hydraulic threshold as laminar flow gives way to turbulent; a transport threshold as required velocities are attained for the throughput of insoluble residues and clastic sediment; and a kinetic threshold as conduit waters become undersaturated. From White (1988).

the principal cause of the development of large conduit karst in the quartzite Sarisarinama Plateau of Venezuela (Szczerban & Urbani 1974). Its conduit initiation may have begun in the Late Cretaceous or even earlier.

Once the conduit extension is completed, conduit enlargement may proceed comparatively rapidly because turbulent flow will soon come to prevail, at least in flood conditions when the solvent capacity may also be maximized. Conduits can expand to diameters of 1–10 m in a few thousand years (Fig. 5.20), or even in a few hundred years in high relief, wet terrains such as many alpine regions.

5.6 Classification and characteristics of karst aquifers

Significant characteristics that might be incorporated into a classification of karst aquifers include: flow media, flow type, style of recharge, conduit network topology, stores and storage capacity, and outflow response to recharge. Depending on the purpose of the classification, one or more of these may be emphasized or neglected.

An important starting point for the more useful general classifications is the conceptualization of flow type. Burdon & Papakis (1963) first drew attention to this, distinguishing *diffuse* circulation from *concentrated* (or localized) circulation and diffuse infiltration from concentrated infiltration.

Table 5.8 Hydrologic classification of carbonate aquifers. After White (1969).

Flow type	Hydrological control	Associated cave type
Diffuse flow	Gross lithology. Shaley limestones; crystalline dolomites; high primary porosity	Caves rare, small, have irregular patterns
Free flow	Thick, massive soluble rocks.	Integrated conduit cave systems
Perched	Karst system underlain by impervious rocks near or above base level	Cave streams perched – often have free air surface
Open	Soluble rocks extend upward to level surface	Sinkhole inputs: heavy sediment load; short channel morphology caves
Capped	Aquifer overlain by impervious rock	Vertical shaft inputs; lateral flow under capping beds; long integrated caves
Deep	Karst system extends to considerable depth below base level	Flow is through submerged conduits
Open	Soluble rocks extend to land surface	Short tubular abandoned caves likely to be sediment-choked
Capped	Aquifer overlain by impervious rocks	Long, integrated conduits under caprock. Active level of system inundated
Confined flow	Structural and stratigraphic controls.	
Artesian	Impervious beds which force flow below regional base level	Inclined 3-D network caves
Sandwich	Thin beds of soluble rock between impervious beds	Horizontal 2-D network caves

They also pointed out that the style of recharge may not precondition the type of flow in the saturated zone: thus concentrated circulation can occur even if recharge (infiltration) was diffuse and vice versa. White (1969) termed these flow styles ‘diffuse’ and ‘conduit’. He also suggested how the occurrence of these contrasting forms of circulation might be deduced from readily observable geological variables. Thus his well known classification of carbonate aquifers (Table 5.8) is important both for identifying the range of aquifer types that occur and for assisting in the interpretation of their groundwater flow characteristics.

Work in France by Mangin (1975) and Bakalowicz (1977) has brought closer attention to the structure and transfer functions of drainage systems within karst (see section 6.5). Spring hydrograph response to recharge has been used as a measure of aquifer karstification (Bakalowicz & Mangin 1980), four groups of karst systems being recognized that range from aquifers with extremely well developed speleological networks to those carbonate terrains which can barely be considered karstified.

Chemical contrasts in groundwaters draining from aquifers that are dominantly porous, fissured or cavernous (conduit) have encouraged the view (Bakalowicz 1977) that to consider carbonate flow media as being

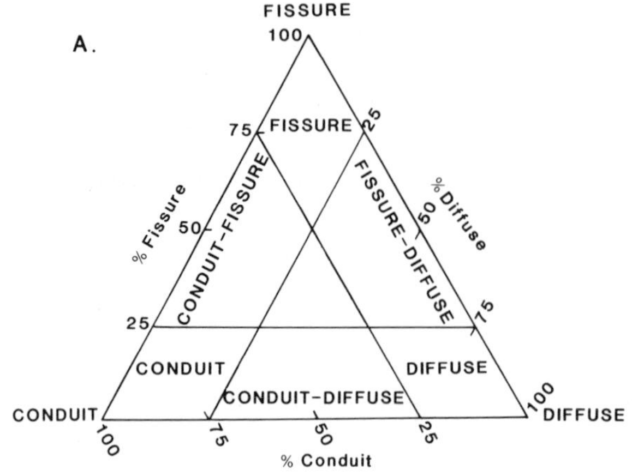

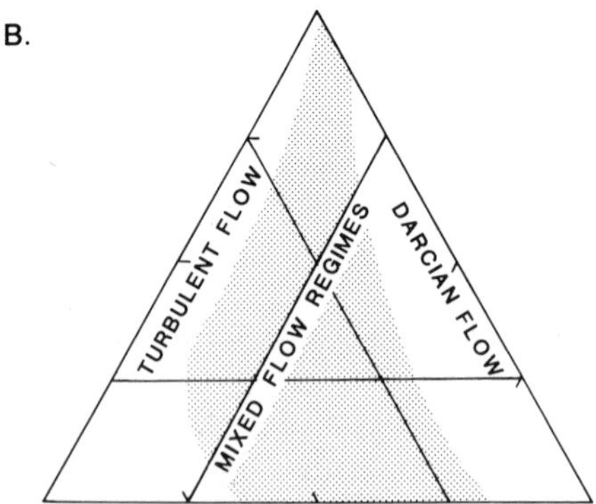

Figure 5.21 (A) A conceptual classification of karst aquifers and (B) their presumed relationship to predominant flow regimes. From Atkinson (1985).

essentially bimodal is perhaps too simple. Thus Atkinson (1985), also citing data from China (Yuan 1981, 1983), suggests that a ternary or three end-member spectrum may be a more appropriate way of visualizing the older concept of granular, fracture and conduit aquifers (Fig. 5.21A). This conceptual classification of flow media is then related to presumed phreatic flow regimes (Fig. 5.21B). Hobbs & Smart (1986) elaborate this approach. They propose a model in which the three fundamental attributes of recharge, storage and transmission are ranged between end members; thus giving a three-dimensional field into which carbonate aquifers may be plotted.

An alternative method of qualitative and quantitative classification is to use a systems approach. An appropriate starting point is precipitation

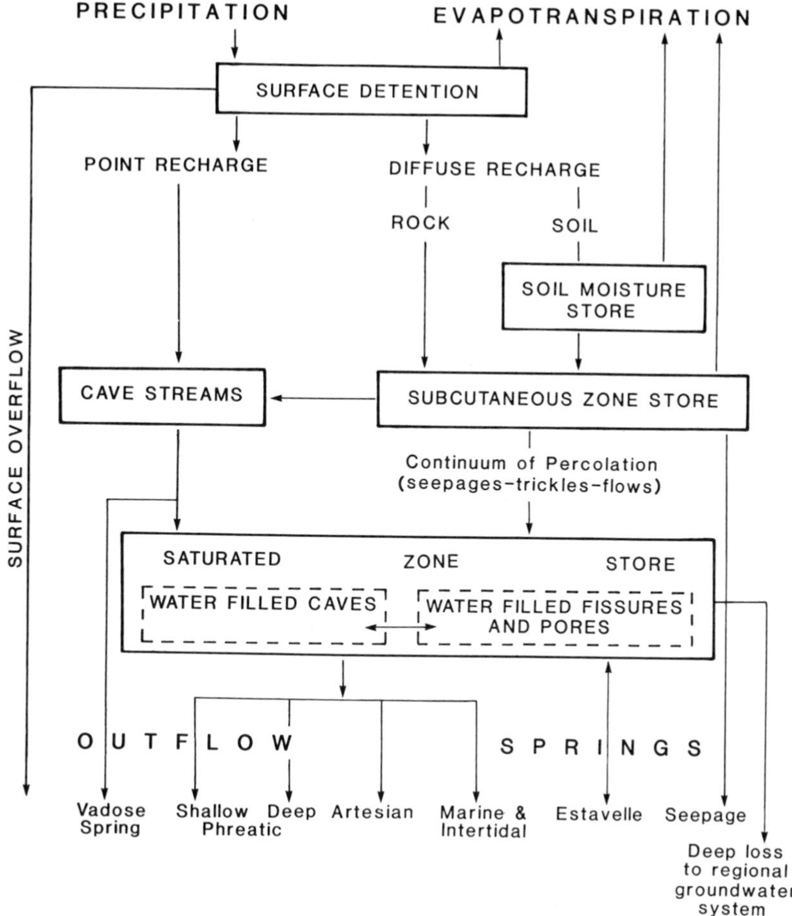

Figure 5.22 Stores and linkages in a karst drainage system.

leading to recharge. Burdon & Papakis' (1963) notion of diffuse and concentrated recharge has been termed percolation input and allogenic water (or swallet stream) input by Drew (1970) and Smith *et al.* (1976), or 'infiltration retardee' or 'rapide' by Mangin (1975). Subsequent downstream linkages and storage components in the recharge – throughput – discharge system have been conceptualized by them. Smith *et al.* used data from Scotland, Jamaica and Mendip to interpret the relative significance of the flow routes identified. Halliwell (1981) has done the same for north Yorkshire. Thus some important characteristics of different aquifers can be convincingly compared. However, recognition of the epikarstic aquifer (Mangin 1973, 1975) with its subcutaneous flow system (Williams 1972a, 1978, 1983) requires refinement of this conceptualization of subterranean

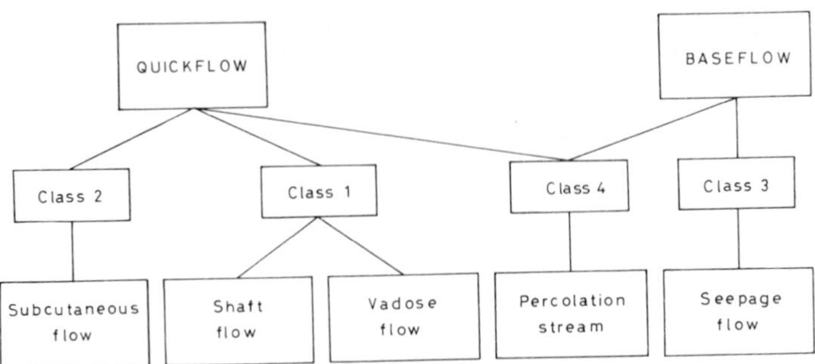

Figure 5.23 Empirical classification of autogenic percolation waters. From Friederich & Smart (1982).

stores and linkages. Figure 5.22 expresses our present understanding of the structure of the system, but further modifications will be inevitable as parts of it become understood in greater detail, e.g. the autogenic percolation subsystem (Fig. 5.23) as analysed by Friederich & Smart (1982).

6 Analysis of karst drainage systems

6.1 The 'grey box' nature of karst

Determination of the structure and main physical properties of an anisotropic and heterogeneous karst aquifer poses severe practical problems. Yet it is essential for water resources estimation, planning and management to be able to answer such questions as how much water can be used, where is it coming from, and what are the physical parameters characterizing the aquifer?

Karst aquifers pose more problems to the hydrologist than any other, because their characteristics are poorly defined and water flow in them is of a very particular type. It is also difficult to generalize about karst, because many different kinds of aquifers exist (section 5.6 & Table 5.8). But at least there is sometimes the opportunity to make direct observations underground, in caves, even though accessible passages penetrate only a small part of any given karst terrain.

With direct observations limited to caves, boreholes, inputs and outputs, the rest of the aquifer characteristics must necessarily be deduced. Sometimes a choice is made when modelling the aquifer function in terms of rainfall-response relationships to treat the system as a 'black box' (Knisel 1972, Dreiss 1982), but such treatments are often fairly far removed from physical reality and tell us little about the structure of the aquifer and how it really operates. More realistic is a 'grey box' approach that uses such information as is available on subterranean conditions to clarify the structure of the system and to help explain its observed operation. Although each karst aquifer is unique in its individual characteristics, some structural components are widely found (Fig. 5.22), although they vary in relative significance in different systems.

Comprehensive analysis of karst drainage systems involves determining the following:

- the areal and vertical extent of the system
- its boundary conditions
- input and output sites and volumes
- the interior structure of linkages and stores
- the capacities and physical characteristics of the stores
- the relative importance of the linkages

Table 6.1 Assumptions and decisions made about the nature of a karst aquifer and the most appropriate method for its analysis.

Flow conditions	Boundary conditions	Aquifer characteriterics	Scale & state	Form of analysis
Linear–Laminar (diffuse Darcian)	Infinite areal extent	Confined/unconfined		Borehole dilution
			Site specific	Borehole recharge
	Impermeable/leaky upper and lower boundaries	Constant/variable thickness		Borehole pumping
			Local	Recharge-response modelling
Mixed laminar and turbulent	Spatially uniform/variable vertical recharge	Homogeneous/heterogeneous	Regional	Water budget
			Groundwater basin	Spring hydrograph and chemograph
	Constant/variable potential recharge boundary	Isotropic/anisotropic	Steady state	Network linkages
Turbulent (conduit flow)	Constant/variable potential discharge boundary		Transient state	
	Fixed/mobile phreatic divides			

- throughput rates
- the response of storage and output to recharge
- the system's response under different flow conditions.

The information required can be obtained by taking four complementary approaches: water balance estimation, borehole analysis, spring hydrograph analysis, and water tracing. The first two apply conventional water resource survey techniques to karst, whereas spring hydrograph analysis and water tracing have been mainly developed for karst and are little used in other hydrogeological contexts.

Decisions are often made about the most appropriate form of analysis that depend on certain initial assumptions about flow and aquifer characteristics (Table 6.1). Such assumptions, for example that the aquifer is isotropic and flow is laminar, should be recognized as such and treated as working hypotheses that may be modified in the light of field results.

This chapter elaborates these points and discusses methods available for exploration, survey, data analysis and interpretation – the clarification of the grey box.

6.2 Exploration and survey techniques

Practical methods for the evaluation of groundwater resources in general are explained in a wide variety of excellent publications, for example by Brown *et al.* (1972), Freeze & Cherry (1979), Todd (1980), Lloyd (1981a), the United States Department of the Interior (1981), Castany (1982), and UNESCO/IAHS (1983). Repetition of material presented there is unnecessary, but evaluation of its applicability to karst is essential, because most groundwater texts deal inadequately with karst.

Defining the limits of the system

All aquifers have boundaries that modify flow conditions. Examples of boundaries are the confining beds of an artesian aquifer, the water table and lower limit of karstification in an unconfined aquifer. These factors determine vertical boundary conditions. Also very important are those limiting the horizontal extent of an aquifer; and in this context discharge and impermeable boundaries should be distinguished from recharge boundaries.

In non-karstic terrains, groundwater divides are sometimes assumed to directly underlie surface topographic divides as determined from contour maps or aerial photographs. This approach is acceptable in a karst context only as an initial working hypothesis, because experience in innumerable karst catchments has often shown phreatic and vadose divides to deviate significantly in plan position from overlying surface watersheds. Furthermore, phreatic divides may migrate laterally according to prevailing water flow conditions. A groundwater divide determined at low water table levels may not be valid when the piezometric surface is higher.

Within a simple karst aquifer there may be several groundwater basins with minimal hydraulic connection, each draining to a different spring (or set of springs). In an unconfined aquifer, the limits of each system can be determined:

(a) by mapping piezometric contours and thus establishing regions of divergence of flow (groundwater divides); and
(b) by water tracing, perhaps using fluorescent dyes or isotopes (see section 6.10).

In a recent study in Kentucky, Thrailkill (1985) has shown that dye tracing can reveal narrow groundwater basins that cannot be identified by interpretation of groundwater contour patterns. This indicates that water tracing should be used to verify groundwater divides determined from water table maps.

Groundwater divides may also be deduced from the potentiometric surface in confined aquifers, but since flow through times are likely to be much longer than under unconfined conditions, water tracing by dyes may not be practical. Instead, the provenance of groundwaters may have to be determined using environmental isotopes or by pulse train analysis (section 6.10).

The impervious basin area sustaining allogenic inputs along a recharge boundary can be accurately determined by conventional plan mapping of surface watersheds. However, the swallow holes of sinking streams are less easily located, especially if the sites are small or under forest. Influent rivers that gradually lose their flow over several kilometres are also difficult to detect, especially if not all of the flow is lost. Field mapping and discharge measurement is the only sure way of obtaining accurate information, but aerial photograph interpretation is of considerable assistance.

Remote sensing using multispectral and geophysical techniques

Remote sensing using conventional stereoscopic aerial photographs as well as multispectral imagery from aircraft and satellites is an important reconnaissance tool in groundwater hydrology investigations (Moore 1980, Farnsworth *et al.* 1984). For example, Parizek (1976) presents a convincing review of its value in mapping fracture traces and lineaments in carbonate terrains; and Brown (1972a), Harvey *et al.* (1977) and LaMoreaux & Wilson (1984) show the particular importance of thermal infrared imagery in identifying recharge and especially discharge points, including submarine springs (Gandino & Tonelli 1983). Brook (1983) illustrates the value of

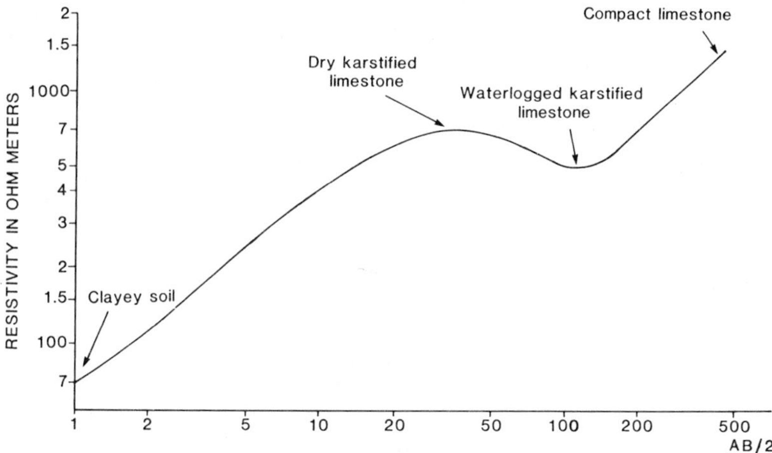

Figure 6.1 Typical electrical resistivity sounding in a calcareous zone. The depth of rock mass investigated increases with the electrode spacing AB/2. From Astier (1984).

Landsat imagery for flood studies in the remote subarctic Nahanni karst of northern Canada. A general review of multispectral remote sensing in karst is provided by Milanovic (1981).

Remote sensing using geophysical techniques is a long established and widely accepted tool in groundwater hydrology. Milanovic (1981), Arandjelovic (1984) and Astier (1984) explain the general application of the methods to karst. Electrical resistivity surveys have proved to be particularly important for establishing the vertical dimensions of a karst aquifer, because the method can sometimes distinguish between compact limestone, water saturated karstified limestone, and dry karstified limestone (Fig. 6.1). Arandjelovic's (1966) work in Yugoslavia, interpreting the base of karstification in the Trebisnjica valley (Fig. 6.2a & b) is an especially good illustration of the place of resistivity surveys. From geophysical and other data, Milanovic (1981) considers the base of karstification in the Dinaric region to be usually no deeper than 250 m. Moore & Stewart (1983) demonstrate that several surface geophysical techniques can be used to delineate zones of higher water yield within the wider surface expressions of the kind of photolinears discussed by Parizek (1976). Nevertheless, a cautionary note is appropriate here. In irregular to rugged topography and where the depth of soil or clastic detritus is variable, results must be treated with extreme caution and only be accepted for planning purposes where confirmed by independent methods such as drilling.

There are wide variations in the depths to which karstified rock is encountered. Yuan (1981) mentions a case in the Sichuan basin in China, where a 4.45 m high cave was found by drilling 2400 m below the surface;

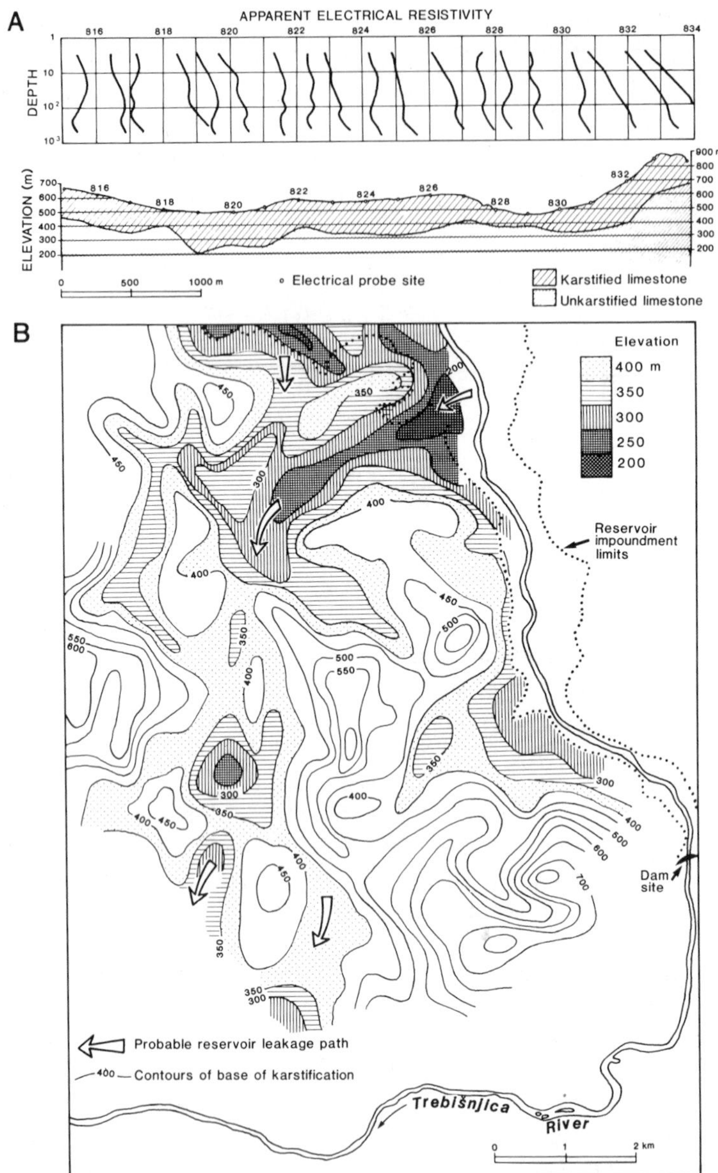

Figure 6.2 (A) Results of an electrical resistivity investigation along a cross-section in the right abutment of the Bileca reservoir and (B) contour map of the base of karstification beneath the Trebisnjica valley, Yugoslavia. From Arandjelovic (1966).

Bogli (1980) gives several examples where bore holes intersected water filled cavities at depths of 2000 m to almost 3000 m below the surface; and Milanovic (1981) states that karstified rock has been found 2236 m below the surface (1600 m below sea level) in the Dinaric region. These particularly deep occurrences probably represent foundered paleokarst, a result of tectonic subsidence.

In coastal aquifers another problem of aquifer boundary delimitation is encountered that is particularly amenable to geophysical study: that of determining the depth to the salt water/fresh water interface (Mijatovic 1984a). Since its position cannot be very accurately located using the Ghyben–Herzberg principle (Eqn 5.10 and Fig. 5.9), electrical resistivity surveys are often conducted to estimate the distance to the interface (Jacobson & Hill 1980, Arandjelovic 1984). This technique is appropriate because in a saturated rock of given density and porosity, the resistivity is largely dependent on the salinity of the saturating fluid. In a survey of Niue Island, an uplifted coral atoll, Jacobson & Hill (1980) determined the interface depth by resistivity, conductivity profiles down deep bore holes confirming its general location, but showing the existence of a broad transition zone rather than a sharply defined boundary. Seismic refraction, resistivity and salinity profiles were also used successfully to determine the base of the fresh water lens on Pingelap Atoll by Ayers & Vacher (1986).

6.3 Aquifer zonation and thickness

Having determined the extent and boundary conditions of an aquifer, problems often arise in deriving an appropriate figure for the aquifer thickness to be used in transmissivity (Eqn 5.7) calculations. It is not necessarily the entire thickness of the phreatic zone to the base of karstification that should be used, because only the upper part of the phreatic zone may be involved in active groundwater circulation over the span of a few years. Atkinson's (1977b) solution to this problem in the Mendip Hills, England, was to estimate the depth of active circulation from the elevation range of the phreatic looping of a trunk conduit leading to Wookey Hole spring – one of the main resurgences in the region (Figure 7.16). However, information of this kind is available only if springs have been dived for long distances. Alternative solutions are possible.

Active aquifer thickness varies spatially and temporally. The depth of active circulation may be less close to the spring than further into the aquifer and it may also be less under baseflow than under flood conditions (Fig. 6.3). Where there is a stagnant phreatic zone, the karstified rock may have a high permeability and storage capacity, but zero specific discharge if there is no hydraulic head to drive the flow. Hence total aquifer thickness

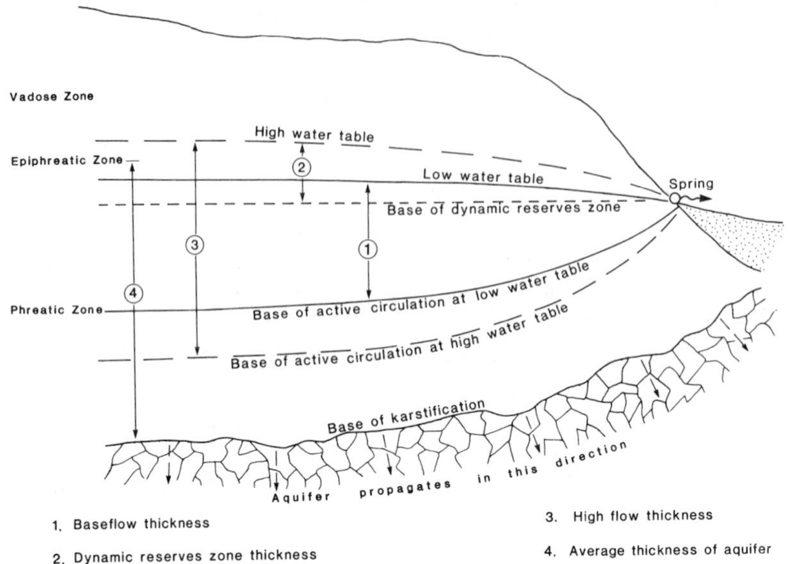

Figure 6.3 Alternative approaches to estimating the thickness of an unconfined karst aquifer. For the calculation of reserves, the maximum dynamic volume is dependent on the thickness of zone 2.

can be appreciated to be a sum of several parts; the different dynamic zones of the aquifer.

Transmissivity calculations for the entire phreatic zone will take into account the aquifer thickness from the water table to the base of karstification. However, since hydraulic conductivity diminishes with depth through the aquifer, average transmissivity can be envisaged as comprising the sum of the thickness weighted transmissivities of different subzones within the phreas (Fig. 6.3).

6.4 Borehole analysis

Data may be obtained from borehole analysis on (a) the location of relatively permeable zones within the karstified rock; (b) the hydraulic conductivity of a relatively small volume of rock near the measurement point; and (c) the specific storage of part of the aquifer.

Using the results from 146 borehole tests in the Dinaric karst, Milanovic (1981) substantiates previous less formal evidence that karstification decreases exponentially with depth (Fig. 6.4). The borehole data can be interpreted as indices of permeability (and consequently of hydraulic conductivity). Thus he suggests that in general the permeability at 300 m

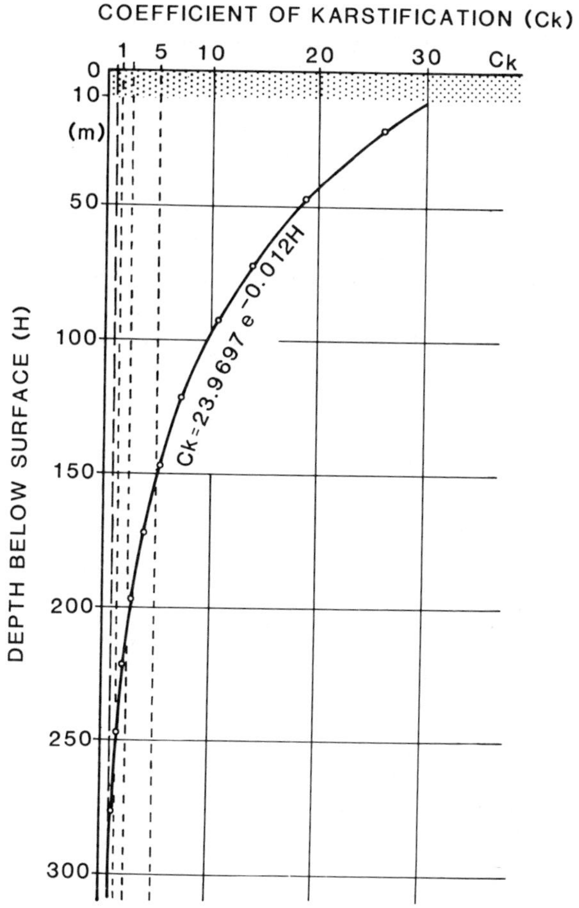

COEFFICIENT OF KARSTIFICATION (Ck)

1 5 10 20 30 Ck

0 10 (m) 50 100 150 200 250 300

DEPTH BELOW SURFACE (H)

$Ck = 23.9697\,e^{-0.012H}$

Figure 6.4 Generalized relationship between karstification and depth based on data from 146 boreholes in the Dinaric karst. From Milanovic (1981).

below the ground surface is only about one-tenth of that at 100 m and one-thirtieth of that at 10 m.

Castany (1984b) summarizes appropriate field techniques for the determination of hydraulic conductivity in karst. Since the permeability of a fissured rock is a function of scale (Fig. 5.4) and the geometry of the fissure network, a detailed structural field study is essential prior to the evaluation of permeability results. Fissures are commonly oriented in one, two or three sets of directions. If the spatial distribution can be defined, the permeability K_1, K_2, and K_3 associated with each set may be measured, and a weighted estimated of the hydraulic conductivity of the field area can be derived.

The average hydraulic conductivity K of a fissured rock mass, assuming linear–laminar flow conditions, is given by Castany (1984b) as

$$K = \frac{W}{n} K_f + K_m \tag{6.1}$$

where W is the average width of opening of the fissure set under consideration; n is the average spacing of fissures in the same set; K_f is the fissure set hydraulic conductivity; and K_m is the hydraulic conductivity of the intervening unfissured rock (the rock matrix). In crystalline karstified limestones, K_m is negligible compared to K_f. Hence it is a reasonable approximation to assume that

$$K = \frac{W}{n} K_f \tag{6.2}$$

Snow (1968) showed that for a parallel joint set with N fissures per unit distance and a fracture porosity $n_f = NW$, the hydraulic conductivity of the set can be evaluated from

$$K_f = \left(\frac{\rho g}{\mu}\right)\left(\frac{NW^3}{12}\right) \tag{6.3}$$

provided that it is applied to a volume of rock of sufficient size that it acts as a Darcian continuum. Snow also concluded that a cubic system of similar fractures creates an isotropic net with a porosity $n_f = 3NW$ and with a

Table 6.2 Borehole recharge and pumping methods and their applicability to field problems. From Castany (1984b).

Distribution of fissures	Investigation	Size of area under investigation	Test method
Irregular	Determination of average horizontal permeability; foundations; construction works; mine drainage; pollution, etc.	In the order few km^2	Standard Lugeon test
	Determination of the horizontal permeability for groundwater development; water resources evaluation.	Greater than 100 km^2	Pumping test
1, 2, or 3 fissured sets	Determination of the average horizontal permeability; foundations, construction works, mine drainage, pollution, etc.	In the order of few km^2	Modified Lugeon test

permeability k that is double that of any one of its fissure sets. Thus in the cubic system

$$k = NW^3/6 \tag{6.4}$$

whereas in an array of parallel joints comprising one set

$$k = NW^3/12 \tag{6.5}$$

However, Bocker (in press) finds it unnecessary to discriminate between fracture sets in the intensively studied karst of the Transdanubian Mountains of Hungary. He determined a statistically significant log : log relationship to hold for the region

$$\log W = 1.087 \log N - 3.37 \tag{6.6}$$

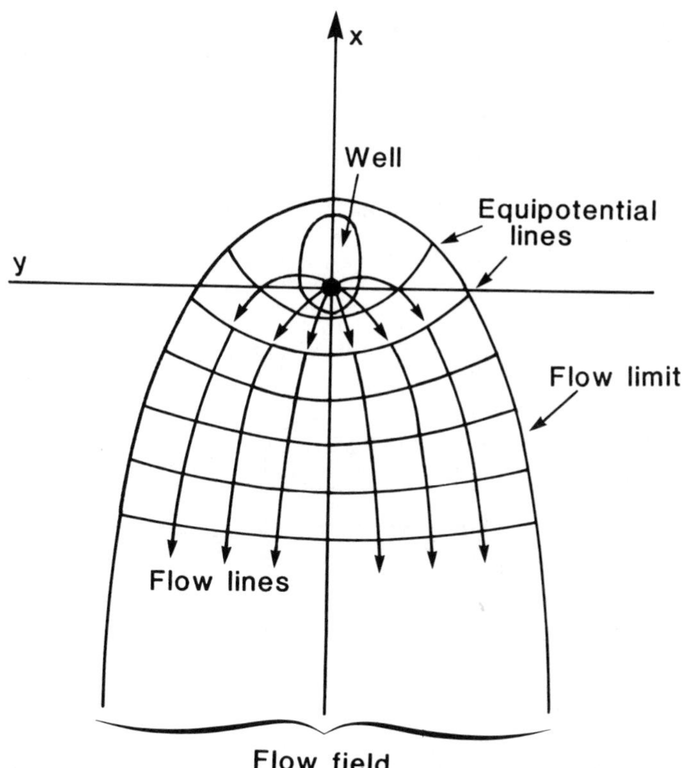

Figure 6.5 Equipotentials and flowlines in a borehole undergoing a recharge test. From Castany (1984b).

where N is the number of fractures <0.4 m in width, where 70% of the porosity is contained in openings <0.001 m.

In the field, hydraulic conductivity is usually determined by borehole *recharge tests*, sometimes called Lugeon tests, or by borehole *pumping tests*. The appropriate technique depends upon the purpose and scale of the investigation (Table 6.2). However, it may also be obtained from in hole tracer dilution (see section 6.10).

Recharge and pumping tests both distort the initial potentiometric surface into cones of recharge or abstraction, depending on the direction of induced water movement. The resulting flow pattern can be represented by an orthogonal network of equipotentials and streamlines (Fig. 6.5).

Borehole recharge tests

Borehole recharge tests for permeability are of three types: pressure tests, constant head gravity tests, and falling head gravity tests, (US Dept Interior 1981).

Castany (1984b) notes that three kinds of pressure injection tests are in common use in karst terrains. They are performed in uncased boreholes:

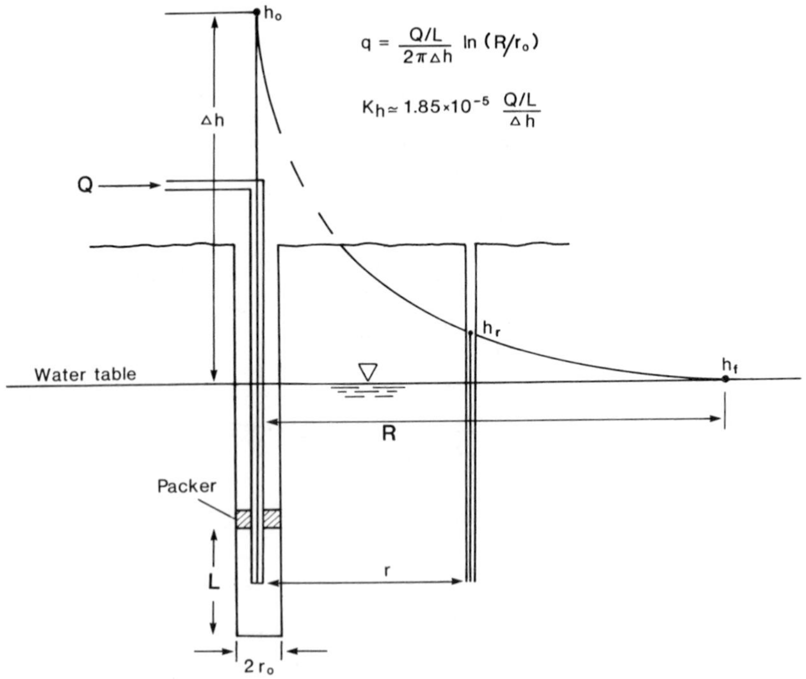

Figure 6.6 Definitions required to interpret the results of a standard borehole recharge test. From Castany (1984b).

(a) The standard Lugeon test, which yields an average local horizontal hydraulic conductivity without taking into account the anisotropy of the rock formation (Fig. 6.6).
(b) The modified Lugeon test, by which a directional hydraulic conductivity on the basis of relative orientation of the test hole to the system of fissures is determined (Fig. 6.7).
(c) A triple hydraulic rig test, which gives directional hydraulic conductivities directly.

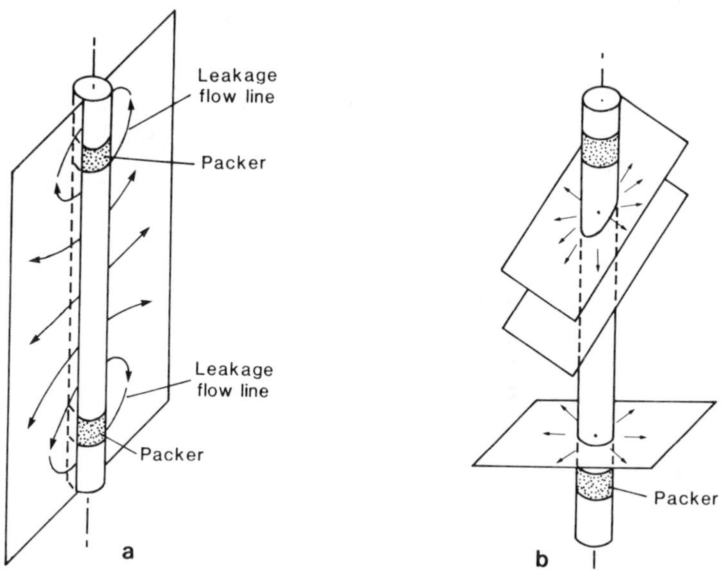

Figure 6.7 Influence of the direction of fissures on flow during a recharge test: (a) fissures parallel to bore and (b) fissures transverse to bore. From Castany (1984b)

In the standard recharge test, water is pumped into the borehole to form a recharge cone. It is performed for a selected zone of length L above the bottom of a borehole (Fig. 6.6). Injection is organized in fixed steps at prescribed time intervals. The water pressure in the bore is first increased gradually in steps such as 2,4,6,8,10 bars (1 bar = 10.2 m head of water; 1 m head = 9.8 KN/m^2 = 9.8 kPa), and then is allowed to decrease in a reverse series of similar steps. Test lengths within a borehole are characteristically 5 m, but may be reduced to 1 m, using isolating packers to identify highly permeable zones (Fig. 6.8).

The permeability of the rock is determined from the pressures and measured water quantities injected. It can be expressed as *specific permeability* q, defined as the volume of water that can percolate through the karstified rock per unit length of borehole under a pressure of 1 m of

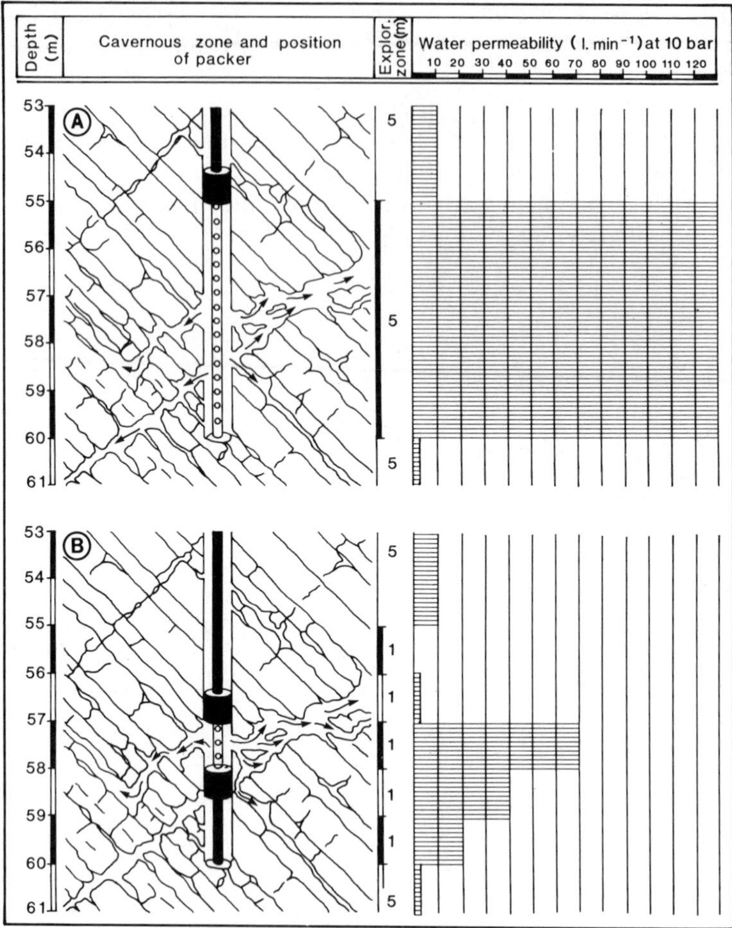

Figure 6.8 The use of variable packer spacing to isolate highly permeable zones within a borehole. From Milanovic (1981).

water (0.1 atmospheres) over a duration of 1 min; or can be referred to in Lugeon units (1 Lu = 1 l min^{-1} m^{-1} under 10 atmospheres. Milanovic finds seven categories of specific permeability suitable for describing permeability in the Dinaric karst (Table 6.3).

Results are plotted on graphs relating the quantities of water injected to pressure at each step (Fig. 6.9). An acceptable test for analysis is one that produces a matching reverse cycle. The flow pattern around the test zone depends upon the characteristics of the fissure systems (Fig. 6.7). Where fissures are essentially horizontal, the analysis of the test is possible using

Table 6.3 Classification of rocks according to their specific permeability. From Milanovic (1981) after Trupak (1956).

Category no.	Specific permeability ($l \min^{-1}$)	Rock category
1	0.001	Impermeable
2	0.001–0.01	Low permeability
3	0.01–0.1	Permeable
4	0.1–1	Medium permeability
5	1–10	High permeability
6	10–100	Very high permeability
7	100–1000	Exceptionally high permeability

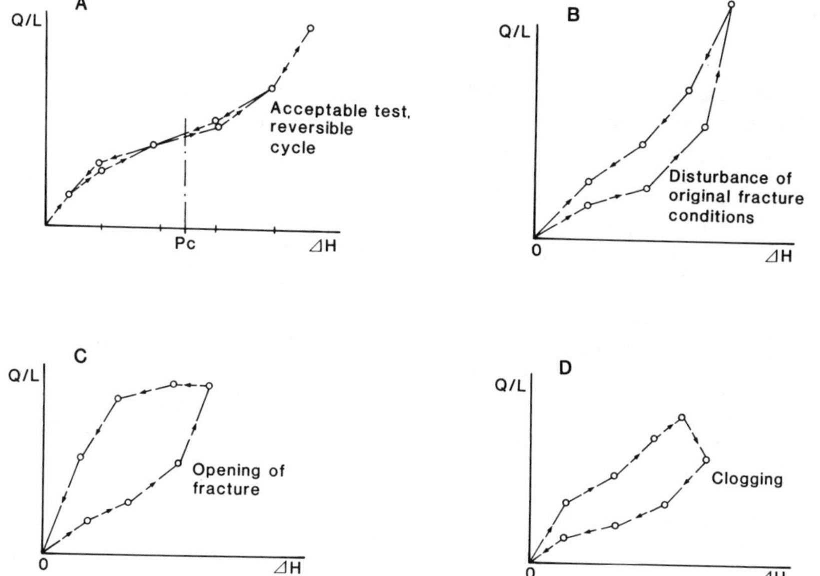

Figure 6.9 Four characteristic curves for recharge tests. Modified from Castany (1984b) after Louis (1969). (Pc refers to the head loss due to the packer).

cylindrical flow principles (Castany 1984b), and the following relationship holds

$$q \text{ (Lu)} = \frac{100\ Q/L}{\Delta h} \qquad (6.7)$$

where Q/L is a volume rate of flow per unit test length; and Δh is the difference between head in the test hole h_o and that of the natural rest level of the water table h_f beyond the radius of influence R of the test (Fig. 6.6). The specific permeability q can also be derived as follows

$$q = \frac{Q/L}{2\pi\Delta h}\ln\,(R/r_o) \tag{6.8}$$

where r_o is the diameter of the borehole.

Castany (1984b) suggests that since $\frac{\ln\,(R/r_o)}{2\pi}$ is roughly constant and because R/r_o varies very slightly and can be assumed close to 7, then for numerical computations

$$q^* = 1.85 \times 10^{-5}\frac{Q/L}{\Delta h} \tag{6.9}$$

Here the quantity q^* is in m s^{-1} and thus has the same dimensions as hydraulic conductivity.

Milanovic (1981) discusses further the relationships between specific permeability and hydraulic conductivity. If a linear relationship between the two can be assumed then

$$K = j\,q \tag{6.10}$$

For three media models (the unsaturated zone, the saturated zone, and a combination of each), j was calculated for a bore of radius 0.02 m and a test length of 5 m to vary from 1.2 to 2.3, being least in the saturated model and greatest in the mixed case. Milanovic (1981) suggests that a reasonable approximation will be derived from

$$K = 1.7 \times 10^{-5} q \tag{6.11}$$

where K is in m s^{-1} and q is in l min^{-1} m^{-1} under 0.1 atmospheres.

Borehole pumping tests

The removal of water from a well by pumping produces a cone-shaped zone of depression in the water table. The drawdown cone (or cone of depression) is unique in shape and lateral extent, depending on the hydraulic characteristics of the aquifer and the rate and duration of pumping (Fig. 6.10).

Two general types of analyses are available for these tests:

A. Plan

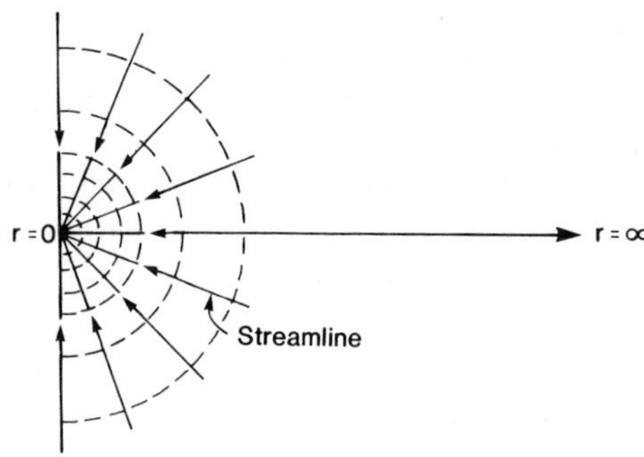

B. Confined Aquifer

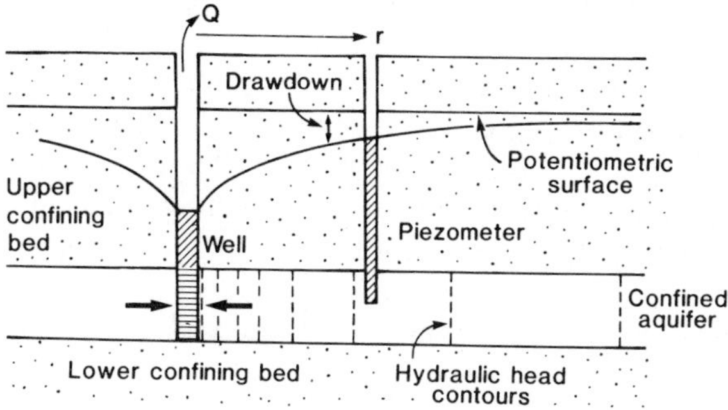

C. Unconfined Aquifer

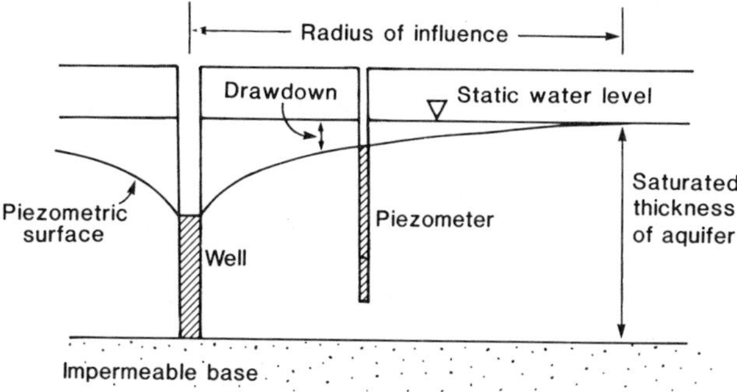

Figure 6.10 Development of drawdown cone and streamlines in confined and unconfined aquifers. Modified from Freeze & Cherry (1979).

(a) steady-state or equilibrium methods, which yield values of transmissivity T and storativity S; and
(b) transient or non-equilibrium methods from which can also be derived values of storativity and information on boundary conditions (Brown *et al.* 1972, US Dept Interior 1981). The latter permits analysis of groundwater conditions that change with time and involve storage, whereas the steady-state methods do not.

We emphasize that all of these methods were developed for porous media in which Darcian conditions hold. Experience shows that the techniques still produce acceptable results in karst, provided the constraints of scale (Fig. 5.4) are not overlooked.

The boundary conditions of aquifers may vary. Some can be assumed for practical purposes to have infinite areal extent. Others are limited horizontally, e.g. by impermeable rocks, a fault, or by a marked reduction in aquifer thickness. A constant potential recharge boundary can be provided by rivers or lakes, whereas the ocean may provide a constant potential discharge boundary. Impervious confining beds in artesian aquifers preclude vertical recharge and discharge, but semi-permeable beds permit leakage.

The simplest aquifer configuration for analysis is that which is:

(a) horizontal and of infinite extent,
(b) confined between impermeable beds,
(c) of constant thickness, and
(d) homogeneous and isotropic.

In such a case the well fully penetrates the aquifer and at a constant pumping rate Q there is no drawdown in hydraulic head at the infinite boundary. The relationship of drawdown in the well to the aquifer properties was first elucidated by Theis (1935). Castany (1984b) expresses it as

$$\Delta h = \frac{0.183Q}{T} \log \frac{2.25\ Tt}{r^2 S} \tag{6.12}$$

where Δh is the drawdown (m) of the potentiometric surface in an observation well, T is the transmissivity ($m^2\ s^{-1}$), S is storativity (dimensionless), Q is the constant rate of pumping ($m^3\ s^{-1}$)), t is the time elapsed in seconds since the start of pumping, and r is the distance from the pumped well to the observation well. The formula is considered valid to within 5% when $t \geqslant 10\ r^2S/4T$.

The recovery to the original level of the potentiometric surface is given by

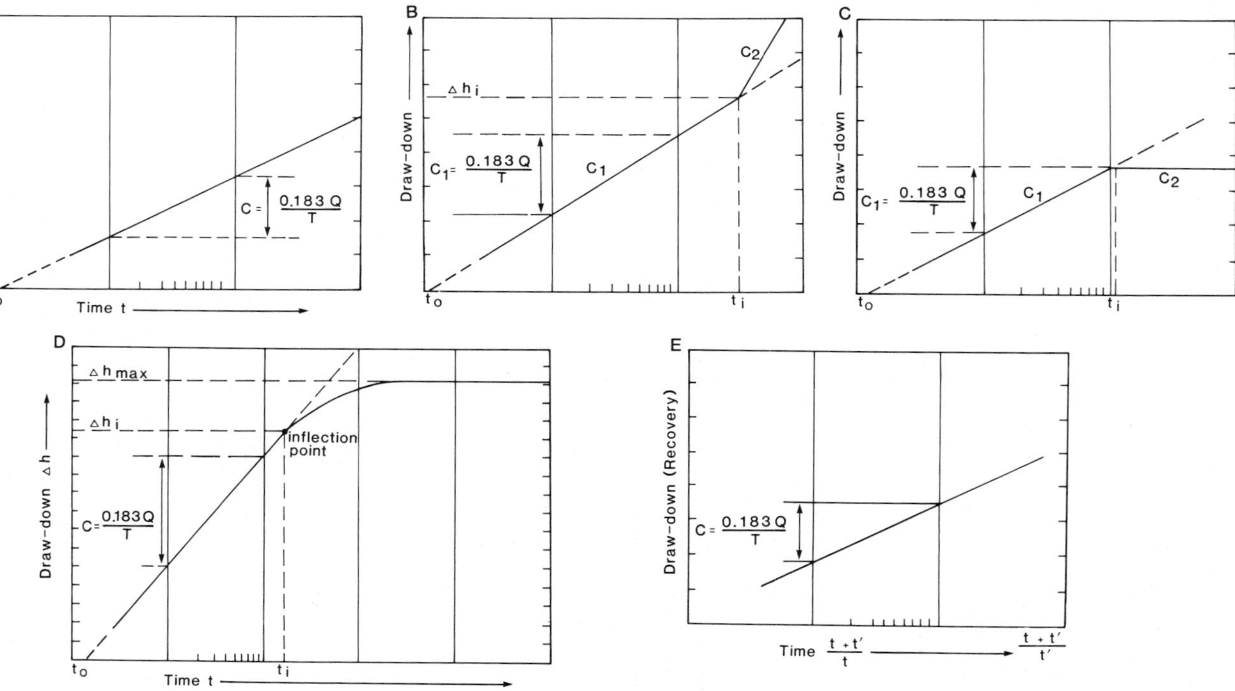

Figure 6.11 Semi-logarithmic plots of drawdown versus time for well pumping and recovery tests in confined aquifers with various boundary conditions; A infinite areal extent; B neighbouring lateral barrier; C neighbouring constant potential recharge boundary; D leaky confining beds wth time lag in recharge transfer; and E recovery straight line plot, C, C_1 or C_2 defines the slope of the curves. Pumping is at a constant rate. From Castany (1984b).

$$\Delta h = \frac{0.183Q}{T} \,.\, \log \frac{t + t'}{t'} \qquad (6.13)$$

where t' is the time elapsed in seconds since pumping stopped.

The drawdown values at certain times during the pumping test (or recovery test) are recorded as in Figure 6.11. A straight line drawn through the plotted points after 1 to 2 hours of pumping is regarded as representative of well behaviour. The slope of the line C is termed Jacob's logarithmic approximation

$$\frac{0.183Q}{T} C \qquad (6.14)$$

It corresponds to the change of drawdown in one logarithmic cycle on the time scale. Extrapolation of the fitted line to intersect the time axis yields t_o, the time at zero drawdown.

Simple rearrangement of Equation 6.14 permits evaluation of transmissivity and, if the thickness of the aquifer b is known, the hydraulic conductivity may be calculated by rearrangement of Equation 5.7. However, the generalized value of K may not be very meaningful in karst given the characteristic variability of permeability mentioned earlier.

In addition, storativity can be derived from

$$S = \frac{2.25 \; T \; t_o}{r^2} \qquad (6.15)$$

In aquifers limited by vertical barriers (Fig. 6.11b), the first straight line segment corresponds to the infinite aquifer, but the second has a considerably steeper slope and is represented by the revised expression

$$\Delta h = \frac{0.183Q}{T} \,.\, \log \frac{2.25 \; Tt}{r^2 2lS} \qquad (6.16)$$

where l is the distance from the pumped well to the impermeable boundary. The distance may be determined from

$$l = 0.5r. \; (t_i/t_o)^{0.5} \qquad (6.17)$$

In the case where the drawdown cone reaches a recharge boundary of constant potential (Fig. 6.11c), stabilization of drawdown results because the external recharge (for example from a flooded conduit) introduces a steady-state condition. The secondary straight line (horizontal) segment is represented by a linear function

$$\Delta h = \frac{0.183Q}{T} \cdot \log \frac{2l}{r} \tag{6.18}$$

l being the distance from the pumped well to the constant potential recharge boundary, and also determined from Equation 6.17. Castany (1986b) reports how leakage factors can also be determined (Fig. 6.11). Other borehole analysis methods are discussed by the US Department of the Interior (1981) and Brown *et al.* (1972).

Borehole logging

Further information can be derived on aquifer characteristics by various measurement techniques, mainly geophysical, down boreholes. General details are provided by US Department of the Interior (1981) and Robinson & Oliver (1981); application to karst is discussed by Astier (1984) and Milanovic (1981).

Borehole logging provides *in situ* measurements of parameters related to the physical characteristics of the rock formation, the fluid in it and the borehole. It is most productive when different methods are employed in appropriate combination with each other, as illustrated by Maclay & Small's (1983) study of the Edwards Limestone aquifer in Texas (Fig. 6.12). Electrical, radioactive, caliper, geothermal, television, and stereophoto techniques are available for logging and have been used with varying success.

Electrical logging is of two sorts: spontaneous potential (or self potential) and resistivity. An electrode is lowered down a borehole and in the first case the naturally occurring potential difference at various depths between a surface electrode and the borehole electrode is measured. In the second case, a source of current is connected and the potential difference is measured at different depths for a given current strength. This leads to a log of apparent resistivity versus depth. Both properties are measured in uncased wells, and can be interpreted in favourable cases to distinguish rock unit thickness and stratigraphic sequence, i.e. formation characteristics.

Temperature logs measuring fluid properties are often run in association with electrical logs. They use a sonde with a thermocouple and record temperature variations with depth. This is useful because of the known relationship between temperature and electrical conductivity and also because temperature variations can indicate discrete groundwater bodies sometimes related to water movement from different source areas.

Caliper logging may also be used in uncased boreholes. This technique measures the variation of well diameter with depth, which helps to identify and correlate solution openings, bedding planes and zones that may need grouting.

However, Milanovic (1981) warns that the use of a caliper will not

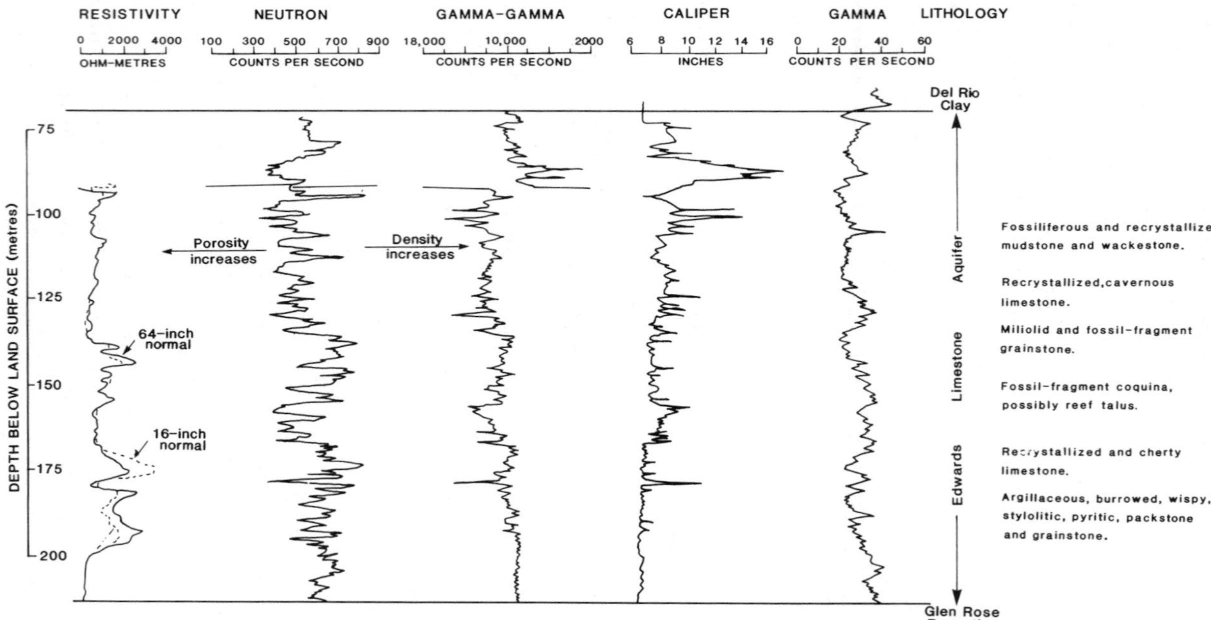

Figure 6.12 Hydro-stratigraphic subdivision of the Edwards aquifer, Texas, using borehole logging techniques. From Maclay & Small (1983).

produce meaningful results if caverns over ten cubic metres in volume are encountered – as commonly found in the Dinaric Karst – and that 'current experience shows that lowering probes into uncased boreholes should be avoided'. This is because rubble may cause jamming and the loss of the instrument. Nevertheless, an 84 mm diameter television camera was lowered successfully down bores in Yugoslavia, although lighting problems provided severe limitations when observation distances exceeded 30 cm.

Radioactive logging methods vary according to their sources and detectors (Astier 1984). The recording of natural radioactivity is referred to as gamma logging, gamma radiation having a penetration distance of about 30 cm. The radiation flux is measured by a scintillation counter lowered down the borehole. Variations in flux are assumed to relate to rock unit boundaries, with clays and shales being commonly several times more radioactive than sandstones, limestones and dolostones.

Gamma–gamma and neutron–gamma logging techniques require artificial radiation sources and detectors, and hence are less frequently used. Gamma–gamma logging has been successfully applied (Milanovic 1981, Maclay & Small 1983) in the determination of rock density variations down boreholes (Fig. 6.12). Neutron–gamma logging is a standard technique for soil moisture measurement, but is less frequently used in boreholes. It provides a measure of the hydrogen abundance per unit volume of rock, and so is related to the abundance of water and to porosity (Fig. 6.12).

The further the aquifer deviates from the behaviour of an ideal or high density fracture medium, the less likely a given borehole will intercept characteristic conditions. Hence in karst it is often much more productive to focus attention on springs, because these *must* be integrating significant conditions. To these we now turn.

6.5 Spring hydrograph analysis

The shape of the outflow hydrograph recorded at a spring is a unique reflection of the response of the aquifer to recharge. The form and rate of recession, in particular, provide significant information on the storage and structural characteristics of the aquifer system sustaining the spring. For these reasons, the analysis of spring hydrographs offers considerable potential insight into the nature and operation of karst drainage system. The information that can be obtained is considerably improved if water quantity variations (hydrographs) are analysed alongside corresponding water quality variations (chemographs).

The duration and intensity of precipitation strongly influences the form of flood hydrographs in surface rivers. But it is well known that basin characteristics such as shape, size, slope, drainage density, lithology, and vegetation modify the runoff response. In addition, antecedent conditions of storage strongly influence the proportion of the rainfall input that runs off and the lag between the input event and the output response. As a consequence, flood hydrograph form and baseflow recession characteristics show considerable variety. But for a particular climatic regime, lithology emerges as one of the dominant controls on hydrograph form. Impermeable rocks yield strongly peaked hydrographs because of little storage and rapid runoff, whereas basins composed of highly permeable formations such as limestones and basalts tend to have flatter, broader and more delayed responses. There is a continuum of rainfall – response functions in nature, with karst terrains with diffuse flow characteristics lying near the permeable end of the spectrum.

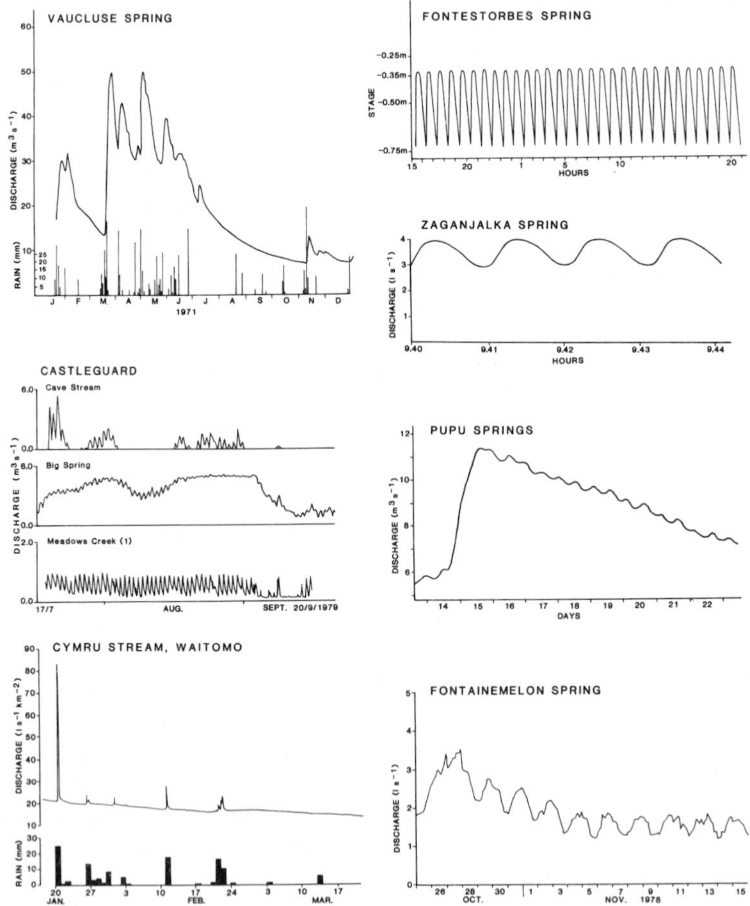

Figure 6.13 Hydrograph forms encountered in karst springs and streams. Data from Gavrilovic (1970), Mangin (1969a), Durozoy & Paloc (1973), Gunn (1978), Siegenthaler *et al.* (1984), C. Smart (1983b).

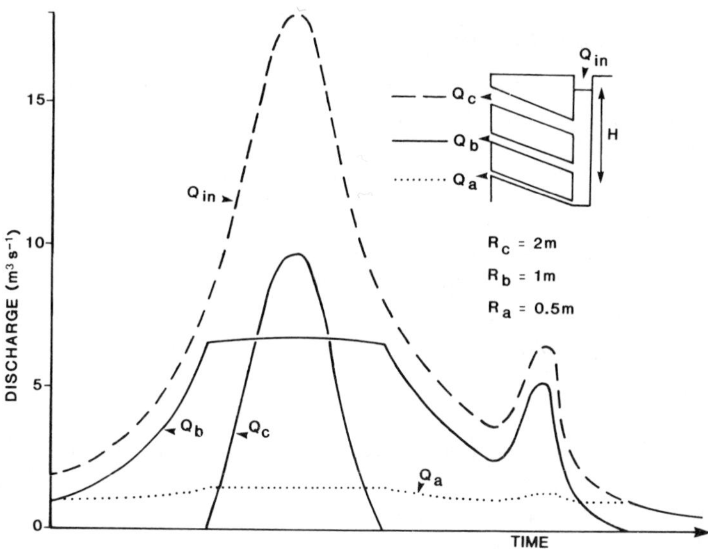

Figure 6.14 Static system model (no storage considered) for a three component underflow – overflow spring system responding to an arbitrary input hydrograph. Q signifies discharge and R conduit radius. From C. Smart (1983a)

The flood hydrographs of vadose cave streams tend to be peaked and similar to surface rivers. But should cave streams flow into a flooded phreatic zone before their waters later emerge at a spring, then the influence of their inflow hydrographs on the composite outflow hydrograph of a spring is similar to that of tributaries flowing into a lake. The overflow or output response is a delayed, muted reflection of the input. In karst systems there is considerable variety in the extent and degree of groundwater reservoir development, as well as various mixes of autogenic and allogenic inputs; thus there are many variations in the outflow responses of karst springs (Fig. 6.13). Some outflow hydrographs are highly peaked, others are oscillatory, and many are broad and relatively flat. Sometimes a flat-topped apparently truncated hydrograph is an indication that it is an underflow spring in Smart's (1983a & b) sense, the 'missing' peaked remainder of the hydrograph emerging at an intermittent high water overflow spring draining the same aquifer (Fig. 6.14). Thus in interpreting hydrograph form, it is important to know if the whole output is being dealt with or only part of it. The discussion that follows assumes that the entire output from a karst drainage system is being monitored.

Simple hydrograph recessions

Given precipitation leading to a discrete recharge event over a karst basin, the output spring will show certain important discharge responses, characterized by:

(a) a lag time before response occurs;
(b) a rate of rise to peak output (the 'rising limb');
(c) a rate of recession as spring discharge returns towards its pre-storm outflow (the 'falling limb'); and
(d) small perturbations or 'bumps' on either limb although best seen on the recession.

When the hydrograph is at its peak, storage in the karst system is at its maximum. The rate of withdrawal of water from that storage is indicated by the slope of the subsequent recession curve.

Quantitative analysis of hydrograph recession derives particularly from the work of Maillet (1905), who proposed that the discharge of a spring is a function of the volume of water held in storage and described it by the simple exponential relation

$$Q_t = Q_o\, e^{-\alpha t} \qquad (6.19)$$

where Q_t is the discharge ($m^3\ s^{-1}$) at time t; Q_o is the previous discharge at time zero; t is the time elapsed (usually expressed in days) between Q_t and Q_o; e is the base of the Napierian logarithms; and α of dimension $[T^{-1}]$ is termed the *recession coefficient*. If this curve is plotted on semi-logarithmic graph paper, it is represented as a straight line with slope $-\alpha$.

The logarithmic form of Equation 6.19 is

$$\log Q_t = \log Q_o - 0.4343 t\alpha \qquad (6.20)$$

from which α may be evaluated as

$$\alpha = \frac{\log Q_1 - \log Q_2}{0.4343\ (t_2 - t_1)} \qquad (6.21)$$

Since $e^{-\alpha}$ in Equation 6.19 is a constant, it is sometimes replaced by a term β, the *recession constant*, and the equation is written

$$Q_t = Q_o \beta^t \qquad (6.22)$$

The recession constant may be evaluated from the logarithmic expression

$$\log \beta = \frac{\log Q_t - \log Q_o}{t} \tag{6.23}$$

On constructing a karst hydrograph recession curve on semi-logarithmic graph paper, it is often found to be non-linear at least in part. Hence other expressions have sometimes been used to improve the fit to the data. These include the double expontential

$$Q_t = Q_o{}^{e-\alpha t^n} \tag{6.24}$$

which in logarithmic form is written

$$\log\left(\log \frac{Q_o}{Q_t}\right) = n \log t + \log x - 0.36222 \tag{6.25}$$

Alternatively, a better fit may be provided by the hyperbola

$$Q_t = \frac{Q_o}{(1 + ct)^n} \tag{6.26}$$

or expressed in logarithmic form

$$\log Q_t = \log Q_o - n \log (1 + ct) \tag{6.27}$$

The parameter c can be evaluated from

$$c = \frac{1}{t}\left(\frac{(Q_o)^{1/n}}{Q_t} - 1\right) \tag{6.28}$$

and the recession constant from

$$\beta = c/Q_o{}^{1/n} \tag{6.29}$$

The exponent n usually lies in the range 0.5 to 2. Other expressions applicable to karst systems are discussed by Mangin (1975).

With regard to the baseflow recession of surface streams, Martin (1973) has pointed out that the recession constant values (Eqn 6.23) range from zero to unity, but usually lie between 0.500 and 1.000 with a very distinct bunching as β approaches unity, and that because of the compressed range of likely values, the recession constant is a relatively insensitive measure of the recession rate. Furthermore, it has no clear physical meaning: the rate of recession giving rise to a particular value of β is not easily appreciated in practical terms. Martin therefore suggests that a concept equivalent to the half-life in nuclear physics would be more appropriate.

A *half-flow period* $t_{0.5}$ can be defined as the time required for the baseflow of the stream (or spring) to halve, hence $2Q\ t_{0.5} = Q_o$ by definition. Then by substitution into Equation 6.22

$$Q_{t_{0.5}} = 2Q_{t_{0.5}}\beta^{t_{0.5}} \tag{6.30}$$

hence

$$1/2 = \beta^{t_{0.5}} \tag{6.31}$$

and

$$t_{0.5} = \text{constant}/\log\beta \tag{6.32}$$

The parameter $t_{0.5}$ has the following properties (Martin 1973):

(a) it is independent of Q_o and Q_t and of the time elapsed between them;
(b) it is sensitive to change and can take values in the range zero to infinity;
(c) it can be easily evaluated from Equation 6.32 and is simply related to β by means of that equation; and
(d) it is a direct measure of the rate of recession and therefore can be used as a means of characterizing exponential baseflow recessions.

The value of the recession coefficient α in Equation 6.19 derives from the hydrogeological characteristics of the aquifer, especially effective porosity and transmissivity. When α or β are large and when $t_{0.5}$ is small the recession is steep, indicating rapid drainage of conduits and little underground storage. If no recharge is occurring but α and β are small and $t_{0.5}$ is large, then very slow drainage of the karst aquifer is indicated, probably from an extensive fissure or porous network with a large storage capacity and high resistance to recharge throughout.

Composite hydrograph recessions

Semi-logarithmic plots of karst spring recession data generally reveal the recession to consist of two or more segments, one or more of which may be linear. In these cases, the data are best described by separate expressions for the different segments identified. Such curves reflect the complex hydrological characteristics of karst aquifers and the various stores and linkages (Fig. 5.22) that may be involved in the drainage system.

Where a complex recession curve occurs, consisting of two or more linear

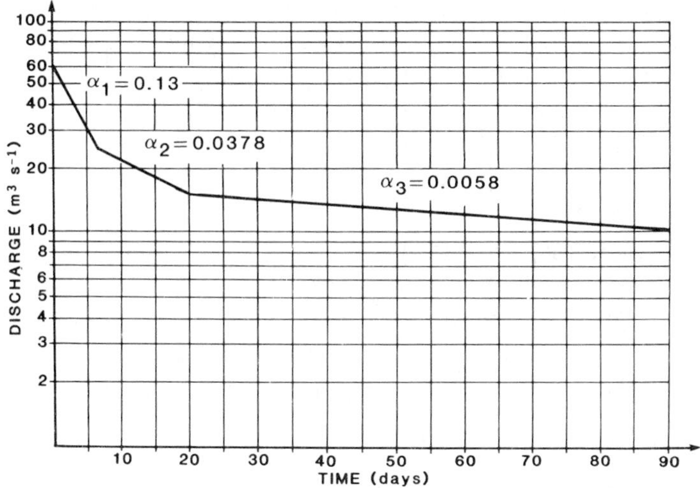

Figure 6.15 Composite hydrograph recession of Ombla spring, Yugoslavia. Note that the three recession coefficients are of different orders of magnitude. From Milanovic (1981).

segments, it can be represented by an equation of the form

$$Q_t = Q_{0_1}\, e^{-\alpha_1 t} + Q_{0_2}\, e^{-\alpha_2 t} + \ldots + Q_{0_n}\, e^{-\alpha_n t} \qquad (6.33)$$

used by Torbarov (1976) and Milanovic (1976) in the analysis of flow from the Bileca and Ombla springs respectively, in Yugoslavia. Figure 6.15 illustrates the Ombla outflow regime. From this Milanovic interprets its aquifer as being characterized by three types of porosity, represented by the three recession coefficients of successive orders of magnitude. He suggests that α_1 is a reflection of rapid outflow from caves and channels, the large volume of water that filled these conduits emptying in about seven days. Coefficient α_2 is interpreted as characterizing the outflow of a system of well integrated karstified fissures, the drainage of which lasts about 13 days. Finally, α_3 is considered to be a response to the drainage of water from pores and narrow fissures including that in rocks and soils above the water table, as well as from sand and clay deposits in caves.

When the various sub-regimes yielding flow to a spring cannot be adequately represented by separate linear segments on semi-logarithmic plots, other approaches must be used. For example, in an analysis of the discharge of Cheddar spring, England, Atkinson (1977b) constructed a master recession curve by superimposing recession segments of all the winter and spring floods of the 1969–1970 water year (Fig. 6.16). He found that the overall recession could be described well by an equation of the form

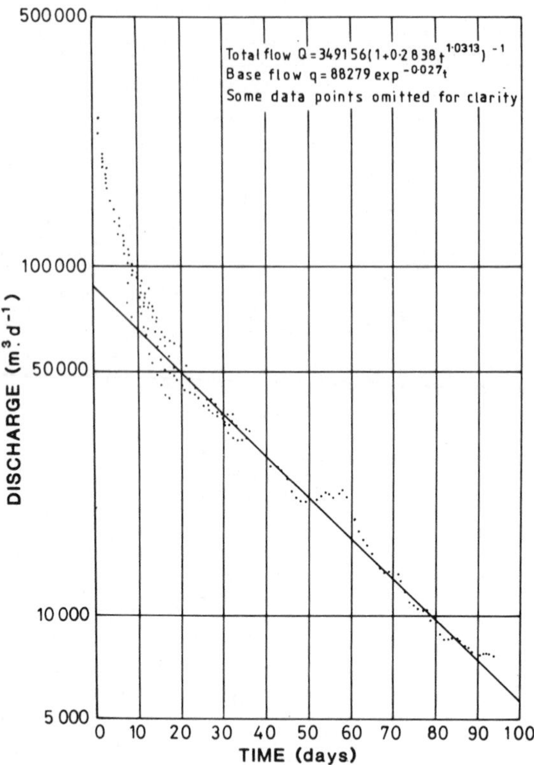

Figure 6.16 Master flow recession at Cheddar spring, England. Note the linear and non-linear segments. From Atinson (1977b).

$$Q_t = Q_o\ (1 + xt^y)^{-1} \tag{6.34}$$

but after 25 days of runoff of the highest flows, the best fit for the remainder of the recession was provided by a simple exponential relation of the type discussed earlier.

This experience reaffirms the important conclusion reached by Mangin (1975) who, in the most thorough analysis of karst spring hydrographs made to date, found that the baseflow phase in all the cases he studied is accurately represented by Maillet's expression (Eqn 6.19). Mangin considers that two basic hydrological entities are distinguishable in the interior of a karst drainage system, one being the non-saturated zone with a non-linear flood recession, which he represents with the function ψ_t and the other being the saturated zone with a linear baseflow recession represented by the function ϕ_t.

Thus $$Q_t = \phi_t + \psi_t \tag{6.35}$$

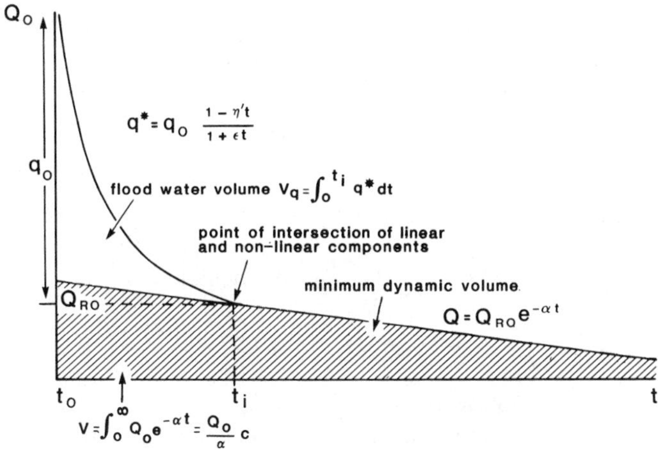

Figure 6.17 Karst spring recession curve analysis following Mangin's (1975) approach.

By comparison to the flood hydrographs of surface rivers, ψ_t corresponds to the quickflow part of the recession and ϕ_t to the delayed flow segment. In the case of karst, ϕ_t relates essentially to the saturated zone and to its relatively low transmissivity, whereas ψ_t translates the effect of surface recharge to the spring: it is an infiltration function modulated by its transfer through the saturated zone. ϕ_t can be described by Maillet's formula (Eqn 6.19), whereas ψ_t is, above all, an empirical function that Mangin considers is best expressed by

$$q^* = q_o \frac{1 - \eta' t}{1 + \epsilon t} \tag{6.36}$$

where q_o corresponds to the difference between the total outflow Q_o at the spring at $t = o$ and baseflow component termed Q_{Ro} (Fig. 6.17). The function is defined between $t = 0$ and $t = \frac{1}{\eta'}$, which is the duration of the flood recession. Whether one takes $\eta' = 1/t_i$ or $\eta' = q_o/t_i$ does not change the interpretations, but the first case is preferred by Mangin as a form that facilitates comparisons.

The coefficient ϵ characterizes the importance of the concavity of the flood recession curve, termed by Mangin the *coefficient of heterogeneity*. A high value for the coefficient corresponds to a recession that falls steeply at first but rapidly levels out. It is evaluated from the expression

$$\epsilon = \frac{q_o - q^*}{q^* t} - \frac{\eta' q_o}{q^*} \tag{6.37}$$

Mangin notes that the form of the flood recession can be described by the function

$$y = \frac{1 - \eta' t}{1 + \epsilon t} \tag{6.38}$$

which being independent of discharge permits comparison of different karst systems. After commencement of the flood recession, the time required for the initial flow to diminish by 50% can be calculated from

$$t_{0.5} = (\epsilon + 2\eta')^{-1} \tag{6.39}$$

Recalling from Equation 6.35 that $Q_t = \phi_t + \psi_t$ and with reference to Figure 6.17, the entire recession curve may be represented by the expression

$$Q_t = Q_{Ro}\, e^{-\alpha t} + Q_o \frac{1 - \eta' t}{1 + \epsilon t} \tag{6.40}$$

In practice, the baseflow segment is first calculated using Equation 6.21 to determine the recession coefficient. Points Q_1 and Q_2 should be as distant from each other as possible, but on a straight line segment. Their choice is critical for the accuracy of α and the reliable estimation of Q_{Ro}. The point of intersection of the linear and non-linear components determines t_i, when the flood recession is assumed essentially zero, i.e. $q^* = 0$.

The volume of water in storage in the saturated zone above the level of the outflow spring (Fig. 6.3) is termed the *dynamic volume V*. Under baseflow conditions this can be found by integrating Equation 6.19. Hence

$$V = \int_0^\infty Q_o\, e^{-\alpha t} = \frac{Q_o}{\alpha} c \tag{6.41}$$

where c is a constant that takes the units used into account (when Q_o is in m^3s^{-1} and α in days then $c = 86\,400$). Therefore from the initial volume to time t, the volume of outflow is

$$V_t = \int_o^t Q_o e^{-\alpha t} = \frac{Q_o}{\alpha} c\, (1 - e^{-\alpha t}) \tag{6.42}$$

Mangin (1975) points out that with the aid of these formulae, the percentage of the dynamic volume that has flowed out can be calculated as a function of time

$$\%V = (1 - e^{-\alpha t})\, 100 \tag{6.43}$$

and he provides a useful diagram that shows $\% V$ as a function of time for various values of α.

Given a complex recession curve of the type shown in Figure 6.15 where there are several α values, then the aquifer volume is the sum of each of the component volumes. Milanovic (1981) discusses the procedure.

The appropriateness of calculating V by using Q_o as the starting point is questioned by Mangin (1975), although it is a widespread practice. He contends that only when baseflow actually begins is the dynamic volume determined a reality, and that to take Q_o at the height of a flood leads to a fictitious value. Hence the preferred departure point for his calculations is the moment when baseflow becomes a certainty, at time t_i with discharge Q_{Ro} as on Figure 6.17. The minimum dynamic volume thus calculated should be regarded as an *index* of aquifer conditions.

The difference between baseflow reserves and the total volume discharged during a flood recession yields the flood water volume V_q. Approaches to its calculation vary. With reference to a situation of the type analysed by Milanovic (1976) and depicted in Figure 6.15 and Equation 6.33 the flood water volume can be estimated by using the first (steepest) recession coefficient over the appropriate duration (seven days in this case).

Alternatively, it can be obtained by subtracting the baseflow component from the total runoff volume. This was the method used by Atkinson (1977b) in the case illustrated in Figure 6.16. Mangin's (1975) technique is to integrate equation 6.36

$$V_q = \int_o^{t_i} q^* \, dt$$

$$= \frac{c\, q_o\, [2.30\eta' \,(1+\epsilon/\eta') \log\,(1+\epsilon/\eta')-\epsilon]}{\epsilon^2} \qquad (6.44)$$

It is interesting to note that karst spring hydrograph separation focuses on the recession and works backwards from the baseflow component, whereas flood hydrograph separation techniques developed for surface streams usually work forward in time from the start of the hydrograph rise, but is usually arbitrary in its separation – for example, 'quick flow' and 'delayed flow' (Hewlett & Hibbert 1967). Both approaches can be improved, but the necessity depends on the purpose of the separation. A distinction should be made between flow separation to determine resources available for exploitation and separation used to elucidate the structure and functioning of the system. In the latter case, chemical data also help.

6.6 Spring chemograph interpretation

Discharge variations at a spring are often accompanied by changes in water quality. Characteristics that may vary include ions in solution, electrical conductivity, environmental isotopes, pH, suspended sediment and temperature. Although some of these are physical rather than chemical attributes of water, it is convenient to consider plots of any of these water quality aspects against time as 'chemographs'.

Jakucs (1959) showed that the chemical quality of karst spring water may vary with time (as well as discharge), and he appears to have been the first to recognize that the response of a spring to rainfall depends on the nature of the recharge, whether autogenic via karstified joints or allogenic via stream–sinks (Fig. 6.18). Thus the transfer function of the system was perceived as conditioned by the flow network.

Shuster & White (1971) substantially reinforced these ideas when they argued that whether a karst aquifer is characterized by a diffuse flow type of circulation or is dominated by conduit flow can be determined by the chemograph. From an analysis of 14 carbonate springs in the central Appalachians, they concluded that the characteristics Ca, Mg, HCO_3, pH, °C of conduit fed springs displayed great variability throughout the year compared to the assumed diffuse flow springs which had relatively constant characteristics. They found the variation in total hardness (expressed as $CaCo_3$ mg l^{-1}) to be a better index of aquifer type than hardness itself; the coefficient of variation of hardness being usually <5% in the case of diffuse flow systems. The latter were also close to saturation, whereas conduit fed springs were undersaturated by factors of 2 to 5. Drake & Harmon (1973) were the first to apply rigorous statistical methods (in this case stepwise linear discriminant analysis) to determine which characteristics varied significantly, i.e. to discriminate objectively between different types of water. Bakalowicz (1979, 1984) has used principal components analysis to describe statistically the quality of karst waters from springs and other sites.

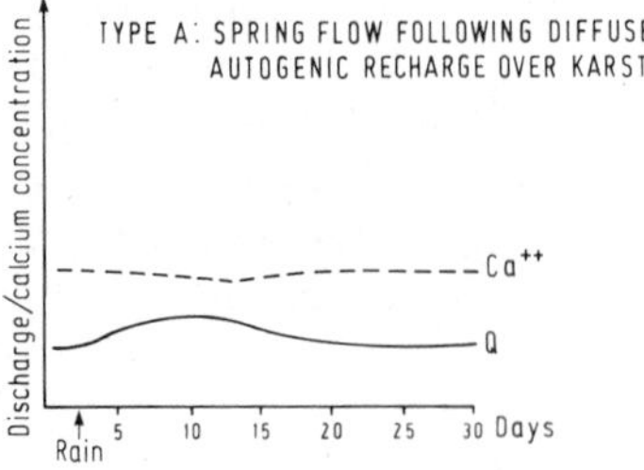

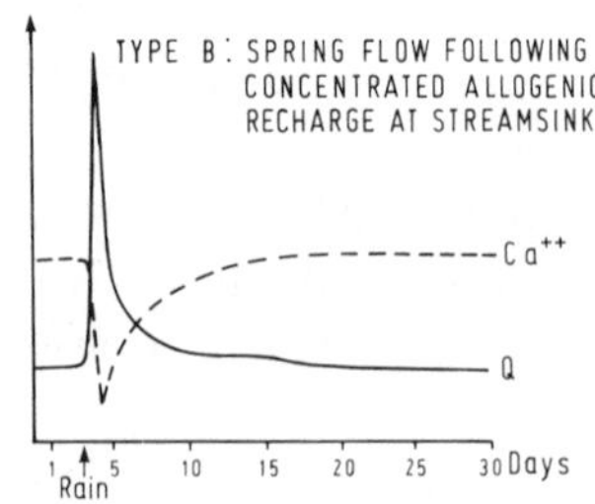

Figure 6.18 Hydrographs and chemographs of karst springs following (A) diffuse autogenic recharge and (B) concentrated allogenic recharge. From Jakucs (1959).

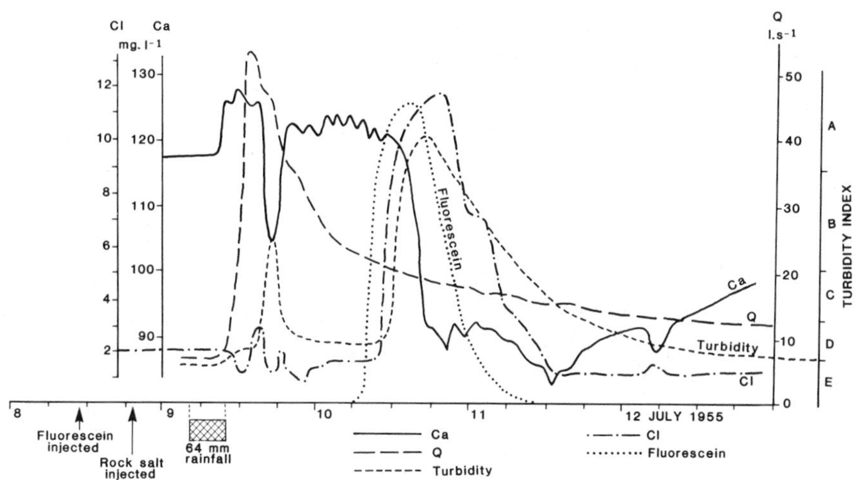

Figure 6.19 Hydrograph and chemograph of the Komlos spring, Hungary, during a water tracing experiment punctuated by rainfall. Note (a) that the discharge pulse arrives before the introduced chemical tracers, in spite of the rain occurring after their injection, (b) that the Ca concentration increases prior to its dilution by floodwaters, and (c) that the turbidity peak lags the discharge peak. From Jakucs (1959).

An insight into the response to recharge of a conduit fed spring was also provided in Jakucs' 1959 paper. In this he presented details of a dye tracing experiment in which the introduction of the dye at a stream-sink was closely followed by a heavy rain storm. Analytical data for samples taken at the resurgence are reproduced in Figure 6.19. They clearly show that the recharge pulse from the rain arrived at the spring well before the dye, despite the latter having been introduced first. These data were interpreted by Ashton (1966) in terms of pulses of water from different input points passing through essentially discrete sections of the phreas. He explained the increase in Ca concentration at the start of the flood as due to the flushing out of water with a long residence time in the deeper phreatic zone; recharge forces water out by a piston-like process. Thus spring chemographs have a pattern composed of sequential and sometimes superimposed pulses of water of different quality and quantity from different stores and tributary inputs. Hence, if water quality and quantity data are combined, it becomes possible to calculate the volume of stores flushed by the recharge. This enables the hydrograph to be separated into different components.

The reality of 'old' water in storage being pushed out by new recharge water was confirmed by the work of Bakalowicz *et al.* (1974) and Eberentz (1976) (cited by Bakalowicz & Mangin 1980) in France. Using ^{18}O variations in three karst springs, it was shown that a precipitation event can cause (old) water in different stores in a karst system to be flushed out and mixed in different proportions, thus giving rise to a fluctuating pattern of $\delta^{18}O$ at the

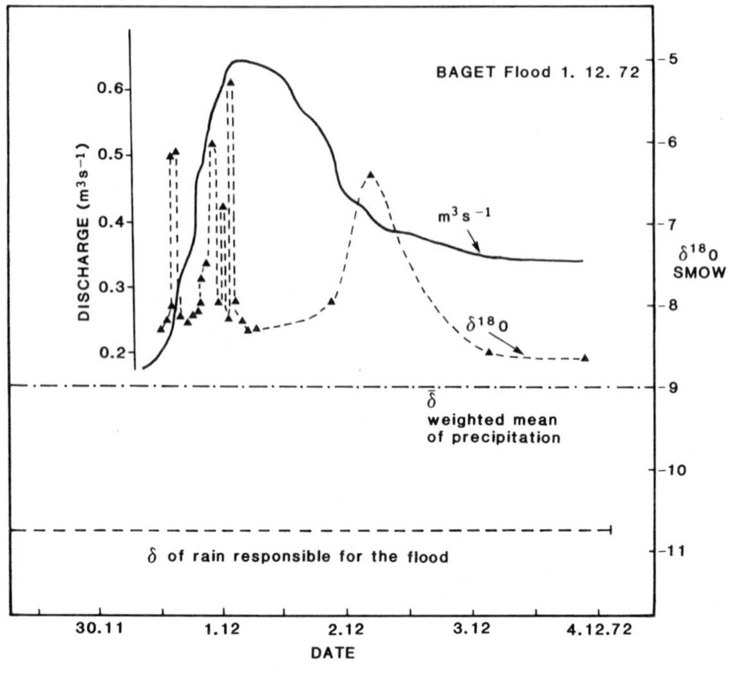

Figure 6.20 Oxygen–18 variations with discharge during a flood event at the Baget spring, France. From Fritz & Fontes (1980).

spring that is different to the pattern of the storm event (Fig. 6.20). Similar conclusions have been reached about the contributions of stored water from previous events to flood hydrographs in surface streams, also using ^{18}O as a label (Sklash *et al.* 1976). Other chemical indicators used to assist in the study of the groundwater component of streamflow include K, Na, Ca, Mg, Cl, SO_2, HCO_3, and electrical conductance (Pinder & Jones 1969; Brown *et al.* 1972).

The separation of karst spring hydrograph and chemograph data is necessarily guided by the conceptual model of the karst system being used. Thus Ashton (1966), Smith *et al.* (1976) and Atkinson (1977b) did not recognize the subcutaneous zone (the epikarst) as a store and possible contributor of displaced water, since its significance was only just beginning to be appreciated in the mid-1970s. Consequently, the volumes they ascribed to the phreatic store are correspondingly too large. The model of stores and linkages in a karst system adopted here is presented in Figure 5.22. An idealized separation of spring data is shown in Figure 6.21, although each component separated is in reality a mixture in which the component is dominant but not usually the only source of the water. A further point is that while the separation model may be a reasonable

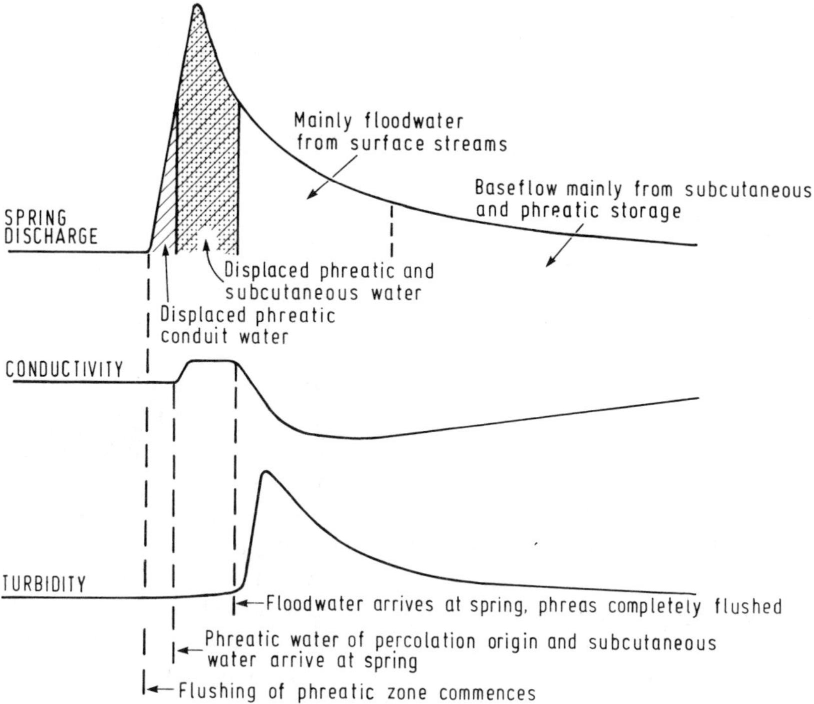

Figure 6.21 Interpretation of a karst spring hydrograph and chemograph. From Williams (1983).

approximation for most springs it requires modification for estavelles that may first discharge invaded flood waters.

At any given instant, the mass–balance of constituents in spring water can be expressed as the sum of the discharge components and their respective concentrations

$$Q_r\,C_r = Q_sC_s + Q_eC_e + Q_pC_p \qquad (6.45)$$

where Q_r is the discharge of the resurgence and Q_s, Q_e and Q_p represent the contributions to it made by sinking surface streams (allogenic), percolation from epikarstic storage, and withdrawal from phreatic storage respectively. The concentration of a given water quality characteristic at the resurgence is C_r and C_s, C_e and C_p are the contributions made to it by the various components. In practice it is difficult to obtain all the component values for Q, although chemical concentration data are relatively easily acquired; thus a combined approach using graphical hydrograph separation and calculation is necessary to derive a realistic assessment of the proportions involved.

For example the allogenic surface water volume may be derived from a contraction of Equation 6.45 in which the only distinction made is between that and karst water; Q_e and Q_p being lumped together as Q_k and C_e and C_p as C_k. Hence

$$Q_r C_r = Q_s C_s + Q_k C_k \tag{6.46}$$

and consequently

$$Q_s = Q_r \frac{C_r - C_k}{C_s - C_k} \tag{6.47}$$

Thus the dominant components may be identified using the graphical approach illustrated in Figure 6.21, but the actual proportions involved in each sector may be estimated from equations of the form of Equation 6.47.

A particular difficulty is encountered in the recognition of subcutaneous water as opposed to phreatic water, because in a simple autogenic system there will be little distinction in their chemical quality. However, in mixed autogenic–allogenic systems, phreatic conduit water will be influenced by allogenic stream inputs and hence will normally have lower concentrations of Ca than water derived solely from the subcutaneous zone. In addition, evapotranspiration losses from the soil in summer tend to elevate Cl concentrations in the soil water store and hence in the immediately underlying epikarstic aquifer. A storm which breaks a summer drought will thus tend to displace the Cl-rich water, and its signal is often encountered at a karst spring as a marked Cl peak. Under such circumstances, the soil water plus epikarstic component can be recognized.

Since at the end of a long dry period subcutaneous storage is at a minimum, a water balance calculation can also be used to estimate the volume of the epikarstic aquifer V_e. The difference between storm precipitation P and flood discharge Q at the spring, less other losses to evapotranspiration ET and to storage S in the soil and phreatic zone, yields a rough estimate of V_e

$$V_2 = [P - (S + ET)] - Q \tag{6.48}$$

Over a heavy storm event ET will be negligible, but S will present difficulties of estimation with correspondingly large uncertainties conveyed to V_e. Estimates of this sort are best made with data from free-draining springs (section 5.3) where phreatic storage is minimal.

The above discussion assumes a monomineralic karst aquifer. Chemically complex springs that discharge water from a variety of aquifers, e.g. bicarbonate, sulphate, chloride, thermal, etc., require a modification of this approach. Assume a spring has two main water quality components, termed

Table 6.4 Flood dilution mixing model calculations for four karst springs in Mexico. From Fish (1977).

Spring	Date	$Q_{r(\max)}$ m^3s^1	$\frac{Q_B}{Q_{r(\max)}} \times 100$	Predicted SO$_4$ mg l^{-1}	Measured SO$_4$ mg l^{-1}
Choy	7 July 72	56	1.16	32	30
	15 June 72	48	1.35	37	35
Mante	17 Aug 71	~ 28	10.7	220	180
	11 July 72	~ 23	13.0	225	240
Frio	27 Dec 71	13.2+	4.17	90	92
Coy	20 June 72	<118	>5.93	>112	25

Note: Q_B was not sampled directly

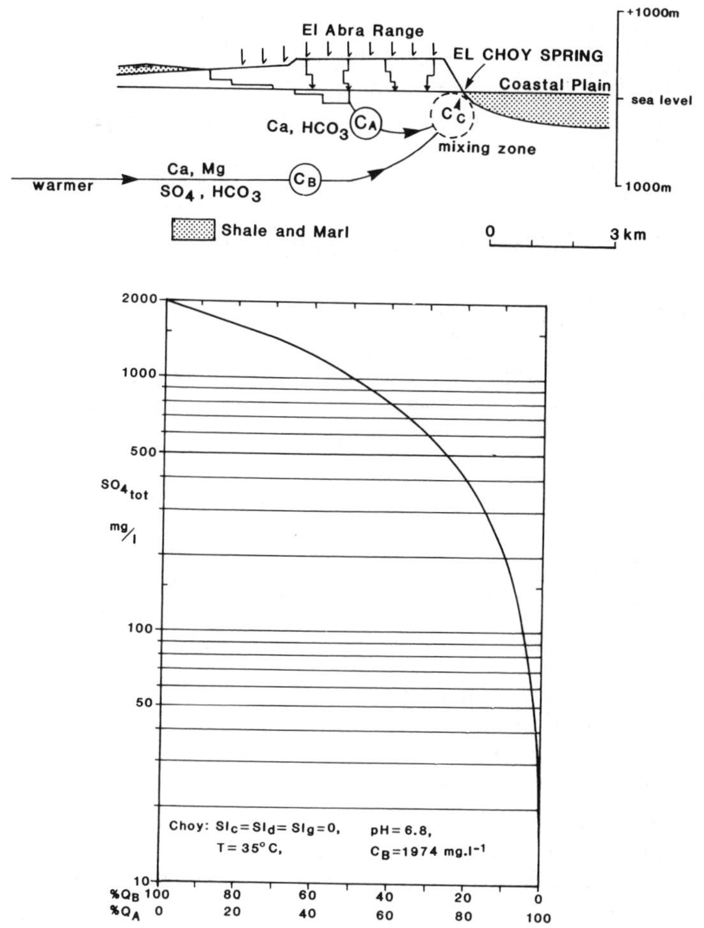

Figure 6.22 Two source mixing model for the Choy spring, Mexico. From Fish (1977).

A and B, perhaps from limestone and gypsum respectively. The basic mixing equation has the form of Eqn 6.46

$$Q_r C_r = Q_A C_A + Q_B C_B \qquad (6.49)$$

Component B may have a deep, inaccessible source, and we may wish to estimate its proportional contribution to the discharge at the resurgence. Dividing through by Q_r, the expression becomes

$$C_r = Q'_A C_A + Q'_B C_B \qquad (6.50)$$

where $Q'_A + Q'_B = 1$. The value of C_A can be assessed from measured values of monominerallic spring waters elsewhere in the region. Knowing the water balance of these other springs, it may also be possible to obtain a reasonable figure for Q_A. Model compositions for C_B are estimated and then tested in high stage and low stage conditions where C_A, C_r and Q_r are known, to arrive at an acceptable estimate of the true values of C_B and Q_A. Table 6.4 and Fig. 6.22 from Fish (1977) illustrate results from an application of this method.

6.7 Interpretation of the degree of organization of a karst aquifer

In a most important recent conrtribution to karst hydrology, Bakalowicz & Mangin (1980) show how the degree of organization of karst aquifers may be deduced from hydrograph and chemograph data. They point out that the karst aquifer appears very heterogeneous, but that it is not the result of a random juxtaposition of different types of voids (primary pores, fissures and karstic voids) because it stems from the ordered distribution of voids around a drainage axis, according to a certain hierarchy. The principal drain receives water from tributaries, the epikarst, and 'systemes annexes' – namely ancient flooded voids and networks of fissures – as explained in Chapter 7. Thus the aquifer in a karst basin can be considered structured with reference to the drainage. This has certain consequences:

(a) Water flows are increasingly organized as the structure develops. The point of departure for this evolution is a porous or fissured aquifer with a diffuse flow system, while the end point is an aquifer with flow limited essentially to conduits (these are the two end members described by Shuster & White 1971). Karst aquifers can include all possible combinations and stages of drainage evolution.
(b) Because of the existence of a particular drainage structure, the scale appropriate to the resolution of any problem concerning the aquifer has to be precise. For example, at a scale of 1–100 m^3 the existence of

randomness can be recognized in the aquifer – a drill hole can go through a karst cavity or can miss it by a few metres, without being able to predict it one way or the other. On the other hand, at the scale of the karst massif an organized drainage structure can be recognized. Thus at one scale the aquifer appears random, while at the other it is structured (see section 5.3 and Fig. 5.4). Our interest is usually in the regional scale, and at his level a reference unit that encompasses the whole structure is the karst system (i.e. drainage basin).

Bakalowicz & Mangin (1980) regard the karst drainage system as being characterized by an impulse response function that transforms the input pulses from precipitation to hydrograph responses at springs, and they contend that by analysis of this function it is possible to identify particular features of the aquifer and its degree of organization. Their approach focuses on flood and baseflow recession analysis techniques discussed in section 6.5 (Eqns 6.35–6.44) supported by hydrochemial data.

Information on the importance of the phreatic zone is obtained from calculation of the baseflow reserves. In order to facilitate comparison of the relative significance of the permanently saturated zone in different karsts, a dimensionless ratio $\underline{\varkappa}$ is employed, defined by Mangin (1975) as the ratio of the maximum value fo the baseflow reserves to the average transit volume of the flood event(s) considered. This coefficient has a value between 0 and 1 and is a measure of the degree of karstification of the system.

Mangin suggests that corresponding information on the unsaturated zone can be obtained from the flood flow recession curve (Eqn 6.36). For this purpose he defines an index i, as the value of y (Eqn 6.38) for $t = 2$ days. The choice is a convention that he found could take account of the different cases studied in France. When i values are small, flood hydrographs are sharply peaked, as opposed to being broad with a relatively gentle recession when values are high. However, systems cannot be described using this index if flood recession is complete within two days (Fig. 6.13), as is often the case with steep gradient and free draining springs (Ecock 1984). The variability of the index over several events and several years may also be of interest in its own right.

Recalling our earlier discussion on the half-flow period applying to the baseflow recession (Eqn 6.30) and the time that the flood recession requires to diminish by 50% (Eqn 6.39), a better basis for comparing linear and non-linear sectors of the karst spring hydrograph may be the values of $t_{0.5}$ for each segment.

On the basis of the values of $\underline{\varkappa}$ and i, Bakalowicz & Mangin (1980) propose a classification of karst aquifers, but since the index i requires further evaluation and possible replacement, caution should be exercised in its use, even though the principles are helpful.

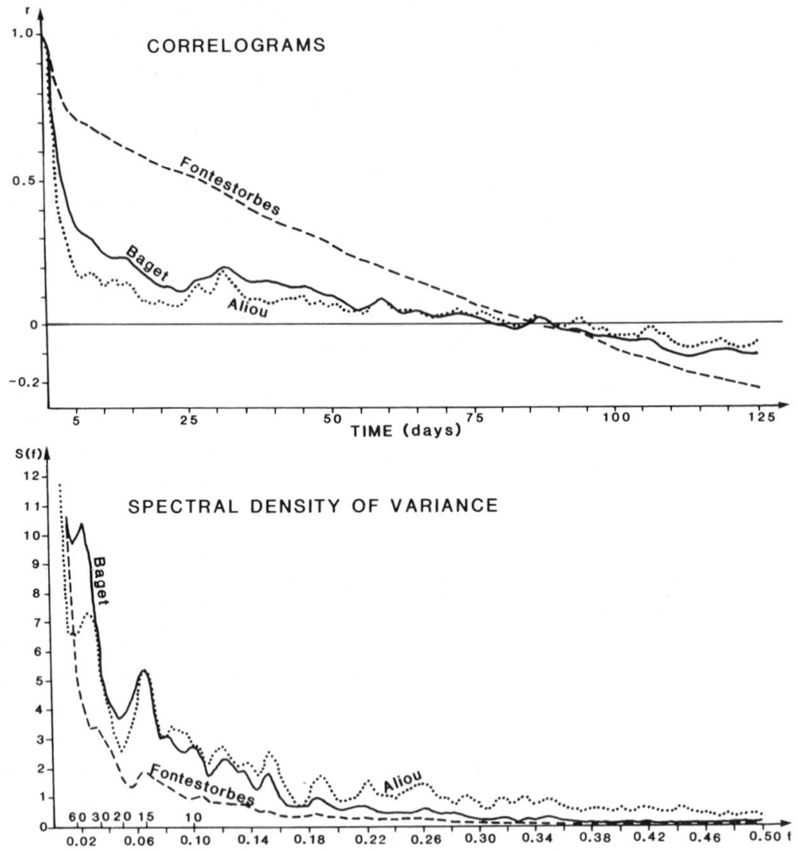

Figure 6.23 Time series analysis of the discharge of three karst springs. From Mangin (1981a).

In a subsequent analysis of rainfall and spring discharge data using autocorrelation and spectral analysis, Mangin (1981a & b, 1984a) added further depth to his interpretation of the Pyreneean karst aquifers. The contrasting reactions of three systems (Aliou, Baget and Fontestorbes) to recharge is illustrated in Fig. 6.23, which shows that the speed of decline of the three correlogrammes is different: most rapid for Aliou and much more even and gradual for Fontestorbes. The memory effect can thus be gauged to be very great in the Fontstorbes system, but short in the case of Aliou. The spectral density curves translate the same phenomena. In another paper (Mangin 1984b), he suggests that karst aquifers may also be classified according to these characteristics (Table 6.5). Mangin interprets this memory as a reflection of storage in the system at the time of recharge, undrained reserves being considerable in the Fontestorbes basin but slight in the case of Aliou. This may be partly a function of basin size

Table 6.5 Classification of karst aquifers. From Mangin (1984a).

Types	Memory effect (days)	Spectral band	Regulation time (days)	Unit hydrograph
Aliou	Reduced 5	Very large 0.30	10–15	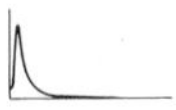
Baget	Small 5–10	Large 0.20	20–30	
Fontestorbes	Large 50–60	Narrow 0.10	50	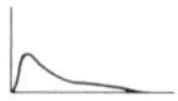
Torcal	Considerable 70	Very narrow 0.05	70	

(A = 11.9 km^2, B = 13.2 km^2, F = 85 km^2) as well as extent of karstic organization. Other studies employing time series analysis in karst hydrology include those by Knisel (1972), Brown (1973), Jakeman *et al.* (1984) and Ozis & Keloglu (1976).

Bakalowicz & Mangin (1980) suggest that if $\underline{\varkappa}$ values are close to or greater than 0.5, then it is possible to analyse the aquifer using the classical methods for porous and fissured media. But they caution that if $\underline{\varkappa}$ values lie between 0.2 and 0.4 then the use of such methods becomes arguable, and that if they should be < 0.1 then interpretations based on porous media concepts are absolutely erroneous. They also suggest that when i values are small a simple system is indicated without recharge delays, but when the index is large a complex system is involved. Thus they attribute the shape of the flood flow recession to the relative complexity of the system rather than to the importance of karstification.

Chemical data reinforce and complement the information derived from the spring hydrographs, although according to Bakalowicz (1977) the structure of a karst aquifer cannot be defined from the coefficient of variation of chemical variables describing springwater, as was suggested by Shuster & White (1971), because the distribution of the values is usually multimodal rather than normal. Atkinson (1977) also noted that the very small range of $CaCO_3$ in resurgances from the Mendip Hills, England, would suggest that they are diffuse flow springs according to Shuster & White's criteria, yet they are known to be fed largely by conduit flow. Bakalowicz & Mangin (1980) demonstrate from French examples that the frequency distribution curve of a hydrochemical variable such as conductivity provides a more accurate description of the aquifer. Figure 6.24 shows that waters issuing from the Baget aquifer originate from three distance populations, each having undergone a unique geochemical evolution, and also illustrates the different forms of frequency distribution that can be encountered. Bakalowicz & Mangin place the following interpretation on the curves:

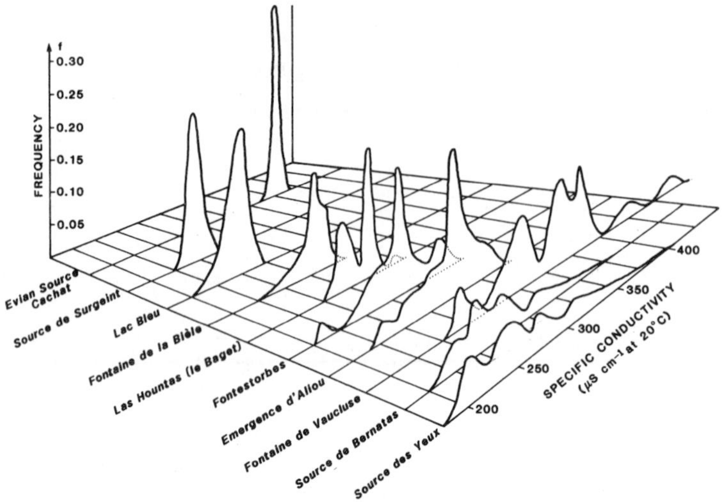

Figure 6.24 Frequency distribution of electrical conductivity of waters from different springs in France. From Bakalowicz & Mangin (1980).

Porous aquifers – unimodal, relatively high electrical conductivity
Fissured aquifers – unimodal, relatively low electrical conductivity
Karstified aquifers – multimodal, wide range of electrical conductivities.

The development of a preferred conduit drainage axis through the saturated zone permits easy transfer of water and therefore transmits a not too distorted image of the recharge process. But if such a hydrological organization is not developed, then homogenization of infilitration waters occurs. This point will be evident from the geochemical data acquired from even just one flood event, which in fact will yield most of the information required to characterize the aquifer. Figure 6.20 illustrates this point perfectly. The succession at the spring of water with different ^{18}O composition would only be possible if a preferred drainage axis exists that minimizes homogenization by the operation of a piston-like throughput process.

6.8 Polje hydrograph analysis

Poljes are large flat-floored basins with interior drainage (Fig. 6.25). They are usually the largest of the enclosed basins in a karst region and may measure many square kilometres in area. Their geomorphology is discussed in detail in section 9.8. A very important hydrological feature common to most poljes is their periodic inundation. In the context of their water

A

B

C

Figure 6.25 A. Popovo Polje, Yugoslavia, with the Trebisnjica River incised into its planed rock floor. The residual limestone hill in the background is beside the hamlet of Hum.
B. Planinsko Polje in Slovenia, northern Yugoslavia, flooded in the wet season.
C. A polje in the Longgongdong karst area near Anshun, Guizhou Province, China.

regime, Ristic (1976) classifies them into four simple types: wholly enclosed, open upstream, open downstream, and open both upstream and downstream. Flooding characteristics involved range from periodic lakes at one extreme to inundated floodplains at the other. Since this does not emphasize the essential quality of poljes – that they are enclosed internally drained basins – the hydrologically more restrictive but geomorphologically more acceptable classification of section 9.8 (Fig. 9.30) is preferred here.

When the sum of the various inflows Q_i into a karst polje exceeds the outflow Q_o then the excess volumes $+\Delta V$ for a time interval Δt is stored in the polje; but when $Q_i < Q_o$ water is withdrawn from storage $-\Delta V$ and flooding recedes. The process is described by a simple water budget equation

$$Q_i - Q_o = \pm \frac{\Delta V}{\Delta t} \tag{6.51}$$

Figure 6.26A illustrates the idealized flooding process in a polje and Fig. 6.26B is an example of the more complex real situation.

Inflows into a polje may come from surface streams, springs and estavelles. However, measurement of the output of flooded springs and estavelles is particularly difficult. Estimation of the total swallow hole capacity is also problematic, but Zibret & Simunic (1976) show how an

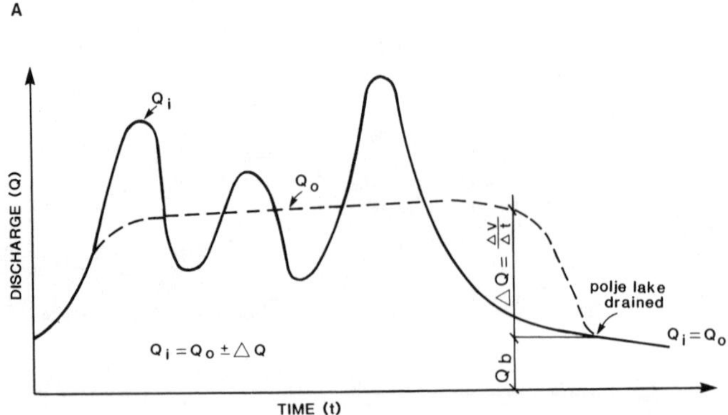

Figure 6.26A Schematic inflow (Q_i) and outflow (Q_o) hydrographs and changes in storage volume (V) in a flooded karst polje. From Ristic (1976).

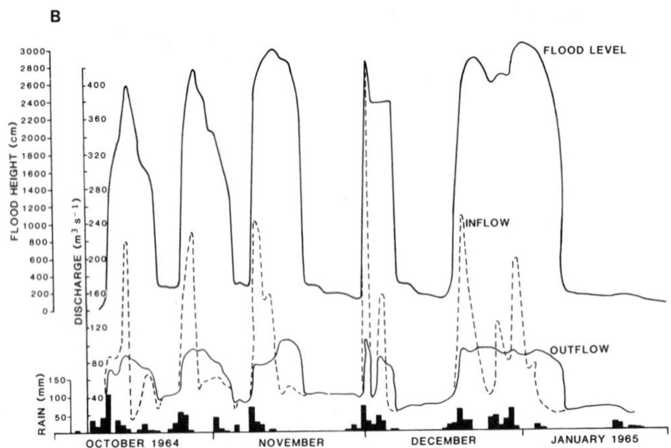

Figure 6.26B Rainfall, hydrographs and flood levels during the inundation of Nevesinjsko polje, Yugoslavia. From Zibret & Simunic (1976).

order of magnitude value may be obtained from the final dewatering phase of an inundation, provided that the reduced inflow Q_i has attained a nearly constant low discharge; the sum of the inflow and the volume reduction $\Delta V/\Delta t$ during the emptying phase then corresponds to the total stream–sink capacity Q_o for a given inundation stage in the polje (Fig. 6.26a). As the lake level reduces the hydraulic head diminishes and the swallow holes may absorb less water.

Although the volume that can be absorbed by swallow holes varies with the depth of flooding, it also depends on underground piezometric pressures. The deeper the water table is below the polje floor, the longer the

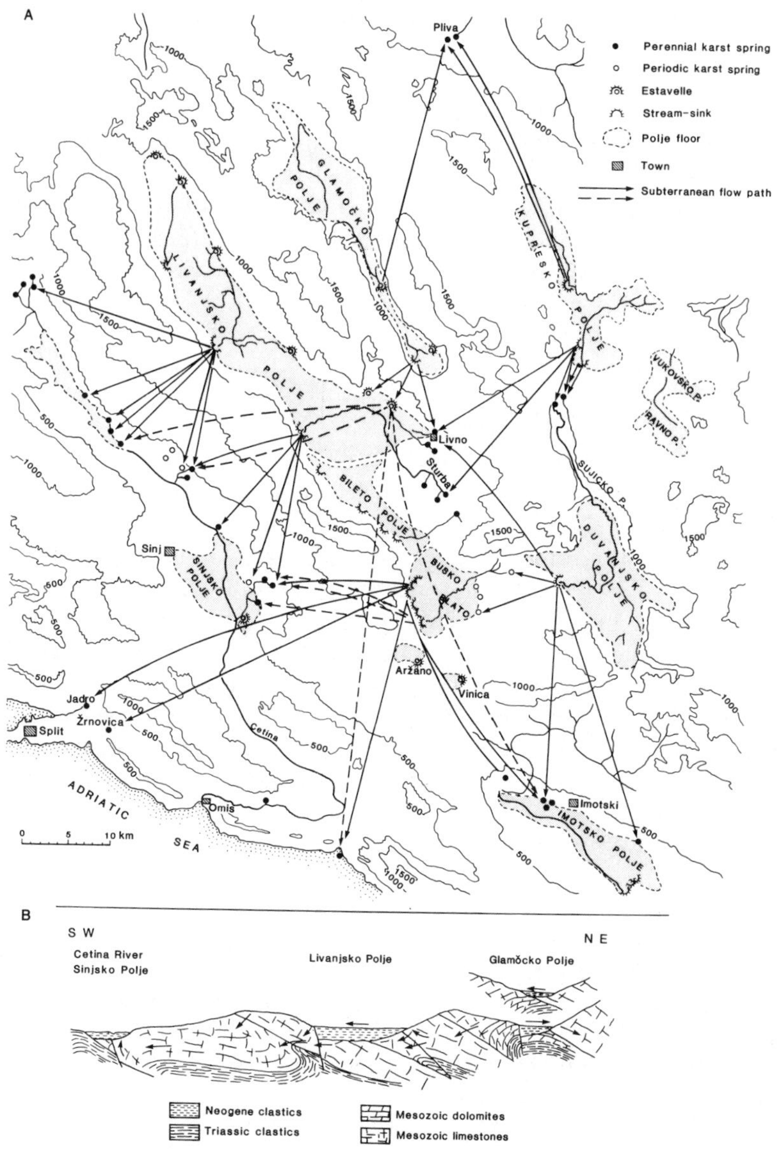

Figure 6.27 The hydrogeological system of poljes, springs and their traced interconnections in southern Yugoslavia. A from Zötl (1974) and B from Herak (1972).

stream–sink will be able to operate at full capacity. Hence antecedent rainfall conditions are influential in determining quickflow runoff and the magnitude of Q_i, while seasonal effects influence subterranean storage and the height of the water table. Antecedent rainfall, through its influence on the generation of quickflow runoff, is a particularly important variable in the flooding of border poljes, whereas seasonal phreatic zone storage is a critical factor in the inundation of baselevel poljes (Fig. 9.30).

The individuality of flooding behaviour in different Yugoslav poljes is well illustrated by a number of excellent case studies: Popovo polje (Milanovic 1981); Cerknisko polje (Kranjc 1985, Gospodarič & Habič 1978); Planina polje (Gams 1980); and Kocevsko polje (Kranjc & Lovrencak 1981).

Poljes are best seen as subsystems within a broader karst drainage system, the storage and overflow from one polje often affecting the inundation behaviour of others in the series (Fig. 6.27). Hydrological interconnections between poljes can often be elucidated by water tracing experiments, a topic discussed in section 6.10.

6.9 Water balance estimation

Hydrograph recession analysis and chemograph separation techniques produce values for components of aquifer storage. Water balance calculation provides an order of magnitude estimate of reserves and storage changes, as we have just seen in regard to polje flooding.

A hydrologic budget is a quantitative evaluation of inflows, outflows and changes in storage over a specified time interval, usually a hydrological year, although it can be applied to longer or shorter periods. A hydrological year runs from dry season to dry season, i.e. it starts and ends when storage is at a minimum, and thus often does not correspond to a calendar year. With respect to surface streams, the water balance equation in its simplest form is usually written

$$Q = P - E \pm \Delta S \tag{6.52}$$

where Q is runoff, P is precipitation, E is evapotranspiration and $\pm\Delta S$ represents withdrawal from or replenishment of storage. Dunne & Leopold (1978) expand on this approach. But in a groundwater context the perspective differs; thus the US Department of the Interior (1981) suggest that the hydrologic budget may be summarized as

$$\Delta S = P - E \pm R \pm U \tag{6.53}$$

where R is the difference between stream outflow (−) and inflow (+), and U is the difference between deep groundwater outflow (−) and inflow (+); ΔS being the resultant change in storage in groundwater, soil moisture, channels and reservoirs. The groundwater component of this is ΔSg and can be estimated from

$$\Delta Sg = G - D \tag{6.54}$$

in which G represents recharge to the aquifer and D the discharge from it.

Both G and D are themselves comprised of several components discussed by the US Department of the Interior (1981). Brown *et al.* (1972) also explain how various components of the water balance may be estimated.

A good example of large scale karst water balance estimation is given by Bocker (in press). The aim was to investigate the environmental impacts of coal and bauxite mine dewatering operations in the 15 000 km^2 of the Transdanubian Mountains, Hungary. A finite element model using 4 km^2 elements was developed. Infiltration into each element was estimated from an algorithm derived from a 15 year field experiment, and data from 480 waterworks, 93 mines, 270 observations wells and 155 meteorological stations were included in this very major computation.

6.10 Water tracing techniques

Water tracing is a well developed, powerful tool of the karst hydrologist, enabling catchment boundaries to be delimited, groundwater flow velocities to be estimated, areas of recharge to be determined, and sources of pollution to be identified. Chemicals such as common salt, ammonium sulphate and fluorescein have been used for water tracing for more than a century. Back & Zotl (1975) review the full range of modern techniques from naturally occurring labels to artificially introduced radio isotopes. Recent reviews by Quinlan (1986a) and the US Environmental Protection Agency & US Geological Survey (1988) also provide much useful information and comprehensive bibliographies.

Three classes of water tracing agents are available:

Natural labels	(a)	flora and fauna, principally microorganisms
	(b)	ions in solution
	(c)	environmental isotopes
Pulses	(a)	natural pulses of discharge, solutes and sediment
	(b)	artificially generated pulses
Artificial labels	(a)	radiometrically detectable substances
	(b)	dyes
	(c)	salts
	(d)	spores

The use of naturally occurring labels

(a) NATURALLY OCCURRING MICROORGANISMS

Bacteriological and virological examination of karst spring waters can be undertaken to establish the hygienic quality of the water and, if contaminated, to help trace possible sources of pollution. Karst aquifers are notorously bad water filters; thus transmission of microorganisms is to be expected. There is an extensive literature on the movement of bacteria through porous media, reviewed by Romero (1970), Gerba *et al.* (1975), and contributors to Gospodarič & Habič (1976). Bacterial species used for groundwater tracing include *Serratia marcescens*, *Chromobacterium violaceum* and *Bacillis subtilis*. A virus (bacteriophage) has also been used as a tracer (Martin & Thomas 1974). Further information on the use of organisms in water tracing is available in Moeschler *et al.* (1982).

(b) IONS IN SOLUTION

Chloride has long been used as a natural tracer to determine the fresh water–salt water interface in coastal aquifers (section 5.1) and to detect possible intrusion of sea water into bore water supplies near the sea. Variations in concentration of the calcium ion have also been used to determine flow through times (Pitty 1968b) and flow through routes (Gunn 1981c, 1983) especially in the vadose zone, although Friederich & Smart (1982) remark that it does not provide an unambiguous variable for the hydrological classification of percolation waters because hardness is controlled by non-hydrological factors. The provenance of water emerging at springs can also be determined by concentrations and changes in concentration of different water quality characteristics. This is the basis of spring chemograph separation (section 6.7). The nature of the flow system sustaining the spring is also interpretable from such data (Shuster & White 1971, Bakalowicz 1977).

(c) ENVIRONMENTAL ISOTOPES

The use of stable and radioactive isotopes in groundwater hydrology has made enormous advances in recent years (International Atomic Energy Agency 1981, 1983, 1984, Fritz & Fontes 1980, Lloyd 1981b). The term 'environmental isotopes' is used to describe isotopes that occur naturally in the hydrological cycle. Hydrogen possesses ^{1}H, ^{2}H (also called deuterium, D) and ^{3}H (termed tritium, T), the last being radioactive. Oxygen isotopes include two of interest in hydrology: ^{16}O and ^{18}O, both being stable. These constitute the water molecule, so are the best conceivable tracers of water. They occur in various combinations, such as $H_2{}^{16}O$, $H_2{}^{18}O$, HDO and HTO. Other important environmental isotopes commonly used in groundwater research are ^{12}C, ^{13}C and ^{14}C. Less frequently used are isotopes of argon, chlorine, helium, krypton, nitrogen, radium, radon, silicon, sulphur, thorium and uranium (IAEA 1983, 1984).

The main applications of environmental isotopes to groundwater hydrology are

(i) to provide a signature of a particular groundwater type that can be related to its area of origin;
(ii) to identify mixing of waters of different provenance;

(iii) to provide information about throughflow velocities and directions; and

(iv) to provide data on the underground residence time (age or age spectrum) of the water.

In order to be able to use natural isotopes effectively, the reasons for and empirical laws of their distribution in natural waters must be understood. These points are discussed in Fritz & Fontes (1980) to which readers are referred for further detail.

Radioactive isotopes

Radioactive isotopes are unstable and undergo nuclear transformation emitting radioactivity. Their decay is spontaneous and unaffected by external influence. It occurs at a unique rate for each radioisotope, defined by its half-life $T_{1/2}$, which is the time required for one half of the radioactive atoms to decay following an exponential law

$$N = N_o e^{-\lambda t} \tag{6.55}$$

where N is the number of radioactive atoms present at time t, N_o was the number present at the commencement of decay, and λ is the half-life or decay constant. The half-life for ^3H is 12.43 years and for ^{14}C is 5730 years. Consequently ^3H is useful for dating waters up to about 50 years in age, whereas ^{14}C has an upper limit of approximately 35 000 years. Fontes (1983) explains how groundwaters can be dated and discusses environmental isotopes suitable for that purpose.

Both ^3H and ^{14}C are produced naturally in the atmosphere, but atmospheric thermonuclear explosions since 1952 increased the abundance of these isotopes in precipitation. The subsequent moratorium on atmospheric thermonuclear testing has resulted in considerable diminution of these isotopes in rainfall to the extent that in the Southern Hemisphere ^3H levels have practically returned to natural background concentrations, virtually eliminating the value there of ^3H for dating. Tritium concentrations are expressed in Tritium Units (TU), defined as one atom of ^3H per 10^{18} atoms of ^1H. Natural background levels are 4–25 TU, whereas ^3H concentration in rainfall attained a peak of 8000 TU in the Northern Hemisphere in 1963. The Southern Hemisphere experienced a lower peak a year or two later.

Quite apart from thermonuclear effects, the amount of ^3H in precipitation also shows some important natural variations (Erickson 1983). For example there is a marked latitude dependence, with values lower by a factor of 5 or so in the tropics (Gat 1980). Tritium values also slowly increase inland, doubling in central Europe over a distance of 1000 km. Superimposed upon this is a seasonal variation with a maximum in late spring and summer 2.5 to

6 times higher than in winter. These seasonal variations are potentially important in karst water investigations (Back & Zotl 1975), because they may provide a signature for the season of recharge.

Interpretation of radioisotope data at a spring depends very much on the model of flow and mixing adopted (Yurtsever 1983). Several alternatives have been suggested with the *piston flow* model at one extreme and the *completely mixed reservoir* model (sometimes termed the exponential or one-box model) at the other. The first is similar to the conduit flow or 'perfect pipe' end member commonly applied to karst (sections 6.6 and 6.7). Piston flow assumes that recharge occurs as a point injection and that a discrete slug of tracer moves through the groundwater system. In nature this never occurs perfectly because of mixing and dispersion effects but, nevertheless, the flow behaviour of well developed karsts is sufficiently similar to it (Fig. 6.20) for such a model to be taken as an acceptable first approximation. By contrast, the completely mixed reservoir model assumes uniformity at all times, each episode of recharge mixing instantaneously. It also assumes stationary conditions with respect to reservoir volume, discharge and infiltration rate.

Actual tracer behaviour in most groundwater systems lies between these two extreme cases. *Dispersive* models have been proposed to describe intermediate situations, taking account of mixing and dispersion within the system but assuming that pulse variations in tracer output can be related back to concentration variations in recharge events. A cascade of mixing compartments connected by linear channels has been used to approximate this situation, being termed a *finite state mixing cell* model. Individual cells are assumed to behave like well mixed reservoirs but may be of different volume and concentration. Fontes (1983) and Yurtsever (1983) discuss applications and theoretical considerations.

Stewart & Downes (1982) provide an instructive illustration of the different interpretation that emerges depending on the model chosen. They deal with a karstic artesian aquifer in the Takaka valley, New Zealand (Fig. 11.1). that drains to the Waikoropupu Springs (Williams 1977, Stewart & Williams 1981). Tritium measurements were made at the springs (mean discharge 15 $m^3 s^{-1}$, Table 5.6) in 1966, 1972 and 1976 and were compared with tritium values in local allogenic recharge and rainfall (Table 6.6). Tritium increased rapidly in local rainfall to about 40TU in 1964/65. The result for the Main Spring in 1966 shows the input of this young tritium enriched water, whereas the 1972 figures reveal the presence of some low tritium (older) water. Values in rainfall declined steadily after 1971, the 1976 results indicating that not much high tritium water remains. Stewart & Downes thus conclude that the springs' outflow has a spread of ages, but that young and old components predominate, the best estimate of throughput time depending on the model adopted and the year of sampling (Table 6.6). Geological circumstances in a partly capped artesian aquifer

Table 6.6 Interpretation of tritium data for the Waikoropupu Springs, New Zealand. From Stewart & Downes (1982).

	Tritium content (TU)			Flow times (a)	Turnover times (a)
Date	Annual weighted mean rainfall	Local runoff recharging aquifer	Main spring	Piston flow model	One-box model
27/5/66	34		14 ± 0.9	3–4	7–8
29/7/72	15	20.1 ± 1.2	15.3 ± 1.9	< 1 or 8–10	10–12
4/5/76	8	11.9 ± 2.0	11.2 ± 1.2	2–4 or 12–14	0–20

(Fig. 11.1) suggest a piston flow model to be more realistic than a one-box model, but detailed stable isotope data demonstrate the aquifer to be very well mixed (Stewart & Williams 1981). It is likely that the mean age of the outflow increases during low discharge conditions.

Our conclusion is that in many karst aquifers the water may have components of different residence times and so to assign one mean age may be misleading. This point was well made by Siegenthaler *et al.* (1984) in regard to a spring in the Swiss Jura. Strong evidence showed that in periods of baseflow relatively homogeneous old water sustains the spring, whereas after a storm or meltwater event rapid runoff of new water is also important. 'Consequently, it would not make sense to talk of *one* mean age of the spring water.' The main interest is in the average residence time of the older reservoir water.

Tritium concentration has been largely thought of as a means of dating water, but in view of the currently decreasing ^3H values in karst reservoirs it is becoming less important for that purpose. Rozanski & Florkowski (1979) and Salvamoser (1984) suggest that as a consequence the use of environmental krypton-85 ($T_{1/2}$ = 10.8a) should be considered for dating. In several respects it is an ideal tracer: its amount is still increasing, it shows no marked seasonal variations, it has virtually no sinks outside of radioactive decay, and its atmospheric concentration in the Southern Hemisphere is only 10–20% lower than in the north.

Stable isotopes

Stable isotopes undergo no radioactive decay. D, ^{18}O and ^{16}O occur in the oceans in concentrations of about 320 mg l^{-1} HDO, 2000 mg l^{-1} $H_2{}^{18}O$ and 997 680 mg l^{-1} $H_2{}^{16}O$. Variations in these proportions are measured by mass spectrometry and compared to the composition of a standard termed SMOW (standard mean ocean water). The ratios D/H and $^{18}O/^{16}O$ are expressed in delta units, which are parts per thousand (per mil) deviations of the isotopic ratio from the standard

$$\delta‰ = \frac{R_{\text{sample}} - R_{\text{standard}}}{R_{\text{standard}}} \times 1000 \qquad (6.56)$$

where R is the isotopic ratio of interest.
Thus a sample with $\delta^{18}O$ = + 10‰ is enriched in ^{18}O by 10‰ (i.e. 1%) relative to SMOW. Values for δD usually fall in the range −50 to −300‰ and for $\delta^{18}O$ in the range +5‰ to −50‰. The accuracy of measurement is approximately ± 2‰ for δD and ± 0.2‰ for $\delta^{18}O$.

Differences in the isotopic composition of water samples reflect fractionation processes that occur in the hydrological cycle. The heavy isotope molecules (HDO, $H_2{}^{18}O$) have slightly lower saturation vapour pressures than the ordinary water molecule ($H_2{}^{16}O$), hence when changes of state occur during evaporation and condensation, a slight fractionation takes place. For example, when evaporation occurs from an open water surface the vapour is depleted in heavy isotopes in relation to the remaining unevaporated water; whereas when condensation occurs the initial precipitation is slightly enriched so that later precipitation becomes increasingly depleted with respect to SMOW.

Temperature is an important factor in fractionation: the lower the temperature the greater the depletion in heavy isotopes. This influence is reflected in both latitude and altitude differences in heavy isotopic composition of precipitation (Dansgaard 1964; and see also section 8.6). Back & Zotl (1975) identify four general rules for stable isotope fractionation:

(a) in regions of moderate climate precipitation is isotopically lighter in winter than in summer;
(b) precipitation becomes isotopically lighter with increasing latitude and altitude;
(c) D and ^{18}O in precipitation show a good linear correlation (Eqn 6.57); and
(d) enrichment of D and ^{18}O occurs in lakes because of evaporation.

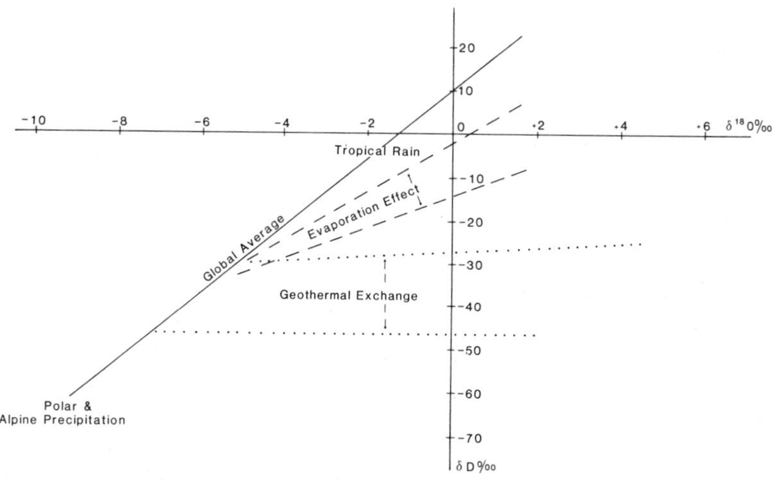

Figure 6.28 Oxygen–18 and deuterium relationships in natural waters. From Fontes (1980).

The relationships which underlie these rules and result in the natural labelling of water in stable isotopes can be summarized as in Fig. 6.28. Generalizing global data, a linear relationship (the 'meteoric water line') emerges (Craig 1961, Gat 1980)

$$\delta D = (8.17 \pm 0.08)\ \delta^{18}O + (10.56 \pm 0.64)‰ \quad (6.57)$$

with a correlation coefficient of $r = 0.997$ and a standard error of estimate of ± 3.3‰. Regional variants may also be identified – for example a Mediterranean curve.

The relationships that exist in the natural distribution of stable isotopes in water have been valuable in determining the recharge areas and recharge seasons of groundwaters, especially where allogenic inputs are involved. Autogenic recharge is usually so well mixed in the epikarst that seasonal patterns and recharge events are difficult to identify (Yonge *et al.* 1985, Even *et al.* 1986). Mixing of groundwater bodies and leakage of water from lakes and reservoirs can also be discerned (Fontes 1980). Illustrations of the use of these methods in karst aquifers can be found in numerous publications, for example by Dincer & Payne (1971), Dincer *et al.* 1972, Bakalowicz *et al.* (1974), Zotl (1974), Moser *et al.* (1976), Stewart & Williams (1981) and Celico *et al.* (1984). An interesting application of isotope hydrology to elucidate the complex relations between confined karst groundwaters, geothermal water, and groundwaters in shallow basaltic and alluvial aquifers is that published by Issar *et al.* (1984). However, Margrita *et al.* (1984) point out that the concepts of theoreticians do not always appear to be well understood by the users of isotopic tracers, and they identify the kinds of errors that may appear in interpretation of results.

Pulses as tracers

(a) NATURAL PULSES

A pulse is a significant variation in water quantity or quality. A storm is the usual natural trigger mechanism for its generation, although snow and glacier melt also produce pulse waves (e.g. Meadows Creek in Fig. 6.13). Pulses were being used as karst tracers by the mid-19th century (e.g. Tate 1879), although their potential to discriminate details of conduit network geometry was formally developed by Ashton (1966). He showed how input pulses combine to produce complex output signals and explained how the network geometry might be determined by their interpretation (Fig. 6.29). In practice, it has not yet proved possible to interpret spring hydrographs with confidence in the detail that Ashton's theoretical arguments suggest, but pulse train analysis as a means of water tracing has proved strikingly successful in a system too difficult for conventional dye tracing (Williams 1977). Further elaboration of pulse train techniques can be found in papers by Wilcock (1968), Brown (1972b, 1973), Christopher (1980) and Smart & Hodge (1980).

Given an input pulse, the related output pulse will vary according to the transfer function of the system, as we have seen in sections 6.5 and 6.6.

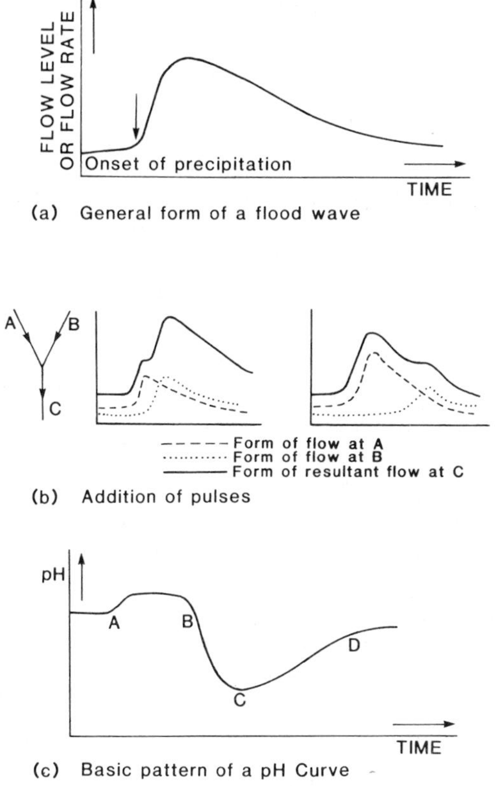

Figure 6.29 Pulse generation and pulse amalgamation. From Ashton (1966).

Whereas natural and artificial labels such as isotopes and dyes approximately measure the rate of flow of the water being traced, pulses almost always exceed it by a significant amount. A flood wave (discharge pulse) will travel as a *kinematic wave* down an open vadose passage, and as a *pressure pulse* through a phreatic conduit. Large kinematic waves travel faster than smaller ones; both travel more quickly than the water itself, especially through pools. Transmission of a pressure pulse through a flooded tube is almost instantaneous (at the speed of sound). It is necessary therefore to distinguish between the *pulse through time* (i.e. hydraulic response time) and the *flow through time* of the system. For example, the Komlos spring pulse through time is about six hours, whereas the flow through time of a tracer dye is roughly 40 hours (Fig. 6.19).

Brown's (1972b, 1973) work on the Maligne basin, Canada (Table 5.6), is particularly fine illustration of the application and development of these techniques. Wheres Ashton employed aperiodic and unique pulses, Brown used time series analysis on long term discharge data. The Maligne valley contains Medicine Lake, an intermittent internally draining lake that in winter reduces to small ponds, and overflows about one year in two for a few weeks in summer. The lake is at 1500 m asl and drains to a line of springs 16 km distant and 410 m lower in altitude. Brown undertook a cross-covariance analysis of lake input and springs output data and identified a negative cross-covariance peak at −20h. He explained this as the output of daily snow melt cycles occurring in the lower one-third of the basin (i.e. closer to the springs than the input from the lake up valley). A subtle secondary 'peaking' at +70 to +124 h was considered to be the response to known flow inputs from the lake, and was later confirmed by a dye trace that appeared +80 to +130 h after injection, peaking at +90 h. Since the travel time of the dye was of the same order as the pulse through time of the river, Brown was able to conclude that open channel (vadose) conditions prevailed underground at the time of the experiments. Later work by Kruse (1980) showed the system's response to vary with discharge conditions.

(b) ARTIFICIALLY GENERATED PULSES

Ashton (1966) discussed the use of artificial flood waves produced by collapsing small temporary dams at stream-sinks. Periodic water releases in the course of hydroelectric power generation yield much larger pulses that have been traced for a considerable distance. In the Takaka valley, New Zealand, Williams (1977) showed by cross-correlation analysis that kinematic waves produced by releases from a hydroelectric dam take 5 h to travel about 15 km down an open natural channel to enter a marble aquifer and then 9–11 h to travel a further 20.2 km through it to an artesian spring with a mean discharge of 15 $m^3\ s^{-1}$. The karst section consists of a recharge zone where the river looses flow into its gravelly bed for about 10 km and then a confined zone where pressure pulse transmission may be assumed. Whereas the tritium concentration of the spring water suggests a minimum flow through time of 2–4 years (Table 6.5), the pulse through time is less than half a day; and whereas the gradient in the Maligne system analysed by Brown is about 26 $m\ km^{-1}$, in the Takaka valley it is a maximum of 2.7 $m\ km^{-1}$. Thus pulse train analysis is an effective means of demonstrating point-to-point connections over long distances with low hydraulic gradients and through large flooded zones (the Takaka marble aquifer probably has a phreatic water volume of 1.5–3.8 km^3).

Pulse travel times in these investigations ranged from 3 $km\ h^{-1}$ for kinematic waves travelling down the steep channel of the upper Takaka river to approximately 2 $km\ h^{-1}$ through the marble aquifer of the valley. This compares to 0.2–1.45 $km\ h^{-1}$ in the much steeper Maligne system under vadose conditions (Table 5.7).

Artificial labels

One of the shortcomings in the use of natural labels is that they seldom provide information on specific point-to-point connections. Hence artificial tracers are widely used especially to establish stream–sink to resurgence linkages.

To be suitable for water tracing experiments artificial tracers should be:

(1) non-toxic to the handlers, to the karstic ecosystem and to potential consumers of the labelled water;
(2) soluble in water, in the case of chemical substances, with the resulting solution having approximately the same density as water;
(3) neutral in buoyancy and in the case of particulate tracers, sufficiently fine to avoid significant losses by natural filtration;
(4) unambiguously detectable in very small concentrations;
(5) resistant to adsorptive loss, cation exchange and photochemical decay, and to quenching by natural effects such as pH change and temperature variation;
(6) susceptible to quantitative analysis;
(7) quick to administer and technologically simple to detect; and
(8) inexpensive and readily obtainable.

Few tracers satisfy all of these criteria, but several do to a significant extent. Non-toxicity is the most important characteristic.

(a) RADIOMETRICALLY DETECTABLE SUBSTANCES

The International Atomic Energy Agency (1984) gives guidance on the choice of artificial isotopes for water tracing:

(i) The isotope should have a life comparable to the presumed duration of the observations. Unnecessarily long-lived isotopes will create pollution, a health hazard, and interference if the experiment needs repeating.
(ii) The isotope should be resistant to adsorption by soils and rock.
(iii) For most problems, it is desirable to be able to measure the radioactivity in the field; thus γ emitters are preferred in general, although β emitters are also suitable.
(iv) the isotope must be readily available at reasonable cost.

In our earlier discussion on environmental isotopes, we identified ^{3}H as a perfect tracer because it is part of the water molecule itself. It is also identifiable in very low concentrations, to $1 : 10^{15}$ or even more dilute. However, using the first criterion above, ^{3}H is immediately eliminated. *In situ* detection (criterion iii) is also not possible.

With respect to adsorption, Lallemand & Grison (1970) noted ^{3}H, ^{82}Br as bromide and ^{35}S as sulphate to be the best of 12 isotopes considered. Ramljak *et al.* (1976) undertook field experiments with ^{51}Cr, ^{46}Sc, ^{140}La, ^{131}I and ^{82}Br and concluded that the most suitable were ^{46}Sc as a chelate with EDTA and ^{51}Cr in the form of chromate–bichromate. To trace 10^6 m^3 of water, they calculated that 40 kg of fluorescein dye would be needed compared to less than 2 g ^{82}Br and about 25 g^{51}Cr; an impressive difference.

The public health disadvantages of using radioactive isotopes may be overcome for some species by the post-sampling activation analysis technique, although it has the severe drawback of requiring very specialized facilities. Principles of the method are dealt with by Kruger (1971). Buchtela (1970), Schmotzer *et al.* (1973) and Burin *et al.* (1976) have applied the technique to karst, and a general discussion of results in that context is provided by Zotl (1974). The method involves injection of an initially non-radioactive tracer that is later sampled and irradiated with neutrons in a reactor. If the tracer is present in the sample it is detectable by its activity. Buchtela used an EDTA complex of indium (detectable to $1 : 10^7$), whereas Schmotzer *et al.* and Burin *et al.* found bromide (detectable to $2 : 10^8$) to be the best ion, applied either as sodium, potassium or ammonium bromide. As with the radioisotope tracers, chelation with EDTA forms a negatively charged complex ion that minimizes adsorption losses. Added advantages of the activation technique are that it makes possible the simultaneous use of several tracers (e.g. Br, In, La) and problems of half-life can be disregarded.

Nevertheless, despite the advances of the last 25 years or so, Burdon & Papakis's (1963) recommendation remains true, that before turning to radioactive tracers, which are costly, hazardous and need skilled handling by personnel supported by atomic laboratories, it is better to try to solve the problem with coloured or chemical tracers. Back and Zotl (1975) reach a similar conclusion.

(b) DYES AS WATER TRACERS

Artificial dyes are the principal and most successful karst water tracers at the present time. Dyes have been used to trace underground water for more than a century. The green dye fluorescein ($C_{20}H_{12}O_5$) was discovered in 1871, and was first used as a karst water tracer at the stream-sinks of the upper Danube river in 1877 (Kass 1967). Its disodium salt under the trade name 'uranine' ($C_{20}H_{10}O_5Na_2$) was introduced a few years later. For many years detection of fluorescein and uranine depended on visual observation at a resurgence. Thus if the dye peak has passed, the chances are that no unambiguous visual trace of the dye would remain. Tracing with fluorescein therefore advanced considerably when it was found (Dunn 1957) that the dye is adsorbed by charcoal grains from which it may later be released by an elutant of potassium hydroxide (5%) in ethanol. The field requirement is simply to suspend small mesh bags containing a few grams of granular activated carbon in a moderate current in the monitored springs. Wittwen *et al.* (1971) adsorbed on charcoal sulphorhodamine B and sulphorhodamine G-extra, two orange dyes, which they eluted with various alcohols (see also Perlega 1976). Rhodamine WT is also adsorbed and may be eluted with a warm solution of 10% ammonium hydroxide in 50% aqueous 1-propanol (Smart & Brown 1973). The great advantage of this technique is that detector bags can be changed when convenient and many sites may be easily monitored. There are no time constraints on when analysis must be undertaken, although air drying of detector bags is required if more than a few days will elapse between collection and examination. Charcoal bag detectors also work effectively in the sea should resurgences be submarine or intertidal. With careful elution and the use of a fluorometer, the charcoal method may be used semi-quantitatively (Smart & Friederich 1982).

The most recent important development in dye tracing techniques has been the advent of quantitative fluorometric procedures following the work of Kass (1967) in Europe and Wilson (1968) in North America. Kass examined two green dyes (uranine and eosine) and two orange dyes (rhodamine B and sulphorhodamine B), and showed uranine to possess the strongest fluorescence of the four and to be detectable in solutions down to 0.01 mg m^{-3}). He subsequently participated in a major experiment tracing Danube River water to the Aach spring in Germany, where fluorometric techniques were used successfully to identify sulphorhodamine G-extra as a tracer and to confirm uranine in the elutant from activated charcoal detectors (Batsch *et al.* 1970). This showed that orange and green fluroescent dyes could be separately identified when mixed.

Fluorescent substances emit light immediately upon irradiation from an external source. The emitted or fluoresced energy usually has longer wavelengths and lower frequencies than that absorbed during irradiation. This property of dual spectra is utilized to make fluorometry an accurate and sensitive analytical tool, because each fluorescent substance has a different combination of excitation and emission spectra (Wilson *et al.* 1986). Since some naturally occurring substances such as plant leachates, phytoplankton and some algae also fluoresce, it is important to know the background fluorescence of the water being traced before any experiments are conducted. Industrial and domestic wastes also introduce background problems. Natural substances tend to fluoresce in the green wave band; thus the use of dyes with an orange emission overcome problems of possible misidentification.

Numerous fluorescent dyes are commercially available and many have been used in groundwater tracing (Table 6.7). Because different manufacturers often assign different product names to what in fact is the same dye, confusion sometimes occurs in comparing results of tracers by different people. Hence we endorse the recommendation of Quinlan & Smart (1976)

Table 6.7 Fluorescent dyes used in water tracing with an indication of their properties.

Fluorescent dye	Colour index general name (number)	Peak excitation nm	Peak emission nm	Minimum detectability $\mu g l^{-1}$	pH Effect	Adsorptive dye loss
Optical brightners						
Photine CU	Fluorescent brightner 15	345	447	0.36	emission reduced below pH7	High on organics
Photine CSP	Fluorescent brightner 145	352	455		emission reduced below pH7	
Fluolite BW	Fluorescent brightner 23	355	435		emission reduced below pH7	
Leukophor C	Fluorescent brightener 231	349	442		emission reduced below pH7	
Tinopal CBS-X		355	430			
Blankophor REU	Fluroescent brightener 119	340	425			
Blankphor DBS	Fluorescent brightner 192	375	450			
Blue dye – intermediate						
Amino G acid		355	445	0.51	emission maximum at pH6–8.5	Low
Green dyes						
Fluorescein LT	Acid yellow 73 (CI 45350)	490	520	0.29	emission reduced below pH7 to trough at pH3	Moderate
Pyranine	Solvent green 7 (CI 59040)	455	515	0.09	emission increased above pH7 but reduced below pH7	Low
Lissamine FF	Acid yellow 7 (CI 56205)	420	515	0.29	emission maximum at pH4–10	Low
Diphenyl brilliant Flavine 7GFF	Direct yellow 96	392	510			
Eosine FA	Acid red 87 (CI 45380)	515	535	0.05–0.1		
Orange dyes						
Rhodamine B	Basic violet 10 (CI 45170)	555	580	0.01	emission maximum at pH 5–9	Very high
Sulphorhodamine B'	Acid red 52 (CI 45100)	565	590	0.06	emission maximum at pH 4–10	High on organics
Rhodamine WT	Acid red 388	555	580	0.1	emission reduced below pH 5.5	Moderate
Aminorhodamine G-extra		535	552			

that publications reporting the use of dyes should record where possible the Colour Index generic name and Colour Index constitution number as well as the manufacturer's name for the dye. The name and location of the manufacturer is also helpful especially in the case of new products. More than one fluorescent dye may be injected simultaneously into a karst drainage basin and yet be mutually distinguishable at the resurgence. Although the absorbance and emission spectra of the many dyes overlap, three groups of them have sufficiently minimal interference to permit their simultaneous use: i.e. those that fluoresce in the blue, green and orange wavelengths. The most suitable dye from each group depends on other characteristics.

Photochemical decay rate	Temperature sensitivity	Toxicity to fish	Background fluorescence interference factors	Water tracing suitability: quantitative	qualitative	tropics	sea water
Very high	Moderate inverse	Very low	Domestic detergents organic leachates	N	Y	Y	N?
	Moderate inverse		Domestic detergents organic leachates				
	Moderate inverse		Domestic detergents organic leachates				
	Moderate inverse		Domestic detergents organic leachates				
High	Low inverse						
High	Low inverse	Low	Organic leachates	Y?	Y	Y	Y
Very high	Low inverse	Low?	Organic leachates	N	Y	Y	
Low	Low inverse	Low?	Organic leachates	Y	Y	Y	N
					Y		
Low	High inverse	Moderate	None significant	N	Y	Y	Y
Low	High inverse	Moderate	None significant	Y	Y	Y	
Low	High inverse	Low	None significant	Y	Y	Y	Y

Data sources: (1) Auckland University Geography Dept.; (2) Smart & Laidlaw 1977; (3) Smart 1976a & b; (4) Gospodarič and Habič 1976

Notes:
(i) dyes produced by different manufacturers may show variations in excitation and emission peaks by ± 5 nm
(ii) secondary excitation and emission peaks often occur but are not noted here
(iii) detectability in distilled water at pH7
(iv) Y = yes; N = no

In a complex water tracing experiment in Slovenia (Gospodarič & Habič 1976), uranine, eosine, amidorhodamine G-extra, rhodamine FB and tinopal CBS-X were injected and analysed separately in two laboratories. All traces were successful but the evaluation of the percentage recovered varied considerably. Amidorhodamine in particular showed a poor return of only 11–17%.

Smart & Laidlaw (1977) assessed the relative merits of eight fluorescent dyes (amino G acid, photine CU, fluorescein, lissamine FF, pyranine, rhodamine B, rhodamine WT, and sulphorhodamine B). They considered their sensitivity, minimum detectability, effects of water chemistry, photochemical and biological decay rates, adsorption losses, toxicity and cost. They concluded that the orange dyes are more useful than the others because of lower background fluorescence, which permits higher sensitivities to be obtained. They also showed the fluorescence of some dyes, such as pyranine, to be strongly affected by pH of the water (see also Launay *et al.* 1980) which precludes its quantitative use. Other dyes have little resistance to adsorption losses, the dye adhering to organic and inorganic surfaces – including the inside of plastic sampling equipment! Some also decolourize in sunlight or decay in the fluorometer when exposed to ultraviolet irradiation during measurement. Smart & Laidlaw recommended rhodamine WT (orange), lissamine FF (green) and amino G acid (blue) as the three dyes most suitable for simultaneous injection. Their spectral properties are shown in Fig. 6.30 and other details in Table 6.7.

Smart & Laidlaw (1977) and Smart (1982, 1984) have reviewed the toxicity of fluorescent dyes used in water tracing. Tests on mammals indicate a low level of both acute and chronic toxicity for the 12 dyes investigated by Smart (1982), with rhodamine B (the only cationic dye) showing the highest toxic effects; indeed 'Studies of carcinogenicity and mutagenicity often yielded conflicting results: the optical brighteners, Fluorescein and Eosin, were however judged nonmutagenic, while the Rhodamine dyes were suspect'. It is apparent that rhodamine B is carcinogenic and more toxic to aquatic organisms than either rhodamine WT or fluorescein; so it should not be used as a water tracer. Sulphorhodamine B may be mutagenic. Further testing of dyes for mutagenicity is clearly required.

Another dye used with spectacular success in water tracing is Direct Yellow 96 (Diphenyl Brilliant Flavine 7GFF). Its use was introduced in the Sinkhole Plain–Mammoth Cave area of Kentucky (Fig. 6.31) where hundreds of dye traces have since been conducted (Quinlan & Ray 1981, Quinlan & Ewers 1981, Quinlan *et al.* 1983). The dye fluoresces in the green wave band and thus like fluorescein is not very suitable for quantitative work because of background problems, even though detectability is high in distilled water. The sampling technique used follows the method developed by Glover (1972) and discussed by Smart (1976a & b) for the detection of optical brighteners (fluorescent whitening agents that emit in the blue wave bands). Since Direct Yellow 96 has a high affinity to cellulose, it can be adsorbed onto cotton (bleached calico is preferred for optical brighteners). In the field cotton wool hanks are used as dye detectors, being suspended in the flow in the same way as charcoal bags. Under light from an ultra-violet lamp optical brighteners exhibit a characteristic light blue fluorescence, whereas Direct Yellow 96 is typically pale yellow. It is important to use

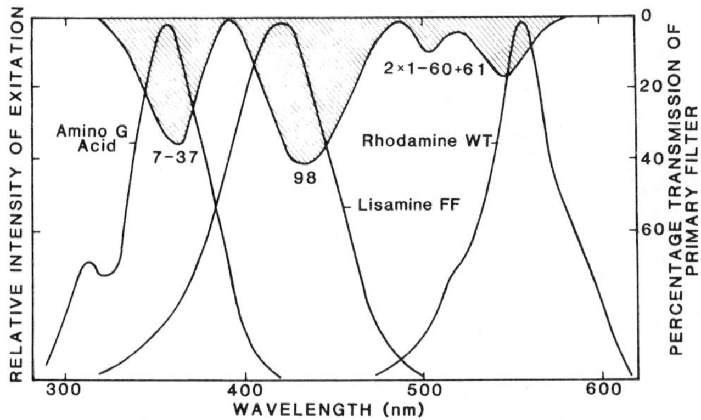

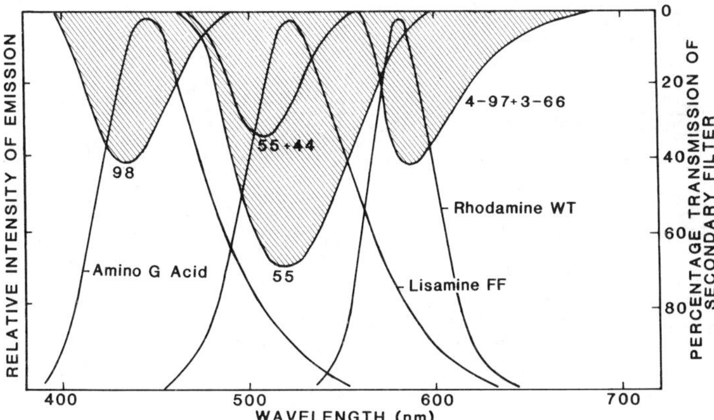

Figure 6.30 Absorption and emission spectra of blue (amino G acid), green (lissamine FF) and orange (rhodamine WT) fluorescent dyes. From Smart & Laidlaw (1977).

surgical quality cotton that has not been previously treated by any brightening agent. If the spring becomes discoloured and the cotton wool very dirty, it can be vigorously rinsed under a tap and the dyes will still be detectable. Serious disadvantages in the use of optical brighteners are that they decolourize rapidly in sunlight and, secondly, are indistinguishable from brighteners in domestic detergents that may have infiltrated the aquifer. These problems are less severe with Direct Yellow 96. On the other hand a major advantage of optical brighteners is that they are colourless in solution and are the least toxic of all dyes used in water tracing. Direct Yellow 96 is bright yellow in solution and of unquantified though probably low toxicity.

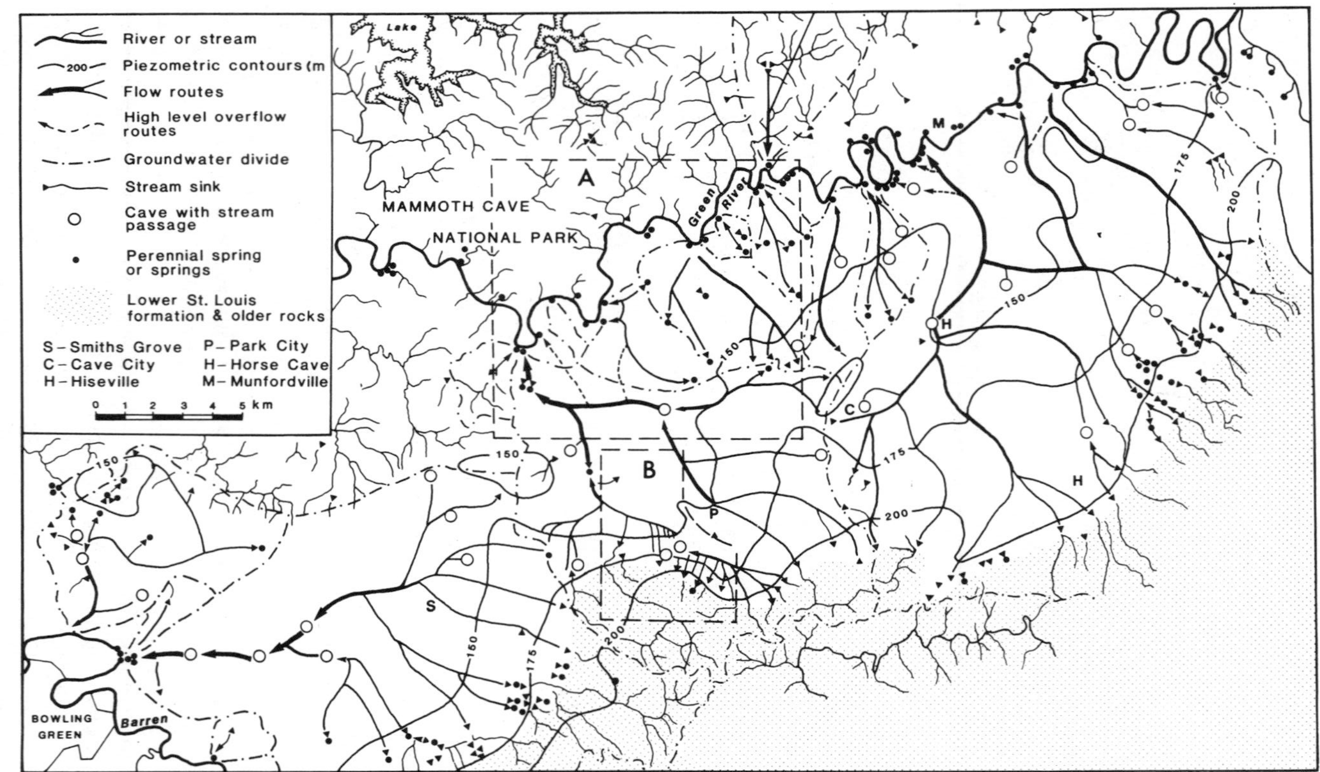

Figure 6.31 Groundwater basins, piezometric surface, subsurface flow routes, and surface drainage in the Sinkhole Plain – Mammoth Cave region, Kentucky (simplified from Quinlan & Ewers 1981). Boxes A and B show the locations of Figures 7.12 and 9.45 respectively.

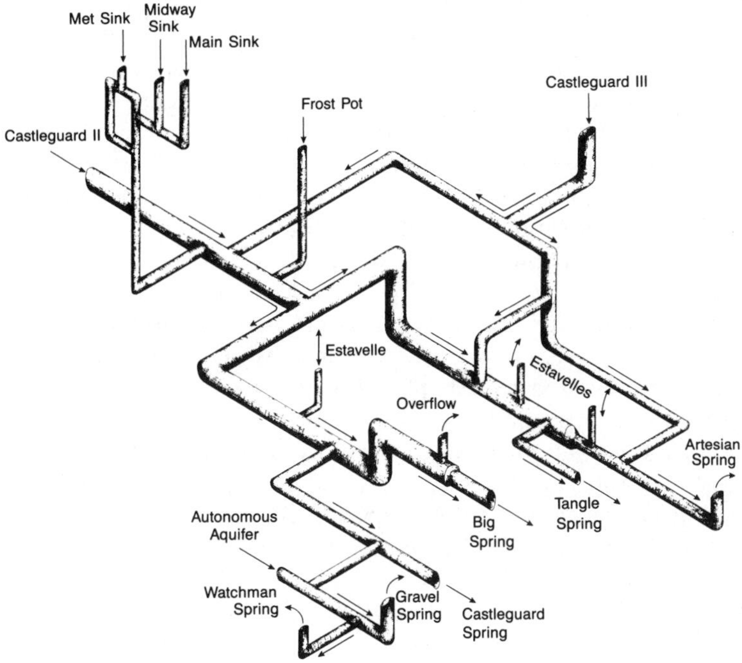

Figure 6.32 Proportional geometric model of Castleguard II conduit system, Canada. From C. Smart (1983b)

Although elutants from charcoal may be used in a fluorometer much more significant dye concentration data are obtained from direct sampling of the spring waters. These can be taken by hand sampling, or by automatic water samplers collecting at fixed intervals or taking time-integrated samples.

The most accurate quantitative water tracing at the time of writing is obtained by continuous fluorometry, i.e. passing water from a spring or other sampling point directly and continuously through an adapted fluorometer. The shape of the dye discharge pulse is then accurately compared to other continuously monitored variables such as discharge, conductivity and temperature. Fluorometers powered by 12-volt batteries are now available, making work possible in the most remote field areas (see C. Smart 1984b, for details).

C. Smart's (1983a) proportional geometric model of the inaccessible Castleguard II conduit system is an outstanding example of the use of combined water tracing technology (three fluorescent dyes with repeated traces, continuous fluorometry, natural discharge pulse analysis, isotope and chemograph analysis) in a remote alpine location (Fig. 6.32).

Borehole dye dilution tests

Dye testing is mainly used to establish point-to-point connections, including borehole to spring and borehole to borehole linkages, but it also has important applications in determining aquifer characteristics in the vicinity of bores. The estimation of hydraulic conductivity by tracer dilution is well established in groundwater hydrology in granular aquifers. It normally employs radioisotope tracers and the monitoring of isotope activity by a scintillation probe down the borehole (Drost & Klotz 1983), but salt dilution has also been used (Brown *et al.* 1972). Fluorometric techniques were first employed by Lewis *et al.* (1966), although not in a karst context. With their recent advances, fluorometric procedures are now the preferred method in karst and other rocks, having the advantages of ease and accuracy of application and low toxicity (Smith & Greenaway 1983).

Determination of groundwater flow velocity can be achieved either by injection of a tracer up hydraulic gradient from an observation well or from measuring the rate of dye dilution in a given bore. The relationships between tracer travel time t, well spacing R, effective rock porosity n_e, hydraulic conductivity K, and hydraulic gradient $\frac{dh}{dl} = i$ may be approximated as follows

$$t = \frac{R\,n_e}{K\,i} \tag{6.58}$$

In karstic rocks an injection quantity of fluorescein of 2–10 g per 10 m of flow path is recommended by Brown *et al.* (1972). Travel time is taken to be the time to peak concentration in the observation well, rather than the time to first arrival of the dye.

We saw in Chapter 5 that under Darcian conditions groundwater movement can be described by

$$u = -K\frac{dh}{dl} \tag{5.2}$$

where $u = Q/a$ and is variously termed specific discharge, macroscopic velocity or filtration velocity. The actual microscopic velocity u^* of the water through the pores of the aquifer is obtained by dividing the macroscopic velocity by the effective porosity of the rock

$$u^* = \frac{u}{n_e} = \frac{Ki}{n_e} \tag{6.59}$$

Lewis *et al.* (1966) discuss equations that may be used to derive u from in-hole dye dilution. To achieve valid results various conditions should be satisfied: steady-state conditions should prevail; there should be uniform groundwater flow and tracer and distribution; and the diminution of tracer concentration over time should be only due to dilution by horizontal groundwater flow. With these points satisfied, Lewis *et al.* (1966) found that the results may be satisfactorily evaluated by plotting against time the logarithm of the ratio C_0/C. The slope of the semi-logarithmic plot is then evaluated and a value for u is determined

$$\log (C_0/C) = 1.106\ ut/2r \tag{6.60}$$

where C_0 is the initial tracer concentration, C is its concentration at time t, and r is the internal radius of the borehole. Smith & Greenaway (1983) have refined this technique. In bores with well screens, Drost & Klotz (1983) suggest velocity is best derived from

$$u = (\pi r/2\alpha \mathrm{t})\ ln\ (C_0/C) \tag{6.61}$$

The term α corrects for borehole disturbance of the flow field and is usually taken to have a value of 2. Drost *et al.* (1968) show how it can be determined, but it usually lies in the range of 0.5 to 4.0.

If Darcy's law can be assumed valid (and there might be *serious* doubts about this in a karst context – see section 5.2) then the hydraulic conductivity can be calculated from

$$K = u/i \tag{6.62}$$

In comparing such a result with that obtained from a pumping test, it should be recognized that tracer dilution involves a much smaller rock volume. Estimation of K by this method is considered by Lewis *et al.* (1966) to be more economical from the standpoint of time, cost and repeatability than conventional pumping tests, but we consider it unlikely to be valid in well karstified aquifers.

(c) SALTS AND OTHER WATER LABELS

Sodium chloride is one of the earliest used artificial water tracers, three tons of it having been used in 1899 to trace water sinking at Malham Tarn in Yorkshire, England. It also featured amongst the 13 different substances used to trace the waters of the upper R. Danube to the Aach spring in the catchment of the R. Rhine in Germany. In that case 50 tonnes were used (Batshe *et al.* 1970). Lithium and potassium chlorides have also been widely used in water tracing, as have non fluorescent dyes such as Congo Red and Methylene Blue (Brown *et al.* 1972).

Dyes dependent on visual detection have been superseded by fluorescent tracers which can be objectively detected by instruments, but some salts are still used as a supplement to other tracers. Lithium, for example, is relatively rare in nature and so with low background values present is a useful tracer, although uncertainty exists regarding its acceptability on toxicity grounds. Behrens & Zupan (1976) provide a recent example of the results of karst water tracing with lithium chloride.

(d) SPORES AND YEAST AS WATER TRACERS

The use of spores to trace water was introduced by Mayr (1953) and was developed by Maurin & Zotl (1959). The method is fully described by Drew & Smith (1969) and Zotl (1974). Spores of the club moss *Lycopodium clavatum* have a diameter of 30–35 μm and a buoyancy only slightly greater

Table 6.8 A comparison of *Lycopodium* and fluorescent dye tracer techniques. From Smart & Smith (1976) with additions.

Lycopodium spores	Fluorescent dyes
(1) limited to a maximum of six simultan- inputs	(1) limited to three simultaneous inputs
(2) requires only periodic sampling	(2) requires frequent sampling
(3) sampling requires the use of special plankton nets, easily damaged in floods	(3) sampling requires no special equipment and is possible using an automatic water sampler
(4) qualitative	(4) quantitative
(5) cost of capital equipment (microscope, centrifuge) moderate	(5) cost of capital equipment (fluorometer) high
(6) cost of non-capital equipment (nets, glassware, etc.) high	(6) cost of non-capital equipment (glass-ware) low
(7) pre-treatment to colour spores time consuming and moderately expensive	(7) no pre-treatment required
(8) post-collection treatment, time consuming	(8) no post-collection treatment
(9) analysis time consuming and requires skilled personnel	(9) analysis straightforward and fast, requires no skilled personnel
(10) immediate field analysis not possible	(10) immediate field analysis possible
(11) cost of tracers expensive	(11) cost of tracers moderate
(12) unaffected by water chemistry and pollutants	(12) perhaps detrimentally affected by water chemistry and pollutants
(13) affected by high sediment concentrations	(13) affected only at extremely high sediment concentrations
(14) contamination problems are virtually insuperable and cast grave doubts on the results	(14) contamination short term
(15) once injected, an aquifer is contaminated with spores for a very long time; so tests cannot be repeated with confidence	(15) tests are repeatable after a short interval

than water. This permits them to drift in turbulent flow through karstified conduits without excessive loss by filtration. They are recovered at resurgences by 25 μm mesh conical plankton nets suspended in the current.

Spores dyed blue, red, violet, green and brown can be used simultaneously to inject different stream-sinks, and their presence in samples can be detected under a microscope. If the sample filtered by the plankton net contains large amounts of other organic material it can be treated by digesting in warm carbamide, formaldehyde and potassium hydroxide (Dechant 1967).

Smart & Smith (1976) review tracing work in tropical waters using spores and fluorescent dyes, and summarize the relative merits of the methods (Table 6.8). They point out that the largest single cost in tracing work is that of the time and transport involved in collecting samples. Consequently, a potential advantage of using *Lycopodium* is that up to six input sites can be traced simultaneously whereas only three can be dealt with using fluorescent dyes. However, some operators cannot discriminate the different coloured spores, and likely significant operator error has never been adequately evaluated. Once injected, the aquifer is contaminated by spores for a long period. Moreover, the quantitative work possible with dyes is precluded with spores because of sampling difficulties. Thus since the early 1970s spore tracing has given way to the superior fluorometric techniques (Smart *et al.* 1986).

A relatively untested but potentially useful microorganism for water tracing is baker's yeast, *Saccharomyces cerevisiae*. Wood & Ehrlich (1978) maintain that it is more suitable for this function than other microbal species, because yeast is readily available in quantity, is not naturally present in any aquifer, and is easy to enumerate over a wide range of concentrations. Yeast cells are also roughly the same size as the pathogenic bacterial cells (2–3 μm) that they may be being used to trace. The effect of filtering can therefore be assessed.

7 Cave systems

7.1 Classifying cave systems

In Chapter 5 it was shown that patterns of interconnected solution conduits determine the characteristic behaviour of well developed karst aquifers. These patterns are systems of caves and proto-caves. Their development is the subject of this chapter.

Solution cave systems are among the most complex of all landforms. This is because they ramify in a great variety of three-dimensional patterns in rock masses that exert many different influences upon their organization, extent, and shape. They are further affected by hydrochemical factors dependent upon petrologic, tectonic, climatic, biotic and pedologic conditions, and by external baselevel controls. They may survive in the rock as active or relict features after these conditions have ceased to apply, and perhaps be altered under radically different conditions.

Being so varied, there are many different ways in which solution caves are described and interpreted. No single theory of genesis has been able to encompass them all except at a trivial level of explanation and there is no one classification that accommodates all the needs of geomorphologists, geohydrologists, economic geologists, etc.

Definitions of caves

The definition adopted by most dictionaries and by the International Speleological Union is that a cave is a natural underground opening in rock that is large enough for human entry. This definition has merit because investigators can obtain direct information only from such caves, but it is not a genetic definition. From Chapters 3 and 5, we define a karst cave as a solutional opening that is greater than 5–15 mm in diameter or width. This is the effective minimum aperture for turbulent flow (section 5.1), although that will not necessarily occur as soon as the minimum is attained.

Isolated caves are voids that are not and were not connected to any water input or output points by conduits of these minimum dimensions. Such *non-integrated* caves range from vugs to, possibly, some of the large rooms occasionally encountered in mining and drilling.

Proto-caves extend from an input or an output point and may connect them, but are not yet enlarged to cave dimensions.

Where a conduit of 5–15 mm diameter extends continuously between the input points and output points of a karst rock it constitutes *an integrated cave system*. Most enterable caves are parts of such systems. In this chapter we are concerned principally with the building of integrated cave systems but for brevity will refer to them simply as 'caves'.

Caves and their classification

Tens of thousands of solution caves have been explored, in part at least, and many thousands of them are accurately mapped. Table 7.1 lists some approaches to classifying them. The longest and deepest caves (as at April 1988) are given in Table 7.2 and Figure 7.1. Their distributions are highly skewed. This is primarily a function of exploration difficulties. A majority of caves have been explored for less than 1 km and penetrated to less than 100 m in depth. The number of greater examples increases by dozens or hundreds every year but the form of the distribution curves remains the same.

Mammoth–Flint Ridge–Roppel–Procter System in Kentucky (Fig. 7.12) has maintained its position as the world's longest set of connected passageways for most of the years since the 1840s, when some 25 km were known. It is developed in no more than 100 m of limestones but they are

Table 7.1 Some classifications of solution caves.

(A) By internal characteristics

(1) *By size*: aggregate length or depth or volume.
(2) *By measure of vertical or horizontal dimensions*.
(3) *By plan form*: entrance or niche (abri), chamber (room), linear passage, branchwork, network, anastomosis, spongework, multiphase branchworks, rectilinear combinations.
(4) *By passage cross-section form*: circular or elliptical, canyon, breakdown, compound.
(5) *By relation to a regional water table*: vadose, water table cave, phreatic, compound, relict.
(6) *By categories of deposits*: speleothem cave, gypsum (crystal) cave, sand cave, ice cave, archeologic site, etc.

(B) In relation to external factors

(1) *Modes of geological control*: rock type (limestone, gypsum, etc); joint-guided, fault-guided, etc; horizontal strata, steeply dipping, folded, etc.
(2) *By topographic setting*: mountain caves, plateau caves, etc.
(3) *By relation to topography*: underdrain valley or valley flank, meander cut off, connect poljes, foot cave, etc.
(4) *By role in fluvial system*: allogenic river caves, holokarst drains, shortcut caves, combinations, sea cave, etc.
(5) *By aquifer type*: ideal pipe cave → continuum → perfect spongework cave.
(6) *By role in geomorphic and hydrologic cycles*: active cave → episodic → relict cave (preserved, intercepted, truncated, destroyed).
(7) *By climatic setting*: humid tropical, semi-arid, mediterranean, temperate, alpine, arctic, etc.

Table 7.2 The longest and deepest caves as at April 1988.

	The longest	km			The deepest	m	
1	Flint Mammoth Cave System	530.00	USA	1	Reseau Jean Bernard	1535	France
2	Optimisticeskaya	157.00	USSR	2	Vjaceslav Pantjukhin	1465	USSR
3	Holloch	133.50	Switzerland	3	Puerta de Illamina	1408	Spain
4	Jewel Cave	120.73	USA	4	Sistema del Trave	1380	Spain
5	Siebenhengstehohlensystem	110.00	Switzerland	5	Snezhnaya Meznonnogo	1370	USSR
6	Ozernaya	105.30	USSR	6	Sistema Huautla	1353	Mexico
7	Ojo Guarena	89.07	Spain	7	Reseau de la Pierre St Martin	1342	France
8	Reseau de la Coumo d'Hyouemede	87.00	France	8	Reseau Berger	1241	France
9	Zolushka	82.00	USSR	9	V. Iljukhina	1220	USSR
10	Wind Cave	80.46	USA	10	Schwersystem	1219	Austria
11	Sistema Purificacion	71.60	Mexico	11	Complesso Fighera Corchia	1215	Italy
12	Fisher Ridge Cave System	71.50	USA	12	Sistema Aranonera	1185	Spain
13	Friar's Hole Cave System	68.82	USA	13	Dachsteinmammuthohle	1180	Austria
14	Organ Cave System	60.51	USA	14	Jubileumschacht	1173	Austria
15	Reseau de l'Alpe	53.67	France	15	Sima 56	1169	Spain
16	Reseau de la Dent de Crolles	53.20	France	16	Anou Ifflis	1159	Algeria
17	Res del Silencio	53.00	Sarawak	17	Sistema Badalona	1149	Spain
18	Ease Gill Cave System	52.50	Great Britain	18	Sistema del Xitu	1148	Spain
19	Sistema Huautla	52.10	Mexico	19	Riviere de Soudet	1137	France
20	Mamo Kananda	51.82	Papua New Guinea	20	Axematic	1130	Mexico
21	Gua Terangair	51.66	Sarawak	21	Kujbyshevskaja	1110	USSR
22	Reseau de la Pierre St Martin	51.20	France	22	Schneeloch	1101	Austria
23	Kap–Kutan/Promezutocnaja	46.10	USSR	23	Sima GESM	1098	Spain
24	Crevice Cave	45.38	USA	24	Jagerbrunnentrog system	1078	Austria
25	Complesso Fighera–Corchia	45.00	Italy	25	Sistema Ocotempa	1063	Mexico

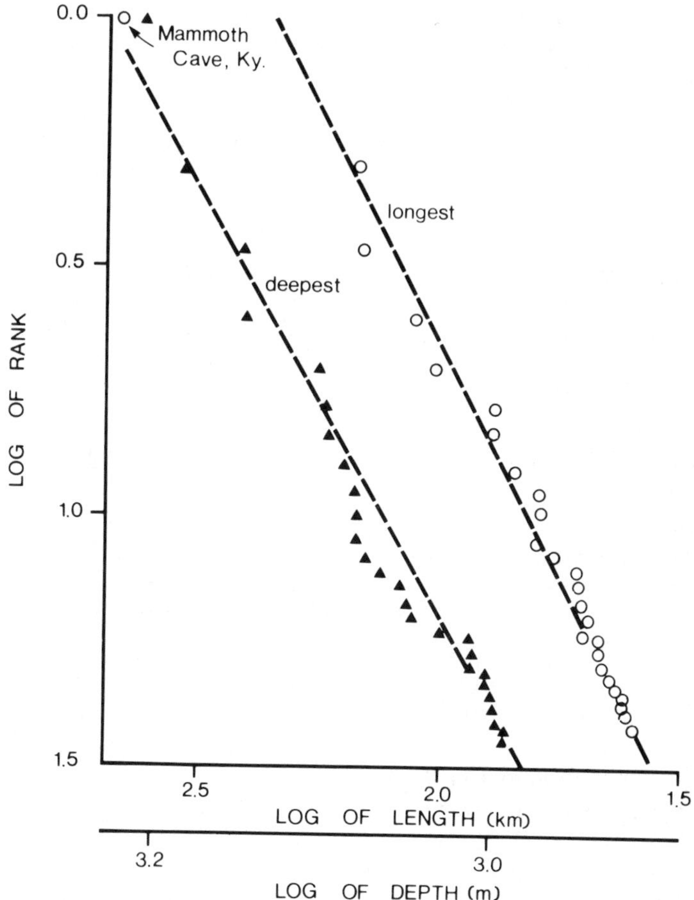

Figure 7.1 Rank/size correlation plots (Zipf Plots) of the 26 longest and deepest known cave systems. According to the rank/size rule, the Mammoth–Flint Ridge–Roppel–Procter cave system, Kentucky, should be 225 km in length; it exceeds 500 km. Lengths of other caves and depths of caves obey the rule quite closely.

nearly flatlying and the caves extend as multiphase branchworks at several levels beneath sandstone ridges that preserve them. Optimisticeskaya (Fig. 7.24) and other great caves in the Soviet Union are rectlinear mazes in gypsum strata only 12–30 m thick but again flatlying in low plateaus. Holloch (No. 3) is a mountain cave in steeply dipping limestones, as are Nos. 6, 7 and 8. Jewel Cave (No. 4) and Wind Cave (No. 10) are three-dimensional mazes in gently domed limestone hills. No. 13, Friar's Hole, lies beneath a deep valley in shales and sandstones where the limestone is scarcely exposed at all. These longest caves thus display a wide range of physical type. They have in common the indefatigability of their explorers.

There is less variety amongst the deepest caverns. It is infeasible to dive below ~ 200 m in waterfilled caves at present; therefore, the deepest known caverns occur in mountain massifs where the drained depth will be greatest. Most are in the alpine ranges of Europe because these are the most intensively explored. Recently, exploration in the highlands of Mexico has added three tropical contenders. All are alike in being systems of shafts and steeply descending stream canyons that terminate at siphons or breakdown barriers.

When actually underground most cave scientists categorize caves by the vadose, phreatic or breakdown form that they see about them, or by their deposits. But extensive cave systems may display all of the differing forms and a wide variety of deposits, so that these characteristics are not well suited for general classification.

A majority of karst researchers are concerned with surface landforms or hydrologic studies and so classify the unseen plumbing in terms of appropriate external factors, as in Chapter 5 and Table 7.1B. Some geological factors will be stressed later in this chapter. Relationships with topography and with the fluvial systems are of particular relevance to geomorphologists. Some caves simply underdrain valleys, others abstract water across topographic divides. A very common category is the subterranean piracy that forms a shortcut across the neck of an incised river meander, through a spur or bypassing a knickpoint. Some of these are 'ideal pipes', being single straight conduits that neither gain nor lose notable quantities of water during comparatively short passes underground.

Caves are most frequent in the wettest climates. However, large caves do occur in the desert Nullarbor Plain of Australia (section 10.2). There is little relationship between climate and most aspects of cave form although the biggest river cave passages are created by the biggest rivers. These are usually allogenic, as in the case of the magnificent Nanxu Cave (Fig. 7.2) but Nare Cave and other recent discoveries in New Britain are probably autogenic. Some very big passages such as Carlsbad Caverns (Fig. 7.28) display no relation to the modern topography, fluvial channels or presently active groundwater systems at all. They are known because surface erosion has chanced to intercept them.

Classifying caves by these external factors does not explain the structures that the systems adopt or the form of their component parts. Table 7.3 presents a simple classification that is generically based. It is not well balanced. Three quarters or more of caves that have been adequately described and mapped must be placed in the first class. These are caves created by meteoric waters circulating in karst rocks without any unusual constraints such as confinement below aquiclude strata. These we call 'common caves'. Other classes cover those caves formed under unusual geological constraints or where unusual waters are present. They are discussed in the later sections of the chapter.

Figure 7.2 The river cave at Nanxu, Guangxi Province, China. (Photo by Andy Eavis, with permission).

Table 7.3 Classification of karst solution caves

A	Normal meteoric waters	Unconfined circulation in karst rocks = hypergene caves	1	Common caves (80% of known caves?)
		Confined circulation in karst rocks, or partial circulation in non-karstic rocks; includes some hypogene caves	2	Maze caves and outlet basal injection caves
			3	Combinations of types 1 & 2
B	Deep enriched waters	Enriched by exhalative CO_2 (normally, thermal waters); hypogene caves	4	Hydrothermal caves (~ 10% of known caves?)
		Enriched in H_2S, etc. (basin waters, connate waters)	5	Carlsbad-type cavities and gypsum replacement caves
C	Brackish waters	Chiefly marine & fresh waters mixing	6	Coastal mixing zone cavities
D		Combinations of B or C with A, developing in sequence	7	Hybrid caves

Many caves display multiple phases of development. One phase ends in a cave and another begins: (1) when a spring position is shifted upwards or downwards sufficiently to compel the creation of extensive new passages; or (2) when there is an externally-caused change of water quality or quantity that results in net cave infilling where net erosion prevailed before or *vice versa*. Section 7.5 discusses effects of the shift of springs, i.e. local or regional base level change. Net erosion–deposition changes are considered in Chapter 8 because they involve cave deposits.

System information

Cave systems are the functional equivalent of river networks in fluvial geomorphology. Study of the latter has seen major academic advances in the past 35 years from morphometric analyses of channel and basin properties. There has been little of this in speleology and it cannot yet provide the framework for genetic explanations of phenomena of interest. This is because comprehensiveness is lacking in cave system information in almost every case. Many passages are too small to enter; others are sealed by breakdown, sand, etc. Either it is not known where they originate or where they terminate. In a majority of the great systems much of the passage is waterfilled. Cave diving is hazardous and only a few active phreatic caves have been well mapped. Most morphometric success so far has been obtained with small features such as erosional scallops and alluvial sediment samples, working in local sections of caves. These are summarized at the end of this chapter and in the next. There is lacking any substantial body of quantitative results that can link these small-scale findings to the highly generalized system descriptions obtained from the dye tracing and hydrograph analysis discussed in Chapter 6. That is one of the major challenges facing future students of caves.

7.2 Formation of plan patterns of common caves

Initial conditions

In Section 5.3 we examined the controls on the development of karst aquifers and concluded that the orientation of the groundwater flow will reflect a balance between the direction in which resistance to flow is least (i.e. where hydraulic conductivity is maximized) and the direction in which the rate of energy loss is maximized (i.e. the shortest and steepest route). In this section we examine the principles that govern the propagation of solution conduits through fissures and down the gradient. The rocks normally are dense carbonates with no significant permeability outside the planes of penetrable fissures, i.e. some bedding planes, joints and faults. Before karst solution begins, minimum apertures of connected voids in the fissures are small, < 10 μm to 1 mm and their aggregate volume is also small. Available runoff thus is readily able to fill them and the water table is consequently at or near the ground surface. These are simplified conditions that, because of diagenetic or tectonic history, will not always prevail but

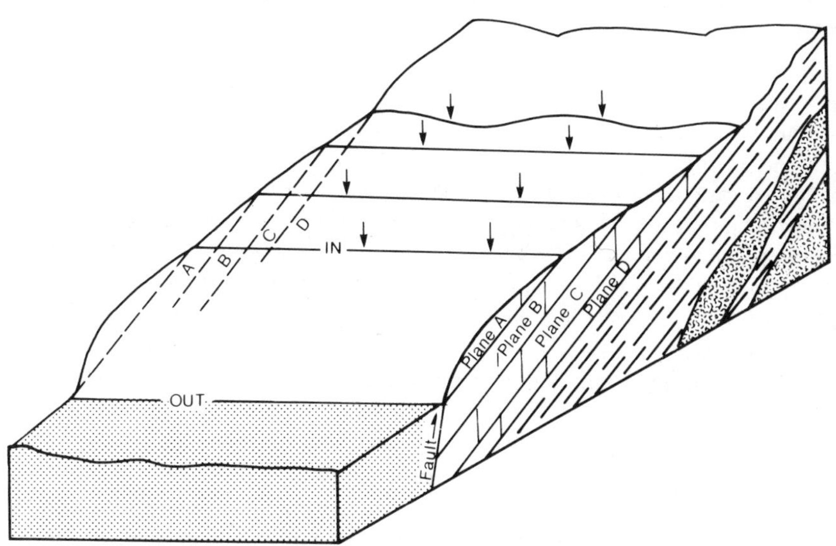

Figure 7.3 Conceptual model structure to explain the development of common cave systems in penetrable bedding planes A to D and joint systems in between them.

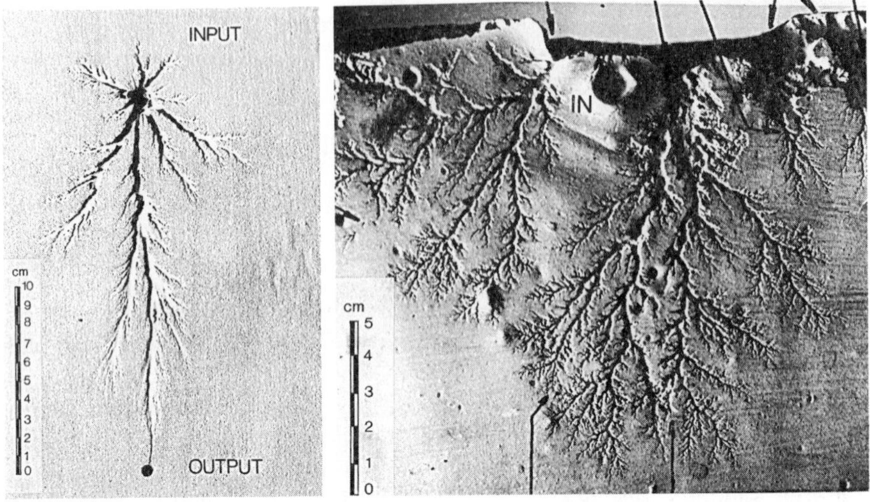

Figure 7.4 Experimental studies of the propagation of solution conduits through fissures, using plaster of paris (gypsum) as the medium and with a constant head applied. A. In this experiment with a low transmissibility fissure, a proto-cave has extended from the input to connect to an output point. This will enlarge to become a cave. Many subsidiary conduits have also been created; their enlargement is attenuated once the protocave is connected. B. This experiment shows us what happens when a fissure has too low transmissibility for effective (connecting) conduit propagation. Flow is dispersed into ever smaller tributaries, creating a fractal tree. Both experiments display 'fractal fragmentation' – the repetitive subdivision of geometric figures (Mandelbrot 1983, Curl 1986). In Case A, this natural process was attenuated by the connection. Where fissures display high initial transmissibility it may not even commence. (Experiments by Ewers 1982, and Waterman 1975).

they form an acceptable starting point for our discussion.

Figure 7.3 illustrates a hypothetical situation that provides a hydrogeological framework for this section and the next. It is drawn for the case of steep stratal dip with regional flow in the direction of that dip because this is easy to illustrate. With the mind's eye, readers may flatten the dip to the horizontal or even reverse it away from the output boundary. Strata may be folded, or the output boundary (potential spring line) may be shifted to the strike on one or both sides of the model. The analysis remains the same.

The model and analysis are for bedding planes because these are the most continuous entities permitting groundwater flow. For conduit propagation in normal vertical joints, tip the model on edge.

In this section of the chapter we are concerned only with plan patterns, that is with propagation in plane A, because this is where the first caves will develop. Connections with planes B, C, D, etc. introduce the depth dimension and are dealt with in the following section. The analysis is based upon model simulations and field studies by Ewers (1972, 1982) and field studies by Ford (1965, 1968, 1971). Ewers investigated a great variety of initial situations using electrical and sand model flowfield analogs, and direct solutional simulations with plaster of Paris and salt (Figure 7.4). Comprehensive details appear in Ewers (1982).

Conduit propagation with a single input

This is the simplest situation (Fig. 7.5). Length of the fissure between input and output may be no more than 1 m (in which case we are considering the karren or epikarst scale of aquifer development) or as great as 10 km. The pressure head ranges from the thickness of a single limestone bed to hundreds of metres. The flow envelope (Fig. 7.5 and c.f. Fig. 6.5) indicates the initial mode of flow; it is laminar within the plane and can be treated as Darcian.

Distributary patterns of solutional micro-conduits develop that extend

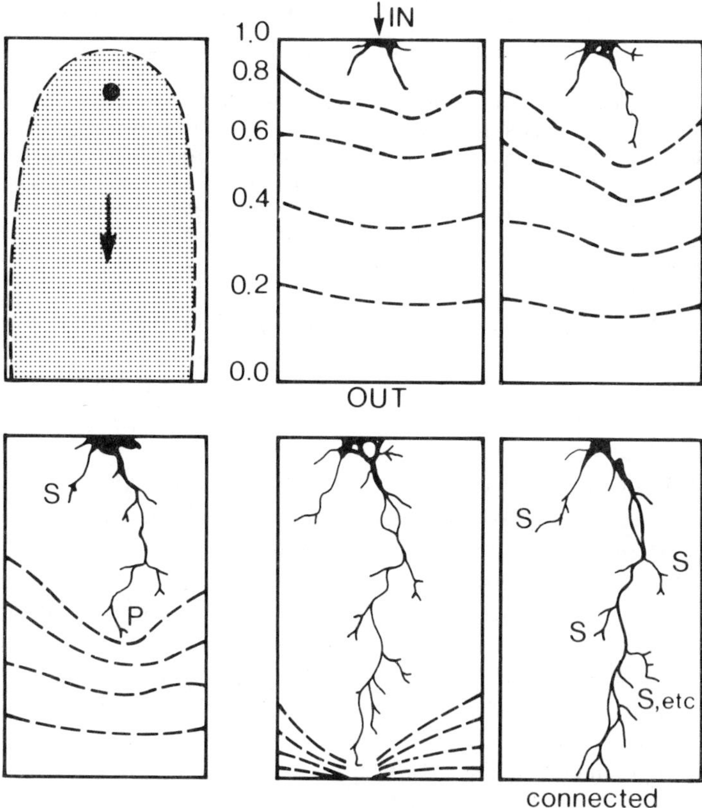

Figure 7.5 The propagation of a proto-cave from a single input to an output boundary in plane A. In shading, the flow field or envelope at the start of dissolution. Dashed lines are equipotentials. P=principal (or victor) tube. S=subsidiary tubes. Adapted from Ewers (1982).

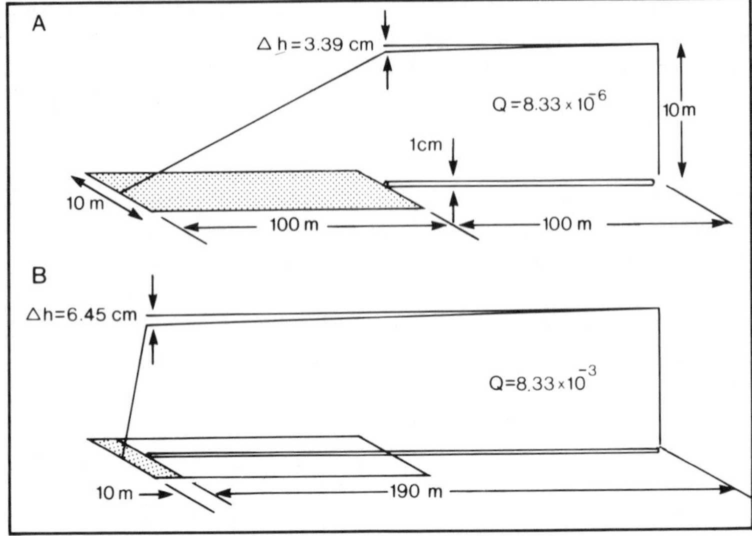

Figure 7.6 This important figure emphasises that resistance to groundwater flow through a fissure in which a cave is propagating remains very high until the moment that the principal tube is connected to the output boundary. Here the principal tube is straight and 1 cm diameter. The head is 10 m. Headfall is only 6.45 cm when the tube has propagated across 95% of the distance to the output (lower figure). The fall of head will be 9.9 m when the tube has crossed the remaining 5%. Cave systems are built in a cascading sequence, flow fields reorienting drastically each time that a connection (with its abrupt fall of head) occurs. From Ewers (1982).

preferentially in the direction of the hydraulic gradient. Their rates of extension are determined by solvent penetration distance (section 3.8) and the factors that control it. Actual courses adopted, metre by metre, are random. They depend upon geologic micro-features of the plane.

In electrical analogue terms, all tubes are connected in series with the solutionally unmodified plane beyond them (Fig. 7.6). The latter maintains high resistance so that flow is slow and small in amount. *Variations in cross-section and other properties of the tubes are insignificant while this high resistance element remains.*

In every experiment one tube grows ahead of others (Fig. 7.5). It deforms the equipotential field, reducing the gradients at the solution fronts of its competitors ('subsidiaries') and thus slowing their rates of advance. When this 'principal' or 'victor' tube attains the output boundary, the high resistance to flow is destroyed. Darcy flow conditions no longer apply because the connected tube is an inhomogeneity. The equipotential field is re-oriented onto it because of its high conductivity and consequently low head.

A proto-cave system now extends throughout the fissure. From Ewers' experiments, its diameter is 1 mm or more and, because flow rate is much increased by the loss of resistance, region 2 solution kinetics (Fig. 3.15) probably apply. The conduit may expand comparatively rapidly to minimum cave (turbulent flow) dimensions and even to enterable dimensions, as discussed in section 5.3. In proportional terms, the longest timespan in speleogenesis is that required for proto-cave extension to the output. When Jennings (1983a, 1985) described *nothephreatic* conditions he referred to that time.

In past writing (Ford 1965a, 1971b, etc.) these early tubes were termed 'dip tubes', because in steeply dipping strata where they were first studied they are usually oriented close to stratal dip. In reality they are gradient tubes because they are broadly oriented down the hydraulic gradient. Here we shall name them *primary tubes* because they are the first conduits to develop in the fissure.

The more open the fissure is before solution begins, the straighter and less branched is its pattern of proto-caves. If it is a bedding plane with joints terminating at it or passing through it, distributary proto-caves preferentially extend along the intersections. However, these will only become 'victor' tubes if they are favourably oriented with respect to the hydraulic gradient. Bogli (1980, p. 152) stresses this need for favourable orientation where (normally small) joints or joint-bedding plane intersections are used. (His additional contention that fissure intersection is essential because it furnishes the chemically different solutions for mixing corrosion we believe to be an overstatement (see Chapter 3.8); conduits may propagate without mixing for great distances in single fissures, including bedding planes. Mixing is not necessary, but will assist if present).

When principal tubes enlarge following their connection much of the subsidiary branchwork is swallowed up. However, provided a pressure head is maintained, surviving elements can continue to extend slowly and may play an important role where later passages develop to create a multi-phase cave (section 7.5). In other instances the subsidiary net may be sealed by clay and rendered inert.

Where the output boundary is a line or zone, rather than a single spring point, it is common for several downstream distributaries to connect to it and then expand, forming a small distributary network.

This model describes development of the most simple cave, a single passage following a single bedding plane, joint or fault. Many shortcut caves are of this type.

Multiple inputs in a single rank

Here we may imagine a number of streams at the input boundary of plane A. If their pressure heads are equal, the initial groundwater flow envelopes are as in Figure 7.7A. Experimentation shows that there is zero potential between envelopes: this means that there can be no mixing of

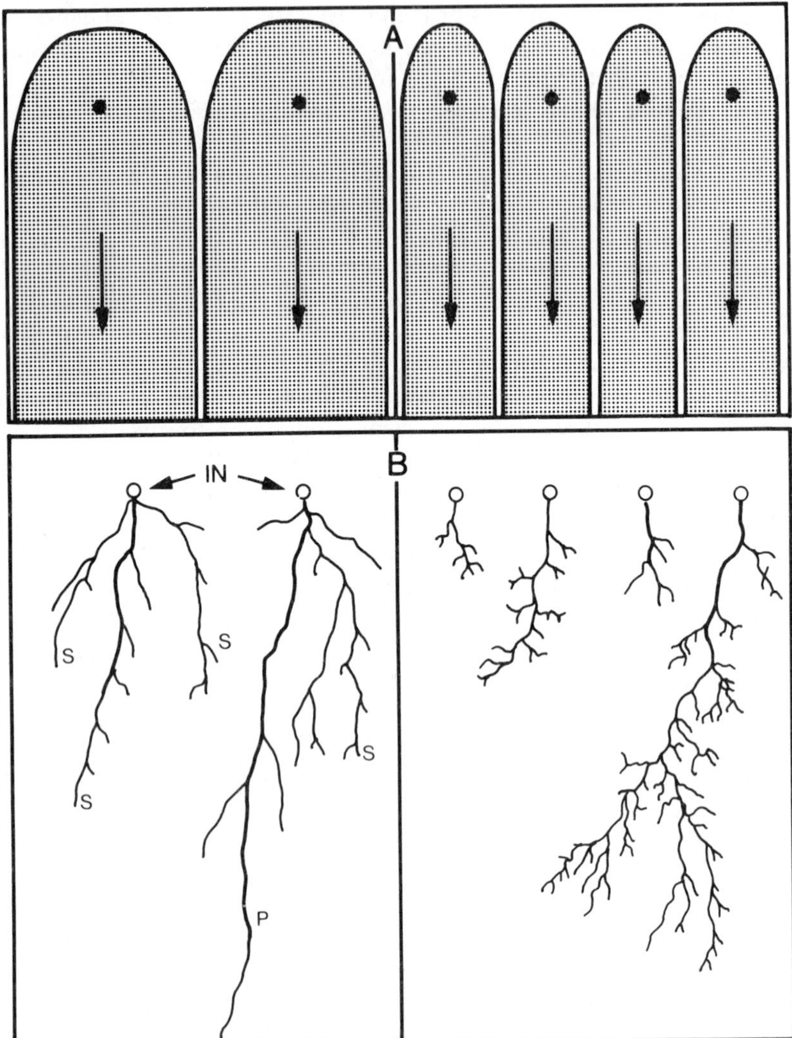

Figure 7.7 The competitive development of proto-cave systems where there are multiple inputs in Plane A, drawn for the cases of two inputs and four inputs. Heads are equal, thus flow fields are equal (A). In B it is seen that competitive branching has occurred, as in the single input cases shown in Figures 7.4 and 7.5. Adapted from Ewers (1982).

water from the different inputs (no lateral mixing corrosion) while the high resistance conditions are maintained.

Typical competitive development is seen in 7.7B. Because resistance is not isotropic in the fissure, because pressure heads never will be precisely equal, or because some inputs are initiated before others, one or more of them extend preferentially. Their flow envelope spaces are increased and those of competitors are diminished.

The closer that parallel inputs such as these are spaced, the greater must be the competition. The steeper the hydraulic gradient, the greater will be the number of inputs whose principal tubes are able to reach the output boundary and so establish separate caves. Close spacing models the initial conditions for what will become diffuse flow in the epikarst zone described in Chapter 5.

When one proto-cave becomes connected, the flow envelopes (or equipotential fields) of its near neighbours are re-oriented towards it. Unless already close to the output boundary, the near neighbour proto-caves will be captured as tributaries. The principles are illustrated as Figure 7.8. This is drawn for the case of a single point outlet (one spring) that is to strike with respect to the near neighbours. Two different conditions are envisaged. In Figure 7.8 a–c the host fissure is tight, with high resistance. The conduit links that are established, first, between victor tube 1 and tube 2, then between tubes 2–3 (and so forth) are quite random in their position on the plane. Subsidiaries of adjoining tubes that happen to be nearest each other

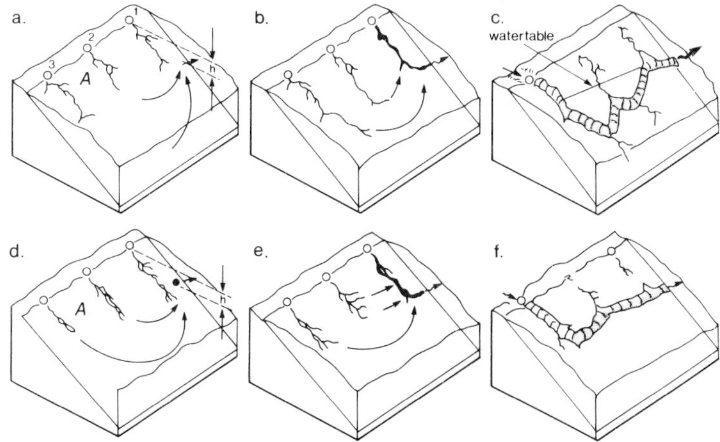

Figure 7.8 Illustrating the principles governing lateral connections between multiple inputs in a rank, where drainage is to the strike. a–c show the sequence of development where resistance in the fissure is high. Single connections at random points occur between adjacent input proto-caves. d–f show the sequence of development where resistance is low. Multiple connections are achieved, increasing the likelihood that a horizontal water table passage will be created as the cave enlarges. Modified from Ewers (1982).

when the equipotential field is re-oriented by establishing connections will be the most likely to link up. In Figure 7.8d–f the fissure offers low resistance. Many alternative linkages can occur at the proto-cave and micro-cave scales. Those which offer the most direct route to the spring will win, creating a rather straight cave. In both cases the link-up routes are those of least resistance.

We are now in a position to understand the basic structure of the Holloch (Switzerland), the second longest limestone cave that is known (Fig. 7.9). It is a multi-phase sequence of trunk conduits draining along the strike in a major, faulted bedding plane that dips at 12° to 20°. The sinuous, highly irregular pattern on the plan occurs because the trunk conduits ascend or descend subsidiaries and angle across the plane between them in the manner of Figure 7.8a–c. Link lengths between subsidiaries of different tubes range from 50 to 250 m. Most of the illustrated passages are now relict. A similar trunk conveys the modern water along the bedding plane at lower elevation. We may presume that subsidiaries are extending slowly below the active trunk, setting up the situation for a new trunk when external erosion lowers the spring elevation once again.

Strike passages like this are common structural components of caves though rarely are they as predominant as in Holloch. They are termed *irregular strike passages* when displaying the high resistance form of Figure 7.9a–c, and *regular strike passages* in the low resistance form. They develop in inclined joint or fault planes as well as in bedding planes. Figure 7.10 presents a second spectacular example.

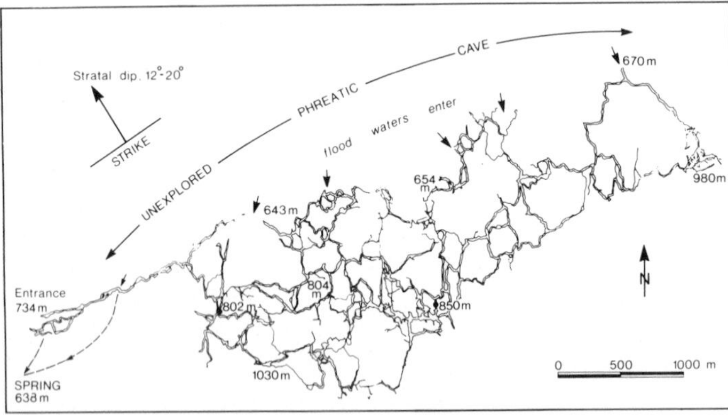

Figure 7.9 Principal passages of the main system of the Hölloch, Muotatal, Switzerland. The system is contained in one major bedding plane dipping at 12–20° NW. Drainage is to the strike. This is a multi phase system with 400 m of relief. It is seen that principal passages are connected up in segments in the manner of the high resistance model, Figure 7.8 a–c. Subsidiary tubes extend between principal passages of different phases. Drawn from the cave map in Bogli (1970).

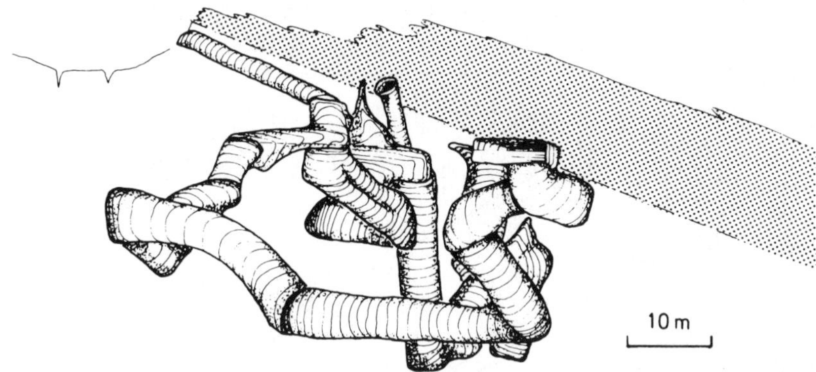

Figure 7.10 A 'head-on' view of the underground outlet of Lake Glomdal, Svartisen, Norway. This shows a complex of 500 m of irregular strike passages developed in steeply dipping marbles. All except the highest passages remain waterfilled today; they were explored and mapped by divers. Drawing by Stein-Erik Lauritzen, reproduced by permission.

Inputs in multiple ranks

This is the situation most frequently encountered in karst systems. As presented in the highly schematic form of Figure 7.11, it will be best understood if the reader now imagines that plane A is horizontal and that the successive ranks of inputs descend parallel joints. The analysis, therefore, is particularly appropriate for clint and grike topography (Fig. 9.6), which is small in scale. It also applies to the lengthiest cave system that is known.

The new element that is introduced in this situation is readily understood. The flow envelopes of further input ranks (Rank 2 in Figure 7.11) are obstructed by those closest to the output boundary. Proto-caves of further ranks can scarcely connect via joints to plane A until some near inputs have connected to the output, reducing their resistance and steepening the hydraulic gradient in their rear (Fig. 7.11A–C). Lateral connections in near ranks and first connections (principal tubes) from further ranks then proceed simultaneously. High or low resistance rules apply as in the single rank situation discussed above. The cave systems will link together headwards and laterally until all of the available karst rock area is incorporated or some minimum hydraulic gradient (for the remaining resistance of the systems) is attained. As systems expand to enterable or greater dimensions their resistance becomes very low so that, given sufficient time and water, most areas of adequately pure karst rocks will become drained by caves.

Figure 7.11 depicts a model with uniform conditions (e.g. of pressure head at the inputs) that cannot be expected in reality. Geological, topographic and hydrologic variations will always distort it. Nevertheless,

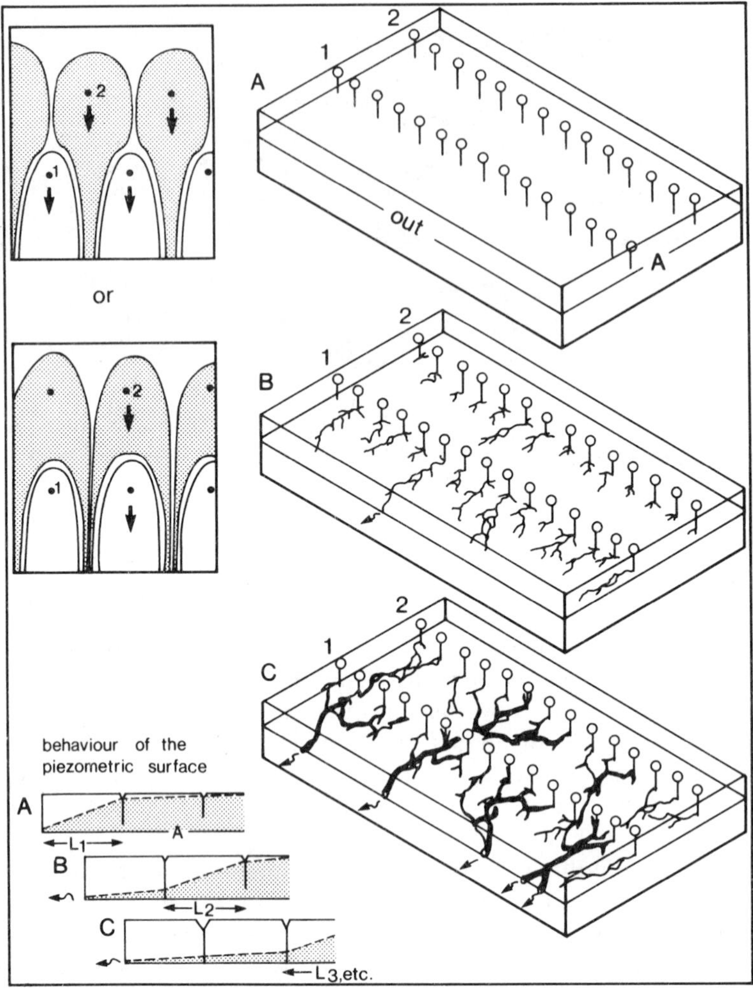

Figure 7.11 The initiation of cave systems with multiple ranks of inputs, drawn for the case of two ranks. Flowfield configurations limit rear rank access to an output boundary. The cave network is connected up in a headward sequence, with recession of the zone of steep hydraulic gradient (lower left).

many cave systems display patterns of systematic headward and lateral connection such as drawn here. There are obvious similarities to the building of Hortonian stream channel nets, but Horton's laws have not been successful when tested on caves because of geologic distortions and incompleteness of information.

The structure of karst basins and cave systems in the Central Kentucky Karst appears in Figure 6.31. Figure 7.12 shows a small part of it. Spring

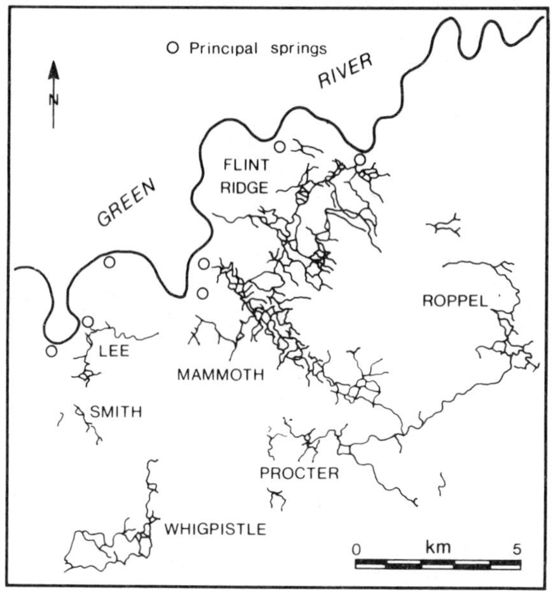

Figure 7.12 Mammoth–Flint Ridge–Roppel–Procter Cave System and nearby caves, part of the Kentucky karst shown in Figure 6.31. This is the most extensive explored cave. Only major passages are shown. Although it is a multiphase complex with additional invasion passages the basic structure is of a multirank system draining down dip to Green River.

points have shifted downwards and laterally many times in response to entrenchment by the Green River. As a consequence, the pattern of cave passages is very complicated because much of it belongs to earlier phases and is now relict. However, it is possible to discern incomplete patterns of near and far ranks connecting to the springs. A particularly strong feature is a recent amalgamation along the strike of a far rank (Procter and Roppel Caves) that beheads Mammoth Cave drainage. Detailed analysis of Mammoth–Flint Ridge–Roppel–Procter System is given in Palmer (1981).

The restricted input case

This case completes the range of significantly different karst input–output configurations that can occur. Inputs are restricted to a line or narrow zone where the karst rock is exposed, most often a valley floor. The spring point is usually at the lower end of the zone. Flowfield geometry constrains the far inputs (Fig. 7.13B), and so the initial cave in plane A is built in headward steps like single inputs (Fig. 7.5).

This is a common type of cave. The example shown is complicated because, once again, the network displays more than a single phase of

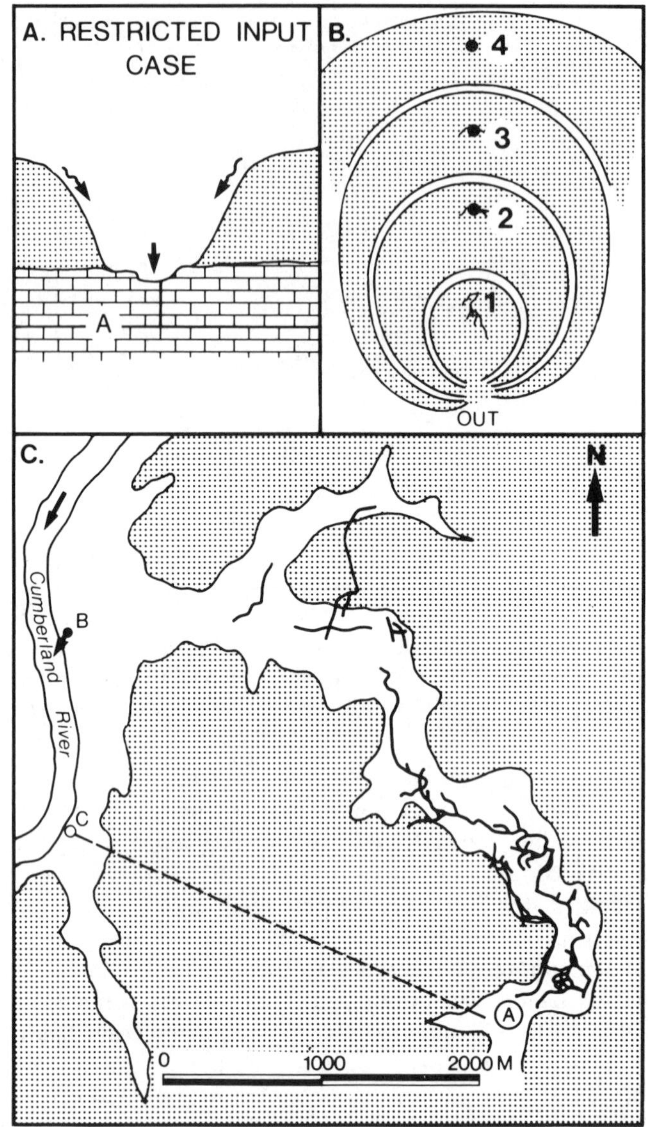

Figure 7.13 The restricted input case of cave system initiation. The example shown is Cave Creek, Ky. (from Ewers 1982). Cover strata are thick shales. Known caves underlie the valley floor and drain to springs at B. A–C is a new groundwater route of inaccessible dimensions.

development and it is incompletely explored. Entrenchment in the Cumberland River has exposed new, lower output points so that, in the latest phase, a new input A in the headward part of the cave has abandoned the sub-valley course and propagated beneath the divide toward spring C. These new galleries are not yet enterable in size.

Cave systems and General Systems Theory

In terms of General Systems Theory applied to geomorphology (Chorley & Kennedy 1971), caves are a type of *cascading system*. There is a cascading event each time that a proto-cave connects to a spring or to a pre-existing cave system. In that event, the local flow field and hydraulic gradient are reoriented. Mathematical modelling of the pattern development of common caves has been unsuccessful hitherto because it has not incorporated the cascade events.

7.3 Development of common cave systems in depth

This section completes the formal analysis of the construction of cave systems by investigating the third dimension. We are now considering the constraints upon conduits propagating in all planes of the general model (Fig. 7.3), or equivalent multi-fissure sequences in other possible geometric–structural arrangements. Sequences of passages may display a predominant vadose or phreatic morphology or they may appear to have developed along a piezometric surface.

The construction of phreatic and water table types of caves – the 'Four-state Model'

The differentiation of phreatic and water table caves is indicated in Figure 7.14. In each case all principal tubes from inputs have connected to outputs. The spring position is stable. It will be appreciated that the frequency of connecting initial fissures that are penetrable by water varies from very few in some karst strata to a great many in others. The real variation is as a continuum but its effect is to permit four distinct phreatic or water table cave geometries to occur.

A bathyphreatic cave (state 1) makes just a single pass beneath the stabilized water table because the frequency of available fissures is too low for any alternative. It has the highest hydraulic resistance. The pass drawn in the figure happens to be quite complex with several looping parts. Successive tubes propagated in planes A, B, C, etc. and then connected as in the multi-rank input case already discussed. Before any of the others had expanded more than a few centimetres, the tube in plane C (having the

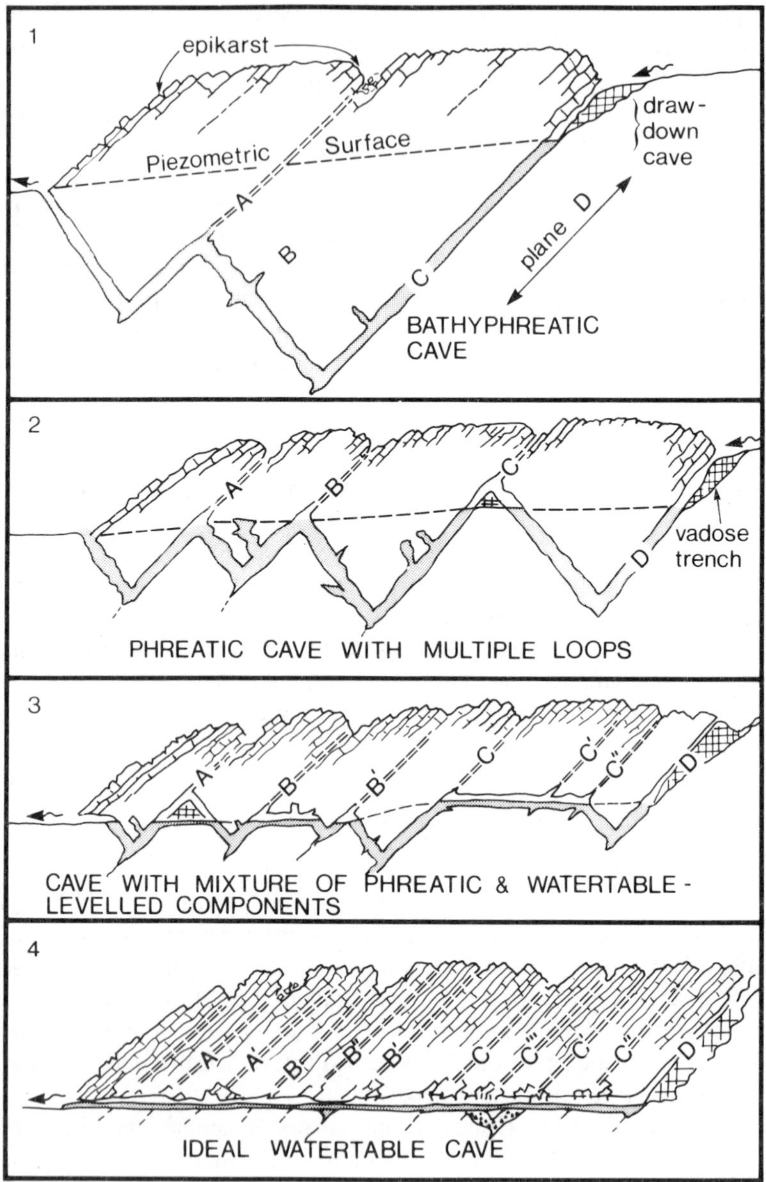

Figure 7.14 The Four-state Model that differentiates the basic types of phreatic and water table caves, drawn to correspond to the conceptual framework of Figure 7.3.

headward advantage) captured most available runoff with the consequence that it alone expanded to enterable size.

Some examples of active and relict bathyphreatic caves are given in Figure 7.15. Note that system information is particularly incomplete. Active (filled) bathyphreatic caves are difficult or impossible to explore. If drained and abandoned, they are most frequently obstructed by detritus towards the base of loops. We do not know the maximum depth attained by bathyphreatic caves but it is certainly greater than 300 m at Vaucluse (Fig. 7.15). Exploration drilling in several areas has tapped caves filled with young flowing water at depths to as great as 3000 m. Some may be bathyphreatic as shown here, but it is likely that many deep interceptions are of shallower types of caves that have been lowered by tectonic activity.

Multiple loop phreatic caves (Fig. 7.14 state 2) are created where there is significantly higher fissure frequency. It must be understood that it is the elevation of the tops of higher loops that defines the stable position of the piezometric surface and not vice versa. Until the system is quite enlarged, the piezometric surface is higher. As the tubes expand in volume it falls until arrested locally at the tops of loops.

Holloch is an excellent example where the phreatic loops are built of irregular strike passages. The vertical amplitude of its looping is ~ 100 m, rising to 180 m during the greatest recorded floods. Other caves loop down the dip and up joints, being built of primary conduits, as in the model of Fig. 7.14. They include many in the Kentucky Sinkhole Plain where the amplitude of looping does not appear to be greater than ~ 40 m.

Caves that are a mixture of shorter, shallower loops and quasi-horizontal canal (i.e. water table) passages represent a third, higher state of fissure frequency and diminishing resistance. The horizontal segments exploit major joints or propagate along the strike in bedding planes as shown in Fig. 7.8. Swildon's Hole–Wookey system in the Mendip Hills of England is an outstanding example. Exploration from the upstream end has passed eleven consecutive shallow phreatic loops and is arrested at a twelfth, deeper loop. At least eight loops have been overcome from the downstream end (Wookey Hole, Fig. 7.15) where exploration is halted at −80 m in a ninth and greater loop.

In state 4 (Fig. 7.14), fissure frequency is so high or resistance is so low that low gradient, most direct, routes are readily constructed through successive ranks of inputs behind the spring points. When sufficiently enlarged these can absorb all runoff; thus, the piezometric surface is lowered into them. They are 'ideal' water table caves.

Ideal water table caves that are wading or swimming canal passaages with low roofs are very common. Hundreds of short examples pass through residual limestone towers on alluvial plains in southern China, Vietnam, Malaysia, Cuba, etc. and many longer systems are known. One instance is the Caves Branch system, Belize, with ~ 10 km of river passages that

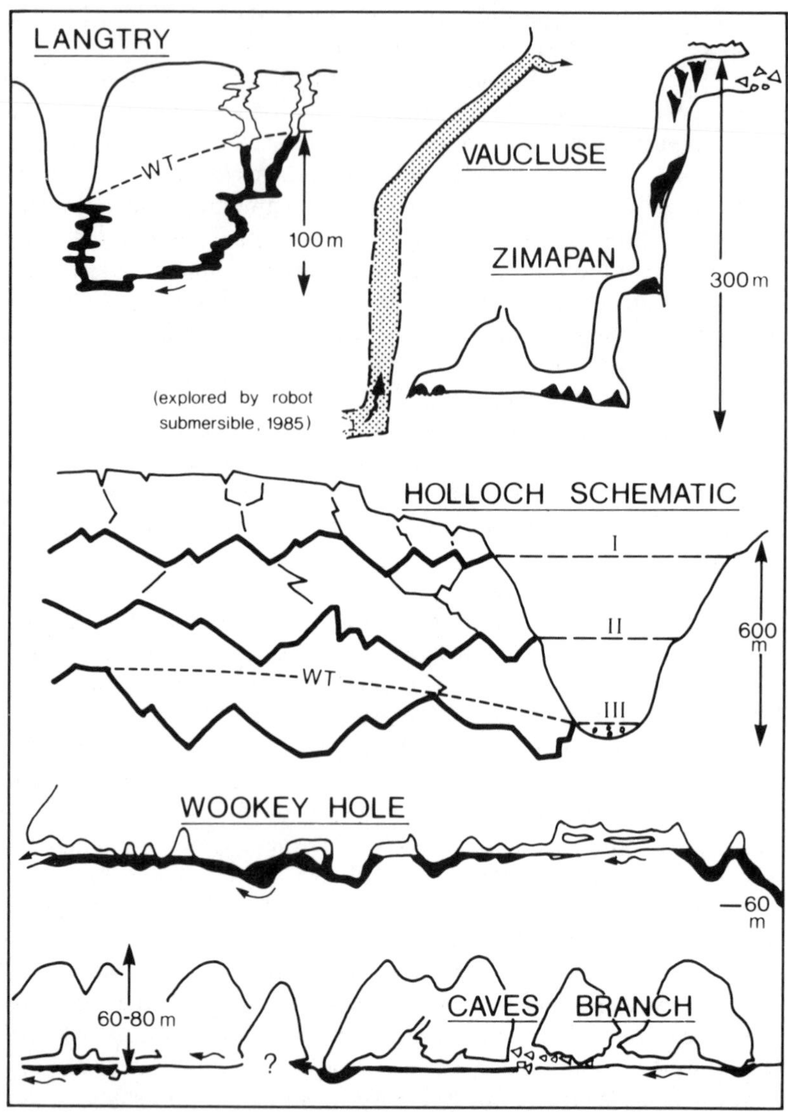

Figure 7.15 Examples of phreatic and water table caves. Langtry Cave, Texas, is in horizontally bedded strata (after Kastning 1983). Vaucluse, France is a phreatic lifting passage where water ascends 315 m in a gallery 10–20 m in diameter. It is the outlet end of a bathyphreatic system. La Hoya de Zimapan, El Abra, Mexico, is another outlet with at least 300 m of lift but it is now relict. The base is infilled with clay. Upper galleries are 20–30 m in diameter and contain massive speleothem. Das Hölloch is an excellent example of a multi-loop phreatic cave, with three relict 'levels'; (from Bogli 1980). Wookey Hole, England, is a mixture of phreatic and water table elements. Caves Branch, Belize, is a great water table cave developed along a polje margin. From Miller (1981).

develop in thoroughly brecciated rock (Miller 1981).

This Four-state Model to differentiate phreatic and water table caves (Ford 1977) is simplified and idealized as it is drawn in Fig. 7.14. In reality there is a greater mixing of the types. Ideal water table passages rarely extend for more than one km before being interrupted by shallow phreatic loops where local geological conditions change. A cave may be assigned to one category or another according to its predominant characteristics.

Many speleologists have contended that caves develop preferentially in the epiphreatic or intermittently saturated zone that is inundated seasonally or in storms by fast-flowing (and chemically aggressive) flood waters. As we have explained it here, such a zone is only of significance to cave genesis where fissure frequency is state 3 or 4.

Measures of fissure frequency

There can be no simple assignment of fissure frequencies to the four states because of differing resistances within the fissures. In a situation of low frequency of penetrable fissures there may also be low resistance so that state 3 systems evolve, or even state 4 where caves are short as in meander cut-offs. High frequency but high resistance may yield state 2, as in parts of Vancouver Island, Canada. Fissure frequency that is measured at the natural surface or exposed in quarries is a poor guide to the effective frequency below the epikarst zone.

In the Mendip Hills of Britain fissure frequency has been linked to the effective porosity (Eqn 5.6). It is found that all four states of cave systems have developed where the effective porosity is $\leq 1\%$ (Atkinson 1977b, Ford & Ewers 1978); hence this measure is not sufficiently discriminating for our purposes.

Increase of effective fissure frequency with time

At the onset of karstification, initial frequency of penetrable fissures varies within and between formations. With passage of time (and solvent waters) after that it tends to be increased, as pointed out in Chapter 5. As a result later caves in a multiphase complex tend to be of a higher state. This is illustrated in Figure 7.16 which models the general situation in the Mendip Hills (SW England) over, it is believed, Pliocene and Pleistocene time. In each of the six steps of the model fissure frequency diminishes with depth. In the earliest situation it was low even close to the surface and state 2 phreatic systems developed that were composed of few loops with great vertical amplitude (Fig. 7.16b). While these aged (Fig. 6.16c) primary tubes extended below them and some intervening bedding planes and joints of higher resistance were connected into the system for the first time, increasing the frequency. When spring positions were lowered by allogenic

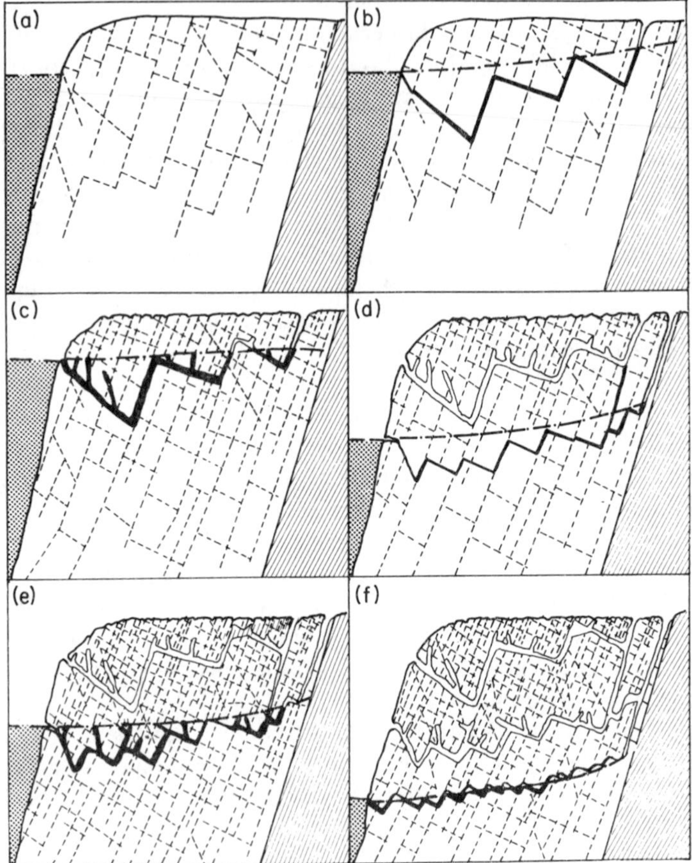

Figure 7.16 The increase of penetrated fissures in a karst rock with passage of time after the onset of karstification affects the geometry of successive caves developed in a multiphase system. Here the first generation of caves comprises a few deep loops (Fig. 7.16c, state 2). Later caves display more loops and pass to state 3 form (Fig. 7.16f).

processes (Fig. 7.16d) this higher frequency yielded state 2 systems composed of many loops of lesser amplitude. Repetition (Fig. 7.16e, f) yielded systems where gradational processes (outlined in the next section) could produce a state 3 system. In one example the amplitude of early loops was > 40 m, diminishing to ~ 15 m in a second phase, and to < 10 m in a third phase.

This pattern of development can be recognized in the cave fragments that are preserved in many tower karsts. The massive bedding that is necessary to sustain the verticality of the tower walls also favours low fissure frequency. Higher, ancient caves display state 2 features; the modern caves of the floodplain are often state 4. The great tropical river cave systems of Selminum Tem, Papua New Guinea, and Mulu, Sarawak, have similar histories (Brook 1976, Waltham & Brook 1980, Gillieson 1985). However, Holloch has conserved its high amplitude, state 2 form through at least three successive phases, so this generalization is not always true.

Some effects of geologic structure

Where strata dip quite steeply (2–5° or more) the bedding planes tend to entrain groundwater to great depth. Their gradients are steeper than most initial hydraulic gradients and so there is a quasi-artesian trapping effect, as Glennie (1954) noted. As a consequence the deeper types of phreatic caves are favoured.

Where strata are quasi-horizontal, bedding planes (the most continuous form of fissuring) may extend to the perimeter of the karst, where they outcrop to offer many potential spring points. It is only joints or faults that can guide water to great depth. In addition, conditions are most favourable for perching groundwater streams upon aquitards such as dolomite beds, thick shale bands, etc. States 3 and 4 caves are particularly favoured as a consequence. Very large water table passages of the Nullarbor Plain, Australia, provide excellent examples such as Cocklebiddy Cave (Grodzicki 1985). Deep phreatic systems can develop where there is strong jointing, however, e.g. Yorkshire Dales, England (Waltham 1970).

In tightly folded rocks (where cave systems may extend over several or many folds) fissure frequency is generally high because of the high stressing. Water table caves are consequently common, as in the folded Appalachians, USA (Davies 1960).

Frequency of penetrable fissures may increase where strata are unloaded. This helps to explain why many caves draining plateaus ramify into complex distributary networks close to the springs where the overburden is reduced (Renault 1968), and why state 3 or 4 caves can be well developed in valley flanks (e.g. Droppa 1966).

Hydration effects in anhydrite inhibit deep phreatic loops; hence gypsum caves tend to be states 3 or 4 where the strata are not confined.

Differentiating vadose caves

As primary tube systems extend, connect and then enlarge, the reservoir capacity of the rock increases. The piezometric surface falls until stabilized at some minimum gradient to the springs. The two basic types of vadose caves that may develop in the drained rock are shown in Figure 7.17.

Drawdown vadose caves are those guided by the early networks of primary tubes. They tend to be the first type of vadose cave to appear in the rock. Although 90%+ of their volume may have been created by erosion in

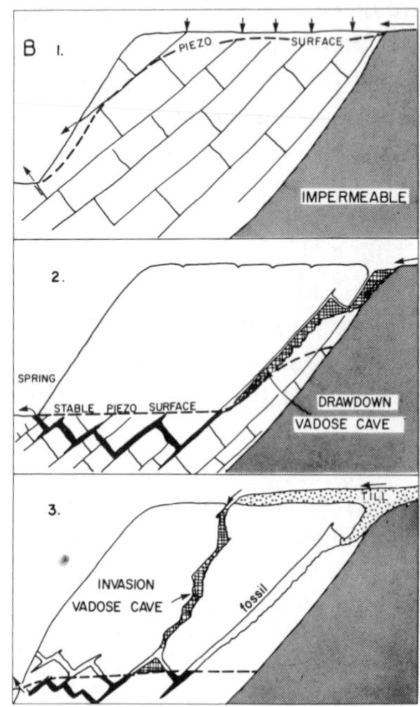

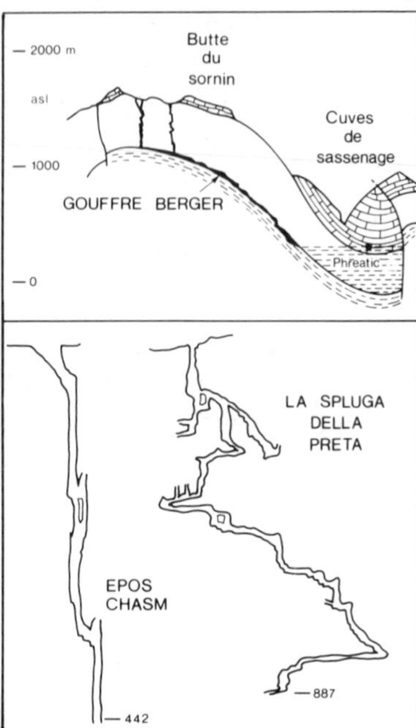

Figure 7.17 The development of the drawdown and invasion types of vadose caves. Gouffre Berger, France, is predominantly of the drawdown type. Epos Chasm, Greece, and La Spluga della Preta, Italy, are examples of the very steep shaft systems associated with invasion vadose caves.

vadose conditions, the passage plan pattern and skeleton is largely phreatic. If preserved at all, the phreatic morphology is usually in the roof where it may be difficult to inspect. Gouffre Jean Bernard, currently the deepest explored cave, appears to be predominantly of the drawdown vadose type with shaft drains conveying water to a trunk vadose canyon (Delannoy & Maire 1984) although it contains some invasion vadose and phreatic multiphase elements.

Invasion vadose caves are so named because, in many instances, they are created by new streams invading rock already drained during one or more previous phases of speleogenesis. As a consequence they lack a sustained primary tube network: the phreatic element in their morphology is either very restricted and rough, or absent. In dipping strata these caves tend to be steeper than the drawdown type, as illustrated in Figure 7.17. Invasion vadose caves can be the first type to develop where the effective porosity is very high or where resistance in fissures is very low. Bogli (1980) terms the latter caves 'primary vadose caves' and suggests that they develop where initial apertures are $\geqslant$ 1 mm. Such apertures might be expected where there is rapid tectonic uplift accompanied by some deformation, i.e. in young mountain systems. Systems of vertical shafts with short basal drains are generally of invasion vadose origin. The 'domepits' of Kentucky are an excellent example (section 7.11).

The invasion is often a result of karstification of surface streams as they are lowered from an eroding caprock (Fig. 9.25). It often follows from the clogging and sealing of former sinkpoints, the most potent agency of such infilling being regional glaciation (Fig. 7.17). Invasion vadose caves perhaps predominate in alpine mountain regions because of the combination of uplift, deformation and repeated glacial disturbance of drainage.

Short sections of phreatic cave often occur within vadose caves. They are due to lithologic perching or local phreatic lifting, often at faults.

Extent and magnitude of vadose cave development

The extent of vadose caves (either type) is a function of the depth of the vadose zone and of many lithologic and other effects tending to divert groundwater from a simple, vertical descent. Topographic relief above the springs is most important; the deepest vadose caves occur in mountain areas. Variations of water table gradient are also significant. Where vadose caves drain to state 1 or 2 systems variation of resistance in the phreatic passages may cause local variations of tens of metres in the depth of the vadose zone i.e. in the topography of the water table surface.

In general hydrologic models, water in the vadose zone is presumed to drain vertically downwards. In many karst regions this is true only in the sense that it does not flow upwards in the zone. Thousands of kilometres of lateral vadose passages are known.

The magnitude of vadose caves is a product of the size of their streams and the duration of erosion. The largest occur where large allogenic streams have flowed underground for a long while. In many autogenic systems individual streams are tiny because recharge is relatively diffuse and doline frequency is high. Underground, all of the vadose zone (and perhaps more) may be required to effect the amalgamation of streams that is necessary to generate caves of enterable dimensions. Such areas therefore appear to lack vadose caves.

Review

The following conclusions may be drawn from the models developed above: (1) in the first phase of development a common cave system may be composed entirely of one type between its inputs and its outputs e.g. bathyphreatic, drawdown vadose etc.; (2) more frequently, in the first phase it will display a combination of one or both vadose types supplying water to one of the four phreatic or water table types (e.g. Fig. 7.18); (3) in multi-phase caves, two or more of the phreatic/water-table types may be present

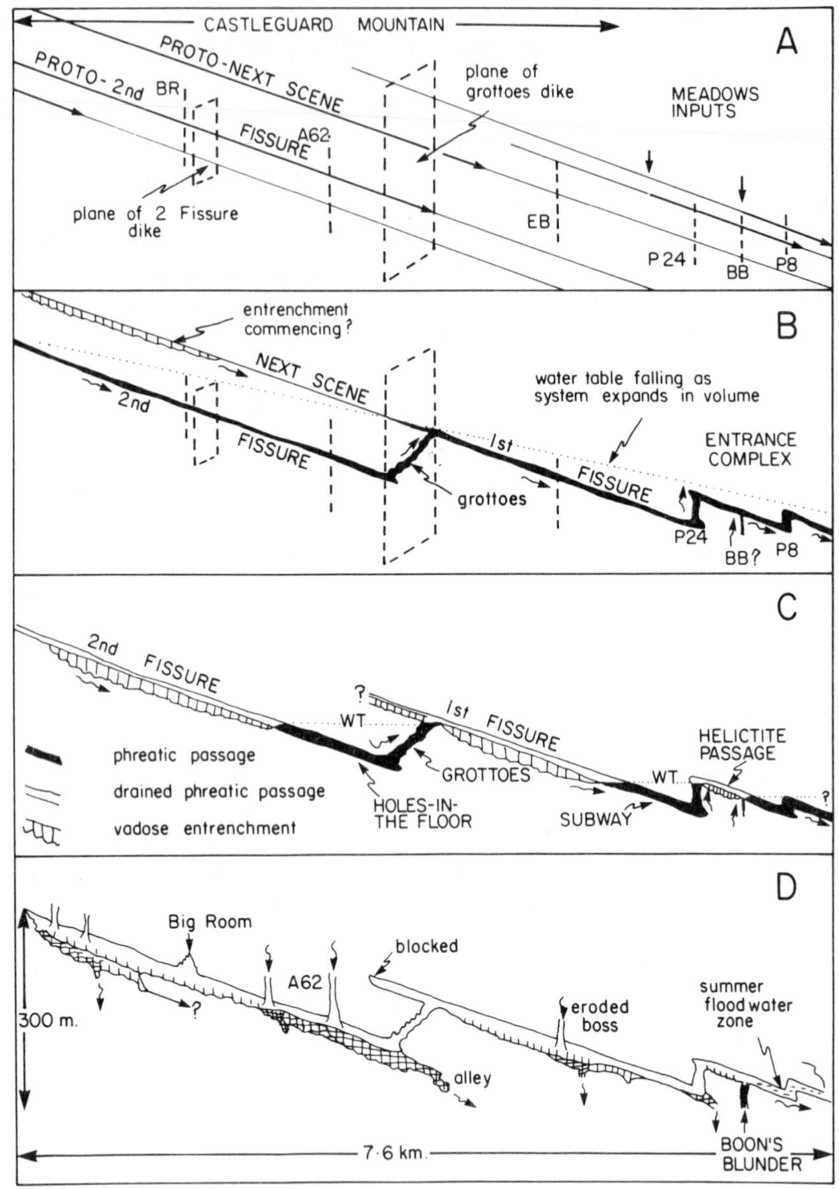

Figure 7.18 The central section of Castleguard Cave, Rocky Mountains, Canada, is an alternation of drawdown vadose passages (canyons) draining to phreatic loops (C). During abandonment (D) underfit streams entrenched bases of the loops. P8, BB, P24 = phreatic lifts of 8, 22 and 24 m straight up faults.

in addition to the vadose types, the later caves tending to be of higher state; and (4) many systems display some admixture of these types that has developed within a single phase.

This set of models resolves a debate that was the focus of much speculation for many years, especially amongst English-speaking researchers. One school of thought contended that major caves develop preferentially within the vadose zone, a second that they were principally of deep phreatic origin, while a third supposed that they developed proximate and parallel to the water table (Martel 1921, Davis 1930, Swinnerton 1932, Rhoades & Sinacori 1941). The arguments are reviewed by Warwick (1953), Thrailkill (1968), Watson & White (1985) and White (1988). We now appreciate that geological and geomorphic conditions, in carbonate rocks and gypsum at least, vary so widely that all of the cave types in contention occur widely and there is no theoretical reason to suppose that any one of them should predominate. In addition, there are the other categories of solution caves still to be discussed.

Generalizing the Four-state Model

In terms of speleogenesis it follows that *six* states may be distinguished in the karst rocks. State 0 is that where fissure frequency is too low (or resistance is too high) for common cave systems to propagate in the available geological time. This is true of some marble formations whose karst, therefore, is restricted to the pit and gutter types of karren that do not require cavernous drainage. States 1 to 4 are as specified. State 5 occurs where fissure frequency is too high; the solutional attack is dispersed in innumerable proto-cave channels. Amalgamation to create cave systems, of enterable dimensions at least, does not occur. This is true of many chalks, some corallian and aeolian limestones and dolomites. It is the ideal diffuse flow end of this spectrum of karst hydrologic states.

7.4 System modifications occurring within a single phase

Here we are concerned with two sets of effects that can significantly change the cave patterns described above within the phase that saw their creation. The more drastic pattern changes brought about by multiple phases are then discussed in the following section.

Gradational features in phreatic systems

The three different effects shown in Figure 7.19 all serve to reduce the irregular looping profiles of state 2 and 3 caves. *Isolated vadose trenches* develop when, as a result of water table drawdown, apices of upward loops

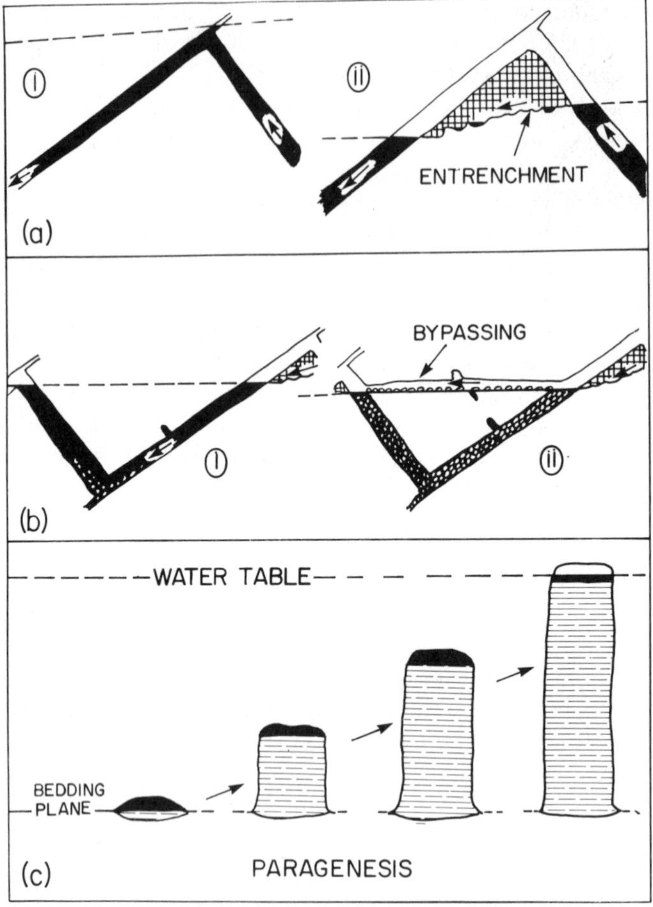

Figure 7.19 Gradational features common in phreatic caves (a) isolated vadose entrenchment of the upward apex of a phreatic loop; (b) development of a bypass tube above a loop clogged with detritus. (c) development of a 'paragenetic' gallery upwards to the water table.

become aerated. Entrenchments 50 m deep and hundreds of metres long are known, e.g. the great river passage of Skocjanske Cave, Slovenia, appears to be at least partly of this origin.

Bypass passages or 'tubes en raccord' (Ford 1965a, Renault 1968) develop where sharp downward loops below a water table become obstructed by alluvial detritus. Normally this will be in upstream parts of a system. If rapid flooding occurs a big head of water may build above the obstruction, greatly steepening the hydraulic gradient across the loop. Local fissures that are impenetrable under normal gradients are then exploited to open one or a series of sub-parallel bypasses over the loop. The latter becomes fully silted and inert; new clastic load is carried through the bypass to clog the next loop. The process propagates downstream and is also transitional to the floodwater maze (below).

As originally used by Renault (1968) a paragenetic passage is any phreatic or water table passage where the erosional cross-section is partly attributable to effects of accumulations of fluvial detritus. As applied by Ford (1971a) and other English-speaking authors, it has meant only the type of passage that has a steadily rising solutional ceiling (Fig. 7.19c); this is the extreme or 'ideal' type of paragenesis. Pasini (1975) perhaps has shown more flair in naming it 'erosione antigravitativa'!

Paragenetic passages originate as enlarged principal tubes. With enlargement, groundwater velocity may be reduced, permitting permanent deposition of a portion of any insoluble suspended or bed load. This protects the bed and lower walls so that solution proceeds upwards on a thickening column of fill. The vertical amplitude of such paragenesis can exceed 50 m. Remarkably flat roofs can be bevelled across dipping strata. The process terminates at the water table.

Floodwater maze passages

Floodwater mazes are evaluated in detail by Palmer (1975). They develop in state 3 or 4 systems or low gradient vadose caves, where fissure frequency is high.

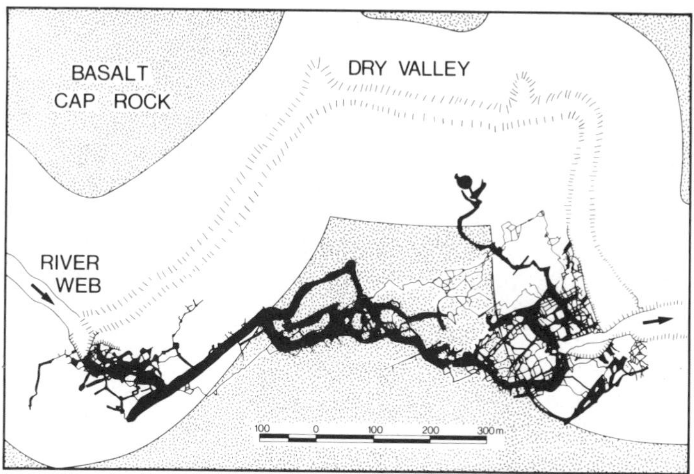

Figure 7.20 Sof Omar Cave, Ethiopia, a spectacular example of flood water maze development. The cave has developed as three successive levels in horizontally bedded limestones beneath a basalt caprock. It is a meander cut-off system that floods frequently during the rainy season. (From cartography by Steve Worthington).

Maze development occurs where allogenic flash flooding invades karst and/or where a trunk passage with large carrying capacity becomes obstructed by clastic or organic debris. Flood heads of many metres can be applied quickly, generating very steep hydraulic gradients as in bypass passages. Interlinked fissures around the obstruction may be penetrated rapidly to relieve the pressure. This creates a maze that is hydraulically inefficient compared to the trunk, but which may function for the remainder of the active life of the cave.

Floodwater maze development is most significant where caves drain large and rugged allogenic catchments (i.e. where big floods are applied rapidly to one point in the karst) and is most prominent at the upstream end of systems e.g. Bullock Creek Caves, New Zealand (Williams 1982a). It may extend throughout the length of comparatively short systems such as meander cut-off caves: e.g. the spectacular Sof Omar Cave, Ethiopia (Fig. 7.20).

7.5 Multi-phase cave systems

Most extensive common caves display multi-phase features due to the negative shift of springs. Multi-phase developments due to positive shifts flood or fill caves with sediment, making them less likely to be explored. At the extreme, the cave system may be de-coupled from all hydrologic throughput to become an inert, buried paleokarst system. Oil well drilling penetrates examples down to depths of 3 km or more, e.g. in Rumania.

Because of net continental erosion, negative shifts predominate in most karst areas and generally create lower series of passages (e.g. in Yorkshire; Sweeting 1950). Most authors describe the two or more series as cave 'levels'. This raises difficulties because the actual passages are frequently not level (horizontal) at all e.g. state 2 caves. 'Storey' has similar connotations, although Gregor (1981) recommends that it be restricted to local perching of passages due to tectonic effects, where there is no connection with erosional base levels. 'Stage' has evolutionary implications that can create misunderstanding. Here we use 'phase', 'level' and 'storey' interchangeably, but readers should bear in mind that actual cave passages of a phase need not be level *sensu stricto*.

Types of network reconstruction

The type and scale of reconstruction that follows the lowering of a spring is varied. At its most simple, the spring shifts downwards only, relocating at a lower point on the same vertical fissure. If it follows this fissure, the cave may display simple vadose entrenchment to the new level, e.g. Grotte St Anne de Tilff (Ek 1961 & Fig. 7.35D). Normally, the entrenchment is

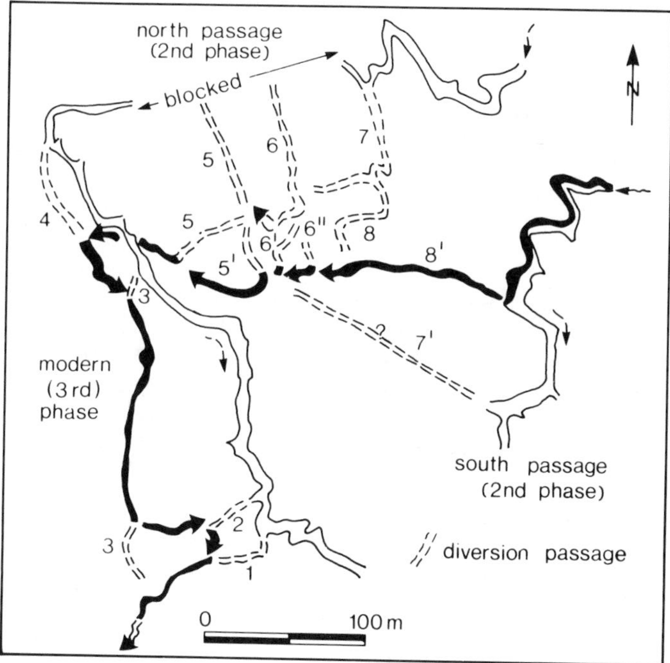

Figure 7.21 The central galleries of Swildon's Hole, Mendip Hills, England. Shown in white are north and south passages of a second genetic phase. In black are passages of the third (modern) phase. Numbered dashed lines are the headward sequence of diversion passages by which the modern route was built beneath that of the second phase.

retrogressive, i.e. a knickpoint. There is little or no change of the plan pattern of the cave. This type of reconstruction occurs particularly where the initial cave is one of the vadose types but has a gentle gradient. However, it is comparatively rare.

In many more instances new series of passages are constructed. They propagate headwards through the system and are similar to the connections of further ranks in a first phase. This is because many primary tube systems of previous phases have survived cave enlargement and, although carrying an insignificant proportion of the discharge, have slowly extended and can be exploited when hydraulic gradients are steepened by downward shift of the spring. In addition, the spring is displaced laterally as well as vertically in a majority of cases, which compels the development of at least some distinct new passages. At Holloch, the lateral shift was of the most simple kind, i.e. by small increments down the edge of a single, inclined plane. Nevertheless, this created several new series of state 2 irregular passages across that plane (Fig. 7.9). At Mammoth Cave, because stratal dip is very gentle, when the Green River (on the down dip, output boundary)

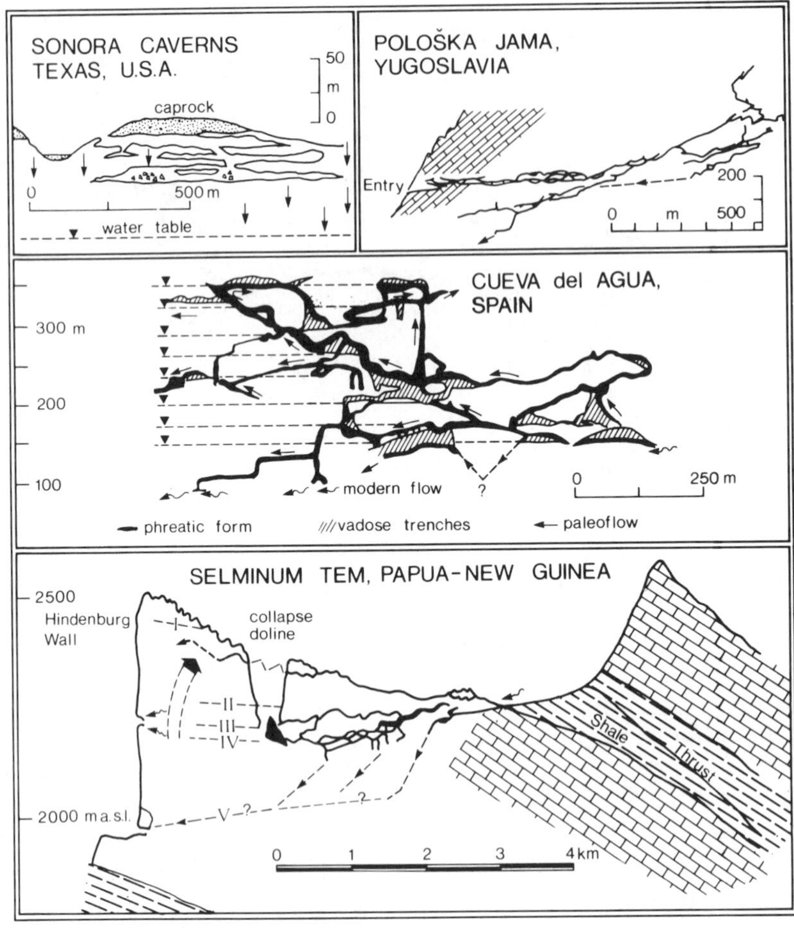

Figure 7.22 Some examples of multiphase caves shown in long section. See text for details.

entrenched its channel a few metres it might create a favourable new outlet position more than 1 km from previous springs.

An excellent example of the passage complexity that can be created by multi-phase reconstructions is given in Figure 7.21. Swildon's Hole displays just three major erosion phases. The phase 2 conduits were irregular strike passages of state 2 type. The stable phase 3 conduit is a river passage 25–30 m lower. The 1000 m of its course that is shown was built piecemeal from at least eight, consecutive and headward, diversion channels from the northern phase 2 tributary, plus two or more diversions from the independent, southern phase 2 tributary. The earlier diversions were wholly phreatic; later ones were vadose at their upper ends, signifying the lowering of the water table that resulted from creation and expansion of this new

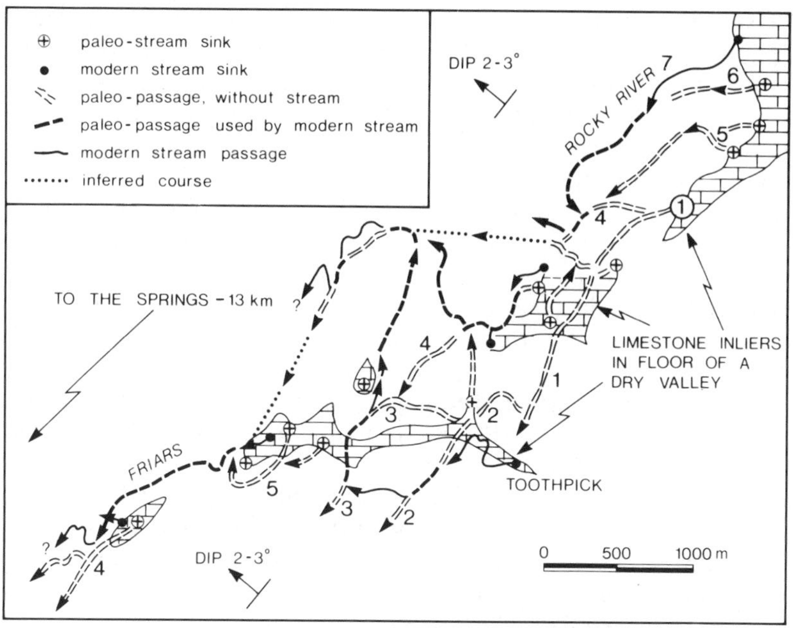

Figure 7.23 The multiphase pattern of Friars Hole System, West Virginia, USA. Numbers indicate the phase of different passages; see text for details. From S. R. H. Worthington (1984).

cave. Parts of each diversion channel were incorporated into the final phase 3 river passage. Other parts were abandoned entirely and, in some cases, remain too small to enter.

Figure 7.22 displays some other multi-phase systems chosen to emphasize the wide variety that occurs. The Caverns of Sonora, Texas (Kastning 1983) display simple lowering of shallow amplitude. Strata are flatlying and some limestone units are quite porous. Three levels occur in 50 m, interconnected by the collapse of roofs. Poloska Jama, in the Julian Alps of Yugoslavia (Gams 1974) has a long series of dip passages that appear to be a multiphase sequence, plus a prominent, more horizontal level that also displays several phases within it and which leads to a paleospring position.

Cueva del Agua, Spain, is shown in the outstanding morphological reconstruction by P.L. Smart (1986). It is a state 2 multiphase complex containing steep-to-vertical phreatic lifting segments that climb 120 m or more. There are many isolated vadose entrenchments that are taken to represent downcutting to eight different levels within a range of 200 m. The levels may represent paleo springs. The modern river cave remains of the state 2 type. The Selminum Tem cave systems of Papua New Guinea (Brook 1976, Gillieson 1985) are different again. Here it appears that five successive cave levels of state 2-becoming-state 3 type propagated against the dip. This is a plate boundary region of vigorous tectonic activity. Brook suggested that strata have been raised and inclined towards the sinkpoints, tilting the oldest caves back on themselves. Doline formation then dissected them. Gillieson prefers the simpler explanation that the cave postdates uplift, its passage levels being controlled by lowering of the spring.

Friars Hole System, West Virginia (Fig. 7.23) offers one final contrast. It is an instance of the restricted case of cave development (Fig. 7.13). Limestone outcrops in only a few narrow strips in the floor of a steep valley of shale and sandstone, yet 68 km of passages are mapped. Stratal dip is $\sim 3°$ westerly and the springs lie down strike to the south. The system displays at least six distinct state 4 channels that follow faults or selected bedding planes to the strike. The highest channel is only 125 m above the modern springs. Later ones shifted downdip as at Holloch. The plan pattern is extremely complex because stream-sinks have been randomly blocked and sealed for long time spans by debris from the surrounding shale slopes. New sinks were created both upstream and downstream of old ones and many underground diversions have appeared. At present three separate rivers flow through different parts of the explored system, at differing elevations and in different stratigraphic positions. They join somewhere before reaching the springs. The lowest river enters a siphon that is 13 km straightline distance from the spring yet only 24 m higher.

7.6 Meteoric water caves developed where there is confined circulation or basal injection of water

Single-storey and multi-storey, reticulate maze caves (Table 7.3, type A2)

These caves are quite frequently encountered, either as separate entities or as anomalous parts of common caves. Their form is everywhere similar. They are reticulate mazes of comparatively small passages that follow joints confined to one or a few rock beds. Passages are phreatic in form. Parallel passages (i.e. in one joint set) are of similar dimensions. Characteristically, their morphology is that created by slowly flowing waters. They lack the high velocity erosional scalloping and sharp cuspate morphology of typical floodwater mazes, although transitional examples do occur.

In essence such mazes develop where strata possessing a high fissure frequency are geologically confined. The trapping may be artesian, e.g. in a syncline beneath a shale aquiclude, or it may be a local 'sandwich' situation where one or a few densely jointed limestone beds are trapped between massive limestones with few joints. Artesian maze caves are very widespread in the north of England (Ryder, 1975) and in Arkansas. Where

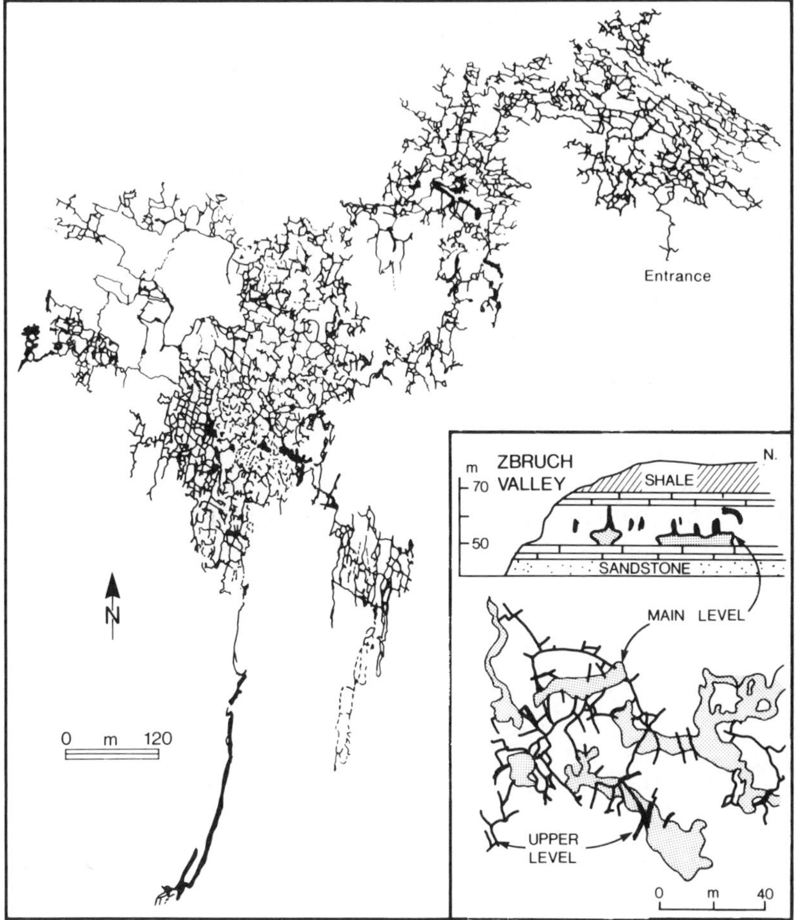

Figure 7.24 Plan of Optimisticeskaya Cave, Podolia, Ukraine (from Courbon & Chabert, 1986). 165 km of galleries were mapped as of October 1988. Inset-section and plan of part of the nearby Atlantida Cave, showing a fragmentary upper joint maze developed from a main cave level at the base of the gypsum; from A. B. Klimchouk, Kiev.

small areas of reticulate maze passages occur within common branchwork caves (e.g. Lummelund Grotta, Sweden; Engh 1980) it is most frequently because a 'sandwich' situation exists.

The mazes are most frequently of a two-dimensional kind i.e. they are guided by one local joint system, so forming a single 'storey' or 'level' of passages. Three-dimensional but one-phase mazes may be created where phreatic waters penetrate *upwards* into overlying joint systems. Usually such joint systems are separated from one another by thin aquitard beds or locally breached aquiclude beds such as clays, shales, etc. The upper passages normally are subordinate to the basal storey; they are smaller and may comprise only discontinuous fragments of maze superimposed upon the larger base e.g. Atlantida Cave, Figure 7.24 inset. Where, however, major outflow to springs passes through them they will be enlarged to full size; they may also be enlarged by local mixing corrosion effects during final drainage. 'Endless Caverns' and other caves in backreef limestones at McKittrick Hill near Carlsbad Caverns, NM are excellent examples of multi-storey, one-phase mazes, as are many of the great Ukrainian gypsum caves described below.

Maze formation by diffusion through an overlying sandstone aquifer

Palmer (1975) investigated a sample of maze caves in the United States and found that 86% of them had developed directly beneath permeable sandstones. Often, passage roofs were of the sandstone. He therefore proposed the model illustrated in Figure 7.25A/B. An equidimensional maze is created because water is introduced to the soluble rock by way of an overlying, insoluble and homogeneous diffusing medium, i.e. sandstone with

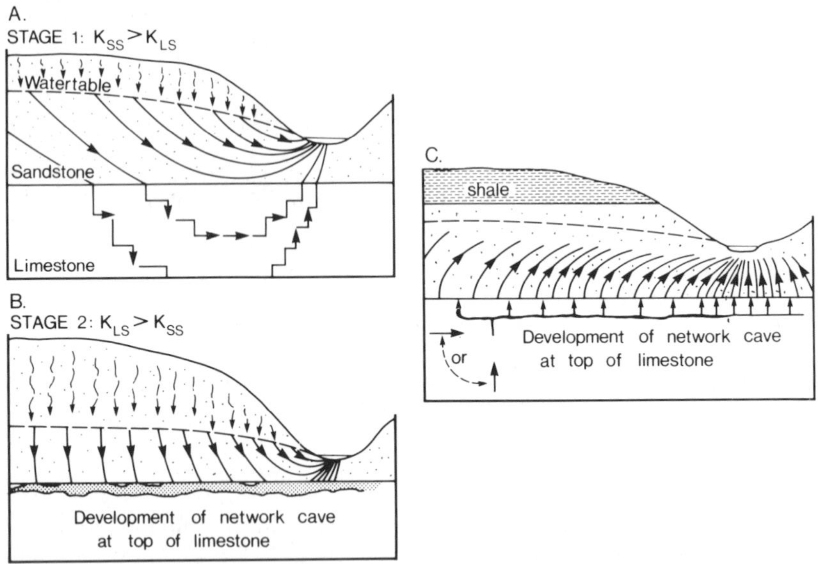

Figure 7.25 A and B. Development of a two-dimensional maze cave in the top of a limestone formation as a result of diffusion of meteoric groundwater *downwards* through an overlying sandstone aquifer. The model of A. N. Palmer (1975). C. An alternative proposition that fits many instances is that groundwater (meteoric or other) is compelled to diffuse from a large conduit into a maze of lesser dissolutional passages in order to disperse into diffuse flow *upwards* through the sandstone granular aquifer. In this example, a shale caprock is superimposed to avoid the added complication of local meteoric water descending through the sandstone.

well developed primary permeability. In stage 1 the hydraulic conductivity of the sandstone is greater than that of the unkarstified limestone beneath, but the situation is reversed by stage 2. This model will explain single-storey mazes directly beneath the sandstone cover but cannot explain development of multi-storey mazes. An alternative possibility, that the groundwaters must disperse into a maze in the karst rock in order to *discharge* through the lower conductivity sandstone (Fig. 7.25C) will permit both single-storey and multi-storey development.

The Gypsum Maze Caves of Podolia and Bukovina, Ukraine

These caves are the greatest known reticulate mazes. They include Optimisticeskaya and Atlantida (Fig. 7.24), Ozernaya, Zolushka (Table 7.2) and many other examples that are also astonishing when viewed on maps.

The caves occur in the flanks and spurs of the Dniester and Prut valleys and their tributaries, and across gentle interfluves. The gypsum is only 10–22 m thick. It rests on underlying limestone containing a major sand aquifer bed. It is overlain by a variety of rocks, predominantly marine clays which function as an aquiclude. The large caves developed when rivers incised the clays and could begin to lose groundwater into the karst strata (Dublyansky 1979, Dublyansky & Lomaev 1980, Klimchouk 1986, Klimchouk & Andrejchouk 1986).

Optimisticeskaya, Podolia, is now drained and relict as a result of continued river entrenchment. Initial penetration by groundwater was via the underlying sand aquifer. Solution upwards towards a thick clay cap created four storeys of passages in four different joint systems in the gypsum, though rarely are all present in one area. There is some late-stage paragenesis above clay filling.

Zolushka Cave, Bukovina, remains phreatic but is being revealed by groundwater pumping in a gypsum quarry. It has two levels of very large passages, separated by a prominent clay layer. The cave was also initiated by flow from underlying limestones and sand. It lies within an interfluve 25 km wide across which groundwater drains from the Dniester valley to the Prut. The maximum hydraulic gradient is 4 m km^{-1}.

The Ukrainian maze caves developed as lateral adjuncts of entrenching rivers. At much smaller scale, such caves can be found locally where densely jointed, sandwiched beds occur in the walls of many canyons and lesser channels in limestone and dolomite. They may be viewed as hybrids of the artesian and floodwater maze processes.

Simple outlet caves created by basal injection of meteoric waters into karst rocks

In an excellent local study, Brod (1964) described small solutional pit and fissure caves scattered for 60 km along the crest of a denuded anticline in eastern Missouri. They extend downwards through 30 m of limestone with basal shale, and then through 30 m of underlying dolomite where the form may change from shaft to a small, basal maze. Beneath the dolomite underlying sandstone is exposed in some cave floors, including detached sandstone blocks partly rounded *in situ*.

The sandstone unit is 40 m thick and very porous. In Brod's interpretation the sandstone aquifer is discharged upwards and out through the impure carbonate aquitard where this is fractured at the anticline. Localization of the fracturing allowed sufficient concentration of waters to dissolve enterable caves. This mode of cave genesis is identical in hydrogeologic terms to that of the Satorkopuszta type of hydrothermal cave (below) but the waters were of the shallow meteoric type.

7.7 Hypogene caves A. Hydrothermal caves associated with CO_2

This section considers solutional caves created by deep waters that are heated, perhaps enriched in a wide range of dissolved gases and minerals, and flow out to the surface in most instances. By definition, thermal waters are those which, at their springs, have a mean annual temperature $\geqslant$4°C higher than that of the region (Schoeller 1962). Temperatures at hot springs in karst rocks are most frequently in the range 20–80°C. Much higher temperatures can occur at depth. Geothermal gradients of 16°C km^{-1} occur in carbonate rocks in the Transdanubian Mountains, Hungary, a region with many hydro thermal caves. In overlying shales they become as steep as 40–60°C km^{-1}, a contrast which emphasizes the role of groundwaters in transporting heat through the karst rocks.

The waters may be juvenile (i.e. volcanic in origin) or connate (trapped sedimentary depositional waters that are being expelled), deeply circulating meteoric waters, or mixtures of these types in any proportions. Shallow meteoric waters (normal karst waters) may then mix with them as they approach the surface. CO_2 and H_2S are the gases most commonly reported. One or both may be important in a given water. Here, we simplify by supposing that one class of these cavities (the more abundant) is created predominantly by CO_2 solution processes, while in the second class (section 7.8) the role of H_2S may be more significant.

When hot, carbonated water rises and cools in carbonate rocks, with or without mixing with shallow meteoric water, there may be net dissolution, incongruent dissolution or net deposition. A cave may straddle several of the different process zones (e.g. dissolving at its base, receiving subaqueous precipitates at the top) or be entirely within one. The zones may migrate through it once (erosion→deposition) or oscillate across it several times. Many suspected hydrothermal caves experience only dissolution by the hot waters. In others the bedrock walls can scarcely be seen at all, being covered by later hydrothermal precipitates.

Criteria for hydrothermal origin or part origin

The most certain criterion is hot water flowing from the cave! However, some common caves have been invaded by hot waters because these will migrate to pre-existing conduits if possible. They offer the easiest exits. Most explorable hydrothermal caves are drained and relict. Presence of hot springs in their vicinity offers a strong hint concerning their origin but cannot be conclusive.

Certain erosional features are diagnostic although, once again, no one of them may be conclusive. First is the form of the entire system. Remarkable tree-form effluent chimneys and rectilinear, multi-storey mazes are characteristic (Muller & Sarvary 1977); they are summarized separately below. Within a system strong index features are deep, rounded and multi-cuspate solution pockets (Dublyansky 1980). These are termed *cupolas* to distinguish them from the simpler pockets of common caves. They are attributed to convectional solution (see section 7.11). In a few instances there are highly corroded patches or rills on walls where steam has condensed above hot pools. Dolomitization of limestone walls has also been reported. In some Soviet caves cherts in the limestone are strongly corroded, which is good evidence of hot alkaline waters. Perhaps as significant is the lack of evidences for moderate to fast water flow found in meteoric caves.

Better indices are provided by the exotic deposits. Most important are uncommon forms and abundances of calcite and aragonite, but a wide range of rare minerals may also be present in vugs, breccia, etc. Layered crusts of nailhead or Iceland spar calcite are common, with thicknesses up to 2 m. These may cover all surfaces (floors, walls and ceilings) indicating subaqueous precipitation. Quite frequently, fine-grained breccia with or without barite ($BaSO_4$) or silica is reported beneath the spars. Large quartz crystals are sometimes found that suggest cooling from a hot environment.

Paleo-waterlines tend to be marked by thick botryoidal or corralloid growths or by fin-like accretions of calcite. There may be abundant debris from the most fragile of all speleothems, rafts of calcite that precipitate about dust grains on the pool surfaces. Calcite geyser stalagmites may signify late, feeble effusions of the hot waters through drained cavern floors. In some instances there is further precipitation overhead from the expelled vapour. Some hollow stalagmites may be subaqueous ('white smokers'). Gypsum as big crystals or extruded 'flowers' and 'whiskers' is common in freshly drained hydrothermal cavities.

All of these calcite, aragonite and gypsum forms can be found in meteoric water caves. What is distinctive in the hot water caves is their unusual abundance and their association together (Hill & Forti 1986). Strong supporting evidence of precipitation from past thermal waters is obtained by measuring ^{12}C and ^{18}O isotope abundancies (see section 8.6).

Monogenetic hydrothermal caves

Budapest is centred upon hotsprings caves in hills overlooking the River Danube. It is not surprising that Hungarian researchers have been foremost in the study of hydrothermal caves (Muller & Sarvary 1977), which they divide into two main categories. The prototype for the first is

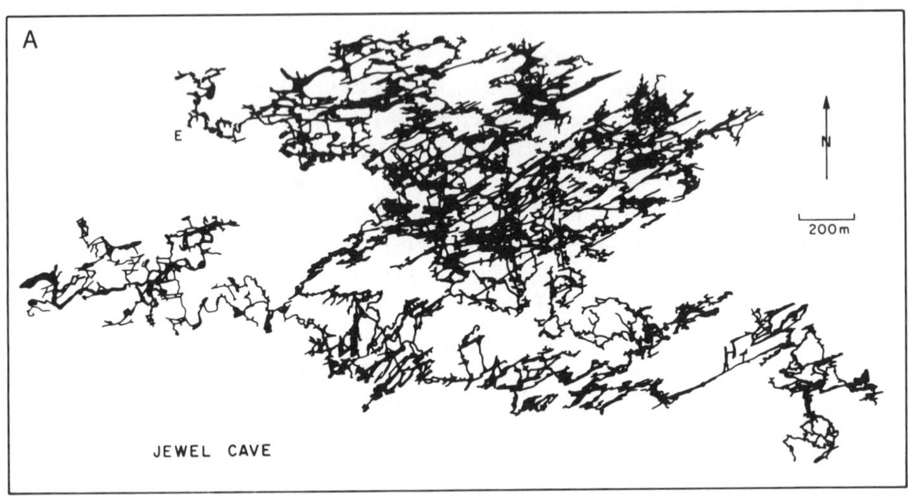

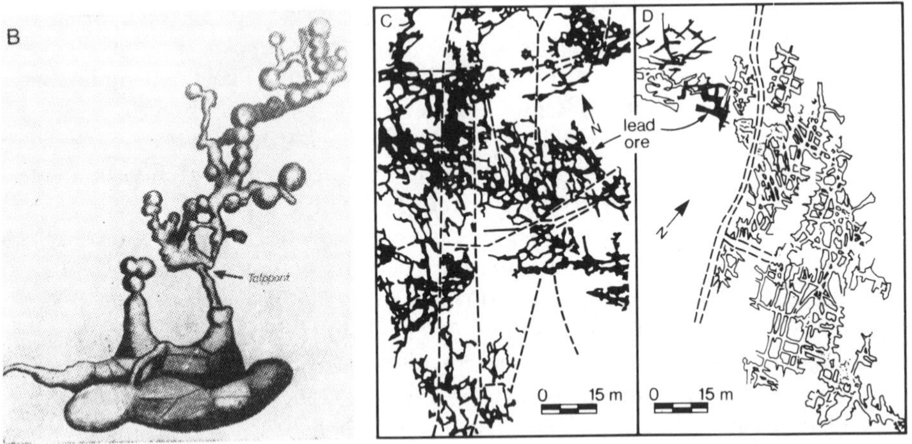

Figure 7.26 Satorkopuszta Cave, Hungary, (B) is an example of a monogenetic hydrothermal outlet cave. Note the cupola-form dissolution pockets (from Muller & Sarvary, 1977). Jewel Cave, S. Dakota, USA is a three-dimensional rectlinear maze with 118 km of mapped passages. Most are lined with 6–15 cm of precipitated calcite spar. (C) Jefferson City Mine, Tennessee, is a zinc/lead mine with the orebodies indicated in black. (D) Devil's Hole Mine, England, is a maze of open passages plus a few ore-filled passages indicated in black. Dashed lines are mine adits. Adapted from Ryder (1975).

Satorkopuszta Cave (Fig. 7.26). This remarkable cavity consists of a basal chamber (the limit of exploration) that may be likened to a magma chamber, from which a branching pattern of rising passages have grown like the trunk and branches of a tree. Their form is heavily modified by convection cupolas, which terminate most of them. One gained the surface and discharged the hot water.

This type of cave appears to be created by the delivery of hot water to a single point at the base of carbonate rocks of low fissure frequency. It is more truly 'dendritic' than any other erosional landform in geomorphology because it grows upwards, branching and terminating.

Hungarian speleologists consider this to be one hydrothermal cave style that is certainly formed by hot waters alone (monogenetic). It is rare.

Phreatic maze caves

This is the second hydrothermal category. Two-dimensional rectilinear mazes are created where rising water is trapped in carbonate rock against relatively impervious cover beds (Fig. 7.25C). Cserszegtomajikut Cave, Hungary, is an excellent example. High joint frequency is required, and perhaps some cooling by admixture of meteoric water. Very stable thermal stratification in saturated rock may also create this effect.

Multi-storey rectilinear mazes that have been created in a single phase are more frequent. All authors stress the necessity for dense fracturing (which may be created by the tectonic forces that also furnish the heat and CO_2) and many stress the evidence for episodes of meteoric water invasions or mixing that is sometimes found (see Muller & Sarvary 1977, Dublyansky 1980, Bakalowicz *et al.* 1987). The extent of the mazes suggests that areal inputs of hot water (rather than point inputs) are converging upon point outlets. It is probable that most of these mazes are not as purely hydrothermal in origin as the Satorkopuszta type.

The caves of the Buda hills are 3–D mazes. Those at the Danube River level discharge highly mineralized water at 42–48°C. Precipitates form at the pool rims, diminishing below. There is strong vertical zoning of precipitates in some higher, relict caves as well. In Eocene times the strata were penetrated by hot volcanic waters which reduced rock along the major fissures to clay (Kovacs & Muller 1980). Modern thermal waters invade them via solutional passages at the sides and sap the clay, creating tall fissures with residual clay roofs. As another example, Stari Trg Mine, Yugoslavia, is a complex of vertical and horizontal galleries and chambers extending to a depth of 600 m. It was discovered during zinc/lead mining. Its lower 400 m was filled with warm water until pumped out (Petrovic 1969).

Jewel Cave and Wind Cave, South Dakota

Jewel Cave (Fig. 7.26) and Wind Cave are the greatest known mazes of this kind and amongst the largest caves (Table 7.2; Bakalowicz *et al.* 1987). They are developed in 90–140 m of well bedded limestones and Ca-rich dolomites uplifted and fractured by Eocene intrusion. There are modern hot springs nearby but at lower elevation.

Jewel Cave is entirely relict. It has no relation to the surface topography and is entered where intersected by gully erosion. Its great features are big passages (up to 20 m in height) largely or entirely covered by nailhead spar. In the highest places the spar has been partly removed in a later condensation corrosion phase. Elsewhere, it is generally 6–15 cm thick and has a silica overgrowth (Deal 1968).

Wind Cave is lower in elevation and nearer to the hot springs. The morphology is similar to that at Jewel Cave but there is little spar encrustation. It, too, is a hydrologic relict but its lowest accessible parts are filled by semi-static waters fed from below. They are cooling backwaters of mixed thermal-meteoric composition. Calcite rafts form upon them.

Boxwork is uniquely well developed in Wind Cave (Fig. 7.27). It occurs in

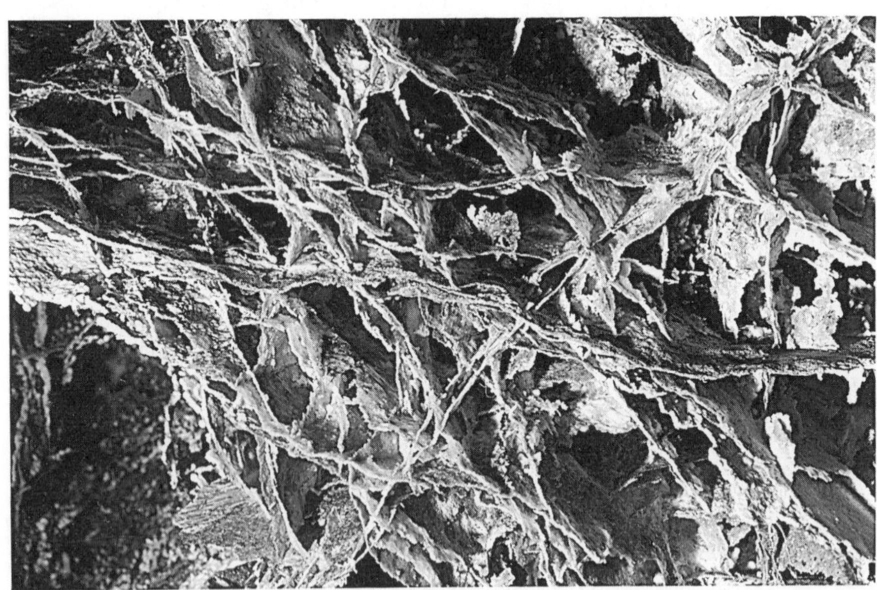

Figure 7.27 'Boxwork' in the ceiling of Wind Cave, South Dakota. The boxes are a few centimetres in diameter and up to 30 cm deep. Their walls are calcite veins preserved when intervening dolomite was preferentially dissolved (='incongruous dissolution', Section 3.4). Calcite overgrowths were deposited onto the veins during and after the dolomite dissolution. Photo by A. N. Palmer.

micro-fractured dolomites where the fractures are filled with calcite. Under incongruent dissolution conditions (section 3.4) the dolomite was dissolved but the calcite veins were preserved. Later, further calcite precipitated onto the veins, which now protrude as much as 100 cm from roofs and walls.

7.8 Hypogene caves B. Caves formed by waters containing H_2S

Formation of $H^+ + HS^-$ and of H_2SO_4 were summarized in Chapter 3. Despite their potency, the quantitative significance in most karst regions will be nearly negligible because the quantities of H_2S obtainable are tiny. However, recent studies are suggesting that H^+ ion or sulphuric acid from basinal fluids or other sources (Eqns 3.67 to 3.71) are important in some very large caves, including the celebrated Carlsbad Cavern.

Caves of the Guadalupe Mountains, New Mexico–Texas

The geology and topography of the southern Guadalupe Mountains is shown in Figure 7.28. The escarpment is ~ 50 km in length and contains more than 30 major caves. These are in or near the reef facies. The region is semi-arid.

The caves are alike in possessing a network form with great rooms linked to higher passages or shafts and underlain by blind pits. Many contain blocks of layered gypsum, up to 7 m in thickness, resting on thinner residual clay deposits. The Big Room of Carlsbad (Fig. 7.43) is representative. It is up to 80 m in height and contains enormous calcite stalagmites and columns. The rooms are interconnected by breakdown or spongework mazes. There

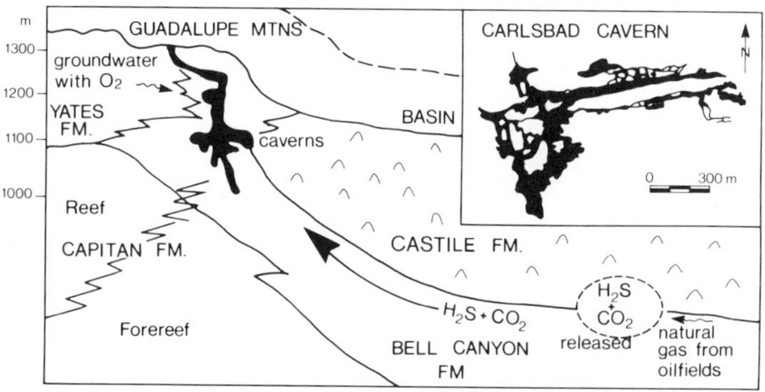

Figure 7.28 Model for the development of Carlsbad Cavern and other large caves of the Guadalupe Mountains, New Mexico, with H_2S (forming H_2SO_4) and CO_2 expelled from Permian oilfields. Modified from C. A. Hill (1987). Capitan Fm. = reef and mound limestones. Bell Canyon Fm. = sandstone and shaly limestone. Castile Fm. = cyclic evaporites, impermeable but soluble. Yates Fm. = silty limestones. Inset of Carlsbad Caverns in plan.

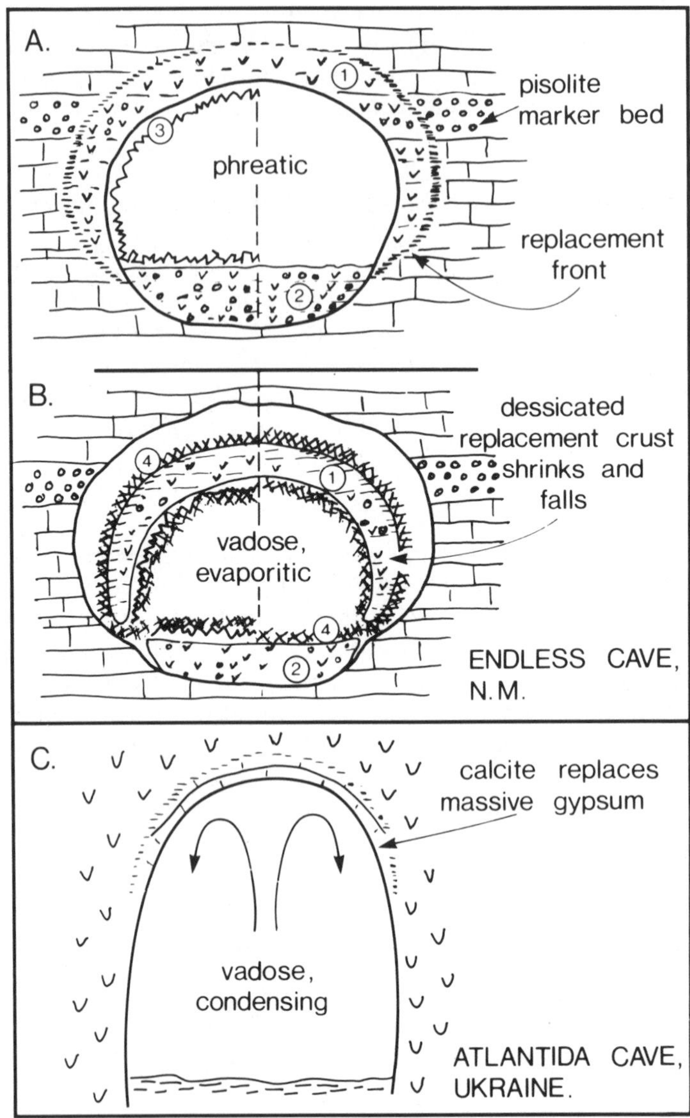

Figure 7.29 Gypsum deposition under phreatic conditions in a New Mexico cave. 1. Gypsum replaces limestone wallrock to depths up to 30 cm. 2. Layered gypsum with residual clay and pisolites accumulates on floor. 3. Possible late stage of slow deposition of large gypsum crystals on all exposed surfaces. B. The cave drains and dries. Gypsum replacement crusts and floor deposits are dessicated and shrink. An overgrowth evaporitic crust of gypsum (4) forms on the dessicated deposits. (From observations by Marcus Buck.) C. In contrast, at Atlantida Cave, Ukraine, standard atmospheric condensation water with dissolved CO_2 is replacing massive gypsum wall rock with ~5 cm of calcite. (From observations by Alexander Klimchouk.)

are few indicators of the direction of paleoflow. The caves are relict, without any relation to modern topography or surface hydrology.

Sulphur in the gypsum is very depleted in the heavy isotope, ^{34}S. This suggests that it derives from H_2S produced by biogenic reduction in the adjoining oil and gas fields. Davis (1980) and Hill (1987) propose that the H_2S migrated up-dip with expelled basin fluids. Approaching the water table it became oxidized (perhaps by some mixing with meteoric waters) to form H_2SO_4. The blind pits are then explained as input points at the base of the oxidizing zone, whereas the big chambers represent the areas of most effective oxidizing. When supersaturation occurred some gypsum was precipitated within them, as in a lagoon. In nearby maze caves there was direct replacement of limestone under phreatic conditions (Fig. 7.29).

We consider Carlsbad Caverns to be of polygenetic origin. In a sequence of phases hypogene water dissolved lifting shafts and strike-oriented exit passages to paleosprings. Both basal input points and spring outputs shifted progressively eastwards. The waters may have been enriched in both CO_2 and H_2S and meteoric water mixing could have played a role. In late stages, as these caves were being abandoned in favour of more easterly, lower springs, there was backwater ponding of H_2S-rich waters, which created the big rooms with their gypsum at Carlsbad.

Grotte di Frasassi, Ancona, Italy and Akhali Atoni Cave, Georgia, USSR are other spectacular systems of large chambers and galleries created by waters rising through them. The latter is in the footslope of the Caucasus Mountains at the edge of the Black Sea. A mixture of cool springs and warm springs with CO_2 and minor H_2S emerge a few hundred metres distant. When they are in flood the lower cave is also back-flooded. It appears to be essentially of the Carlsbad type but with deep CO_2 and warm–cold water mixing corrosion processes playing more of a role than H_2S (Tintilozov 1983). Dublyansky (1980) categorized Akhali Atoni Cave as a 'chamber' type of hydrothermal cave that is common in the USSR.

Limestone caves formed by gypsum replacement and solution

Egemeier (1981) describes a distinctive sub-type of hypogene cave in the Big Horn Basin, Wyoming, USA. It comprises short, horizontal vadose outlet passages in limestone. They taper in dimension inwards where they terminate in springs rising from narrow fissures. The waters are warm and rich in H_2S which is liberated into the air. Walls and roofs are deeply encrusted with gypsum.

It is suggested that H_2S is oxidized to sulphate and hydrogen ion, which reacts to replace the limestone with gypsum (see Eqns 3.68, 3.70 and Fig. 3.29). In the vadose situation the gypsum replacement crusts fail because of their weight plus expansion forces exposing fresh limestone to alteration. Fallen crust is dissolved by the vadose streams. The caves enlarge headwards from original valley floor springs, into the rock. The greatest reported example is 420 m in length and has an average diameter of 14 m.

The same mechanism is proposed for the genesis of Gaurdakskaya Cave, Turkmenia, USSR (A.B. Klimchouk, personal communication, 1986).

7.9 Sea coast mixing zone cavities

It was shown in Chapters 3 and 5 that solution with cavity formation may occur in salt–fresh water mixing zones (Fig. 7.30), or that calcite may be replaced by dolomite there. The two processes may occur in the closest proximity. There is a Boulder Zone of high solution porosity beneath southern Florida, where dolomitization is believed to take place (Hanshaw & Back 1979). Some large cavities have been intercepted by drilling at 200 and 600 m below sea level in northern Florida, where there has been mixing (Leve 1984).

We should expect these caves to differ from all types discussed above, especially in rocks that are not diagenetically mature when the cave formation occurs. There is high effective porosity via pores and vugs. Spongework maze caves extending along the mixing zone are anticipated. If large in size, they will be partly collapsed. In denser rocks, cavern formation follows joints and bedding planes in the mixing zone.

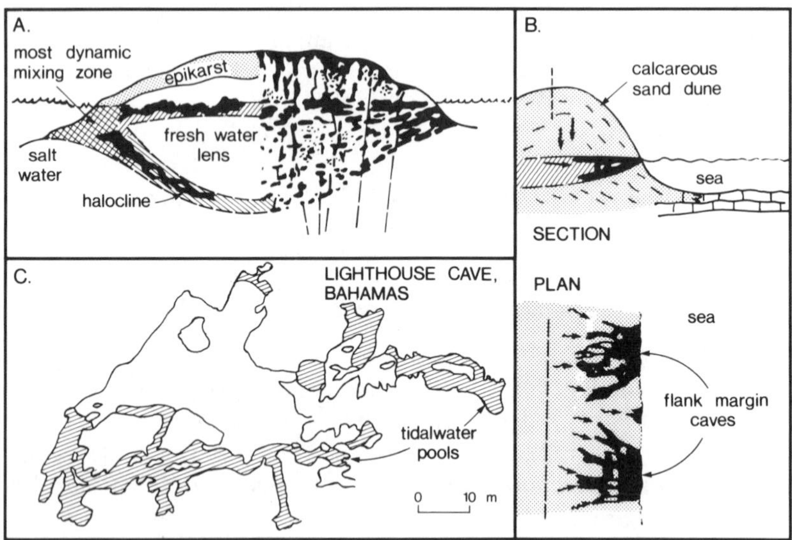

Figure 7.30 A. General model for zones of enhanced solubility in a young carbonate island. To left, development of enterable caves at the water table and along the halocline is stressed. To right, model distribution of cavities in the Yates oilfield buried paleokarst, Texas; adapted from Craig, 1987. B. Model for the development of flank margin caves in aeolianite; from Mylroie (1988). C. Lighthouse Cave, a water table-mixing cave zone on San Salvador Is., Bahamas; from Mylroie (1988).

Back *et al.* (1984) describe a 'Swiss cheese zone' where there is modern mixing on the Yucatan coast of Mexico. Dense honeycomb development within linear zones is described by Gregor (1981) in the British West Indies and by Ollier (1975) in the Trobriand Islands, New Guinea. When very high cavern porosity is created by this or other mechanisms the mixing zone migrates up close to sea level at the coast so that the cavities become sub-horizontal (Palmer 1986).

Most known caves in or close to modern mixing zones appear to be more complex because they are polygenetic. During the Quaternary, global sea level oscillated at least in the range, +5 to −80 m, over comparatively brief time spans (Figs 10.16 & 10.19). Mixing zone caves are modified by phreatic and vadose meteoric solution if sea level descends through them. Meteoric water caves then may be modified by mixing zone solution as sea level rises. The drastic changes of conditions cause much rock collapse. Breakdown morphology pre-dominates in many coastal solution caves, with an abundant overlay of speleothems because of the diffuse permeability.

Perhaps our best examples of mixing zone caves are to be found not in present-day karst but in paleokarst that developed when sea levels were somewhat more stable than during the past 2 million years of the Quaternary. One instance occurs in west Texas where Late Permian mixing zone caves developed within a height range of 45 m appear to be the principal source for a billion barrel oilfield (Craig 1987 and Fig. 7.30A). A further example is discussed in the next section.

7.10 Massive sulphide deposits in karst cavities

The majority of economic lead and zinc ores (PbS, ZnS) occur with pyrite and coarse-grained sparry dolomite as discordant (or 'strata-bound') masses within carbonate rocks. Many evidently formed at comparatively low temperatures (30–180°C) and are known as Mississipi Valley-type deposits (or 'MVT') because they are abundant in that region.

It is clear that many of these were deposits into solution caves. In some instances, the caves may have existed long before ore deposition, but more often there was an important measure of syngenesis (much of the cavity was being dissolved by CO_2 and H_2S solutions as sulphides were precipitated elsewhere in it). The literature on MVT deposits is vast (e.g. Dzulynski & Sass-Gutkiewicz, in press). Here we round off the discussion of cave types with summaries of three sharply contrasting MVT examples.

Jefferson City zinc/lead mine, Tennessee

Jefferson City Mine (Fig. 7.26) is a two-storey maze of the Atlantida type (Fig. 7.24) but with its passages entirely filled with sulphides and sparry dolomite. Lead/zinc maze deposits of this kind are quite common. Because of the equidimensionality of the passages, distinct phases of cave erosion (in diagenetically mature rock) and of sulphide-carbonate precipitation seem necessary. The similarity to Jewel Cave (where the precipitation phase was limited to calcite spar that occupies no more than 1–2% of the void space) is evident.

Pine Point Mines, NWT, Canada

The general geology of the Pine Point structure is shown in Figure 2.9. It is a reef complex of Devonian age, with known lead/zinc ore deposits outcropping along the strike for 30 km or more. There are two distinct modes of mineralization. 'Tabular' ore bodies are low, wide, sub-horizontal solutional passages wandering along the flanks of carbonate mounds. They are filled with channel deposits of silt and sand partly replaced by sulphides. Lesser replacement extends into dolomitized strata between channels and for 100 m or more on their flanks. 'Prismatic' ore bodies are collapse dolines filled with breccia and sulphides (Rhodes *et al.* 1984).

The Pine Point tabular deposits are here interpreted as examples of marine mixing cave genesis. This began with zonal reflux dolomitization. Solution cavities then developed through the zonal cores. Episodes of shallow emergence permitted fresh waters to invade and link these cavities, depositing debris eroded from the vadose zone in them. Later, the ore fluids invaded and were able to fill most voids and partly replace dolomite and sediments. One open room of $\sim 30 \times 10 \times 2$ m has been discovered recently.

This is a paleokarst. The mixing zone caves are of Devonian age. The ore may be younger, but certainly antedates the Tertiary. However, the system has been partly rejuvenated. Doline collapse was locally renewed during the Quaternary; some dolines contain tills of two or more glacial episodes (Fig. 2.9).

Nanisivik Main Ore, Baffin Island, Canada

This last example is again quite different. It is a single great cave of Precambrian age filled with ore and sparry dolomite. The cave is horizontal, slightly sinuous, 100 m wide and 8 m high. It wanders along the strike across a horst block of tight, impermeable dolomites dipping at 15°C. It is truncated by modern erosion at each end. A subsidiary tongue of ore climbs to the cave elevation from the north side where it is deposited in a trap against a roof of shales (Fig. 7.31).

The ore cavity is remarkable. Frail fins of eroded dolomite extend 20 m into it from the walls. They are carved to razorsharp edges, as in some vadose channels. Corrosion notches (section 7.11) filled with layered pyrite are slashed 25 m or more into the walls. One notch-and-sheet fill perhaps

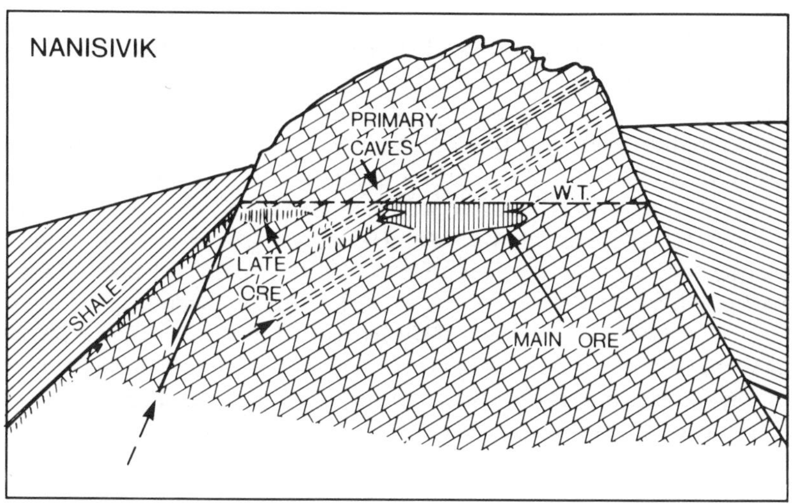

Figure 7.31 The Main Ore deposit at Nanisivik, Baffin Island, Canada. The ore body is 3 km long, ~ 100 m wide and 8 m high. It occupies a paragenetic cavity dissolved by ore fluids targeted onto an earlier, much smaller, primary cave. WT = water table or other control of level during ore deposition.

extends laterally for 400 m across the dipping dolomite. The main sulphide deposits show repeated episodes of channel cut-and-fill and of bank collapse.

Olson (1984) interpreted this as a Mammoth Cave type of passage that was lowered, filled with hydrocarbons and later occupied by the ore. Ford (1986) and Figure 7.31 interprets it as an extreme example of paragenesis close to a water table or other stable control of a fluid interface such as a deep gas–liquid contact, with hot ore fluid substituting for meteoric water. In a preliminary phase, a small common cave system of primary tubes and strike passages developed. Then ore fluids rose on boundary faults and invaded the body of the horst because the caves functioned as a low-pressure target zone. The fluids discharged along the strike. In a series of episodes sulphides were precipitated on the floor as the roof was dissolved upwards. At times of greatest discharge channels were cut through the sulphides and carved fins where they encountered the dolomite walls. There were episodes of very low discharge (pond phases) when the corrosion notches were cut. Dzulynski & Sass-Gutkiewicz (in press) describe similar examples in Poland.

These examples of lead/zinc ore deposition in large cavities are an appropriate conclusion to this summary of cave genesis by dissolution. The three deposits differ radically in their morphology because their genetic circumstances were different. There is a greater range of morphologies amongst the common caves and other types, for the same reasons.

7.11 Passage cross-sections and smaller features of erosional morphology

The erosional form of a cave passage may be attributable entirely to solution in phreatic (pressure flow) conditions or in vadose (free flow) conditions or alternating (floodwater) conditions. Many passages are compound forms displaying, first, phreatic erosion and then vadose erosion. Dissolutional forms of every kind may be modified or destroyed by collapse of walls or roof. Breakdown processes are reviewed in the next section.

There is a great variety of small forms eroded into cave walls, floors, etc. (Bini 1978). There are standard and variant stream channel forms plus many types of karren, some of which also occur above ground. Bretz (1942) gives a comprehensive review that is still pertinent.

Phreatic passage cross-sections

When first connected, phreatic passages (French – 'conduites forcés') have a subcircular cross-section or (if resistance is low) are elongated along the fissure (Fig. 7.32). Diameter or width is a few millimetres. The solutional attack is delivered to all parts of the perimeter.

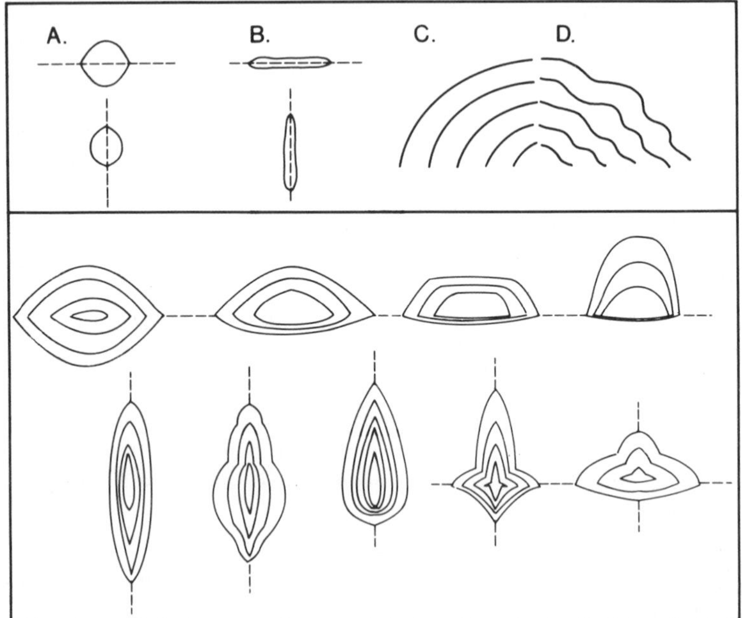

Figure 7.32 The evolution of phreatic passage cross-sections. See text for details.

Figure 7.33 Cross-sections of some phreatic passages. Photos by Ford and (bottom left) A. N. Palmer.

The form that develops as the passage enlarges is a function of the interaction between passive variables (lithologic and structural) and active mass transfer variables (fluid velocity, solution potential, type and abundance of clastic load). Figure 7.32c illustrates the case where geologic properties are isotropic or the active variables are much more significant (normally, flow is very fast). The minimum friction cross-section (a circular pipe) is maintained. Circular cross-sections are common, even at great size, e.g. at La Hoya de Zimapan (Fig. 7.15) diameter is ~ 30m, both where flow is vertically upwards and where it is horizontal.

Isotropic geology with slow mass transfer will similarly permit the simple fissure (Fig. 7.32b) to enlarge uniformly to great size. More often, however, passive variables are anisotropic normal to bedding planes (i.e. properties change significantly from bed to bed). The irregular type of profile (Fig. 7.36d) must then evolve. Note that the extent of irregularity may be a function of size, and thus of the aggregate amount of erosion time.

The variety of phreatic cross-sections is enormous (e.g. Fig. 7.33). All are variations on these themes. Lange (1968) and Sustersic (1979) have attempted formal geometric analyses.

Anastomoses, pendants and half tubes

Anastomoses may be independent forms or, with pendants or half tubes, can constitute a gradational set of features (Fig. 7.34). Independent anastomoses are the subsidiaries of primary tubes. They may continue to extend throughout the phreatic history of a cave. The frequency of their divergence and convergence is a function of properties of the fissure and of its gradient. Vertical joints, steeply-dipping bedding planes, etc. show relatively little anastomosing.

Excellent anastomoses can originate late in the history of a cave where water from an established large passage opens up hitherto impenetrable bedding planes or fractures (i.e. effective fissure frequency is increased). This is best seen where stratal dip is low. Often it appears that the penetration is by floodwater (Palmer 1975).

Pendants (German – 'Deckenkarren') are residual pillars of rock between anastomosing channels. They can develop in bedding planes and joints, where they are gradational from anastomoses. They also appear on unfissured erosional surfaces such as cave walls where they developed at the contact with rather impermeable clastic fill. They may be as much as 1 m long (Kuffner 1986). They can be carved by water draining up, down or along the contact. Diverse and complex patterns can be created.

Ceiling half tubes (French – 'chenaux de voute', German – 'Wirbelkanal') are equally common but more controversial. Some develop where a bed is more resistant. We consider that the great majority develop to carry the surviving flow through a passage that has become choked with sediment i.e.

Figure 7.34 *Top*. Dissolution pockets in passage ceilings. Since the cave drained the pocket at right has begun to fill with calcite speleothems fed from the joint that guided its dissolution. Photos by Ford. *Bottom left.* Ceiling anastomoses. *Bottom right.* 'Boneyard' – intense pocketing in reefal rock at Carlsbad Caverns, N.M., also characteristic of coastal mixing caves. Photos by A. N. Palmer.

they are a stage beyond the ideal paragenetic passage (Fig. 7.19). Although normally found in the apex of a passage, horizontal examples are known in walls, while others climb from wall to apex. In the apex, they often transform into a broader pattern of pendants (and back again), especially where ceilings are horizontal. Lauritzen (1981) has reproduced them experimentally.

If in the apex, half tubes are the last places to fill during a flood. This led Bogli (1956) to propose that they develop because the CO_2 of trapped air is dissolved there, enhancing the local solvent capability. They are found in deep phreatic conditions, so this explanation can be only partial.

Solution pockets or cupolas, and bell holes

Solution pockets are one of the most attractive features of phreatic caves and one which most surprises geomorphologists who are not cave specialists. They may occur in floors and walls but are best developed in the roofs (Fig. 7.34). They are varieties of blind pockets that extend as much as 30 m upwards. Many terminate in a tight joint or micro-joint that they have followed during expansion. They may be single or multiple features. Some are multi-cuspate and transitional to honeycomb structure; these develop in vuggy rock, e.g. reefs. Some are complexly multi-cuspate but have neither vugs nor joints to guide them.

It is probable that the initiation of most solution pockets (those propagating up small joints) is a trigger effect of mixing corrosion. Bogli (1980) and others attribute their excavation entirely to mixing. Where a mere thread of flow descends a tight joint to join a trunk passage in the phreatic zone, however, the mixing ratio is probably less than 1 : 1000. Mixing cannot furnish sufficient solvent to create the entire void but does operate at the junction to open a small void that the trunk waters enlarge when they are chemically aggressive e.g. during floods. Gravitational convection (see discussion of corrosion notches, below) may play a secondary role.

Highly complex, multi-cuspate pockets without joints or vugs to guide them are believed to be created by thermal convection in near-static waters. This 'cupola' type is an index feature of thermal caves (Fig. 7.26B), although examples are known elsewhere. Szunyogh (1984) and Cigna & Forti (1986) suggest that they may develop where steam is condensed above the heated waters.

Bell holes (Wilford 1966) are enigmatic features. They are vertical cylinders up to 1 m in height and ~ 5–20 cm in diameter, i.e. they are comparatively small pockets. Some are fluted and evidently developed as micro-domepits (below). Others are plain and lack any vug or micro-joint at their terminations. They seem too small to be convectional. Bell holes in caves have been reported from the humid tropics. They may be vadose or floodwater forms of biogenic origin. Densely packed, small bell holes occur in limestone and dolomite in the zone of water level oscillation of some fresh water lakes in Ireland and Canada.

Vadose passages, corrosion and potholes

The form of vadose passages is that of entrenchment, with or without widening (Figs 7.35 and 7.36). At its minimum development, the entrenchment is an underfit in the floor of a phreatic passage (Fig. 7.35A). This suggests that regional phreatic flow has been re-routed and the passage now handles only local epikarst drainage. T-form or keyhole compound passages as in Figure 7.35B suggest that the same river has always occupied them, switching from phreatic to vadose conditions. Figure 7.35C and E are

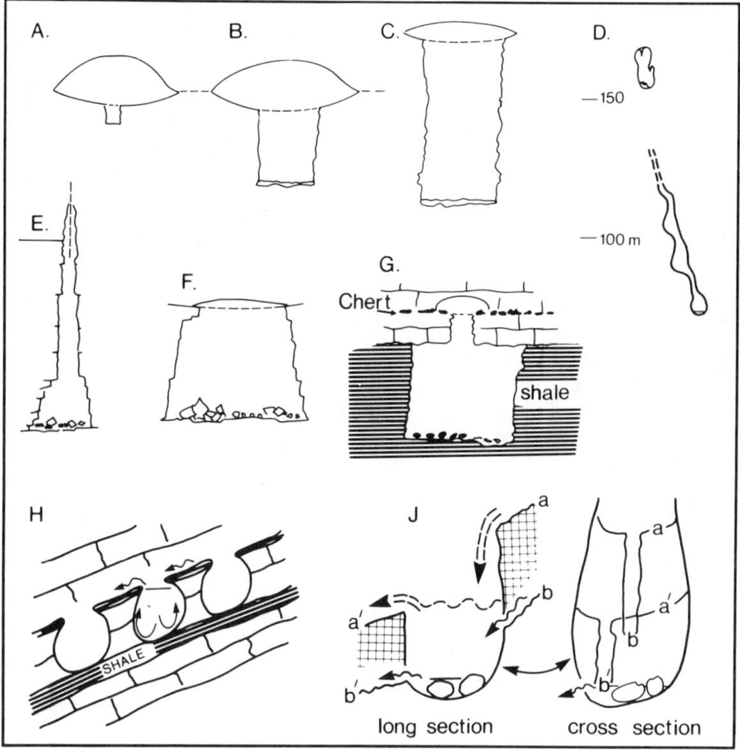

Figure 7.35 Vadose entrenchments: A–F, see text. Cross section D from Grotte Ste. Anne de Tilff, Belgium (Ek 1961). G – McClung's Cave, West Virginia, a typical example of entrenchment from an initial phreatic passage in limestones down into weaker shales. H – purely dissolutional stream potholing in limestone arrested by a basal shale band, and (J) a large pothole eliminated by underfit stream entrenchment which destroys its plunge. From Ford (1965).

Figure 7.36 *Top.* Vadose entrenchment; note dissolutional scallops in left and centre frames. *Bottom.* Basal undercutting to create trapezoid or breakdown forms. Photos by A. N. Palmer, A.C. Waltham and Ford.

characteristic of drawdown vadose morphology; entrenchment predominates but the phreatic part was first and fixes the position of the trench. There may be multiple entrenchments, where each floor represents a former base level (Fig. 7.35D). Single trenches as deep as 100 m are known; they are canyons with a roof.

Many entrenching streams are able to establish long profiles that in fluvial geomorphology would suggest rough equilibrium. The channel becomes armoured by bedload moved only during floods. This encourages channel widening. Cave walls are undermined, and large blocks collapse from them into the channel which deflect flow to the opposite wall, where the process is repeated. By the mixture of undercutting and breakdown a trapezoid cross-section of stable width is achieved (Fig. 7.35F). This is a common form.

Although all categories of vadose channels can be created by solution alone, mechanical erosion (corrasion) is often important (Newson 1971). This is attested by innumerable instances where the entrenchment is largely or entirely in underlying insoluble rocks (Fig. 7.35G). Gouffre Pierre St Martin, one of the world's deepest systems (Table 7.2) is a celebrated example.

Where channel gradients are steep and the rock is hard, stream potholes (French – 'marmites') may develop. These occur in all strong rocks, being drilled when grinder boulders are trapped and spun (rock mills) in any small hole in the channel bed. However they are most frequent and display the most regular form in hard limestones, dolomites and (pre-eminently) in marbles because dissolution in the swirling water reinforces the grinding, and may replace it entirely (Ford 1965, Fig. 7.35H). To maintain and deepen a pothole a minimum plunging force (height times discharge) is necessary. In some caves this is lost, a narrow entrenchment being cut through the pothole so that it is left abandoned in the wall (Fig. 7.35J). There are two successive flights of eliminated potholes in Swildon's Hole, England, and a third flight in the active channel is in the process of elimination. This is apparently an effect of changes in the dominant discharge.

Meandering channels in vadose caves

Three distinct types of meandering channels occur in vadose caves. The 'intrenched' meandering canyon develops where a waterfall (a knickpoint) recedes along a meandering channel. The canyon has simple vertical walls and meanders only in plan view. This type is common where strata are flat lying and there is a deep vadose zone with many invasion streams.

'Ingrowing' meandering canyons are created by meandering entrenchment down into bedrock. There is no waterfall recession. As the channel is cut down it migrates forward (Fig. 7.37A). Ingrown meanders are common in surface canyons (e.g. the famous 'goosenecks' of the San Juan River, Colorado), but they achieve their supreme morphological development in well bedded limestones in caves. Many examples are several tens of metres deep and some kilometres in length, but too narrow for an explorer to pass through most parts.

Smart & Brown (1981) studied this type of meander in the Burren, Ireland and at Waitomo in New Zealand. The most striking result is that the relationship between meander wavelength and channel width is the converse of that found in other meanders, alluvial or bedrock (Fig. 7.37C). In the examples they measured channel width increases downstream (the standard

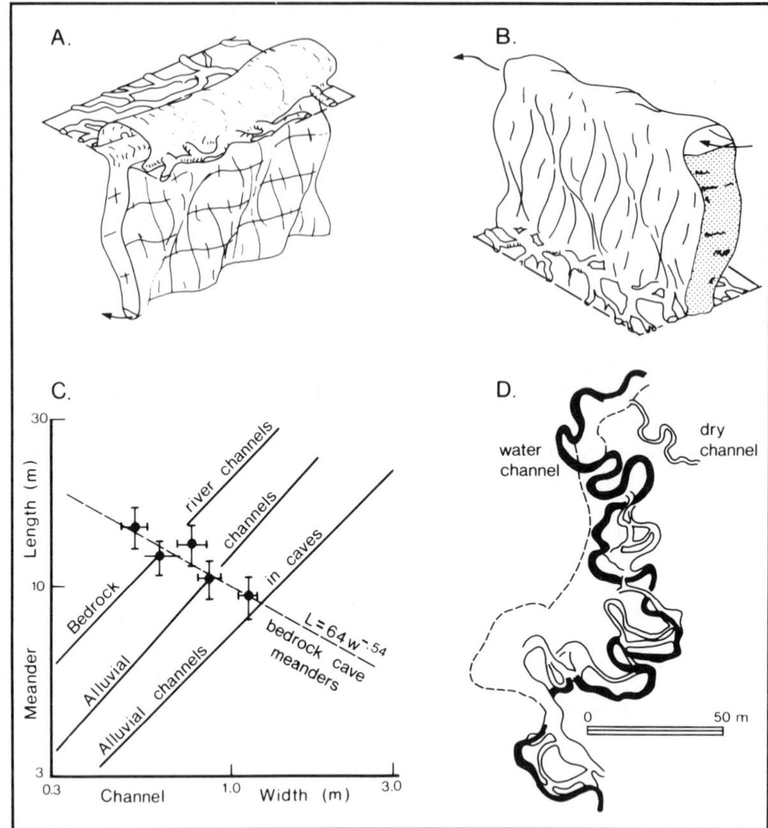

Figure 7.37 Meandering channels in caves. (A) bedrock meanders entrenching below an initial phreatic passage, and (B) paragenetic passage meandering propagating above an initial phreatic passage. From Ewers (1982). (C) Entrenched (vadose) bedrock meanders in caves compared to alluvial channels and surficial bedrock channels. Modified from Smart & Brown (1981). (D) part of the composite (bedrock and alluvial) channel pattern in Hurricane River Cave, Arkansas, USA.

response to increased discharge from addition of tributaries) but canyon height diminishes as the water table is approached.

Alluvial meanders develop where the channel is formed in sand, gravel, etc. on a broad passage floor. Deike & White (1969) showed that examples in Missouri caves behaved morphometrically in the same manner as all surface channel meanders (Fig. 7.37D). There is one major distinction, however; because the rock is soluble the alluvial channel may pass smoothly into the wall (becoming a bedrock channel) and back out again. Where there is slow downcutting (or paragenetic rise) of the channel floor this creates tapered fins of rock (bedrock point bars) projecting into the passage. They are another channel form that is rarely or never found at the surface.

Vadose shafts

Shafts created by falling water are known up to 400 m in depth. Many of the deepest caves are of the invasion type and consist of spacious vertical shafts linked by short sections of constricted meander canyon (Fig. 7.17).

The form of shafts ranges between two extremes. The first is that created by a powerful waterfall. The water mass itself will tend to create a simple circular or elliptical cross-section for its fall but this is often modified. Breccia is swept out of any guiding fault, to form a parallel wall shaft. A plunge pool undercuts the walls which, therefore, tend to display irregular taper rather than parallelism near the base. Spray at all levels attacks weaknesses to produce local block fall. Many shafts are highly irregular in form as a result of these effects.

The other extreme is that of the *domepit* created by a relatively slow and steady flow. This may occur as leakages at the base of the epikarst (Fig. 5.17) or below point recharge depressions (Fig. 9.21), and is able to attack drained joints in the invasion vadose zone. In the ideal condition the flow is never big enough to detach and fall free in the vertical plane. Instead, it is retained against the rock by surface tension. It disperses radially from an input point and carves a set of solutional flutings down the walls. The pit has a symmetric dome at the top (where the first dispersion occurs) and is circular below. This kind of pit is best developed where strata are flat-lying and joints are few and with high resistance. This is the case in the Mammoth Cave area of Kentucky, where the form was first analysed (e.g. Merrill 1960). There is a sandstone caprock there which functions as an additional regulator to maintain a steady, filming flow of aggressive water.

Many shafts show a mixture of the two forms, with waterfall features down the fall line and fluting of farther parts of the perimeter (Fig. 7.38).

Erosional scalloping by solution and sublimation

Erosional *scallops* are spoon-shaped scoops (Figs 7.36, 7.39). They occur in packed patterns so that individuals are overlapping and incomplete. They are common on walls, floors and ceilings in caves. Inspection shows that they are smallest where flow is fastest e.g. in a venturi. Measurement reveals that their length is log-normally distributed, usually with relatively little statistical dispersion. Most scallops are 0.5–20 cm in length but they range up 2 m. Width is ~ 50% of length. In the right conditions it is evident that patterns of scallops of characteristic length extend to colonize all available surfaces. They are the stable form of these surfaces in the prevailing conditions.

Many scallops are strongly asymmetrical in the direction of flow, the perimeter being steeper at the upstream end i.e. facing downstream. '*Flutes*' (Curl 1966) show the same asymmetry and are really scallops of infinite

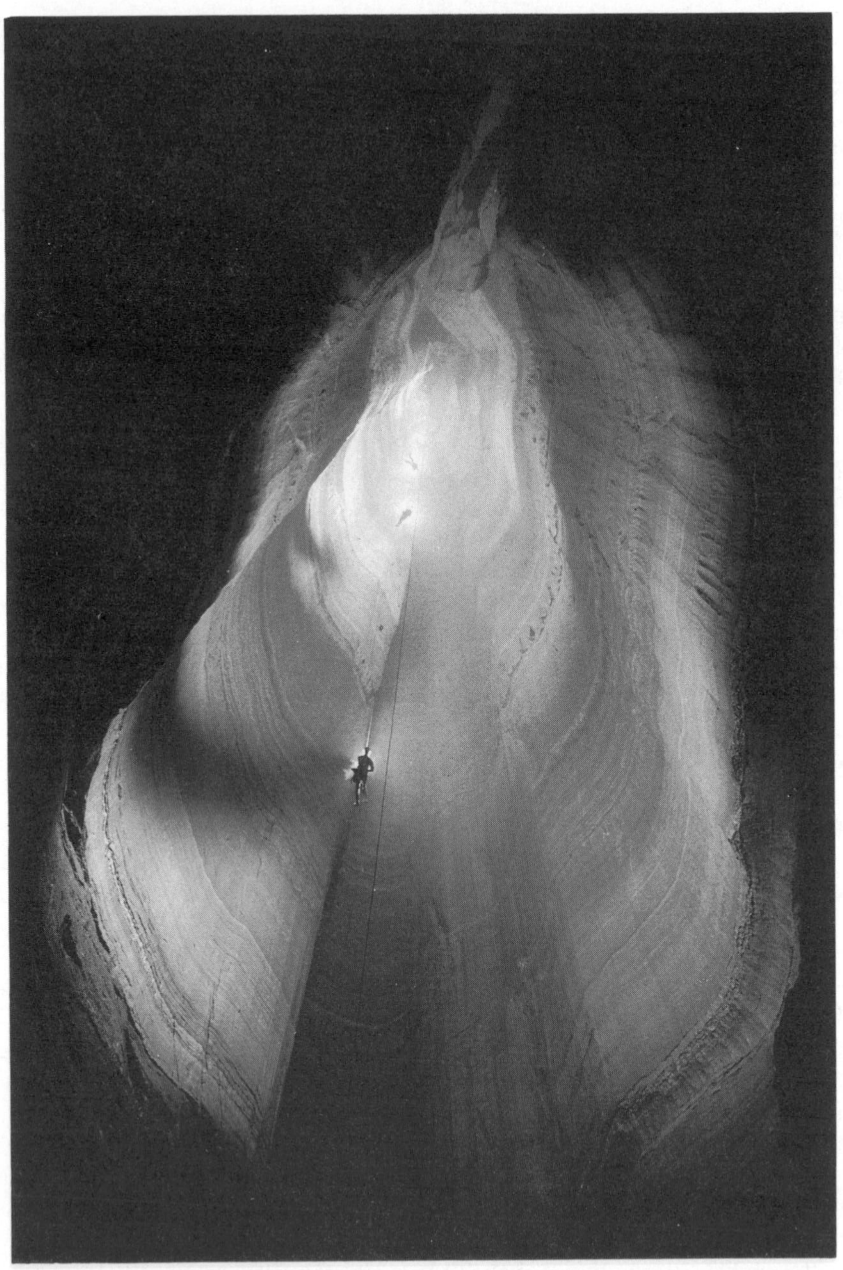

Figure 7.38 Fantastic Pit, Ellison's Cave, Georgia, USA. This is a beautiful example of a fluted domepit. It is 190 m deep, the deepest vertical shaft in the USA; the tie-off point for the rope was 178 m above the floor. The cave is in Mississippian Bangor and Monteagle formations. Photo by A. N. Palmer.

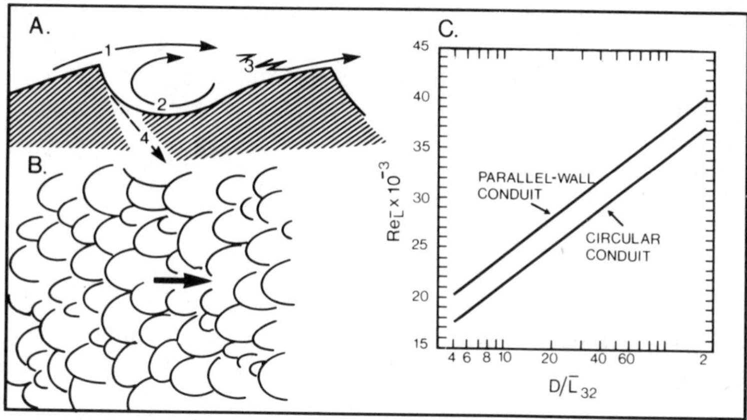

Figure 7.39 Dissolutional scalloping. A. Section through a scallop: (1) the saturated boundary layer detaches; (2) turbulent eddy = locus of maximum dissolution; (3) diffusion, mixing and reattachment; (4) course of the steepest cusp following further dissolution (the scallop migrates downstream). B. The characteristic appearance of a fully developed scallop pattern on a surface; the entire surface is occupied and individuals overlap. C. The predicted relation between Reynolds Number and the ratio, conduit width or diameter (D) to Sauter mean scallop length, L_{32}; from Curl (1974).

width; they are rare. In relict caves, scallops and flutes are potentially important indicators of both the direction and velocity of paleo-flow.

Scalloping is common in limestone, gypsum and salt caves. However, not all limestone regions display it and it is quite rare in dolomite caves. This is because uniform rock grain size is necessary for good scalloping, and a lack of heterogeneities such as insoluble fragments or open pores. Many limestones and dolomites are too heterogeneous to accept scalloping, for example reef rocks.

Scalloping also develops where wind blows over old, well-densified snow or over ice in a glacier cave. It even develops (but without the steeper upstream cusps) on infalling iron meteorites! In these cases the process is mass transfer by sublimation of homogeneous crystalline material. In the karst rocks it is the same, but with solution substituting for sublimation.

Curl (1966) proposed an elegant theory of scallop and flute formation in which there is detachment of the saturated boundary layer at a specified Reynolds Number (Eqn 5.11), occurring in the subcritical turbulent flow regime. Detachment permits aggressive bulk fluids to erode the solid rock directly (Fig. 7.39). The frequency of detachment increases as velocity increases, thus reducing the erosion length available to each individual scallop. Scallop length will be inversely proportional to fluid velocity and to fluid viscosity. Air has a much lower viscosity than water; for a given velocity, scallops are longer (bigger) in ice and snow than on limestone, etc.

Curl's theory has been quite exhaustively tested and confirmed with laboratory simulations and in the field (Goodchild & Ford 1971, Blumberg & Curl 1974, Gale 1984, Lauritzen *et al.* 1986). Mean paleovelocity (v) in a channel may be computed by obtaining Re_L from Curl's graph for circular and parallel wall conduits (Fig. 7.39C) and substituting the value in the equation

$$\mathrm{v} = \nu \frac{\overline{Re_L}}{\overline{L}_{32}} \tag{7.1}$$

Where $\overline{Re_L}$ is a Reynolds No. for scallops, based on the fluid velocity at a distance from the wall equal to $\overline{L}_{32}$ and ν is kinematic viscosity (Table 5.3). $\overline{L}_{32}$ is the Sauter mean length of scallops

$$\overline{L}_{32} = \frac{\Sigma L_i^3}{\Sigma L_i^2} \tag{7.2}$$

L_i being the greatest length (in the direction of flow) of the *i*th scallop. This is used to suppress the statistical significance of the sub-population of very short scallops that occurs in many scallop distributions because of bedrock inhomogeneities.

An alternative, direct calculation for a compromise between circular and parallel wall cases given by Curl is accurate to ~ ± 15%.

$$\mathrm{v} = \frac{\nu}{\overline{L}_{32}}[55 \ 1\mathrm{n} \ (D_\mathrm{h}/\overline{L}_{32}) + 81] \tag{7.3}$$

D_h = hydraulic diameter (four times cross-section area divided by length of the wetted perimeter).

The concept of dominant discharge is important throughout hydrology and fluvial geomorphology because this is the flow assumed to be responsible for channel width in alluvial rivers. It is difficult to establish and may not apply to bedrock channels. In well-scalloped cave passages, however, we may think in terms of one or several 'scallop dominant discharges', i.e. those discharges that create the one or several Sauter mean scallop lengths measured. In examples from some Norwegian caves, Lauritzen *et al.* (1983, 1985) have shown that the modern scallop dominant discharge is approximately equivalent to the annual snowmelt flood, i.e. to the upper 5% of the flow regime, and some three times larger than mean annual discharge.

At very high velocities sand may be entrained as suspended load (Fig. 8.2). Its abrasive effects then may override dissolutional effects so that scallops become highly elongated and polished, resembling 'flute markings' obtained by Allen (1972). This will only occur where there is an abundant supply of hard (i.e. silica) sand and thus is rare in caves. Excellent examples are seen at narrow points in the great gallery of Niaux Cave, Pyrenees, which was subjected to violent glacial melt floods.

Corrosion notches, bevels and facets

Scallops are created by an accelerated erosion mechanism requiring fast, turbulent fluids. This survey of solutional forms in caves is concluded by going to the other extreme and considering a special family of features developed because the water is nearly static, so that the slightest fluid density gradients may establish sharply delimited zones of accelerated erosion.

Corrosion notching is illustrated in Figure 7.40. In a standing pool heavy solute ions and ion pairs sink, driving a cellular convection that carries fresh H^+ ions to the walls at the water surface. A sharp notch is dissolved there, tapering off very steeply below the waterline.

This is the normal form. It is widespread in limestone caves, e.g. at siphon pools where it develops in low stage conditions (Cocean 1975, Serban & Domas 1985). We have seen examples where the notching is as much as 1 m deep, though this is exceptional. It can signify paleo-water table levels very precisely. Some times there are several notches one above another.

This notching becomes much larger in the foot caves of karst towers abutting alluvial floodplains (Figs 7.40B & 9.37). Notching at the seasonal flood level and extending several metres into the rock is common. Because such notches are more extensive and create a flat roof regardless of geologic structure, they may be termed *corrosion bevels* (German – 'laugdecke' of Kempe *et al.* 1975). In examples in some karst towers of southern China, the bevelling extends throughout the length of state 4 (water table cave) passages through them (Fig. 7.40C). These caves have lost their prime hydrologic function because most water now discharges around the towers on the floodplain.

The greatest corrosion notching occurs in the hyper-acid conditions of some paragenetic massive sulphide emplacements, e.g. at Nanisivik mine, Baffin Island, where the greatest notch is ≥ 400 m wide, 1 m deep and filled with syngenetic layered pyrite (Figs 7.31 & 7.40D).

The first theoretical analyses of density gradient corrosion notching were made by Lange (1968) who supposed that, beneath the waterline, the corrosional cut would taper away in a smooth exponential function. Kempe *et al.* (1975) suggested on the basis of physical simulations and examples in gypsum caves of southern Germany, that the taper should be linear. They measured 1–3 mm thick density currents descending the natural gypsum

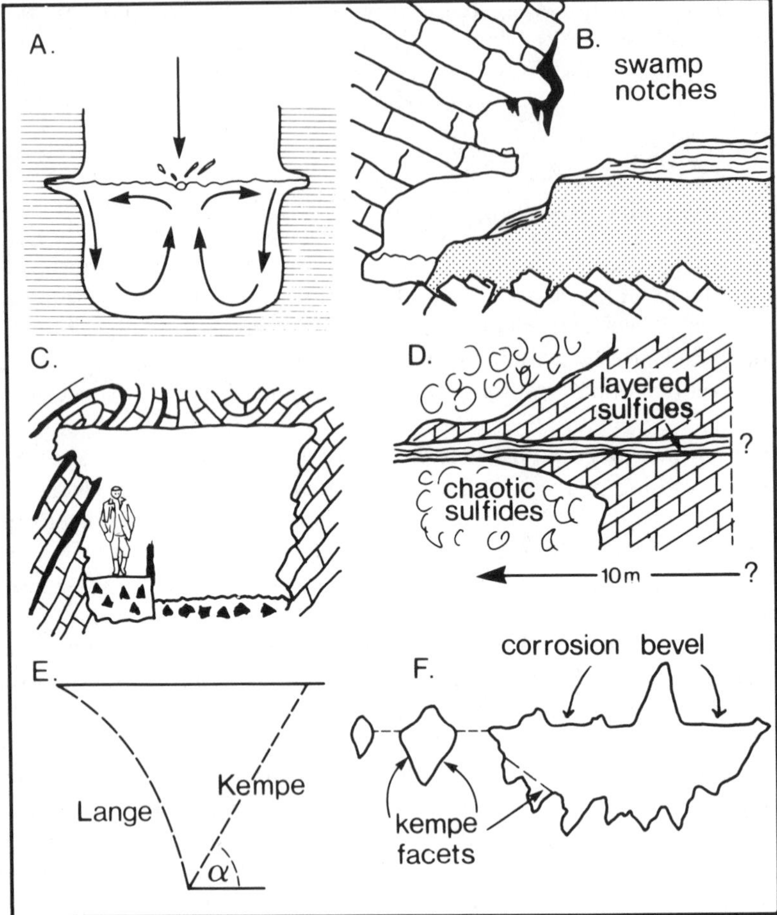

Figure 7.40 A. Corrosion notching at the surface of a pool. B. Foot cave or swamp notches in tower karst. C. Fung Kui Water Cave, Guangdong, China. D. Corrosion notch filled with layered pyrite, Nanisivik zinc/lead mine, Baffin Island, Canada. E. Corrosion taper; the models of Lange (1968) and Kempe *et al.* (1975). F. Na Spicaku Cave, Moravia, Czechoslovakia; corrosion notching is sharpest where passages are largest.

walls, and calculated velocities of ~ 0.5 cm s^{-1}.

These linear corrosion surfaces Kempe *et al.* term 'facets'. They are rare in our experience, presumably because the very still conditions and stable rate of solvent supply, etc., that they require are not often achieved. The sharper 'notch' develops instead. However, they are reported in a number of gypsum caves. In the less soluble limestone, Na Spicaku Cave (Fig. 7.40F) and Ochtinska Cave in Czechoslovakia are spectacular examples. Retreat of a Lange or Kempe facet under stable water table conditions yields a *corrosion bevel.*

Condensation corrosion facetting and runnelling

The condensation of acidic vapour above the water table in hydrothermal caves may cut broad, curving corrosion facets into cave walls or ceilings (Cigna & Forti 1986). The forms appear to be a kind of large, shallow dissolutional scalloping. They are oriented, indicating that there is significant flow of the air upwards or outwards in the cave. Szunyogh (1984) has modelled cupola-form solution pockets as condensation corrosion cells, as noted.

In Carlsbad Caverns and neighbouring caves limestone bedrock plus calcite and gypsum speleothems are facetted by the flow of vapour across them. Small portions of the dissolved speleothem mass are precipitated again at sheltered downwind edges of the facets or in other lee positions. Facetting of speleothems indicates significant changes in relative humidity and in the abundance of reactive gases (chiefly, CO_2 and H_2S) at the site that may be related to episodes of geothermal activity, to climatic changes in semi-desert regions, or to more mundane effects such as the establishment of a bat colony upwind. Condensation corrosion is quantitatively important in semi-desert cave areas of the USSR (A. Maltsev, personal communication, 1986).

Preferred condensation occurs directly above steam vents in thermal water caves and on rock projections in the entrance areas of common caves. This normally runs off to create small, sinuous or irregular wallkarren, as described in section 9.2. Such condensation may considerably enlarge small vertical caves in major cliff faces such as the scarps of the Crimea, USSR, where warm external air is chilled when passing through during the summer season (V.N. Dubljansky, personal communication, 1986).

7.12 Breakdown in caves

Jagged surfaces of rupture in walls and roofs and piles of angular rocks that have fallen from them are found in a majority of caves (Fig. 7.41). They constitute the third basic morphology seen underground, modifying or replacing previous phreatic or vadose solutional forms. This morphology is termed *breakdown* or *collapse* by English-speaking speleologists, *incasion* by many European specialists.

The proximate cause of all breakdown is mechanical failure within or between rock beds or joint-bounded masses. The summary of cave breakdown that follows is also applicable to cliff faces exposed in surface karst, e.g. at stream sinks, pinnacles, towers, etc.

The load on a point in a rock mass may be expressed simply as

$$p = \rho g h \tag{7.4}$$

Figure 7.41 Ceiling breakdown occurring at a passage junction. Photo by A. N. Palmer.

where ρ = rock density, and h = thickness of the rock mass or height of the cliff overhead (c.f. Eqn 5.20). The distribution of this load, as a stress field about a cave cross-section, is given in Figure 7.42A. A tension dome is created in the rock above the passage. Its height is determined principally by passage width. Rock in the dome is subject to sagging and the overlying weight is transferred to the rock adjoining the passage walls, greatly increasing the stress there. There is also tension in the floor, but in natural caves (as opposed to mined cavities) this is of little significance.

The largest cave breakdowns are failures in the tension dome. These are most regular in form where strata are well-bedded and horizontal. As a consequence, this is the situation used in basic analysis (e.g. Davies 1951). Bonding across bedding planes is presumed to be much weaker than strength within the beds; therefore, the beds sag elastically away from each other. Each bed then can be considered to function as a separate beam if it extends the full width of the passage without any fracturing across it, or as a cantilever where it is much fractured (e.g. by a wide central joint) or does not extend the full width. In practice, fractured spans are often much stronger than simple cantilevers. Cliffs at the surface most often fail as cantilevers.

Mechanical rupture and fall of the bed will occur where a critical span is exceeded for a given thickness and strength, and vice versa. A simple equation for a beam is

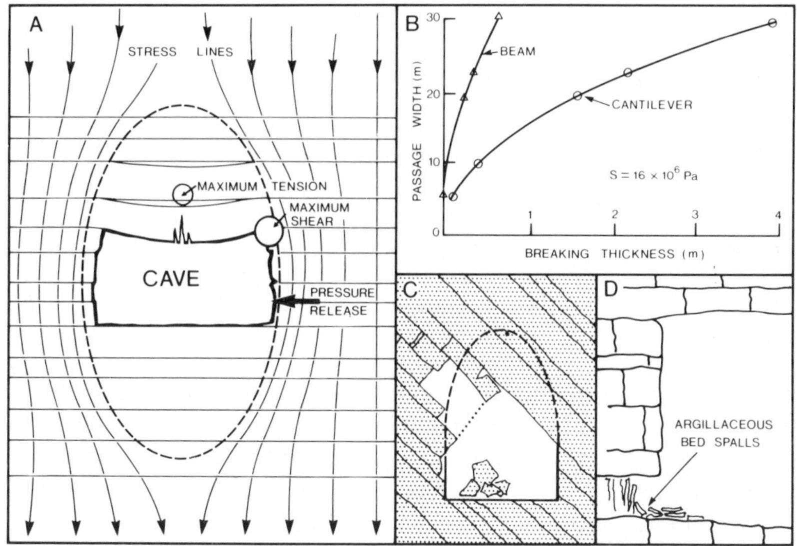

Figure 7.42 A. Distribution of stress lines around a cave and tension dome. B. The relationship between passage width (or span) and the breaking thickness of sample limestone beds; bending stress (S) ~ 16×10^6 Pa. After White & White (1969). C. Deformation of the theoretical breakout dome where stratal dip is steep. D. Pressure release spalling where load is maximum e.g. at a pillar or the foot of a wall.

$$h_{\text{crit}} = \frac{\rho\, \iota^2}{2S} \tag{7.5}$$

where ι = span of the beam (passage width), and S is bending stress (White & White 1969). For a cantilever

$$h_{\text{crit}} = \frac{3\rho\iota^2}{2S} \tag{7.6}$$

i.e. cantilevers are much weaker. Rock density is used here as an approximation for strength: in mining practice, triaxial compressive test values are substituted for it (Brady & Brown 1985).

Bending stress is defined as

$$S = \frac{Mc}{I} \tag{7.7}$$

where M, the maximum moment of the beam, is given by

$$M = bh\rho\iota^2 \tag{7.8}$$

(b = width of the beam, i.e. normal to ι and h)

$$c = \frac{h}{2} \tag{7.9}$$

and I, the moment of inertia of the beam

$$I = bh^3 \tag{7.10}$$

Figure 7.42B illustrates critical spans (passage widths) or breaking thicknesses for some West Virginia limestones whose value of S at failure is approximately 16×10^6 Pa. Medium bedded carbonates of this strength can support roofs up to 20 m wide before failure where they are unbroken beams, and thick bedded carbonates may extend 35 m. There are few cave passages wider than 35 m that do not exhibit at least some breakdown in the tension dome. Where joints are prominent and open in caves, so creating cantilever situations, it is seen that beds as massive as 5 m thick will fail at or before 30 m width.

Failure in the tension dome tends to progress upwards, one or a few successive beds falling at one time. The breakdown stabilizes when the span of the bed newly exposed is less than the critical width for its thickness.

The proportion of failure in the tension dome (partial, or completed to the upwards limit) and its geographical extent varies greatly within and between caves. There may be no breakdown. For example, it is negligible in much of the Big Room at Carlsbad Caverns (Fig. 7.43) although the room is often wider than 50 m. This is because it is located in massive reefal rock. Where the wide passages extend into adjoining backreef beds, breakdown promptly occurs.

Where cavernous strata are generally horizontal and medium- to thick-bedded, it is common to find collapse domes at particular places only. They tend to develop at passage junctions or where a prominent joint intersection creates a cantilever. At the extreme there is partial or complete failure throughout the length of a passage. This is characteristic of most large passages in gypsum (because the rock is weakened by hydration stresses) and in many caves in medium- to thin-bedded limestones. In the Mammoth–Flint Ridge–Roppel System of Kentucky wide passages that are oldest display the most continuous breakdown, indicating that time plays a role.

Where strata dip steeply the same tension dome with a vertical orientation exists in theory (Davies 1951) but it is evident that up-dip walls and roofs are the more unstable, creating asymmetric breakdown (Fig. 7.42C). Here, the adhesion across bedding planes and joint faces (neglected in the analysis given above) becomes an important variable. In mines, adhesion will be measured by laboratory or *in situ* shear box testing (Brady & Brown 1985).

Tension dome breakdown that is completed at a given time may not be stable over the thousands to millions of years that the cave may then survive, especially if it is in a vadose state. Infiltrating waters drain preferentially towards domes and, if chemically aggressive, will attack the rock, reducing S values and converting beams to cantilevers, etc. This renews the process of upwards failure of rock, termed *stoping*. Many karst terrains contain deep breccia pipes or geological organs (section 9.13; Quinlan 1978) that are the debris-filled products of long sequences of tension dome formation and collapse. Frequently, they extend into non-karstic cover strata. Where the latter are mechanically weak but elastic (e.g. clays, many shales), it is common for the progressive fragmentation to be replaced by cylindrical down faulting *en masse*. Stoping upwards has propagated through more than 1000 m of cover rocks at sites in Canada and the USSR.

In mines, rock bursts (explosive release of stress) sometimes shatter walls and floors. This is unknown in natural caves because the slow rate of their excavation permits less violent adjustments to the stresses being imposed. However, it is common to see pressure release spalling of particular beds exposed in cave walls (especially at pillar sites such as passage junctions), or at or near the bases of limestone or dolomite cliffs (Fig. 7.42D). The spalling beds are usually the most argillaceous and/or fine-grained in the sequence. The spalled fragments display platy or conchoidal fracture. Such basal spalling undermines upper walls, widening the tension dome, and so may induce a general collapse at the site.

Davies (1949), Gèze (1964) and others have suggested that continued spalling parallel to the stress field surrounding initial, subcircular conduits created in the phreatic zone ensures that they will be enlarged with a circular to sub-circular cross-section. However, we believe such stress field control to be of lesser significance than local structures and lithologic effects, as noted in section 7.11.

White & White (1969) divide breakdown fragments into three categories:

(a) Block breakdown: rock fragments consisting of more than one bed remaining as a coherent unit.
(b) Slab breakdown: rock fragments consisting of single beds.
(c) Chip breakdown: rock fragments derived from the fragmentation of a bed.

Single blocks with volumes greater than 25 000 m^3 are known, e.g. from a reef–backreef junction at Carlsbad Caverns. Slab breakdown is predominant in thin- to thick-bedded, horizontal strata. There are many forms of chip breakdown because here petrological properties will determine the shape; splinters, flakes, plates, arrowheads, cubes, and saucer-shaped conchoidal fragments are amongst the more common forms.

Mechanical failure is the proximate cause of all breakdown and cave genesis is the ultimate cause. Between these extremes, other factors determine the sites and timing of the breakdown events. Probably, the three most important factors are: (i) draining of phreatic caves. This removes the buoyant force of water, which may effectively increase the load by 30% to 50% in different karst rocks (see section 9.5): (ii) vadose streams, which overwiden a passage beyond its beam or cantilever width limits, either locally at meander undercuts and passage junctions, or generally along a passage that becomes alluviated, (iii) aggressive vadose seepage waters weakening the roofs, as noted above. In very well-aerated caves dissolution by water condensed onto walls may produce much slab and chip breakdown.

In many caves it is suspected that major falls may have been triggered by earthquakes. In the headwater passages at Castleguard Cave, Canada, it is evident that the addition of 200–400 m of glacier ice overhead induced pressure release spalling (Schroeder & Ford 1983): probably, this was a factor in caves in many formerly glaciated terrains. Gypsum precipitating within limestone from SO_4-enriched seepage waters causes widespread chip and slab breakdown at Mammoth Cave and in many other caves around the world; salt plays the same role in some Nullarbor caves.

Frost shattering and frost pockets

Frost shattering is an effective agent of breakdown in limestone and dolomite caves in cool regions. Where there is only one entrance that admits or discharges thermally significant quantities of air or where such entrances are far apart, frost penetration is limited to an entrance cold zone of vigorous temperature fluctuation over the year. This zone is normally a few tens to hundreds of metres in length and is defined by relaxation lengths X_0 (Wigley & Brown 1976) where

$$X_0 = 100D^{1.2}v^{0.2} \tag{7.11}$$

D is passage diameter (m), v is the mean air flow (m s^{-1}). Passage dimension is seen to be the most important control. Effective frost action normally is limited to the first one or two relaxation lengths.

The shattering zone will be longer or frost action more severe where the cave has two or more large entrances at different elevations, as is the case of many relict caves in mountainous regions. The frost debris produced are normally varieties of chips. Slab breakdown is common close to entrances.

Frost shattering on cliffs may create a shallow, shelter-type of cave termed a *frost pocket*. These develop in all stronger rocks where there are wetted fractures, but they appear to be most abundant in the carbonate rocks because they are wetted via solutionally enlarged openings. Solution is the trigger process for larger scale frost action about the point of groundwater emergence (Schroeder 1979a & b). The pockets often expand upwards into the tension dome that they create, forming a rounded arch that may be metres to tens of metres in height. At a distance these cannot be distinguished from the frost-modified entrances to large caves, which has led to many exaggerated reports of solution cave frequency in mountainous regions.

The largest known voids

Great open spaces in caves are termed *rooms* by American cavers and *chambers* by British cavers. Some of the largest that are known are shown in Figure 7.43. Sarawak Chamber, in the Mulu karst, is the greatest. It has a volume of approximately 20×10^6 m^3. Belize Chamber, Salle de la Verna and the Carlsbad Big Room are each in excess of 1×10^6 m^3. Hundreds of rooms are known with volumes of 100 000 to 500 000 m (Gilli 1986).

A majority of these rooms are centred at vadose river passage junctions where undercutting has widened tension domes to produce repeated breakdown. The rivers are able to remove much of the debris in solution, and so maintain open voids rather than breccia pipes. Carlsbad Big Room and some large chambers occurring in hydrothermal caves probably owe most of their excavation to exotic mixing corrosion effects (sections 7.7 & 7.8) and display relatively little breakdown morphology.

Impressive as they are, the greatest cave rooms are minor in scale when compared to the debris-filled volumes of many breccia pipes.

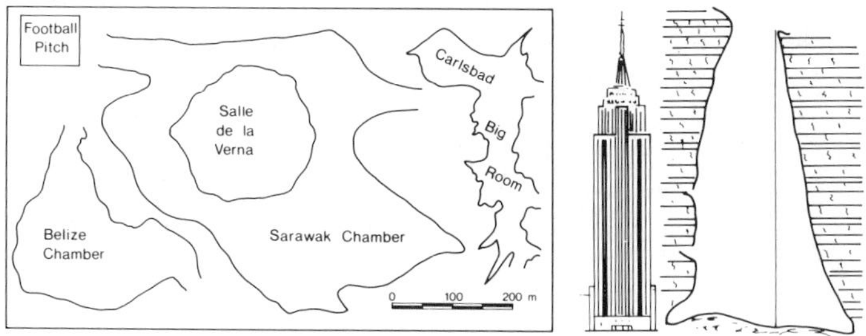

Figure 7.43 Examples of the largest cave chambers or rooms. Sarawak Chamber, Lubang Nasib Bagus (Good Luck Cave), Mulu National Park, Sarawak; Belize Chamber, Actun Tun Kul (Tun Kul Cave), Belize. Big Room, Carlsbad Caverns, USA and Salle de la Verna, Gouffre Pierre St Martin, France. The great shaft of El Sotano, Mexico, is compared to the 430 m Empire State Building, New York.

8 Cave interior deposits

8.1 Introduction

Caves function as giant sediment traps, accumulating samples of the clastic, chemical and organic debris mobile in the natural environment during the life of the cave. They are the most richly varied deposits that form in

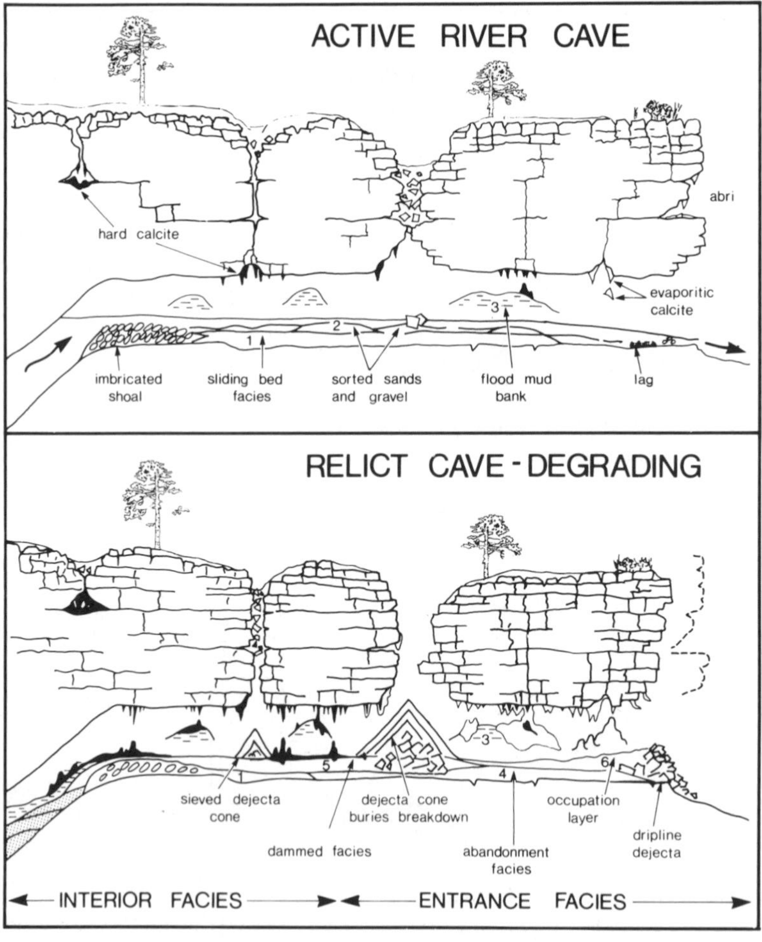

Figure 8.1 Model for the principal categories of clastic and precipitate deposits in hydrologically active or relict caves.

Table 8.1 Cave interior deposits.

A Allochthonous or allogenic			
Clastic	1	Fluvial	– many kinds – *dominant allochthone*
	2	Filtrates	– from seepage. Minor
	3	Lacustrine	– rare
	4	Marine	– beach facies
	5	Eolian	– normally minor except at entrances
	6	Glacial and glacifluvial injecta	– common in glaciated regions
	7	Dejecta, colluvium and mudflows	– normally restricted to entrance areas
	8	Tephric	– volcanic areas; inwashed ash and pumice
Organic	9	Waterborne, windborne, etc.	– scale ranges from spores to tree trunks
	10	Exterior fauna	– sometimes cave-using species – bones, nests and middens, faeces
B Autochthonous or autogenic			
Clastic	11	Breakdown	– mainly by failure; minor thermoclastic
	12	Fluvial	– derived from breakdown or erosion of karst rock
	13	Weathering earths and rinds	
	14	Eolian	– derivatives of 11, 13
Precipitates and evaporites – more than 180 different minerals are known to be generated in caves. Dominant are			
	15	Ice	– as water ice, glacières, frost and glacial injecta
	16	Calcite	– *most significant autochthone*
	17	Other carbonates and hydrated carbonates	
	18	Sulphates and hydrated sulphates	
	19	Halides	
	20	Nitrates and phosphates	
	21	Silica and silicates	
	22	Manganese and hydrated iron oxides	
	23	Ore-associated and miscellaneous minerals	

continental environments and tend to be preserved for greater spans of time than most others do. It is not by chance that a majority of important Lower Paleolithic archaeological sites have been caves.

Cave sediments have received much more study than the cave erosional genesis discussed in Chapter 7, because of their archaeological and paleo-environmental significance and because they can be sampled and analysed like other sedimentary deposits. They are divided into cave entrance plus rock shelter (abris) deposits or facies, and cave interior deposits (Fig. 8.1). This chapter focuses on the latter. Entrance deposits, because of their archaeological importance, can be considered a partly separate subject; see Schmid 1958, Laville 1973, Sutcliffe 1976).

Table 8.1 presents a classification of cave interior sediments. There is a basic division into clastic, organic and precipitated types and into sediment formed outside of a cave and transported into it (allogenic or allochthonous) and that created within a cave (autogenic or autochthonous).

This introduction closes with a warning. Cave sediments can be most complex. The law of superposition (that an upper deposit is younger than a lower deposit on which it rests) is often violated because of shrinkage, slumping, flowstone instrusion, burrowing and other effects. Many facies are diachronous (i.e. they differ in age laterally e.g. Osborne 1984). Variation in rates of deposition can be extreme. Downstream movement of sediment may be much obstructed because breakdown barriers sieve it to varying degrees or dam it entirely for long periods. Undoubtedly, there is much re-working and re-deposition.

8.2 Clastic sediments

Mechanics of aqueous sediment transport

Local breakdown (section 7.12) and allochthonous fluvial sediments are the predominant categories of clastic deposits. Sediment transport in open alluvial channels such as rivers has been intensively studied; e.g. Richards (1982). Vadose caves are broadly equivalent to them (White & White 1968) but are usually rock-bound channels, hence width is fixed during an event, the channels are particularly rough and they usually lack flood plains. Transport of solids in slurries through liquid-filled pipes has been the subject of much engineering experiment (Newitt *et al.* 1955). Phreatic and floodwater caves again are approximately equivalent, but with much higher friction due to their irregularity.

Sediment begins to be moved when the shear stress at the bed (boundary) exerted by the specific stream power exceeds a critical value (Eqns 5.17 & 5.18). A general expression for boundary shear stress is

$$\tau_o = \rho g \, \theta \, \frac{a}{2d+w} \tag{8.1}$$

where ρ = fluid density, θ = channel slope, a is cross-sectional area, d = depth and w = width. The critical value increases with particle diameter and density but is also dependent on its shape, packing among other particles, etc. An approximation for spherical particles is

$$\tau_{crit} = 0.06\,(\rho_s - \rho) g D \tag{8.2}$$

D = particle diameter in mm, ρ_s = particle density.

The greatest particle moved by a given flow is estimated by

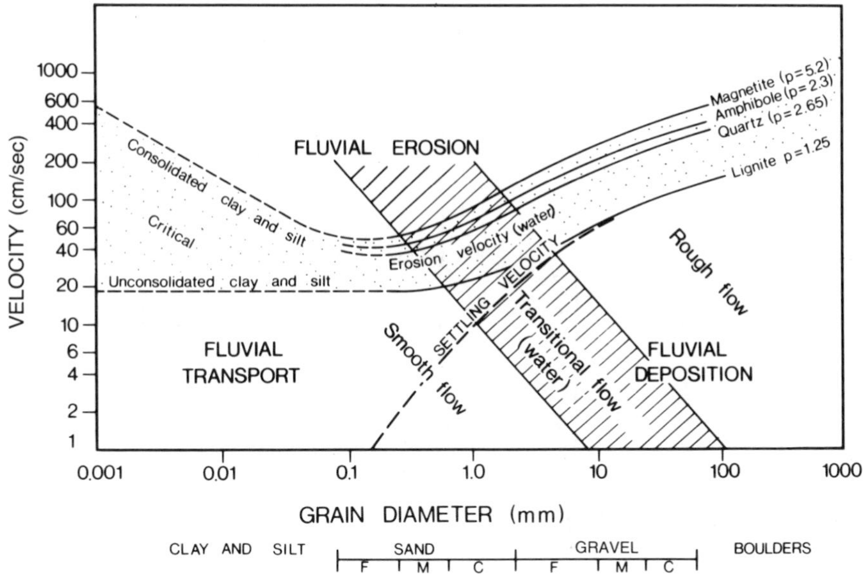

Figure 8.2 Curves showing relations of grain size to critical erosion velocities for flowing water with material of differing densities. Fluvial erosion curves refer to velocity at 1 m above the channel bed. Erosion, transport and deposition regimes are defined. From Sundborg (1956).

$$D_{max} = 65\ \tau_o^{.54} \tag{8.3}$$

τ_o is in kg m^{-2}. Equations estimating bedload transport *en masse* face many problems (Graf 1971). That of Meyer-Peter & Muller (1948), is widely used

$$\text{Rate} = 0.253\ (\tau_o - \tau_{crit})^{3/2} \tag{8.4}$$

For deposition, the terminal settling velocity v_t for a spherical particle is given by *Stokes' law* or similar equations

$$v_t = \frac{1}{18}\ \frac{(\rho_s - \rho)}{\mu}\ gD^2 \tag{8.5}$$

The effects of these relationships in open channels are summarized in Figure 8.2.

For pipefull flow Newitt *et al.* (1955) noted five modes of transport: (a) a slowest mode of rolling grains producing ripples on a stationary bed; (b) saltation of individual grains above the bed; (f) a sliding bed involving at

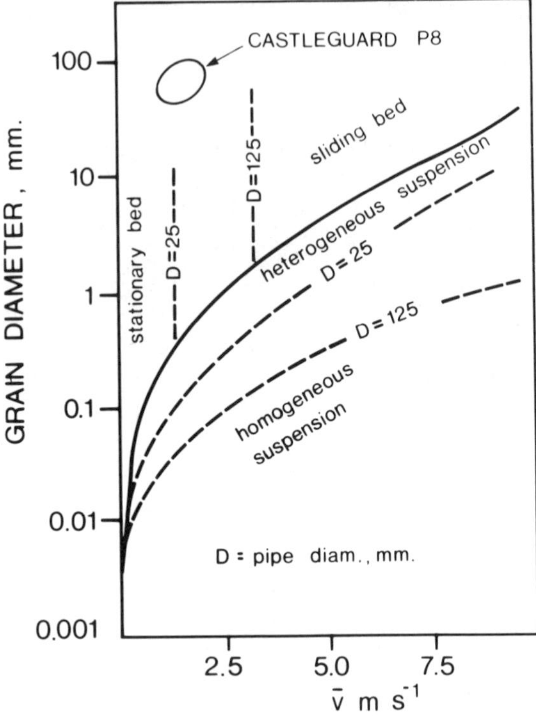

Figure 8.3 Clastic sediment transport regimes under conditions of pipefull water flow, from Newitt *et al.* (1955) with data for Castleguard Cave added. See text for details.

first, the upper part of the bed load but extending to all of it with further increase of velocity; (d) heterogeneous suspension, and (e) homogeneous suspension at the highest velocities (Fig. 8.3). Their experiments were limited to horizontal pipes of circular cross-section, and to grain sizes up to 10 mm. Later experimenters have obtained similar results from larger pipes.

There is no doubt that at least the first four of these transport modes occur in phreatic or flooded caves. There are many instances of pebbles and cobbles being carried straight up vertical shafts and expelled at the top. This represents heterogeneous suspension, at the least. An example from Castleguard Cave is plotted on Figure 8.3; 40–100 mm diameter pebbles of dense limestone ($\rho_s = 2.85$) are carried >7.5 m up a floodwater shaft there. Minimum velocities > 0.80–1.00 m s^{-1} are calculated from the settling laws, in good agreement with mean velocities of 1.15–1.33 m s^{-1} obtained from measured maximum flood discharges. Note that the Castleguard grains plot in the 'stationary bed' domain of Newitt *et al.*'s experiments. This does not invalidate those experiments but does emphasize that grain shape and other factors create great variability in real behaviour.

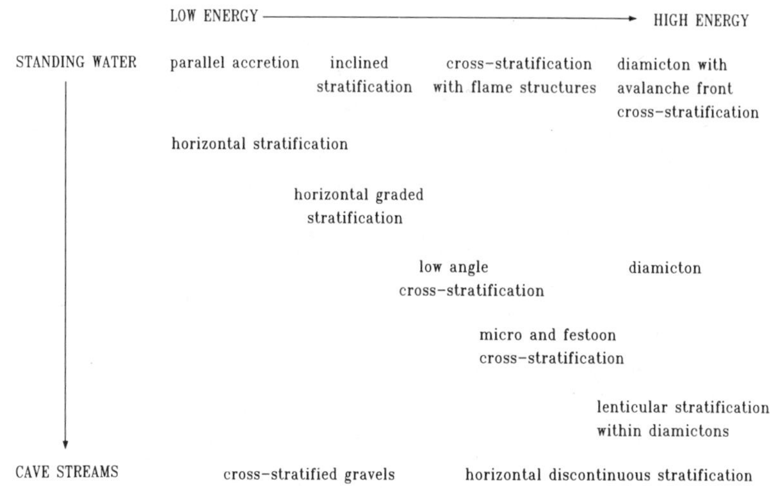

Figure 8.4 Relationships between water flow and depositional energy as expressed in cave clastic sediment structures. From Gillieson (1986).

Figure 8.4 presents a summary of relationships between cave hydraulic conditions and common sediment structures.

Deposits of gravel to boulder-sized material

This is typical bedload in many hilly karst regions if there are extensive allogènic catchments. It is the predominant bedload entering caves in most glaciated terrains, where it is re-worked from till and outwash. It can also be generated from autogenic breakdown.

The extreme of transport is seen in steep floodwater caves such as many alpine systems where the maximum grain size may be determined by minimum passage dimensions. Individual boulders firmly wedged between floor and roof are common.

A facies that is frequently encountered is a poorly sorted to chaotic mixture (a diamicton) of all sizes up to boulders, with patches that are grain-supported and others that are matrix-supported. There is little preferred orientation or other structure. The top of the deposit is an abrupt contact that is often succeeded by well sorted gravel or sand. If in a sequence of deposits it is usually the basal unit. Such deposits in caves we attribute to the pipefull, sliding bed mode i.e. all the mass was in motion, and then deposited simultaneously with sorting only by the dispersive pressure of colliding particles (McDonald & Vincent 1972). Acaroglu & Graf (1968) suggest that this is the equivalent of the antidune mode in alluvial channels. Sliding bed deposits are also recognized in subglacial esker deposits i.e.

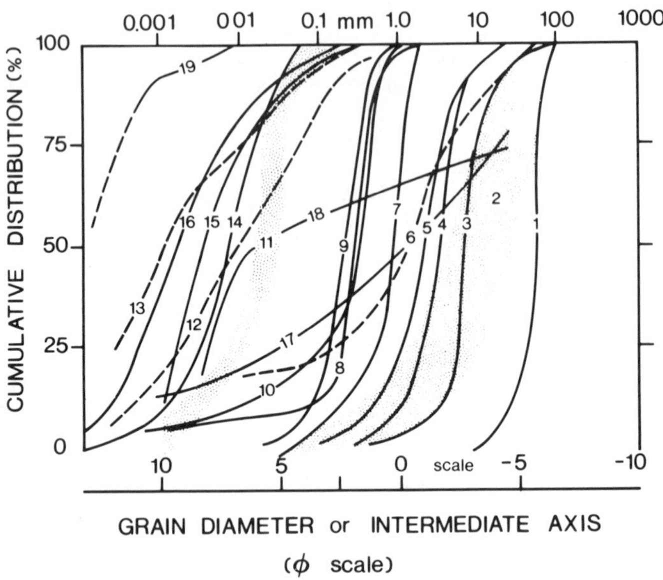

Figure 8.5 Cumulative distribution of grain size curves for 18 different cave deposits plus samples from six eskers. Data from many authors. Details are cited in the text.

floodwater ice caves (Saunderson 1977). Figure 8.5 displays examples of Gillieson's 1986 results (curves 17 & 18) from the very wet caves of New Guinea, and the grain size envelope for six esker samples (envelope 2).

In open cave channels or where pipefull velocities are too low, gravels and boulders move by rolling and sliding. They are deposited as shoals or bars of medium to well-sorted grains, exhibiting lateral, downstream and upwards fining trends. Bar form is best downstream of constrictions such as short siphons. The gravels are often well imbricated. Wolfe (1973) described a paleodeposit in West Virginia where the imbrication indicated flow in the direction opposite to that indicated by solutional scallops on adjoining walls, a consequence of an alluvial stream invading an abandoned phreatic cave.

A distinct facies may occur where a passage is progressively infilled. As the flood cross-sectional area is decreased by filling, velocities increase and thus the deposit coarsens upwards, perhaps terminating with a lag of boulders or cobbles jammed into the roof.

Sands and sand structures

Allochthonous sand is an abundant constituent of cave sediments because of the common occurrence of limestone and sandstone together in basins. The typical sand structures of flumes and alluvial channels (Middleton 1976, Allen 1977) have been reported many times in caves. Sorting ranges from moderate to very good. Fining upwards, laterally or downstream is common, as is coarsening upwards. Sands are often interbedded with silts and clays.

In Figure 8.5, curves 6, 7 and 8 are representative of coarse and medium sands measured in the temperate river caves of Slovenia (Gospodaric 1974, Kranjc 1981); curve 9 is the mean for well sorted sands from the Niaux System, (Pyrenees) where the deposits are proglacial outwash into giant passages (Sorriaux 1982), and curve 10 is sand ejected from a phreatic tube only 67 mm in diameter (Gale 1984).

Suspended load; silts and clays

Silts and clays are the most widespread clastic deposits in caves, as they are elsewhere. Because they are transported in suspension they may coat walls and even ceilings, although most accumulation is on the floors.

Their sources are the most diversified. Allogenic sources include the fluvial and lacustrine, and also filtrates from soils overhead and windborne dust as in Figure 8.5 (curves 14 and 16). There is often a significant autogenic component from the weathering of walls (curve 15) or the winnowing or decomposition of older sediments. Terra rossa at the surface in southern France (curve 12) is winnowed to curve 13 at a depth of 30 m. Mixed allogenic and autogenic fines are often the final deposit in relict caves and are referred to as 'cave earths'.

The clays and silts usually accrete as laminae that are parallel to the depositional surface, which may be horizontal, inclined, vertical or inverted. There may be strong fining upwards within a lamina, which creates a colour change that gives it the appearance of a couplet, e.g. from buff (silt dominant) to grey (fine silt–clay) in many glaciated regions. Thickness of individual laminae are reported to range from ~ 0.2 to >50 mm, but 1–10 mm is most common. Aggregate sections can total many metres.

Some clays lack lamination or fining structure, which suggests that they were decanted from an homogeneous suspension that was steadily renewed. However, this is rare. Lamination and fining (sometimes coarsening) upwards are normal, and represent deposition from waning floodwaters or other pulsed flow.

Typically, silts and clays grade laterally from centres of passages and are finest-grained in deep recesses (curve 19). Floods may repeatedly scour and fill along the main passages but there is only accretion in the recesses which can build up great banks with the most complete sedimentary records. 'Mud mountains' >20 m in height are known in protected places.

Where frequent flooding and draining occur silt and clay banks steeper than ~ 40° may exhibit patterns of rilling (Fig. 8.6); rills are dendritic on slopes between 40° and 70°, becoming parallel where they are steeper (Bogli

Figure 8.6 Some representative cave sediments. *Upper left.* A basal chaotic (probably, sliding bed) layer is terminated by clay at the level of the hand. A sequence of pebbles–silt–clay overlie this and fill the passage to the roof. A later, imbricated fill of pebbles is emplaced in the upper clay by cut-and-fill processes. *Upper right.* Banks of flood silt and clay, Selminum Tem Cave, New Guinea, Photo by A. S. White, by permission. *Lower left.* An 'abandonment suite' of pebbles fining upwards to clays succeeded by silts fining upwards to clays. *Lower right.* Typical flood clays in a backwater passage. Note the dendritic surge marks. Photo by A. N. Palmer, by permission.

1961). These 'surge marks' are depositional, at least in part (Bull 1978a, 1982).

Thin deposits onto vadose cave roofs are often reorganized into stringers and subcircular clusters of clay when the cave dries. They have the appearance of worm trails and are termed *vermiculations* (Bogli 1980). The patterns appear to be produced by a combination of droplet distribution and electrical charge attraction.

In extra-glacial regions clays and silts are normally composed of clay minerals and quartz fines. In glaciated areas carbonate grains typically compose 20–80% or more of the fines. They are re-worked 'glacier flour' produced by basal ice erosion.

Characteristic sedimentary facies and suites

All categories of sediments noted above can be found in common phreatic caves. Allochthonous gravels and even boulders may be transported through a surprising number of deep loops. However, laminated silts and clays usually predominate in the phreatic environment (e.g. Ford & Worley 1977). Where there is a large clastic load paragenetic deposition and roof solution may occur. The paragenetic facies may include sands, granules and even gravels, usually in fining upwards sequences. Laminated fines in small amounts are the chief deposits in thermal water caves and artesian mazes.

The standard suite in gently sloping vadose or shallow phreatic caves is a series of cut and fill deposits with marked lateral fining (Fig. 8.6). There may be a sliding bed facies at the base of any given sequence or this facies may cross-cut remnants of earlier sequences along a central channel.

Where a dam of breakdown is abruptly created in such caves the effects can be highly variable. Most often, there is infilling with parallel beds or delta facies on the upstream side, initially coarse, becoming finer as the barrier becomes increasingly clogged. Downstream, there is re-working or complete scouring of older deposits, plus a varying addition of finer grain sizes passed through the dam or winnowed from its components. If much of the dam is then swept away in a flood, a particularly complex 'pond and sieve, winnow, cut and fill' suite is left.

Where upper passages are progressively abandoned by flowing water distinctive 'abandonment suites' may occur. These are cut and fill sequences with strong lateral fining, that also tend to become finer upwards as the floods become weaker and weaker in their capability. The final deposits are finest at the upstream end (the overspill flood entry point where only suspended load can enter) and coarsen downstream as a result of local reworking. In Figure 8.5, grainsize curves 3, 4 and 5 generalize the progression from perennial channel sediments to the overspill channel and then to the completely abandoned channel in river caves of West Virginia (Wolfe 1973).

Varved clays are rhythmic sequences of laminated silt and clay each appearing as a couplet that is darker and coarser below, finer and lighter above. They form in proglacial lakes, the finer fraction only settling out when the lake is frozen over in winter i.e. each couplet represents one year's accumulation. Thick sequences of varve-like rhythmites are common in caves in glaciated regions. Figure 8.5 curve 11 is the envelope for 14 samples from Castleguard Cave. They may be the predominant deposits, with later channels carrying bedload being cut through them. They represent deposition from pondings beneath or along the flanks or Quaternary glaciers. They have a high carbonate content. It is not yet established whether each couplet is an annual (true varve) deposit or represents some other event; see Bull (1977), Maire & Quinif (1984), Schroeder & Ford (1983), for discussion.

Reams (1968) attributed couplet laminae in caves of Missouri to individual floods generated by thunderstorms.

Other types of clastic deposits

Caves everywhere display cones of dejecta accumulated by a mixture of piecemeal falls and small slurries from steep or vertical openings to the surface (Fig. 8.1). Passages that slope more gently from the surface may channel tongues of colluvium for many tens of metres. This is unsorted debris, usually matrix-supported and with a high organic content. Gillieson (1986) reports the injection of fluidized polymodal mudflow deposits (diamictons) up to 3 km into New Guinea caves.

Glaciers may discharge meltwater and clastic load directly into caves. Typical here is an injection facies; the passage entrance is blocked by wedged boulders and a tail of gravels and sands extends inwards. There are also simpler injecta where an advancing glacier merely bulldozed till or outwash into entrances, filling them entirely.

Caves that terminate in the coastal breaker zone may contain beach sand and shingle deposits. These take the form of berms (barriers) with steep fronts facing the waves and gentle backslopes that can extend inwards for several tens of metres.

Particle shape, changes of shape and size in caves

The shape of a particle is described in terms of form (platy, bladed, etc.), sphericity, and the roundedness of corners (e.g. Sneed & Folk 1958, Cailleux & Tricart 1963). Most studies in caves have been concerned with pebble and cobble sizes. Siffre & Siffre (1961) suggested that limestone pebbles in caves display a tendency to be rounder and flatter than in surface channels, but Bull (1978b) found no systematic downstream changes in the Agen Allwedd caves, Wales. Variation was entirely attributed to original lithological properties. This is not surprising. Caves that offer sufficient accessible lengths of alluvial channel for substantial changes of shape to be expected will be low gradient and likely to contain breakdown barriers that screen out upstream particles and substitute fresh stocks of angular fragments. Excessive roughness of channels also causes much breaking into angular pieces rather than steady attrition over all the surfaces of grains. Figure 8.7 presents an extreme illustration of the problem. Sandstone

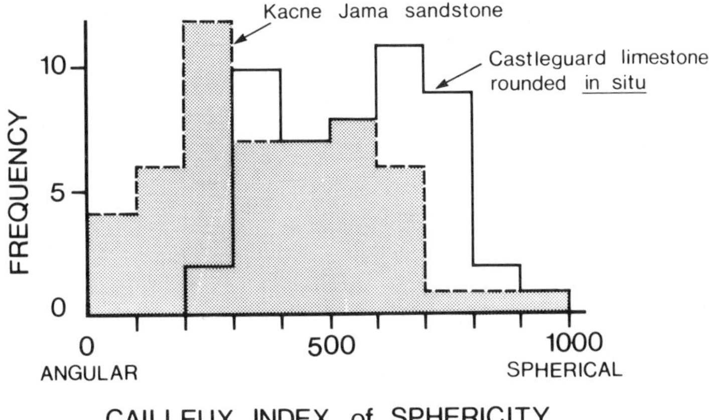

Figure 8.7 The degree of rounding of two pebble samples in caves. At Krizna Jama, Slovenia, sandstone pebbles have been transported through at least 11.5 km of cave passages. At Castleguard Cave, Canada, pebbles are significantly rounder but have experienced zero forward transport, being rounded *in situ*.

pebbles in Kacne Jama, Slovenia, have travelled at least 11.5 km through limestone caves. They display much lower sphericity and rounding than do pebbles of dense, particularly resistant, limestone rounded *in situ* (curve 1, Fig. 8.5) at the bottom of P8 floodwater shaft, Castleguard Cave.

The size of allochthonous pebbles, cobbles and boulders, and their proportion in the total bedload is certainly reduced during transport through caves. Kranjc (1982) obtained a mean diameter loss of 4 mm km^{-1} for the Kacne Jama sandstones, but random samples of all clasts (allochthonous plus autochthonous) yielded no trend there. In big karst basins in West Virginia, Wolfe (1973) found that mean bedload grain size was weakly related to channel length and gradient (Table 8.2) but filtering in breakdown destroyed any trend at the outlet of the longest basin.

The shape of sand–clay size quartz grains in caves has been studied under the electron microscope by Bull (1977) and others. These grains suffer comparatively little damage during underground transport, so the studies determine their origin or provenance. In Agen Allwedd, Bull (1977a) recognized a suite of glacigenic sands modified by fluvial transport and a later suite of diagnetic quartz sands from the local caprock, plus distinct glacial, fluvial or eolian features in the silt and clay grains. Gillieson (1986) recognized two further classes, quartz silts from local soils and from volcanic tephra, in the Papuan caves. Using grain size variations, Bull (1977a) correlated particular groups of silt and clay laminae over distances up to 5 km in Agen Allwedd. In current work he recognizes a fini-glacial loessic phase (component) in capping muds in caves throughout southern Britain (P.A. Bull, personal communication, 1985).

Table 8.2 Sediment characteristics in three West Virginia karst basins. From T.E. Wolfe (1973).

Basin area, km^2	Carbonate outcrop %	Channel length Allogenic, upstream km	In caves km	Mean cave channel gradient	Mean grain size in cave, mm	Max. grain size discharged from cave, mm	Kaolinite: illite ratio in cave clays
12.4	42	2.4	1.6	0.133	62	140	0.890 active 0.540 relict
90	30	21	5	0.033	8	10	1.019 active 0.488 relict
138	23	23	10	0.025	5	<1	0.930 active 0.424 relict

Composition of particles – provenance studies

Many studies have used pebbles and cobbles of some distinctive lithology as an easy but accurate means of tracing sediment sources and flow paths. Normally they will be allogenic.

In fine fractions heavy mineral abundancies may differentiate the sources. Other studies have emphasized clay minerals and their ratios (Frank 1965). Although there is much variation, abundant kaolinite tends to indicate warm conditions either in the source rock or during its weathering. Illite and chlorite are usually the most prominent clay minerals in glaciated regions, increasing montmorillonite suggests a drier climate, etc. Wolfe (1973 & Table 8.2) found a kaolinite : illite ratio of ~ 3 : 1 in Carboniferous sources rocks. This reduced to 1 : 1 in modern sediments entering the caves and towards 0.6 : 1 as they were traced 8–12 km underground. The ratio in abandoned passages also decreased downstream, from 0.6 to < 0.4, suggesting that there was further alteration of kaolinite as the ancient deposits weathered a little in the caves.

Diagenesis of cave sediments

The nature and amount of diagenesis occurring in cave sediments is much less significant than that in original limestones (Chapter 2) and has been little studied. Physical processes include effects such as loading (load casts or deformation), drying out, abrupt re-wetting, etc. Shrinking away from walls is common; it permits later sediment to intrude. Polygonal dessication cracking may extend to depths greater than 1 m. Bull (1975) described 'birdseye' or vesicular structures attributed to organic decay, shrinkage or rapid wetting.

Burrowing by animals (*bioturbation*) is a second physical category. Cave-specific (hypogean) fauna in the interior are small and weak. They do little but disturb the uppermost laminae of soft sediments, which may suffice to ruin their paleomagnetic signals (see below). Burrowing by large mammals such as badgers (and humans) is important in many entrance deposits.

Drying permits oxidation, which extends throughout most well-drained sediments. Periodic wetting encourages further hydrolysis of aluminosilicate minerals and the alteration of clay minerals, as noted. Such chemical weathering takes place in most cave sediments (Gèze and Moinereau 1967), but normally at very low rates.

Chemical precipitation within sediments is the most important process. Calcite is the principal precipitate. Its mode and abundance range from a few loose crystallites in pores to comprehensive cementation that converts the loose sediment into a solid rock of low porosity. Between these extremes, patchy, layered or graded cementation are all common (Frank 1965).

Other precipitates include surface veneers of gypsum and other crystals where enriched solutions in the original sediment have been drawn out by capillary suction as it dried. In parts of the Mammoth–Flint Ridge cave system silts and clays have suffered *evapoturbation* to depths of 1 m or more i.e. gypsum precipitated between the laminae breaks them up. In quartz sand deposits Bull (1976) reports deposition of amorphous silica (crystallites) on grain surfaces, though this will rarely cement them. Nitrate and phosphate minerals also accumulate in cave earths.

8.3 Calcite, aragonite and other carbonate precipiates

Calcite is the principal secondary precipitate in caves. General terms for it include *travertine*, *sinter* and *speleothem*, although the two latter terms may

Figure 8.8 Red Cave, Crimea, USSR, an excellent example of a river passage that is profusely decorated with calcite stalactites, stalagmites, draperies and flowstones. Photo by G. Semyenov and V. Kisseljov.

apply to other cave minerals as well. In most karst regions the calcite mass far exceeds that of all other secondary minerals combined (Fig. 8.8). Aragonite is more widespread than has generally been supposed and is possibly the second mineral in abundance, gypsum being the third. The other carbonates and hydrated carbonates are much less significant; they form small but attractive decorations in a minority of caves.

Lengthier reviews are to be found in Kunsky (1954), Gèze (1965), White (1976), Cabrol (1978), Bogli (1980) and Bull (1983). Hill & Forti (1986) present the most complete account.

Calcite and aragonite crystal growth

In the evaporitic conditions of most cave entrances and seasonally dry caves, much calcite deposition is by the rapid formation of separate crystallites. This creates a soft calcite of globular or irregular clumps of tiny crystals, earthy in texture, with high porosity and often pasty to the touch, at least on first deposition. Hard calcite is precipitated by slow exsolution of CO_2 from solution and becomes predominant in humid cave interiors (Figs 8.1 and 8.8). Some speleothems are composed of sequences of hard and soft calcite, or have soft calcite on upwind sides.

Most hard calcite grows by enlargement and coalescence of crystallites deposited as syntaxial (axially aligned) overgrowths on previous crystals (Kendall & Broughton 1978, Broughton 1983a). The c-axis is oriented in the

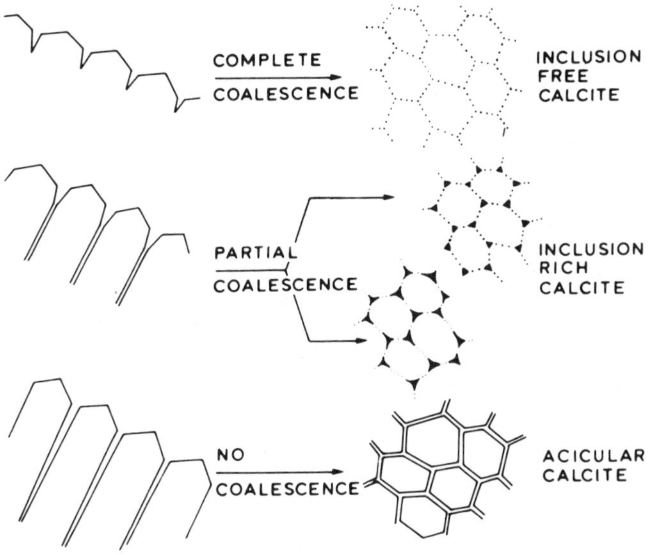

Figure 8.9 Different patterns of the growth of length-fast calcite in speleothems. From Hill & Forti (1986) by permission.

direction of growth, creating 'length-fast' calcite. Crystals are columnar, and parallel or radiating slightly ('palisade' patterns). They may coalesce laterally so that the entire speleothem constitutes a single crystal, normally translucent (Fig. 8.9).

Spaces left between growing columns are sealed off by later deposition and constitute inclusions, normally filled with fluid. Variation in the abundance and scale of inclusions creates the smallest features that can be distinguished by eye or hand lens, which are faint, parallel growth layers. They reflect variation in rates of deposition. Stronger layering is created by temporary cessation and drying, by periods of dissolving at the growth front (re-solution) or other erosion, or by deposition of foreign particulates. Syntaxial crystal growth may extend through such breaks, or new generations of crystals may start upon them with greater or lesser bonding across the break. At the extreme, there is no bonding; the speleothem can fall apart at the hiatus.

Folk & Assereto (1976) found finely laminated layers above certain breaks, with c-axes subparallel to the growth front = 'length-slow calcite'. A final crystalline type is an irregular mosaic, created by inversion from aragonite.

Aragonite in caves displays a principal habit as radiating clusters of needles, termed 'anthodites'. They grow from bedrock or calcite speleothems. There is also massive, acicular aragonite in regular stalagmites and flowstones. It may be interlaminated with calcite or display a patchy

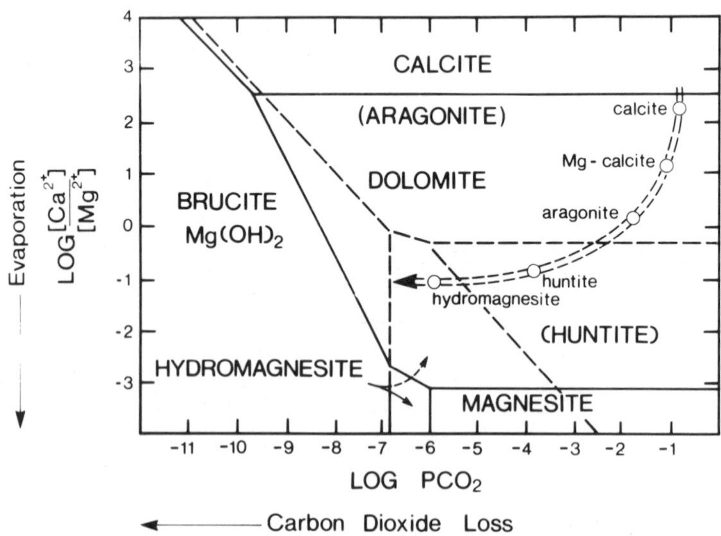

Figure 8.10 Stable and metastable (dashed lines) phase relationships in the system Ca–Mg–CO_2–H_2O showing a model evolutionary path for a hypothetical water.

appearance where inversion to calcite is occurring.

There has been much discussion of why aragonite is found in some caves (or parts of them) and not others. Figure 8.10 indicates that it is meta-stable in the domain of calcite and dolomite. The latter is virtually unknown as a primary precipitate in caves, so it is likely that depletion of Ca ion in Mg-rich solutions is the principal factor, e.g. by prior deposition of gypsum. Other suggestions include ion substitution or 'poisoning' (e.g. by Sr) or unspecified effects of organics or other seed nucleii when growth is initiated e.g. at Grotte de la Deveze, France, aragonite is deposited only where the host rock contains >0.25% Fe, >7.5% Mg, and displays micro-fissuring (Cabrol 1978).

Stalactites and draperies

The principal forms of speleothems are shown in Figures 8.8, 8.11 and 8.12. The fundamental form is the soda straw stalactite (French – 'fistuleuse'), a single sheath of crystal enclosing a feedwater canal. Growth occurs at the tip and the c-axis is oriented down the straw. Andrieux (1965) suggested that straw stalactites require a slow, steady feedwater without suspended fines or organics to block the canal. M. Roda (personal communication 1988) finds that a water supply between 1000 and 2000 ml a^{-1} at the tip is optimal. Some straws grow to 3–6 m in length before breaking under their own weight.

Leakage through the walls of straws may overplate and thicken the sheath. Partial or complete blockage of the canal compels leakage, usually at the stem to form the common tapered or carrot-shaped stalactite. C-axes can be subparallel to the canal, radiate from it, or be randomly oriented.

Curl (1973a) suggested that because of *g* forces and surface tension effects the minimum diameter of straw stalactites must be ~ 5 mm. Hill & Forti (1986) cite diameters of 2–9 mm. The more complex forms of tapered stalactites have not been analysed. Because of accelerated deposition at points of water extrusion they can grow protuberances in a myriad of forms, including crenulations and corbels with drapes or subsidiary stalactites. Maximum dimensions are determined by the strength of attachment to the roof, strength of roof rock, or dimensions of the cave. It is rare to see free-hanging stalactites longer than ~ 10 m and greater than 1–2 m in diameter.

Aragonite soda straws are rare. Somewhat more common is a clumped 'spathite' or 'chapelets de boules' form where the narrow necks may represent temporary halts in growth.

Draperies or curtains develop where the feedwater trickles down an inclined wall or beneath a tapering stalactite. Deposition is along the trickle course with the c-axis perpendicular to the growth edges so that scalenohedral crystal patterns normal to it are well displayed. The form may

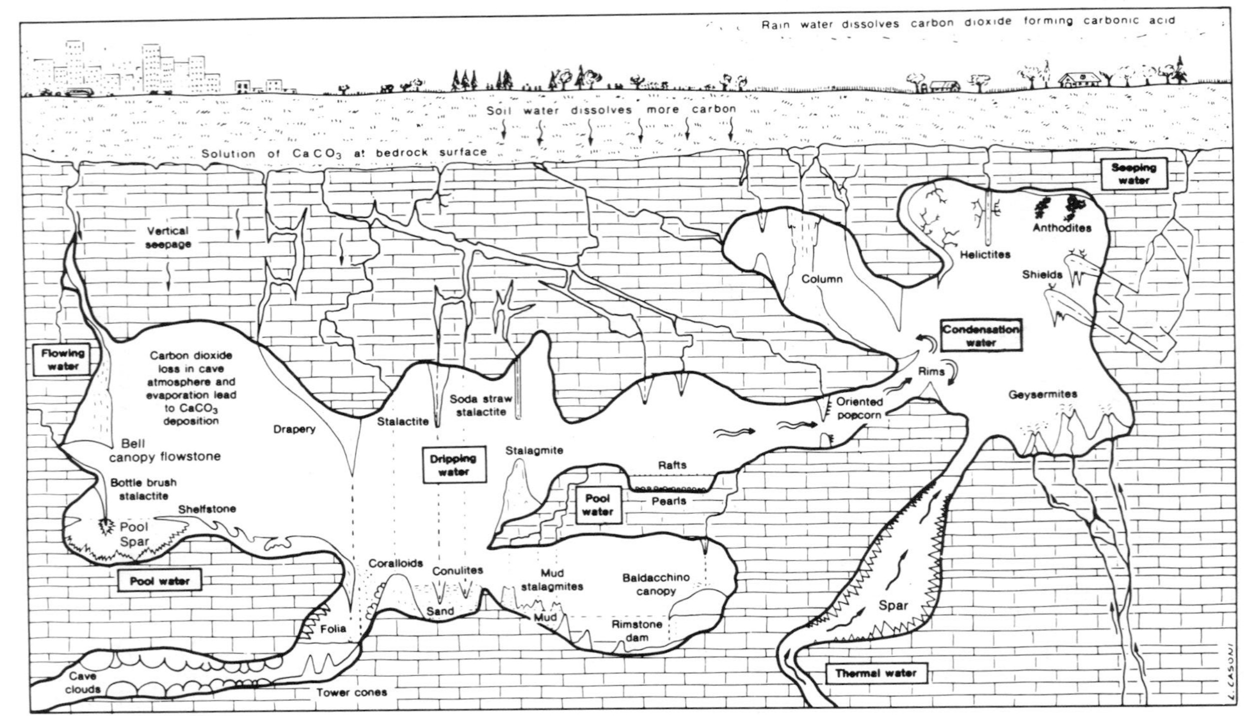

Figure 8.11 Model for types of calcite and aragonite deposits in caves. From Hill & Forti (1986).

Figure 8.12 *Upper left*. Soda straw stalactites, stalagmites and flowstone. *Lower left*. Stalactites and stalagmites aligned along joints. Photos by A. C. Waltham, by permission. *Right* Giant stalagmites and columns in Ogle Cave, New Mexico. The large scale and highly ornamented form are typical of warmer caves with strong evaporation. Photo by A. N. Palmer, by permission.

be sinuous. Overgrowth may not occur because the feedwater is on the lower edges; thus, drapery tends to have the width of a single crystal and is translucent. Colour banded varieties are known as 'bacon'.

Stalagmites and flowstones

The forms of stalagmites are similarly varied. They also grade into general floor and wall coverings termed *flowstone*.

Franke (1961, 1965) divides stalagmites into three categories. Those of uniform diameter add almost all new growth at the top (not down the sides) and are considered to be created by a uniform supply of feedwater with uniform solute concentrations. Curl (1973b) specified their diameter by the relation

$$D = 2\,\frac{C_o Q}{\Pi \dot{z}} \qquad (8.6)$$

where C_o = $CaCO_3$ available for deposition ($C_s - C$ of Eqn 3.78), Q = flow rate and $\dot{z}$ = rate of growth in height. The minimum diameter is about 3 cm. Increase in diameter is broadly proportional to Q. Gams (1981) shows that it also increases with drip fall height, a first complicating factor. 'Broomstick' stalagmites of this type are quite common.

Stalagmites with terraced or corbelled thickening are Franke's second type. At the extreme they possess leaf-like protuberances and are likened to piled plates or palm tree trunks (e.g. at Ludi Cave, Guilin). In between these growths the stem has a uniform diameter. Franke attributed them to periodic variations in growth rate and to greater fall height causing greater splash.

The third, most common, type is the conical or tapered form. Franke attributed it to a decreasing growth rate, Gams to increasing fall height. It is evidently due to there being significant calcite deposition from film flow on the sides. There is a transition to shallow bosses and flowstone sheets.

The greatest reported heights of broomstick and corbelled stalagmites are ~30 m (e.g. Mas d'Azil, France). Tapered stalagmites can be higher. Some bosses attain diameters approaching 50 m.

Flowstones are deposited from a rather uniform flow and accrete roughly parallel to the host surface. They are most common on floors or gentle slopes but occur on vertical walls where they are transitional to stalactites, drapery, etc. Maximum thicknesses up to a few metres are known underground. Flowstones are the dominant deposits at karst springs where, because of rapid growth due to evaporation and incorporation of vegetation, thicknesses of tens of metres can occur (section 9.11). Flowstones may cover a cave floor for tens or hundreds of metres downstream of a single source of feedwater.

Growth layering and hiatuses achieve their maximum development in flowstones. Many sediment sections in caves display flowstones interbedded with fluvial deposits, etc.

Excentric speleothems

Excentric or erratic deposits occur in most well decorated caves. They are termed *excentric* because they grow outwards from a wall or earlier formation, seeming to defy gravity. 'The form develops when crystal growth forces are dominant over the hydraulic forces of vertically moving water. Specifically, this implies that the flow to the excentric must be sufficiently slow to prevent drops forming' (White 1976). If drops form, a regular straw stalactite will grow downwards from the excentric.

There are innumerable varieties. Most excentrics are small, individuals rarely being >1 m and commonly <10 cm in length. These can be divided into (a) the helictites or linear forms (spicular, sinuous and/or branching) and (b) globular or semispherical forms. In addition, there is one large calcite erratic form, the shield or palette.

Hill & Forti (1986) recognize four types of helictites:

(a) filiform or thread-like growths of ~ 0.2 to 1 mm in diameter; aragonite and calcite compositions are both common;
(b) vermiform – sinuous to twisting, often bifurcating calcite sheaths of 1–10 mm diameter. This is the most common type. In a tangled mass on a wall it looks like the hair of the Medusa;
(c) spathite, the aragonite, beaded variety, which also curves and branches, and
(d) twig-like helictites which develop more angular branching.

All these types possess a central capillary. Branching is due to blockage by crystal growth during drier periods, compelling diversion or bifurcation of flow. The cause of the curving–spiralling deformation is controversial (Moore 1954). Foreign solutions or solid impurities twisting the c-axis have been proposed (Broughton 1983b), but experimental helictites have been grown from pure solutions. Cser & Maucha (1966) suggest that the lower or outer edge at the tip receives more $CaCO_3$ molecules, systematically displacing the axis. Andrieux (1965) suggests that stable microclimate and hydraulic regime are important factors. From the preferred distribution of helictites this appears to be true in many caves, but others negate it. Kempe & Spaeth (1977) showed that sample capillaries widen and narrow, appearing like pearls on a string; they suggested a seasonal effect that might also drive the deformation.

Cser & Maucha (1966) noted two further types:

(e) straight monocrystals, 0.2–3.0 mm diameter. They lack a capillary and are believed to form from aerosol $CaCO_3$ accreting at the tip;
(f) straight or tapered monocrystals $\geqslant 2$ mm in diameter and often lacking a capillary. Most often this is a wind flag or 'anemolite', growing into oncoming air from an extrusion in a stalactite.

The globular family contains many varieties. Names given to them include 'cave coral', 'popcorn', 'globulite', 'botryoidal calcite'; 'boules' in French, etc. Occurrence ranges from single individuals growing from a wall or regular speleothem, to lines or patches one globule deep and, finally, great clusters like bunches of grapes. Clusters may be tightly packed together and many layers deep.

Figure 8.13 gives a comprehensive model for the subaerial type. Feedwater is extruded from micro-fissures and evaporated slowly from the globule surface i.e. it is an optimum evaporation form. Depletion of Ca ion may permit acicular aragonite to grow outwards from the globule, tipped by

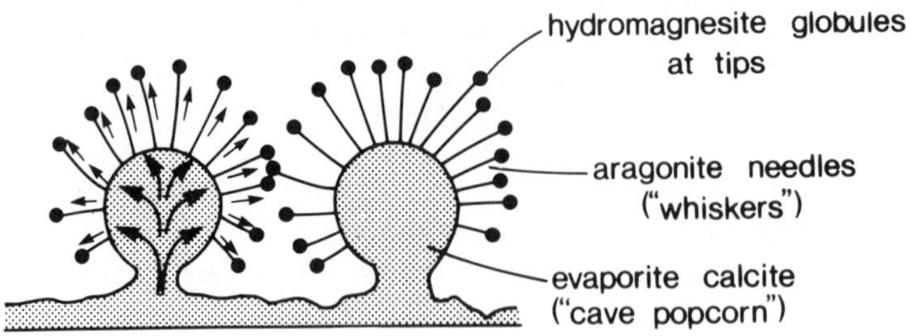

Figure 8.13 Development of the 'popcorn–whiskers–globules' or 'calcite–aragonite–hydromagnesite' sequence common in many caves where there is slow evaporation.

hydromagnesite (see Fig. 8.10). In most caves, only the calcite is present. Globular deposits also form underwater in saturated pools.

Shields or palettes are composed of an upper and a lower plate, both displaying concentric layering, jutting from a wall (Fig. 8.11). Kunsky (1950) suggested that they form where water is discharged under hydrostatic pressure from a fissure inclined upwards. The water diffuses radially outwards between the plates and new calcite accretes to their edges. There is usually deposition of draperies underneath. Shields are up to 5 m in diameter, 4–10 cm thick (White 1976).

Pisoliths or cave pearls

Cave pearls are regular accretions of radial calcite about a foreign nucleus such as a sand grain (Fig. 8.14). They grow in groups of a few to thousands in shallow pools agitated by falling drops of feedwater. Homann (1969) showed that gentle oscillation (rather than repeated rolling) is all that is needed to cause regular spherical accretion. Ideal cave pearls are spheres that range ~ 0.2 to 10–15 mm in diameter. Those in a given nest are similar in size. If the nucleus is elongated, bladed or roller shapes develop. Irregular shapes form about larger granules. There are a few known instances of edge-rounded cubical pearls (Roberge & Caron 1983) and one example of nested hexagonal pearls, the optimum packing pattern (J. Roberge, personal communication, 1986).

Growth of cave pearls calls for a delicate balance in the supply of water. If too little, the pearls cement together to form the core of a new stalagmite; if too much, they are displaced from the pool. Nevertheless, the balance is attained often enough. Nests of cave pearls are known in caves everywhere between the arctic and the tropics. Irregular examples develop rapidly in old mines.

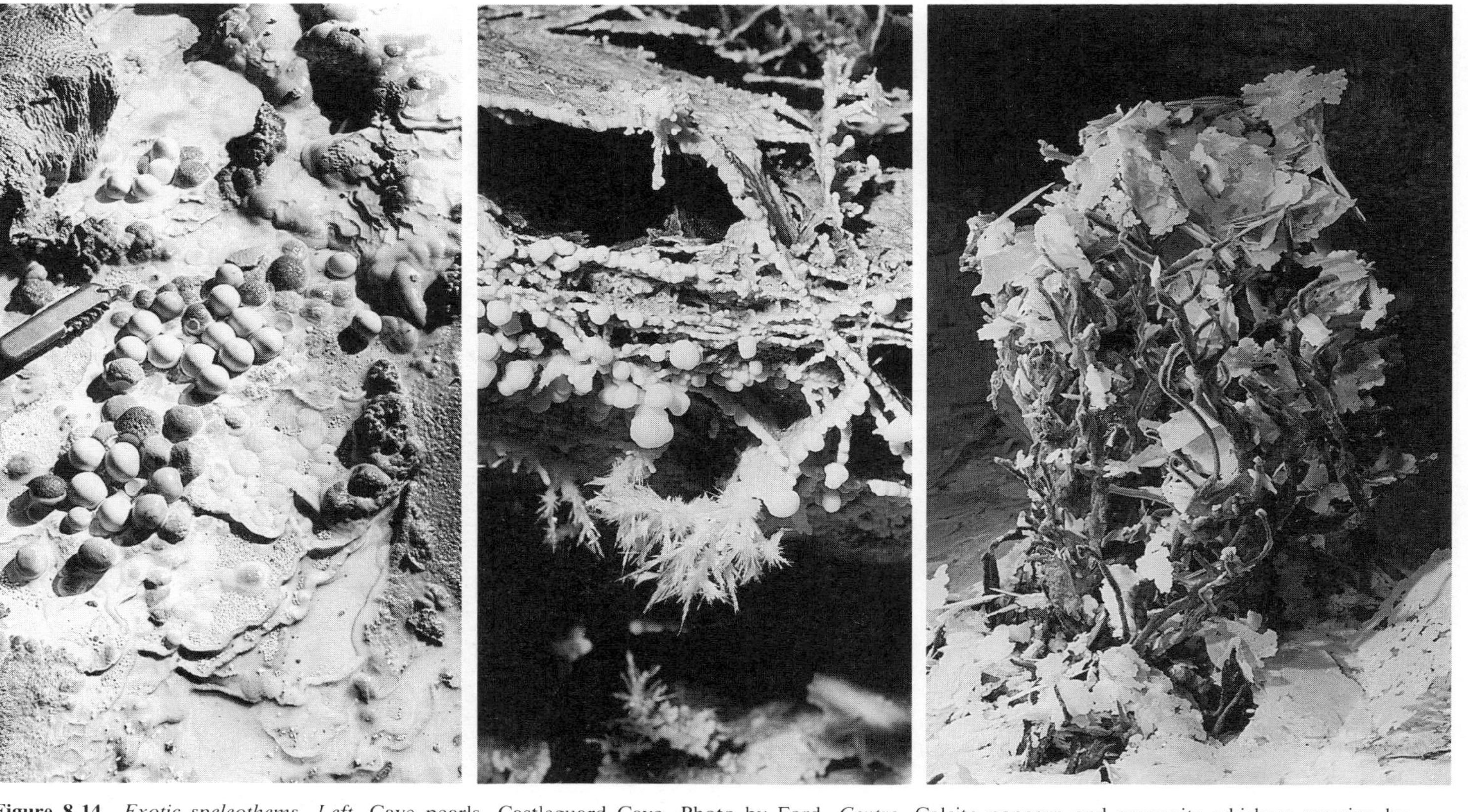

Figure 8.14 *Exotic speleothems. Left.* Cave pearls, Castleguard Cave. Photo by Ford. *Centre.* Calcite popcorn and aragonite whiskers growing by evaporation. *Right.* A 'heligmite' bush that is probably of subaqueous origin, draped by sunken calcite rafts, Wind Cave, South Dakota. Photos by A. N. Palmer, by permission.

Rimstone dams or gours

These features build up in channels or on flowstones. The height ranges from millimetres (microgours, often on the downstream faces of larger dams) to many metres. They may be single, or interlocking to create a staircase of pools. Rims are straight, curved or crenulated. The form is also common at tufa springs; e.g. Plitvice Lakes, Yugoslavia (Fig. 9.40) and hotsprings, such as those of Yellowstone Park, USA.

Rimstone dams originate at any irregularities that will thin a film of flow. They are the best example of positive feedback operating in speleothems. Calcite precipitation is fastest at the rim because the water layer is thinnest there and CO_2 loss is greatest. Transitions to turbulent or to supercritical flow may help accelerate the process. Cavitation (detachment of flow from a bed, to form temporarily free sheets or drops) may explain dense patterns of microgours on steep flowstone slopes.

Moonmilk

This term applies to white, amorphous masses of crystals that are pasty or plastic when wet, powdery when dry. They occur on rock and clastic sediment surfaces or as overgrowths on earlier speleothems, in irregular patches varying from centimetres to many metres in area. They are often bulbous, like cauliflower heads. Thickness may be several tens of centimetres, though that is exceptional.

Moonmilk has been reported from caves in all climatic regions. It is more prominent in cold or dry caves because there is often a lack of other types of speleothems in them.

Moonmilk crystals have an average thickness of only 1 μm or so. They display needle, branched or helicoidal forms. In very humid regions the crystals are entirely or predominantly of calcite or aragonite. Where there is effective evaporation, the hydrated carbonate minerals and gypsum may be added, even huntite, dolomite, some phosphates and silicates.

Moonmilk has been intensively studied (Bernasconi 1976). The presence of loose crystals of evaporite minerals is readily understood as being due to late stage deposition from minute quantities of concentrated solutions. But this habit is quite anomalous for calcite. It is suggested that it is of bacterial origin; organisms break down ordinary calcite crystals and re-deposit them as micro-fibres (Williams 1959).

Subaqueous deposits

A characteristic suite of precipitates develops in and on the surfaces of pools of supersaturated meteoric water, normally rimstone pools (Andrieux 1963). The basic form is a wall encrustation of scalenohedral calcite (dogtooth

Figure 8.15 Typical passage in Jewel Cave, South Dakota. Walls are covered by 6–15 cm or more of nailhead spar precipitated from thermal phreatic waters. Photo by A. N. Palmer, by permission.

spar) with projecting faces. Sometimes there are botryoidal crusts instead. At the water surface calcite may precipitate on dust particles and float as 'calcite ice' or 'rafts' supported by surface tension until too thick, when it sinks. Raft material and shelfstone grow outwards from the pool edges at the water surface. It may also grow radially from any projecting rocks in the centre, creating lily pad patterns. Calcite even encrusts bubbles on pool surfaces, a testimony to the extreme physical stability that can pertain in some cave environments.

The greatest subaqueous deposits are associated with the degassing of thermal waters enriched in exhalative CO_2. At the extreme, all surfaces become coated with dogtooth or nailhead (rhombohedral) calcite spar. These are 'crystal caves'. In Jewel Cave (Fig. 7.26) surfaces are covered with 6–15 cm of nailhead spar in sheets and bulbous masses (Fig. 8.15), except the highest places where it has been removed by re-solution. Other caves nearby contain sheets $\geqslant$ 1 m thick.

Continuous calcite precipitation over a vertical range of >130 m as at Jewel Cave is exceptional. It is more common in thermal caves to see strong outward growth of calcite fins or curving folia at and immediately below the water surface, with botryoidal or dogtooth encrustations below them that rapidly diminish with depth. In some parts of the Budapest caves there are no deposits below −6 m except sunken raft detritus. Wind Cave, South Dakota, displays thin, discontinuous crusts to −30 m or more. Deeper precipitation must be associated with very slow renewal of water and slow, steady de-gassing.

Deposits of other carbonate minerals

These are the Mg-rich carbonates plus hydrated carbonates (Table 8.3). They are all tiny in amount, products of concentrated solutions created by evaporation or by previous deposition of calcite and aragonite. They are difficult to identify in the field and so have been reported from comparatively few caves. Most have been recognized only as components in moonmilk. For example, in the cold and humid centre of Castleguard Cave (+ 3°C and relative humidity > 95%) small nodular growths of aragonite or of calcite plus huntite are surrounded by aureoles of pure hydromagnesite moonmilk (Harmon *et al.* 1983).

Colour in speleothems

Pure, inclusion-free calcites and aragonites may be colourless and translucent. High frequencies of fluid inclusions make them white and opaque. Flood muds sealed in by overgrowth appear to colour the entire deposit although they are only minor, discrete layers: black, grey, brown and red are the most common mud colourings.

Much colour may be contained within the growth layers. Often there is change of colour or tone between successive layers. Kral (1971) showed that optically continuous calcite may appear yellow. Many speleologists attribute intracrystalline colour to trace elements in the lattice. Pb, Zn, Fe and Mn may produce pale yellow colours when in low concentrations. Cu is associated with green and blue in aragonite. High concentrations of Mn, Fe and of Ni may produce black and yellow–green colours respectively.

Gascoyne (1977, 1979) investigated strongly coloured samples from tropical, temperate, arctic and alpine environments and found poor or no correlation between kind or strength of colour and type or quantity of trace elements, up to 15 000 ppm of the latter at least. Most colour was due to organic matter, presumably carried down in the feedwater. White (1981) confirmed these results independently and Lauritzen *et al.* (1986) showed that the organic matter was predominantly humic and fulvic acids of high molecular weight, produced by decomposition in soil. The strongest, darkest

Table 8.3 The principal minerals deposited in caves

Carbonates		Phosphates and nitrates	
Calcite	$CaCO_3$	Whitlockite	$Ca_3(PO_4)_2$
Aragonite	$CaCO_3$	Monetite	$CaHPO_4$
Magnesite	$MgCO_3$	Hydroxyapatite	$Ca_5(PO_4)_3.OH$
Huntite	$CaMg_3(CO_3)_4$	Carbonate-apatite	$Ca_{10}(PO_4)6CO_3.H_2O$
		Crandallite	$CaAl_3.(PO_4)_2.(OH)_5.H_2O$
Hydrated carbonates		Taranakite	$(K.NH_4)Al_3(PO_4)_3(OH).9H_2O$
Mono-hydrocalcite	$CaCO_3.H_2O$	Potassium nitre	
Tri-hydrocalcite	$CaCO_3.3H_2O$	(saltpetre)	KNO_3
Nesquehonite	$MgCO_3.3H_2O$	Soda nitre	$NaNO_3$
Hydromagnesite and others	$4MgCO_3.Mg(OH)_2.4H_2O$	Nitrocalcite and others	$Ca(NO_3)_3.4H_2O$
Sulphates and hydrated sulphates; halides		Iron, manganese and aluminum oxides	
Anhydrite	$CaSO_4$	Geothite	Fe OOH
Barite	$BaSO_4$	Hematite	Fe_2O_3
Celestite	$SrSO_4$	Limonite	$Fe_2O_3.nH_2O$
Thenardite	Na_2SO_4	Birnessite	MnO_2
Halite	NaCl	Hollandite	$BaMn_8O_{16}$
Bassanite	$CaSO_4.H_2O$	Psilomelane	$(Ba.H_2O)2Mn_5O_{10}$
Gypsum	$CaSO_4.2H_2O$	Todorokite	$(Na.Ca.K.Mn.Mg)_6O_{12}.3H_2O$
Epsomite	$MgSO_4.7H_2O$	Bauxite	$Al_2O_3.3H_2O$
Hexahydrite	$MgSO_4.6H_2O$	Boehmite	AlO(OH)
Mirabilite	$Na_2SO_4.10H_2O$	Gibbsite and others	$Al(OH)_3$
Bloedite and others	$Mg_2SO_4.Na_2SO_4.4H_2O$		
Sulphides		Silica	
Pyrite	FeS	Quartz, chalcedony	
Marcasite	FeS	and cristobalite	
Galena	PbS	(opal)	SiO_2
Sphalerite	Zn.S		
plus fluorite (fluorspar)	CaF_2	Ice	H_2O

colours appear in arctic and alpine samples because breakdown to smaller molecules is retarded there.

Hill & Forti (1986) conclude that the question of colour in speleothems is complex: 'Some of the contradiction in results may stem from the difficulty in distinguishing between colour due to chromospheres (metal ions that are actually in solid solution in the lattices of minerals), that due to pigments (metals concentrated on grain boundaries of minerals), and that due to organic stains' (1986, p. 166).

Rates of growth of calcite speleothems

Most visitors to caves are interested to know how fast the speleothems grow. This is a difficult question because the rate control environment is complex. Table 8.4 cites fourteen different conditions that a given deposit may experience. Some re-solution (net erosion by waters from the original

Table 8.4 Conditions of calcite speleothem growth, destruction or decay

		state of
A	The speleothem feedwater is the dominant control	
1	Continuous deposition	= net deposition
2	Periodic deposition	= net deposition
3	Periodic deposition, periodic erosion	= net deposition
4	Periodic erosion, periodic deposition	= net erosion
5	Periodic erosion, no deposition	= net erosion
6	Continuous erosion	= net erosion
B	Other waters are the dominant control (usually floodwater)	
7	One of A may pertain but body (floodwater) re-solution predominates	= net erosion
8	One of A may pertain but mechanical erosion predominates	= net erosion
9	One of A may pertain but burial by clastic sediments predominates	= burial
10	Permanently inundated by fresh water or seawater	= no change, or net erosion, or marine boring and overgrowths
C	All effective water flow has ceased	
11	Very slow weathering with minor accumulation of dust and aerosols; loss of lustre	= weathering
12	Corrosion by acidic waters condensed from flowing, CO_2– or H_2S-rich air	= net erosion
13	Speleothem freezes	= shattering
14	Speleothem buried by weathering earth, organic matter, etc.	= burial

depositional source or by flood waters, etc.) is seen in most well decorated caves.

Rates of growth are usually quoted in terms of the extension of a given form rather than its mass accumulation. Straw stalactites 'grow' fastest because they have the greatest extension per unit of mass deposited. Other dripstones and flowstones grow especially rapidly where they are tufaceous i.e. incorporate vegetal matter. In show caves, rates between 0.2 and 20 mm a^{-1} are quoted for straw stalactites, between < 0.1 and 3 mm a^{-1} for carrot stalactites, and between < 0.005 and 0.7 mm a^{-1} for stalagmites. The sampling is often biassed in favour of deposits that appear to be growing the fastest.

Significant seasonal variations in rate of deposition have been detected even where deposition is continuous. At Postojna Cave, Slovenia, Gams (1965) showed that the rate approximately doubled when summer soil water reached the speleothems in the autumn months.

A different approach has been to calculate long-term mean rates in

Table 8.5 Representative mean extension rates for sample stalagmites (candlestick and tapered types) that have been established by two or more radiometric dates per sample

Site	Age	Mean rate of extension mm a^{-1}
A ^{14}C dating method		
Central Europe[1]	Holocene	0.4
Central Europe[1]	Holocene	0.09
Central Europe[1]	Holocene	0.26
Central Europe; post-glacial mean of Franke & Geyh[1] =		0.10
Central Europe; interstadial mean of Franke and Geyh[1] =		0.01
B U-series dating methods		
Yorkshire, England	Late Glacial–Holocene	0.1–0.2
Yorkshire, England	Late Glacial–Holocene	0.09–0.14
Yorkshire, England	Late Glacial–Holocene	0.02–0.04
Yorkshire, England	Interstadial	0.002
Yorkshire, England	Last Interglacial	0.03
West Virginia, USA	Holocene	0.042
West Virginia, USA	Interstadial	0.002–0.02
West Virginia, USA	Glacial stade	0.33
Mexico[3]	Holocene	0.36
Jamaica	Glacial–Holocene	0.02
Tasmania, Australia[2]	Interglacial–Glacial	0.02–0.08
	Postglacial	0.02–0.26
Wind Cave, USA	Phreatic crust from thermal waters	0.001–0.0001

[1] From Franke and Geyh (1971). [2] From Goede & Harmon (1983). Other data from McMaster University results. [3] Sample DAS2, illustrated in Figure 8.26.

stalagmites and flowstones that have been dated by radiometric methods (section 8.6). Table 8.5 presents some examples of stalagmite extension rates. There is a range of about 2–3 orders of magnitude within this small sample, although the stalagmites are restricted in their morphologic type and are largely from humid temperate regions. This is to emphasize that growth rates are site-specific to a considerable degree; they can vary a lot within one cave.

In summary, in humid cave interiors, carrot stalactites probably grow up to 10 mm a^{-1} and straw stalactites may grow several times faster. Stalagmites rarely extend more than 1 mm a^{-1} and mean rates as low as 0.001 mm a^{-1} are calculated from dated samples. Flowstone thickening rates are generally less than stalagmite extension rates, perhaps by as much as one order of magnitude.

Patterns of distribution of calcite speleothems

This subject has attracted comparatively little discussion, which is surprising because distribution patterns can reveal much concerning effective fissuration, the epikarst and environmental controls above a cave or in its region (section 4.4). The two extremes are readily identified. Many caves have no speleothems. Others are largely or entirely filled with them.

Absence of speleothems in a given region or cave or particular passage is due to four principal causes:

(a) although vadose, all of the passage is in a net erosional state. This is typical of young caves in any region, of the 'active' (usually lowest) passages in a multi-phase system, etc. The first speleothems become established in sheltered recesses in the roof when the flood frequency begins to drop.

(b) There is insufficient soil CO_2 so that infiltrating waters are not supersaturated with respect to the cave air. This is the major factor in explaining the small number and size of speleothems in many arctic and alpine caves and some semi-arid caves. However, lack of well developed soils does not always imply their absence. Castleguard Cave is covered by glacier ice yet some sections of it are profusely decorated, as a consequence of suspected exotic acid effects (Atkinson 1983).

(c) A non-carbonate caprock prohibits infiltration of feedwaters, or changes their ionic compostiion so that calcite precipitation cannot occur. This is true of many middle and upper level passages in the Mammoth–Flint Ridge–Roppel System. Underneath the caprocks there is some precipitation of gypsum (section 7.12) but calcite only appears in profusion if the galleries pass under valleys where the capping has been removed.

(d) There is no hydraulic connection between the epikarst zone and the passage, or connections are intercepted by overlying galleries which capture all precipitates. It is rare to find an entire, extensive cave that does not possess effective linkages to the epikarst, but particular passages remain unconnected or intercepted in many systems. Partial interception is common, reducing the number of feedwaters entering lower passages. These effects are most important where fissure frequency is low.

Where speleothems are abundant their distribution varies with characteristics of the fissures transmitting the feedwater. At one extreme is limestone possessing high primary permeability. Feedwater emerges at points a few millimetres to centimetres apart, tending to create a dense and uniform cover of straw or carrot stalactites in a roof. This is characteristic of shallow caves in Quaternary limestones, but particular beds or lesser patches of rock in older strata also display it. At the other extreme, feedwater supply is limited to a single fissure and the stalactites, etc. are aligned along it. In most grottoes the supply is from a variety of alignments.

Often the type and scale of precipitation are functions of the hydraulic efficiency of the final fissures that supply the water. The largest fissures support the largest stalactites, stalagmites, etc. Small fissures support only straw stalactites and a few tapered examples plus thin calcite crusts, while the smallest may yield only excentrics and evaporite minerals.

Microclimatic factors may be locally significant in determining the type, scale or density of speleothems. This is most obvious in the cases of evaporites and anemolites; they grow best where air currents are strongest. Many hard calcite speleothems have soft calcite overgrowths on the upwind side, especially near entrances. When there is a sequence of hard and soft calcite layers oscillation of the cave climate about some critical value of relative humidity may be implied. Some caves display strong vertical zonation of temperature or humidity, with the consequence that excentrics (particularly the cave coral types) are seen to terminate abruptly at some upper or lower limit on the walls.

8.4 Other cave minerals

Hill & Forti (1986) report more than 180 non-carbonate minerals in caves. Some are alteration products such as the standard clay minerals. However, a majority are precipitated from solutions. Some of these are rarely found except in caves (i.e. they can almost be described as 'cave specific'). Many others attain their greatest or purest development in the stable, unchanging environments of deep but inactive caves. Here, we summarize the most commonly encountered types (Table 8.3).

The sulphates and halides

Gypsum is believed to be the third cave mineral by volume and frequency. Its source may be gypsum strata, inclusions, etc. in the feedwater course, or it may be complexed from oxidized sulphides and dissolved calcite as explained in Equation 3.70.

Gypsum is deposited in three principal modes in caves. The first is as evaporite inclusions in rock or clastic sediments. The inclusions are typically coarse, tubular or fibrous crystals in lenses up to a few cm thick. They disrupt the rock or sediment (*evapoturbation*).

The second mode is as speleothems growing from rock, sediment or calcite speleothems. Hill & Forti (1986) list 18 forms including dripstones, flowstones and crusts. However, more frequent are gypsum flowers ('oulopholites'), needles ('selenite') and 'whiskers' or 'hair'. Flowers are curving, branching bundles of fibrous crystals as long as 50 cm. Epsomite and mirabilite, the most common Mg and Na cave sulphates, also display this form. Gypsum needles usually grow in dense clusters from sediment, like needle ice in freezing soils. Whiskers and hair are the finest, monocrystalline strands. Massive, sword-like crystals of gypsum that may exceed 1 m are also known.

The third mode of deposition is as regularly bedded deposits or wall encrustations from evaporating lakes or pools, etc. (Fig. 7.29).

Gypsum, plus the minor but frequent epsomite and mirabilite, is most abundant in temperate to tropical caves that are quite dry, at least at some seasons. They may also suffer significant condensation corrosion in such caves. They form in lesser amounts in cold, humid caves such as Castleguard and in arctic caves where temperatures are -1 to -3°C.

The other sulphate and halide minerals in Table 8.3 are much rarer and are known only in warm, dry-to-arid caves. Epsomite loses one water of hydration to form hexahydrite. Mirabilite dries to powdery thenardite. Anhydrite, bassanite and bloedite are known in polymineral crusts or inclusions on or in drying cave earths. Halite forms crusts, single crystals or flowers in desert limestone and gypsum caves and is abundant in caves in salt.

Celestite is occasionally reported as crusts or distinctively blue needle clusters on gypsum, and barite is reported as inclusions in cave earths. Both minerals occur in hydrothermal precipitates (Ozoray 1961).

The sulphide minerals and fluorite (fluorspar) are primarily deposited from metalliferous thermal or other waters invading caves, with or without contemporaneous dissolution (section 7.10). Occasionally, there is solution and re-precipitation of these minerals as small stalactites in later caves.

Phosphate and nitrate minerals

Most of these minerals are produced by urine, faeces and other decaying animal matter reacting with cave rock, speleothems or clastic sediment. Bat guano is the principal source of reactants. The minerals therefore are most common in temperate and tropical cave areas with abundant bat populations. They attain significant bulk as micro-crystalline aggregates within cave earths and also appear as thin, discontinuous alteration crusts on walls and speleothems.

The cave phosphate system is outlined in Figure 8.16. Reactions between guano, etc. and calcite normally produce hydroxyapatite. Other calcium and mangesium phosphates are rare. Alumino-phosphate minerals such as crandallite and taranakite form within permeable silts and clays on cave floors.

Nitrate minerals are of unusual interest because they have been extracted

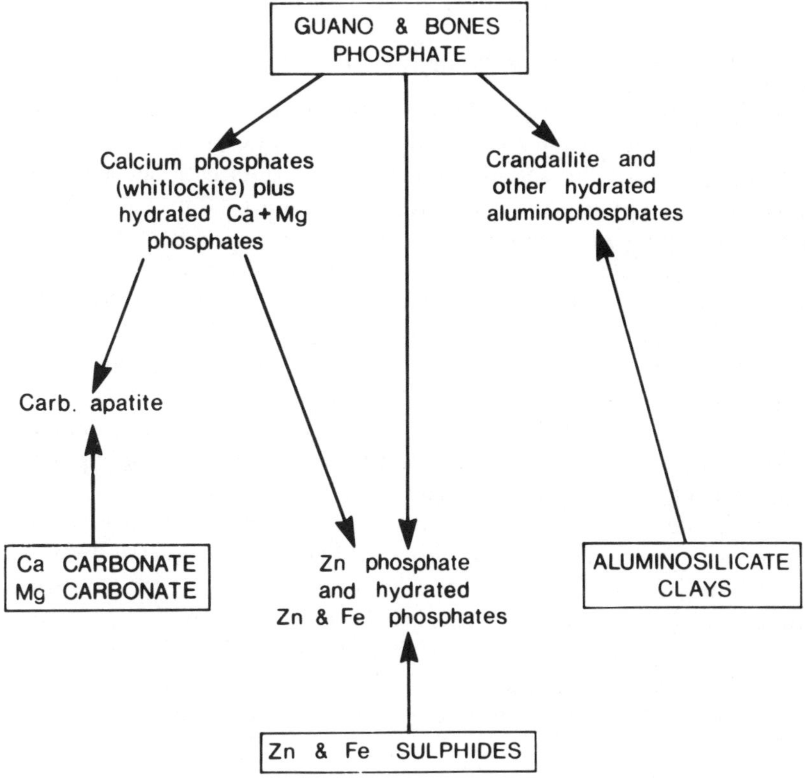

Figure 8.16 Phosphate minerals derived from the reaction of guano, bone, etc. decay products with wall rocks, clay fillings or sulphide ores in caves.

from cave earths for millenia, to use as medicines and in ceramics. There was widespread mining in US caves during wars of the 18th and 19th centuries to manufacture saltpetre (KNO_3) by addition of potash (wood ash) mixed with water. River caves with deep clastic fills in their upper levels were therefore favoured and in the Appalachian region few of these escaped attention. The excavations, tools and vats are still well preserved in many of them; e.g. in the Historic section, Mammoth Cave.

The subject is comprehensively reviewed by Hill (1981). Where bats are abundant, the nitrates derive from nitrous guano. There are few bats in most Appalachian 'saltpetre caves', however, and their typical nitrate earths are alkaline and low in organic matter. They contain 0.01–4% nitrates by weight, mainly in the top few metres. These are soil nitrates carried down in solution in seepage waters.

Nitrates only crystallize where caves are dry and warm (>12°C). KNO_3 and $NaNO_3$ occur as thin wall crusts and small stalactites, and darapskite ($Na_3(NO_3)(SO_4) \cdot (H_2O)$), occurs as small flowers, hair, dripstones and flowstones.

Aluminium, iron and manganese coatings

Most aluminium and iron oxides are carried into caves in suspension but some form *in situ*. Alumina (in bauxite, boehmite or gibbsite; Table 8.3) mostly occurs as thin encrustations in caverns directly beneath bauxite sediments filling karst or paleokarst depressions (Bardossy 1982). Iron oxides occur chiefly as limonite coatings on walls and sediments. Some thick limonite within sediments is reported, and a few instances of limonitic stalactites. In thermal caves or very dry caves hematite is formed. Tiny crystals of magnetite precipitate within the calcite of many ordinary speleothems, where they carry a chemical remanent magnetism (Latham 1983).

In many caves, dark brown–black stains or coatings are prominent on cobbles and boulders in stream channels and may extend up walls or speleothems as obvious deposits from ponded floodwaters. They are usually shiny and soft. These are manganese oxides, chiefly birnessite but also psilomelane, hollandite and todorokite. They are deposited from highly oxidized (e.g. turbulent vadose) waters carrying Mn, Ba, etc. ions from the solution of impure limestones (Moore 1981). They may be associated also with hydrated iron oxides. Thick deposits of pure managanese in Jewel Cave, South Dakota, appear to be residuals of local limestone solution also, in this case settled out of slowly circulating thermal waters (Hill 1982).

Silica

Most silica precipitated in caves occurs as thin coatings of euhedral quartz in thermal waters. Fibrous, length-slow chalcedony replaces gypsum on some speleothems (Broughton 1971).

Opal (cristobalite) is a very fine-grained silica that occurs as thin interlayers in some calcite and gypsum speleothems. It is precipitated by high silica super-saturation attributable to evaporation.

8.5 Ice in caves

Ice caves are those containing seasonal or perennial ice or both. Ice caves may be 'static' or 'dynamic' in their meterological behaviour (Fig. 8.17). The static cave is a simple deep hole with no air drainage out of the bottom; thus, it traps cold, dense air and excludes warmer air. Such static caves extend permanent ice to lower latitudes and altitudes than other natural cold traps. Most dynamic caves have two or more effective entrances for airflow and distinct entrance zones where the temperature varies > 1°C over the year, as explained in section 7.12.

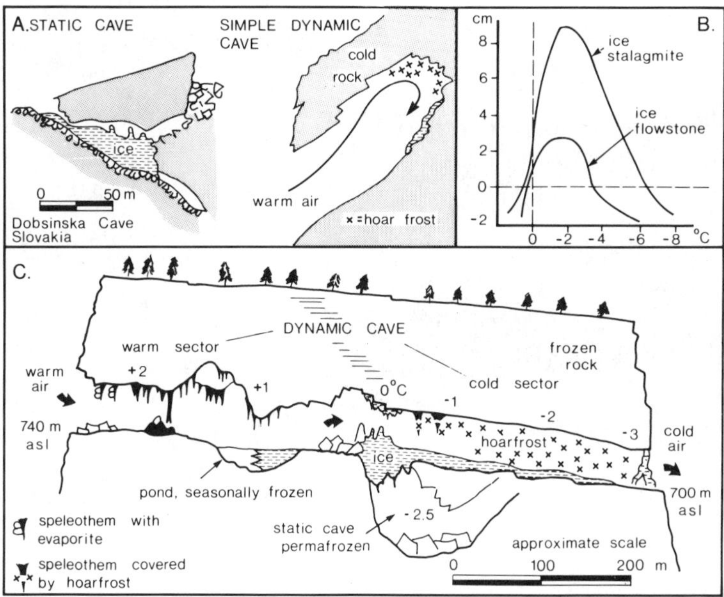

Figure 8.17 A. The structure of static and simple (one entrance) dynamic ice caves. B. Relationship between daily rates of growth of sample ice stalagmites and flowstones and air temperature. From Pulinowa & Pulina (1972). C. Features of Grotte Valerie, a complex dynamic ice cave in the subarctic Nahanni karst, Canada. Vertical scale is much exaggerated.

Types of ice

At least seven different types of ice are known in caves. The most common consists of dripstones and flowstones formed by freezing of infiltrating waters (Fig. 8.18). These appear every winter in many cave entrances in

Figure 8.18 Ice in caves. *Top left.* Ice curtains, stalagmites and flowstone, photographed in mid-winter. There is late hoarfrost on the curtains and the stalagmites are ablating. *Top right.* A grotto of hexagonal ice crystals grown by condensation and freezing. *Bottom left*, a cold trap. The ice in the foreground traps a lake of very cold, dry air in the rear. Hoarfrost is seen above the lake surface. Grotte Valerie, NWT. *Bottom right.* Phreatic water frozen in Coulthard Cave, Alberta, and now ablating slowly in cold dry air. Ablating surfaces are large scallops. Photos by Ford.

temperate regions. They are often predominant in perennial ice caves. The morphologies are similar to those of calcite dripstones and flowstones but with less variety and ornamentation. The ice is clear to opaque and polycrystalline; see Pulinowa & Pulina (1972) for details. Eisriesenwelt, at 1660 m asl in the Austrian Alps, is a perennial ice cave with outstanding ice of this type. It includes an entrance zone of flowstone 650 m in length and up to 12 m thick.

The second type is ice created by the densification and recrystallization of old snow (firn or neve) accumulated in cave trap sites. It is perennial and always contains a substantial proportion of infiltrating water ice (Balch 1900). It usually displays distinct dirt layers formed by concentration during superficial melt in the summer. The ice is opaque and bluish in section. It occurs as irregular masses conforming to the topography plus snow or water supply patterns of the host cave. This is true glacier ice in many respects but there are no authenticated cases of it flowing as plastic glacier ice. Dobsinska Ice Cave at 970 m asl in Slovakia is an excellent example (Skrivanek & Rubin 1973; Fig. 8.17A). The ice mass is 145 000 m^3, with a mean depth ~ 13 m. Ice stalagmites 10 m in height are superimposed on it. Ice temperatures range between −0.2 and −1.0°C. This is delicately balanced; melting problems were encountered in the early years of tourist display, and were solved by modifying the natural air circulation. The mean annual external air temperature at the site is +6°C.

The third type is formed by freezing of ponded, static water. It may be seasonal or perennial. Freezing takes place from the top downwards and tends to occur at steady rates once begun. This creates coarse polycrystalline ice with horizontal 'C' axes (Marshall & Brown 1974). Normally, it is the most transparent of all types of ice, although containing spectacular bubble patterns. At Coulthard Cave, Rocky Mountains, water-filled phreatic passages are now completely frozen up (Fig. 8.18). It is possible to view along them through the clear ice until the passages turn away at an apparent optical depth of many metres. The invariant air temperature is approximately −2.5°C and the ice is sublimating at about 3 mm a^{-1}, forming large scallops.

Hoarfrost is the fourth type, deposited from water vapour onto rock or ice surfaces below 0°C. There are at least four different kinds:

(a) a mixture of small needles, rosettes and hexagonal crystals that develop rapidly at temperatures close to 0°C. They form densely packed crusts of clear to opaque ice growing into oncoming drafts.
(b) the most spectacular are larger, hexagonal plate crystals of transparent ice that grow by accretion at the edges. Individual plate diameters as great as 50 cm have been measured and many interlocked, sequential plates may extend up to 2 m as stalactitic masses (Fig. 8.18). Accretion rates of 5 mm a^{-1} have been measured in a Canadian cave, implying

that large individual crystals take about 50–60 years to form and that the stalactitic masses are more than 500 years old. These deposits require formation temperatures of −1 to −5°C and very slow, gentle air circulation to supply vapour.

(c) are fused, regelation masses of type (a) or (b) hoarfrost. They are opaque and bulbous and signify an increase of temperature.

(d) are regularly tapering, prismatic crystals of hexagonal cross-section. These are rare and small.

Extrusion ice forms curving fibrous crystals similar to gypsum flowers. Lengths up to 20 cm are recorded in Canadian caves. This ice probably forms where super-cooled water is forced under pressure through micro-fissures in frozen rock. It freezes where pressure is relaxed at the point of emergence. It is rare.

Intrusive ice is true glacier ice at the pressure melting point, intruded by plastic flow into open caves beneath glaciers. The only known examples fill passages at the head of Castleguard Cave, which is situated beneath ~ 280 m of temperate glacier ice. The face of the principal intrusion remained constant in its position, neither advancing nor receding by melt or sublimation, between 1974 and 1986.

The final category is ground ice in clastic sediments. Where there is rapid freezing it forms thick, irregular masses. Slow freezing creates segregated ice, consisting of thin, scattered lenses. At intermediate rates needle ice or pipkrake is formed, and may extrude as it does in ordinary soils; it is perhaps the most common type reported in caves. Pulinowa & Pulina (1972) and Schroeder (1979) discuss Polish and Canadian examples in detail, and also describe their effects on the sediments.

Distribution patterns of cave ice

In static caves and simple dynamic examples (Fig. 8.17A) the ice may extend to fill all of the cold trap, seasonal or perennial. This is rarely more than ~ 100 m in length (e.g. Dobsinska) and for hoarfrost it is normally only a few tens of metres.

In multi-entrance dynamic caves patterns become more complex, especially where entrances are at widely differing altitudes. Wigley & Brown (1977) and Bogli (1980) discuss the relevant meteorological effects. However, most ice will form in the entrance dynamic zones, as defined in section 7.12. The greatest quoted extent of ice into cave interiors is 1300 m at the Eisriesenwelt, but this is something of an overestimate because sub-parallel galleries are being added together. The usual maximum is ≤ 400 m.

Perhaps the most complex and unexpected patterns occur in permafrost regions. The example of Grotte Valerie, Northwest Territories of Canada is given in Figs 8.17C, 8.18. The site is at 61.5°N and the mean annual temperature is about −8°C. The ice zone is a composite of dripstone and flowstone ice plus hoarfrost. It extends throughout what is, in the summer season, the interior or downwind half of the dynamic cave. This is because the upwind half (i.e. the conventional dynamic entrance zone) is thawed by warm exterior air that is drawn in. A static cave below the dynamic cave remains a cold trap at all times; it is permafrozen, arid and dusty, without ice.

Seasonal patterns of ice accumulation and ablation

In a majority of caves, the ice accumulates in the autumn and early winter and melts in the spring and early summer. In colder mountain regions there may be more complex sequences (Fig. 8.17B). In alpine Canada dripstone ice accumulates during the autumn, at a decreasing rate as soil–water sources freeze up. Hoarfrost accretes to the dripstone in late autumn–early winter. It may continue to form all winter where there are flowing allogenic streams to provide some humidity; otherwise, the balance swings to net sublimation in the depth of winter if cave entrances remain open to permit cold, dry air to enter. A brief burst of renewed hoarfrost formation in early spring is succeeded by general melting that normally proceeds from the entrance inwards.

Racovitza (1972) studied rates of net accumulation of perennial type 1 (dripstone) and 2 (firn) ice at Scarisoara Cave, Bulgaria, over periods ranging from a few years to the last 370 years. He correlated them to variations in the severity of central European winters and concluded that a period of net ice build-up began around 1700 AD and terminated about 1920 AD.

In the Austrian caves, Trimmel (1968) has suggested that the major perennial ice masses began to accumulate after the Boreal Optimum (the Holocene warmest phase) ended about 4000 years BP. Using oxygen isotope evidence, Marshall & Brown (1974) concluded that type 3 (pond) ice masses in Coulthard Cave may be of this age. In colder regions some ice probably survives from the last (Würm–Wisconsin) glaciation.

8.6 Dating and paleo–environmental analysis of calcite speleothems and other interior deposits

During the past 25 years there have been substantial advances in dating and paleo-environmental reconstruction with cave interior deposits, especially speleothems. This section briefly reviews the principal developments. It is an area of research where new analytic and interpretive methods are developing at a rapid pace today. A general review of Quaternary dating methods is available in Mahaney (1984).

Absolute dating: (A) Carbon-14

The absolute dating methods rest upon the decay of natural radioisotopes in a statistically random manner such that the half life (λ) is a constant for a given isotopic species. Thus the decay is exponential (Eqn 6.55).

^{14}C is created in the atmosphere when cosmic radiation interacts with ^{14}N. The half life is 5730 ± 40 years. Although age estimates up to 75 000 years BP (75 ka) are occasionally published, most workers place the reliable limits of the method between 38 and 45 ka (i.e. seven or eight half lives).

When calcite is precipitated in a cave it is presumed that the system is 'closed', i.e. carbon atoms in the crystal lattice cannot exchange with those in any later HCO_3 passing over or through the calcite. Radioisotope decay thus measures the time elapsed since precipitation occurred.

A major problem with ^{14}C dating of calcite is that of 'dead' carbon. From the reaction

$$CaCO_3 + H_2O + CO_2 = Ca + 2HCO_3 \tag{8.7}$$

half of the carbon atoms in the HCO_3 should derive from bedrock $CaCO_3$ (where ^{14}C is absent) and half from soil air or other sources of modern CO_2. However, during deposition heavier isotopes such as ^{14}C may pass preferentially into the solid phase, slightly enriching it. Researchers have generally assumed 'live' carbon proportions of 50 or 60%. From a study of Holocene speleothems in Belgium, Bastin & Gewellt (1986) place it at 80–85%. Gascoyne & Nelson (1983) obtained a proportion of only 30–35% from Castleguard Cave, where there is no soil CO_2 source and thus ^{14}C may be especially depleted. Because of the 'dead' carbon problem and the limited time range of the ^{14}C method it has seen comparatively little application to speleothems. The principal studies are by Franke *et al.* (1958), Hendy (1970) and Franke & Geyh (1971).

The method has been extensively used to date organic matter (wood, bone, etc.) in cave entrance deposits. In the interior deposits of humid caves the frequent wetting and draining favour rapid organic decay and the system may also be open to addition of ^{14}C *post mortem*. It has been little used.

Absolute dating: (B) Uranium series methods

Disequilibria in the U species are widely used for dating, for the study of weathering systems and groundwater flow. They are the principal methods of dating speleothems at the present time. Comprehensive accounts are given in Cherdyntsev (1971) and Ivanovich & Harmon (1982). Shorter technical accounts especially concerned with cave deposits include Gascoyne *et al.* (1978), Hennig *et al.* (1980) and Schwarcz (1980).

There are two natural parent isotopes, ^{238}U ($\lambda = 4.47 \times 10^9$y) and ^{235}U

($\lambda = 7.13 \times 10^8$y). With such long half lives they and their daughter isotopes survive as common trace elements in igneous and derived rocks, especially black shales. They decay by emission of α-particles (helium nuclei) and β particles to produce stable ^{206}Pb and ^{207}Pb. The intermediate daughters, ^{234}U ($\lambda = 2.48 \times 10^5$y).^{230}Th ($\lambda = 7.52 \times 10^4$y), ^{226}Ra ($\lambda = 1602$y) and ^{231}Pa ($\lambda = 3.43 \times 10^4$y) are also suitable for dating because of their comparatively long half lives.

When a rock containing U is weathered, a higher proportion of ^{234}U atoms are mobilized than of ^{238}U or ^{235}U atoms i.e. there is 'daughter excess' (Fig. 8.19). This is because many of the ^{234}U atoms are loosened in their lattice position when emitting an α-particle. All three species are readily oxidized and transported in solution in bicarbonate waters as the complexed ions, $UO_2\ (CO_3)_2^{\ 2-}$ and $UO_2\ (CO_3)_3^{\ 4-}$. They may then be co-precipitated in calcite or aragonite. Their long lived daughters, ^{231}Pa and ^{230}Th, are essentially insoluble. When detached by weathering these bond to clay or other particles. Therefore, they are not precipitated in the calcite. In an ideal closed system they will accumulate there only as a function of the decay of the parent U species. One gram of calcite with a trace U content of 1.0 ppm contains 10^{15} atoms of uranium available for spontaneous decay.

The chief dating method uses the decay of excess ^{234}U to ^{230}Th (Fig. 8.20). It can be used to a normal limit of 350 ka (or approximately ten times the ^{14}C range) by the standard method of counting α disintegrations. Substitution of atom counting by mass spectrometry may permit extension

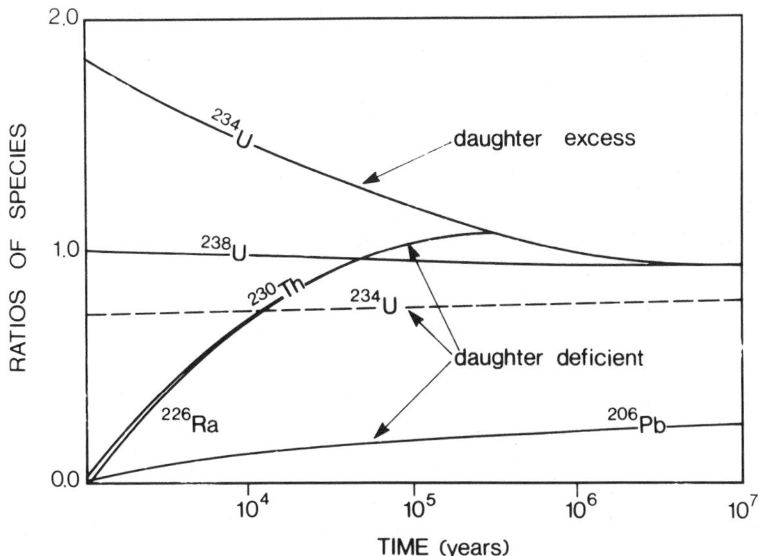

Figure 8.19 Evolution of the longer lived radioisotopes in the series, ^{238}U–^{206}Pb, to display the various dating schemes. Diagram is illustrative only; gradients are not to true scale.

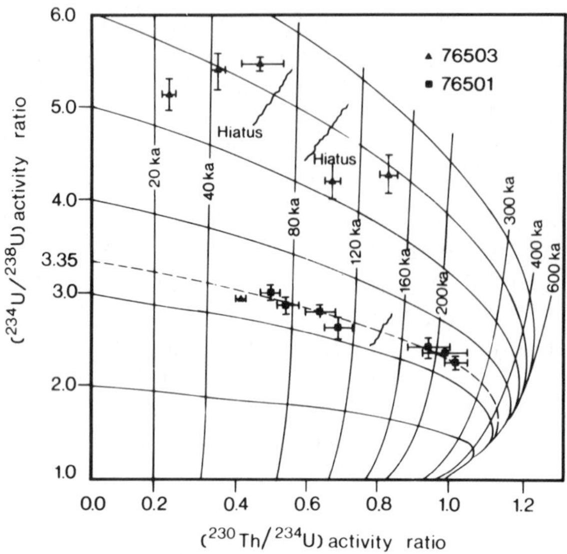

Figure 8.20 Graphical illustration of the ^{230}Th:^{234}U dating method. Most clean speleothems are deposited with an initial ^{234}U:^{238}U activity ratio greater than 1.0 and a ^{230}Th:^{234}U ratio of 0.0. With passage of time, ratios evolve to the right. Sample 76501 grew between 250 ka and 50 ka (with one hiatus) with an initial ^{234}U:^{238}U ratio always close to 3.35. The initial ratio in sample 76503 varied between 5.3 and 6.3.

of 500 ka. More than 1200 ^{230}Th/^{234}U dates of speleothems have been published.

The ^{231}Pa/^{235}U method permits dating only to ~ 200 ka because the half life of ^{231}Pa is shorter than that of ^{230}Th. A major problem is that the natural abundance of ^{235}U is low; ^{238}U : ^{235}U = 137.9 : 1. As a consequence this method has been used only to check ^{230}Th/^{234}U results in samples where U concentration is greater than 1.0 ppm.

Decay of ^{230}Th to ^{226}Ra permits dating to ~ 10 ka. Because of this short timespan the method is rarely used. Latham *et al.* (1986) applied it when studying the magnetic record of a Mexican stalagmite that had apparently grown to a height of 72 cm in the past 2000 years.

The flow of groundwater through U source rocks that have already been highly leached may lead to deposition of calcite with a deficiency of ^{234}U rather than an excess with respect to ^{238}U. This offers an absolute dating method if the initial deficiency can be determined, and establishes the duration of growth of a deficient calcite where it cannot. The only certain example of daughter-deficient calcite known to us is some nailhead spar precipitated from thermal water in Jewel Cave, S. Dakota (section 7.7).

Decay of ^{238}U to ^{234}U may be differentiated for 5–6 half lives of the daughter, i.e. to 1.25–1.50 Ma (million years). Unfortunately, the initial

$^{234}U/^{238}U$ ratio cannot be determined analytically unless ^{230}Th is also in disequilibrium, i.e. the calcite is younger than 350 ka. $^{234}U/^{238}U$ disequilibrium only indicates that a sample is younger than 1.25–1.50 Ma. In an early study, G. Thomson *et al.* (1975) assumed that the initial $^{234}U/^{238}U$ ratio measured in the part of a stalagmite that was <350 ka in age also applied where it was >350 ka. Harmon *et al.* (1975) showed that these initial ratios vary considerably within a cave and even within a single stalagmite; strictly speaking, G. Thomson's assumption is not justified. However, Ford (in Gascoyne *et al.* 1983) suggested that where the initial ratio has been measured at many places and over the whole range, 0–350 ka, in a given small region, the mean ratio might be applied if the standard deviation is low. This method is termed RUBE dating, for 'Regional Uranium Best Estimate'. It permits informed guesses of the true age in the timespan, 350 ka to 1.50 Ma.

Finally, the decay of parent uranium to stable lead in principle permits dating of calcite back to the origin of the Earth itself. In practice, discriminating radiogenic ^{206}Pb or ^{207}Pb from background traces of lead accumulated in a speleothem will be analytically difficult and, at present, appears to be infeasible where the sample is younger than ~ 2.6 Ma.

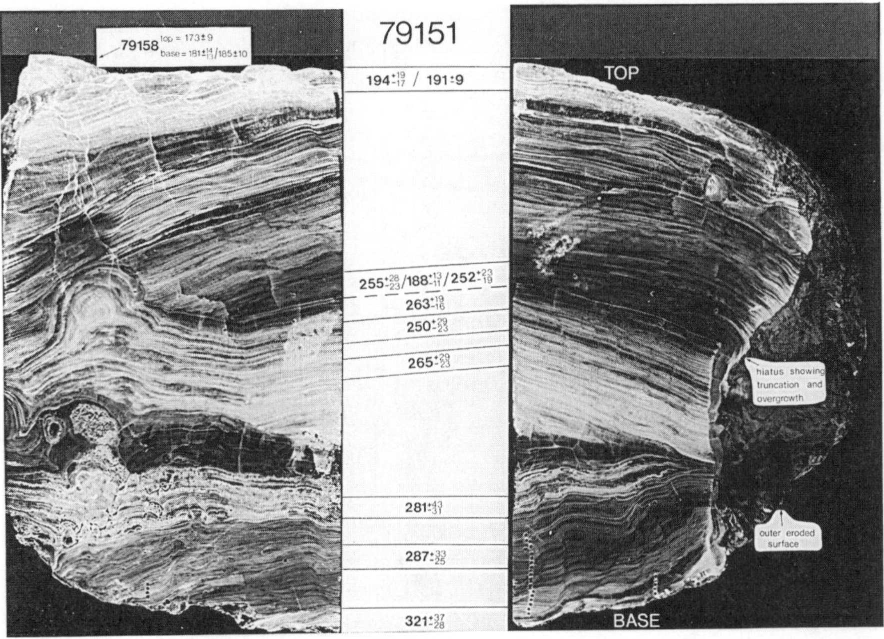

Figure 8.21 $^{230}Th/^{234}U$ dating of flowstone sample 25 cm in thickness, Victoria Cave, Yorks, England. Growth commenced ~ 320 ka. It was interrupted by lengthy erosion phases around 280 ka and 250–260 ka before growth ceased at ~ 175 ka BP. Growth was probably halted by cold phases. From Gascoyne *et al.* (1983).

There are three basic requirements for all U series methods: (1) that the calcite contain sufficient U. Measured concentrations range <0.01 ppm to >120 ppm. 0.01 ppm is the current minimum for $^{230}Th/^{234}U$ dating by α spectrometry; at that concentration, 50 g of sample will be necessary. Ninety per cent of assayed speleothems contain more than this minimum concentration. (2) is that the system be closed after co-precipitation of U and calcite. Often this will be violated. Many speleothems are partly or wholly re-crystallized. Others are porous so that water flows freely through them and may preferentially leach ^{234}U. This results in too great an age being calculated. For this reason stalactites (with their central feedwater canals) and porous tufa deposits are to be avoided, if possible. The most important requirement (3) is that no ^{230}Th or ^{231}Pa be deposited in the calcite. In fact, most calcite contains a proportion of these species and of ^{232}Th ($\lambda = 1.39 \times 10^{10} a^{-1}$) bonded to clay deposited in the speleothem. They are contaminants. As they increase, reliability of calculated dates deteriorates. In practice, where the ratio, $^{230}Th/^{232}Th$, is >20 it is presumed that radiogenic ^{230}Th completely predominates and that contamination is not significant. For highly contaminated deposits ($^{230}Th/^{232}Th < 5.0$) three or more determinations are recommended in order to calculate the isochron, $^{230}Th/^{232}Th$ v. $^{234}U/^{232}Th$ (Schwarcz 1980). If possible samples that are visibly dirty should be avoided. Unfortunately, those are often the most interesting, e.g. spring travertines or flowstones in cave entrances.

U series methods are also applied to bones and shells. Because of *post*

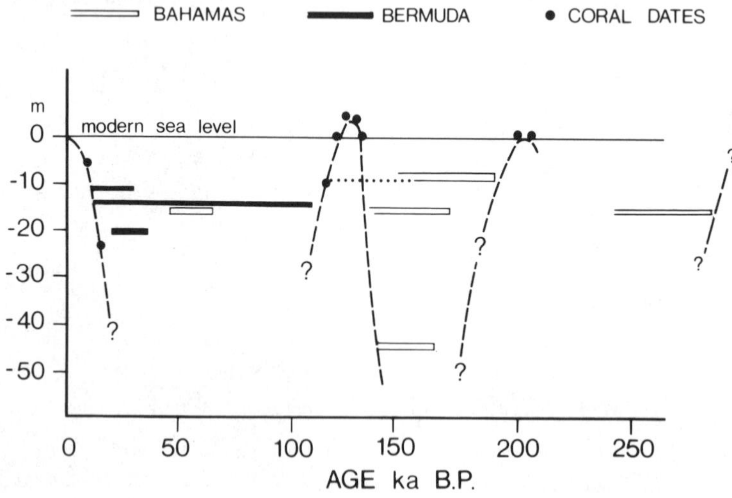

Figure 8.22 Dating of subaerial speleothems submerged to depths of −11 to −45 m in Bermuda and the Bahamas, plus corals that grew at high sea levels, helps to determine the oscillation of global sea level during the Quaternary. Data from Gascoyne *et al.* 1979, Harmon *et al.* 1983 and Ford unpublished.

mortem uptake of ^{234}U and other factors the analytical difficulties are considerable (Schwarcz 1980).

More than 1200 examples of U series dating of speleothems have been published (Gascoyne 1984). The coverage is worldwide. Dates are used to determine mean rates of growth of stalagmites and flowstones (Table 8.5, Fig. 8.21). More fundamentally, they give the minimum ages for the cutting of the vadose trenches or draining of the phreatic passages that they now

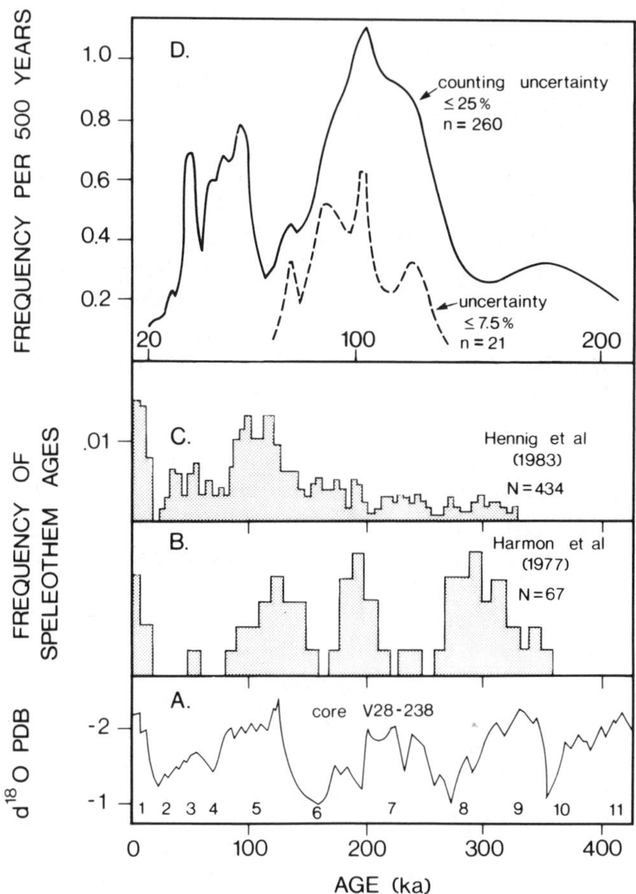

Figure 8.23 Frequency of $^{230}Th/^{234}U$ speleothem dates from formerly glaciated and marginal regions being used to date glacial episodes. A. The tropical oceanic foraminifera core V28–238, that is believed to record the glacial cycles (Shackleton & Opdyke 1973, plus numbered isotopic stages defined by Emiliani 1966). B. The simple distribution of 67 speleothem dates from the Rocky and Mackenzie Mountains, Canada (Harmon *et al.* 1977). Class interval=10 ka. C. Simple distribution of 434 speleothem dates from Europe and North America. Class interval = 5 ka (Hennig *et al.* 1983). D. The distribution of 260 speleothem dates from the British Isles by weighted frequency. (See Gordon *et al.* (1989) for details).

occupy (Fig. 4.13). By extension, mean maximum rates of channel entrenchment can be computed (e.g. section 4.4) and the rates at which river valleys or glacial valleys have been entrenched below paleospring positions; the principles are set out in Ford *et al.* (1981). Ford (1973) used the dated drainage of phreatic caves on a rising anticline to estimate the age of antecedent canyons along the South Nahanni River, Canada. Williams (1982b) dated the tectonic uplift of coastal terraces in New Zealand by the same means. A feature of most work of this kind is that the caves or their draining have proved to be much older than was previously supposed.

Speleothem ages are used to date episodes of clastic sedimentation and erosion in cave interior facies e.g. Sorriaux (1982), Gascoyne *et al.* (1983) and Milske *et al.* (1983). More obviously, the methods are applied to date entrance facies or spring travertines containing bone, artefacts or other remains of early Man, e.g. Caune de la'Arago and La Chaise-de-Vouthon, France (Turekian & Nelson 1976, Blackwell *et al.* 1983), Bilzingsleben Quarry, GDR (Glazek *et al.* 1980), Pontnewydd Cave, Wales (Green *et al.* 1981), and the famous Peking Man site, Zhoukoudian, in China (Yan *et al.* 1984). All of these encountered contaminant thorium problems.

Speleothems in caves of the Bahamas and Bermuda that are now submerged supplied the first absolute dates for Quaternary global low sea levels (Fig. 8.22; Spalding & Matthews 1972, Harmon *et al.* 1978b, 1983, Gascoyne *et al.* 1979).

Finally, speleothem growth in glaciated regions and their peripheries tends to cease during periods of greatest cold (e.g. Fig. 8.21). Harmon *et al.* (1977) and later workers used this feature to broadly date interglacial and interstadial periods (Fig. 8.23). Much speleothem growth ceased at ~ 26 ka in Britain, the onset of the Late Devensian glaciation, and began again 10–12.5 ka (Gascoyne *et al.* 1983, Atkinson *et al.* 1986). However, there are always exceptions. Castleguard Cave is the one cave that is known beneath an extensive *modern* glacier; it has growing calcite speleothems, attributed to sulphate complexing in the feedwater (Atkinson 1983).

Absolute dating: (C) thermoluminescence (TL) and electron spin resonance decay (ESR)

These methods rest on the principle that environmental radiation both from radionuclide decay (alpha, beta and gamma radiation) and cosmic sources may dislodge electrons from their initial sites in a crystal lattice and so give rise to electron deficiencies. The displaced electrons may themselves be trapped elsewhere in the lattice, becoming surplus electrons. These defects (or 'traps') accumulate at a rate proportional to the environmental radiation until they reach a certain density, when the rate of spontaneous decay is equal to the rate of formation. The crystal is then said to be 'saturated'. In an unsaturated sample, age may be determined from

$$\text{age } [a] = \frac{\text{Accumulated dose } (AD) \text{ [Gy]*}}{\text{environmental dose rate } (DR \text{ (Gy a}^{-1}))} \quad (8.8)$$

Because of the natural variation in the response of natural substances to radiation, the accumulated dose is determined by the additive technique, i.e. samples are given stepwise additional radiation (usually from a gamma source). Their response is measured, extrapolated back to zero and thus the pre-irradiation accumulated dose (AD) is determined. In TL dating the response is measured in the luminescent glow curves obtained on heating, whereas in ESR dating the response is measured by ESR spectroscopy (Fig. 8.24) which is based on the ability of unpaired electrons to absorb microwave radiation. The environmental dose rate is estimated from concentrations of radionuclides (U, Th, K) in the sample and by γ-ray measurement at the site (Hennig & Grun 1983, Debenham & Aitken 1984, Grun 1985, Ikeya 1985, Lyons *et al.* 1985).

These methods consume less sample calcite than do the U series methods, and have the potential of applying to a wider age range up to 1 million years, but determination of dose rates throughout the lifetime of the sample is inherently difficult. As a speleothem grows the dose rate need not be constant. Considerable differences of dose rate have been recorded a few centimetres apart in cave deposits (Debenham & Aitken 1984). Correlation with U series ages is sometimes excellent, sometimes poor (Fig. 8.24B). Furthermore, great uncertainty still exists in the reliable determination of both AD and DR, with the consequence that meaningful error limits cannot yet be quoted for the estimated ages. While recognizing the considerable potential of ESR dating in particular, especially for speleothems beyond the

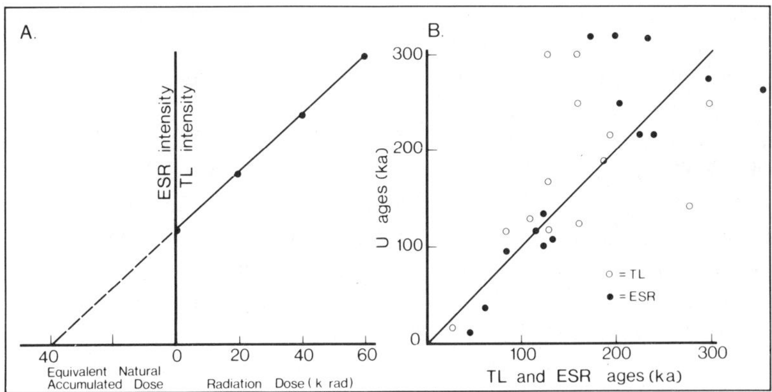

Figure 8.24 A. The additive radiation dose method of determining the accumulated natural dose of radiation in a crystal by means of TL or ESR signals. B. TL and ESR dates of calcite speleothems compared to ^{230}TH/^{234}U dates. Data from Debenham & Aitken (1984), Grun (1985) unpublished.

limit of the ^{230}Th : ^{234}U method, we stress that all TL and ESR ages on speleothems published so far must be treated with considerable caution.

TL and ESR methods have been applied also to shells, teeth, bones, loess, sand and tephra. Traps in the loess or sand grains are emptied when they are exposed to sunlight. Saturation occurs ~ 100 ka after burial (Wintle & Huntley 1982). Quinif (1981) used TL measurements of quartz grains from a Belgian cave to show that the quartz derived from ancient plateau gravels, i.e. the method was applied as a tool in provenance studies.

Comparative dating methods: (A) Paleomagnetism

The Earth's magnetic field displays secular variations in its polar declination and inclination and in its intensity. The variations are generally small in amount and irregular. Complete reversals of the field occur at 10^5–10^6 year intervals, termed 'magnetic epochs' or 'chrons'. The present Brunhes epoch commenced ~ 730 ka ago and is defined as 'normal'. It was preceded by the Matuyama 'reversed' epoch, 0.73–2.5 Ma; that contained some normal episodes, termed 'events' or 'subchrons' (IUGS 1979).

Use of records of ancient variations or reversals as a dating tool relies on matching the curves of declination and inclination (and perhaps intensity) in a given deposit with established curves that have been dated by independent methods. For the Pliocene and Pleistocene the established curves derive chiefly from lake and ocean floor sediments dated by ^{14}C, U series or ^{10}Be methods, plus some lavas dated by the K/Ar method.

In caves paleomagnetic studies have been applied principally to deposits of laminated clays and silts (e.g. Turner & Lyons 1987). Schmidt (1982) shows that sands may also be used where they have remained moist and stable (show no obvious deformation structures). Williams *et al.* (1986) report some success with calcite-cemented clays and even with layers of cemented cave pearls. The carriers of the remanent magnetism are usually detrital grains of magnetite.

Early work was confident. For example, Kopper & Creer (1973) dated the base of a Majorcan cave clay to 12±0.4 ka by fitting the declination and inclination curves to a ^{14}C controlled record from the bottom of Lake Windermere, England. There followed recognition that clastic sediments can suffer much post-depositional alteration of the D and I signals, especially if they have drained (Verosub 1977, Noel 1986). Bioturbation may also be a major problem e.g. at the Mulu Caves, Sarawak (Noel & Bull 1982). As a consequence, cave sediments are now studied chiefly to establish whether their declinations are normal or reversed, the latter implying that they are probably >730 ka in age.

One of the foremost studies is by Schmidt (1982). He collected more than 500 samples of layered clays, silts and fine sands from active and relict passages in Mammoth Cave–Flint Ridge system (Fig. 8.25). Modern

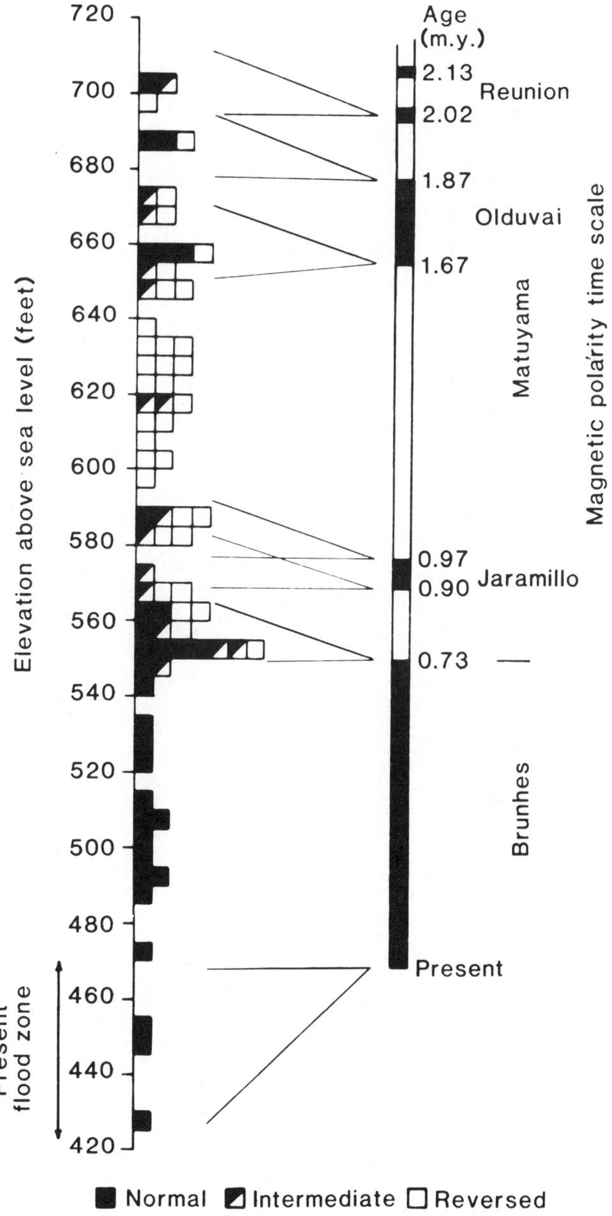

Figure 8.25 The record of normal and reversed magnetic declinations obtained from 500 samples of clays, silts and sands deposited throughout the height range of the Mammoth–Flint Ridge–Roppel–Procter cave system, Kentucky. The record is correlated with the magnetic polarity timescale to estimate that the oldest deposits are 2.0 to 2.1 Ma in age. From Schmidt (1982) by permission.

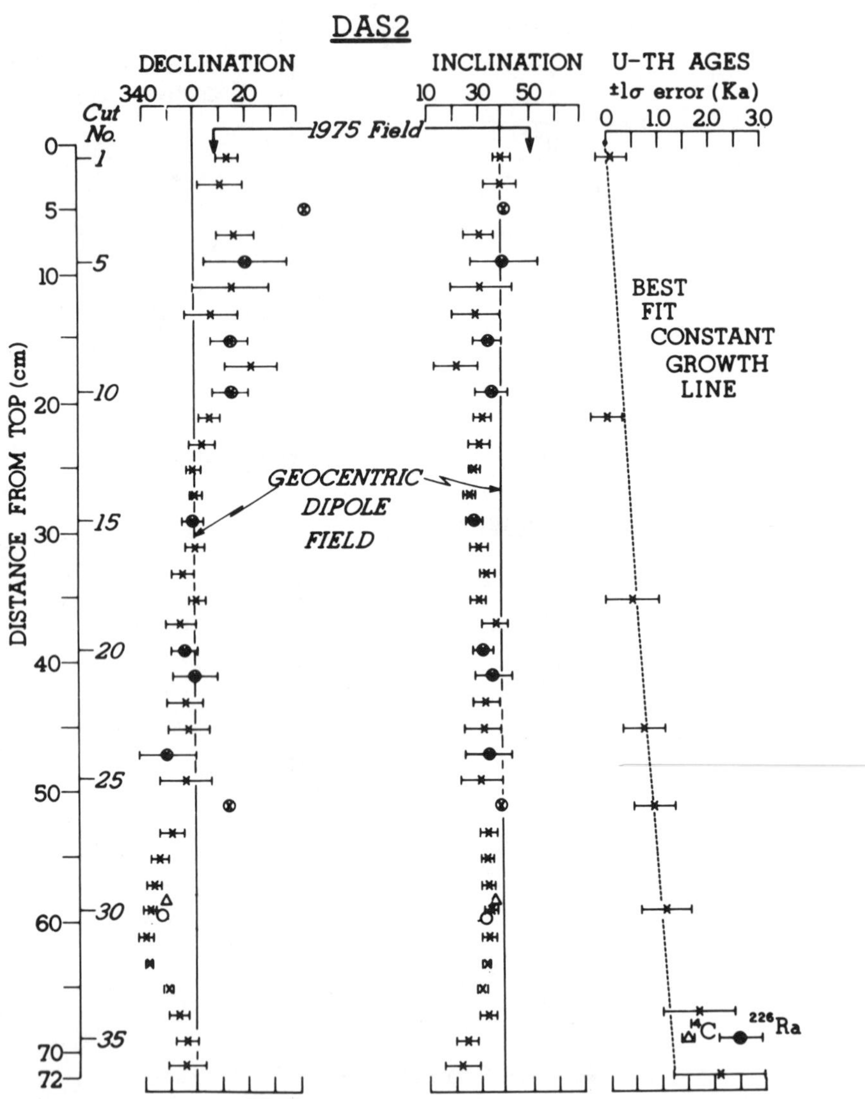

Figure 8.26 The magnetic declination and inclination of Mexican stalagmite, DAS2, and its growth curve from 1200 a^{-1} BP as determined by $^{230}Th/^{234}U$ and ^{14}C dating. The site is Sotano del Arroyo, at Lat. 22.08°N, Long. 99–0°W. From Latham *et al.* (1986).

deposits extend 30 m above the water table at 130 m asl and display normal polarity, as do relict deposits immediately above them. A magnetic reversal occurs in most deposits sampled between 35–40 m above the water table. This was correlated with the Brunhes–Matuyama reversal at 730 ka. Normal polarity was measured in many sediments in the highest passages (190–210 m) and is tentatively correlated with normal events within the Matuyama epoch. This implies that the earliest sediments accumulated during periods between 1.7 and 2.1 Ma.

Post-depositional alteration cannot occur if the magnetic grains are cemented inside calcite, and their preserved records of magnetic variation then may be dated independently by U series methods, etc. Latham *et al.* (1979) recognized this possibility and showed that many stalagmites and flowstones carry natural remanent magnetism either as a chemical precipitate (CRM) or as floodwater or filtrate detrital grains (DRM). Magnetite is the carrier in both cases. The signal is weak, requiring a high sensitivity magnetometer to measure it. Full technical details are given in Latham (1983).

Paleomagnetism of speleothems has been used; (a) to test for normality or reversal where a sample is known to be older than 350 ka (^{230}Th/^{234}U method) or older than ~ 1.5 ma (^{234}U/^{238}U method) and (b) to obtain dated, high resolution curves of recent secular variation e.g. Figure 8.26.

Comparative dating methods: (B) Biostratigraphic methods; fauna, flora and pollen, humic and fulvic acids

Here deposits may be used for approximate dating by correlation with external type sections, and also in the reconstruction of past environments such as successive ecological assemblages above a cave.

Troglobitic flora and fauna (i.e. exclusively cave dwelling, Jefferson 1976) are too small in number and volume to be significant in most instances. Animals that roost or nest in caves but forage outside (trogloxenes) are more important. Rodent nests, bones and faeces are often found in the furthest interior parts of shallow cave systems. More striking in many caves of middle Europe are remains of an extinct bear, *Ursus spelaeus*, that is conventionally dated to the last interglacial and the lower half of the Wurm glacial period. Little dating usage has been made of other trogloxenes. Extinct fauna are common in cave entrance facies, where they have been intensively studied.

Flora carried into cave interiors are generally small in amount and prone to rapid decay. Most attention has focused on pollen and spores. Many clastic deposits are barren or contain only degraded grains because of oxidation. There have been recoveries of well preserved pollen from laminated silts and clays in some Belgian and French caves, in Kentucky and a few other places (Barrett 1963, Leroi-Gourham 1967, Renault-Miskovsky 1972, Damblon 1974, Peterson 1976).

Geurts (1976) and Bastin (1978) pioneered the extraction of pollen from spring travertines, stalagmites and flowstones. Comparatively large volumes of sample are usually needed to obtain significant pollen counts e.g. 100–200 g or more. This means that the time resolution will be poor unless the host deposit grew rapidly. However, this is offset by the ability to date the calcite independently by the absolute methods.

The scope for Quaternary palynology with speleothems appears to be good. Bastin has concentrated on Holocene deposits in Belgium and shown that in samples taken close to cave entrances the sub-Boreal to Boreal, and other floral transitions of western Europe can be recognized (Bastin & Gewellt 1986). Work in progress in Canada has recovered arboreal pollen from Late Tertiary speleothems in subarctic caves.

Caution is needed in interpreting pollen assemblages from cave deposits because there are three potential, distinct sources of supply: (a) eolian – which presumably gives a valid sample of the contemporary regional pollen, (b) speleothem feedwater or other infiltration; pollen grains range 0.5–100 μm in diameter so that where infiltration is an important source of them, species represented by larger grains will probably be screened out, (c) floodwaters, in which much or all of the pollen may be reworked from older deposits. Young stalagmites out of range of floods and close to entrances so that the eolian source of supply is maximum have been selected by Bastin. No stalagmite further than 650 m from an entrance has been reported upon as yet.

Humic and fulvic acids (long chain organic molecules) and possibly terpenes (smaller organic molecules) are present and well preserved in many stalagmites, where they supply brown–dark brown colour (Lauritzen *et al.* 1986). Potentially they will permit broad identification of soil types at paleo-feedwater sources.

Stable isotopic studies of speleothems

The deep interiors of perhaps the majority of extensive caves are very stable climatically. Away from major streams their temperatures vary by less than 1°C and are close to the mean annual external air temperature. Relative humidity is similarly invariant, often being close to 100%. Further, Yonge *et al.* (1985) have shown from a study across North America that the stable isotopic composition of feedwaters to speleothems is also essentially invariant over the climatic year. Seasonal contrasts in precipitation or those occurring within individual storms (section 6.9) are homogenized during the flow of these tiny threads of water through the epikarst and to the speleothems beneath it. Therefore, cave interior speleothems are potentially excellent subjects for study of the long term changes of continental mean temperatures and perhaps other climatic parameters.

The isotopes of interest are ^{16}O, ^{18}O, ^{12}C and ^{13}C. Oxygen is the more readily fractionated because C is held at the centres of CO_3 groups in calcite. As a consequence, most study has focused on the enrichment or depletion of ^{18}O with respect to ^{16}O. The basic principle is that the heavier isotope is preferentially concentrated or retained in the denser phase. The amount of such fractionation when calcite is deposited may be determined by the ambient temperature (termed 'equilibrium fractionation') or by a combination of temperature plus evaporation ('kinetic fractionation'). The criterion for equilibrium fractionation is that there is no correlation between the trends of ^{18}O and ^{13}C measured along a calcite growth layer (Hendy & Wilson 1968).

Data are expressed in permil or 'delta' (δ) notation as explained in section 6.10. Fractionations in water are expressed against standard mean ocean water (SMOW), those in calcite against a standard based on a fossil belemnite, PDB.

$$\delta^{18}O_{SMOW} = 1.03086\delta^{18}O_{PDB} + 30.86 \tag{8.9}$$

The equilibrium fractionation factor, calcite–water, is expressed as

$$1000 \ 1n \ \alpha_{c\text{–}w} = \delta^{18}O_c - \delta^{18}O_w \tag{8.10}$$

The temperature relationship (O'Neill *et al.* 1975) is

$$T = 16.9 - 4.38 \ (\delta^{18}O_c - \delta^{18}O_w) + 0.10 \ (\delta^{18}O_c - \delta^{18}O_w) \tag{8.11}$$

where T = °C. In a majority of applications published thus far, only $\delta^{18}O$‰ of the calcite has been determined.

As $\delta^{18}O_c$ varies in equilibrium calcite it indicates changes to warmer or cooler cave temperatures. To determine the absolute temperature change that occurs it is necessary to measure $\delta^{18}O$ of the formation water as well. One of the great advantages of the dense, crystalline calcite typical of humid cave interiors, when compared to most other calcareous substances used in paleo-environmental reconstructions, is that it may retain the formation water in fluid inclusions (section 8.3). The modern $\delta^{18}O$ in an inclusion may not be valid because it can exchange with the calcite lattice if there is a change of temperature. However, the ^2H : ^1H ratio (D/H ratio) in the inclusion is stable, and the formation $\delta^{18}O$ value is obtained from the meteoric water line (section 6.10)

$$\delta D_{inclusion} = 8\delta^{18}O_w + 10 \tag{8.12}$$

This relation has been proven to apply in modern karst waters ranging from

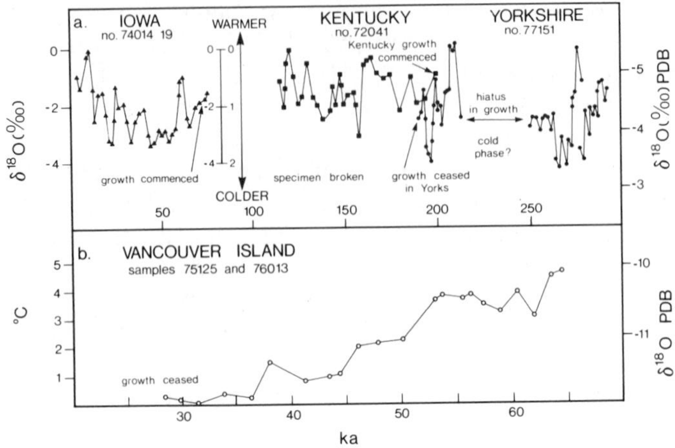

Figure 8.27 (a) The interpretation of equilibrium ^{18}O changes in U-series dated speleothems from Yorkshire, England, Kentucky and Iowa, USA. $\delta^{18}O$‰ scales on left are normalized to zero points at the peaks (inferred warmest positions) to facilitate comparison. As temperatures became colder the calcite became 'heavier'. From the work of Harmon (1975) and Gascoyne (1979). (b) $\delta^{18}O$ trend and computed paleotemperature of speleothems from Vancouver Island, where the oceanic effect overrides the cave temperature effect dominant in (a). As temperatures became colder the calcite became 'lighter'. From Gascoyne *et al.* (1981).

subarctic to tropical locales (Schwarcz *et al.* 1976).

From the use of fluid inclusions, it has been suggested that cave temperatures shifted as much as 8°C between glacial and interglacial times at sites south of the ice limits in the interior United States (Harmon *et al.* 1978a). However, the method has seen comparatively little use because it requires access to a D/H mass spectrometer and because of problems encountered in the fluid extraction. Yonge (1982) presents a full technical review.

The relationships between regional or global temperature changes and the proportion of ^{18}O isotopes deposited in equilibrium calcite are not simple. As noted, when a cave interior cools in response to external cooling more ^{18}O will be deposited; i.e. the calcite is said to become 'heavier'. This is the 'cave temperature' effect. However, the ultimate sources of the feedwater are the oceans. Temperate and polar oceans are cooled during glacial phases so that precipitation derived from them is depleted in ^{18}O i.e. becomes 'lighter'. This is the 'oceanic effect'. Intertropical oceans are believed to have experienced little cooling.

The two effects are opposed to each other, and further fractionation effects may intervene between ocean source and cave sink. Isotopic characteristics in each karst region therefore need to be studied carefully before ^{18}O trends are interpreted as paleotemperature changes. At a majority of modern temperate and alpine sites the cave temperature effect predominates i.e. equilibrium calcite becomes 'heavier' as the temperature falls (Fig. 8.27). But at Vancouver Island, a perhumid ocean margin site in the temperate belt, the oceanic effect prevailed during the last glaciation. This permitted direct computation of paleotemperatures from $\delta^{18}O_c$ (Gascoyne *et al.* 1981). Between 64 ka and 28 ka (U series dates) the mean annual temperature in the sample cave fell from +4.5°C to ~ 0°C, when calcite deposition ceased (Fig. 8.27b). The modern temperature is 8°C. Goede *et al.* (1986) obtained a similar result from a Tasmanian speleothem.

There has been little study of equilibrium speleothems from intertropical regions. Harmon (1975) found negligible change of ^{18}O as some Mexcian stalagmites grew into the cold phases of the last glaciation.

It is now recognized that speleothems growing entirely in equilibrium fractionation conditions are comparatively rare. They are to be sought in recesses or closed chambers away from vigorous air flow, and where relative humidity is always high.

Some speleothems that have grown continuously over long spans of time display an alternation of equilibrium and kinetic fractionation conditions. These are interpreted as changes between no evaporation and effective evaporation, that may be correlated with Pleistocene climatic oscillations.

There has been little interpretive study of $^{13}C : ^{12}C$ trends in speleothems. $\delta^{13}C = 0 \pm 5‰$ PDB in most carbonate rocks. HCO_3 ions in soil water display ^{13}C ranging −16 to −24‰, depending upon the plant association. As with the case of ^{14}C discussed above, the isotopes subsequently precipitated in speleothems may be drawn equally from each of these sources but differing enrichment or depletion effects can distort the proportions. For example, thermal water calcites are comparatively enriched in ^{13}C due to high temperature leaching of limestone along the flowpaths (Bakalowicz *et al.* 1987). Where such special effects may be discounted, a change of several ‰ over time in an equilibrium speleothem is to be interpreted as a major change in the type or extent of vegetation cover at the feedwater source.

Seismospeleology

In many caves a proportion of the calcite speleothems has been fractured by natural causes and has fallen from its growth position. Straw stalactites break under their own weight and larger carrot stalactites fall when their weight overcomes the bonding to the ceiling. Stalagmites and flowstone masses built upon unconsolidated sediments may topple because the sediment foundation fails under the increasing load. However, there are instances where columnar stalagmites are broken off above a base that appears not to have shifted. More than 100 years ago European researchers suggested that such breakage might be caused by earthquakes; see Schillat (1965) for a review of early work.

Most attention has focused upon the position and orientation of the broken columns. For example, a strong earthquake shock propagated from due East might be expected to place the broken section on the West side of its base, with a westerly orientation. Directions to epicentres of some past earthquakes have been determined by this means. Moser & Geyer (1979) studied the relationship between basal diameter and length of the broken fragment in an attempt to determine the magnitude of the vibration in Austrian alpine caves. In recent work in tectonically active regions of Italy, Forti & Postpischl (1984, 1985) have investigated the changes in long axis orientation that some stalagmites display, arguing that these are caused by slight tilting of the cave due to tectonism. In one example, 21 small shifts of the axis were measured in a specimen 36 cm in height. Directions of shifts were shown to fit the main tectonic trends in the cave areas.

The ages of axial displacements can be measured by U series dating, as can the date of a breakage if calcite growth is renewed on the stump. However, both breakage and axial displacement may be caused by cave processes alone, as indicated. Correlated results from a number of speleothem samples in a given region will be necessary to establish that there is a common, external cause that is probably seismic.

8.7 Mass flux through a cave system; the example of Friars Hole, W. Virginia

It is instructive to close this chapter with an estimate of the flux of all matter through a cave system during its lifetime. Few such estimates have been attempted because of the evident difficulties in devising them. Many quantities may be in error by at least one order of magnitude.

One set of estimates (Table 8.6) has been prepared by Worthington from his 1984 study in one of the world's most extensive caves, Friars Hole. This is something of an extreme example because only 3% of the surface catchment basin consists of the host limestones (Fig. 7.23). These outcrop as inliers in narrow valley floors, into which flow the detritus from steep slopes of shale, sandstone and argillaceous limestone, i.e. it is a situation favouring the maximum flux of allogenic debris through a system.

Extrapolating from U series, RUBE and paleomagnetic results, the earliest passages are 4 million years or a little greater in age. The mean solute mass flux from the host limestone has approximated 15 000 m^3 ka^{-1} over that period. Note that only 4% of this flux is represented by dissolution in the known cave. Generous estimates of the additional volume of unknown caves in the system will increase this value to 15–25%, i.e. 75%+ of net dissolution has occurred in the epikarst developed in the small windows or inliers.

Thirty-seven per cent of the volume of the known cave was created by

Table 8.6 Estimated dimensions and mass fluxes through the Friars Hole Cave System, West Virginia. (Adapted from computations by Worthington, from the data of Worthington, 1984).

Contributing drainage basin			–	85.7 km^2
including				
(Host limestone exposed at surface			–	2.6 km^2)
Friars Hole Cave System				
Explored length			–	68.12 km
Total volume of known cave			–	2700 × 10^3 m^3
(of which – now open			–	1800 × 10^3 m^3
now infilled with detritus			–	900 × 10^3 m^3)
Age of the oldest passages			–	>4.0 ma
Mass fluxes		Total (10^3 m^3)	In cave now (10^3 m^3)	
Dissolved host limestone				
(1) from the input surface	–	57 300	trace	
(2) from the known cave	–	2 400	trace	
Cave breakdown	–	1 000	280	
Authigenic fluvial detritus (mostly from breakdown)	–	400	20	
Allogenic fluvial detritus (mostly from siliciclastic rocks)	–	3 000 000	600	
Eolian deposits		<1?	<0.001	
Organic matter – all sources	–	100	0.001	
Calcite speleothems	–	~ 1	0.15	
Gypsum and other precipitates	–	0.001	0.001	

mechanical breakdown. Seventy per cent of the clasts produced have been removed, chiefly in solution.

Clastic rocks outcrop over 97% of the area of the basin and are estimated to have furnished 95–98% of the total mass flux in the system during its history. This component overwhelms all others. Only 0.2% of the aggregate detrital flux is in transit through the caves at the present time, yet this suffices to infill about 22% of their volume. The mean underground transit time of a clast is ~ 80 000 years. Flowpath lengths (sink to spring) will be between 15 and 60 km. Effective hydraulic gradients have been 0.006 or lower.

Although it has a few grottoes with large and abundant speleothems, Friars Hole is not a well decorated cave by world standards. It is estimated that only the tiny proportion of 0.0016% of solutes in transit from the epikarst have been intercepted and precipitated as cave calcite during its history.

9 Karst landform development in humid regions

9.1 Coupled hydrological and geochemical systems

It was explained in Chapter 4 that most dissolution is expended near the surface, in the epikarst. We now consider the landforms and assemblages of landforms that are created there. These vary from small features such as karren to large scale landforms measured in kilometres e.g. poljes. Within the dynamic karst system they can also be classified as input, output or residual features (Fig. 1.1). This chapter is organized to discuss input forms first, proceeding from smaller to larger, and then output and residual features. It concludes with a discussion of landform sequences in carbonate terrains and a review of the special features associated with evaporite rocks.

We introduced karst hydrology (Chs 5 & 6) and cave development (Ch. 7) before discussing surface landforms because the essence of karst is that its drainage is subterranean, although only a small proportion of the dissolution takes place deep underground. The initiation of karst plumbing is an essential pre-condition for the early development of medium to large scale surface landforms.

Karst landforms result from processes operating in coupled hydrological and geochemical systems. Essentially the same processes can operate over a very wide range of environments, but limiting conditions are provided by aridity and extreme cold. Karst is therefore characteristic of humid regions, where water normally occurs in its liquid phase. In this section we examine 'normal' karst development in humid areas, leaving consideration of karstification under extreme climatic conditions until the next chapter.

In endeavouring to understand the relation between processes, karst rocks and resultant landforms the following points are important:

(a) Hydrological processes determine the general location of erosion within karst lithologies and hence are usually the principal control on landform development. In particular, the nature of hydrological recharge, whether autogenic, allogenic or mixed, has considerable morphological significance because of its influence on the horizontal and vertical distribution of corrosion and corrasion.

(b) Lithology and structure can be so important as to dominate landform development although, in general, geology influences karst development through its control of (i) the provision of solute pathways; (ii) rock strength; and (iii) susceptibility to corrosion and corrasion.

(c) Different amounts of runoff in various humid regions influence annual karst erosion and hence the rate of landscape evolution, but not necessarily the morphological style of karst topography that is developed.

(d) Temperature variation is significant to morphological development mainly through its influence on (i) the water balance (via evapotranspiration), (ii) the rate of chemical reactions and hence the vertical distribution of corrosion, and (iii) biochemical processes leading to the acidification of infiltrating water. Depositional landforms are also influenced by temperature via evaporation and biological processes.

9.2 Small scale solution sculpture – microkarren and karren

The German term 'Karren' and the French term 'lapies' are widely used to describe small scale dissolution pit, groove and channel forms at the surface and underground. Here we anglicize the German and define *microkarren* as features whose greatest dimension or characteristic dimension (length, width, diameter, depth, etc.) is normally less than 1 cm. Karren range from 1 cm–10 m in greatest dimension in most instances, though kluftkarren and some solution channels can be longer. Assemblages of many individual karren, termed *Karrenfeld*, may cover much larger areas.

Karren develop upon the carbonate and sulphate rocks and dominate all outcrops of salt. They are also the dissolutional landforms most frequently encountered on other rocks such as sandstone, quartzite and granite. Lithologic properties are of great importance; many specific karren forms develop only if rocks are homogeneous and fine grained.

There is a vast range of karren features. Bogli (1980) wrote 'The multiplicity of possible karren forms makes a morphological system endless, while a genetic one allows a meaningful collection'. In 1960 he proposed a classification based primarily upon whether the host rock was bare ('free karren'), partly covered ('half free'), or entirely covered by soil or dense vegetation ('covered karren'). His scheme and the German names he gave to particular classes of karren have been adopted by most later reviewers (Sweeting 1972, Perna & Sauro 1978, Jennings 1985, Trudgill 1985).

We sympathize with the principle that a genetic classification is to be preferred to a morphologic one, but believe that the genesis of many karren is not sufficiently understood to warrant a wholly genetic basis at this time. In particular, much of the variety in karren occurs because two or more differing processes combine to produce a polygenetic form. The classification

Table 9.1 Classification of karren forms.

A Circular plan forms

Micropits and etched surfaces – wide variety of pitting and differential etching forms commonly less than 1.0 cm in characteristic dimension.
Pits – circular, oval, irregular plan forms, with rounded or tapering floors, > 1.0 cm in diameter.
Pans – rounded, elliptical, to highly irregular plan forms; planar, usually horizontal floors in bedrock or fill, > 1.0 cm in diameter.
Heelprints or trittkarren – arcuate headwall, flat floor, open in downslope direction. Normally 10–30 cm diameter.
Shafts or wells – connected at bottom to proto caves/small caves draining into epikarst. Great range of form.

B Linear forms – fracture controlled

Microfissures – microjoint guided, normally tapering with depth. May be several centimetres long but rarely more than 1.0 cm deep. Transitional to
Splitkarren – joint-, stylolite- or vein-guided solution fissures. Taper with depth unless adapted for channel flow. From centimetres to several metres in length, centimetres deep. Closed type terminates on fracture at both ends. Open type terminates in other karren at one or both ends.
Grikes or kluftkarren. Major joint- or fault-guided solutional clefts. Normally 1–10 m in length. Master features in most karren assemblages, segregating *clint* blocks (Flachkarren) between them. Scale up to karst bogaz, corridors, streets, etc. Subsoil forms are termed cutters.

C Linear forms – hydrodynamically controlled

Microrills – as on *rillenstein*. Rill width is ~ 1.0 mm. Flow is controlled by capillary forces and/or gravity and/or wind.
Gravitomorphic solution channels

1. *Rillenkarren* – packed channels commencing at crest of slope; 1–3 cm wide. Extinguish downslope. Rainfall-generated, no decantation.
2. *Solution runnels* – Hortonian channels commencing below a belt of no channelled erosion. Sharp-rimmed on bare rock (Rinnenkarren), rounded if subsoil (Rundkarren). Channels enlarge downslope. Normally, 3–30 cm wide, 1–10 m long. Linear, dendritic or centripetal channel patterns.
3. *Decantation runnels*. Solvent is released from an upslope, point-located store. Channels reduce in size downslope. Many varieties and scales up to 100 m in length, e.g. wall karren (Wandkarren), Maanderkarren.
4. *Decantation flutings* – solvent is released from a diffuse source upslope. Channels are packed; may reduce downslope. 1–50 cm wide.
5. *Fluted scallops or solution ripples* – ripple-like flutes oriented normal to direction of flow. A variety of scallop. Prominent as a component of *cockling patterns* on steep, bare slopes.

D Polygenetic forms

Mixtures of solution channels with pits, pans, wells and splitkarren. Subsequent development of *Hohlkarren*, *Spitzkarren* and subsoil *pinnacles*. Superimposition of small forms (microrills, Rillenkarren, small pits) upon larger forms.

Assemblages of karren

Karrenfeld – general term for exposed tracts of karren.
Limestone pavement – a type of karrenfeld dominated by regular clints (flachkarren) and grikes (Kluftkarren). Stepped pavements (*Schichttreppenkarst*) when benched.
Pinnacle karst – pinnacle topography on karst rocks, sometimes exposed by soil erosion.

continued

Table 9.1 *continued*

Arete-and pinnacle, stone forest, etc., with pinnacles to 45 m high and 20 m wide at base. *Ruiniform karst* – wide grike and degrading clint assemblage exposed by soil erosion. Transitional to *tors*.
Corridor karst – (or *labyrinth karst, giant grikeland*); scaled-up clint-and-grike terrains with grikes several metres or more in width and up to 1 km in length.
Coastal karren – distinctive coastal and lucustrine solutional topography on limestone or dolomite. Boring and grazing marine organisms may contribute. Includes intertidal and subtidal notches, and dense development of pits, pans and micropits.

adopted here (Table 9.1) is based on morphology, with subdivision that incorporates genetic factors. Bogli's (1960) nomenclature is retained wherever possible. Perna & Sauro (1978) give the equivalent names in many other European languages. It is emphasized that our classification stresses the comparatively simple end member forms. In reality there is a great mixture of forms created by factors of lithologic variation and poly-genesis combining together.

Microkarren

All the microforms distinguished in Table 9.1 are considered together here. Dissolutional topography can be recognized at a scale of a few micrometres under the electron microscope, but where the relief is less than 1 mm or so it is convenient to consider the surface to be smooth. Exposed karst rock surfaces generally display relief greater than 1 mm unless they are being subjected to vigorous scouring or polishing action. This relief can develop upon limestones within a few decades.

Many apparently bare carbonate surfaces are partly or entirely covered by bacteria, fungi, green algae, blue-green algae, or lichens. These may contribute to the preferential etching of weaker grains and to the development of micropits (section 4.4). Most attention has focused on the activity of blue-green algae (cyanophytes) since Folk *et al.* (1973) suggested that they produce much of the relief of coastal phytokarst. Most species are surface dwellers (epilithic), but in ecologically stressful environments some cyanophytes bore into rocks to depths of ~1.0 mm while others dwell in vacated borings or other micro-cavities. Borers create pits directly; other species may contribute to their creation or enlargement by way of the organic acids or CO_2 that they excrete (e.g. Verges 1985). Viles (1987) could not relate algal activity to specific microkarst except on sea coasts, however, so that their role remains uncertain. It is widely agreed that, once established, small pits and fissures may be preferentially deepened if fungi, lichens or mosses can establish in them and excrete CO_2. Cyanobacteria-induced pits have been measured to 14 mm (Fig. 4.10).

Figure 9.1 *Top left* – microrills on the walls of shallow solution pit in micrite. Coin is 2 cm in diameter. *Top right* – dissolution pan. *Lower left* – meandering micro-rills from stem flow into rund karren on micrite. *Lower right* – a pattern of multiple trittkarren (heel prints) in marble, with rillenkarren developing upon residual ridges. Can is 12 cm in length.

Microrills are typically 1 mm wide, roundbottomed dissolutional channels that are tightly packed together (Fig. 9.1). They are sinuous or anastomizing on gentle slopes, becoming straighter on steep slopes. Lengths are up to a few centimetres. Most reports of them are upon fine grained to aphanitic limestone but they also occur on gypsum. They are known in most climates. Clasts with microrills are *rillenstein* (Laudermilk & Woodford 1932).

Some microrills are created by waters flowing down surfaces (e.g. from acid stemflow over clasts; Trudgill 1985). But in the majority of instances the rilling takes place when waters move upwards, drawn by capillary tension exerted at an evaporating front (Laudermilk & Woodford 1932, Ford & Lundberg 1987). Capillary flow is believed to explain much of their characteristic sinuosity.

Solution pits, pans, heel-prints (tritt), shafts or wells, cavernous weathering

Solution pits are *round bottomed or tapering* forms that are circular, elliptical or irregular in plan view. Diameters greater than 1 m are rare; the form merges to a pan as that scale is approached. Pits can occur singly, aligned, clustered or packed. They may drain by evaporation, by overspill, and/or by basal seepage via primary pores or tight microfissures. Together with shafts, they are the most widespread karren form, both on bare rocks and beneath soil. They are predominant where the rocks are very heterogeneous (e.g. many limestone, and most dolomite, reefs).

Many pits are located along small joints, taper down into them and are transitional to shafts or to split karren (below). Others have developed at a cluster of primary pores or a vug, or where an insoluble fossil has fallen out. Deeper pits are often colonized by moss that appears to have accelerated the deepening of algae, etc. Some pits display raised rims, where water has precipitated calcite upon evaporation, armouring them. At an experimentally cleared limestone site in Yorkshire, England, Sweeting (1966) noted that pits 3–5 cm deep developed in ten years. The water was peaty i.e. enriched by organic acids.

Solution pans display a flat or nearly flat bottom that is usually horizontal. This may be created by an organic or clastic filling of a round bottomed pit, but most often it is a dissolutional bevel in the bedrock with an organic or other veneer. Walls are steepened by undercutting and may display a basal corrosion notch. Overflow channels are common. Individual pans attain diameters of several metres and depths greater than one metre. Amalgamation of adjoining pans is common, creating larger features of cuspate or irregular form. Pans develop well on limestone, dolomite, gypsum, quartzites, granites and well cemented sandstones. They are termed *solution basins*, *kamenitze* and *tinajitas* by other authors.

Solution pans occur on bare or lightly vegetated rock. They appear to be rare or absent beneath a soil cover, whereas solution pits are abundant there. This emphasizes that pans develop where a pool may form with a floor that becomes partially armoured by detritus, focusing dissolution around the perimeter. Pans cease to function when the floors, lowered by dissolution, intercept a penetrable bedding plane or other fissure.

Trittkarren or heelprints (Fig. 9.1) are comparatively rare. They occur on bare limestone and dolomite surfaces that are gently inclined or shallowly stepped. Each tritt comprises a planar corrosion bevel (the heel) open in the downslope direction. The bevel is usually horizontal, 10 to 30 cm in diameter. Upslope it is enclosed by a steep, cirque-like backwall a few centimetres in height. This may be indented by rillenkarren. The contact between backwall and bevel is sharp but not undercut. Tritt occur singly or adjoining one another where they indent a step.

Some trittkarren are modified solution pans but the majority appear to be of different, though related origin. Bogli (1960) ascribed their development to the accelerated dissolution that might occur where a film of flowing water is thinned upon descending a pre-existing step. Possibly, the early process is a boundary layer detachment as in dissolution scallops, though this cannot be true once rillenkarren are established. Haserodt (1965) attributed trittkarren to micro-snowdrift formation.

Trittkarren we have seen have been limited to homogeneous, fine grained to aphanitic limestone, dolomite or marble. They are also limited to surfaces where scouring agencies (chiefly glaciers, but also waves and flood waters) generate microscarps such as chatter marks.

Karren shafts or wells are very short caves draining into the epikarst. Most are guided by bedding planes, joints or calcite veins. More sinuous examples may follow primary porosity. They may be vertical, horizontal or inclined. Length (depth) ranges from a few centimetres to 2–3 m. Cross-sections tend to be circular or elliptical and up to 1 m in diameter, but there is great variety.

These features develop from proto-caves as described in section 7.2. In addition, pits and pans are converted into shafts where their floors intersect bedding planes. Many grikes are initiated by shaft development into an underlying bedding plane. Karren shaft forms and assemblages can be complex and variable where they develop beneath a deep, periodically saturated soil cover because the dominant condition is epiphreatic; it becomes a 'bone yard' morphology that is much favoured for ornamental rockeries throughout the world (Fig. 9.2).

Cavernous weathering refers to the bone yard type of cave and pit morphology described above. It can also describe individual blind caves or clusters of caves produced by weathering back into steep faces. The latter are common on some dolomites, and in sandstones, quartzites, conglomerates and granitic rocks where salt weathering and/or hydrolysis may play a role. In this kind of weathering the water does not pass via the cave into the epikarst; it is a superficial phenomenon.

Fissure-controlled linear karren

The term *splitkarren* was proposed by Pluhar & Ford (1970) and is extended here to encompass all karren depressions that are elongated along minor joints, veins, stylolites, or micro-fractures such as may develop normal to stylolite seams. Such features range in length from one or a few centimetres to several metres. Length : maximum width ratios are greater than ~ 3 : 1 and depth is usually much less than length. Unless adapted by a channelled flow, the features taper sharply with depth and thus appear to be splitting the rock. Closed splitkarren terminate on the host fissure. Open split karren terminate at one or both ends in other karren (e.g. a grike).

Figure 9.2 A. subsoil pit and shaft development in a limestone block set up as an ornament in the garden of the Imperial Winter Palace, China. B. subsoil clint-and-grike (or 'pinnacle-and-cutter') exposed in a quarry in thick bedded limestones, Kentucky. Arrow indicates persons for scale. Photo by A. N. Palmer.

Splitkarren may be transitional to pits, karren shafts or grikes. On slopes they are often intermingled with the larger gravity-controlled karren. Where a rock is densely fractured they may display a bewildering variety of orientations and intersections; development of other karren types then is prohibited by their density, with the exception of grikes and some shafts.

Grikes or kluftkarren are the master karren features in most karren assemblages. They are the principal drains, either to the deeper epikarst, or to dolines or to surface discharge such as river channels. They develop along the major joint sets or systems and thus tend to intersect at angles of 60°, 90° and 120° (shear and tension systems). Blocks isolated between them are termed *clints* (flachkarren) and host the smaller forms of karren. In bedded rocks most grikes terminate at a penetrable bedding place at depths of one half to a few metres. A small minority may extend down to deeper bedding planes and receive the drainage of the shallower members.

Grike length is inversely proportional to the density of major joints. In most karsts it ranges from one metre to a few tens of metres. Widths at the top range from ~ 1 cm to ~ 1 m where rocks are bare or thinly vegetated. Rose & Vincent (1986) reported a range from 1.5 to 25 cm on samples of glaciated limestone pavement in Lancashire. Where the distribution of widths was bimodal they recognized a class of wider grikes created before the last glaciation and of narrower grikes that are postglacial in age. At a given site grikes tend to be longest, widest and deepest near escarpment edges where jointing is expanded by tension unloading. This is where they will develop first.

Grike walls may be parallel or taper with increasing depth. They are often indented with cavernous weathering, rillenkarren or cockling, or dissected by splitkarren, rinnen or rundkarren. Many grikes have been created by the amalgamation of earlier shafts developed at intervals along the joint. This creates a pattern of widenings and narrowings. Alternatively, one wall of a grike may project and the opposite wall display a matching recess. This indicates a mechanical dislocation rather than dissolution.

Beneath a deep soil, grikes become much widened at the top and taper with depth. These forms are termed *cutters* by American authors (Howard 1963). Intervening clint tops are reduced in areas and sharpened by runnel cutting to form *subsoil pinnacles* (Fig. 9.2).

In many young reef rocks grikes are the only linear karren form that develops because the others are prohibited by the textural heterogeneity.

Hydraulically controlled linear forms – dissolution channels

Channel karren have received more study than the other types because of their similarities to erosional channels and, therefore, their supposed amenability to an hydraulic explanation.

Rillenkarren are perhaps the most striking because they appear to be the

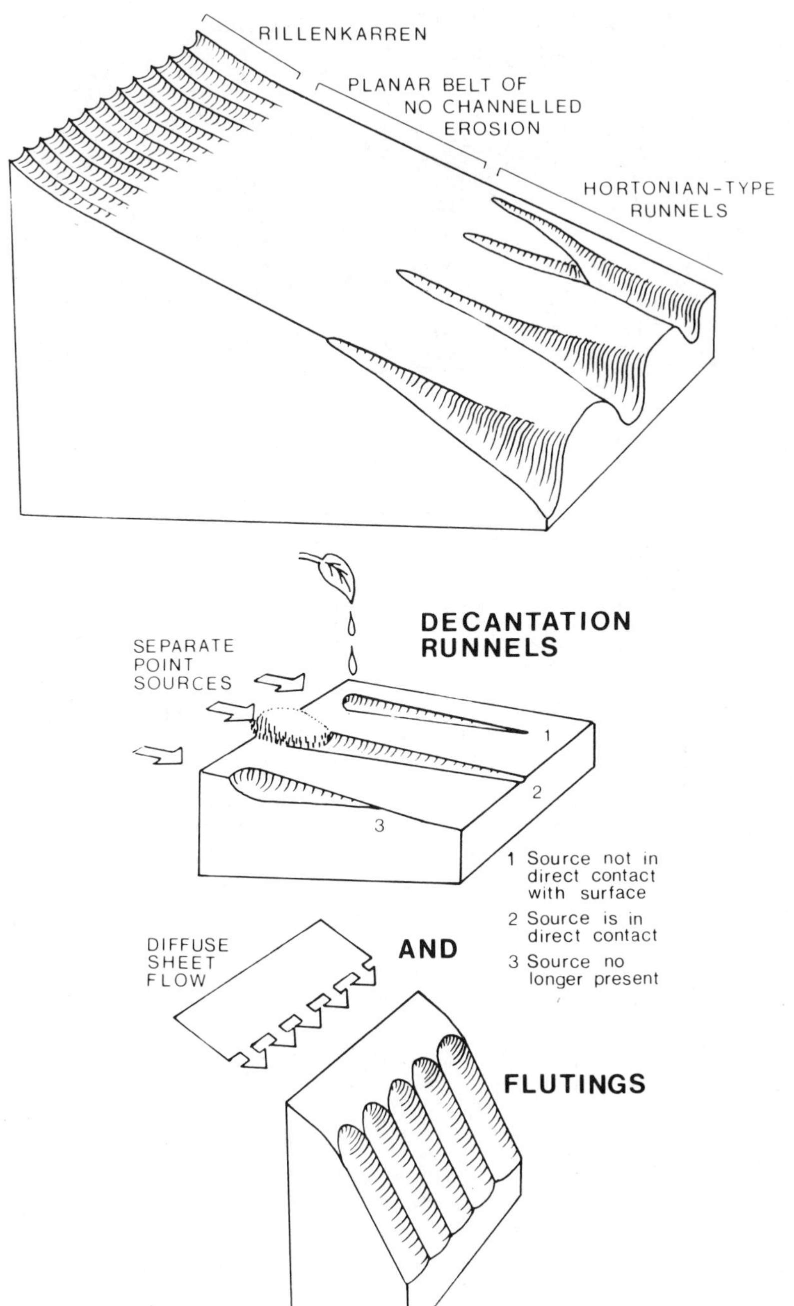

Figure 9.3 Rillenkarren, Hortonian-type dissolution channels and decantation channels, as defined in Table 9.1. Drawings by J. A. Lundberg.

Figure 9.4 Rillenkarren on limestone (*left*) and salt (*right*). Relationships between the length of rillenkarren and the angle of slope. Open circles and the linear relationship are from the controlled experiments of Glew & Ford (1980) with plaster of Paris and a rain machine. Rocky Mountain data are from different blocks in a Jasper Park landslide pile. This was an ideal site to study rillen karren because there is no inheritance effect; nevertheless, it is seen that correlation with the experimental results is poor.

antithesis of the normal (or Hortonian) erosional rills that are generated by runoff on soil, etc. (Fig. 9.3). Rillenkarren head at the crest of a bare slope and diminish in depth down slope (Fig. 9.4) until they are replaced by a planar solution surface, *Ausgleichflache* (Bogli 1960). Dunkerley (1983) and Jennings (1985) use the term 'solution flute' as a synonym for rillenkarren, but confusion may arise because Pohl (1955) used *fluting* to describe decantation channels on shaft walls.

Hortonian rills start below a belt with no channelled erosion and are uniformly separated by interfluves on a simple surface. Rillenkarren are uniformly packed together and at a given site, they display only two or three characteristic widths. They do not develop on gentle slopes and on the steepest slopes degenerate into *cockling patterns* – a mixture of scallops, fluted scallops or ripples (below), and discontinuous rills. Rillenkarren must be the product of direct rainfall because there is no other feasible source of water.

Rillenkarren develop well upon fine grained, homogeneous limestones, dolomites and marbles, with dominant channel widths usually between 1.0 and 2.5 cm. On coarse-grained or more heterogeneous carbonates rilling is only partial or is absent entirely. They develop well upon gypsum and are the predominant karren form on salt outcrops; characteristic widths appear to be a little greater than on the carbonates but this is not confirmed. Channels that also head at crests but are notably wider (> 4–5 cm) are varieties of the decantation flutings discussed below. Admixture of the two types occurs in nature, so they are readily confused.

At many sites it appears that, setting aside textural factors, length of rillenkarren increases with an increase of gradient. However, a wide variety of field studies in Europe (Gerstenhauer & Pfeffer 1966, Miotke 1968), Australia (Lundberg 1977, Dunkerley 1983) and South Africa (Marker 1985) failed to obtain statistically significant relationships. Glew & Ford (1980) tackled the problem by hardware simulation, exposing texturally uniform plaster of Paris slabs at differing inclinations to constant rainfall at 25°C (Fig. 9.4).

The simulations demonstrated that rillenkarren propagate from the crest downslope until a stable length is reached. Both rillen and the ausgleichflache are then removed by parallel retreat. Formation begins as many short, shallow rills that deepen and lengthen, coalescing laterally to achieve a characteristic width. Rills at lateral margins are longer because thin flow is maintained there. The rill cross section approaches the parabolic, which focuses rain splash into the centre. Short rills appear between 5 and 10° (depending upon texture). Mean rill lengths increase with slope, being 25–30 cm at 60° in the experiments. Greater lengths can occur on steeper slopes but there is degeneration into cockling.

Rillenkarren are produced by a hydraulic 'rim effect'. At the crest of a slope raindrops penetrate the fluid boundary layer (section 3.8), permitting turbulent reaction at the mineral surface. Depth of flow increases downslope to some critical value (0.15 mm in the experiments) where drops cannot impact the surface directly; uniform mass transfer then creates the ausgleichflache (Glew & Ford 1980, Ford & Lundberg 1987).

As in the models, natural rillenkarren develop towards an equilibrium width at a site, but lengths are most irregular. This is because the dynamic controls (temperature, droplet masses, velocities and intensities) and the passive controls of slope and texture are all variable. Variation is greatest where the rillenkarren are superimposed upon earlier, complex karren surfaces; this invalidates many published results.

Solution runnels are normal Hortonian channels, heading where sheetflow or wash on a slope breaks down into linear threads (Fig. 9.3). On steeper slopes the channels are parallel; on gentle slopes there may be dendritic confluence or centripetal orientation into a karren shaft or grike. *Rinnenkarren* display sharp rims and flat or rounded bottoms (Fig. 9.5). They develop on bare slopes. *Rundkarren* have more rounded cross sections because they develop beneath vegetation or soil.

These are conventional stream channels that gain discharge downstream; thus they normally widen and deepen downstream. For example, rinnen and rundkarren developed upon 35–40° dip slopes on Vancouver Island, Canada, increased in mean width from 4 to 8 cm along lengths of 3–5 m and maintained width : depth ratios between 1.5 and 2.5 (Gladysz 1987). After 3 to 5 m all examples amalgamated with decantation forms (i.e. became composite or polygenetic) or were intercepted by split karren, shafts or grikes. Sweeeting (1972) quotes rinnenkarren depths up to 50 cm and lengths up to 20 m but these are exceptional.

Rinnenkarren occur on slopes as low as 3°. Their courses may be sinuous but full meandering (sinuosity ratio > 1.5) is rare. Rundkarren can extend headwards across horizontal clint surfaces from grike edges, draining soil percolines. Rinnenkarren have been measured on slopes as steep as 48° in the Dolomites (Fig. 9.5) but that is unusual. Rundkarren can propagate on slopes that are nearly vertical because cavitation rarely occurs beneath soil.

Rinnen and rundkarren often accumulate moss or excess soil at places along their channels. These may be widened to create a segment with overhanging walls (*hohlkarren*), or deepened to create a locally reversed gradient or even a flat pan bottom. Overdeepened segments occur in most rundkarren we have seen, making them more complex forms.

Rinnenkarren and rundkarren develop on most carbonate rocks but are best formed where they are homogeneous and medium- to fine-grained. They also develop well on gypsum, basalt, granite and sandstones but are not known on salt.

Decantation runnels and flutings are classes proposed by Ford & Lundberg (1987) to differentiate channels created where water is released steadily from an upslope store, from the rillenkarren and Hortonian types that are created during episodes of storm-generated runoff. Decantation runnels occur principally upon bare and partly covered surfaces, but also beneath continuous soil or vegetation covers if small surfaces are temporarily freed by soil piping or root decay. The diagnostic characteristic of pure decantation forms is that, because they do not collect additional acidic water downslope, their cross-sections are largest at or close to the input point and diminish downstream.

Decantation runnels are approximately equivalent to the *wandkarren* type

Figure 9.5 *Upper left* – spectacular rinnenkarren on a landslide surface, Dolomites. *Lower left* – wallkarren on marble, Norway. *Upper right* – Rundkarren exposed by modern deforestation, Vancouver Island. *Lower right* – long denuded rundkarren are sharpened into rinnenkarren, Pyrenees.

of Bogli (1960). Each individual is supplied from a point store such as a patch of moss or the stem of a tree. The dimensions of the runnel are in proportion to the volume of water released from storage and its acidity. At the two extremes, tiny pits along the crests of rillenkarren slopes overspill to enlarge particular rills below them, while perennial snow banks may support runnels 50–80 cm wide and deep that are up to 100 m in length. A majority of wandkarren have widths and depths in the range 1–10 cm and are extinguished or have amalgamated with other karren types within 10 m of their source.

The slow release of water permits meandering. Most *Mäandekarren* described by previous authors are point–source decantation forms.

Decantation stores accumulate within earlier split, rinnen and rundkarren to produce composite or polygenetic forms, as noted. These are more abundant than the pure forms at most sites; in a Vancouver Island site Gladysz (1987) measured 66 rinnenkarren, 27 decantation runnels and 423 composite runnels.

Decantation flutings are adjoining, shallow channels formed where water is released from a linear source such as a bedding plane or a soil mat at the top of a cliff. They develop best where slopes are steep to overhanging, as on grike walls or in vadose shafts (section 7.11). Channel widths typically are 5 to 25 cm, and lengths are up to 25 m. Depths and depth : width ratios have not been reported, but the features appear to be shallow in proportion to their width when compared to rillen, rinnen and decantation runnels.

On steep surfaces fluting requires that the film flow be thin enough to be retained on the rock by surface tension rather than detaching as happens when cockling is formed. Such films develop 'parting lineations' oriented in the direction of flow (Allen 1977). These probably establish the flutings in the rock; their separation (flute width) is inversely proportional to the velocity of flow.

Fluted scallops resemble transverse ripples in sand. They are oriented normal to the direction of flow and may have an assymetric cross-section, being slightly steeper on the upstream side. Where fully developed they adjoin one another and extend across a wall or a cave roof. Curl (1966) defined them as an ideal end member of the class of dissolutional scallops (section 7.11) and termed them 'flutes'. Jennings (1985) termed them 'solution ripples'. They develop only partially on many steep walls such as grikes, to form a prominent element in cockling patterns.

The composite nature of karren

We re-emphasize that there is an immense variety of karren forms. Simple or monogenetic end members have been stressed but many karren are composite, being both polygenetic and also varying with lithology. Chief among the lithologic factors are chemical purity, grain size, textural homogeneity (including pores), bedding thickness and joint frequency. Greatest density of karren occurs where strata are thin, closely jointed and heterogeneous in composition but types will be limited largely to grikes, split karren, pits and shafts. The greatest variety of karren and their best developed form is associated with massively bedded, fine grained and homogeneous limestones and marbles.

Some authors have sought to relate particular karren types or scale of development to specific climatic conditions, but this encounters many problems. The *rate* of their development is greatest where it is wettest (section 4.1) and types that are limited to bare rock (principally tritt and rillenkarren) are less common where the climate can support forest. For the other types it appears to us that variations in lithology, hydraulic gradient and the duration of the karst denudation outweigh climatic control.

Assemblages of karren and giant karren

Great areas of karren evolve beneath complete soil and vegetation cover. They are ignored in most regional morphologic studies. Bare or partly bare exposures of karren are widespread above the treeline in alpine areas, on desert slopes, or where there has been deforestation and soil erosion. *Karrenfeld* is the general term for such assemblages. There are contiguous areas of many square kilometres in Yugoslavia and elsewhere.

Where strata are flatlying or gently dipping, karrenfeld are termed *limestone pavements* (Williams 1966a) if they are dominated by patterns of regular clints and grikes so that they appear like artificial paving (Fig. 9.6). These are the most studied type of karren assemblage (e.g. Goldie 1981). In addition to limestone they develop well on dolomite, and on some well bedded sandstones and quartzites.

Areas of the individual clints are typically 1–10 m^2. Where bare or lightly vegetated these are indented by pits, pans with decantation runnels, shafts and splitkarren. Beneath dense vegetation or an acidic soil rundkarren tend to become predominant.

Pavements develop best upon thick to massively bedded strata. Where beds are thin or medium, clints are readily broken up by mechanical processes such as frost wedging, root wedging and fire. Massive beds eventually break up along subsidiary (previously impenetrable) sedimentation planes also, degrading the clint surfaces to rubble (or *shillow*). As a consequence, optimum pavement development occurs where an agency periodically can scour off the shillow and dissected upper beds to restore pristine surfaces. The chief scouring agents have been Quaternary glaciers. The most extensive pavements are exposed in areas subject to the last (Wisconsinian–Wurm) glaciation and so have developed largely during the 10 000–15 000 years of postglacial time. Lesser pavements occur where the scour is by wave action, river floods, or even sheetfloods on pediments.

Figure 9.6 Clint-and-grike topographies or 'limestone pavement'. *Upper left* – at a low relief, low energy coast, Quebec, *upper right* – a rugged staircase (schichttreppen) at tree line in the Pyrenees. *Lower left* – 'classical' clint-and-grike at Malham Cove, Yorks; subsoil rundkarren dissection dominates in

Where soils are deep but acidic there is clean dissolution. Clints become tapered by rundkarren entrenchment to form subsoil pinnacles or cutters. If their tops are exposed as a consequence of soil erosion, they become sharpened – depending on the extent of vegetation cover – by rillen, rinnen, wandkarren and decantation flutings, especially the latter. They therefore emerge as a form of *pinnacle karst*.

The most celebrated pinnacle karst of this type is the 'Stone Forest' in Yunnan, China (e.g. Chen *et al.* 1986). A rugged tor-and-pediment topography in pure, massive limestones was buried by Tertiary 'red beds' (cover sands and clays). Smooth, rounded pinnacle development occurred beneath them, etched into the old topography which is now being exhumed and sharpened, exposing tracts of pinnacles over an area of 30 000 hectares. Characteristic heights range 1 to 35 m and diameters, 1 to 20 m (Fig. 9.7).

The spectacular arete-and-pinnacle terrains of Mt Kaijende, Papua New Guinea (Williams 1971) and Mt Api, Sarawak (Osmaston & Sweeting 1982) have pinnacle forms in excess of 45 m high, penetrating and rising well above tropical rain forests (Fig. 11.15). These appear to have developed directly from dissolution of the limestone without previous burial and exhumation phases. Similar but much less sharp forms occur in the semi-arid karsts of northern Australia (Jennings & Sweeting 1963). Essential prerequisites are dense rocks, very massive bedding and well spaced major jointing.

Pinnacle karst or *Schlotten* develop quickly in massive gypsum exposed to high hydraulic gradients. The rounded or fluted pinnacles are segregated by deep karren shafts or small dolines centred at intervals of 2 to 5 m. Gypsum pinnacle karst with a relief of 20 m has developed during the Holocene in Nova Scotia and Newfoundland. Exposed salt flats and slopes may be dissected into smaller schlotten with rilled and pitted walls within a few years.

Ruiniform (Perna & Sauro 1978) describes terrains where soil has been removed from exceptionally deep and wide grikes but where the clints are not sharpened into pinnacles. Instead, they stand out like miniature city blocks in a ruined townscape. Ruiniform tracts are common on gentle hillslopes as in the French Causses that have suffered deforestation and substantial soil erosion. On hill crests they are transitional to *tors*, which develop particularly well on massive, coarsely crystalline dolomites.

Grikes may be expanded to larger scales to create aligned or intersecting corridor topographies. Types of large grikes are termed *bogaz* by Cvijic (1893), *corridors* (Jennings & Sweeting 1963), *zanjones* (Monroe 1968) and *streets* (Brook & Ford 1978). Squared valleys and closed depressions formed by grike wall recession are *box valleys* and *platea*, respectively. The assemblages have been termed *giant grikelands*, *corridor karst* or *labyrinth karst*. Examples are known in tropical and temperate rainforests, and desert and semi-desert areas (e.g. section 10.2). Grikelands develop on

Figure 9.7 The Stone Forest of Yunnan. *Lower left* – first exposure of subsoil pinnacles. *Lower right* – emerging. *Upper right* fully emerged and with basal planation on alluvium. *Upper left* – a typical group. Arrow indicates people for scale.

sandstones in dry mountain ranges of the Australian and US deserts, also. Small grikelands are common in limestones, dolomites and sandstones along crestlines in the subpolar, periglacial Mackenzie Mountains of Canada and are expanded to a spectacular labyrinth with individual streets longer than 1 km and deeper than 50 m in the Nạhanni limestone karst in that region (Fig. 9.8). The prime requirements for such assemblages appear to be very

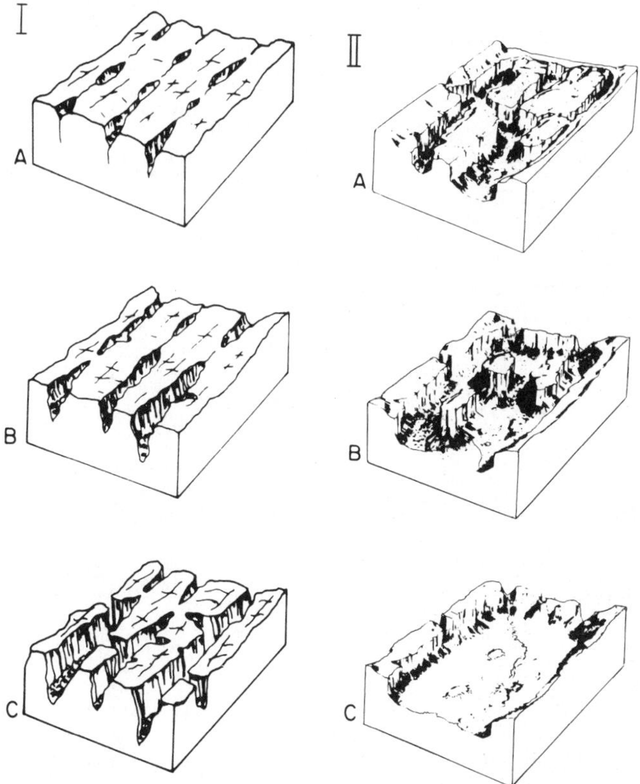

Figure 9.8 Model for the development of the Nahanni labyrinth karst or giant grikeland. Mackenzie Mountains, NWT, Canada. Grikes are created by shaft formation and elision along master joints (I). A combination of further dissolution and frost shattering widens them into 'streets' (IIA), reducing clints to residual towers (IIB) and, finally, creates large closed depressions or 'platea' (IIC). Grikes are up to 10 000 m in length and 50 m deep. From Brook & Ford (1977).

major joint sets (often, small faults with lateral displacements), massive strata, climatic or topographic conditions that permit a deep water table, and a long duration of sustained dissolution.

Littoral karren

A variety of karren forms and assemblages occurs on limestone and dolomite sea coasts and around lakes. They have been intensively studied because of relationships to modern limestone accretion and diagenesis. There is a considerable literature which cannot be adequately reviewed here; see Trudgill (1985). Section 10.5 discusses effects of changing sea levels on karst development in coastal regions.

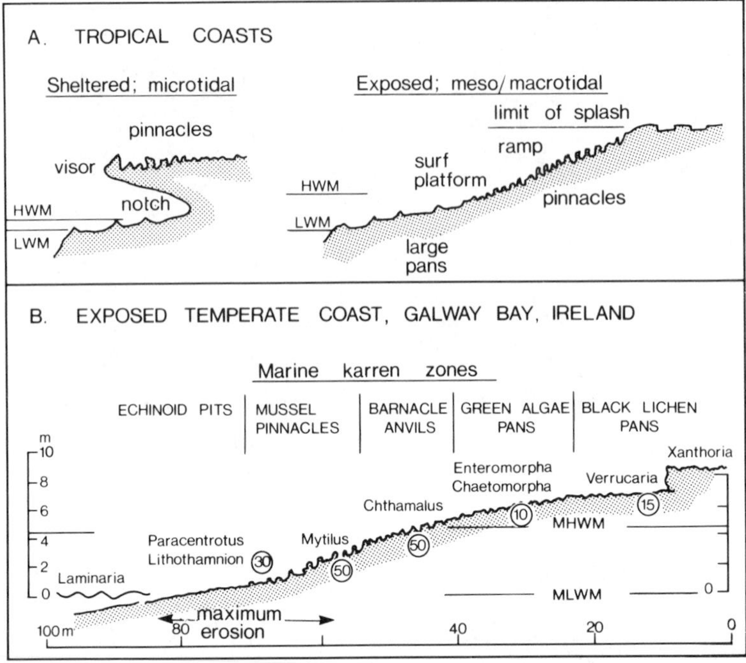

Figure 9.9 A. Zonal features of tropical limestone coasts; generalized models to display the contrasts between sheltered and exposed, micro- and macro-tidal settings. B. Transect of an exposed limestone coast in Galway Bay Ireland. Latin names = principal colonizing species. Circled numbers are local relief in centimetres.

In addition to physico–chemical dissolution and bio-erosion, carbonate coasts may be sculpted by wave action, wetting and drying, salt weathering and hydration. The relative effectiveness of these competing processes is determined by many different factors, chief of which appear to be (a) wave energy, (b) tidal range, and (c) variations of lithology and structure, and (d) climate. Combining the range of processes and factors, two extreme (or end member) types of eroding carbonate coasts can be recognized. The first is that dominated by mechanical erosion; dissolution forms are few or absent. This will tend to be a high energy (or exposed) coast with low tidal range and weak strata. At the other extreme is the low wave energy coast where a high tidal range exposes wide intertidal flats upon mechanically resistant rocks. Karst studies have favoured this latter type for obvious reasons; even there, structural factors such as dipping strata often blur other karst zonation (Ley 1979).

Karst sea coasts display two distinctive features, *notches* and a high density of pits and pans (Figs 9.9 & 9.10). On protected coasts gypsum cliffs display sharp intertidal notching in all climates. Notches (or *nips*) in

Figure 9.10 *Upper left* – phytokarst, Puerto Rico. *Lower left* – notch and phytokarsted platform, Okinawa. *Upper right* – coastal dissolution pans modified to serve as sea salt evaporation pans, near Bari, Italy. *Lower right* – freshwater micro-pitting, Lake Huron shore; some pits are aligned along glacial striations.

limestone or dolomite are confined largely to tropical and warm temperate waters. A prominent notch develops in the intertidal zone, and a lesser notch may be present below the low tide mark on steep coasts. The intertidal notch is cut partly by boring organisms (algae and sponges) and molluscs that graze upon them, rasping the limestone to obtain their prey (Neumann 1968). On Aldabra Atoll, Trudgill (1976) measured notch recession rates as high as 1.0–1.25 mm a^{-1}, with grazing contributing 0.45–0.60 mm. The sharpest notches occur on the most protected, microtidal coasts, where the depth of the undercut may be 2 m or more. On exposed coasts the intertidal notch is replaced by an eroding ramp.

Pitting attributable to boring and grazing organisms is sharp and may be densely packed or coalescing. This reduces the divides between pits to sharp crests and pinnacles sometimes termed *lacework*; coastal pit topography may be extremely rough. Pit depths range from < 1 cm to ~ 1 m, and diameters can be as great as several metres. Width : depth ratios tend to fall between 1 : 2 and 6 : 1.

The interaction of marine, biological and chemical processes may produce a karren zonation on limestone shores that varies with climate and exposure. It is best developed where unbroken beds dip gently towards the sea in areas of high tidal range, e.g. on the Aran Islands, western Ireland (Fig. 9.9B).

There has been little study of the freshwater littoral. In our experience development usually is restricted to formation of small pits and pans in a narrow zone about the waterline. They may develop to high densities, and coalesce. Many are occupied by blue-green algae, which possibly initiate them. We have noted pitting that extends vertically *upwards* from opened bedding planes in a zone of seasonal lake level oscillation in Lough Mask, Ireland, and below the waterline in Lake Huron. These pits are narrow cylinders like bell holes (section 7.11) but densely packed.

9.3 Dolines – the 'diagnostic' karst landform?

Karst geomorphologists have always accorded special importance to dolines since Cvijic (1893) identified them as giving karst topography its particular character. Grund (1914) considered their place in the karst landscape to be similar to that of valleys in fluvial terrain. These features then are basic or index landforms of karst, although their often simple basin form may belie a complex origin. They are called *sinkholes* in North American literature (e.g. Beck 1984). In this chapter we use *doline* for any small to intermediate enclosed karst depression regardless of genesis or climatic context. It is derived from *dolina*, an everyday Slovenian expression for any depression in the landscape (Gams 1973a). Sweeting (1972) discusses other local terms. Roglic (1972) reviews early ideas on their origin.

Dolines are usually circular to subcircular in plan form, and vary in

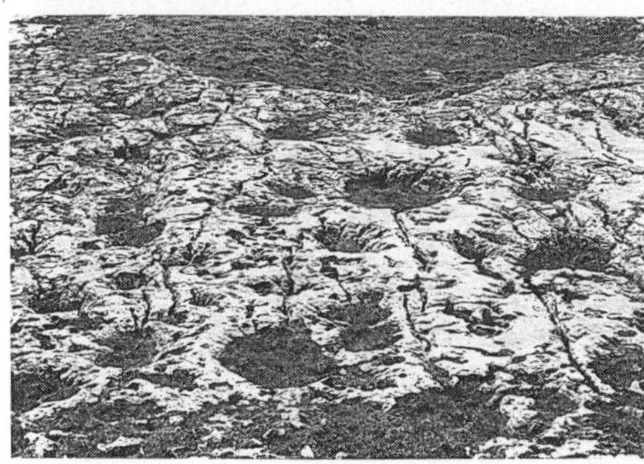

Figure 9.11 *Top left* – An isolated solution doline, Causse Mejean, France. The floor of the basin has been infilled by slopewash and periglacial debris and levelled for agriculture. *Top right* – An aerial view of polygonal karst terrain in southern Yugoslavia (photo by I. Gams). *Lower left* – Small solution depressions that have developed on a glaciated limestone pavement since deglaciation about 12 000 years ago. Burren plateau, County Clare, Ireland.

diameter from a few metres to about one kilometre. Their sides range from gently sloping to vertical and they vary from a few to several hundred metres deep. This yields a spectrum of features from saucer shaped hollows, to funnels to cylindrical pits. In the landscape they may occur as isolated individuals or as densely packed groups that totally pock the terrain (Fig. 9.11).

Cvijic (1893) recognized that solution and collapse account for the formation of most dolines, although he considered the majority to have a predominantly solutional origin. He was one of the first geomorphologists to apply morphometry to landform description, when on the basis of depth : diameter ratios from numerous field measurements he distinguished (a) shallow trough or bowl shaped basins with flattish floors from (b) steeper, deeper funnel shaped dolines and (c) shaft-like dolines in which the breadth is usually less than the depth (see Sweeting (1972) for discussion of this).

Thus early ideas on the nature of dolines were drawn mainly from field experience in Europe, although Danes (1908, 1910) also investigated humid tropical karsts in Jamaica and Java. He and Grund (1914) were of the opinion that tropical dolines, or *cockpits* as they are known in Jamaica, develop mainly by solution in much the same way as dolines in the temperate zone, although some early researchers believed them to be collapse features. Later work by Lehmann (1936) indicated tropical cockpits to be morphologically distinct from most temperate dolines (Fig. 9.12),

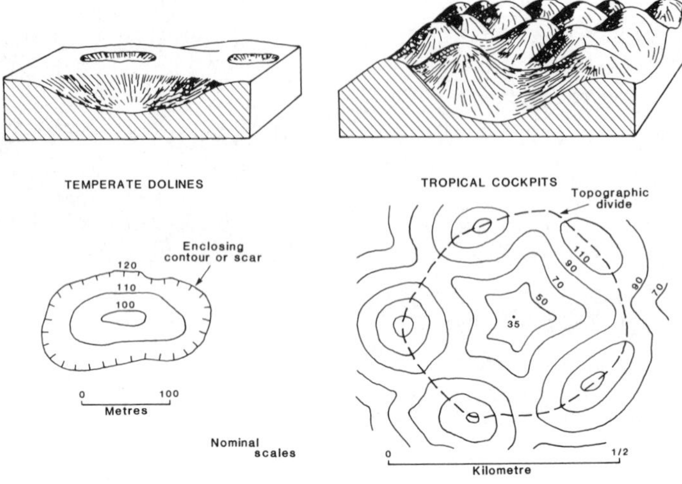

Figure 9.12 Delimitation of temperate and tropical closed depressions. From Williams (1969).

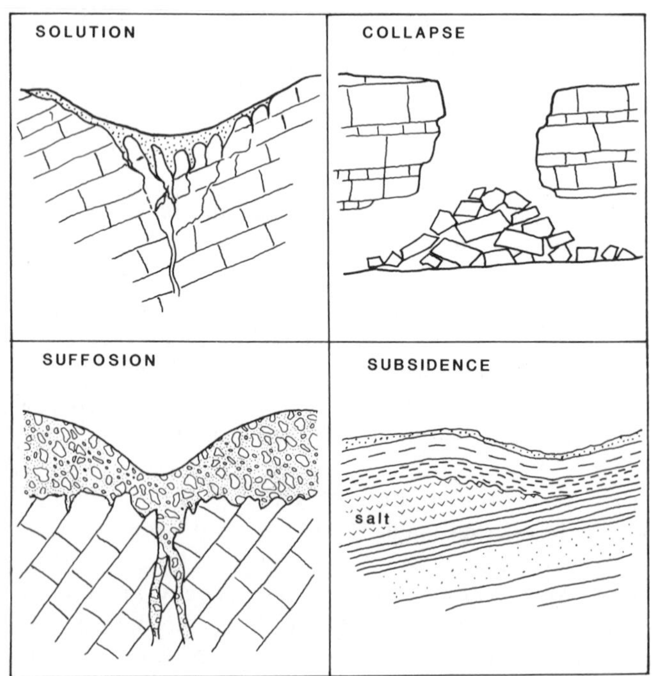

Figure 9.13 Principal genetic classes of dolines.

although he supported an origin by solution processes.

The basic genetic distinction between solution and collapse dolines was elaborated by Cramer (1941) in an extremely thorough paper on their development and morphology. He examined topographic maps of karst regions from many parts of the world and made a most comprehensive morphometric description of the doline terrains concerned. His study revealed a picture of variations in the density of doline fields in different areas.

From this early work emerged an awareness that there are several types of dolines developed in different ways and in different materials but with a convergence of form. The end members are illustrated in Figure 9.13. We distinguish *collapse*, which involves fracturing and rupture of rock and soil, from *subsidence* which is a more gradual process involving sagging or settling of the surface without obvious breaking of the soil. Natural dolines of a purely subsidence origin are rare, being found where there is interstratal solution of evaporite rocks at shallow depth (artificial subsidence dolines are often created during salt mining). The actual form of most dolines is of polygenetic origin, plotting within a ternary diagram of end members; but it is strongly biased towards one or another of the end points, thus we discuss them under those headings.

9.4 The origin and development of solution dolines

The shape of solution dolines indicates that a greater mass of rock has been removed from their centres than from around their sides. This implies there to be a common natural process that focuses corrosion. Since the mass transport in solution is the product of solute concentration and runoff, variations in either or both factors could explain the focused corrosion within dolines. If local variability in solute concentration alone were sufficient to explain them, then they would be found on every type of limestone in a given climatic zone, but this is not the case as is illustrated by comparison of the Cretaceous, Jurassic, Carboniferous and Devonian limestones of England, for example. If the solute concentration factor is eliminated, it follows that sufficient local inequalities in corrosion to form dolines must be created principally by spatial variations in water flow. How does water flow through the rock commence and how can it be focused? Here we must refer back to Chapter 7 and specifically to the initiation of proto-conduits (Fig. 7.11).

Initiation of point recharge dolines

In an undeformed sequence of limestone interbedded between clastic rocks, fluvial erosion of the cover beds will reveal inliers of limestone. These provide an input boundary for a developing karst, while the valley of a main stream cut deeply into the limestones will provide an output boundary at a lower elevation. Input to output connections are established as explained in section 7.2. Dolines can only commence to form when proto-caves connecting recharge points to a spring (or springs) are developed. Until such links are established, resistance to flow through the rock is too high to permit removal of enough limestone to create a doline-sized depression, but once connected the focus of drainage (and solute) removal is a surface funnel and the karst pipe that drains it.

Dolines formed in this way can be termed *point recharge depressions* in recognition of the process responsible for focusing corrosional attack (Fig. 5.17(I)). Often only the bottom of the enclosed basin is composed of limestone, the flanks being relatively impermeable cover beds. These dolines act as centripetal drainage points for allogenic runoff and, clearly, the larger their contributing catchments the bigger these depressions may become.

As more clastic rocks are eroded, so more limestone inliers are exposed. A second generation of point recharge sites may be envisaged, not necessarily in the stylized ranks of Fig. 7.11 but nevertheless with the same implications for subsurface conduit development and its feedback effects permitting doline growth. Continued caprock erosion and more doline initiation further increases the frequency of point recharge sites. The drainage areas of individual dolines are reduced as a result. Where caprocks have long been entirely removed and soils are thin or discontinuous, the high density fissuring of weathered bedrock, sometimes exposed as karren fields, may begin to substitute for dolines as drainage routes. Under such circumstances recharge has moved to the diffuse end of the spectrum (see section 5.3) and accompanying corrosion is also spatially diffuse; thus unless flow is focused lower down in the subcutaneous zone the form of any dolines that have been initiated may not be maintained.

Initiation of drawdown dolines

Some karst landscapes have been developed in limestones without caprocks, either because none was deposited or because morphological development followed the uplift of a baselevelled erosion surface. Solution dolines are often found in such terrains although corrosion cannot have been focused by point recharge. In other karsts where caprocks have long gone, there are apparently active, rather than degenerate, doline landscapes.

Any uplifted erosion surface subjected to renewed karstic attack will possess a vestigial conduit network developed beneath it by phreatic corrosion in the previous phase. It can be assumed that some input to output connections already exist, although they will certainly develop further. Recharge from rainfall will be relatively diffuse, depending on soil cover, and percolating water will accomplish 50–80% of its solutional work within about 10 m of the surface (see section 4.4). Hence fissures that are considerably widened by corrosion beneath the soil close rapidly with depth (Fig. 9.2B). As a result, infiltration into the top of this highly corroded subcutaneous zone is much easier than drainage out of it (Williams 1983). The bottleneck effect results in much storage of water in this zone after heavy rain, constituting a perched epikarstic aquifer (Mangin 1973, 1975) with a base that is essentially a leaky capillary barrier (Fig. 9.14) as discussed in Chapter 5. Because of initial spatial variability in fissure frequency and permeability arising from tectonic and lithological influences, preferred (low resistance) vertical leakage paths develop down connected pipes at the base of this aquifer. These paths are enlarged by solution with

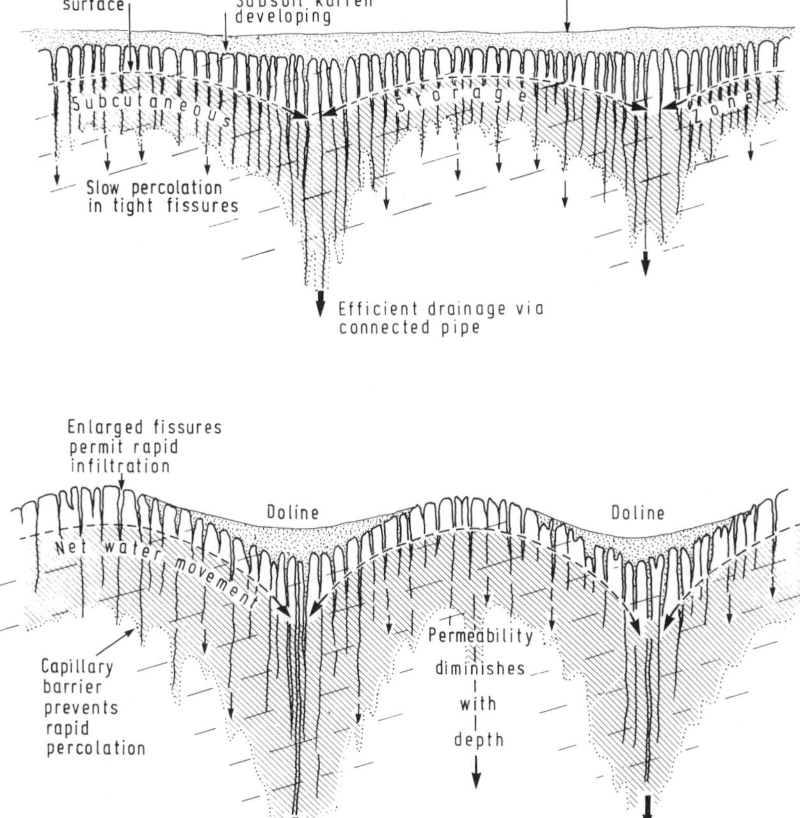

Figure 9.14 Drawdown doline initiation in the subcutaneous zone. From Williams (1983).

the result that a depression develops in the overlying subcutaneous water table similar to the cone of depression around a pumped well. Flow paths then adjust in the epikarstic aquifer to converge on the dominant leakage route. The extra flow encourages more solution and with it the enhancement of vertical permeability. The zone of influence of the leakage route(s) widens according to the radius of the cone of depression in the subcutaneous water table. The dimension of this radius depends on the hydraulic conductivity of the epikarst and the rate of water loss down the leakage path at its base.

These processes, then, can explain the focusing of corrosion where caprock point recharge is not important. As the surface lowers the more intensely corroded zones begin to obtain topographic expression as solution dolines, their diameters being controlled by the radius of the subcutaneous drawdown cone. The theoretical relationships between surface doline topography, underlying relief on the subcutaneous water table, and vertical hydraulic conductivity near the base of the epikarstic aquifer are illustrated in Fig. 5.16.

From this dicussion we can see the necessity to distinguish (a) between doline initiation where there has been no proto-cave development and that where a ready made, permeable, vadose zone is inherited from an earlier phase of karstification, and (b) between doline corrosion focused by point recharge as opposed to that focused by epikarstic drawdown. We can also appreciate that different generations of dolines (termed primary and secondary or parent and daughter) can develop either from successive caprock stripping or from secondary drawdown within a major drawdown cone (Drake & Ford 1972, Kemmerly 1982).

The occurrence and enlargement of solution dolines

Once a doline is established, positive feedback will encourage its further development because of the centripetal focusing of flow and hence corrosion (Fig. 9.15). The agressiveness of the water may be enhanced by greater biogenic CO_2 production in the thick soils that tend to accumulate in the bottoms of depressions. Such soils may also be damper longer because of drainage accumulation and lingering snow melt; thus the duration of active corrosion may also increase. Further, with efficient vertical drainage encouraged by the corrosional enlargement of rifts and shafts, the increasing average velocity of water flow leads to a corresponding growth in mechanical transport of down-washed soil and rock, evacuating them underground. The freer vertical drainage permits much greater leakage from the basin and so steepens the epikarstic hydraulic gradient, stimulating further drawdown in the subcutaneous water table, and encouraging expansion of the radius of influence of the centripetal drainage system.

Although the general tendency during solution doline development is for

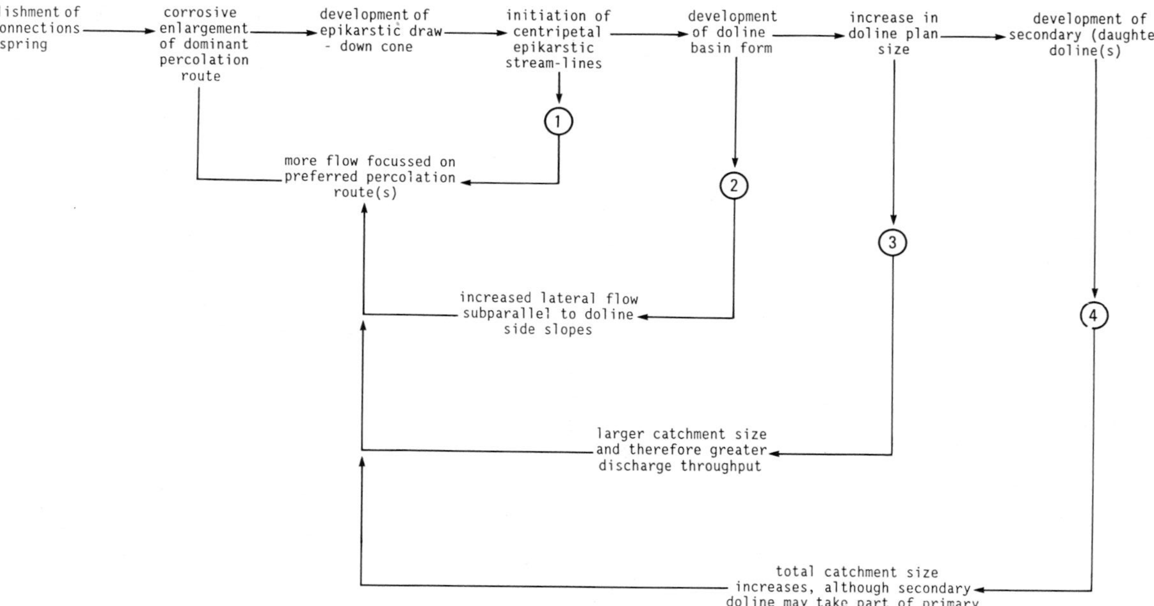

Figure 9.15 Positive feedback loops in the development of solution dolines. Modified from Williams (1985).

self-reinforcement, some effects have negative feedback influences. For instance, the soil covering some karsts may have a lower saturated hydraulic conductivity than the underlying rock. Depending on its composition it could range from 10^{-2} to 10^{-5} m day^{-1}, whereas in karstified limestone conductivity may be 10^{-1} to 10^{3} m day^{-1} or more. The soil therefore commonly acts as an infiltration regulator and, where it has a high proportion of clay, it may even be an infiltration inhibitor. Hence thick soils mantling doline bottoms can result in the ponding of runoff and thus in the reduction of peak throughput following storms. It can even create semi-permanent ponds – doline ponds – that reduce storm surcharges of flow underground to zero.

Although solution dolines are a common feature of karst in humid areas not all carbonate rocks support them. Williams (1985) suggests that they will not develop at all (a) if vertical hydraulic conductivity is so great throughout the vadose zone that little or only short-lived subcutaneous storage occurs, as is the case in rocks with high primary permeability such as aeolian calcarenites (Jennings 1968); (b) if vertical permeability is always so spatially uniform and sufficiently dense that no drawdown depressions develop in the subcutaneous water table, as may be the case within Cretaceous chalk terrain of England and northern France as well as on some raised coral atolls; and (c) on steep hillsides, sloping more than about 20°, where the dominant subcutaneous hydraulic gradient is sub-parallel to the topographic slope.

Factors (a) and (b) that exclude dolines are products of primary rock control on hydraulic variables (Fig. 5.15). With respect to (c), Williams (1972a) noted that closed depressions incised into moderately sloping hillsides in Papua New Guinea are asymmetrical, being elongated down-slope and having their deepest point off centre close to the down-slope margin. It is interesting that the shape of drawdown cones around pumped wells penetrating a sloping water table have an identical asymmetry; and so it is to be expected that rapid leakage beneath a sloping epikarstic aquifer would produce a similar drawdown effect and create the observed asymmetry of solution dolines.

The development of solution dolines leads both to increased permeability (and hydraulic conductivity) and to increased spatial variation in permeability (Fig. 9.16) in the upper vadose zone. In the vertical plane there is a difference of several orders of magnitude between hydraulic conductivity in the epikarstic aquifer and that in the remaining underlying part of the vadose zone (see Figs 5.16 & 6.4). In the horizontal plane, major variations occur in hydraulic conductivity near the base of the subcutaneous zone, and it is these that determine the relief on the epikarstic water table. It is probable that the subcutaneous processes responsible for the genesis of solution dolines become less relevant as the topography develops, because of positive feedback mechanisms strongly enhancing the topography. The

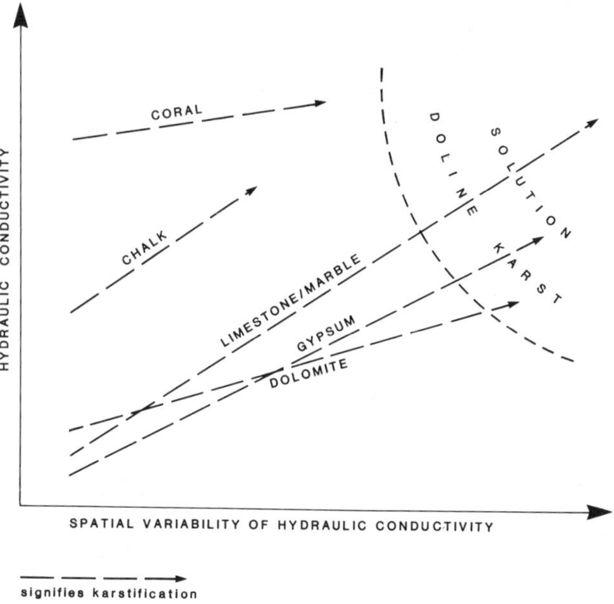

Figure 9.16 Theoretical relationship between hydraulic conductivity, its spatial variation, and karstification giving rise to solution dolines. From Williams (1985).

sharp contrast that sometimes exists between shallow temperature zone dolines and tropical cockpits with larger relief may be a direct consequence of the variable strength of those feedback factors under the differing environmental conditions.

9.5 The origin of collapse and subsidence depressions

Collapse dolines

Collapse dolines are usually steeper-sided than solution dolines and of smaller plan area. But as their side slopes degrade and bottoms infill with slope wash, or other detritus, their surface form may assume the bowl shape of solution dolines, with which they can then be confused. Only excavation will reveal their true origin. Nevertheless, when newly collapsed or still actively developing there is no mistaking some collapse dolines. The deepest open example reported is the chasm containing Crveno Jezero (Red Lake) in Yugoslavia. It is 421 m deep on its lowest rim and 518 m on its highest; the diameter at lake level being about 300 m (Roglic 1976). Filled breccia pipes above deep evaporites may be much larger but lack either prominent or any topographic expression (section 9.13).

1. SOLUTION FROM ABOVE

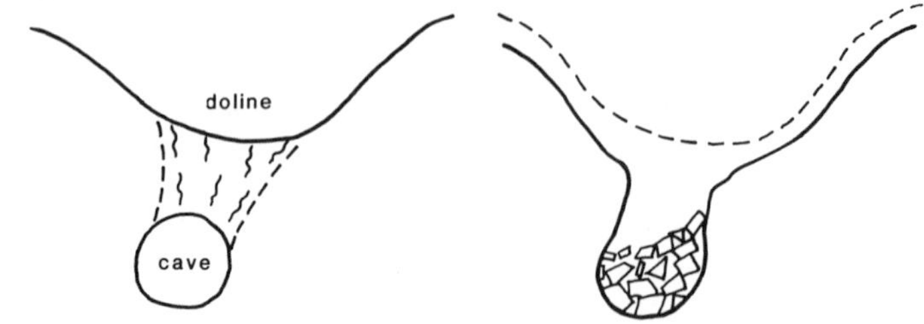

2. UNDERMINING FROM BELOW

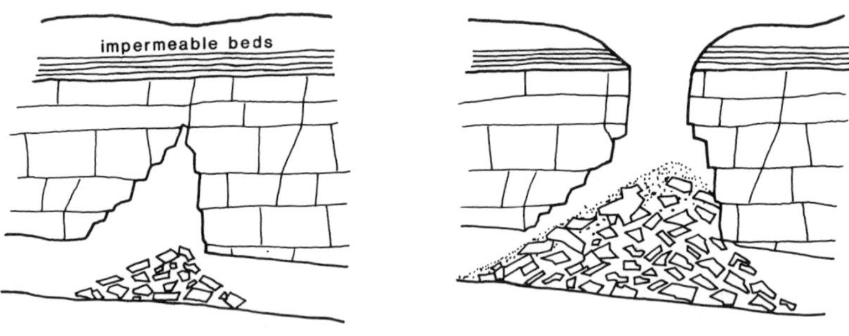

3. REMOVAL OF BUOYANT SUPPORT

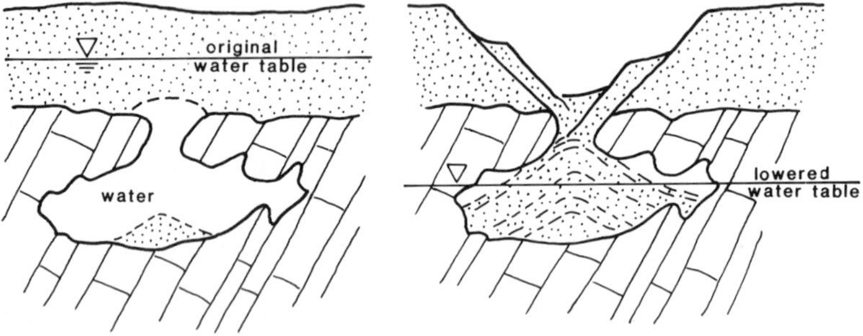

Figure 9.17 Mechanisms giving rise to the development of collapse dolines.

Three main mechanisms are responsible for collapse dolines (section 7.12 and Fig. 9.17):

(a) solution from above that weakens the span of a cave roof;
(b) collapse from below that widens and progressively weakens the span of a cave roof; and

(c) removal of buoyant support by water table lowering, that increases the effective weight on the span so that its strength is exceeded.

In practice these mechanisms often interact. Corrosion by water percolating down a fissure combines with upwards stoping of a cave roof along the same plane of weakness. Failure then occurs because of weakening both from above and below.

The nature of rock strength and the stresses to which it is subjected when associated with underground openings such as caves are discussed in sections 2.8 and 7.12. No single criterion defines rock strength, but unconfined compressive strength, shearing resistance and tensile strength are important elements (Selby 1982), in underground environments compressive stresses being the most significant. If stress exceeds the rock's strength then the cave roof will fail. This will be by fracturing in the case of competent brittle rock or by deforming in the case of incompetent lithologies. The first gives rise to collapse depressions and the second to subsidence depressions.

The study of subsidence and roof failure by mining engineers provides a useful model that is sometimes applicable to karst. The collapse of a horizontal mined gallery can produce a depression at the surface (Fig. 9.18A) similar in morphology to a collapse doline. The process involves lateral as well as vertical movement (Rellensmann 1957). All lateral surface displacements are inwards towards a point over the centre of the collapsing subterranean cavity, at which point lateral movement is zero. Vertical movement is at a maximum at that point and extends well beyond the lateral limits of the cavern. Within the plan confines of the cavity lateral surface strain is compressive, but it is tensile beyond these limits. Observed angles of break are about 85° for limestones and hard sandstones, reducing to 40–45° in unconsolidated materials.

In reality in karst terrains many natural factors usually combine to modify the simple model presented in Figure 9.18. The rocks above the cave are unlikely to be homogeneous in their strength characteristics. Strata of contrasting thickness and composition will have differential susceptibility to failure, and even in otherwise relatively uniform rocks weathering zones are likely to guide failure paths. The model also assumes that movement affects the entire rock mass enclosed by the plane of draw. This is acceptable for relatively incompetent materials but not for strong karst rocks. Thus the model is appropriate for collapse in relatively weak sedimentary formations and for the interstratal removal of support in evaporite sequences as in Figure 9.18B, but is less valuable for understanding the processes that lead to collapse in strong carbonate rocks.

Both the stresses imposed on a cave roof and the strength of the rock constituting the roof are not constant. Over time, surface denudation reduces the loading on a passage roof because of the thinning of the

A

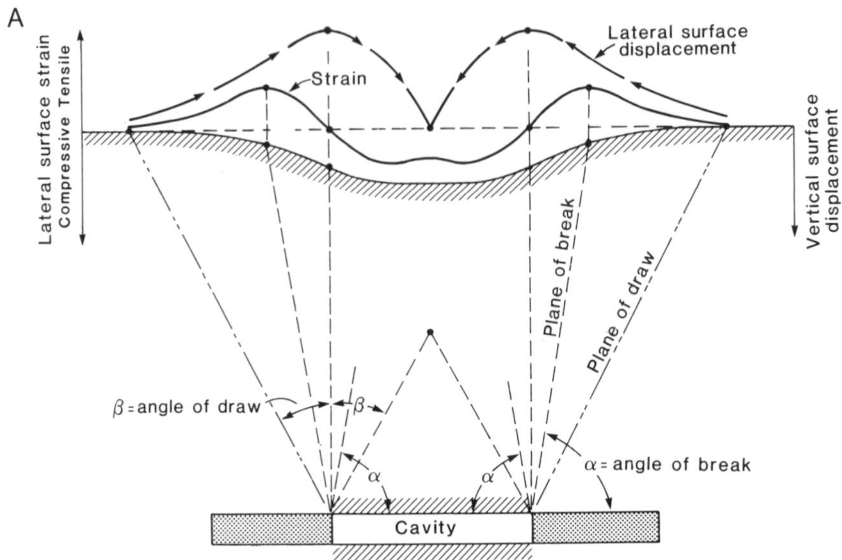

B

Figure 9.18 A. Collapse above a horizontal mined gallery (modified from Obert & Duvall (1976) after Rellensmann (1957). B. Vermilion Creek doline, near Norman Wells, Northwest Territories, Canada; Lat. 65°N. This spectacular feature measures 180 × 100 m and is approximately 40 m deep to the waterline. It is a collapse of Holocene age through calcareous shales overlying gypsum. Photo by R. O. van Everdingen.

overlying rock layer; this reduces compressive stress. But the same denudation processes encourage the development of unloading joints, thus permitting more effective corrosion of the fissure system; this reduces rock strength. Increasing cave passage size results in growing instability, because as size increases the probability of intersecting a mechanical defect in the rock also increases, i.e. it is the strength of the rock rather than the stress affecting it that is influenced by the size of the passage opening.

As karstification proceeds the regional water table is gradually lowered and phreatic conduits are de-watered. Oscillating water tables between wet and dry seasons and temporary backing up of water in conduits due to flash floods also produce rapid changes in stress patterns in a karstified massif. Seasonal water table fluctuations of up to 100 m or more are known, so large volumes of rock can be affected. Groundwater withdrawal can be as abrupt where water resources are heavily exploited by pumping. In a fully saturated medium, the buoyant force of water is 1 t m^{-3} and if the water table is lowered 30 m, the increase in effective stress on the rocks is 30 t m^{-3} (Hunt 1984). If unconsolidated overlying materials are affected by such de-watering, compression occurs and the surface subsides (Fig. 9.17(3)). Compression in sands is essentially immediate, but cohesive materials exhibit a time delay as they drain. Thus the amount of subsidence is a function of the magnitude of the decrease in the water table, which determines the increase in overburden pressures, and the strength and compressibility of the strata. The abruptness of the process is a function of the rapidity of de-watering and the nature of the materials.

Long-term collapse can cause thick non-karst cover beds to be breached (Fig. 9.17(2)). Depressions formed by collapse penetrating caprocks were termed *subjacent collapse dolines* by Jennings (1971, 1985). Examples of this type in carbonate rocks are described from South Wales by Thomas (1974) and Bull (1977b). They are much more important above evaporite rocks (Figs 9.18B and 9.47).

Illustrations of depressions formed by a combination of the first two mechanisms of Figure 9.17 are the huge collapse dolines of the Nakanai Mountains in New Britain, Papua New Guinea (Maire 1981a). They are in polygonal karst terrain and originate from the coincidence of solution from the surface and upwards stoping of the cave roof above large underground rivers with low flow discharges of the order of 15–20 m^3s^{-1}. The bottoms of several solution depressions have collapsed into the river caves below, the large shafts produced ranging in diameter from 100–700 m and in depth from 150–400 m. Corrosion focused near the bottoms of the solution dolines above a cave reduces the rock strength more than the diminution of the rock load reduces the applied stress there. Many examples of this kind of polygenetic doline are found above river caves running through cockpit karsts. They are frequently encountered just upstream of spring heads and just downstream of stream-sinks.

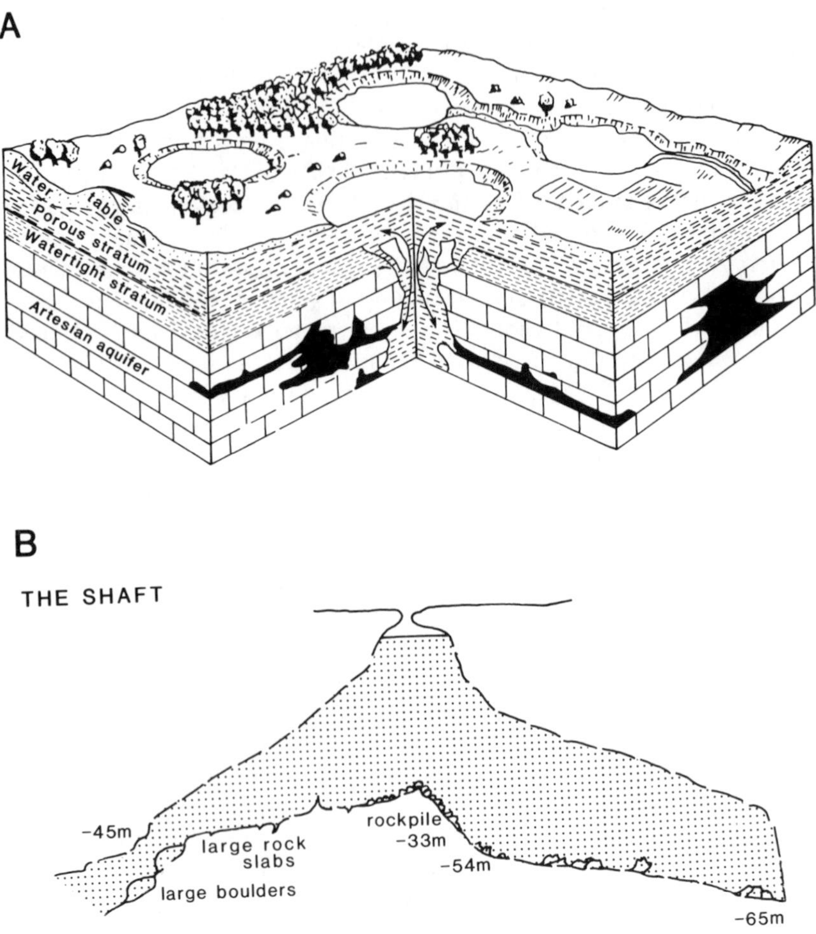

Figure 9.19 A. Drowned collapse dolines in Florida formed by a combination of undermining from below and periodic loss of buoyant support. From Davies & LeGrand (1972) after Cooper & Kenner (1953). B. Flooded, cenote-style collapse doline in southeast Australia formed by a combination of undermining from below and periodic loss of buoyant support. From Lewis & Stace (1980).

Examples of dolines formed mainly by a combination of the second and third mechanisms are shown in Figures 9.19A & 11.9. Johnson *et al.* (1986) provide various case studies. In Florida repeated glacio-eustatic sea level changes caused numerous major fluctuations in the level of the water table. Buoyant support was removed on numerous occasions and consequently cover beds have been penetrated by collapse in many locations. The present high stand of sea level results in most of the depressions being water filled, giving a landscape of doline ponds.

Glacio-eustatic oscillations probably also made a major contribution to the development of flooded collapse depressions in southeast Australia (Marker 1976) and Yucatan (Gerstenhauer 1968), where they are known as *cenotes*. They appear superficially to be water-filled shafts abruptly set into a relatively featureless limestone plain, but diving has shown those in southeast Australia at least often to be the top of enormous flooded bell-shaped cavities with disconcertingly thin roofs (Fig. 9.19B). Shaft-like depressions deeply flooded by sea water in low lying coral islands are known as *blue holes*. They are common features in the Caribbean and are also found on the Great Barrier Reef (Backshall *et al.* 1979, Gascoyne *et al.* 1979, Hopley 1982, Smart 1984c;. They resemble cenotes at the surface because of their usually circular opening. However, some are produced along massive shear fractures formed by steep reef edge foundering following removal of buoyant ocean support during low glacial sea levels. Thus not all blue holes are the product of karst processes, although the cavities are highly decorated with drowned speleothems (Palmer & Heath 1985).

In recent years in Florida, pumping of the carbonate aquifer for water supplies has artificially reduced buoyant support again. Beck (1984) notes that 650 new sinkholes have been recorded in the area, most probably being related to this although building construction on the surface also contributes through loading and runoff effects. The most infamous case of collapse doline development triggered by human activity is reported by Brink (1984) from South Africa, where a major pumping programme was initiated to de-water dolomite and dolomitic limestone over gold-bearing conglomerates. Pumping commenced in 1960, and between 1962 and 1966 eight collapse dolines exceeding 50 m wide and 30 m deep formed in the mining area. One large example developed under a mine building, engulfing the entire structure and taking 29 lives. According to Brink after 25 years of pumping 38 people have now lost their lives in collapsing sinks, and damage to buildings and structures amounts to tens of millions of rand. The depressions are the result of progressive arch collapse in the regolith overlying solution widened joints which lead downwards into large caverns (Fig. 11.10).

Subsidence depressions

In the ideal subsidence depression the ground and underlying beds gradually sag downwards with no significant faulting of the rocks, although folding always occurs (Fig. 9.48). This may be compared with the collapse depression where rupture is a characteristic feature. In practice there is a continuum in nature, many predominantly subsidence depressions having an element of fault displacement in them, depending largely on the competence of the lithologies involved.

Figure 9.20 Suffosion dolines in glacial deposits on Mt Owen, New Zealand. These dolines have developed in the last 14 000 years since deglaciation.

Trough subsidence above underground cavities can take place at different scales. At a small scale and at shallow depth it can create subsidence dolines where the cavity is at the rock–soil interface. Artificial subsidence depressions up to a few hundred hectares in extent have been created by injection mining of shallow salt deposits. At much larger scales and sometimes involving considerable depths are processes leading to development of depressions by solution sag. These are limited to the evaporite rocks and are discussed separately in section 9.13.

Suffosion dolines

Seepage through thick unconsolidated regolith or allochthonous detritus over karst rocks can generate suffosion depressions; suffosion being an evacuation of fines by a combination of solution and downwashing. It is a category of piping. Infiltrating water beneath regolith creates sub-soil karren and widened joints connected with deeper cavities. Suffosion then causes a dimpling of the surface with a multitude of small dolines (Fig. 9.20A). Very often inheritance plays a part, the allochthonous veneer of loess or glacial till, etc. having been deposited onto a well developed karst. Dolines formed in the thick alluvial veneer that commonly floors blind valleys and poljes were termed 'alluvial dolines' by Cvijic (1893), and another class was called 'alluvial stream-sink dolines' by Cramer (1941).

9.6 Morphometric analysis of dolines

Objective description of doline styles and patterns

Dolines can occur as scattered isolated individuals, as scattered clusters of individuals, as densely packed groups, or as irregularly spaced chains along allogenic recharge margins and dry valley systems. The topography is termed *polygonal karst* where enclosed depressions occupy all the available space (Figs 9.11 *Top right* & 9.21), because when tightly packed the common

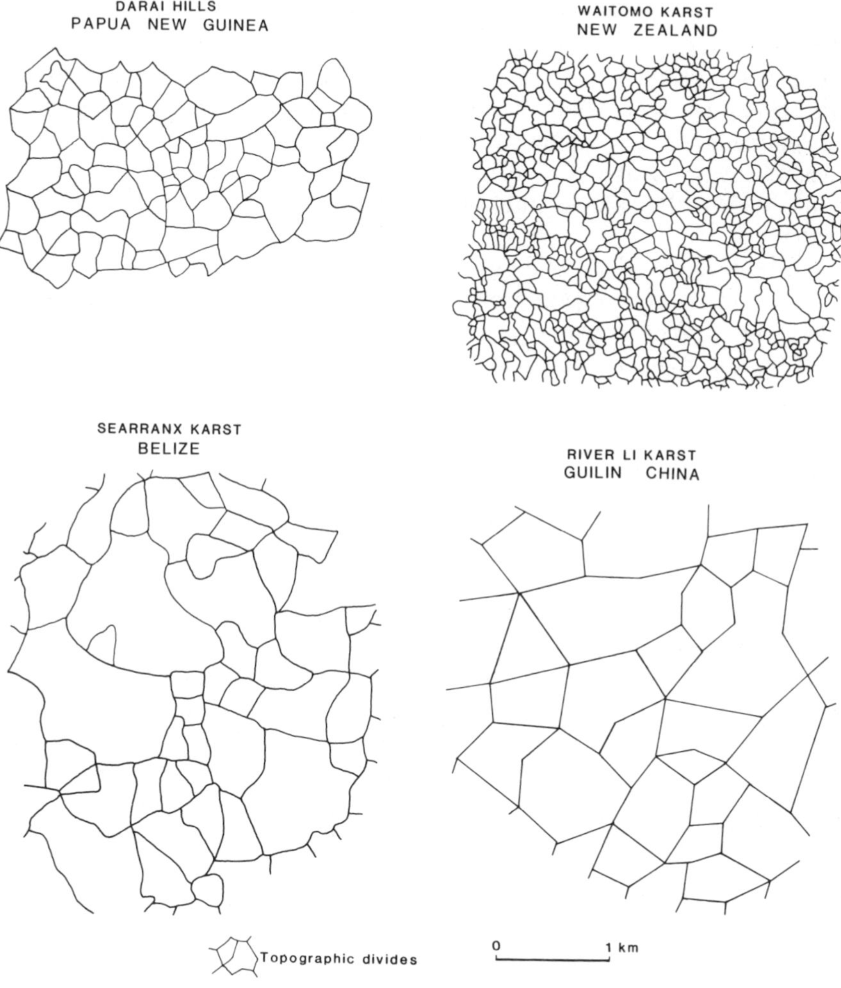

Figure 9.21 Polygonal karst networks in Belize, China, New Zealand and Papua New Guinea. Modified from Miller (1982), Pringle (1973), Zhu (1982) and Williams (1971).

drainage divides of neighbouring dolines form cellular networks in plan (Williams 1971). The mesh sizes of the tessellations may vary widely from one polygonal karst to another (Fig. 9.21) and, depending on the internal morphology of the inosculating basins, the residual hills that they isolate take on a variety of shapes in the range of hemispherical–conical–pyramidal (Balazs 1973, Day 1978). An 'egg box' type of relief is thus produced. The development of such topography is discussed more fully later.

The differing styles of dolines and residual hills encountered impart a unique character to the terrains in which they are found. It has been claimed by Lehmann (1954), Jakucs (1977) and others that morphologically distinct assemblages develop in different climatic zones, but it is also evident that there is a considerable range of karst landscapes even within individual climatic zones (see section 10.1). Questions arise whether the perceived topographic styles can be verified objectively, and whether the landscape differences between climatic zones are greater than those within them. Morphometry is a technique that has been used to help to resolve such questions (e.g. Drake & Ford 1972, Williams 1972a, Day 1976, White & White 1979). Its aim is the objective and quantitative description of landforms.

The application of morphometry to karst in fact commenced long before it became widely used in geomorphology as a whole. Cvijic (1893) and Cramer (1941) in particular were pioneers in this field. Their work and later contributions by many others is discussed by Williams (1972b) and Jennings (1985).

Methods used in karst morphometry

The most accurate data for morphometric analysis is obtained from field survey (e.g. Jennings 1975). However, this is time consuming and only permits coverage of a small area. The most practical medium for morphometric analysis of karst is generally found to be large scale (say 1 : 15 000) aerial photographs viewed stereoscopically under magnification. These are preferable to topographic maps, because even when maps are of large scale and small contour interval, significant terrain information is lost, especially when dolines are shallow. In Barbados, Day (1983) used maps as large as 1 : 10 000, but found their representation of dolines to be arbitrary, underestimating depression numbers by up to 54% when compared to field surveys. The major problem with maps is their variable quality and density of information, e.g. Troester *et al.* (1984) endeavoured to compare results from 1 : 50 000 maps of 20 m contour interval with those of 1 : 24 000 with ~ 1.5 m interval. Problems are not entirely escaped when using aerial photographs because heavy forest or shadows mask hydrographic and topographic detail.

On detailed maps, the contour pattern of deep and large dolines

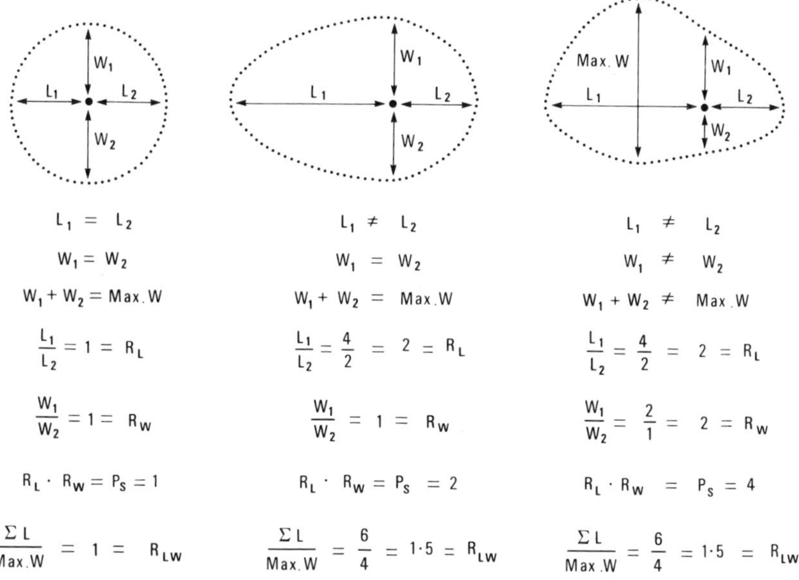

Figure 9.22 Measurements made to estimate the two-dimensional geometry of karst depressions. From Williams (1971).

(cockpits) in the humid tropics is often star shaped (Fig. 9.12). This arises from the small and usually dry valleys that incise the depression slopes and focus on the bottom. Aerial photographs reveal this even more clearly and also show that a similar internal dissection often characterizes the commonly smaller dolines of the temperate zone. The spatial location of a doline is therefore usually represented by its lowest point (the drainage focus), which may not necessarily coincide with the geometric centre of the basin; and it is normally delimited by its topographic drainage divide (Fig. 9.21). This approach applies to any type of doline from any climatic setting. The plan geometry of the individual features thus defined can be expressed by various length : width ratios (Fig. 9.22). Internal shape with respect to the drainage focus can be assessed with the *product of symmetry*, which is the product of length and width ratios measured through the lowest point of the basin (Table 9.2). Because the short valleys are morphologically similar to low order valley systems in a fluvial landscape, Williams (1971, 1972a) ranked depressions hierarchically, employing Strahler's (1957) stream ordering procedure. Subsequent experience has shown Shreve (1966) magnitude ranking to discriminate doline differences more effectively, because it takes better account of the many lower order valleys that usually occur.

Relief attributes of dolines have received less attention than plan form because sufficiently detailed height data are not readily obtainable from

Table 9.2 Some morphometric measures applied to karst depressions.

Measure	Definition	Units
Mean doline area $\bar{A}$	$\Sigma A_d/N_d$	$L^2{}_{(m^2)}$
Depression density D	N_d/A_k	$L^{-2}{}_{(km^{-2})}$
Index of pitting P_i	$A_k/\Sigma A_d$	Dimensionless
Length : width ratio R_{LW}	$(L_1 + L_2)/W_{max}$	Dimensionless
Product of symmetry P_s	$(L_1/L_2) . (W_1/W_2)$	Dimensionless
Depth : diameter ratio R_d	$H.\left(\frac{L_1+L_2+W_{max}}{2}\right)^{-1}$	Dimensionless

A_d area of individual depression
A_k area of karst
H depth of depression (usually bottom to lowest point on rim)
L_1 and L_2 longest and shortest parts respectively of the depression long axis on each side of the lowest point
N_d number of depressions
W_1 and W_2 longest and shortest parts respectively of the depression width axis on each side of the lowest point, measured at right angles to the length axis
W_{max} maximum width axis measured at right angles to the long axis but not necessarily passing through the lowest point

Table 9.3 Average depth and density of karst depressions by region. From Troester *et al.* (1984).

	Average depth (m)	Characteristic depth (m)	Carbonate outcrop area (km^2)	Number of depressions	Depression density (km^{-2})
Temperate karst regions					
Appalachians					
Dolines on Miss. Ls.	7.3	4.62	1510	2182	1.45
Dolines on Ord. Ls.	8.3	4.62	1150	1506	1.31
Dolines on Ord. Dolo.	8.0	4.18	1474	1472	1.00
All Dolines	7.8	4.48	4134	5160	1.25
Kentucky	5.4	4.02	153	830	5.41
Missouri	6.8	3.23	*	2217	*
Florida	*	0.85	427	3395	7.94
Tropical karst regions					
Puerto Rico	19	11.37	799	4308	5.39
Dominican Republic	23	8.93	1262	7205	5.71

maps and field measurement is time consuming. Doline depth is usually taken to be the difference in elevation between the lowest point in the basin and a bounding contour or characteristic height on the polygonal divide. Depth information is important because as Troester *et al.* (1984) remark, the most obvious difference between temperate and tropical karst is often the amount of relief in the landforms (Table 9.3). Terrain roughness is an associated attribute of interest, because it also can be used to help

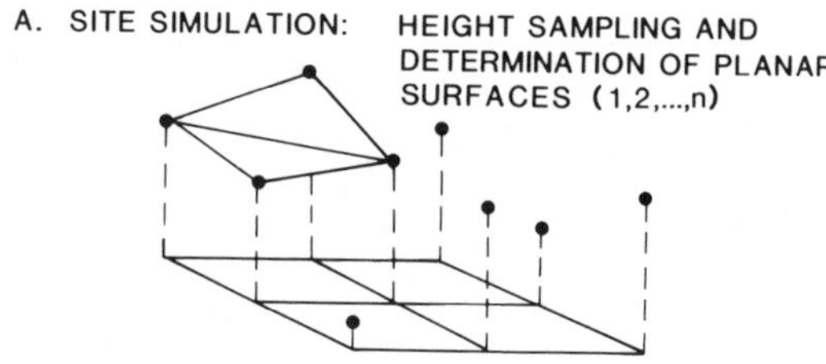

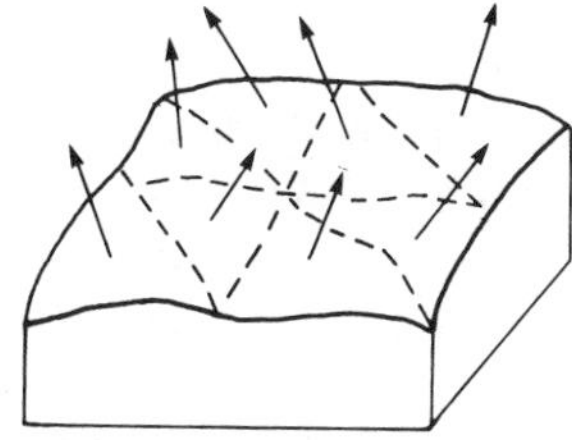

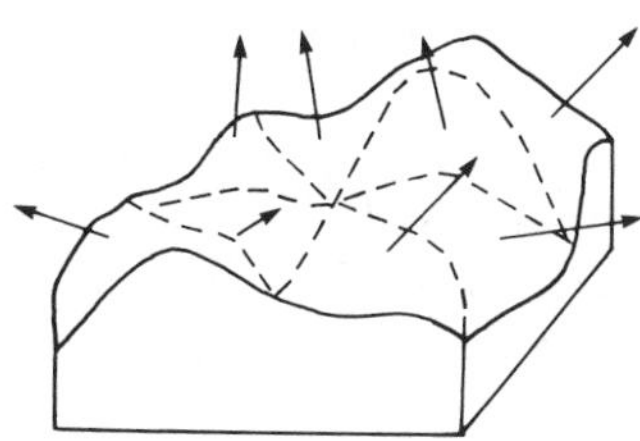

Figure 9.23 Definition of 'smooth' and 'rough' topography using the criteria of vector dispersion and strength. From Day (1979b).

discriminate between karst landscape styles (Day 1977, 1979b). 'Smooth' and 'rough' topography are illustrated in Figure 9.23. Vector orientation, strength and dispersion are calculated, and roughness (K) is defined by:

$$K = \frac{N - 1}{N - Rl} \tag{9.1}$$

where N is the number of observations (triangles in Fig. 9.23) and Rl is the vector strength. In smooth terrain K approaches infinity and in rough terrain it tends to unity. Mathematical models simulating tropical karst terrain with different degrees of roughness have been developed by Brook (1981).

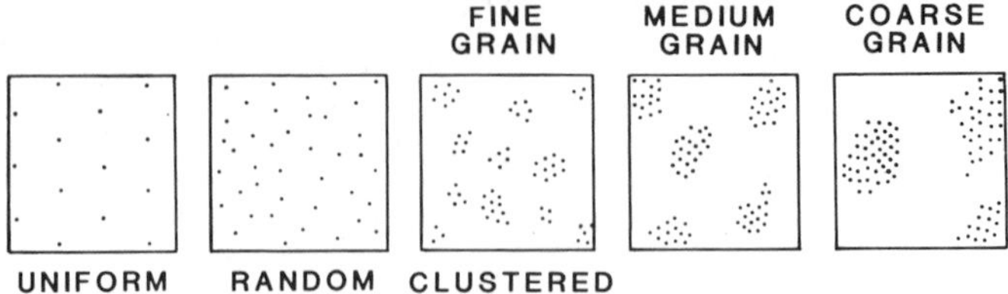

Figure 9.24 Illustrations of dispersion patterns with different intensities and grains. From Jarvis (1981) after Pielou (1969).

Analysis of the distribution patterns of dolines is discussed by Williams (1972b). The most common approaches follow methods first developed by plant ecologists and since extended by geographers. A spatial pattern can have two distinct aspects, termed 'intensity' and 'grain'. Intensity is the extent to which density varies from place to place, whereas grain is independent of intensity and concerns the spacing and areas of high and low density patches in a dispersion pattern (Fig. 9.24). Most geomorphic interest in pattern analysis focuses on intensity, and the question of particular interest in the investigation of doline fields is whether the pattern is best described as random, clustered or regular. Two methods have been used to assess this, quadrat analysis and nearest neighbour analysis (e.g. by Drake & Ford 1972, McConnell & Horn 1972, Williams 1972a & b).

Quadrat analysis compares the number of dolines occurring per cell with expectation according to a model distribution. Thus, for example, if the actual occurrence is not signfiicantly different from that described by the negative binomial frequency distribution, then the pattern can be described as clustered.

The basis of nearest neighbour analysis has until recently been Clark & Evans' (1954) test. It assesses pattern with a nearest neighbour index that compares the average actual distance ($\bar{L}_a$) between points in a spatial distribution with the average expected distance ($\bar{L}_e$) if the points were randomly disposed

$$\bar{L}_e = \frac{1}{2\sqrt{D}} \tag{9.2}$$

where D is the doline density. The nearest neighbour index $\bar{R} = \bar{L}_a/\bar{L}_e$ varies from 0 for a dispersion with maximum aggregation or clustering, through 1 for a random case, to 2.149 for a regular pattern that is as evenly and widely spaced as possible. A summary of doline density and nearest neighbour statistics for various karsts is provided in Table 9.4. Vincent (1987) provides an improved method of investigating spatial dispersion by

Table 9.4 Depression density and nearest neighbour statistics for various polygonal karsts.

Area	Number of depressions	Depression density (km^{-2})	Nearest neighbour index	Pattern	Source
Papua New Guinea (eight areas)	1228	10–22.1	1.091–1.404	Near random* to approaching uniform	Williams (1972a)
Waitomo, New Zealand	1930	55.3	1.1236	Near random*	Pringle (1973)
Yucatan (Carrillo Puerto Formation)	100	3.52	1.362	Approaching uniform	Day (1978)
Yucatan (Chichen Itza Formation)	25	3.15	0.987	Near random	Day (1978)
Barbados	360	3.5–13.9	0.874	Tending to cluster	Day (1978)
Antigua	45	0.39	0.533	Clustered	Day (1978)
Guatemala	524	13.1	1.217	Approaching uniform	Day (1978)
Belize	203	9.7	1.193	Approaching uniform	Day (1978)
Guadelope	123	11.2	1.154	Near random*	Day (1978)
Jamaica (Browns Town–Walderston Formation)	301	12.5	1.246	Approaching uniform	Day (1978)
Jamaica (Swanswick Fm.)	273	12.4	1.275	Approaching uniform	Day (1978)
Puerto Rico (Lares Fm.)	459	15.3	1.141	Near random*	Day (1978)
Puerto Rico (Aguada Fm.)	122	8.7	1.124	Near random*	Day (1978)
Guangxi, China (three areas)	566	1.96–6.51	1.60–1.67	Approaching uniform	Fang (1984)
Spain, Sierra de Segura	817	18–80	1.66–2.14	Near uniform	Lopez Lima (1986)

* significantly different from random expectation at the 0.05 level, i.e. although near random there is a tendency to uniformity of distribution.

calculating an empirical distribution function of nearest neighbour distances. Complete spatial randomness can then be tested for by using the Poisson distribution. He illustrates the method using Williams' (1972a) data from New Guinea, broadly confirming previous conclusions for seven of the eight sites.

Morphometric techniques have also been used to address the question of karstic evolution. Three approaches have been adopted: one analyses growth patterns, the second compares change over time, and the third substitutes space for time.

Drake & Ford's (1972) study illustrates the first. In earlier work Ford subjectively classified dolines in the Mendip Hills into two groups termed 'mother' and 'daughter'. By means of quadrat analysis this was shown to be justified. It was concluded that there is a constant cluster consisting of (on average) four daughter features about one mother feature, the mothers themselves being distributed in random clusters. This indicates that there is some justification for thinking in terms of two 'generations' of dolines on the Mendip Hills, with a definite spatial and causal relationship between them. The features concerned are predominantly solution dolines. Later work by Kemmerly (1982) gives further support to multigeneration evolution, although evidence for this was not found by Day (1983) in Barbados.

Kemmerly & Towe (1978) studied change over time, based on growth of 18 dolines during an interval of 35 years. They used sets of aerial photographs flown in 1937 and 1972, supplemented by field survey. The dolines were mainly of suffosional origin, being developed in drift deposits up to about 10 m thick, with no carbonate rock outcrops showing. Their conclusions were that: (a) depression length and width axes enlarged over the 35 year period, length usually growing more rapidly than width; (b) doline growth is a function of surficial geologic setting, areal growth rates averaging 40, 70, and 100 m^2 per century for loessic, clayey residual, and silty colluvial surface material respectively; and (c) estimates of doline age on the three surficial materials, based on linear growth rates, varied from 25 ka (silty colluvium), through 38 ka (clayey residuum) to 65 ka (loess). These ages were compatible with independent biological evidence.

Substitution of space for time to obtain information on landform evolution is a well known strategy in geomorphology. Its validity rests on the assumption that evolution has followed a similar pattern in the places being compared; hence a weakness is that process rates and combinations may have varied significantly through Pleistocene climatic changes. Yet despite this problem useful results have been obtained. For example, Day (1983) studied solution dolines on a series of raised reef terraces in Barbados, the interest being to measure how doline morphology changes with duration of evolution, i.e. on surfaces of successively greater age. Doline densities were found to increase on terraces up to an elevation of 150 m, but to decline above that, whereas doline dimensions reach a

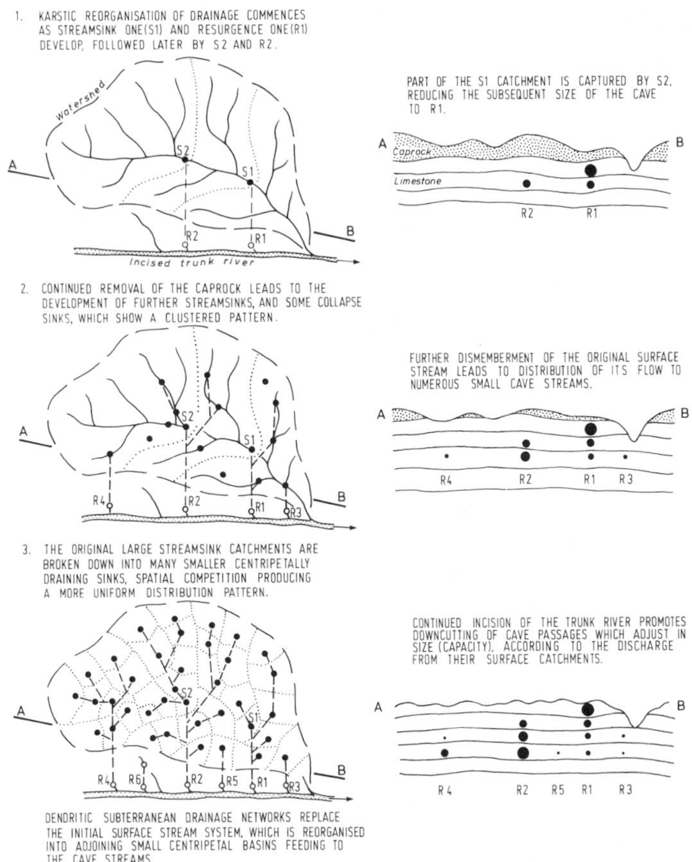

Figure 9.25 A model of the reorganization of drainage by karstification. From Williams (1982a).

maximum at about 225 m. Doline patterns tended to be clustered and two doline sub-populations could be identified: larger dolines in interfluve areas and smaller dolines in valleys. It was concluded that doline complexes evolve through time partly by expansion, interference and coalescence and partly by subdivision. The development of dry valleys appeared to compete with the dolines and to exert some control on their spatial distribution, as noted by Fermor (1972).

The karstification of drainage when a normal dendritic stream pattern is lowered from cover beds onto limestone has also been studied morphometrically using this space–time substitution approach. Feeney (1977) and Williams (1982a) describe results from Waitomo, New Zealand. The karst is developed in Oligocene limestones that rest unconformably on an impervious basement and are overlain by relatively impermeable silty and sandy clastic beds. Every degree of stripping of the caprock and limestone is represented. Feeney examined four sample areas where erosion of the cover beds had exposed 0, 30, 80 and 100% of the limestone, and a fifth area where 60% of the limestone itself had been removed to yield a 40% exposure of the impervious basement. The doline pattern was shown to progress from a clustered dispersion when karstification of drainage commences to a distribution on the regular side of random when the polygonal karst is at its maximum development (Fig. 9.25). Coarse grained clusters of dolines mark the end stages of the evolutionary sequence as blocks of karst are isolated and reduced by denudation and as surface drainage resumes and expands across the extending exposure of basement rocks.

Principal conclusions from karst morphometry

One of the great values of morphometry is that detailed scrutiny of a landscape throws up unexpected observations and stimulates fresh hypotheses. For example, for a long time it was supposed that karsts are chaotic, their landforms being a random jumble of collapse and solution features. Morphometric research has shown this not to be so (Fig. 9.26). Many karsts in widely separated parts of the world have similar spatial organization,

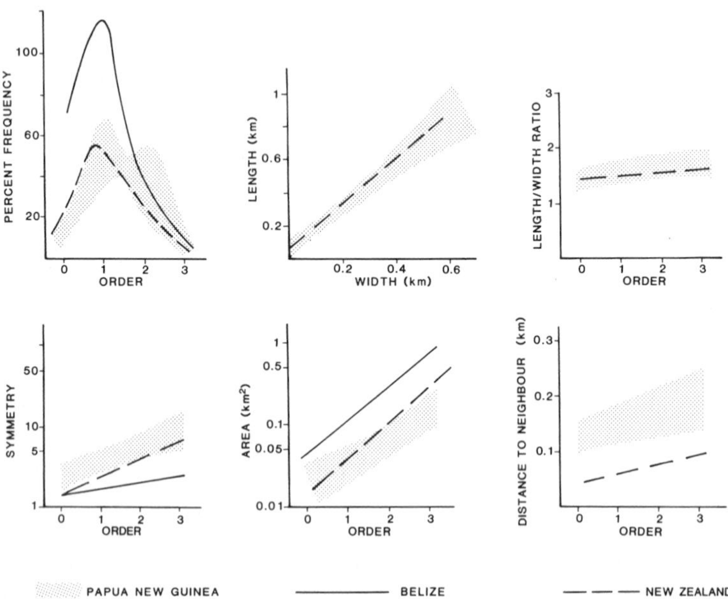

Figure 9.26 Relationships between morphometric characteristics of karst solution depressions in Papua New Guinea, Belize, and New Zealand. Data from Williams (1972a) Miller (1982) and Pringle (1973).

being characterized by a cellular mosaic, e.g. polygonal karsts in America, Asia, Australasia and Europe. Furthermore, their dolines are often dispersed in patterns that are significantly different from random, tending towards uniformity of distribution (Table 9.4). This has been explained as being a result of spatial competition, attainment of perfectly uniform distribution being hindered by distorting factors such as general topographic slope, regional dip and fissure patterns which orient corrosion by directing runoff and infiltration (Williams 1972a). Ever since Penck's (1900) views were known on the importance of a fluvial phase before karstification commenced, there has been interest in the effect of lowering a normal drainage pattern from impermeable cover beds onto underlying karst rocks. The results of morphometry show that such karstification does not *disorganize* the stream pattern as has commonly been assumed. It *reorganizes* it (Fig. 9.25). H. Lehmann (1936) postulated a fluvial phase in Java prior to the development of Kegelkarst, although at that site it was not lowered from cover beds but developed as consequent streams on an uplifted erosion surface. Monroe (1974) and Miller (1982) have kept this idea alive from karsts in Central America and the Caribbean. Miller's morphometric work has enabled reconstruction of paleodrainage lines in Belize and Guatemala, and has convincingly shown their imprint on polygonal karst terrains. However, the paleodrainage (fluviokarst) impression across polygonal karst in Papua New Guinea is almost certainly ascribable to lowering from a caprock (Williams 1972a).

Troester *et al.* (1984) concluded from their map analysis of 23 000 dolines that the average depth of depressions in their tropical Caribbean examples is approximately three times greater than that of US temperate zone examples (Table 9.3). Low relief, subtropical Florida has shallower dolines. Such information is of morphogenetic interest when the same genetic types of doline are being compared, e.g. drawdown solution dolines in each area, but is clearly of little value if the classes are mixed, as would be the case if those types illustrated in Figs 9.14 and 9.19 were compared for example.

9.7 Landforms associated with allogenic inputs

In section 5.3 we discussed input controls on the development of karst aquifers. The landforms associated with allogenic inputs depend upon five factors:

(a) input discharge;
(b) hydraulic conductivity;
(c) hydraulic gradient;
(d) location of input (lateral or vertical); and
(e) time.

Through valleys

When allogenic discharge into a karst is considerable it may exceed the capacity of the karst to absorb it underground. Such rivers maintain their flow at the surface and may entirely cross the karst to the output boundary. Whether they do so also depends partly upon the hydraulic gradient; the steeper it is the greater the tendency for drainage loss underground. This is because for a given hydraulic conductivity throughput discharge can increase if hydraulic gradient steepens (section 5.1). The morphological consequence of a large allogenic river entering a karst region with little elevation difference between input and output boundaries normally is that a through valley will be incised across the karst. A through gorge will form (a) where the height difference is greater but discharge is still sufficient to maintain competent surface flow or (b) where uplift occurs at a rate that does not exceed the river's capacity to incise (an antecedent gorge in this case, e.g. Ford (1973)). Gorges formed in either of these two ways should not be confused with those developed by cavern collapse.

Perennial *effluent* allogenic rivers traversing a karst have an important morphological function because they act as the regional base level. The Green River in Kentucky and the Krka River in Yugoslavia are examples. Such rivers gain in discharge from tributary karst springs. They therefore

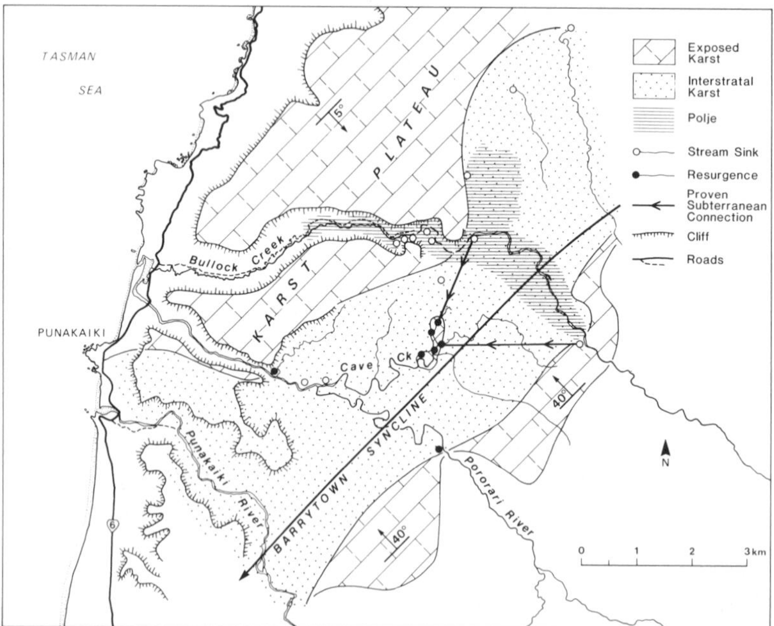

Figure 9.27 Bullock Creek on the west coast of the South Island of New Zealand was once a through valley, but uplift caused steeper hydraulic gradients and the stream has been captured underground. In flood, overflow reoccupies the antecedent gorge. Modified from Williams (1987b) with extra hydrological data from S. Crawford.

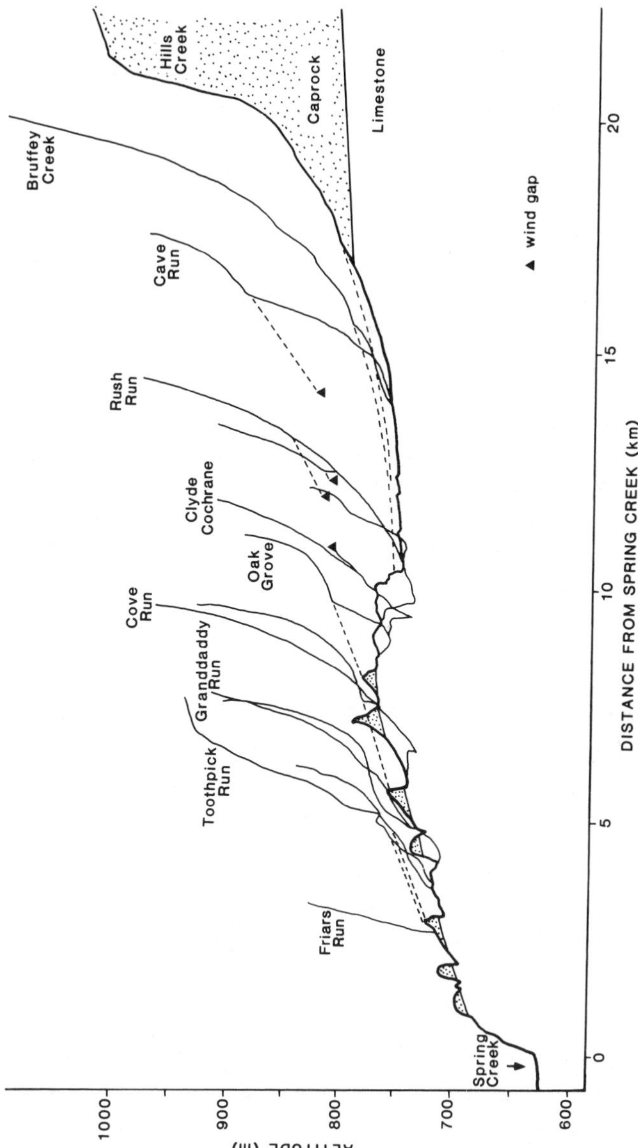

Figure 9.28 Dissected valley above Friars Hole cave system, West Virginia. The multiple blind valley has developed by headwater stripping of the caprock. Wind gaps and caprock residuals mark the level of the former active valley. Numerous stream-sink points pock the valley floor. Modified from Worthington (1984).

contrast with *influent* allogenic rivers that gradually lose flow, in part or totally, but still maintain a continuous channel downstream. The upper River Danube in Germany and the Takaka River in New Zealand (Fig. 11.1) are good examples. Their valleys display no obvious karst morphology, but leakage from the channel into karst drainage is well developed and spread along many kilometres of channel.

As hydraulic gradient steepens and the average flows of incoming allogenic streams reduce, sink points become more localized and morphologically more distinct. Such valleys are characterized by one or more permanent stream-sink depressions that sometimes overflow, the channel downstream being interrupted by further sinkholes (Fig. 9.27). Given time, and especially where drainage has been lowered from above, stream-sinks can incise to the extent that all overflow ceases and the former channel downstream becomes dissected by dolines. We then have a blind valley with an abandoned higher level valley continuing downstream. There are many instances in Appalachian America e.g. Friar's Hole dry valley (Fig. 9.28) which overlies the cave of Figure 7.26. A modification of this kind of situation is encountered as an 'island' of caprock is stripped back. Upstream retreat of a sink point may leave a line of abandoned stream-sinks along a normally dry valley downstream, some of which may be periodically reactivated during floods.

Blind valleys

Karstification of drainage lowered from above can lead to the development of innumerable small stream engulfment points or blind valleys. Flow is directed into the karst either laterally or vertically, depending largely on whether runoff is from an inlier or a caprock (Fig. 5.17). Blind valleys consequently vary in form from normal elongate valleys with a cliff foot sink point at the downstream 'blind' end to those which are circular in plan with drainage focused centripetally into a stream-sink where a window of the underlying karst rock is exposed. The latter features are often scattered in clusters across a surface, whereas the conventional elongate blind valleys are disposed linearly along a lithological contact.

As caprock is eroded, the centripetal blind valleys subdivide into smaller features best regarded as point recharge dolines, although clearly there is a spectrum of form. Eventually all the caprock is removed and only dolines remain. By contrast, the development of conventional blind valleys that receive recharge laterally can involve deepening, widening or both. Incision is most likely when the hydraulic gradient is steep, otherwise lateral enlargement occurs. This may lead to the coalescence of neighbouring blind valleys, to produce a large enclosed depression. If it has a relatively smooth, extensive, alluviated floor (the coalesced floodplains of allogenic streams), it can then be termed a *border polje* (Fig. 9.30).

Sometimes aggradation buries stream-sinks and reduces their absorbance capacity. More frequent overflow then occurs and the valley is considered semi-blind. The semi-blind condition can therefore arise both in the early stages of evolution before subterranean conduits are fully developed and at a later stage if aggradation and fill reduces conduit throughput capacity.

Recharge response

Recharge is a process and valley development a morphological response. Other things being equal, the greater the lateral point recharge the larger is

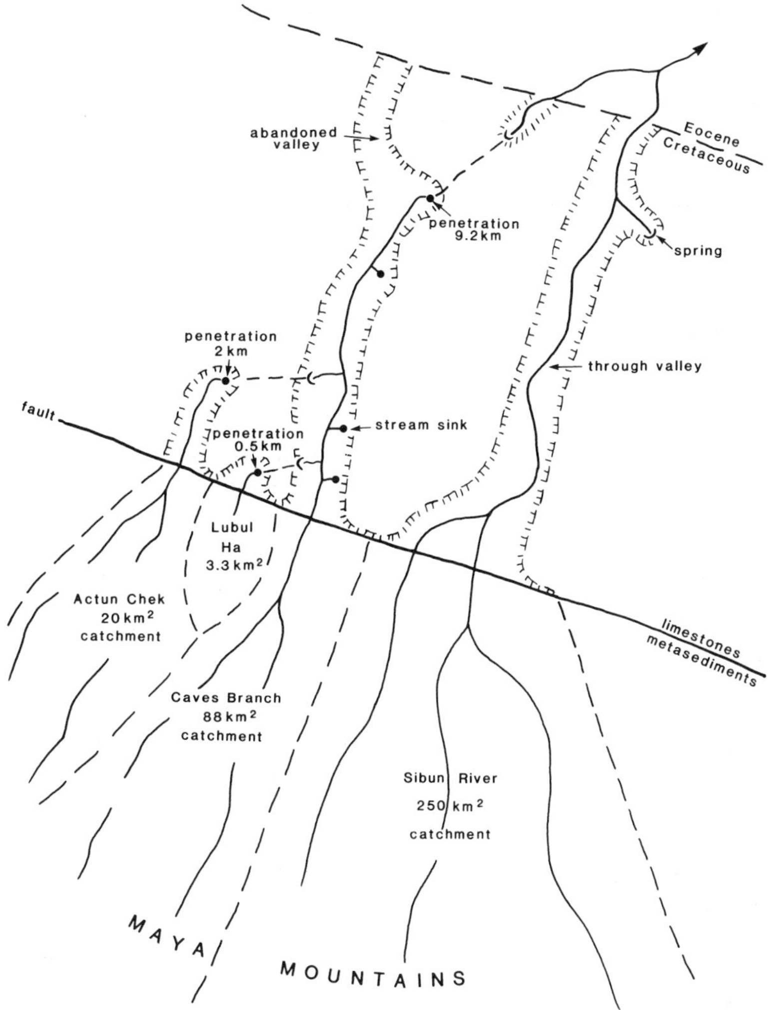

Figure 9.29 Variation in penetration distance of allogenic streams according to catchment size, Belize, central America. Modified from Miller (1982).

the valley penetration into the karst, to the extent that very large recharge provokes complete traversing of the karst by the allogenic river. This general principle is illustrated in Figure 9.29, where rivers in Belize flow from an inlier of non-karst Palaeozoic rocks across a faulted boundary into a polygonal karst in Cretaceous limestones. Lubul Ha blind valley drains 3.3 km^2 of impervious rocks and penetrates merely 0.5 km into the karst; Actun Chek catchment covers 20 km^2 and penetrates 2 km; Caves Branch has an 88 km^2 allogenic basin and penetrates 9.2 km when in flood; whereas the Sibun River draws its large discharge from 250 km^2 of metasediments and completely traverses the karst in a through valley.

Morphological response to recharge is also seen underground. An illustration based on the Waitomo karst is provided in Figure 9.25. A few relatively large blind valleys individually provided large input discharges. Caves of commensurate size developed to transmit the throughput volumes involved. As the caprock was eroded, the same recharge was distributed between a greater number of smaller blind valleys. Their individually smaller throughput volumes produced correspondingly smaller conduits.

9.8 Karst poljes

Definition

Poljes are large, flat floored enclosed depressions in karst terrains (Fig. 6.25). They are landforms associated with the input or throughput of water, and in many respects can be considered inliers of a normal fluvial landscape (Fig. 6.27). The word 'polje' signifies a field, and it is still widely used in Slav languages without particular reference to the terrain, which need not necessarily be karstic (Sweeting 1972, Roglic 1974). However, polje has acquired a special usage in technical karst literature, particularly since the writings of Cvijic (1893, 1901) and Grund (1903). Similar landforms are termed 'plans' in France, 'campo' in Italy, 'wangs' in Malaysia and 'hojos' in Cuba. Although they are written about mainly in tropical and temperate contexts, they are also reported in the subarctic zone (Brook & Ford 1980). The most thoroughly studied are in Yugoslavia (Gams 1978, Mijatovic 1984b).

Gams (1978) considers the many geomorphic definitions of polje that have been published and concludes that uncertainties still remain as to what a polje really is. Nevertheless, several elements are common to most definitions. Gams identifies three criteria that must be met for a depression to be classified as a polje:

(a) flat floor in rock (which can also be terraced) or in unconsolidated sediments such as alluvium;

(b) a closed basin with a steeply rising marginal slope at least on one side; and
(c) karstic drainage.

He also suggests that the flat floor should be at least 400 m wide, but this is arbitrary. Cvijic (1893) took 1 km as a lower limit. In fact poljes vary considerably in size. The most extensive in the Dinaric karst is Lika Polje with its flat floor covering 474 km^2 (Table 9.5).

Regardless of individual particularities all poljes have a common hydrological factor in their history: their development occurred close to the local water table, even though it may be perched in some cases and even though subsequent events (e.g. uplift and karstification) may since have

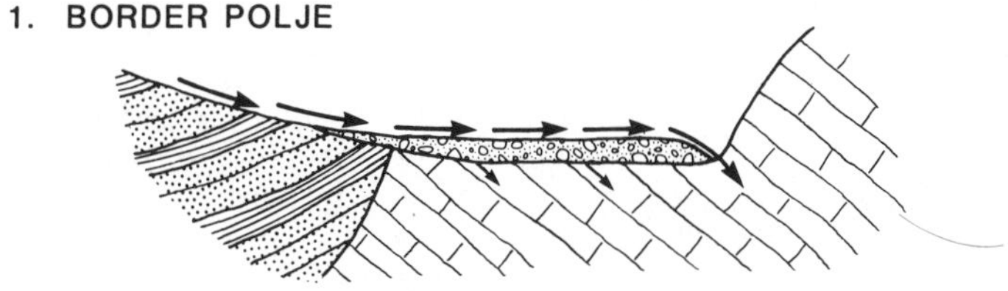

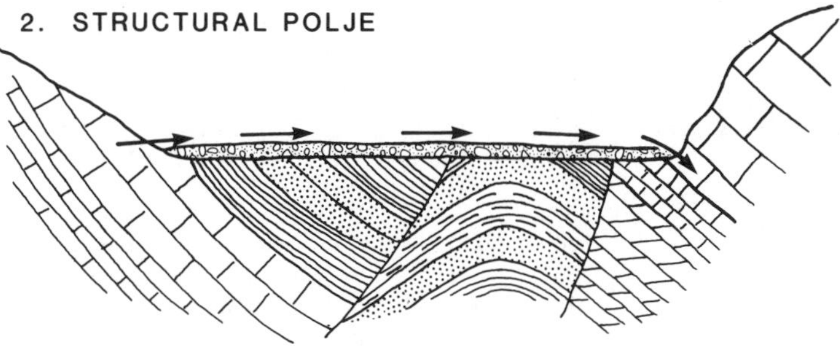

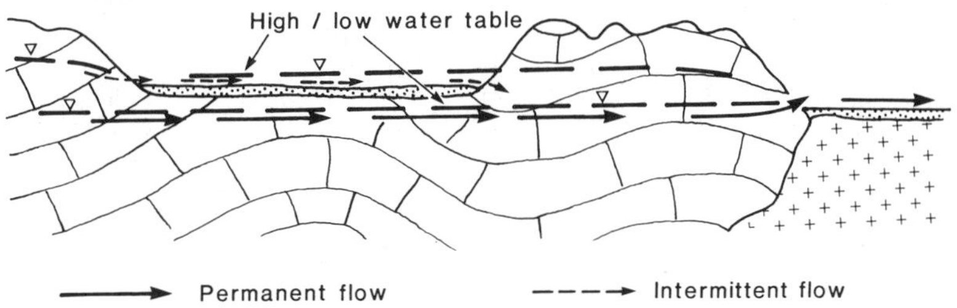

Figure 9.30 Basic types of polje.

Table 9.5 Selection of the most distinctive poljes of the Dinaric karst. From Gams (1978).

Name	Mean altitude m	Size of flat bottom km^2	Water bearing sediments or dam built of	Type of polje according to Gams
Planinsko polje	450	9	Dol	Overflow
Cerkniško polje	550	45	Dol	Overfl+Border
Loško polje	575	12	Dol	Overflow
Bloško polje	720	9	Dol	Border
Postojnsko polje	550	65	Pal	Peripher
Rakitna	780	6.5	Dol	Border
Grosupeljsko-Radensko p.	330	14	Dol	Border
Dobro polje	410	19.5	Dol+Per+Car	Border, Overflow
Ribniško polje	500	29	Dol+Per+Car	Border+Overflow
Kočevsko polje	460	72	Dol+Neo	Periph+Border
Globodol	200	3	(Dol)	Piezom
Polje of Lika	550	474	Pa	Overfl+Border
Gacko polje	450	86	Pa	Border-Overflow
Drežniško polje	470	10	Dol+(Dol)	Border
Ogulinsko polje	320	60	Dol+Pal	Border
Krbava	630	70	Dol, Neo	Border
Koreničko polje	660	19	Dol	Border
Dugo polje	480	2	(Pal?)	Border
Vojničko polje	380	16	Dol	Border
Grahovsko polje	780	25	Dol	Piedmont+Border
Livanjsko polje	700	385	Neo+Paleo	Overflow
Glamočko polje	890	129	Dol+Neo	Border+Periph
Kupreško polje	1130	94	Neo	Peripher
Duvanjsko polje	870	122	Neo+Paleo	Peripher+Border
Podrasničko polje	750	17	Dol	Border
Ravansko polje	1135	22	Dol	Border
Dugo polje	1185	11	Neo+Paleo, Dol	Piedmont+Border
Imotsko polje	260	92	Neo	Border, Overflow
Jezero polje	30	31	Pal	Piezom+Overflow
Mostarsko blato	230	32	Pa, (Dol)	Border
Popovo polje	255	185	Dol + (Dol)	Border
Ljubijsko polje	420	5	Pal	Border
Dabarsko polje	470	30	Pal	Peripher
Nevesinjsko polje	850	188	Pal	Peripher+Border
Fatničko polje	460	27	Pal	Overflow
Ljubomirsko polje	550	12	(Dol)	Border
Grabovo	650	10	(Dol)	Peripher
Dvrsno polje	645	7	(Dol)	Piedmont
Grahovsko polje	720	14	(Dol)	Piedmont
Gatačko polje	940	61	Pal	Peripher
Nikšičko polje	350	48	Pal, Dol	Overflow+Border
Cetinjsko polje	670	7	Dol	Border
mean values	581	61		

Dol = dolomite, mostly Triassic age, (Dol) = dolomitic inliers in the limestone, Pal = Paleogene, Neo = Neogene sediments, Per+Car = Permocarboniferous sediments. For the polje types: Border = Borderpolje, Overflow = Overflow polje, Peripher = Peripheral polje, Piedmont = Piedmont polje, Piezom = Piezometric polje, in the zone of oscillation of the water table.

separated the polje floor from the contemporary position of the water table. Where a low gradient water table is close to the surface, lateral fluvial planation (corrosion and corrasion) and deposition processes are more important than incision; hence plains are created rather than deep valleys.

Five types of polje are recognized by Gams (1973b, 1978):

(a) *Border polje*. Located at a contact across which it receives surface runoff. The flat floor partly truncates karst rocks and is veneered with alluvial deposits. Water escapes through stream-sinks (ponors). This is the 'open upstream' type of Ristic (1976) – see section 6.8.
(b) *Piedmont polje*. An alluviated valley usually located downslope of a glaciated terrain from which much debris has been received. The relatively impermeable polje floor is largely attributable to Pleistocene deposits. Contact with non-karst rocks is not a prerequisite as in the border polje case.
(c) *Peripheral polje*. An enclosed basin that receives surface runoff from a large internal area of impermeable rocks, which are drained radially to stream-sinks around the periphery of the inlier.
(d) *Overflow polje*. A large enclosed depression underlain by a belt of relatively impermeable rocks that act as a hydrological barrier to water emerging at springs on one side of the polje floor. Water overflows or flows around the barrier and escapes down stream-sinks on the other side of the basin. This is the 'enclosed karst polje' of Ristic (1976).
(e) *Baselevel polje* or 'polje in the piezometric level'. The polje floor is cut entirely across karst rock but is located in the epiphreatic zone and consequently is inundated at high stages of the water table.

We consider that these can be reduced to the three basic types illustrated in Figure 9.30.

Border poljes

Border poljes are allogenic input dominated. They develop where the zone of water table fluctuation in non-karst rocks extends onto the limestone. This ensures that allogenic fluvial activity is kept at the surface and that lateral planation and alluviation dominate valley incision; otherwise blind valleys form. Floodplain deposits may partly seal the underlying limestone and encourage water to stay near the surface, although leakage may be widespread upstream of the final point of engulfment. Gams' first two types are variants on the blind valley and are characteristic of recharge margins; thus they can be grouped together.

Structural poljes

Structural poljes are dominated by geological control. They are often associated with graben or fault-angle depressions and with inliers of impervious rocks. Their depression form is elongated with the structural grain, although their tectonic boundaries may be modified by extensive planation across karst rocks. Gams' third and fourth types are variants within this category. Structural poljes are very important features, giving rise to the largest karst depressions in the world and being the dominant class of karst polje in the Dinaric karst (Fig. 6.27). They are inliers of a normal fluvial landscape (a floodplain, often with terraces) within an otherwise karstic terrain, the local water table being near the surface because of the impervious enclave. Water escapes from the basin where the hydraulic gradient steepens, usually on the karst rock side of a bounding fault. Numerous boreholes through Yugoslavian poljes of this type have shown their flat floors to be mainly the result of aggradation of Neogene terrestrial and lacustrine deposits that frequently bury an irregular topography (Mijatovic 1984b). In Duvanjsko Polje, for example, such sediment reaches 2000 m in thickness.

Baselevel poljes

These are water table dominated and occur where the regional epiphreatic zone intersects the surface. They are windows on the water table, and are typically located on the outflow side of karst. Because they do not depend on allogenic inputs or geological control, they can be considered the purest kind of polje. They can develop in an entirely autogenic context. Water table control extends inland from the output boundary, where the sea or an impermeable formation may act as a dam or threshold. If long-term denudation finally erodes part of the terrain down to the piezometric level subterranean streams are exposed and flow across the surface. Their lateral rather than vertical activity expands the window on the water table, creating an interior fluvial plain that may be cut entirely across carbonate rocks, although veneered with alluvium. Positive shifts in the water table result in their aggradation.

9.9 Corrosional plains and shifts in baselevel

Steepheads and pocket valleys

Long-term denudation reduces the relief to baselevel, and may produce a plain that expands headwards by spring sapping and lateral corrosion planation. Such features are known as *karst margin plains*. Springs occur on the output margin of karst regions. They are controlled in elevation by the regional or local water table which may be determined by lakes or the sea or by an impermeable rock threshold on the outflow side (German *Vorfluter*). Springs can retreat either by gravitational undermining and slumping of a slope, a process termed spring sapping, or by irregular collapse of a cave roof above a subterranean river, the collapse depressions ultimately coalescing to yield a gorge with a spring at its upstream end. The amphitheatre-like valleys produced by either mechanism are known as *steepheads* or *pocket valleys*. A particularly good example is Malham Cove in England (Sweeting 1972). Still larger features, ascribed to the emergence of groundwaters and measuring kilometres across, are termed *makhteshim* (erosion cirques) in the Negev and Sinai deserts of Israel and Egypt (Issar 1983).

Baselevelled corrosion plains

Corrosion plains (German *Karstrandebene*) have been described in many parts of the world, for example by Kayser (1955), Gerstenhauer (1960), Sunartadirdja & Lehmann (1960), Pfeffer (1973, 1976) and Rathjens (1954). Although most often associated with the output side of a karst terrain, they can also be produced after a long period of denudation on the input margin, when a border polje becomes connected at the surface to the outflow side (the 'open upstream and downstream' poljes of Ristic (1976), see section 6.8). Very extensive corrosion surfaces of low relief are thus formed, the Gort Lowland in western Ireland being an especially good example (Figs 9.31 & 9.32A).

Because of their low elevation, corrosion plains are commonly veneered with alluvium. Beneath this is a surface cut across the karst rock regardless of structure. When uplifted, glaciated or strip mined for placer deposits, removal of the clastic veneer reveals an impressively planar rock floor that in the tropics, in particular, can sometimes be very rugged in detail because of etching down joints, but has little or no perceptible slope and can extend across many square kilometres. Corrosional plains of this type develop by solutional removal of irregularities down to a surface controlled by the water table. Since mechanical work is not involved, except where there are significant insoluble residues, slopes can be very low indeed ($<0.1°$). Once the topography is corroded down to the level of frequent inundation (the epiphreatic zone), the plain expands by gradual retreat of adjoining karst uplands and by elimination of residual hills. Even on a perfect plain corrosion can be expected to occur beneath the water table because of the continuous supply of aggressive water from rain, but most activity will be restricted to a shallow depth because of the low hydraulic gradient.

The complex of processes producing baselevelled plains in karst has been called *lateral solution planation*, but in fact *corrosion planation* is a better

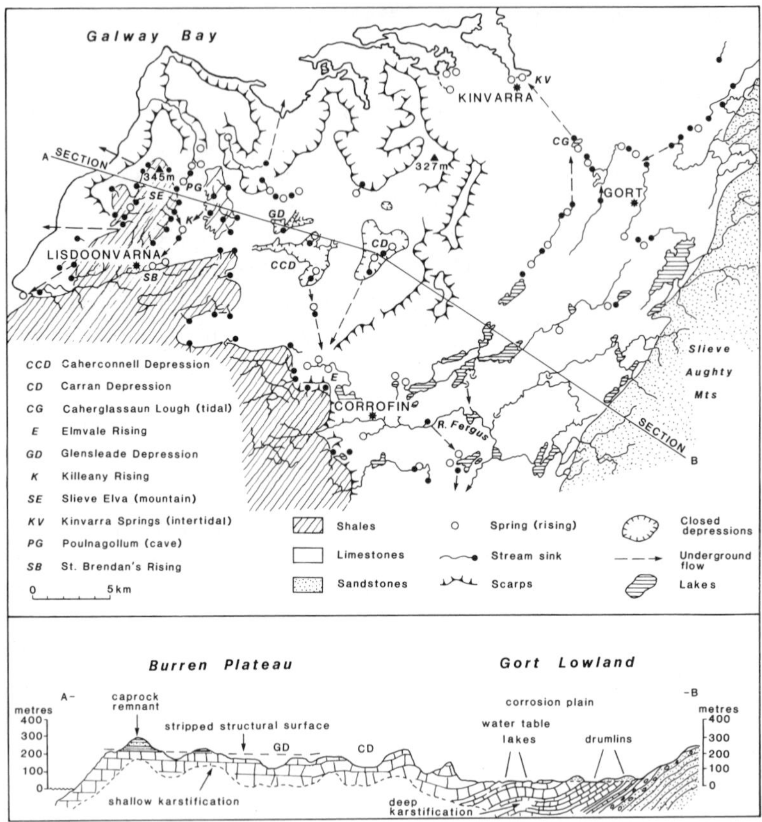

Figure 9.31 A corrosion plain in western Ireland in the vicinity of Gort. Modified from Williams (1970). See also Fig. 9.32A.

term because it involves (a) vertical solution of upstanding remnants by direct rainfall, (b) lateral undercutting of hill sides by accelerated corrosion in swampy zones at their base, and (c) spring head sapping. The relative importance of these three activities depends upon hydrogeological and biogeographical circumstances that control the aggressiveness of water, as discussed in Chapter 4, and the deposition of alluvium that influences (im)permeability of the surface.

Corrosion plains are not *peneplains* in the sense of W.M. Davis (1899), because they are not undulating surface of low relief and are not roughly the same age everywhere across the surface, as is the assumption with peneplains. Corrosion plains are as near perfectly flat as can be produced by natural degradational process (Fig. 9.32A) and are worn back rather than worn down. Thus they are youngest at the foot of the retreating uplands which they replace. In this respect they resemble *pediplains*.

A

Figure 9.32 A. An ancient corrosion plain with residual hills on the Guizhou plateau, China, at 1200 m above sea level.

B

Figure 9.32 B. The Gort Lowland, and extensive baselevelled corrosion plain cutting across folded Carboniferous limestones in western Ireland. The view is to the northwest with the Burren Plateau in the distance.

Field evidence often cited for the effectiveness of lateral corrosion is the occurrence of swamp slots or notches (Fig. 9.33A) and cliff-foot caves, particularly in tropical karsts. Jennings (1976) and McDonald (1976a) mapped in detail the distribution of foot caves and swamp slots around isolated limestone towers in Selangor, Malaysia, and Sulawesi, Indonesia, respectively. Jennings concluded that their frequency of occurrence was compatible with the supposed importance of lateral solution undercutting. McDonald mapped 12 km of hill bases and found 31% to be composed of foot caves, some being relict, and 59% to be characterized by foot slopes that owe their origin to processes other than solutional undermining. These include bedrock inclines (27%) that ascend up to 100 m and also alluvial

A

Figure 9.33 A. Swamp notches at the base of an isolated karst tower in Guilin, China. The different notches reflect changes in local water table level.

B

Figure 9.33 B. The sheer sided karst tower at the base of which are the notches of the upper photograph.

foot slopes (32%). He concluded (p. 89) that the 'erosion of hillslope support and the retreat of limestone hillslopes appears not to take place uniformly . . . but takes place in scattered locations along hillsides'. Our observations elsewhere indicate that the relative significance of lateral solution planation depends on the location of the residual hill: if beside a river or on a floodplain, swamp slot corrosion is very important, but if on a terrace above the reach of contemporary inundation then other slope processes become comparatively more significant.

It has been claimed by Rathjens (1954), for example, that corrosion plains are best formed under seasonally wet, warm environments and that those found in the Dinaric karst are therefore legacies of a previous warmer period. Warm conditions do not necessarily favour corrosion (see Chs 3 & 4) and therefore cannot be a prerequisite. More important is a long stable period under conditions sufficiently humid to permit denudation of the relief down to the epiphreatic zone.

Baselevel control

In all the schemes of karst denudation that have been proposed, the ultimate baselevel of karst erosion recognized is either sea level or underlying impermeable rocks, if they occur above sea level. However, it is well established that karst springs can resurge well below sea level and that active groundwater circulation can extend to greater depths. Since the karst hydrochemical system can operate well below sea level and can corrode large phreatic caves, clearly sea level is not the ultimate base for karst dissolution. The question of note then is: what is the depth of the lowest active conduit?

The baselevel issue is best understood by referring back to our explanations of the controls on aquifer development (section 5.3) and cave evolution (section 7.3). It was pointed out that a karst circulation system undergoes a continuous process of self-adjustment, depending in detail on fissure frequency and hydraulic potential. Where fissuring supports a state 3 or 4 conduit system, a corrosion plain will develop near the level of the water table with only shallow groundwater circulation beneath it. Should the fissure frequency initially have been low, the increase of fissuring with age that appears to occur in a majority of states 1 and 2 systems also tends to favour generally shallower circulation, as will the declining hydraulic gradient. However, there will be exceptions, and deep phreatic loops once formed will maintain their flow unless they are sealed by clastic detritus. Thrailkill's (1968) analysis indicates that for a given pipe network under constant head there is no significant hydrodynamic advantage to a shallower course whether flow is laminar or turbulent.

Therefore we conclude: (a) that the baselevel for corrosion plain development is the water table and (b) that the base of active corrosion varies (i) with the state of the system, being deepest in states 1 and 2, and (ii) with age (if fissure frequency increases it will tend to be shallower) and (iii) with the extent of any clastic infilling. Fresh water : salt water interface effects are ignored for the purpose of this discussion, but it should nevertheless be recognized that considerable corrosion is likely in the mixing zone.

Rejuvenation

A downwards shift in baselevel resulting from tectonic uplift is a common cause of rejuvenation of erosional activity. Uplift of a well developed karst has three principal effects: (a) major allogenic through rivers entrench to produce deep valleys; (b) the vadose zone expands but may be perched above valley bottoms, depending on the rate of valley entrenchment; and (c) the phreatic zone moves downwards into less karstified rock. Once the water table drops beneath an uplifted corrosion plain, percolating rainwater can escape vertically again instead of running off laterally. Solution dolines can incise and suffosion dolines appear on the alluvial plains. Any residual hills become incorporated into divides between developing dolines. Figure 9.34 schematically represents rejuvenation of a tower karst that had reached water table level. A wave of morphogenesis of this sort could be expected to progress inland from the coast or from an incised river. Rejuvenation underground (i.e. development of a multiphase cave) is a prerequisite for rejuvenation (Yang 1982) of the surface topography.

Figure 9.32B shows an uplifted corrosion plain with residual hills in the Guizhou Plateau, in an area still unaffected by the current wave of rejuvenation. Superb examples of uplifted stripped karst margin corrosion plains are found beside the lower courses of the Cetina and Krka rivers, Yugoslavia (Fig. 9.35A).

Submergence

Quaternary sea level history reveals that the level of the ocean has been well beneath its present height for most of the last several hundred thousand years (Fig. 10.16B). Karsts near the sea have therefore generally evolved in response to lower baselevels than at present. Consequently, the positive shifts of baselevel during the postglacial transgressions of the multiple glaciations have given rise to: (a) contracting vadose zones; (b) expanding phreatic zones, possibly with deeper stagnant sections being removed from the layer of active groundwater circulation; (c) upward displacement of the salt water–fresh water interface on the coast and hence of the associated zone of enhanced corrosion; (d) drowned coastal springs, some of which are too deeply submerged to operate; and (e) aggradation of coastal lowlands

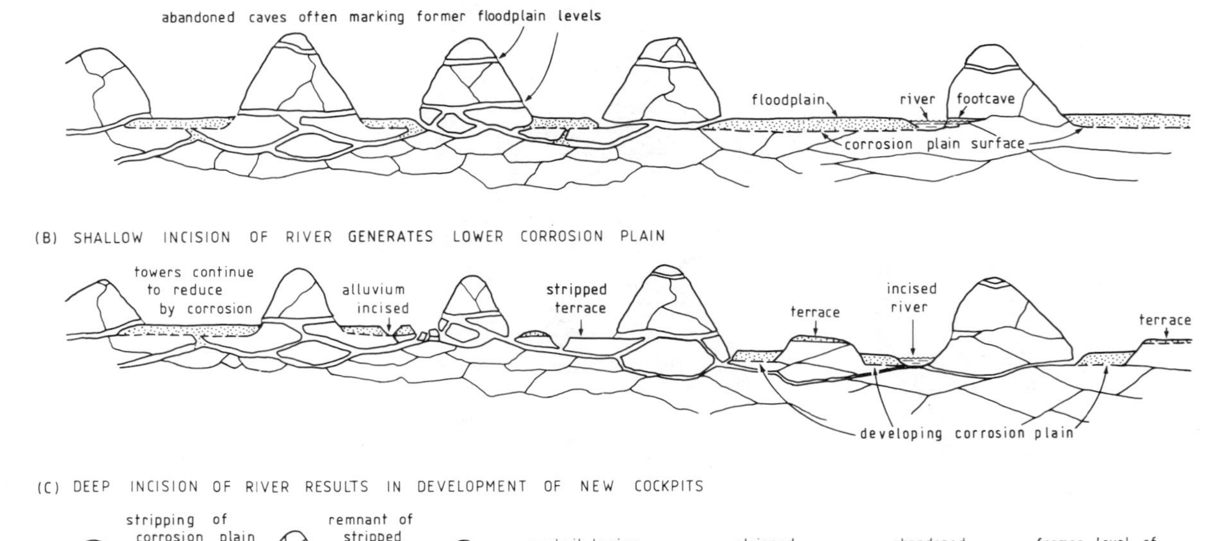

Figure 9.34 Redevelopment of polygonal karst from rejuvenation of tower karst.

A

Figure 9.35 A. An uplifted corrosion plain in Krka valley, Yugolavia. The plain truncates the structure and now forms a rock terrace above the river.

B

Figure 9.35 B. Philippine conical 'towers' in the Thousand Hills area of Bohol Island. Photo by R. Wasson.

and infilling of caves by water-borne sediments. The Florida and Yucatan peninsulas provide outstanding examples.

9.10 Residual hills on karst plains

Final stages of polygonal karst evolution

As a polygonal karst dissolves downwards through a thick carbonate mass, minor adjustments to the doline pattern will occur as beds or zones of different permeability are encountered, but otherwise the network geometry will remain until the water table is reached. When this occurs, the epikarst aquifer merges with the phreatic zone and free vertical drainage is no longer possible. Consequently vertical deepening of dolines rapidly decelerates and their floors begin to widen in the epiphreatic zone. The location and form of residual hills is inherited from the position and shape of hills around the enclosed basins of an earlier stage. Rainfall reduces these residual hills vertically by corrosion, so that, even if lateral solution planation is ineffective, a corrosion plain must ultimately develop. Where allogenic rivers cross the karst, they may considerably aid the planation process by periodically introducing large volumes of aggressive floodwaters. Corrasional planation is also encouraged by gravelly bedload. In addition, overbank deposition of alluvium may stimulate biochemical activity on the floodplain and so enhance soil water aggressivity. We conclude, therefore, that invasion by allogenic flood waters considerably accelerates the production of corrosion plains, but is not a necessary prerequisite for their formation.

Tower karst landscapes

A landscape of residual carbonate hills scattered across a plain is usually referred to as *tower karst*, even though the 'towers' may not necessarily be steep. The residual hills display a variety of shapes from tall sheer sided towers (Fig. 9.33) to cones or even hemispheres (Fig. 9.35B). Others are asymmetric, reflecting the influence of dip or erosional processes. Although some rise directly from the plain, many surmount pedestals. Some towers are isolated and others are in groups rising from a common base. The term tower karst, therefore, subsumes a myriad of forms. Nevertheless, it is associated with some very impressive features, the tower karsts of southern China constituting what must be considered one of the world's great landscapes (Figs 9.32B & 9.33B).

The international literature suggests that there are four main genetic types of tower karst (Fig. 9.36):

(a) residual hills protruding from a planed carbonate surface veneered by alluvium;
(b) residual hills emerging from carbonate inliers in a planed surface cut mainly across non-carbonate rock;
(c) carbonate hills protruding through an aggraded surface of clastic sediments that buries the underlying karst topography; and
(d) isolated carbonate towers rising from steeply sloping pedestal bases of various lithologies.

It is evident from this that the plain between the towers need not be cut entirely across carbonate lithologies, even though this often appears to be the case. Tower karsts sometimes incorporate planed surfaces in other rocks, those in the Sierra de los Organos in Cuba (Panos & Stelcl 1968) and at Chillagoe, Queensland, Australia (Jennings 1982a) being examples. Where deep residual deposits cover the plain between the towers, as is found at the coast in the Bay of Ha Long, Vietnam (von Wissmann 1954, Pham 1985) or in the blanket sand plains between the *mogotes* of Puerto

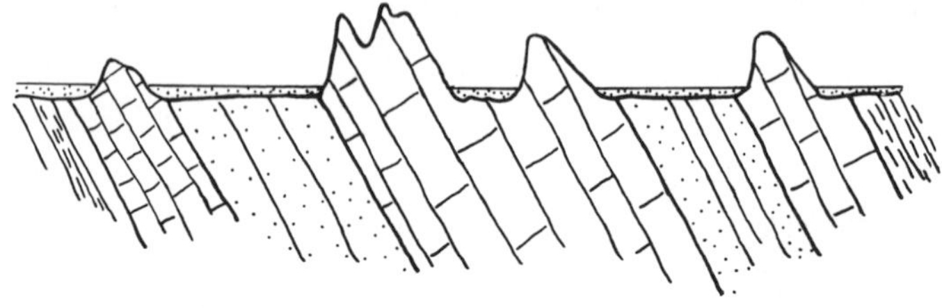

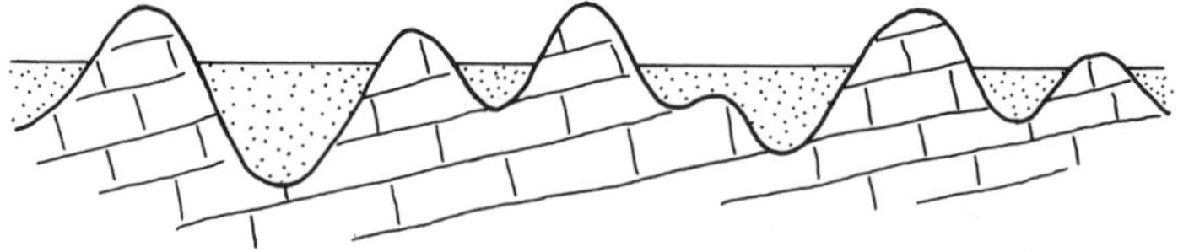

Figure 9.36 Types of tower karst. From Williams 1987a.

Rico (Monroe 1968), the third type of terrain may exist or perhaps aggraded versions of types (a) or (b). An added complexity in all of these cases is that frequently the intervening plain is terraced. The terraces are often developed in the veneer of superficial deposits, as in the tower karst near Guilin, although there is no theoretical reason why they should not also be in bedrock.

Evolution of a karst tower

Chinese geomorphologists distinguish between isolated towers on a plain (termed *fenglin* or 'peak forest') and groups of residual hills emerging from a common bedrock base (termed *fengcong* or 'peak cluster'). Fengcong usually has closed depressions among the peaks; so the combination is sometimes called 'peak cluster depression', but when residual hills are separated by dry valley networks instead of closed depressions the landscape is termed 'fengcong-valley' karst.

Williams *et al.* (1986) investigated the evolution of a tower at the edge of a fengcong group at Guilin. The hill rises abruptly from the floodplain of the River Li. A reconnaissance examination of the paleomagnetism of deposits from caves at different levels in the tower suggests that sediments up to 23 m above the floodplain possess normal geomagnetic polarity, but that some deposits above this have a reversed magnetism. Other evidence from the lower caves shows the foot of the tower to have been buried by fluvial aggradation and then to have been re-exposed by partial excavation of the deposits. The combined information indicates the tower to have grown by the lowering of its base at a net rate not exceeding 23 mm/1000 years during

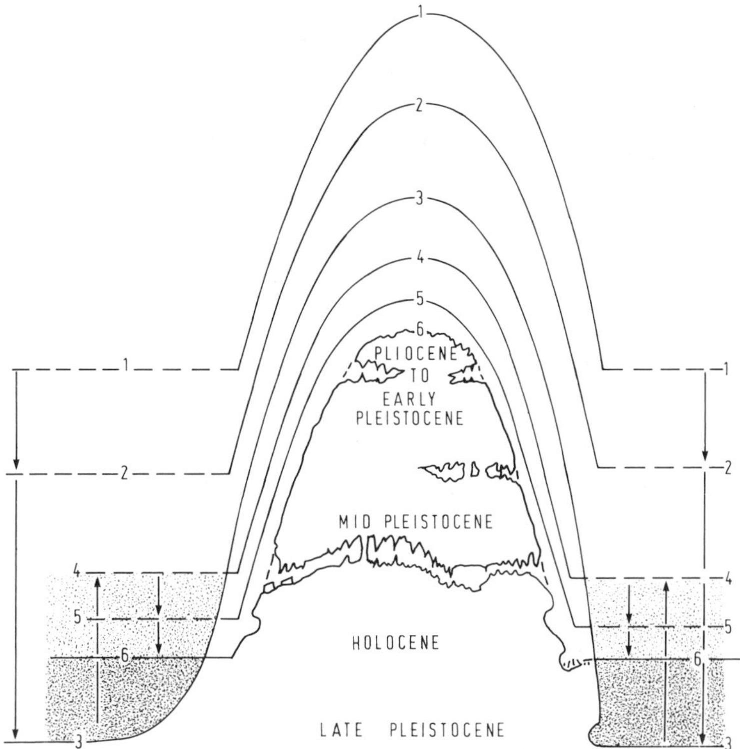

Figure 9.37 Model of karst tower evolution near Guilin, illustrating episodes of burial and exhumation that occur simultaneously with tower reduction by solution. Abandoned caves from earlier phases penetrate the tower and newer swamp notches and foot caves mark present and previous floodplain levels.

the past 1 Ma. This evidence reinforces that already long established from the fossil record in other towers in southern China, where Pliocene to mid-Pleistocene vertebrates of the Stegodon–Ailuropoda faunal complex have been recovered from high level caves (Kowalski 1965). The inescapable conclusion is that karst towers in this region are time-transgressive landforms, being considerably older near their summits than at their base, although showing an oscillatory age pattern in detail because of episodes of partial burial and re-excavation as morphological evolution progressed (Fig. 9.37).

In Belize, McDonald (1979) demonstrated that the morphogenesis of towers located on low interfluves is governed by the progressive lowering of the alluvial plains by erosion caused by overland runoff. As the interfluvial plains are lowered bedrock slopes are formed around the towers, with angles between 20 and 60°. Undermining of towers only seems to occur where rivers actually flow against their bases. Field evidence points towards similar conclusions around Guilin.

A different style of tower karst evolution has been proposed by Brook & Ford (1978) for the giant grikeland or labyrinth karst in the Nahanni region of northern Canada (Fig. 9.8). As the labyrinth of widened grikes, closed depressions and small poljes is expanded, steep residual towers to 50 m high are left emerging from uneven karst margin plains. The entire development probably occurred in the 300 000–400 000 years since the region was last glaciated. Tower karst of this style is also found in seasonally arid karsts in tropical north Australia (section 10.2).

Case-hardening of residual hills and limestone surfaces

Calcareous weathering crusts are often observed on towers and outcrops in tropical to warm temperate environments. Their significance for karst was first appreciated in the Caribbean region, where the phenomenon is referred to as *case-hardening*. It is particularly important in the development of karst on highly porous, mechanically weak, and diagenetically immature lithologies, because it invests the rock with much more strength and resistance to erosion and collapse, as well as rendering it very much less permeable. Porosity can be reduced by a factor of ten or more. Thus it is a type of vadose diagenesis (section 2.3). There is also a transition between development of case-hardening on limestones and formation of calcrete on surface materials that contain carbonate but are not limestones.

The importance of case-hardening for the morphology of mogotes in Puerto Rico was recognized by Monroe (1964, 1966). In nearby Cuba, Panos & Stelcl (1968) found it on practically all bare limestone surfaces, the thickest crusts developing on the most porous rocks, especially where the surfaces are relatively old (pre-Pleistocene). Monroe considered case-hardening to be thickest on the windward sides of mogotes, encouraged by their more frequent wetting and drying. This explained the apparent asymmetry of mogotes observed by Thorp (1934). Recent work sheds further light on the question. The Aymamon Limestone outcrops in an east–west belt about 125 km long and is characterized by mogote karst parallel with the strike. Day (1978) showed that the asymmetry does not have a simple windward–leeward pattern. Some north–south asymmetry probably relates to a gentle northward dip of the limestones and stream–sinks are often responsible for undermining and steepening mogote sides. Ireland (1979) indicated case-hardening to be much more uniform than previously suggested, the zone of induration closely following the present topography (Fig. 9.38). On average the case-hardened zone is 1 to 2 m thick, but it can vary from 0.5 to 10 m.

Ivanovich & Ireland (1984) proposed a model for case-hardening with two main diagenetic processes. In limestones composed of more than 50% fossils the dominant process is precipitation into solution cavities. In rocks with less than about 20% fossils the main process is aggrading neomorphism, a wet recrystallization process resulting in the progressive increase of microspar. Precipitation is predominantly caused by CO_2 degassing. During these processes the total carbonate porosity is reduced from about 30% to 5% or less. They suggested that formation of a 1 m thick indurated layer could occur within 10 000 to 20 000 years, assuming a constant denudation rate of between 50 and 100 mm/1000 years. Fifteen samples of case-hardened limestone dated by $^{230}Th/^{234}U$ methods yielded ages ranging from 220 to >350 ka BP, although it is difficult to interpret these dates as the diagenetic process must still be continuing.

Case-hardening is well developed elsewhere. For example, it occurs on Pleistocene calcareous aeolianites in southwestern Australia, where dense indurated crusts up to at least 2 m thick encase relatively friable dune sands. Indurated layers within the deposits mark paleosol horizons, i.e. case-hardening episodes from the past. Jennings (1968) recognized that in these calcareous dunes lithification and karstification are likely to occur simultaneously, for the same agents are responsible for both. He therefore proposed the concept of *syngenetic* karst development. However, it is point recharge by allogenic streams on the inland side of the dune fields that is responsible for the numerous caves and collapse dolines, whereas it is rain falling directly onto the dune sands that is responsible for their case-hardening.

Induration in calcareous aeolianites is also characterizerd by vertical piping. The mechanisms involved are not clear but it is possible that pipes are not solely solutional in origin; tree roots may play a part in some cases. A feature of some pipes in dune sands is that they are lined with an indurated skin. Should the soil profile be eroded, residual pinnacles between the pipes stand proud, giving rise to impressive 'tombstone' terrains (Fig. 9.39) that have been mistaken for petrified forests.

A more subtle form of case-hardening occurs on emerged coral reefs.

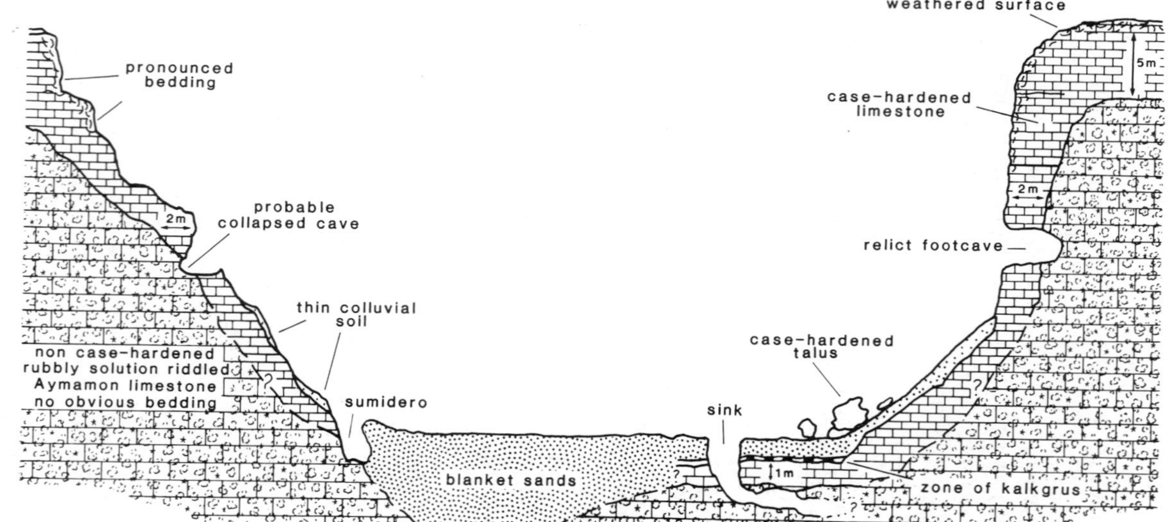

Figure 9.38 Variation in thickness of case-hardening on mogotes from north-central Puerto Rico. Note the extension of case hardening beneath the blanket sands, indicating that the sands were deposited after case hardening had already occurred. From Ireland (1979).

Figure 9.39 The 'Pinnacles' north of Perth, Western Australia. These are solutional residuals developed between vertical soil pipes in indurate calcareous aeolianite, the overlying soil having been stripped off, mainly by wind. Their formation is similar to that of pinnacle-and-cutter relief of Figure 9.2B, but in a weaker more porous rock. The climate is seasonally arid and the location is coastal.

These rocks are mechanically much stronger than dune sands or the chalky limestones of Puerto Rico, but can also be extremely porous, depending on the facies. Primary pore spaces within coral tend to be very large, typically centimetres or decimetres across. Solution near the surface leads to carbonate precipitation in the voids a few metres down the profile, banded flowstone and silt (from soil) being common deposits. In this way the reef rock in the upper vadose zone is rendered less permeable as the terrain is corrosionally denuded.

9.11 Depositional and constructional karst features

Calcrete

When carbonate-rich waters infuse soil, alluvium or weathered rock in regions where potential evaporation exceeds rainfall, chemical precipitation may occur in the profile. This sometimes also involves solution of carbonate clasts near the top of the deposit and precipitation lower down. A material is produced that is predominantly composed of calcium carbonate, in various states ranging from powdery to nodular to indurated. It is termed *calcrete* (or caliche or kunkur). Goudie (1983) explains its origin and distribution. There is a transition from calcrete, which is usually associated with soils, to tufa, associated with springs and waterfalls, to tufaceous alluvium, associated with floodplain sediments.

Tufa deposits

Springs, waterfalls and outflowing rivers in karst are often associated with the precipitates known as *tufa* or *travertine* (sections 2.2 and 8.3). Calcareous tufa is mixed with p!ant remains and the relative importance of organic and inorganic processes in its deposition has for many years been a subject of debate. Recent work demonstrate that both inorganic and organic deposition occur, but that organic processes are much more important than previously assumed. Casanova (1981) defines the tufa-depositing environment of southwest France as continental stromatolitic and shows that mineralization is often located around algal filaments. Chafetz & Folk (1984) provide convincing evidence that bacterially precipitated calcite forms a large percentage of the carbonate in many tufa accumulations in Italy and the USA, exceeding 90% of the framework grains comprising some lake-fill deposits. Their investigations found individual accumulations to range up to 85 m thick and to cover hundreds of square kilometres. They concluded that harsh environmental situations (e.g. hot geothermal water) favour inorganic deposits, while increasingly more moderate conditions favour organically precipitated material.

Tufa dams, terraces, waterfalls and mound springs

Five morphological variations of surface tufa deposits were recognized by Chafetz & Folk (1984): (a) waterfall, (b) lake-fill, (2) sloping mound or fan, (d) terraced mound and (e) fissure ridge. Waterfall or cascade deposits accumulate at the loci of both increased agitation and a place where algae and mosses can readily attach and grow. Tufa accumulation at such sites can produce dams and pond substantial lakes.

Figure 9.40 Longitudinal profile of the Plitvice Lakes, Yugoslavia. These lakes in the Korana valley are impounded by tufa dams. From Roglic (1981) after Petrik (1958).

One of the world's best known landforms associated with impoundment of water by tufa dams is the Plitvice lakes, Yugoslavia (Roglic, 1981), where 14 lakes have been formed along a 6.5 km reach of the upper Korana valley (Fig. 9.40). They lie in a gorge immediately downstream of the confluence of two main tributaries, one flowing from dolomite and already containing tufa, and the other from limestone and without calcareous deposition. An increase in Mg in already saturated carbonate waters is known to induce supersaturation of calcite when its concentration exceeds about 7% (Picknett 1972) because of the common ion effect (section 3.6). Thus the mixing of the water from these two streams could be responsible for much of the deposition, although an ecosystem in which mosses play an important part is also significant. The tufa barriers are up to 30 m high, with compound dams holding back lakes as deep as 46 m. In the largest lake (Kozjak, 0.8 km^2), a drowned dam is found 4.6 m beneath the water surface; presumably a consequence of a downstream tufa dam growing upwards at a greater rate. Photosynthetic sulphur oxidizing bacteria may be responsible for laminated carbonate accumulations in such lakes.

Springs can produce stepped mound deposits, the water flowing through radially disposed pools dammed by tufa barriers similar to rimstone pools in caves. Sulphate and carbonate deposits often occur around artesian springs in arid regions, the spring water emerging at the top of a mound of its own construction; hence the term *mound spring*. Fissure ridges form where spring waters upwell through fissures running along mound crests. By contrast, tufaceous waterfall deposits can have a tapered dome form acquired from the trajectory of cascading water (Weisrock 1981a & b).

Since many tufa deposits are associated with springs, their occurrence may be related to local or regional water table levels. Because such tufa deposits are sometimes datable radiometrically or by means of their fossil fauna and flora assemblages, their occurrence on river terraces assumes a particular importance both in terms of chronology and paleoclimatology, although interpretation may be difficult. Figure 9.41 from Ambert (1981) illustrates

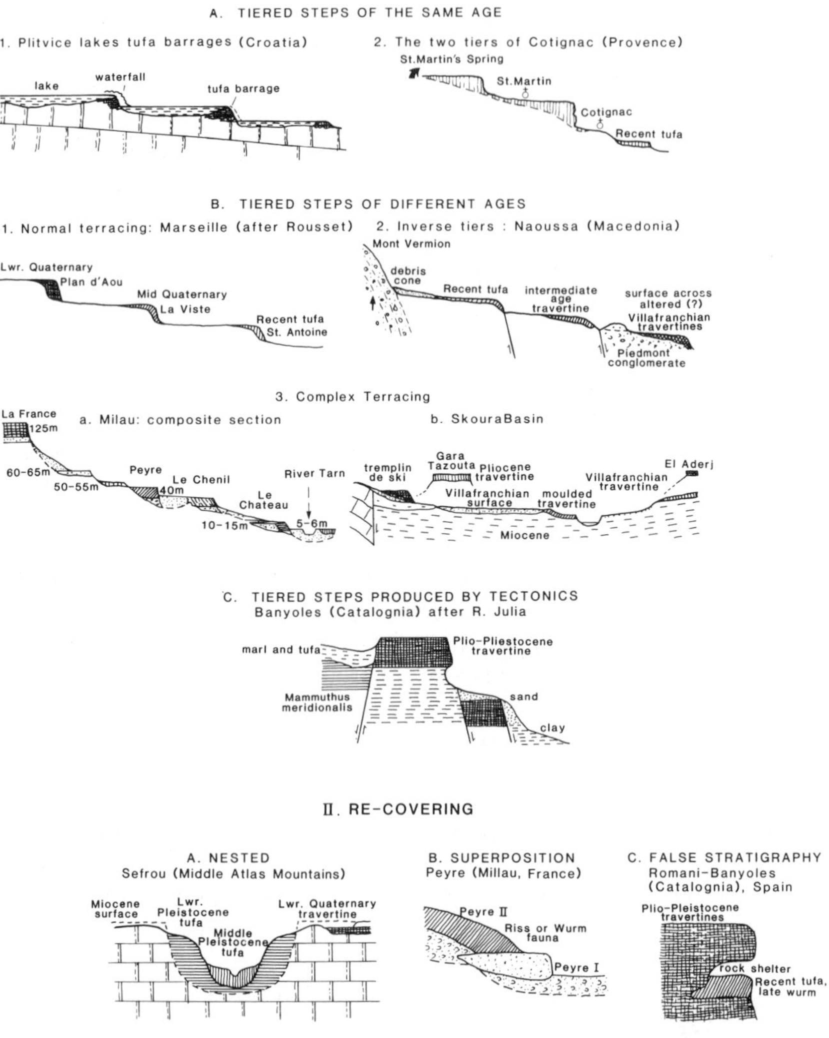

Figure 9.41 Complex field relationships encountered in tufa deposits. From Ambert (1981).

the range of field relationships that can be encountered.

Many of the deposits and forms discussed arise partly through the intervention of biological activity, as do some erosional forms. Such features are termed *biokarst* by Viles (1984) whose suggested typology is presented in Table 9.6.

Table 9.6 A tentative typology of biokarst forms. From Viles (1984).

Erosional forms	**Depositional forms**
Phytokarst (Folk *et al.* 1973)	Tufas and travertines
Directed phytokarst (Bull & Laverty 1982)	Directed speleothems
Coastal biokarst	Moonmilk
	Stromatolites
Root grooves	Reefs
'Zookarst'	
Mixed erosional and depositional forms	
Calcareous crusts	
Case-hardening	
Lake crust and furrow systems	
Degraded tufas	

9.12 Sequences of carbonate karst evolution in humid terrains

Early ideas

At the turn of this century the influential ideas of the American geomorphologist W.M. Davis on the cycle of erosion in 'normal' landscapes were reverberating through the world of geomorphology. In Vienna, one of the best known European geomorphologists, Albrecht Penck, had amongst his students Jovan Cvijic – considered by many to be the father of modern karst geomorphology. Vienna is on the doorstep of the Dinaric karst. Hence, the landforms of what we now regard as the 'classical' karst provided the main source of inspiration from which concepts of karst landscape evolution emerged. Imagine the excitement and stimulation of a fieldtrip in 1899, when Penck and his students accompanied by Davis set out to investigate Bosnia and Hercegovina, in what is now central Yugoslavia. Roglic (1972) considers this meeting in the karst area of two outstanding masters of geomorphology to have been of decisive importance for the further development of geomorphological concepts, although we should not forget that arguably the most influential work ever written on karst had already been published by Cvijic in 1893.

Penck and Davis both contended that karstification is preceded by an episode of fluvial erosion, an idea with which Cvijic concurred. The problem then remained to identify the erosional stages through which the karst landscape progressed. Although Richter (1907), Sawicki (1909) and Beede (1911) first offered solutions, a scheme proposed by Grund (1914) claimed most interest. His theoretical sequence of landscape evolution took into account both the Dinaric region, with which he had first hand experience, and the humid tropical karsts of Jamaica and Java, which he visualized from the writings of Danes (1908, 1910). His scheme is shown in Figure 9.42. It

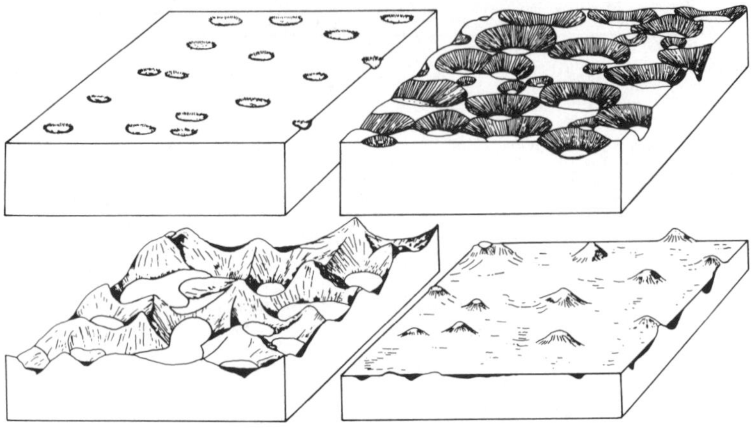

Figure 9.42 The karst cycle according to Grund (1914).

depicts a doline karst of several generations in which individuals enlarge, coalesce and gradually consume intervening residual hills. The terrain is ultimately reduced to a corrosional plain, fluvial activity playing a significant part in its final modelling.

Cvijic's own thinking evolved significantly on the question of karstic evolution. In his 1893 monograph he suggested a genetic sequence involving amalgamation from dolines to uvulas to poljes, but it was not until 1918 that he published his considered opinion on the morphological evolution of karst and its relationship to subterranean hydrology. In this work he drew attention to hydrographic zones within karst and pointed out that subterranean evolution can proceed without the intervention of baselevel change, because karstification itself leads to the lowering of the hydrographic zones as permeability progressively increases at depth. Whereas Grund attempted to produce a universal model, Cvijic's scheme of karstic evolution (Fig. 9.43) was proposed for the Dinaric karst only; and whereas Grund's model assumed an indefinite thickness of limestone, with clear implications for uplift and rejuvenation, Cvijic's focused on a sequential development that terminates on impermeable underlying beds. The role of lithology became a more explicit theme in some of Cvijic's later work, best expressed in his 1925 paper in which he introduced the terms *holokarst* and *merokarst*. Holokarst is pure karst uninfluenced by other rocks, and is developed on limestones extending well below baselevel. It contains the full range of karst landforms and it is exemplified by the Dinaric area. By contrast, merokarst (or half karst) is developed on thin sequences of limestones interbedded with other rocks, as well as on less pure carbonate formations. The landscape thus contains both fluvial and karstic elements and may be thickly covered with insoluble residues. Karsts in Britain were cited as examples. Cvijic also identified transitional types, such as found in the

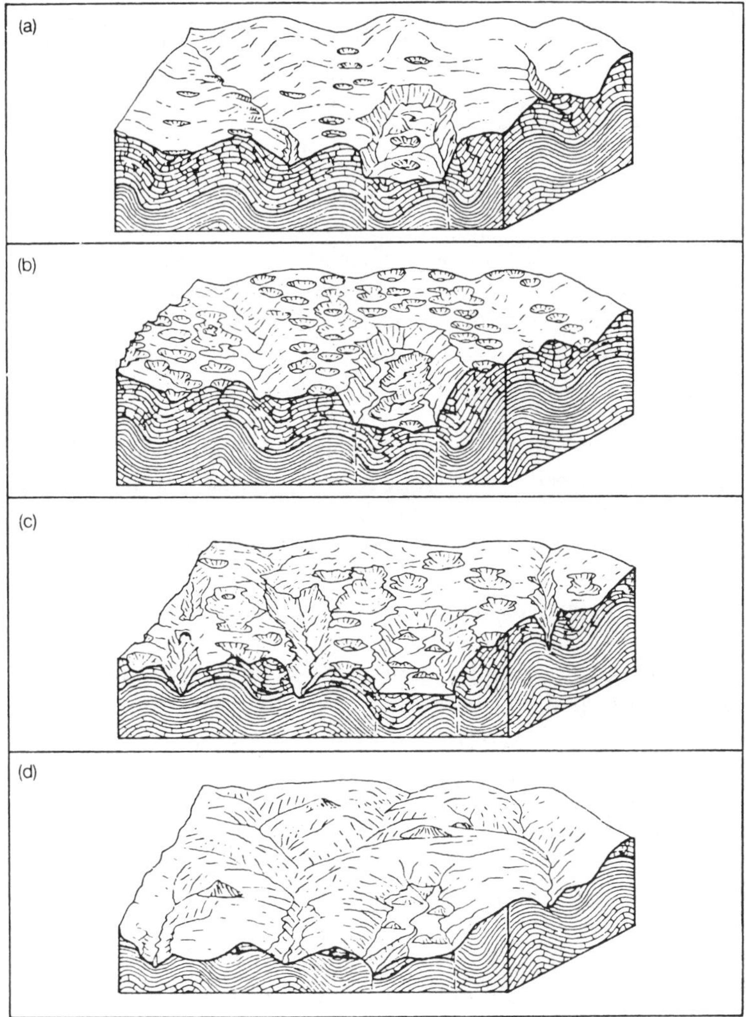

Figure 9.43 The sequence of karstic evolution according to Cvijic (1918).

French Causses, where there is extensive karstic development above underlying impermeable rocks.

These early ideas are discussed more fully by Sweeting (1972), Roglic (1972) and Jennings (1985). A translation into English of Grund's 1914 paper is available in Sweeting (1981), together with Sanders' (1921) translation of Cvijic's 1918 work.

The conceptual framework provided by these early geomorphologists persisted essentially unchallenged until the emergence of ideas on the critical influence of climate on landform evolution. This commenced in 1936 with publication of H. Lehmann's observations in Java and thereafter followed a series of perceived schemes of karst landscape evolution, each tailored to the specific climatic zone in which it occurred. This is discussed more fully in the next chapter.

Alternative conceptual models

One of the major difficulties with models of karst evolution that have been proposed is that they do not adequately accommodate the range of hydrogeological and geomorphological circumstances that occur. Karstic evolution may be envisaged as commencing from one of three main starting points:

(1) uplifted unkarstified dense rock protected by impervious cover beds;
(2) uplifted unkarstified rock of high primary porosity with no cover beds; and
(3) uplifted rock karstified in a previous erosion phase.

There are two important variants of the first case (Fig. 9.44):

(1a) Stratification is horizontal or dips upstream and the impervious cover beds are stripped down and back from the spring (output) boundary.
(1b) The strata dip downstream and the caprock is stripped down and towards the spring boundary.

Highly tectonized terrain with complex geology can often be subdivided into more simple sectors such as the above two, although continuing uplift and tilting during karstification presents special problems.

CASE 1A

The principles regarding the initiation of karstification of dense carbonate rocks beneath impervious cover beds were dealt iwth in Chapter 7. The important point here is that surface karst landforms cannot evolve until subterranean connections have been established from input to output boundaries. The relevance of this for doline development was discussed in section 9.4. Having acquired a hydraulic potential and output boundary by deep incision of an allogenic trunk stream, denudation of the caprock begins to expose the carbonate formation beneath. A first rank of point recharge depressions develops at the input points where the first throughput connections are made (Fig. 5.17 case (1)). As the caprock retreats upstream, successive ranks of new connections form, each supporting new dolines along the lines depicted in Figure 7.11. By this stage the karst has become one of multiple inputs and multiple ranks, with caves at different

CASE 1A HORIZONTAL STRIPPING

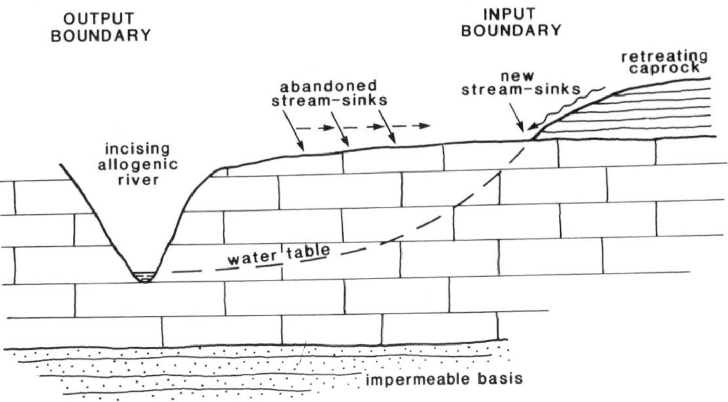

CASE 1B DOWN-DIP STRIPPING

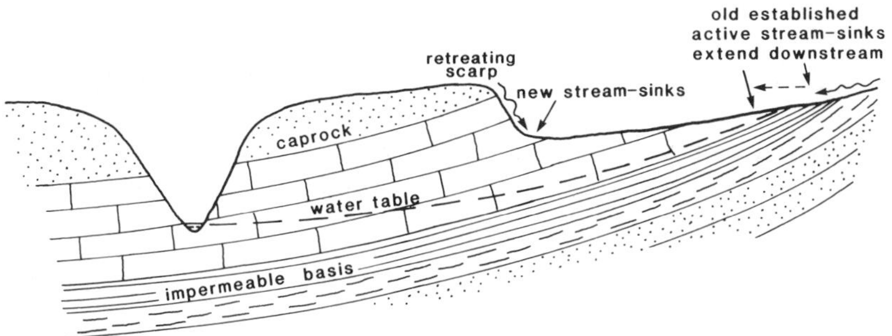

Figure 9.44 Karstic evolution in uplifted dense carbonate rock protected by cover beds: 1A horizonal beds; 1B strata dip downstream.

levels if local baselevel has also changed. During the following complete stripping of the cover beds, the epikarst will develop. Where fissure frequency is relatively low, the surface will be characterized by drawdown depressions. The initial dolines will grow in plan size, but may be limited in their extension or subdivided by development of secondary (or 'daughter') dolines above new leakage routes in the subcutaneous zone (Fig. 9.15). Where fissure frequency is high, diffuse autogenic recharge down innumerable fissures coupled with the absence of a significant capillary barrier (Fig. 9.14) in the epikarst promotes the development of karren field morphology rather than a doline karst.

CASE 1B

The same principles apply, but with some important variations. Because strata dip downstream, the underlying limestone is first exposed at its upstream boundary. Successive ranks of inputs then migrate downdip and downstream with the contracting caprock. Stripping is therefore towards the spring into a previously established deep and well drained vadose zone i.e. invasion vadose development. Any underlying impervious beds are also first exposed at the upstream boundary; thus the karst area contracts as the input boundary migrates downdip. In case 1a the oldest surface landforms are closest to the output margin, whereas in case 1b they are closest to the input margin.

Field examples of the first case are found in Yorkshire and Derbyshire, England; Fermanagh, Ireland; the Dordogne, France; eastern Kentucky, Tennessee and West Virginia, USA. The most celebrated instance of the second case is the Mammoth Cave–Sinkhole Plain region of Kentucky (Figs 6.31 & 7.12). The change in landform texture across part of this area is depicted in Figure 9.45. The newest surface karst features are the recharge depressions of The Knobs, where the caprock is still being removed. These large basins are subdivided to form the finely tessellated texture of the Sinkhole Plain, and these dolines in turn are consumed by the downdip migration of the Sinking Streams boundary. An intermediate case is provided by the schematic Waitomo model of Figure 9.25, depicting lateral capture to a trunk stream. In all of these instances, landscape evolution follows a sequence rather than a cycle, because the effective thickness of carbonates is limited and no further karst development is possible once erosion has removed them.

CASE 2

Here we may imagine the extensive surfaces exposed on coral reefs during Pleistocene glacio-eustatic low sea levels. The rock possesses extremely high ‘fissure frequency’ in terms of Chapter 7, i.e. high density of openings of all types that may provide *ready made* hydrological connections from the recharge zone to the output boundary. Hydraulic conductivity is high and its spatial variability is comparatively small (Fig. 9.16). Diffuse autogenic recharge promotes the development of rough karren at the surface rather than doline karst. Saturated groundwaters inhibit speleogenesis, although field and theoretical evidence indicates that some caves form at the water table and at the fresh water : salt water interface. Outflow sites where stream lines converge and where the mixing zone rises to the water table may be especially important. Glacio-eustatic fluctuations have forced this activity to be located at different levels. Solutional lowering of the surface causes chance interception of caves and the creation of collapse dolines. Autogenic karst of this type never becomes highly developed because of very high primary porosities of 20% or more. However, where denudation exposes an impermeable inlier, then point recharge of allogenic streams transforms the situation with the development of large state 4 caves

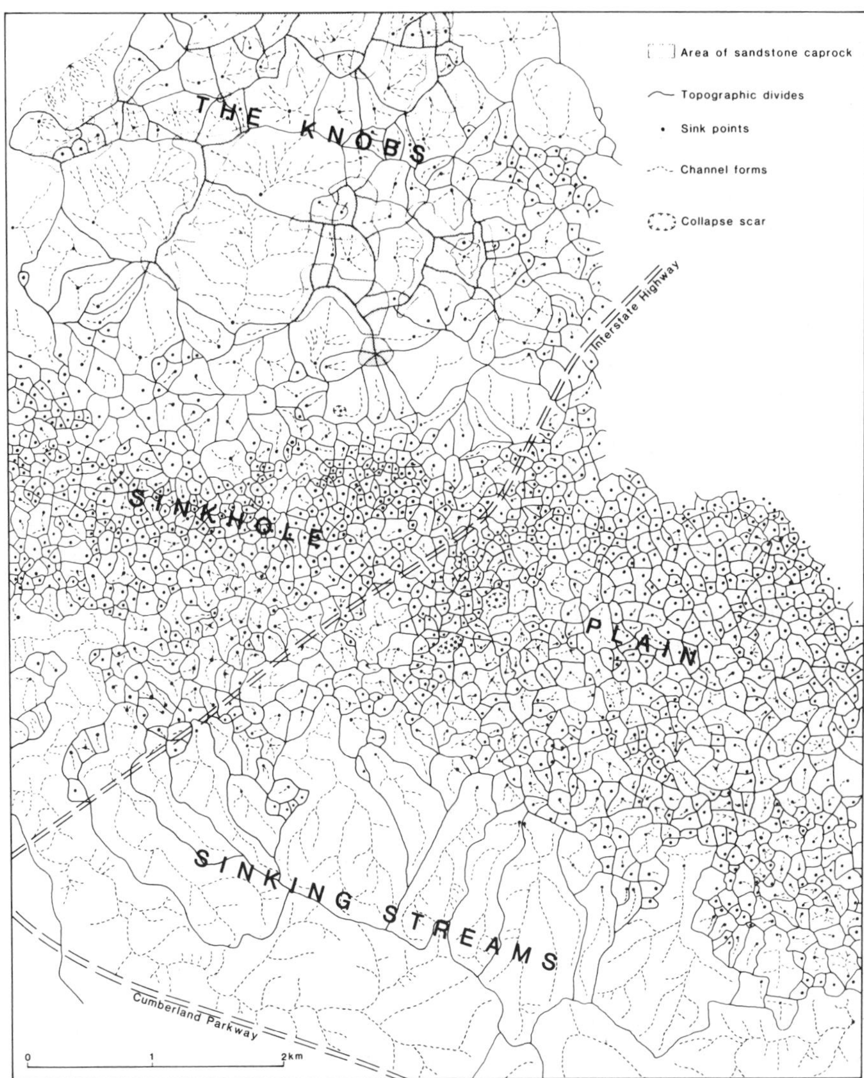

Figure 9.45 Morphological map of part of the Mammoth Cave–Sinkhole Plain district showing the change in closed depression texture from the youngest point-recharge depressions amongst the Knobs to the oldest features at the up-dip Sinking Streams boundary (from Williams & Quinlan, unpublished). The situation is as depicted in Figure 9.44 (1B).

(Fig. 10.19). At the input boundary, blind valleys merge to form interior lowlands (small poljes). Collapse above caves becomes more common, but solution dolines remain rare. An instructive insight into karstification of a coral reef is provided by Stoddart *et al.* (1985), for Mangaia in the Cook Is.

CASE 3

The course of karstic evolution following uplift of a previously karstified surface is strongly influenced by inheritance. The uplifted surface may be a baselevelled plain or more rugged (Fig. 9.32), but in each case the drained phreas provides an instant vadose zone with ready-made connections from input to output boundaries. Any inherited topography will guide runoff underground and hence guide solution in the epikarst. Residual hills, become incorporated into the topographic divides of the emerging, rejuvenated karst (Fig. 9.34). The landscape that develops also depends upon the amount of uplift and comparative rate of river incision. Minor uplift results in development of a new corrosion plain with dissected fragments of the former surface left as terraces. Major uplift and deep gorge entrenchment by allogenic trunk rivers may lead to incision below the base of karstification in the former phreatic zone. This leaves new vadose tributaries suspended above the trunk river. The long profile of these underground streams consequently steepen downstream as the gorge is approached, and subterranean knick point recession works upstream. When the water table eventually lowers upstream, surface incision can commence. Hence a wave of surface rejuvenation follows underground rejuvenation. Water table lowering occurs in two stages: initially by gravity drainage of the former phreatic zone during incision of the trunk river and later as secondary permeability develops in previously unkarstified rock. The margins of the Guizhou plateau, China, offer outstanding examples of this pattern of development (Yang 1982, Song 1986, Smart *et al.* 1986).

9.13 Special features of evaporite terrains

Dolines and deeper, interstratal solution features

Because of their great solubility salt rocks are exposed only in the most arid or cold regions such as Death Valley, the Dead Sea, the Qinghai (Tibetan) plateau and the high Arctic islands of Canada. Even there, individual outcrops are limited to a few square kilometres at the most. Interstratal salt deposits are widespread (Fig. 1.3) and support a wide range of karst features in all climates. Gypsum is much more stable in outcrop but its karst is also better expressed where mean annual precipitation is rather low. Gypsum karst is widespread in the dry midwest–southwest of the United States, in the northern interior of Canada, the Archangel, Bashkir and Perm regions of the USSR, across the Middle East and in northeastern China. Outcrops may amount to thousands of square kilometres. There are smaller outcrops in many other countries. The principal studies of gypsum karst have been conducted in the Soviet Union where Gorbunova (1979), Pechorkin & Bolotov (1983) and Pechorkin (1986) provide recent reviews. Nicod (1976) and Forti (1986) present other European work. Quinlan *et al.* (1986) review gypsum karst in the United States, and Quinlan (1978) provides the most comprehensive English-language discussion of salt and gypsum/anhydrite interstratal solution.

Evaporite terrains display many of the landforms typical of carbonate karst, including varieties of karren, dolines, blind valleys and poljes. Gorbunova (1979) asserts that dolines are the most widespread features in the gypsum karsts of the USSR, and this is true in Canada also. Collapse and suffosion processes are more prominent in doline development than in most carbonate karsts, the former because interstratal dissolution is of much greater extent (Fig. 9.18B) and the latter because, in the USSR and Canada, great tracts are veneered with glacial debris or loess. Nevertheless, wholly solutional dolines are also common, e.g. in Italy (Belloni *et al.* 1972, Burri 1986), in Spain (Gutierrez *et al.* 1985) and in the Pecos Valley karst of Texas and New Mexico (Gustavson *et al.* 1982, Quinlan *et al.* 1986). Shallow, bowl-like forms are prominent, with diameters of 100–500 m and depths of only 10–20 m. In parts of the Pecos Valley they are packed to form a polygonal karst.

In the Perm and Bashkir karsts Gorbunova (1979) reports doline densities of 32 and 10 km^{-2} respectively, although densities up to 1000 km^{-2} sometimes occur at the crown of folds or at contacts with other lithologies. Densities of 1100–1500 km^{-2} occur in the Italian Alps where hydraulic gradients are high (Belloni *et al.* 1972). Dolines there have a mean diameter of only 5 m. The densely packed schlotten type of depression or large karren shaft (section 9.2) have diameters of 2–5 m and depths of 1–8 m in Nova Scotia, and extrapolated densities of 10 000 km^{-2} (Fig. 9.46). They do not extend over any one area as large as 1×1 km, however, being limited to escarpment edges.

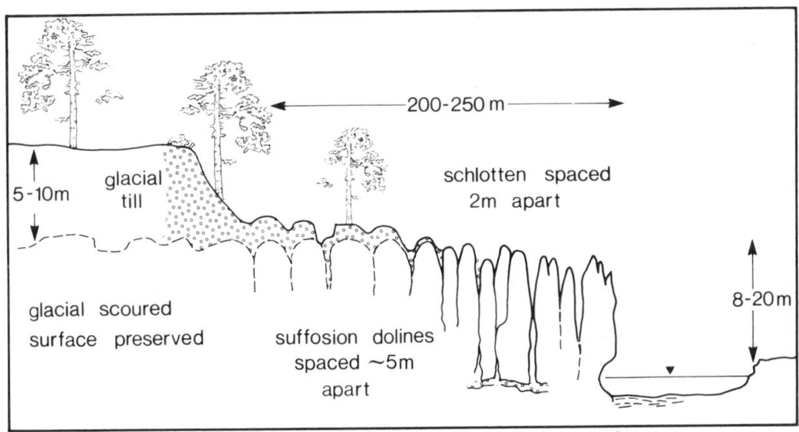

Figure 9.46 Model for the progressive development of schlotten topography in massive gypsum near Windsor, Nova Scotia, Canada. The features are of Holocene age.

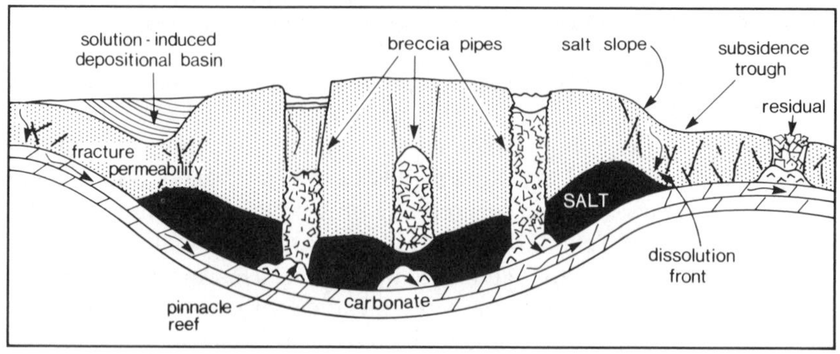

Figure 9.47 Model for the development of breccia pipes and residuals, subsidence troughs and solution-induced subsidence basins.

At larger scale are the breccia pipe type of features which are created by progressive stoping above caves or sites of interstratal dissolution (sections 7.12, 9.5). Although well developed in carbonate rocks they are most abundant and largest above gypsum/anhydrite and salt (Fig. 9.47). Quinlan (1978) reviewed the nature and distribution of an estimated 5000 breccia pipes over salt or gypsum in the United States. They ranged in diameter from 1 to 1000 m and in depth up to 500 m. Quinif (in press) describes pits containing 'cyclopean breccia' that interrupt coal mine workings in Hainaut, Belgium. Typical examples are 50 to 100 m in diameter and have depths of several hundred metres. Similar features interrupt the potash mine workings of Saskatchewan (Fig. 9.48) where they may propagate from depths as great as 1200 m (i.e. beneath 1000 m or more of cover strata).

In some breccia pipes the salt or gypsum (or both) are entirely removed at the base of the breccia pile. In other examples, part of the soluble strata remain. In many pipes all overlying strata are brecciated. In others, a lower brecciated zone is succeeded by an upper zone in which the strata failed as a block, settling downwards inside a cylindrical pattern of steep to vertical faults; downthrow as great as 200 m is reported. Elastic subsidence of upper strata also occurs.

Breccia pipes may exhibit one of four dynamic/topographic states: (a) active, and propagating upwards towards the surface but not yet expressed there; (b) active or inactive, expressed at the surface as a closed depression or a depression with a surface outflow channel; (c) inactive, and buried by later strata (= paleokarst); (d) inactive and standing up as a positive relief feature because the breccia (probably cemented) is more resistant than the upper cover strata.

Solution subsidence troughs (Olive 1957) are elongated depressions created by interstratal solution (Fig. 9.47). Many small examples occur in

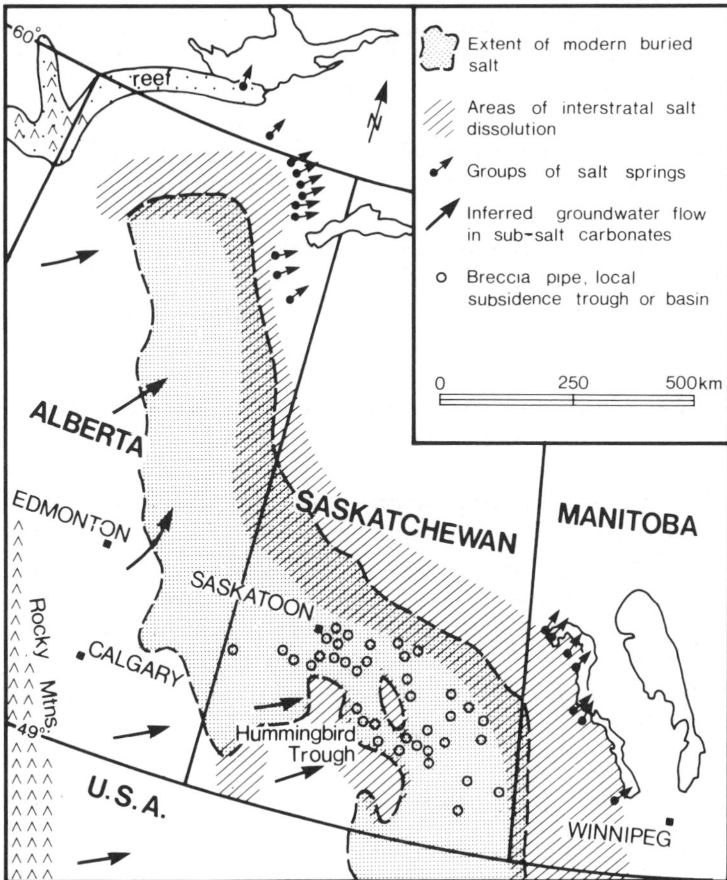

Figure 9.48 Features of the interstratal dissolution of the Elk Point (Devonian) salt deposits beneath the Prairie Provinces of Canada.

the gypsum plain south of Carlsbad Caverns, being 0.7 to 15 km in length, 100 to 1500 m wide but no more than 5–10 m deep (Quinlan *et al.* 1986). Larger troughs tend to be infilled by terrigenous or other sediments and so lack topographic expression in most instances. Quinlan (1978) terms them *solution-induced depositional basins*. They are noted in the paleokarst reports of many nations (e.g. Bosak & Horacek 1981). Many continue to be active but do not appear as strong topographic depressions because the rate of sedimentation approximately equals the rate of subsidence. Intermediate scale examples (5–100 km long, 5–250 km wide and with 100 to 500 m of subsidence and sedimentation) are reported in Canada (Tsui & Cruden 1984), New Mexico and Texas (Bachman 1976, 1984) and the Perm region of the USSR (Tsykin, in press). Pleistocene and Holocene rates of subsidence in active US and Canadian examples are estimated at 5 to 10 cm/1000 years. In the Uralian troughs, which are paleokarst, slightly lower rates of 0.3 to 3 cm/1000 years are obtained.

The largest solution-induced depositional basins tend to occur along the margins of the great interstratal salt deposits, creating a solution front that may be represented by a shallow *salt slope* at the surface (Fig. 9.47). Dissolution can begin as soon as the salt is buried, and the immediately overlying strata (usually, dolomites, gypsum/anhydrite or redbeds) may be comprehensively brecciated. Fig. 9.48 shows the Elk Point Formation of Canada, a sequence of salts with lesser gypsum and redbeds that accumulated to thicknesses of 50 to 500 m in a lagoon barred by the Presqu'ile Reef (Fig. 2.12) during the Devonian period. It is now at a depth of 200 to 2400 m beneath later carbonate and clastic rocks. Where it is most shallowly buried (the eastern edge), the dissolution front has receded an average of 130 km along a distance of 1600 km. More than 50 large breccia pipes are known over the salt in southern Saskatchewan. The Hummingbird Trough (Fig. 9.48) is a re-entrant in the deeply buried SW side where 200–300 m of salt has been entirely removed over an area of 25 000 km^2 (De Mille *et al.* 1964). The Trough is now inert, a paleokarst feature, but the eastern dissolution front continues to recede. Its mean rate of recession (averaged over the 365 Ma that have elapsed since the Upper Devonian) is 36 cm/1000 years. This is scarcely a meaningful average, however, because dissolution will have been strongly episodic.

Positive relief features created by diapiric or hydration processes

Salt has a low specific gravity (2.16 g cm^{-3}) so that when considerable deposits of it are buried by denser strata it deforms in a cellular manner and flows upwards into them. Talbot & Jackson (1987) graphically describe the process. The displaced salt approaches or emerges at the surface in the form of stocks (diapirs), dikes and sills (canopies). Intercalated or overlying gypsum and anhydrite beds are deformed and may flow with the salt. Diapirs are initiated beneath 2000–10 000 m of cover strata. They rise episodically and, in general, very slowly; rates calculated in the Gulf of Mexico are only 0.1 to 1.0 mm a^{-1}. The most active modern diapir formation occurs in the Zagros Mountains of Iran, where convergence of Eurasion and Arabian crustal plates is squeezing deep salt and probably extruding it more rapidly.

Emerged salt diapirs generally range from 2 to 20 km in diameter. In wetter climates they are decapitated by groundwater dissolution but may still create fractured, dome-shaped hills up to 100 m in height because of the displacement of insoluble superficial rocks. In dry and cold regions relief of 500 m or more can be created, with an exposed core of salt (Fig. 9.49).

Where the extrusion is slow the salt displays rillenkarren, wallkarren,

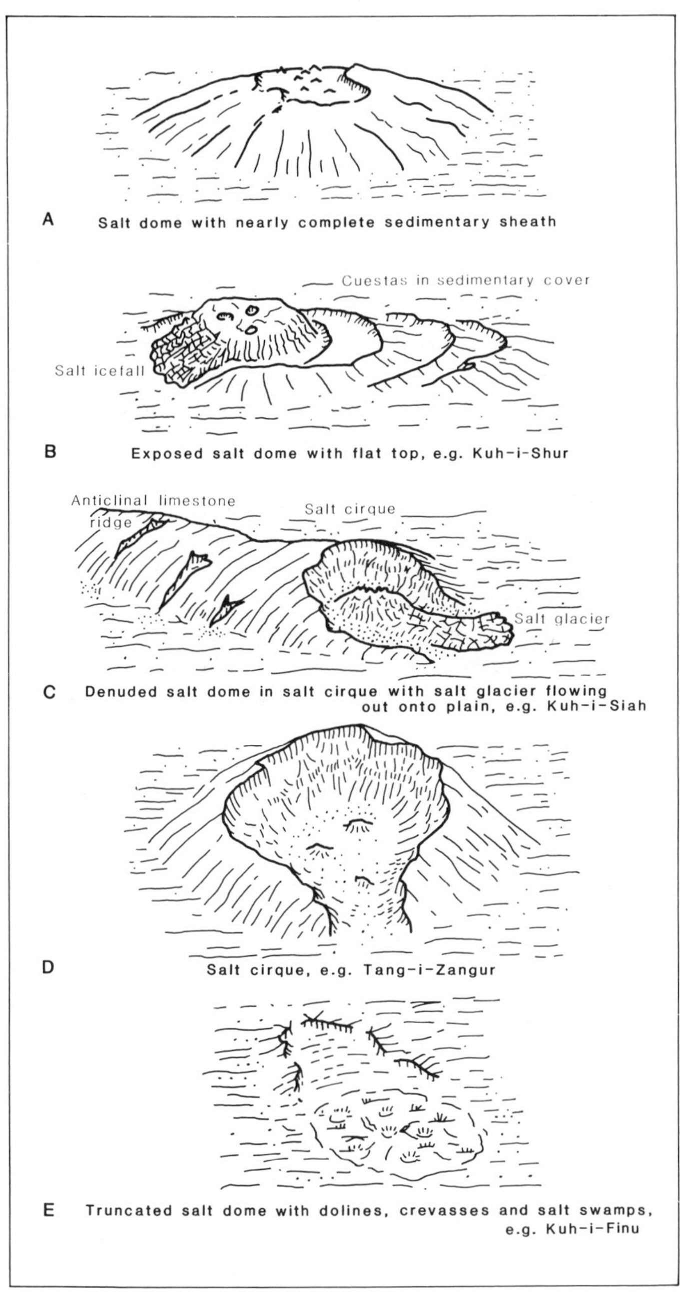

Figure 9.49 Surface expression of salt domes in the Laristan Desert, Iran. From Jennings (1985).

schlotten and pinnacle karst. Where it is more rapid it flows like macrocrystalline ice to create salt glaciers or *namakiers* (Talbot & Jackson 1987). These display standard glacier features such as crevasses, icefalls and ogives (overthrust ridges). Salt glacier flow rates in the Zagros Mountains average a few metres per year, i.e. one or two orders of magnitude slower than in conventional mountain glaciers formed of ice. The salt flow is episodic, resulting from recrystallization following the addition of water (Urai *et al.* 1986). A water content as low as 0.1% by weight can induce the process; thus it may operate sporadically in the driest climates.

Anhydrite is hydrated to gypsum by a process of dissolution and reprecipitation. Under laboratory closed system conditions this is accompanied by increases of volume between 30% and 67% (section 2.4). It is believed by many that the forces created by the expansion may raise and rupture overlying gypsum and thin cover strata. Gorbunova (1979) and Pechorkin (1986) report that the hydration front will be no deeper than 150–200 m and is very irregular because it preferentially follows joints, intercalated shale or dolomite beds, etc. They suggest that many landforms attributed to hydration expansion are caused instead by the viscous flow of the newly formed gypsum, i.e. like the diapiric intrusion of salt. In its own saturated solution, gypsum has a viscosity lower than that of dry salt, and subject to a load of 50 kg cm^{-2} may deform by 20% in one year. We believe that both processes may operate; gypsum tumuli (below) appear to be attributable to hydration alone, while undoubted gypsum intrusions are commonplace in highly deformed redbed and dolomite sections.

Gypsum tumuli (or bubbles or tents) are hollow domes of freshly formed gypsum that are round or elliptical in plan (Breish & Wefer 1981, Pulido-Bosch 1986). Small examples are a few tens of centimetres in diameter but most described in the literature are 2–10 m in diameter and up to 2.5 m in height. These are hydration features created by compressive and shear stresses that separate the gypsum from underlying anhydrite if the hydration front is horizontal or nearly so.

Karst domes are described by Bachman (1984, 1987) in the New Mexico gypsum karst. They are up to 200 m in diameter and 10 m in height. The cores are of gypsum and insoluble residue, with disturbed dolomite or clastic beds or calcrete crusts draped around their annular rims. Bachman (1987) suggests that they are the surficial remnants of pervasive near-surface salt dissolution that has caused general subsidence with projections remaining above insoluble remnants; the domes being part of those remnants.

More dramatic are domes and anticlines in the gypsum terrains of northern Canada described by van Everdingen (1981), Tsui & Cruden (1984), and similar features in the Archangel gypsum karst of the USSR (Korotkov 1974). These features range 10 to 1000 m or more in length or diameter, and up to 25 m in height in a majority of cases. Many are highly fractured, with individual blocks being displaced by heaving and sliding. At the extreme they become a mega-breccia, an upthrust jumble of large blocks. The greatest reported Canadian example is a steep-limbed anticline that extends along the shore of a lake for a distance of 30 km and is up to 175 m in height. Its crest is marked by 'chaotic structure and trench-like lineaments' (Aitken & Cook 1969).

Tsui & Cruden (1984) attributed the examples that they studied in Wood Buffalo National Park (Lat. 59–60°N) to hydration processes operating during the postglacial period. Given the association with recent glaciation in both the Canadian and Soviet sites, however, it is possible that they are attributable to diapiric injection of gypsum during times of rapidly changing glacial ice loading, as discussed in section 10.3. These are regions of widespread permafrost, in addition, so that accumulation of ground ice in initial fractures probably contributes to the heaving and other displacement.

The final category of positive relief features associated with evaporite karsts are cemented sinkhole or breccia pipe fillings that come to be exposed as residual hills when strata surrounding them are preferentially eroded, as noted above. Breccia pipe residuals are normally a few tens or hundreds of metres in diameter and rise 5–40 m above the general land surface. More than one thousand such hills are mapped in the plains of western Oklahoma (Fay & Hart 1978). Shallow hills with a doughnut form (a central depression) in that region are believed to be doline fillings, rather than breccia pipe fillings (Myers 1962).

Castiles are steeper, irregular masses of secondary calcite rising 3 to 30 m above the Gypsum Plain south of Carlsbad Caverns, New Mexico (Kirkland & Evans 1980). These formed to replace gypsum locally at the base of the Castile Formation at the start of the H_2S-generating process described in section 7.8 and Fig. 7.32. They are exposed by the preferential dissolution of the remaining gypsum along the margins of the Plain.

10 The influence of climate, climatic change and other environmental factors on karst development

10.1 The precepts of climatic geomorphology

During a study in Java, H. Lehmann (1936) recognized the coincidence of Kegelkarst relief with the humid tropics. This reinforced a growing belief amongst geomorphologists generally that landform assemblages are strongly influenced by climate through its control of natural processes. There then began a quest to identify the singular relief styles that were expected theoretically in the various climatic zones, and maps of morphoclimatic regions were produced (e.g. Tricart & Cailleux 1972, Budel 1982). Karst geomorphologists were particularly active in this (e.g. Lehmann 1954). Descriptions and comparisons of karsts from the tropics to the arctic provided a focus for research for 40 years, the aim being to determine and explain in morphogenetic terms the contrasting landform assemblages that were found. Nevertheless, some detailed work showed lithology and structure to have a larger effect on relief forms than had been appreciated (Verstappen 1964, Panos & Stelcl 1968).

Most geomorphologists now agree that broad landscape differences exist in regions with contrasting climates, while admitting that subtler variations in style had often been claimed than objective scrutiny can justify. A more important criticism of climatic geomorphology is that it has been unable to explain why many of these contrasts occur. For example, it has not revealed why karstic activity in the humid tropics sometimes results in the development of cockpit karst, whereas in the temperate zone doline karst is apparently more typical – even though dolines are also found in the tropics. It appears to us that climato-genetic geomorphology has reached about the limit of its contribution. We should now pass on, but avoid the mistake of failing to recognize the value of its major conclusions.

The availability of water is the key climatic factor in karst development. It is certainly the principal variable controlling total solution denudation (Fig. 4.4), although the targeting of corrosion is determined by the controls on water flow explained in Chapter 5.3. In the last chapter, we explained how 'normal' karst evolves in areas where water is abundant, but we remarked that aridity and extreme cold place constraints on development. Both of these climatic conditions lead to a scarcity of water in its liquid state, thereby limiting solution and permitting other geomorphic processes to dominate morphological evolution. But if the other processes are themselves not very active, then solution effects may persist unaltered for a considerable time. In this chapter we examine the karsts that result.

10.2 The hot arid extreme

We agree with Jennings' (1983b) assessment that 'Less is known about karst in deserts and semideserts than anywhere else except beneath glaciers and permafrost'. Nevertheless, present information indicates that the determinants of karstification are the same in these environments as elsewhere, although there are differences in the relative influence exercised by the various controls. Because the soil is thin and patchy, it is less influential as an infiltration 'governor' and as a moisture store than in more humid regions. It also has reduced significance as a CO_2 source. The duration and intensity of corrosion is correspondingly reduced. Precipitation is delivered typically in short but violent, aperiodic convectional storms. This favours flashflooding, especially in rugged terrains.

The morphological consequences of these factors vary according to the lithology concerned. In the more soluble rocks such as halites and gypsum, the influence of subhumid conditions is less critical than it is for the carbonates (Fig. 10.1). Indeed, we have emphasized that the best expression of evaporite karsts is found in relatively dry environments. If it is too wet they are effaced by solution (section 9.13). In dense, massively jointed crystalline carbonates, the relatively 'naked karst' of arid zones becomes dissected along joint corridors into blocks heavily fluted by karren (Fig. 10.1). Dissection penetrates to the level of neighbouring pediplains, which may truncate rocks of any lithology. The karst surface is efficiently drained down an open network of grikes or corridors (Section 9.2), and few or no draw-down dolines (Fig. 9.14) evolve because epikarstic detention and storage is small (though not absent). In other areas where the limestones have a high primary porosity, rainfall soaks into the rock with minimal passage over it, and so the conditions favouring karren and joint corridor development are largely absent. Solution dolines are rare, because both high porosity and low water surplus are inimical to their development. However, in both porous and dense lithologies impressive collapse dolines may occur above caves.

Speleogenesis in hot arid regions follows the principles explained in

Figure 10.1 A. Detail of solution runnels on towers in the Limestone Ranges of Western Australia. Photo by A. Goede. B. Steep sided blind valley in gypsum karst terminating at a stream-sink in the Chihuahua Desert, New Mexico. The valley takes flash floods after heavy rain. The three people show the positions of present and former valley floors.

Chapter 7, but the size and frequency of caves developed is necessarily more limited than in humid regions. There is preferred development where simple cut-off caves can capture flash-flood waters. For example, Sof Omar Cave (Fig. 7.20) is a floodwater cut-off maze in a semi-arid region. Short vadose shaft-and-drain systems fed by ephemeral streams are common along escarpment edges. However, there is also deep phreatic circulation in mountainous deserts; deep, inter-basin flow is common in the US West and Soviet deserts. Relict phreatic caves, especially of the maze type, are quite abundant.

Soviet researchers emphasize that seasonal condensation waters may play a major role in enlarging relict phreatic caves by processes of breakdown and scaling. Domepit-type shafts and small drains possibly may be produced in escarpments entirely from dew. Castellani & Dragoni (1986) calculate that shafts 0.5 m diameter and 10–15 m deep may develop in 500 ka by this mechanism; their field site, the Hammada de Guir plateau and scarp in Morocco, receives 50–60 mm a^{-1} of conventional precipitation and has a mean annual temperature of 19.6°C.

With little process study accomplished in the hot arid karsts, it is difficult to know if we should ascribe their landscapes to the cumulative effect of repeated if sporadic modern events, or to more humid periods in the past when conditions for karst development may have been more favourable. A complicating factor is the further difficulty of distinguishing between karst and desert elements in arid landscape (Fig. 10.9). Notable dryland carbonate karsts occur in north Africa, central America, Asia, the Middle East and Australia. We illustrate these points with two of the better studied dryland karsts in Australia.

Karst of the Limestone Ranges

The Limestone Ranges are situated in the north of Western Australia. Mean annual temperature is 20°C and rainfall is 450–640 mm distributed over 30–80 days, depending on the distance inland. The monsoonal wet season is short but intense. Following the work of Jennings & Sweeting (1963), the area has become the type-site of semi-arid karst. Their study revealed not only a distinctive landscape, but also a particular sequence of landform evolution since found applicable to other seasonally humid karsts in northern Australia (Williams 1978). Morphogenesis begins with an uplifted Tertiary planation surface in Devonian reef limestones, standing up to 90 m above a neighbouring lower pediment surface also truncating limestone bedrock. The plateau is dissected and reduced to the level of the pediment, mainly by a process of parallel joint-aligned retreat in interfluvial areas. Through rivers provide baselevel control. The main stages identified in this process are:

Figure 10.2 Tower karst rising from a bare limestone pediment near J.K. Yard in the Limestone Ranges of Western Australia. Photo by J. N. Jennings.

(1) The plateau surface is stripped of soil and joints are penetrated by corrosion, producing fissure caves which enlarge to become intersecting sets of closed solution corridors or 'giant grikes' that isolate large bedrock blocks. These corridors are up to 3 m wide, 33 m deep and hundreds of metres long. Vertical wall karren flute their sides, and flat topped blocks are pitted with solution pans and flutes. Modern fissure caves prolong the open grikes underground and link up joints with different orientations. Intersection points of solution corridors sometimes widen into steep-walled closed depressions. Superimposed allogenic streams cut gorges through the karst, although some pass through the plateau in caves. The largest known is 8 km long.

(2) Solution corridors and infrequent closed depressions amalgamate to form integrated valley systems reflecting the joint geometry in plan. Termed 'box valleys', they have rectangular cross-sections with steep walls, flat floors, and plateau like divides. Their long profiles grade to the adjoining pediment. Significant tufa deposition in some valleys may seal their floors.

(3) Plateau remnants are consumed by the widening of box valleys, thereby isolating towers that are scattered across bedrock pediments. The landscape is of a bare fluted tower karst (Fig. 10.2) that is sharp and

abrupt but of comparatively small relief (< 50 m).

(4) Pediplanation results from continued solution of the towers and from direct scarp recession into the margins of the plateau; scarps are fluted by wall runnels up to 20 m long. Ultimately the upper surface is completely replaced by the lower.

This mode of development and the scale and form of landscape produced is very similar to that of the subarctic Nahanni karst (sections 9.2, 10.4) save in one crucial respect. This is that the Nahanni corridor karst is drained via younger generations of caves developed in the floors of the corridors and platea (box valleys), whereas the Limestone Ranges are drained by a surficial pediment.

Karst of the Nullarbor Plain

About 1500 km SSE of the Limestone Ranges is the Nullarbor Plain. It is distinctly more arid. The mean annual temperature is about 18°C and rainfall diminishes inland from 400 to 150 mm a^{-1}. The low plateau covers an area of about 200 000 km^2 and is underlain by an almost undisturbed, pure, Tertiary limestone sequence that extends well below sea level. Judging by the stratigraphic thickness removed, average solutional lowering has proceeded at a rate of only 2–5 mm ka^{-1} since emergence in the mid-Miocene. Its surface now has an elevation of between 75 and 225 m, being lower in the wetter south where it terminates in cliffs. Dry relict river courses have been traced across the plain for up to 130 km although they are incised no more than 10 m below the general surface. They usually start on non-karst rocks to the north and west, indicating an allogenic origin.

Surface relief on the plain is always small and usually assumed to be karstic (Jennings 1983b). In the wetter areas it is characterized by shallow claypan corridors between low rocky ridges or by a lattice of claypans around bedrocks outcrops. In drier parts there is a scatter of shallow, circular depressions locally called ‘dongas’. These basins may be up to 1 km across, but have a depth of only 1.5–6 m. The extensive limestone plains (‘hammadas’) of the Sahara display similar features (Fig. 10.9). Deeper collapse depressions and explored caves in the Nullarbor are largely confined to a southern belt parallel to the coast roughly 75 km wide, although some are known 150 km inland. About 130 collapse depressions have been mapped, ranging up to 240 m diameter and 35 m deep. Some lead into caves, which can be of impressive dimensions. Mullamullang Cave for example has some 5 km of spacious passages up to 30 m wide (Dunkley & Wigley 1967).

The striking karren sculpture of the Limestone Ranges is entirely absent on the Nullarbor, despite induration of the uppermost 15 m or so of the generally chalky limestones. Yet solution flutes are developed at the same latitude some 1000 km to the east in the Flinders Ranges where rainfall is also slight (250 mm), but the rock is a dense Cambrian limestone (Williams 1978). A feature of the relatively moist (400 mm a^{-1}) southwestern corner of the Nullarbor is the occurrence of numerous inliers of crystalline basement rock. Some scarcely rise above the general level of the plain, but one emerges 450 m. The inliers are commonly ringed by annular depressions 50–150 m across and 3–10 m deep, resembling dry moats. They are clearly formed by the corrosive action of centripetal allogenic runoff from the inliers.

Jennings (1983b) considered these moats to be the only distinctive karst landform at the meso- to macro-scale in this climatic zone. However, similar landforms created by the same process surround the volcanic inliers of some coral islands (section 10.5); they are a product of physical juxtaposition that is independent of climatic factors. We suggest that the extension of pediments into some karstlands is the one truly distinctive feature so far recognized in desert karst. The patterns of poorly integrated to non-integrated, claypans and dongas ('dayas' on the Moroccan hamada) are quite distinctive, but similar patterns occur on humid northern plains where the deranging agency is glaciation rather than aridity. We agree with Jennings' general conclusion that carbonate karst declines with precipitation; quite simply, the richness and diversity of landforms diminishes.

10.3 The cold extreme: 1 karst development in glaciated terrains

The Late Tertiary–Quaternary Ice Ages

The cold extreme is represented by land that is ice-covered or that is bare but permafrozen. At the present time (considered to be 'postglacial') 10% of the aggregate continental area is occupied by glaciers and a further 15% is widely or continuously permafrozen. At the maxima of the Quaternary Ice Ages ice cover increased to approximately 30%. Much of the terrain that is now permafrozen was then glaciated. Because the glaciers were so extensive, relationships between the karst system and glacial processes will be discussed first. Karst development in permafrozen terrain is then considered in section 10.4.

More than 95% of the volume of modern glacier ice is in Antarctica. Ice sheets have existed there for at least the last 8 Ma, though not necessarily at the modern scale. Expansion to ~ 30% continental cover during ice age maxima was accomplished by minor extension of the Antarctic sheets, growth of major ice sheets over Canada and over Scandinavia plus the Soviet northwest, and of ice caps and valley glaciers in all high mountain regions. Withdrawal of so much water from the oceans lowered global sea level as much as 130 m. This radically affected the world's coasts (section 10.5) and significantly changed the $^{18}O : ^{16}O$ ratio in the remaining sea water. Study of fluctuations in that ratio, as it is recorded in foraminiferal tests recovered from tropical ocean sediment cores, suggests that there may have been as many as 17 cycles of glacier ice growth and decay within the past 2 million years, i.e. roughly 100 000 years per cycle (Shackleton & Opdyke 1973). These are correlated with changes of net global solar radiation induced by periodic irregularities in the Earth's orbital motion about the Sun. The number of glaciations recognized on the continents is fewer because of destruction of early evidence. In well studied regions at least three or four glaciations separated by warm interglacial conditions such as the present are known. The first glacial perturbations appear in lowland mid-continent sites such as Nebraska at 2.5–3.0 Ma and in the Mediterranean region at 2.4–2.6 Ma. Local glaciations in high mountain areas probably began earlier e.g. 8 Ma in Alaska.

The warm climate peak of the last interglacial occurred about 125 ka BP. This was followed by a worldwide glacial cycle; all glaciers expanded close to their greatest known extent, reaching their maxima in most regions between 25 and 14 ka BP. Recession to the modern sizes or less was completed by about 7 ka BP. There have been at least two other glaciations of this magnitude since ~ 730 ka BP.

The effects of such radical, comparatively rapid changes of conditions upon karst in the glaciated regions are fundamental and complex. *Glaciokarst* relationships range from the perfect preservation of 'preglacial' karst to its complete destruction and from the prohibition of any postglacial karst to its most rapid development.

Relevant conditions in glaciers

Glacier ice flows by intra-crystalline creep and by sliding on its bed (Drewry 1986). With creep, the velocity of flow is proportional to temperature and to the surface slope of the glacier; it varies from 1–2 m a^{-1} in cold, near-horizontal ice sheets to 100–1000 m a^{-1} in steep, 'temperate' glaciers. Sliding over the bed is limited largely to temperate glaciers, where basal ice temperature is at the pressure melting point (0°C or a little below), so that water is present to lubricate the sliding. Velocity is proportional to bed slope. Wet-based glaciers, flowing by both creep and sliding and loaded with rock debris, can scour rock surfaces very effectively. Dry based glaciers are frozen to the rock. They may protect it completely from effects of creep in the overlying ice, or may wrench and remove large blocks as integral parts of a basal flowing mass. Where wrenching is only partial (the rock block is not entirely removed) glacitectonic cavities can be created. Often, these are best developed in carbonate rocks because earlier dissolution weakens bedding plane and joint contacts, thus permitting differential slip (Fig. 10.3).

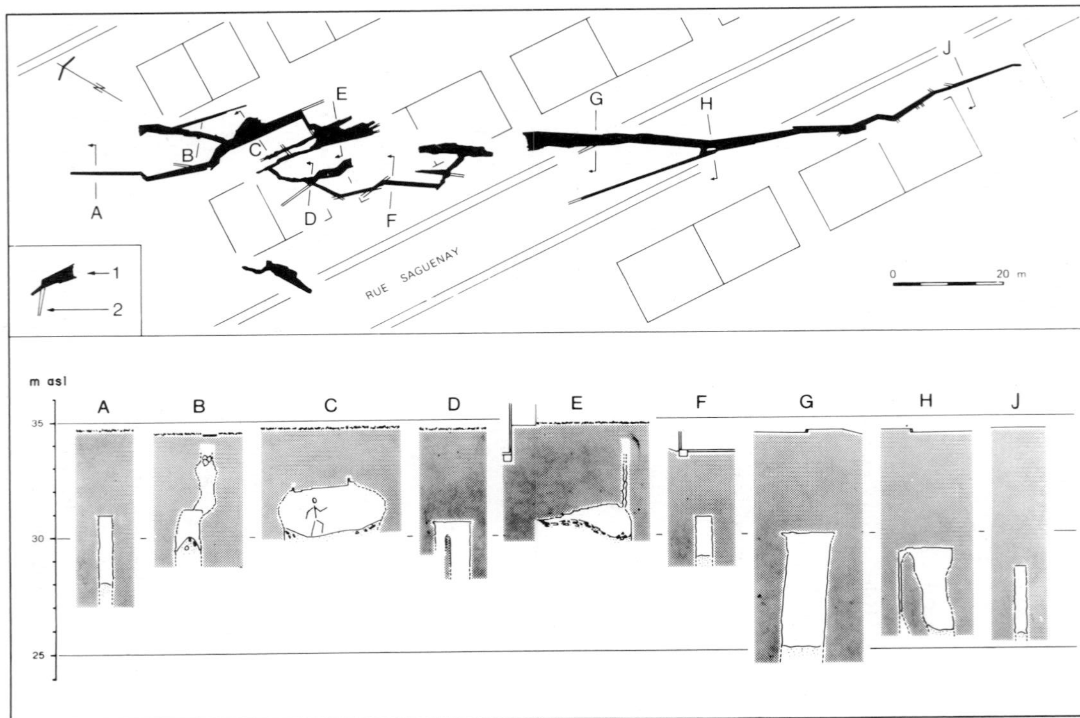

Figure 10.3 Glacitectonic cavities in limestones beneath Montreal, Quebec. From Schroeder *et al.* (1986) by permission.

Water from two different sources may be present at glacier beds. The first source is of melt streams that have descended from the glacier surface or from allogenic terrain (e.g. nunataks). This water will be equilibrated with atmospheric CO_2. It flows in channels melted upwards into the ice (R channels, which may close by annealing during the winter) or in channels partly protected in the bedrock (N channels). A subglacial limestone cave is an N channel fully protected against annealing, although its entry may become obstructed. The second type of water is a thin pressure melt/regelation film present wherever ice and rock are in direct contact (Hallet 1979a & b). It has a complex chemistry but is probably depleted in CO_2 (Ford 1971b, Souchez & Lemmens 1985). It may function as a wholly closed system, melting against an obstruction and freezing again in its lee, or part may escape into the R or N channels or pass down under pressure into the bedrock. Controversy surrounds the question whether the more abundant surface melt waters can accumulate at a glacier base and build up significant pressure heads there (Maire 1977, Ford 1979, C.C. Smart 1984a & b, P.L. Smart 1986). At the least, such accumulation seems inescapable where the ice occupies a closed depression such as an overdeepened cirque (e.g. Lauritzen 1984b), unless karst channels are open already to drain it. It may occur without any closed depressions if R channels are sealed when the melt season begins. There may be high water pressures in wasting ice masses (Fig. 10.4).

Continental glacier conditions exist when all of the topography is buried by flowing ice. The ice is normally 500–5000 m deep. These conditions prevailed over Canadian karst areas east of the cordillera, over most karsts of Scandinavia, northwest USSR and the glaciated British Isles. Alpine glacial conditions prevailed over most other glaciated karsts, where the flowing glaciers were channelled between ice-free ridges and summits that often supplied allogenic streams to the glacier systems.

Alpine karst has received more attention than the continental type because it is well developed in the Alps and Pyrenees, close to the classical centres of karst study (Nicod 1972, Maire 1977, Kunaver 1982). It is important to recognize that two distinct ice–karst relationships have existed in alpine terrains. The first is where glaciers were confined to the highest ground (cirques, summit benches, upper valleys) so that meltwater could discharge underground into lower karst valleys that were ice-free at all times, i.e. the karst inputs are glaciated but not the outputs. This relationship prevailed in most of the Pyrenees and Picos de Europa, much of the Alps and Southern Alps, the Caucasus, the US Rocky Mountains, etc. and may be termed the *Pyrenean type*. In the second relationship, glacier ice occupied all valleys and extended far beyond the output spring positions. This condition prevailed in the Canadian Rocky Mountains and parts of the central Alps and may be termed the *Canadian type*.

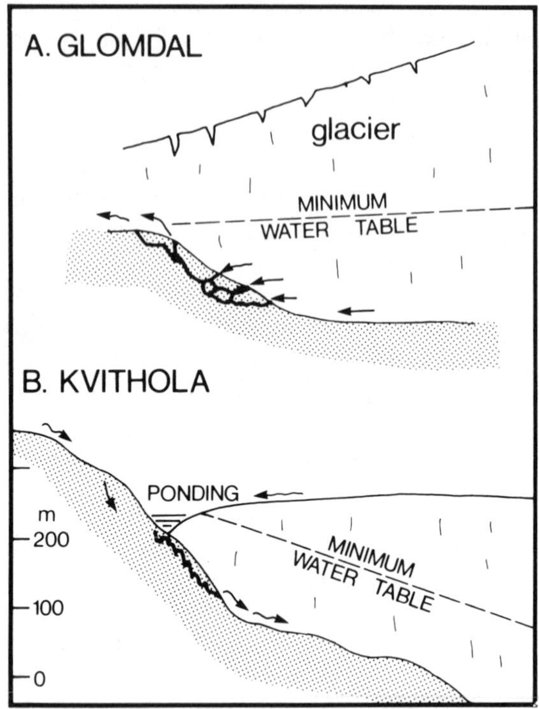

Figure 10.4 Two examples of glacier-induced phreatic cave development from the Norwegian karst. Adapted from Lauritzen (1984b, 1986).

Opportunities for karst develoment are more restrictive in these conditions because all of the karst system is impacted directly by glacial effects.

Surface karst morphology in glaciated terrains

Where there have been multiple glaciations many different relationships between surface karst landforms and glacial action are possible. Those that we have noted are listed in Table 10.1.

Postglacial karst forms are freshest and easiest to categorize. In type 1 (Table 10.1) the form, its dimensions and location owe little or nothing to prior glacial effects. Examples are many types of karren, collapse dolines such as that of Vermilion Creek, NWT. (Fig. 9.18B) and the densely packed schlotten in gypsum (Fig. 9.46).

In type 2 the location and/or recognizable parts of the morphology and dimensions are determined by inherited glacial features. As discussed in section 9.2, the placement and extent of modern limestone pavements are most often determined by prior glacier scour. Suffosion dolines also develop

Table 10.1 Categories of karst landforms in glaciated terrains

Postglacial	1	Simple forms – independent of preceding glacial forms: many karren, collapse dolines.
	2	Suffosion dolines, many solution dolines, limestone pavements, etc. where the location or part of the form is determined by preceding glacial features.
Adapted glacial	3	Glacial grooves, scourholes, potholes, kettle holes, cirques, moraine-dammed valleys, etc. adapted to karstic drainage.
Subglacial and Glacier marginal	4	Normal karst landforms (e.g. solution shafts) occupying anomalous hydrological positions.
	5	Subglacial calcite precipitates; some arctic corridor karst?
Glaciated karst	6	Dolines, etc. subject to one apparent episode of glacial erosion and/or deposition.
Polygenetic forms	7	Most large closed karst depressions in glaciated terrains; created by repeated glacial and karst episodes.
Preglacial karst	8	Some relict caves. (Surface karst features are modified into types 6 or 7).

particularly well because fine-grained glacial detritus is readily piped into surviving epikarst cavities such as grikes (Fig. 9.20).

Probably the majority of postglacial solution dolines belong in this category (Fig. 9.11 *Lower left*). They are located at low points prepared by glacial scour and the form is often irregular because it retains part of the scour morphology. Steep- to vertical-sided dolines on scoured fractures trap and conserve snow, becoming sites of accelerated corrosion termed 'schacht dolinen', schneedolinen' or 'kotlic'.

Solution dolines in glaciated terrains rarely compose regular polygonal karst because of the deranging glacial effects. However, suffosion dolines over buried pavement may yield excellent polygonal drainage.

Type 3 are glacial depressions adapted to karst drainage. Small depressions in bedrock include potholes, grooves and irregular, shallow scours of weaker strata or fracture lines that are up to a few hundred m in length. Larger examples contain karren, small kotlic or suffosion dolines.

Innumerable closed depressions of all shapes and sizes are created during glacial and proglacial deposition of clastic detritus. Amongst those of small to intermediate scale, such as kettle holes, four different conditions may arise (Ford 1979): (a) the depressions cannot drain underground and so fill up as ponds; (b) the depressions drain underground through the glacial detritus without modifying it; (c) the depressions drain underground through drift with a high content of soluble clasts, so that there is dissolution with collapse or piping; (d) the depressions drain underground into bedrock karst; subsidiary suffosion or collapse features may develop. Only this fourth condition belongs to the karst system, but in many continental glaciated terrains covering thousands of square kilometres there is the greatest difficulty in distinguishing it from the others. The 'turloughs' (dry

Figure 10.5 *Upper photos* – postglacial landforms. An erratic block on a pedestal on a glacier-moulded ridge in marble, Pikhauga, Norway. Stream sink in medium-bedded limestones, Anticosti Is., Quebec. *Lower* – 'Moraine Polje', a karst-adapted glacial feature. The photograph is taken from the crest of a terminal moraine which blocks a valley in dolomites (catchment = 90 km^2). The valley floor is seasonally inundated; all waters sink in a cave at the arrow. Mackenzie Mountains, Canada.

lakes) of lowland Ireland provide good examples (Williams 1970).

Moraine-dammed valleys that drain karstically are comparatively rare. Figure 10.5 shows a fine example from the Mackenzie Mountains, NWT, that functions as a seasonally inundated polje.

Karst landforms created beneath the ice include dolines of normal appearance occupying what become anomalous hydrologic positions when the ice is gone. Examples are shafts in the crests of subglacial ridges, created because crevasses formed above them, and others high in valley sides where

Figure 10.6 *Left* – Subglacial calcite precipitates (white material) upon an ice-scoured surface. Ice flow was towards camera. Ice receded approximately 20 years before this picture was taken. *Right* – detail of the calcite deposition around a limestone drumlinoid. A pattern of subglacial micro-rills is seen at the stoss (upstream or near) end of the drumlinoid. Scale in inches. Castleguard karst, Canada.

they swallowed marginal streams from the ice. They are characteristic of alpine karst.

Subglacial calcite precipitates (Fig. 10.6) are a class of karst forms unique to the subglacial environment. They are crusts deposited from the basal pressure melt film as it refreezes (Hallet 1976, Souchez & Lemmens 1985). They are highly streamlined in accordance with local ice flow, including ridge-and-furrow forms and even small, horizontal helictites oriented down-flow. In some instances it is apparent that the feedwater flowed through the bedrock in a miniature epikarst before reaching the freezing site. Crusts are up to a few centimetres thick and may cover 70–80% of a limestone or dolomite surface if it is comparatively regular and free of basal clastic debris. There are a few instances of the precipitates being found on non-carbonate rocks (Hillaire-Marcel *et al.* 1979).

In southern parts of the continuous permafrost zone of Canada there are patterns of solutional corridors (giant grikelands) that are now draped with glacial till, permafrozen and relict. They occur on lowland and plateau surfaces where, despite their northerly situation, the continental ice flow was probably wet-based for at least parts of each glaciation. It is suggested that they may have been created by subglacial dissolution (Ford 1984). They are not glacial erosion forms.

Simple glaciated karst (Table 10.1, type 6) describes karst features subject to one significant episode of glacial action which has modified their form. A small example is shown in Figure 10.7. Such simple forms are comparatively

Figure 10.7 A simple (i.e. one-event) example of a karst form overridden and scoured by warm-based ice. Ice flow was from left to right across this small doline which displays glacial plucking on the upstream (left hand) wall, abrasion and subglacial precipitation on the downstream wall.

rare. Most karst features subject to glaciation must be placed in type 7 – landforms that have suffered several or many later episodes of glacial action plus, in most instances, significant karst solution when ice was absent. They are polygenetic, multiphase features. There are no examples where the precise number of phases has been reliably determined.

A majority of the largest closed depressions now functioning karstically in glaciated regions belong in this category. It includes hundreds, perhaps thousands, of cirques in the carbonate ranges (Maire 1977, Ford 1979, Fig. 10.8). The cirque is the basic alpine glacial landform, normally 0.5 to 5 km in length or diameter. Many are overdeepened by scour into the bedrock i.e. their bases are closed depressions. In others closure is created, partly or wholly, by a moraine barrier. Some contain seasonal or permanent lakes that may overflow at the surface. Others are always dry, with karren and varieties of the postglacial and adapted dolines in their floors. Many drain to springs at the foot of the riegel (bedrock step below the cirque) and thus are local karst hydraulic systems subordinate to the glacial topography. Others contribute to regional aquifers.

Figure 10.8 *Upper* Racehorse Cirque and *Lower* Surprise II Lake. These are two deep, ice-scoured depressions in massive, steeply dipping limestones in the Rocky Mountains of Canada. The depressions drain karstically, their water surfaces being 50 m and 80 m respectively below the bedrock rims. Features such as these are probably produced by many successive episodes of karst and glacial action.

Large depressions also occur in alpine valleys (Fig. 10.8) and on plateaus and plains subject to continental glaciation.

In most glaciated terrains truly pre-glacial karst landforms (i.e. antedating all episodes of glaciation) cannot be recognized because the number of glaciations is unknown. They possibly existed but have been modified into the glaciated and polygenetic types. In the literature, 'preglacial' is often used where it is established only that a feature is older than the last glaciation.

In the alpine karsts of Europe and North America, landforms of types 1, 2 and 3 (Table 10.1) are numerically predominant. Many authors have recognized strong altitudinal zonation of these features, beginning with a doline zone at the tree line, succeeded by a higher karren zone and then a highest zone where frost shattering is predominant (e.g. Rathjens 1954, Bauer 1962, in the Alps). Miotke (1968) in the Picos de Europa and Ford (1979) in the Rocky Mountains emphasized that most alpine karst zonation may be more strongly tied to ice extent, erosion and deposition during the last glaciation than it is to modern floral or climatic zonation. In particular, terrain formerly covered by glaciers displays mixed assemblages of karren and differing types of dolines, varying with slope gradients and with the depth of any detrital cover. Terrain that was not ice covered tends to be dominated by frost debris, regardless of altitude and so does not host karren. It will display kotlic and other dolines if sufficient drainage can be focused.

Effects of glacier action upon karst systems

This section briefly considers the effects that glaciers may have upon the karst system as a whole. The most extensive tracts of formerly glaciated karst rocks occur in Canada where Ford (1983a, 1987) recognized the nine distinctive effects listed in Table 10.2. There may be others awaiting analysis.

Erasure of shallow karst features by glacial scour and plucking is perhaps the most widespread and frequently recognized effect. Positive forms such as spring mounds, pinnacles and hums may be removed entirely, as are micro-karren and the smaller karren such as rillen. Shallow epikarst aquifers such as limestone and dolomite pavements can be stripped away, leaving unbreached, ice-polished surfaces in their stead. However, it is common for the bases of deeper grikes to survive a single glaciation (e.g. Rose & Vincent 1986). They then guide the postglacial renewal of karstification unless they are rendered hydrologically inert by infilling and shielding (below). Most dolines are too deep to be removed by the scouring capacity of a single glaciation although, once again, they may become hydrologically inert.

Glacier ice cannot be abstracted into karst aquifers as invading allogenic

Table 10.2 Effects of glacier action upon karst systems

Destructive, deranging:

(1) Erasure – of karren, and residuals
(2) Dissection – of integrated systems of conduits
(3) Infilling – of karren, dolines and larger input features; aggradation of springs
(4) Injection – of clastic detritus into cave systems

Inhibitive:

(5) Shielding – carbonate- or sulphate-rich drift protects bedrock surfaces from postglacial solution

Preservative:

(6) Sealing – clay-rich deposits seal and confine epikarst equifers

Stimulative:

(7) Focusing inputs, raising hydraulic head – with superimposed glacial streams or aquifers
(8) Lowering spring elevations – by glacial entrenchment
(9) Possible deep injection – of glacial meltwaters/groundwater when bedrocks are being flexed during crustal rebound or depression?

rivers can. Therefore, extensive aquifers may be *dissected* by deeply incised glacial troughs and cirques. The process is most effective in alpine terrains with temperate glaciers. In perhaps the majority of such regions the consequence is that fragments of large, deep phreatic cave systems that are long drained and relict are preserved high in valley sides and even close to the summits of horn peaks, e.g. the famous Eisriesenwelt, near Salzburg, Austria.

Infilling is used to describe the filling of surface karst forms by glacial detritus. In Canada it is known that closed depressions as great as 300 km^2 in area were filled completely during the last glaciation and ceased to exist as topographic entities. It may be presumed that, at the least, tens of thousands of doline-scale depressions have been filled in the world's glaciated regions during the Quaternary.

Filling and burial do not necessarily imply that a doline will be unable to perform its hydrologic role after the ice has receded. If there is an adequate hydraulic gradient the clastic fill may function merely as an infiltration regulator as discussed in section 9.4. With time the depression then is partially re-excavated by dissolution and/or suffosion. The most striking Canadian example is Medicine Lake, where it is estimated that $1.8 \times 10^8 m^3$ of detritus has been removed into the Maligne River conduit aquifer (section 5.4) during the Holocene. This has created a closed depression 6 km in length that is a partial re-excavation of a greater, infilled bedrock depression.

Glacial and glacifluvial debris can be *injected* deep into conduit aquifers.

At Castleguard Cave, the heads of six inlet passages are sealed by intruded glacier ice. This is the only explored cave that extends substantially beneath flowing ice and the intrusions are believed to be short. Normal injecta are clastic detritus of all grain sizes, transported by meltwater (e.g. Hladnik & Kranjc 1977). The streams may be supra-glacial or englacial in position e.g. at the Kvithola cave (Fig. 10.4) but in the majority of studied cases appear to have been subglacial or proglacial. Substantial sections of many deep alpine caves have been filled, nearly or completely, by injecta of varied grain size. Repeated filling is common. However, other sections escape filling so that, during the deglacial or postglacial periods the aquifer can be substantially renewed by a combination of re-excavation of some conduits and opening of new bypasses to others.

Where the relief is intermediate or low, so that it is completely buried by ice, and abundant clay is available from nearby shales or other strata, aquifers may be more comprehensively deranged or rendered quite inert by injection of the clay sealant. The Goose Arm Karst, Newfoundland, is an excellent example of severe derangement (Fig. 10.9; Karolyi & Ford 1983). It is a rugged terrain in massive carbonates, with a local relief of 50–350 m. It contained more than 40 large, bedrock closed depressions (100 to >1000 m diameter), implying the existence of mature conduit aquifers draining to a small number of regional springs. Following clay injection, some depressions were completely blocked and have filled with water to function as normal lakes. Others discharge via short or tiny postglacial conduits to their individual springs. Orientation of the modern flowpaths correlates poorly with both geologic structure and maximum topographic gradient, the two factors that control the directions of flow in normal aquifers (section 5.1).

Postglacial karst development is inhibited where the bedrock is *shielded* from solutional attack because the solvent capacity of the waters is expended on soluble clasts in the cover of glacial detritus. Because they are comparatively weak most carbonate rocks and gypsum yield abundant clasts to scouring glaciers, enriching the local till. Some marbles do not yield significantly and tills immediately downstream of yet weaker rocks such as shale may be deficient in soluble fragments.

At different sites in Ireland, Ontario and Quebec 1–2 m of till or outwash, or no more than 0.25 m of marl, have served to protect underlying limestones and dolomites entirely during the 10 000–14 000 years since the ice receded (Fig. 10.10) despite the fact that an ideally coincident (i.e. optimal) solution system may be operating. At Windsor, Nova Scotia, an ice-scoured surface of gypsum is beautifully preserved beneath 4–6 m of till rich in gypsum fragments (Fig. 9.46).

Much of the terrain underlying sluggish, cold-based glaciers may suffer negligible erosion. In such situations karst features as fragile as densely-fissured pavement may be *preserved* more or less intact. The principal

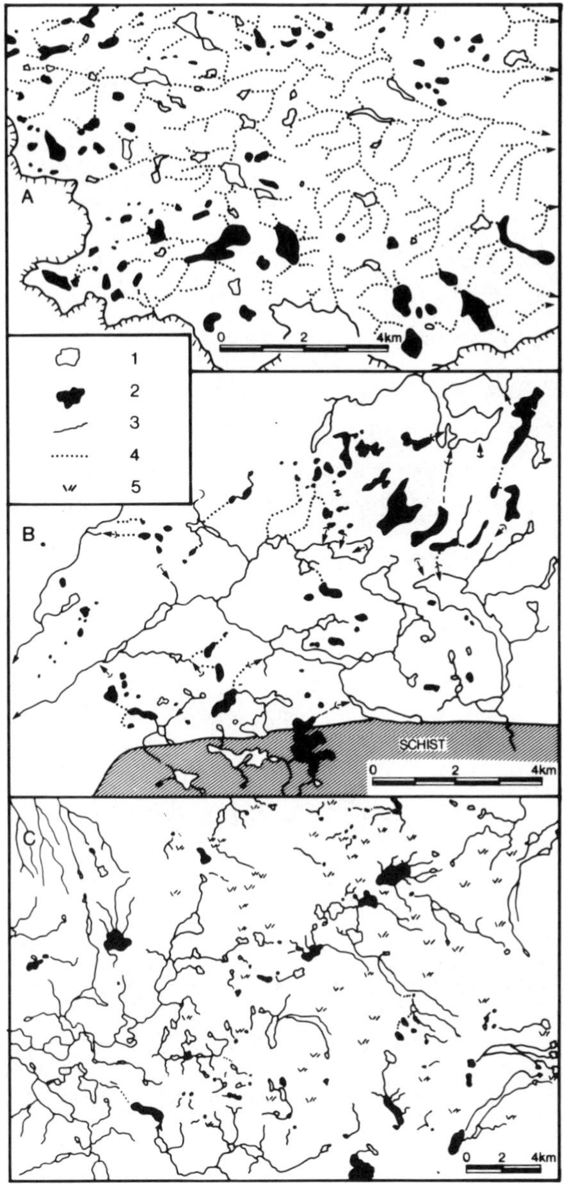

Figure 10.9 Three examples of deranged karstic drainage, drawn from very contrasted climatic environments. A. Hammada de Guir, Morocco low relief limestone desert karst. Adapted from Castellani & Dragoni (1986). B. Goose Arm, Newfoundland – a very rugged carbonate karst subjected to severe glacial scour and injection of detritus. C. Part of the Great Bear Lake karst – a low relief, glaciated dolomite/gypsum terrain at the widespread-to-continuous permafrost transition, arctic Canada. 1. Lake draining normally or, in A, 'dayas' (shallow depressions) that overflow as lakes. 2. Depressions, including dayas, that drain karstically. 3. Perennial or seasonal channels. 4. Ephemeral or meltwater overflow channels. 5. Swamp.

Figure 10.10 Illustrating two extremes of glacial effects upon solutional pavements in dolomites. Both sites were overridden by the (Wisconsinan) Laurentide ice sheet but were close to its margins. *Upper* – Scour beneath warm-based ice led to complete removal of upper beds, plus infilling of grike bases with injected clay, to create a truncated inert paleokarst. Striae are perfectly preserved beneath 1.3 m of diamict rich in dolomite clasts. Hamilton, Ontario. *Lower* – perfect preservation of open clint-and-grike epikarst beneath melt-out tills released from cold-based ice. Winnipeg, Manitoba.

Canadian example is pavement in medium-bedded dolomites that underlies the city of Winnipeg, perhaps extending for as much as 3500 km^2 (Fig. 10.10, Ford 1983a). It is overlain by up to 4 m of melt-out till that is not injected into it significantly. This *sealed* the pavement, converting a quickly drained epikarst into a productive confined aquifer. Such extensive preservation appears to be rare.

Probably the fastest rate of development of ponors and short sections of new cave in carbonate rocks occurs during glacial recession in rugged terrains such as alpine areas. This is because large melt streams can be *focused* on specific input points at a glacier sole or margin for a few tens or hundreds of years, and because the hydraulic head building up in a glacier might be *superimposed* functionally upon that in an underlying karst aquifer. There are many examples of shafts drilled into the crests of subglacial ridges, as noted above, and of others aligned along the outer margins of end moraine ridges where they were evidently fed by supraglacial streams e.g. as described by Glazek *et al.* (1977) in Poland. Lauritzen (1986) considered the Kvithola, a phreatic shaft cave in the wall of a Norwegian fiord, to have been created by a lateral margin stream that was able to maintain a head of at least 250 m in the adjoining glacier. In the modern Columbia Icefield a small melt lake builds up in a depression in the ice until it discharges abruptly through its base, apparently into the underlying Castleguard aquifer i.e. it is a karstic 'jokulhaup' (ice dam burst), an extreme form of superimposition.

Glacial entrenchment also steepens groundwater hydraulic gradients by *lowering* potential spring points. In alpine glacial regions of the Canadian type many springs 'hang' above the floors of limestone valleys because they have not yet adjusted.

The final effect is more speculative than those summarized above. In Canada and elsewhere there is increasing evidence to suggest that there may be very *deep injection of meltwater* into karst aquifers, into interstratal karst, and into paleokarst features (section 10.7) that were buried and inert before the Quaternary glaciations. Injection may be abetted by the crustal flexure that occurs when land is rebounding isostatically from the release of ice load, especially in extensive lowlands.

The most important evidence is that some deep collapse structures have been either initiated or rejuvenated during glaciations. The principal examples are of giant breccia pipes above salt in the plains of Saskatchewan (Fig. 9.48). In Figure 10.11, Howe Lake is a closed depression in which tills of last glacial age were downfaulted 78 m (Christiansen 1971). The depression has been infilling during the Holocene and is now only 10 m deep though it is 300 m in diameter. The 'Saskatoon Low' is more complex. In one or a series of early collapses, Cretaceous and older clastic rocks downdropped at least 190 m to create a principal depression measuring 25 × 40 km (Christiansen 1967). These collapses may have been Quaternary

A.

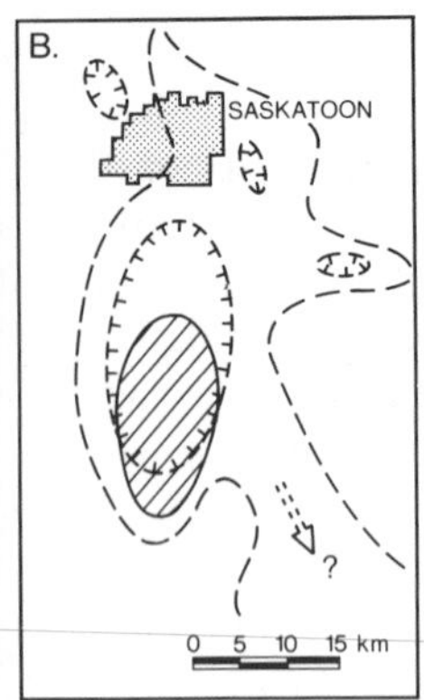

Figure 10.11 A. The surface expression of the Howe Lake collapse structure, Saskatchewan. (E.A. Christiansen, with permission). B. The 'Saskatoon Low'. The shaded feature is a collapse of 70 m depth that is of Late Wisconsinan age and infilled by lake sediments. Earlier depressions (unshaded) are probably Quaternary and glacially triggered but could be earlier. Arrow indicates possible glacial spillway. Note the very large scale of these features. Based on figures in Christiansen (1967).

and glacially triggered but could be earlier. The final collapse occurred at the close of the last (Wisconsinan–Wurm) glaciation, when tills of that age were downfaulted 70 m into the older depression. The collapse propagated from salt approximately 1000 m beneath the surface of the Cretaceous cover rocks. The depression was infilled by postglacial lake sediments and has been inert during the Holocene.

Evidence of glacial rejuvenation of paleokarst of Paleozoic age is being revealed at Pine Point, a zinc/lead mining region on the south shore of Great Slave Lake, Canada. The ores fill paleocaves and collapse dolines in a Devonian barrier reef (Fig. 2.9). Certain dolines display cores of Quaternary till (or multiple tills) passing through the centres of the older breccia and sulphide fillings. In a few cases these cores extend below the paleokarst groundwater circulation base in the carbonates and enter a deeper anhydrite formation (K. Newman, personal communication, 1981). Modern hydraulic gradients are very low and could not support such deep circulation.

This effect, hitherto unrecognized, operates at large spatial and temporal scales. The Saskatoon Low is as large as the greatest modern poljes. The deep injection concept implies that a proportion of the water in some deep aquifers or emerging at regional springs may have been resident underground since the last or earlier glaciations.

To conclude, the development of karst landforms and systems in formerly glaciated terrains is particularly complex. Glacial action may create a wide range of effects within comparatively small regions. The problem of inheritance from previous glacial or interglacial conditions complicates all analysis. In general, destructive or inhibitive effects tend to predominate where the relief is intermediate or low. For example, density and variety of karst features increases markedly to the south of the glacial limits in the US Midwest. Where relief is high and alpine glacial conditions prevailed there is always drainage derangement and complex inheritance present in an extensive karst but individual features or system may develop rapidly and to a spectacular scale.

Nival karst

Nival karst is a term widely used to describe regions where the snowfall is heavy and its seasonal melt contributes a large proportion of the groundwater budget. There are two distinct situations. In the first, the terrain has been glaciated so that the modern nival conditions are superimposed on assemblages of glaciokarst features. In terms of karst form and distribution, the effects of modern snow patch deepening are subordinate to the effects of glacier and karst interaction already discussed.

The second situation occurs where the terrain was not glaciated and was also too warm for deep permafrost to establish during glacial periods. Examples are found among the lower slopes or foothills of most alpine karst regions of the Pyrenean type. The clearest examples of nival effects, however, are found in the Crimea and similar sites that are remote from any former glaciations. The Crimea is a scarpland with limestone escarpments that face the Black Sea and rise to 1400 m asl. V.N. Dubljanskij (personal communication, 1986) has made detailed quantitative studies of the Ai Petru karst basin there. It is a doline karst with an area of 14 km^2. Seventy per cent of mean annual runoff is from snowmelt abetted by spring rains, yielding maximum discharges of 6 $m^2 s^{-1}$ at the springs. Minimum discharges (summer) are only $\frac{1}{1000}$ of this magnitude and derive from condensation. The distribution of dolines is determined by the distribution of snow drifts and cornices, in association with geological controls. Doline density is at a maximum (27 km^{-2}) at the altitudes of maximum snowfall.

10.4 The cold extreme: 2 karst development in permafrozen terrains

The nature and distribution of permafrost

Bedrocks or detrital cover are said to be *permafrozen* if their temperature remains below the freezing point for one year or longer (Brown 1970). A standard classification of permafrost zones is shown in Fig. 10.12. *Glacières* are the distinctive karstic category of permafrost that can form and survive

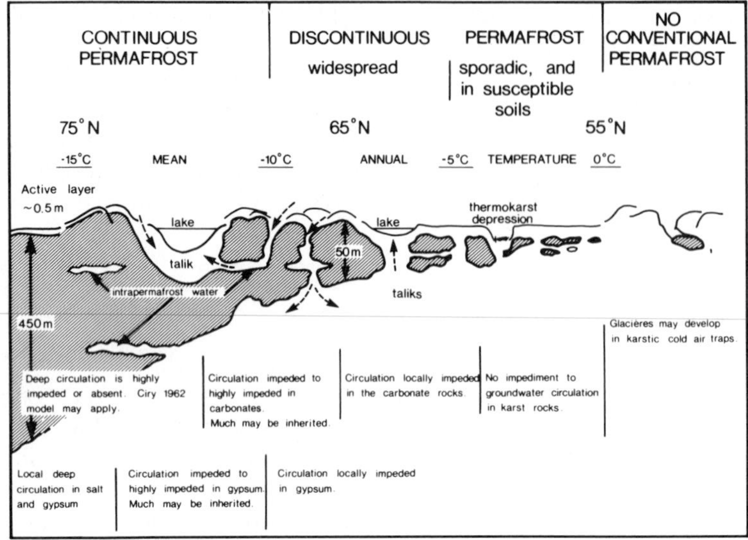

Figure 10.12 Model to depict the general relationships between permafrost and karst activity in terrains of low to intermediate relief. The model is based on conditions in the Interior Platform and arctic islands of Canada.

outside of these zones because of the configuration and large size of some cave cold traps, as explained in section 8.5. *Sporadic* permafrost is confined largely to silts and similar detritus. *Widespread* permafrost extends to all types of rocks and in the *continuous* zone it is present everywhere except beneath lakes, large rivers or in a few special circumstances described below.

Permafrost in alpine regions may be placed in a separate category because of the great local irregularity that it displays in its distribution and depth. Here we treat it under a more general 'rugged terrain' model outlined below.

Groundwater conditions in widespread and continuous permafrost may be most complex. The *active layer* is the top layer, present everywhere, that is thawed and re-frozen seasonally. Its depth diminishes from ~ 2 m at the warm limits of peramfrost to no more than 30 cm at the coldest sites. The *thermoactive layer* (the downward limit of *any* seasonal temperature change) is deeper, ranging −5 m to −30 m at differing sites. *Taliks* are unfrozen areas extending below the active layer. They may reach the base of the permafrost or terminate above it. They are present to some extent beneath all permanent lakes and ponds that are too deep to freeze to their beds. Static or flowing groundwater bodies may be present in taliks or elsewhere within or below the permafrost (Fig. 10.12). In karst rocks most intra-permafrost waters are contained within conduit systems or in particular strata with high diagenetic porosity.

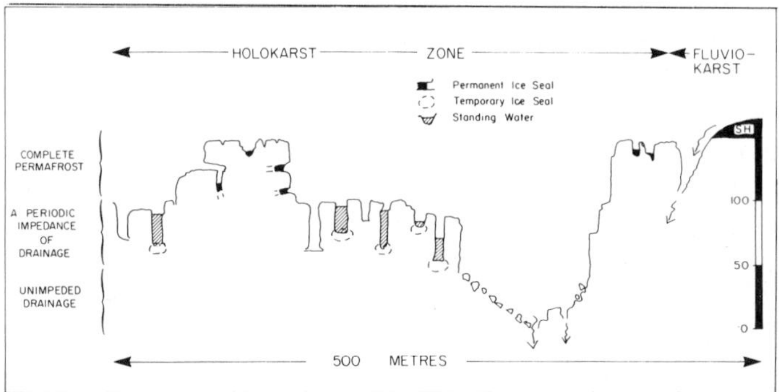

Figure 10.13 A zonal model for the relationships between permafrost and karst drainage in alpine permafrost and for rugged terrain in the widespread-to-continuous permafrost regions. Based upon the doline and corridor karst of Nahanni, Mackenzie Mountains, NWT. SH = shale caprock.

The freezing process is weak because of the release of latent heat as water freezes. As a consequence little energy is required to keep conduits open once flow is established below the thermoactive layer. Maximum discharges into ponors (at the top of that layer) of as little as $5 ls^{-1}$ provide enough heat to keep them perennially open as taliks at the northern limit of the discontinuous permafrost zone in Canada (van Everdingen 1981). Finally, it should be noted that the freezing point of water decreases as its content of dissolved solids increases; thus, flow of saline groundwater may be maintained where fresh waters freeze.

The most extensive tracts of karst rocks in the permafrost zones occur in the USSR and Canada. Most of the Canadian areas were glaciated. Most Soviet karst has not been glaciated, and thus it should provide the best type areas for study of karst-permafrost interactions. We regret that we lack familiarity with detailed studies carried out in the Soviet regions and published in the Russian language. However, the review of Popov *et al.* (1972) and personal discussions with Y.I. Bersenev, V.N. Dublanskij, A.B. Klimchouk and M.V. Mikhailov (1986, 1987) suggest that the models we present from Canadian experience (Figs 10.12 & 10.13) are broadly appropriate in the Soviet Union also.

Karst development in permafrozen terrains

Development of karst systems does not appear to be significantly restricted in the sporadic permafrost zone. Permanent freezing there is confined to soils such as silts that readily accrete ground ice. If these accumulate in traps such as doline bottoms ice growth might temporarily obstruct circulation, as discussed below.

For alpine permafrost conditions and intermediate to high relief in the widespread permafrost zone the Nahanni karst may be taken as a model (Brook 1976, Ford 1984 and Fig. 10.13). It extends between Lat. 61 and 62°N in the Mackenzie Mountains of Canada. Mean annual temperatures range from −6° to −8°C. There is little snow to insulate the ground and winter temperatures fall below −50°C. Summer temperatures can attain 35°C. As a consequence the thermoactive layer is deep. The region was glaciated at some time or times before 350 ka BP but not since then. Topographic and climatic conditions are similar in the unglaciated Lena and Angara highlands of Siberia.

The Nahanni is a mixed doline and corridor karst in limestones (Fig. 10.14). Larger depressions are >100 m deep and up to 1000 m in length. Their walls are frost-shattered and floors are filled with talus. Where allogenic streams supply sand or clay these depressions become alluviated and may function as border poljes (Brook & Ford 1980).

There is unimpeded, highly integrated conduit drainage from all larger depressions to springs at the two ends of the karst belt. Between the depressions the highest karst is permafrozen and relict. An intermediate zone displays a-periodic impedance of the recharge. Its catchments are small, a few hectares at the most. Being in the thermoactive layer, the drains of dolines become sealed by ice or detrital plugs that freeze. Over one or a few melt seasons water accumulates above the seals until, it is supposed, the hydrostatic pressure is sufficient to rupture them; then they drain in the space of a few hours or days. Such behaviour tends to create cenote-form point recharge dolines.

A most important point with respect to the Nahanni karst is that the large depressions and most landforms in the impeded zone have developed since the region was last glaciated. In general, climatic conditions during most of this development will have been more severe than they are at present.

Further north in the Mackenzie Mountains, Tsi-It-Toh-Choh is a similar limestone karst that has escaped any glaciation (Cinq-Mars & Lauriol 1985). Mean annual temperature is −10 to −11°C. Frost-shattered surfaces (*felsenmeer*) are more extensive than in the Nahanni region. The completely permafrozen zone is relatively more extensive and the a-periodic recharge zone is reduced. Relict caves are well decorated with speleothems which (from preliminary U-series studies) are almost entirely of Tertiary age. This suggests that karstic diffuse flow has been eliminated by growth of permafrost during the Quaternary. Nevertheless, over a large area focused drainage to large dolines or valley bottoms permits unimpeded groundwater circulation to perennial springs that are marked by ice buildups ('*aufeis*', German or '*naledi*', Russian). Nearby there is a rugged karst in dolomite breccia. This yields a higher proportion of fines when frost shattered, with the result that soliflucted debris contribute to the a-periodic impedance of the drainage (Fig. 10.15).

Figure 10.14 Aerial oblique views of the Nahanni karst. *Upper* – Raven Lake, a doline that is 150 m deep to the waterline. It drains dry in winter. *Lower* – flood waters rising in a small polje.

Figure 10.15 Solifluction lobes descending into a large doline in the discontinuous-continuous permafrost transition region. Dodo karst, Mackenzie Mountains, Canada.

A-periodic impedance is particularly relevant to the study of temperate highlands that experienced episodes of permafrost but not of glacier cover during the last or earlier glaciations. Examples include the Mendip Hills and Peak District of England, parts of the Ardennes, Jura, Carpathians, etc. At least the smaller threads of drainage from the epikarst base were often halted because speleothem growth in underlying caves was widely arrested (section 8.6). Blockage of dolines is indicated by deep accumulations of layered sediments, that are now dissected and drained by ponors. Occasionally, relict overspill channels cross the doline rims (e.g. Ford & Stanton 1968). The sediments often have a large component of loess. Some dolines became infilled entirely with soliflucted debris.

In Canada, as the continuous permafrost zone is approached in the karst terrains the topographic relief also diminishes. The restraint of low hydraulic gradient is combined with the restraints imposed by permafrost so that the latter are often difficult to distinguish. It appears that modern carbonate karst is poorly developed except along escarpment edges where groundwater gradients are high. Karst systems are widespread where gypsum outcrops or is covered by comparatively thin dolomites or shales. Van Everdingen (1981) mapped more than 1400 closed depressions functioning karstically and 67 perennial springs in gypsum around the western end of Great Bear Lake (e.g. Fig. 10.9C). This is at the discontinuous–continuous permafrost transition. The largest depressions antedate the last glaciation, contain much glacial sediment and function as seasonal lakes that drain karstically.

Further north, active dolines become fewer but are recognized until relief becomes extremely low 200 to 300 km inside the continuous permafrost belt. This northernmost sector was glaciated during the last ice age; much of the karst does not appear to have expanded during the postglacial period and may actually be shrinking in its extent. This last observation introduces the question of *inheritance*.

In the modern permafrost regions it is suspected that many functioning karst systems may have been established under more favourable combinations of hydrological and thermal conditions. They are maintained today because of the comparatively feeble freezing capacity below the thermoactive active, but are essentially inherited. In arctic Canada and other glaciated terrains this implies inheritance from conditions of subglacial, or ice-margin deglacial, thawing of the permafrost. In non-glaciated regions of Siberia researchers suppose conditions to be inherited from warmer interglacial episodes or from Tertiary pre-permafrost times.

It is difficult to prove that a modern groundwater circulation has been inherited. Origin under more favourable conditions is better understood where the inheritance could not be passed on i.e. where permafrost has grown and arrested the circulation during postglacial times. Examples include shallow corridor karst that is quite widely scattered on limestone and dolomite plains between 68 and 73°N in Canada (section 10.3) and large breccia pipe collapses cemented by ground ice known in mines in Spitsbergen and on Baffin Island (Corbel 1957, Ford 1984). Salvigsen & Elgersma (1985) suggest that a doline terrain of low relief in gravels overlying gypsum on the Spitsbergen shore was initiated after it had thawed beneath a marine inundation and then had risen isostatically.

In the extreme conditions prevailing in the northern Canadian islands, Northeast Land, the Siberian arctic islands and Antarctica, mean annual temperatures are below −12°C and the warmest months are cooler than 5°C. As a consequence the thermoactive layer is shallow. In addition, precipitation is generally less than 200 mm. If glaciated, most land remains permafrozen. Karst development on the carbonate rocks is very limited in these conditions. There may be effective groundwater circulation with dissolution to the base of the active layer at 30 to 100 cm. Ciry (1962) suggested that this would favour development of subcutaneous karst such as limestone pavements. Pavement no more than a few metres wide is seen along escarpment edges and channel karren may develop on steep, bare slopes (e.g. Woo & Marsh 1977). However, in most instances the surface is reduced to rubble (felsenmeer) by frost shatter. This is the predominant aspect of carbonate outcrops in arctic Canada (Bird 1967). It appears that shattering is more effective on carbonates than on other comparatively strong rocks because solution opens up cracks to permit veins of ice to form more readily (Ford 1979). Solvent capacity is then dissipated on the rubble. When the water freezes or is evaporated much of the solute carbonate is re-deposited as botryoidal crusts on the undersides of clasts. Some water passes into joints and bedding planes beneath the rubble because small seasonal springs are common along the bases of low escarpments.

Deeper karst systems may develop in salt and gypsum because of the greater solubility of these rocks and the lower freezing point of strong solutions. There are stream-sink dolines on the flanks of salt diapirs in Axel Heiberg Island, 80°N. The groundwater flow paths appear to be no more than a few hundred metres in length. Field studies are lacking. Gypsum solution breccia occurs at 82°N on Ellesmere Island. This may be the most poleward karst but it is not established whether it is actively forming today.

10.5 Sea level changes, tectonic movement and implications for the development of coastal karst

Mean sea level closely approximates the geoid, i.e. the equipotential surface of the gravitational and rotational potentials. Morner (1983) has explained the complex interplay of processes that determine its level and produce changes in its position. The interface between this surface and the land has varied considerably over geological time, particularly in the last 2 million years. This is because there have been world-wide oscillations in ocean level, termed *eustatic* variations, plus regional to continental scale vertical crustal adjustments and linear mountain building or trough-forming disturbances attributable to plate tectonic forces. These have forced repeated vertical displacement of the sites of coastal karst development.

The most important factors producing eustatic changes are of glacial, tectonic and geoidal origin; the former being the most significant in the time scale with which we are concerned. Glacio-eustatic changes stem from the transfer of water mass from the ocean to continental ice sheets during ice ages. Glacial eustasy accounts for sea level movements of up to 130 ± 40 m in periods of 10^3 to 10^5 years.

The world's most important data on sea level variation over time have been obtained from the datable evidence available on limestone coasts. However, Morner emphasizes that since tectono-eustatic and glacio-eustatic rises and falls cannot occur without accompanying geoidal eustatic changes, we cannot expect to find a perfectly parallel history of absolute sea level for different regions of the globe. Curves for some well known Atlantic and Pacific sites illustrate this important point (Fig. 10.16A). Although the major cause of Quaternary sea level variations must be glacial eustasy, Nunn (1986) stresses the danger of assuming negligible geoidal eustatic changes in an ocean with 170 m of movable relief, particularly when the

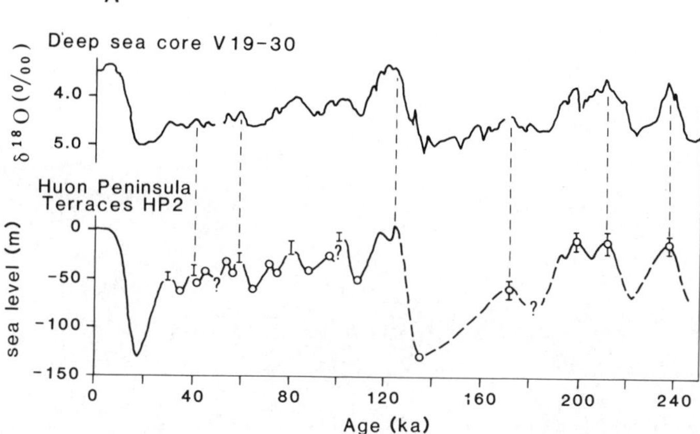

Figure 10.16 A. Huon Peninsula sea level curve correlated with deep sea core V 19–30. From Chappell & Shackleton (1986).

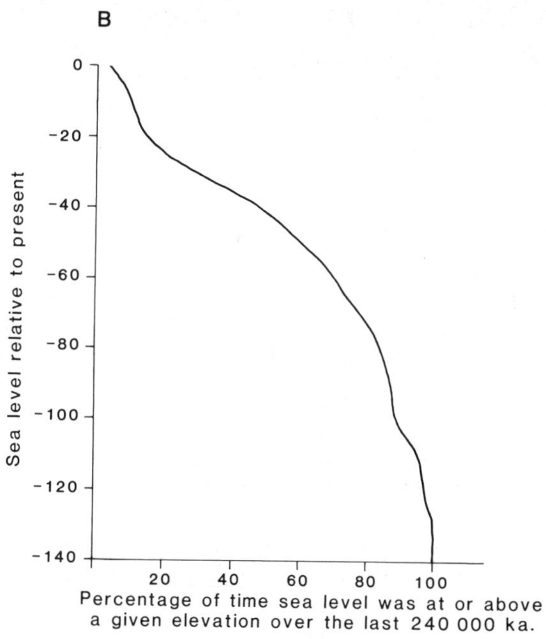

Figure 10.16 B. Percentage of time global mean sea level was at or above a given elevation over the last 240 ka. Data from Fig. 10.16A.

Figure 10.17 Oblique aerial photograph of the Huon Peninsula showing the raised coral reefs used in the construction of the sea level curve of Figure 10.16A. Photo by D. Dunkerley.

above type-sites are in areas of large geoid anomaly, where the probability of change could be correspondingly high.

One of the consequences of variations in sea level is that, even in tectonically relatively stable areas, the present coastline has been within 10 m of its present position for only about 8% of the last 240 000 years. Indeed, it has been at −20 to −50 m for approximately 46% of this time (Fig. 10.16B). The baselevel of erosion has usually been below its present position during the last quarter of a million years, if not longer, and so of course the water table and fresh water/salt water interface zone were also lower.

Many karst landforms have developed in the 10 000 years or so of the Holocene. Hence there was ample time in the Quaternary to produce karst almost anywhere in the zone down to −130 m or more exposed by glacial low sea levels. On tectonically stable coasts, abrasion platforms and coral reefs marking past positions of the shoreline are presumed to develop most extensively during periods of relative stillstand of the ocean. On tectonically active coasts, the corresponding position is at an interval of tangency

Figure 10.18 Dolines on the New Zealand coast, drowned by the Holocene rise of sea level. These are solution dolines that were filled by volcanic deposits, partly removed by suffosion prior to drowning.

between uplift and sea level oscillation. Excellent examples of carbonate coasts where this sort of movement has occurred, although at different rates, include the Huon Peninsula (Fig. 10.17) of Papua New Guinea (Bloom 1974, Chappell 1983), Barbados (Mesolella *et al.* 1969, Fairbanks & Matthews 1978) and Bermuda (Harmon *et al.* 1978, 1983).

On stable coasts, new features may be built on or cut into a base inherited from the past. Thus postglacial morphology may be superimposed onto that developed in the last interglacial some 125 000 years ago, when sea level briefly stood up to 6 m above present. Similarly, landforms developed in coastal zones exposed during low sea levels may have been re-activated and re-submerged several times during their history (Fig. 10.18); thus drowned karst is normal along all carbonate coasts except those subject to very rapid tectonic uplift. We now examine the associated karst features.

Karst phenomena in the coastal zone

The coastal zone in karst is defined by the inland limit of marine tidal influences (Guilcher 1988). In porous, emerged coral islands such as Niue in the Pacific measurable tidal effects extend to the centre of the island (Jacobson & Hill 1980) more than 6 km from the shoreline, and the fresh water–salt water transition occurs about 0.5 km inland. The intrusion of marine influences is much more irregular in dense crystalline carbonates, but penetration along conduits can still be considerable. For instance, in the Gort Lowland (Fig. 9.31), tidal Lough Caherglassaun is 5.3 km inland and nearby Hawkhill Lough 8.75 km inland is also reputed sometimes to show the effect of tides. In this zone a distinction must be made between those features formed at the coast and those that land–sea interface changes have by chance located there. Those formed at the coast mainly comprise littoral karren and notches of the intertidal zone, discussed in Section 9.2. Here we focus predominantly on the other characteristics.

The effect of eustatic variation on karst is best seen on carbonate islands (Fig. 10.19), where the fresh groundwater lens may be so sensitive to sea level change that it readjusts twice daily to the low amplitude variations of marine tides. Emerged coral atolls are dip-sticks of sea level change, recording the relative level of the ocean as it has moved in the past as well as lithospheric flexure (Spencer *et al.* 1987). However, difficulties exist in the interpretation of some morphological features. Thus whereas Darwin (1842) believed barrier reefs around volcanic islands to be due to subsidence, Stoddart *et al.* (1985) explained them as principally the result of karstic erosion of the inner margin of a fringing reef by allogenic runoff, followed by drowning by a rising sea. And whereas numerous authors (e.g. McNutt & Menard 1978) have attributed suites of coral island terraces to uplift, Nunn (1986) has warned that some could alternatively be the result of geoidal eustatic regression.

Negative shifts in baselevel around atolls exposed precipitous fore-reef submarine slopes. With the loss of buoyant support, massive scale cliff failure often followed. Long fracture traces subparallel to the coast now mark the inland margins of foundered blocks. Deep fissures sometimes open at the surface have been confused with karst caves because of their vadose canyon-like morphology and speleothem decoration. But many are simply tectonic features. Good examples of such fracture caves occur in the Bahamas (Palmer 1986) and on Niue Island. Accelerated corrosion at the base of cliffs produces prominent notches in the intertidal zone of tropical seas (Fig. 9.9). Notches have been found to −105 m in the submerged wall of San Salvador Is. in the Bahamas (Carew *et al.* 1984).

Falling sea levels also result in the draw-down of the fresh water lens. This deepens the vadose zone, as well as lowering the level of the fresh water–salt water interface. The zone of mixture corrosion at the halocline and at brackish springs is correspondingly lowered and with it an important horizon of karst cave development. According to Back *et al.* (1984) differential solution generates pocket valleys at spring heads round the coast of Yucatan; their later submergence being responsible for coves and crescentic beaches. The highly porous nature of coral encourages diffuse infiltration and circulation and hence inhibits karst cave development.

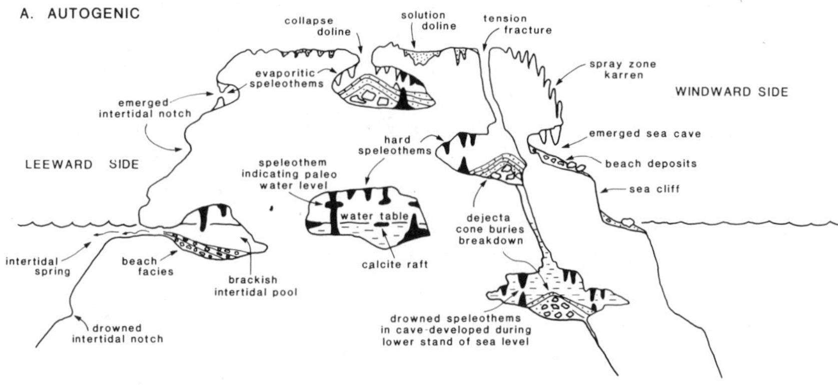

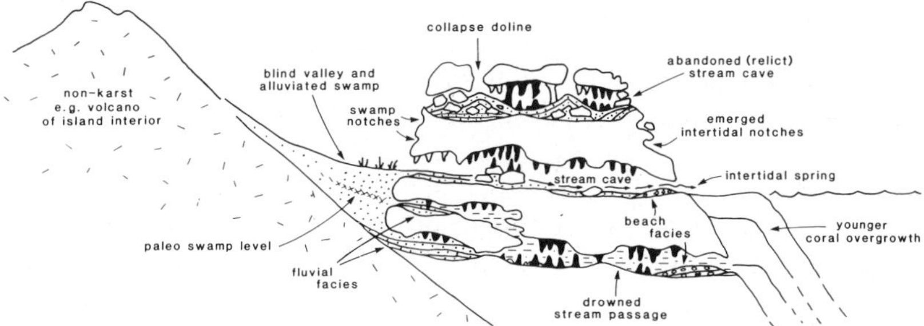

Figure 10.19 Geomorphic and stratigraphic effects of sea level changes observed in coastal cave systems. In the case of A, the tension fracture is a non-karst feature resulting from rock failure due to loss of buoyant support during glacio-eustatic low sea levels. The fracture may nevertheless be exploited by karst circulation. In the case of B, submarine springs often occur where the outlets of drowned stream passages are not blocked by reef overgrowth.

Under such circumstances they are most readily generated where corrosion is focused by allogenic recharge from adjacent non-karst rocks. For example, streams draining radially from the impervious volcanic inliers of some coral islands (e.g. Mangaia Is.) usually terminate in a ring of blind valleys on the inland side of an emerged reef, from where subterranean conduits lead runoff to brackish springs on the coast. Neighbouring blind valleys coalesce to yield a circular moat-like depression with a steep inland facing limestone cliff; the retreating corrosion front. During high stands of the sea, the moat becomes a swamp, lake or lagoon behind the barrier. Dry valleys across the emerged reef mark abandoned stream courses or atoll passes. A few collapse dolines are associated with caves.

The karstification of atolls and fringing reefs during low sea levels has been recognized for many years, although the suites of landforms expected have not been clear. Since hydrogeological prerequisites for the focusing of percolation are generally absent, the generation of solution dolines is uncommon and the evolution of cockpit karst relief is unlikely even in the humid tropics. Hydrological principles also argue against the central lagoons of atolls being drowned solution basins (Purdy 1974), although their form may be accentuated by corrosion. The karst that develops on emerged reefs depends on the hydrochemical controls (Fig. 4.1). Entirely autogenic systems such as Niue and the Trobriand Is. (Ollier 1975) usually support relatively few karst features. But this is more a function of their vadose hydrology than their coastal location. Nevertheless, tropical cyclones ensure the penetration of salt spray influences well inland. Emerged coral islands often have rough corroded surfaces (*makatea*) with the texture of scoriaceous lava rather than sculpted karrenfeld, across which are scattered the occasional small collapse doline, a few giving access to karst caves. A weakly case hardened crust commonly occupies the top few metres of the vadose zone (Hopley 1982). The reef facies is a variable of importance in determining the precise nature of landform response (Bloom 1974), because the extent of lithification and dolomitization affects infiltration. Only rarely, as in Barbados is doline development comparatively extensive (Table 9.4).

By contrast, where there is pronounced point recharge from allogenic streams, exposed reefs may develop a distinctly karstic drainage with large river caves. But the surface topography still need not be particularly karstic, except for collapse features and blind valleys. Suffosion dolines will occur in residuum, but draw-down solution dolines (and cockpits) are still rare. Corrosion plains may form along the input boundary as envisaged by Purdy (1974), but residual towers are not expected unless there was an earlier cockpit phase. In a detailed critical examination of the applicability of the antecedent karst hypothesis to the Great Barrier Reef, Hopley (1982, p. 218) concluded that 'there is no evidence for larger karst features such as marginal plains or towers having evolved. Only the blue holes, and possibly the gorgelike interreefal channels, provide evidence of a major karst process, and both features are very localized'; although as we saw in Section 9.5, some (if not most) blue holes are related to fractures rather than to karst conduits.

The syngenetic karst of Pleistocene calcareous coastal dunes (Jennings 1968) has a number of similarities to that of exposed reefs. Even though aeolian calcarenite is extremely porous, case-hardening welds its surface together and lateral allogenic recharge generates blind valleys and river caves (section 9.10), otherwise few karst attributes are evident.

Calcareous beach sands in tropical and warm temperate regions are often indurated between the limits of high and low spring tides, although sometimes cemenetation extends to the immediate supratidal zone. The resulting coarse grained limestone is termed *beach rock* (Fig. 10.20B). The

A

Figure 10.20 A. Marine molluscs growing part way up a stalactite in a cave on Manus Is., Papua New Guinea. They record precisely the level reached by a former sea level.

B

Figure 10.20 B. Beach rock accumulating on Vava'u Is., Tonga group.

cemented sands are usually identical to unconsolidated materials nearby and retain the seawards dip of the beach bedding. Induration appears to proceed inland, with the stratigraphically lowest of the dipping beds being the most recently cemented. Cementation has been attributed to organic processes, inorganic processes involving fresh water, and inorganic processes from sea water. Hopley (1982) concludes that the latter process is gaining favour because of the frequent occurrence of aragonite cement, as opposed to the calcite spar that would be expected from fresh water. The process can be very rapid, considerable beach rock induration having been recognized to have occurred since World War II.

Where dense crystalline limestones form the coast the features are more clearly karstic, even where drowned, simply because such rocks are more favourable for karst development. For instance, spectacular examples of partly submerged tower karst occur in the Vung Ha Long (Bay of Halong) in Vietnam and near Phuket Is., Thailand. Doline karst affected by high sea level is known along the Adriatic coast and elsewhere (Fig. 10.18).

Submarine springs are common in many coastal karsts, especially around the Mediterranean (section 5.1). Their conduits are presumed usually to have been formed during low sea levels. Activity has been maintained because of sufficient hydraulic head inland. More unusual are sea swallow holes, the best known case being the Sea Mills of Argostoli in the Ionian Is. (Maurin & Zotl 1963). Sea water is drawn into the karst and passes right through the island to springs on the other side. Glanz (1965) (cited in Maurin & Zötl 1963) explained this by the ejector effect of autogenic recharge (Fig. 10.21). Bogli (1980) notes the occurrence of salt water swallow holes elsewhere and the relevance of the Bernouilli equation to their explanation. A more common phenomenon is the reversing flow of coastal subterranean rivers with the ebb and flow of tides (sea-estavelles), particularly in regions with a large tidal range. Tidally generated reversing flow through blue holes is reported from Andros Is. in the Bahamas.

Brackish karst springs are sometimes located above sea level. The main outflow pool of the Waikoropupu springs of New Zealand, for example, is 14 m above sea level but its water is 0.5% salt. Williams (1977) suggested this to be the result of a venturi mixing process drawing in water from the underlying coastal intrusion into the aquifer. This mehcanism is supported by the observation of increasing salinity with spring discharge. Stringfield & LeGrand (1971) and Milanovic (1981) discuss other cases.

Deep submarine karst

Maire (1986) briefly reported the discovery of karren-scale grooves and pitting on steep limestone slopes at depths of 1000 m to 3000 m off the Iberian Peninsula. This is far too deep for any eustatic fluctuation of Quaternary age. He suggests that the features may be dissolved by

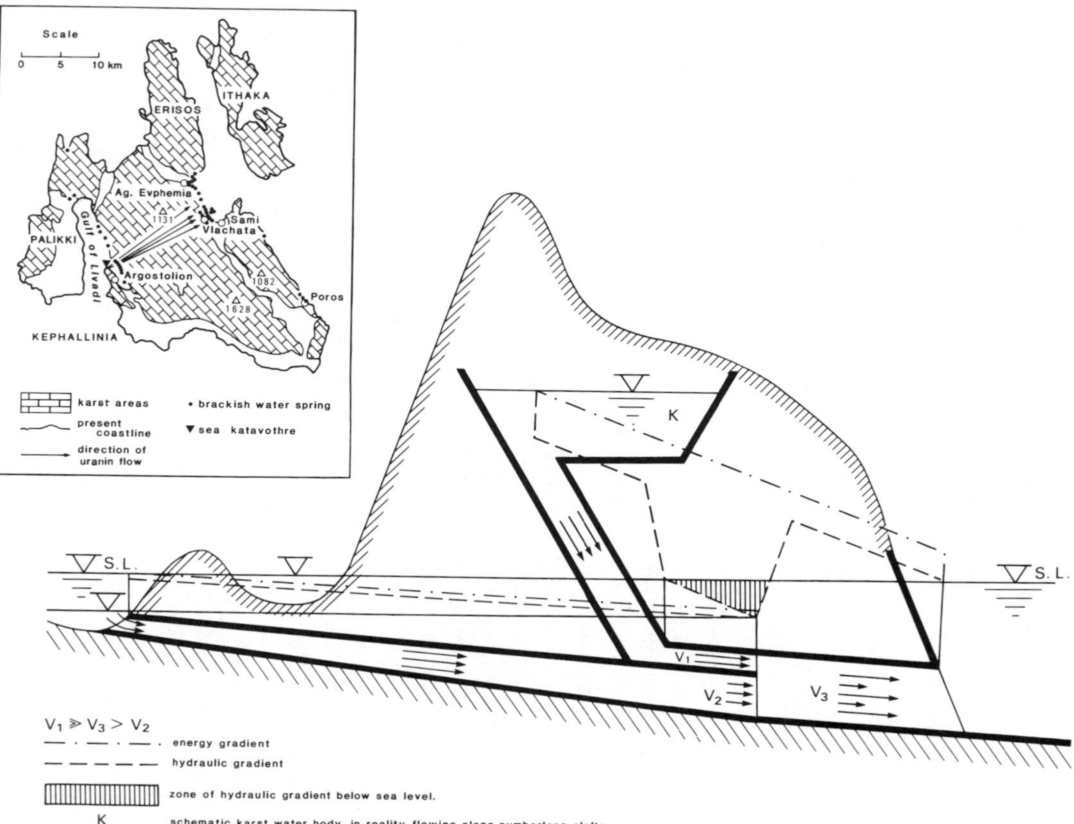

Figure 10.21 The Sea Mills of Argostoli, Greece, showing the hydraulic mechanism for the Sea Mills proposed by Maurin & Zotl (1959, 1967).

descending cold waters, perhaps with thermal or chemical mixing effects. Possibly they are very ancient relict forms, though this seems unlikely.

10.6 Polycyclic and polygenetic karsts

W.M. Davis (1899) maintained that all uplifted land will ultimately be worn down to a surface of low relief termed a peneplain at an elevation close to baselevel, which is usually sea level. This unidirectional erosion process he called the *cycle of erosion*. As the cycle progressed, he considered landforms to pass through an identifiable sequence of stages termed 'youth', 'maturity' and 'old age'. The occurrence of uplift before the completion of the cycle would lead to its interruption; erosion processes would be rejuvenated and a second cycle would ensue, with development proceeding headwards from the sea. Thus landforms of the present cycle would be found downstream of rejuvenation heads (marked by knickpoints in stream profiles), while the still unconsumed morphology of the previous cycle would persist upstream. Landscapes showing evidence of more than one cycle of erosion are termed *polycyclic*. The relationship of stepped erosion surfaces to karsts is discussed in many contexts, for example in the eastern USA by Palmer & Palmer (1975) and in southern China by Song (1981) and Ren *et al.* (1982). In a wide ranging review of karst types in China, Zhang (1980) describes numerous cases of knickpoint recession into such surfaces, which probably range in age back to the Cretaceous.

Grund (1914), Cvijic (1918) and others incorporated Davisian cyclic ideas into their schemes of karst landscape evolution (section 9.12) and Davis himself proposed a two-cycle hypothesis of cave evolution (Davis 1930).

Rejuvenation results from increased erosive energy. This can stem from (a) increased potential energy due to a negative shift in baselevel (following either uplift or sea level fall) or (b) increased erosive power (chemical as well as mechanical) particularly following a climatic change that results in increased runoff. Uplift, if sufficiently great, can also lead to climatic change.

Climatic change does not always stimulate rejuvenation. Often it alters the balance of processes instead, most radically when fluvial erosion gives way to glaciation, as discussed in section 10.3. Landforms modelled under one climato-genetic regime are then remoulded under the next. If modification is incomplete before the climate changes again, then the landscape will bear the imprint of more than one genetic environment and can be termed *polygenetic*. Many variations in climate have occurred during geological time, and in numerous cases the climatic shift persisted long enough to have its geomorphic consequences engraved in the landscape, yet was of insufficient duration to obliterate all traces of previous regimes. As an example, at the height of the Quaternary glaciations both the cold polar and semi-arid mid-latitude zones expanded, with compensating contraction of the intervening Mediterranean and humid tropical zones (Flenley 1979).

Because changes in baselevel and climate have affected large parts of the world, most karsts can be expected to have both polycyclic and polygenetic elements in their landscapes. When the effects of major, repeated climatic changes are superimposed on stage of development and underlying lithological influences, there is clearly considerable potential for great complexity of morphology and landform assemblage. Herein lies the difficulty of interpreting many of the world's great karsts.

10.7 Relict karsts and paleokarsts

Relict karsts are those removed from the situation in which they were developed, although they remain exposed to and are modified by processes operating in the present system. By contrast, *paleokarsts* or *buried karsts* are completely de-coupled from the present hydrogeochemical system; they are fossilized. When stripped of their cover beds they reveal an *exhumed karst*.

Relict karsts can arise in two ways: (a) their hydrographic setting may be changed; or (b) their climatic (morphogenetic) situation may be altered, or both. The former is the more common, although in extreme cases it also leads to the latter. For example, on the Qinghai (Tibetan) Plateau, species represented in fossiliferous deposits discovered in karst terrain at 4 000–5 200 m are now found only under the substantially warmer and more humid conditions that prevail up to 2 500 m above present sea level (Ren *et al.* 1982). Uplift of the order of 3 000 m is indicated since the late Tertiary. Thus the karst has been both drained by uplift and forced out of equilibrium with its process environment. To the southeast in the Guizhou Plateau, where 500–1000 m of uplift has occurred in the Quaternary, at least four erosion surfaces are recognized (Song *et al.* 1983). Figure 9.32B illustrates a karst corrosion surface with residual cones at about 1200 m in Guizhou that is still undissected. The corrosion plain would originally have formed close to the level of the regional water table, which is now being actively lowered as gorges hundreds of metres deep cut headwards into the plateau. Prevailing climatic conditions are probably not sufficiently different to imply morphogenetic disequilibrium.

A lower latitude case where morphogenetic change is certainly not a complicating factor is found in the mogote karst of northern Puerto Rico (Monroe 1966). The surface from which the residual hills rise merges inland with cockpit karst at 130–150 m, the water table in the underlying Aymamon Limestone being just a few centimetres above sea level. Allogenic rivers flowing north from the impervious rocks of the interior cut through the karst in caves and gorges, the mogote karst plateau being sharply incised by valleys up to 150 m deep. The blanket sands which cover

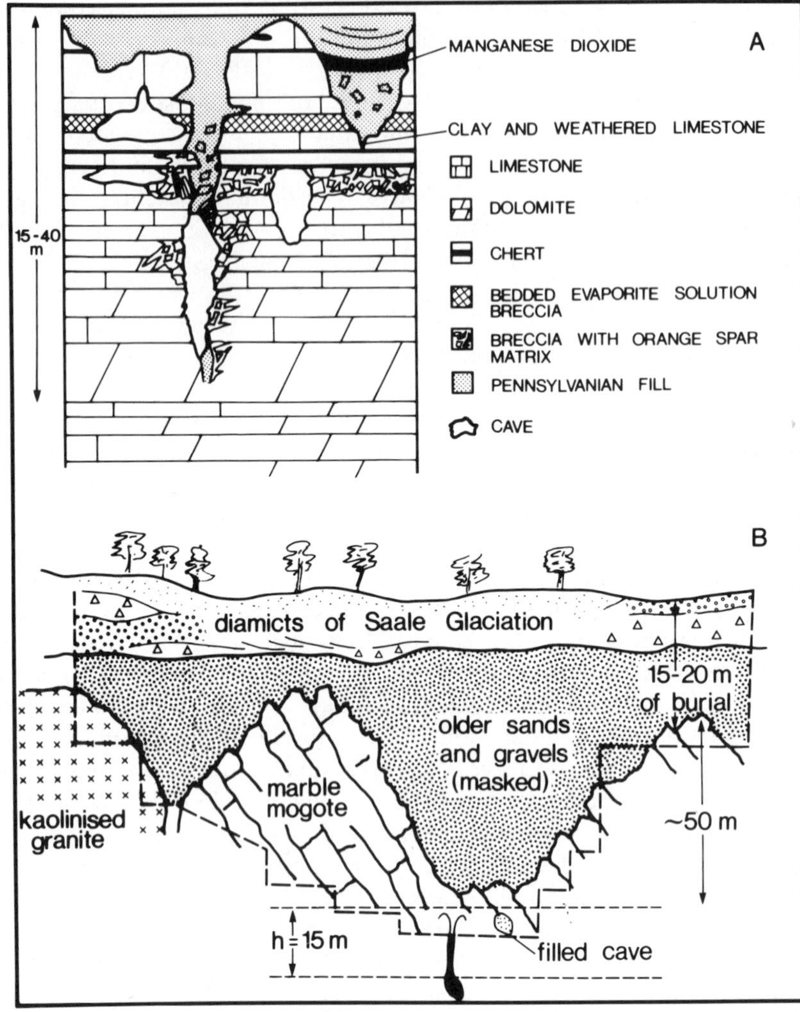

Figure 10.22 A. Fissure paleokarst of post-Kaskaskia (mid-Carboniferous) age in the Black Hills of S. Dakota. From M. V. and A. N. Palmer, with permission. B. Spectacular paleokarst of supposed Mio-Pliocene age exposed in the Supikovice Quarry, Czechoslovakia. Drilling to drain the quarry floor encountered saline waters with a pressure head of 15 m.

the surface between the mogotes are of volcanic origin from the interior and were probably deposited by alluviation of a coastal plain now relatively uplifted. Solution occurs beneath the sands which are being removed by suffosion. The mogote karst is not in equilibrium with prevailing hydrological controls and, as in the Guizhou case above, in the long term will revert to cockpit karst in response to the lowered baselevel.

Examples of relict karst cones and towers thought to have been formed under humid warm temperate to tropical conditions are widely recognized in central Europe (e.g. Fig. 10.22B, Bosak 1981). Gavrilovic (1969) provided evidence for a Tertiary cone karst in the Beljanica Mts of Yugoslavia. Further south, traits of a semi-arid tropical karst of possible Pliocene age are recognized in the Ionian Is. by Zotl (1966).

Landforms that are time-transgressive do not readily fit the classification of relict, however. This difficulty is best appreciated in the case of conventional karst towers, which are younger at their base than at their summit (Fig. 9.37). During the normal course of their development, they will experience de-watering and some will also be exposed to climatic change. Indeed, all relict karsts have been modified in response to conditions that have affected them since their origin. In the Prealps of northern Italy, Magaldi & Sauro (1982) show that early pedogenesis and related doline karst development took place under sub-tropical conditions, but interpret the scree and loess fill of dolines to indicate a subsequent morphogenesis under alternating warm and cold episodes. In China, an eroded limestone topography thickly buried by Eocene continental deposits forms the basis of the famous Stone Forest of Yunnan (Fig. 9.7). Chen *et al.* (1986) explain its development as a result of subjacent solution largely during the long period of burial. Joint enlargement isolated rounded pinnacles, which when exposed were fluted subaerially to yield the present spectacular relief. Thus the gross morphology is pre-Eocene, the pinnacle isolation is mainly mid-Tertiary to Quaternary and the sharp fluting is Pleistocene to Holocene.

Paleokarst (*sensu stricto*) that is buried and inert is known in all continents and is increasingly studied because of its importance in containing or trapping economic deposits of oil and gas, silver, lead and zinc, bauxite and other minerals (James & Choquette 1988, Bosak *et al.* in press). The definition includes karst of interstratal origin and hypogene (paleohydrothermal) origin, which may greatly complicate local sequences. However, the majority of studied occurrences are of the surficial, epikarstic and common cavernous phenomena (T.D. Ford 1977, Bosak in press). These are preserved chiefly in the carbonate rocks because features in the weaker evaporites tend to anneal over the long timespans of burial which are involved. As a particularly rapid example, Iron Age salt mines (~2500 y BP) have closed up completely at Hallstatt, Austria, and are recognized only by tools now trapped inside the massive rock.

The scale of the surficial karst that may be buried ranges across all the scales discussed in this book. At the small extreme is the shallow lagoonal limestone bed briefly exposed to vadose diagenesis and karren pitting before re-submergence and paraconformable deposition of new carbonate upon it. Thin conformable zones of breccia where salt or gypsum has been dissolved are the equivalent in sabkha facies. These are the categories of paleokarst most frequently described by sedimentologists; they are the most abundant but represent only short erosional intervals. At large scale are the wide range of corridor, doline, cockpit, tower or polje karsts with surface features and groundwater solution zones tens to hundreds of metres in their relief. Preserved examples are comparatively rare in the geological record. Intermediate in character and frequency of preservation are karsts of coastal plains or low plateaus, marked by the larger karren, shallow dolines or larger but also shallow depressions containing terrigenous deposits such as bauxite.

Glazek (in press) relates buried karst to plate tectonic history, pointing out that most large carbonate deposits were laid down on the passive margins of continents. Emergence (with little or no deformation) frequently is attributable to major tectonic events deforming the geoid elsewhere. An example is widespread emergence of limestones of upper Jurassic-lowest Cretaceous age (Malm–Neocomian) in central Europe later in the lower Cretaceous.

Preservation of recognizable karst features in the buried rock record is a matter of chance. It is best where karst processes are overwhelmed by rapid deposition, especially of terrigenous rocks, succeeded by prolonged subsidence. Volcanic ash and lavas bury karst most rapidly and may preserve its features most perfectly, as in Papua New Guinea where late Tertiary–early Quaternary karst near the Doma Peaks is buried by deep accumulations including lava dated at 0.85 Ma (Williams *et al.* 1972). Eolian sands, loess, fluviatile fan, estuarine and delta deposits more commonly bury karst, which often suffers moderate fluvial or marine erosion in the process. Particular effects of glacial burial have been noted above (section 10.3).

Much paleokarst has been buried by marine transgressive facies with little or no intervening terrigenous deposition. Such karst is often trimmed by wave action and, if it is of small to intermediate scale, may be removed entirely from localities where wave energy was high. This exacerbates problems of stratigraphic correlation of paleokarst horizons.

Some major paleokarsts

Bosak *et al.* (in press) have attempted the first world review of all paleokarst phenomena. Although preliminary, some major and widespread episodes of karst creation and burial may be recognized in it.

No paleokarst is reported in Archean rocks although, no doubt, detailed study will reveal examples in the future. Marine platform grike and small doline karst is reported from Lower Proterozoic dolomites in Canada, South Africa and the USSR. In the Upper Proterozoic such intermediate scale phenomena are more widely preserved, being reported also in Australia and China. Sililcate karst is quite common in quartzites of that age, perhaps reflecting more alkaline environments. There is a possible cockpit and mogote karst of latest Proterozic–basal Cambrian age in metamorphosed dolomites in Ontario, Canada. This developed before the spread of vascular plants i.e. before the conventional basis for the genesis of abundant soil CO_2 was in place.

In the Cambrian Period karst of intermediate scale and also deep brecciation is widespread in Siberian rocks. It includes early bauxites. The outstanding feature of the Ordovician is a complex, polyphase paleokarst of sub-continental extent over North America, the 'Post-Sauk karst'. It is associated with early Appalachian collisional deformation. A doline, cave and fissure karst with a relief of 15–60 m is well preserved in parts of Georgia and Tennessee and even serves as an important aquifer today. Elsewhere, it was truncated during marine transgression, and in eastern Canada is represented by intermediate to small scale, cyclic platform karst only. These latter types are also predominant in the Silurian of Eurasia and North America.

The Devonian is represented by many paleokarsts that are of major scale but known to extend over comparatively small regions only; one example is the reefal marine-mixing and collapse doline karst of Pine Point, shown in Figure 2.9. Many of these contain economic minerals or are oil and gas hosts.

The Devonian–Carboniferous boundary appears to be one of the greatest global periods of karstification. Major scale paleokarsts are known in European and Soviet Asia, associated with Variscan tectonism. The principal paleokarst of China developed, Lower Paleozoic carbonates being subject to erosion throughout the lower and middle Carboniferous, a timespan of ~40 Ma. As the Lower Carboniferous closed, the 'post-Kaskaskia' or second great paleokarstification of the United States commenced. It is best preserved in the midwest and cordilleran regions where it appears as a doline, fissure and cave karst with much red soil fill that is now indurated (Fig. 10.22).

The Permian record contains major but local paleokarsts in western Europe (Hercynian tectonics), Soviet Asia, China and Canada. Karstification continued into the Triassic in southern England where the Mendip Hills are a famous instance of a karstified semi-arid mountain range that was truncated and buried, and is now substantially exhumed. Karstification commenced in the middle or upper Triassic in eastern Europe and Soviet Asia. It was large scale and widespread in Yugoslavia, generally of lesser scale elsewhere.

The late Jurassic–early Cretaceous (Malm–Neocomian) is another important paleokarst phase, displaying some large-scale buried topography in Europe. Major karst bauxites were deposited in Greece. The middle Cretaceous (Aptian to Turonian in different localities) contains the most significant paleokarst of France and Italy, including the classical bauxites of Les Baux, Midi.

During the Cretaceous post-Gondwana tectonic events raised many platform carbonates into plateau or mountain topographies that have remained emergent. This is the case in much of China, Australia, North America, etc. Because they are emerged, these regions could not develop any further buried paleokarst except that of the most transient kind e.g. buried by ignimbrites (parts of Waitomo, New Zealand), loess (China), or glacial drift (Canada). Much of the European and Siberian platforms and the Mediterranean to Caspian Sea region, however, were subjected to one or more episodes of submergence during the Alpine and Himalayan collisional events. These led to creation of Paleogene and even Neogene (Miocene and Pliocene) paleokarsts of all scales and types that are, perhaps, the best known and studied buried karsts. In Belgium, France, Poland and Hungary, Paleocene doline karsts are buried by Eocene marine clays. On the Adriatic and Black Sea coasts Paleogene cave systems have now foundered to depths as great as −2000 to −3000 m below sea level. Because they contain groundwater (i.e. they are not entirely infilled by detritus) they must be classified as 'relict' rather than 'paleo' karst features. There is active groundwater circulation to at least −1000 m in the Romanian examples.

Bosak (in press) has attempted the first global mapping of Phanerozoic paleokarst, using one palinspastic map per geological period. Most reported paleokarst, was generated between 30°N and 30°S. This is the zone of greatest carbonate rock genesis and so necessarily includes all karst that is syndepositional or developed in very young strata. During the Triassic, Jurassic and Cretaceous karstification extended to 60°N and possibly to 80°S.

11 Karst resources, their exploitation and management

11.1 Karst hydrogeological mapping and water resources assessment

In the previous chapters of this book we have examined mainly the pure science aspects of karst. Here we conclude by discussing some of the important applications of the knowledge acquired. We commence this by considering water resources mapping and assessment.

It is estimated that 25% of the world's population, including many large cities and extensive rural areas depend on karst water supplies to sustain their daily activities. When the more obvious sources of water such as springs are already exploited and sustainable supplies are sometimes exceeded, so the need for scientifically based water resources investigation and management becomes pressing.

The first step in rational resource management and allocation is the objective assessment of the resource in question. A valuable and proven starting point for karst water resource surveys, especially in rocks with high secondary porosity, is speleological investigation (i.e. cave exploration and mapping) with associated water tracing, e.g. Waltham *et al.* (1985). This establishes the basic facts about catchment limits and drainage network geometry as discussed in Chapter 6. But even without this a quantitative estimate is required of annually renewable water resources. This can be derived using a climatic water balance technique (see section 6.9). This approach adopts a reference period, such as a water year (dry season to dry season), and a well defined reference area, such as a representative basin or a karst district with known boundary conditions. Error terms associated with the calculation of the annual water surplus in this way are typically quite large though seldom evaluated; probably of the order of 25% or more. But the estimate can sometimes be cross-checked by a complementary technique based on spring hydrograph analysis (see section 6.5), provided the watershed of the basin feeding the spring is accurately known. The technique usually expresses exploitable resources as a volume in storage or as a mean annual discharge, but can be converted to equivalent depth of runoff if the basin area is known. These and other approaches to the evaluation of karst water resources are presented in volumes edited by

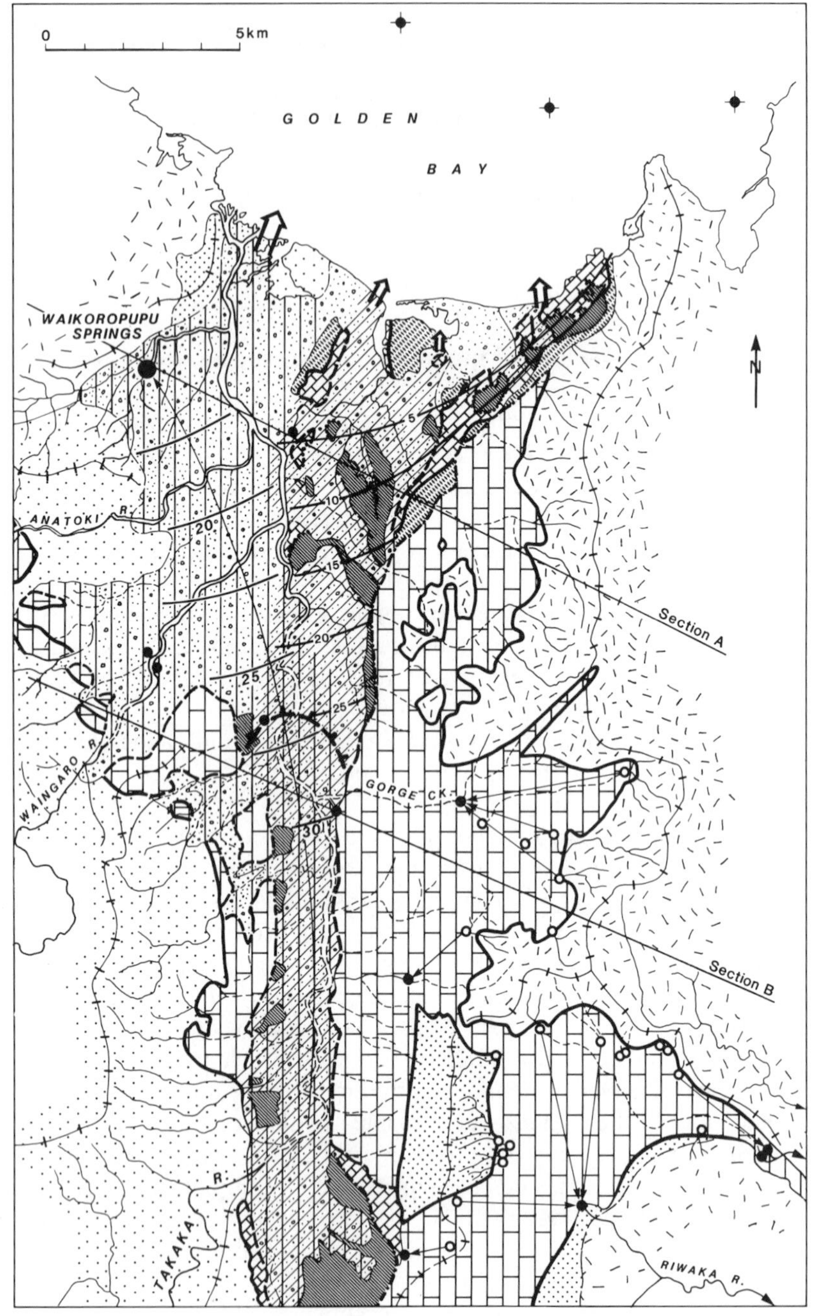

0
5km
GOLDEN
BAY
WAIKOROPUPU
SPRINGS
ANATOKI R.
WAINGARO R.
GORGE CK.
TAKAKA R.
RIWAKA R.
Section A
Section B
N
5
10
15
20
25
30

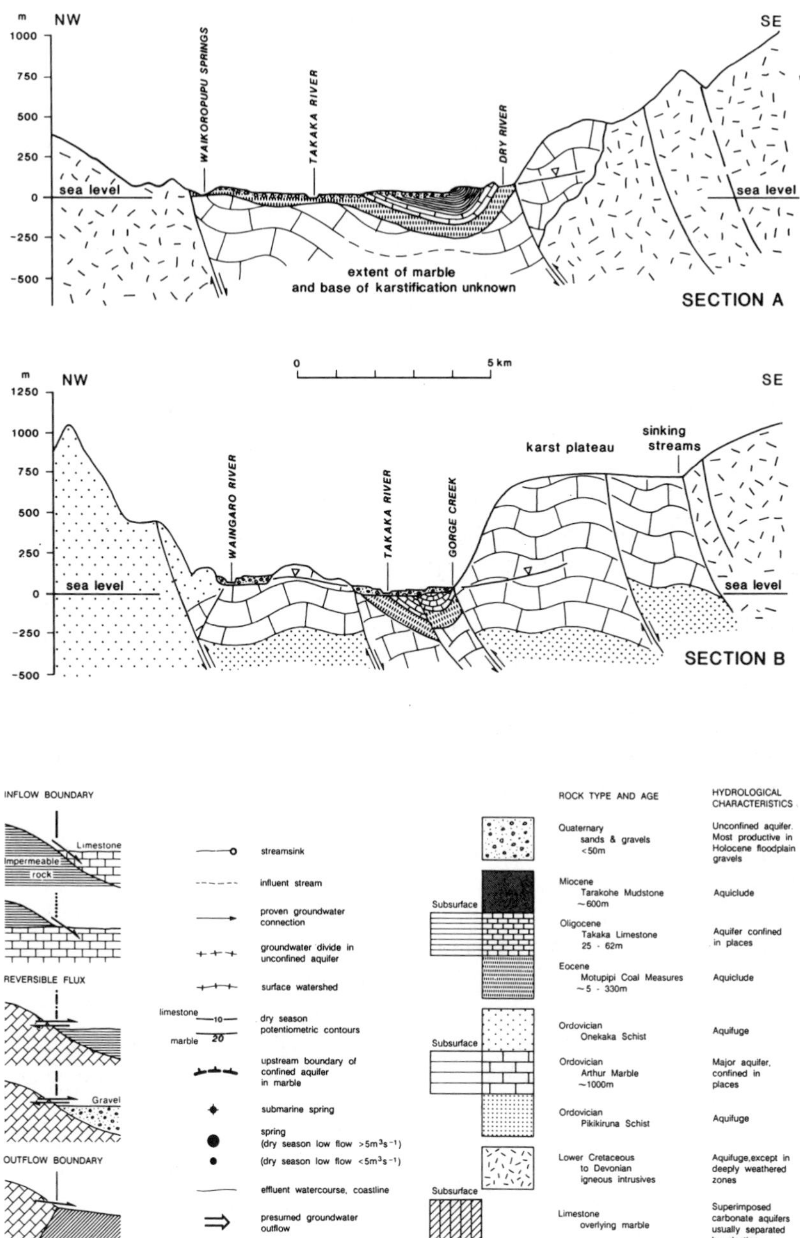

Figure 11.1 *Left and right* – Hydrogeologic map of the Takaka region, South Island of New Zealand. The main aquifer is in marble with a possible total water volume of 3.8 km^3, partly under confined conditions. A second karst aquifer occurs in overlying limestones, also partly confined (data from Grindley 1971, Williams 1977, Mueller 1987).

Burger & Dubertret (1984) and LaMoreaux *et al.* (1984).

Having obtained data of known accuracy on available water resources, its wise allocation next depends largely on the clarity with which scientists communicate this information to decision makers. An efficient way of transmitting this knowledge is through the medium of hydrogeological maps with supporting explanatory notes.

Hydrogeological maps are of various types. They can be prepared at several scales. And they may have a multiplicity of end users, from politicians and planners to hydraulic engineers, well drillers, hydrogeologists and teachers (Paloc 1975, Paloc & Margat 1985). The preparation of any hydrogeological map must therefore take into account the likely variation in scientific expertise of its principal users. Small scale maps are appropriate for atlases, such as the *Hydrogeological Atlas of the People's Republic of China*, which features a map of karst regions in the south of the country. These synthesize and generalize available information. Other excellent examples are of karst water resources in Mediterranean France at about 1 : 760 000 by Drogue *et al.* (1983) and the *Hydrogeological Map of France* at 1 : 1.5 M by Margat (1980). Small scale maps are necessarily imprecise, but are still useful for reconnaissance planning. A range of special purpose maps may present data on water quantity, water quality, pollution and susceptibility, and individual aquifers.

Larger scale maps at 1 : 10 000 to 1 : 200 000 are of greater technical interest. At this scale issues of content and its representation also become more important. Three types of hydrogeological data are basic to all maps of this kind: information on permeability, aquifer geometry, and hydraulic regime. What information can be mapped depends on what is available (Fig. 11.1). And the more abundant the data the larger the scale map needed to convey it clearly. Depending on the end users, it may also be helpful to present other information that complements the hydrogeological data. Relevant material may include topography, rainfall, surface streams, wells, dams, water chemistry, and waste disposal sites. Some of this can be depicted on small scale insets.

Maps are conceptual representations of the real world. Hence the way in which information is depicted is important for the ideas conveyed. This is not just a matter of the symbols used, but of the mapping objectives. Paloc & Margat (1985) point out that two complementary approaches have been proposed, one more *hydrogeolithological* in emphasis and the other one *hydrogeodynamic*. The first is the more common and led to a recommended international legend (UNESCO 1970). The second is being developed and is illustrated by Margat's (1980) map of France. Various colour conventions have also been adopted in these schemes.

The hydrogeolithological approach superimposes three kinds of information:

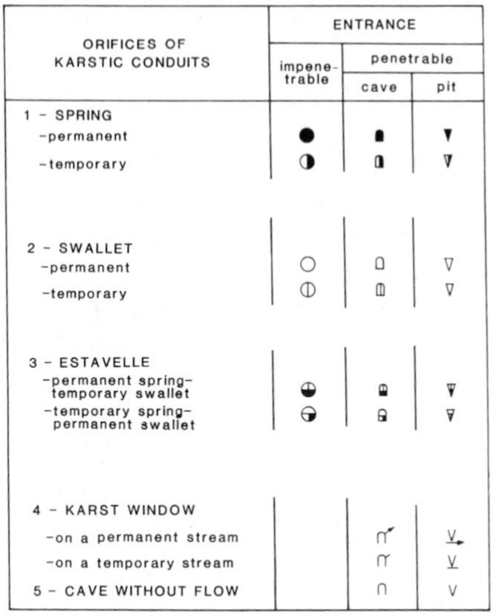

ORIFICES OF KARSTIC CONDUITS	ENTRANCE		
	impene-trable	penetrable	
		cave	pit
1 - SPRING			
-permanent	●	∎	▼
-temporary	◑	◧	▼
2 - SWALLET			
-permanent	○	▯	▽
-temporary	⦶	◫	▽
3 - ESTAVELLE			
-permanent spring-temporary swallet	◒	◒	▼
-temporary spring-permanent swallet	◓	◓	▼
4 - KARST WINDOW			
-on a permanent stream		∩	V
-on a temporary stream		∩	V
5 - CAVE WITHOUT FLOW		∩	V

Figure 11.2 Some symbols used on large scale hydrogeological maps. From Paloc & Margat (1985).

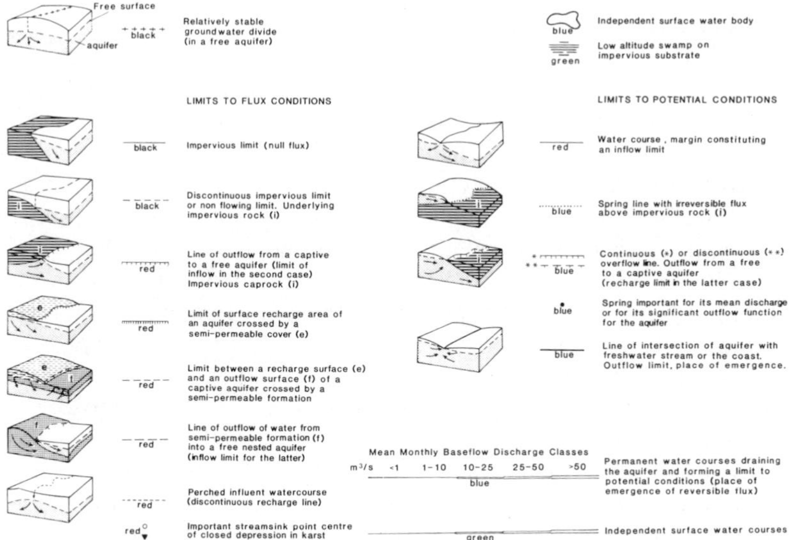

Figure 11.3 Some symbols used in hydrogeological mapping to convey dynamic conditions. From Margat (1980).

(a) lithological types that are assumed to represent permeability classes;
(b) piezometric data from which groundwater flow may be inferred; and
(c) surface hydrography with data on sites of water exploitation.

For karst areas, Paloc & Margat (1985) emphasize the importance of classifying baseflow discharges of springs and surface streams. Some symbols used are illustrated in Fig. 11.2.

The hydrogeodynamic approach conveys information concerning:

(a) the constitution of aquifer systems, based on the distinction between and location of the principal rock bodies (taking into account the degree to which they are water-bearing and their possible layering); and
(b) the boundary conditions of aquifers, distinguishing between
    (i) the direction of water flow (input, output, or static) and
    (ii) flow conditions as opposed to potential conditions.

Suggested symbols are shown in Fig. 11.3.

Other special purpose maps are also frequently used in karst hydrology, especially to convey information on point-to-point connections determined from water tracing experiments. Fig. 6.31 for the Sinkhole Plain in Kentucky is such a case, although the original has much more detail than could be shown here.

11.2 Pollution of karst aquifers

Karst areas are susceptible to a greater range of environmental impact problems than any other terrain, because of the additional set of difficulties associated with highly developed subterranean networks and their associated fragile ecosystems. Unfortunately, in all inhabited karsts, dolines and other sinkpoints are perceived as being particularly suited for the dumping of solid or liquid waste, because it disappears underground and 'out of sight is out of mind!' Yuan's (1983) review of problems of protection of karst in China illustrates the point. The most important of the many impacts encountered is that arising from water pollution.

Karst aquifers are notoriously effective in transmitting rather than treating pollutants. This arises from the unfortunate fact that the relatively large capacity for self-treatment found in many groundwater systems is comparatively poorly developed in karst. The nature of subsurface purification processes is not well known but has been investigated by Golwer (1983), who found it to be composed of the interaction of numerous physical, chemical and biological reactions (Fig. 11.4), and to be significantly affected by transport processes and hydrogeological conditions. The natural treat-

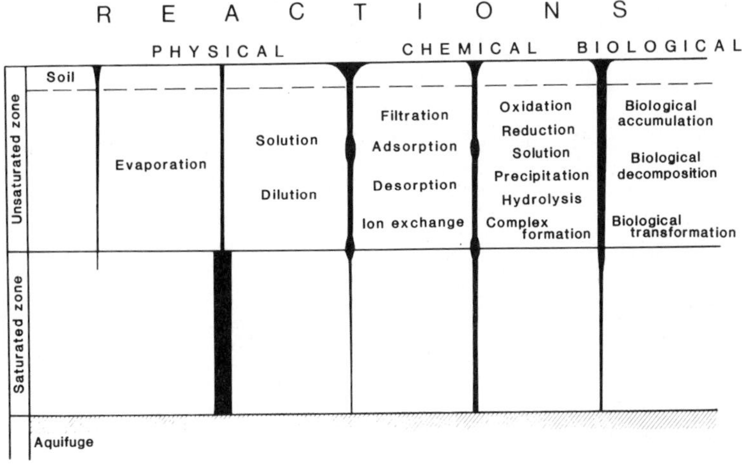

Figure 11.4 Schematic representation of subsurface reactions giving rise to self-purification of infiltrating water. From Golwer (1983).

ment of water-borne contaminants in karst is relatively ineffective, because:

(a) The surface area available for colonization of natural microorganisms as well as for adsorption and ion exchange is much less in dense, fractured karst rocks than in porous clastic sediments.
(b) Rapid infiltration into karst reduces the opportunity for evaporation, a mechanism that is important in the elimination of highly volatile organic compounds, such as solvents and many pesticides.
(c) Physical filtration is relatively ineffective in typically thin karst soils and through rocks with large secondary voids; thus sediment and micro-organisms are readily transported into karst aquifers.
(d) Transmission of particulate matter right through karst systems is assisted by the turbulent flow regime commonly associated with conduit aquifers.
(e) Time-dependent elimination mechanisms (of bacteria and viruses for example) are curtailed in effectiveness because of rapid flow through times in conduits and reduced retardation by adsorption–desorption processes.

In laboratory experiments to assess the survival of nine bacteria in water at 10 ± 1°C, Kaddu-Mulindwa *et al.* (1983) found that *Escherischia coli*, *Salmonella typhimurium*, *Pseudomonas aeruginosa* and other pathogenic or potential pathogenic bacteria survived up to 100 days or even longer in sterilized natural groundwater. Only two bacteria did not survive 10–30 days (Fig. 11.5). Given that the transmission of water through most conduit

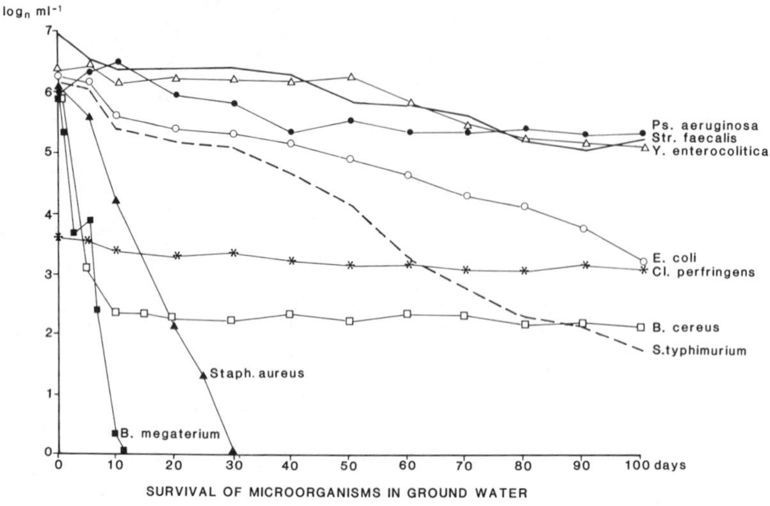

Figure 11.5 Survival of microorganisms in natural groundwater at 10 ± 1°C. From Kaddu-Mulindwa *et al.* (1983).

aquifers is shorter than this (e.g. see Table 5.7 for the flow through times of some of the longest karst systems) and given the neutral buoyancy of bacteria, then the probability of spreading contaminated water through karst is high and potentially serious.

Quinlan (1983) and Quinlan & Ewers (1985) describe how sewage, creamery waste, and heavy metal containing effluent from the Horse Cave area of the Sinkhole Plain in Kentucky (Fig. 6.31) has been spread by a karst distributary system to 56 springs in 16 areas along an 8 km reach of the Green River. In spite of that, none of the 23 wells sampled had contaminated water, presumably because they are up gradient from the conduits or have no direct connection to them. This would give a false sense of security concerning the quality of the total groundwater system, if the condition of discharge through the conduits were not also known. This led the authors to stress that reliable monitoring of water quality in karst must involve springs, not just wells, because they are places where flow converges at basin outfall points (Quinlan & Ewers 1986). This is a vital message in highly karstified fissure aquifers, but becomes of lesser significance in more granular limestones, as the following example shows.

Waterhouse (1984) successfully mapped nitrate contamination in a Tertiary limestone aquifer in the Mount Gambier district of South Australia, using data from 257 wells. Hydraulic conductivities for the bulk aquifer are consistent with coarse gravels, hydraulic gradients are low, and resultant groundwater flow is dominantly intergranular and Darcian, despite the widespread occurrence of fissures and large, flooded collapse dolines

(Fig. 9.19B). Until recently all domestic, industrial and agricultural effluent was disposed of underground. A cheese factory, for instance, discharged 30–50 kl per day of waste into a cave for at least 25 years. A typical polluted groundwater sample from the region has a nitrate concentration of 300 mg l^{-1} compared to uncontaminated water with only 2 mg l^{-1} (concentrations in drinking water are usually considered acceptable if they do not exceed 10–15 mg l^{-1}). In spite of the long term effluent loading, the lower part of the aquifer still remains substantially free of contaminants, showing the effectiveness of the purification mechanisms noted in Figure 11.4 in this relatively porous diffuse flow aquifer.

In view of the above evidence, it appears paradoxical and alarming that karst rocks are being investigated as potential sites for hazardous waste disposal. Regions with anhydrite and rock salt are particularly favoured in these enquiries, the principal argument being that if significant groundwater flow had affected them, then the anhydrite would already be hydrated to gypsum and the highly soluble rock salt would show obvious signs of solution. Hence wastes stored within such rocks are not likely to be dispersed by groundwater. However, in the case of nuclear substances, the time scale is very long and the rate of advance of solution fronts becomes relevant. The US Department of Energy (1982) and Vierhuff (1983) discuss requirements for mined repositories of nuclear wastes. Bachman (1987) describes site investigations in evaporite karst in New Mexico.

11.3 Problems of construction on and in karst rocks – expect the unexpected!

Karst processes and landforms pose many different problems for construction and other economic development. Every nation with karst rocks has its share of embarrassing failures such as collapse of buildings or construction of reservoirs that never held water (e.g. Fig. 11.7). It is probably true to write that the global cost of extra preventive measures or of unanticipated remedial measures in karst terrains now amounts to some thousands of millions of dollars (US) each year.

The problems encountered can be classified by the extent of the impact of construction or other development upon karst features already existing at a site. There may be no impact, as in the case of construction in the path of a potential landslide. There can be small- to large-scale impact where foundations for bridges, buildings, roads and railways are placed upon karst without much effect upon the local water table. Large-scale impact is more common if the water table is raised or lowered; this may become extreme in the construction of dams, tunnels and mines.

Rock slide-avalanche hazards in karst

A *rock slide-avalanche* is the catastrophically rapid fall or slide of large masses of fragmented bedrock such as limestone (Cruden 1985). 'Landslide' is more widely used but is also applied to slides of unconsolidated rocks (Rouse 1984). Rock slides take place at *penetrative discontinuities*, the mechanical engineering term for any kind of surface of failure within a mass.

Carbonate rocks and gypsum are especially prone for two reasons: (a) while faults and joints are the more important penetrative discontinuities in most other rocks, in karst strata there is also major penetration via bedding planes. Often they are particularly favoured as surfaces of failure (Fig. 11.6). (b) large quantities of water may pass rapidly through the rock via its karstic discontinuities to saturate or lubricate underlying weak or impermeable strata. This induces creep or sliding along the contact. For example, such sliding contributed to the Vaiont Dam disaster in Italy.

The forces that resist catastrophic failure within a particular rock are defined by an *internal angle of friction*; see e.g. Brady & Brown (1985). Minimum angles for relatively hard carbonates without shale interbeds range from 14 to 32°. Slab slides are particularly common because they are bedding plane failures. Rotational failures are comparatively rare but there are big ones in dolomites in the Mackenzie Mountains, Canada. The most dangerous situations are the overdip slope and the toppling cliff (Fig. 11.6);

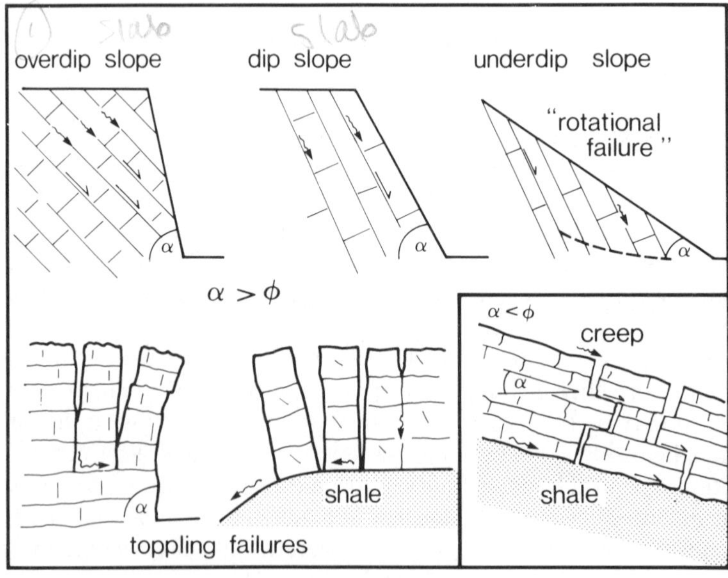

Figure 11.6 Types of rock slide-avalanches or landslides in carbonate rocks. Ø is the internal angle of friction of the rock. Failures on dip and overdip slopes are termed 'slab slides'.

the former usually will involve greater volumes of rock.

Steeply dipping carbonates predominate in the Rocky Mountains of Alberta. On average, there have been one or two rock slides in each 100 km^2 there since regional deglaciation ~ 10 000 years ago (Cruden 1985). The volumes of rock detached and the areas that they bury are found to approximate a Poisson distribution. The largest slide contained 30 million cubic metres of limestone; it buried an area of ~ 3 km^2 to a mean depth of 14 m in a timespan of only 100 s. This was the Frank Slide, which occurred on the night of 29 April 1904, crushing a small town and taking 70 lives.

Setting foundations for buildings, bridges, etc.

Setting foundations where there will be little or no effect upon the natural karst drainage encounters problems with existing cavities. The most important class of cavities are arches in soil that mantles a maturely dissected epikarst (Fig. 11.7). Figure 11.8 illustrates different methods that are used to overcome them. Much has been written on the subject; see Beck (1984), Beck & Wilson (1987) for recent surveys.

Reitz & Eskridge (1977) emphasize that, on flat surfaces where strata are

Figure 11.7 Subsidence of a dwelling house into a collapse sinkhole in soil mantling epikarst, New Jersey. The displacement occurred in three–four hours. Photo by Rick Rader, Philipsburg, NJ, by permission.

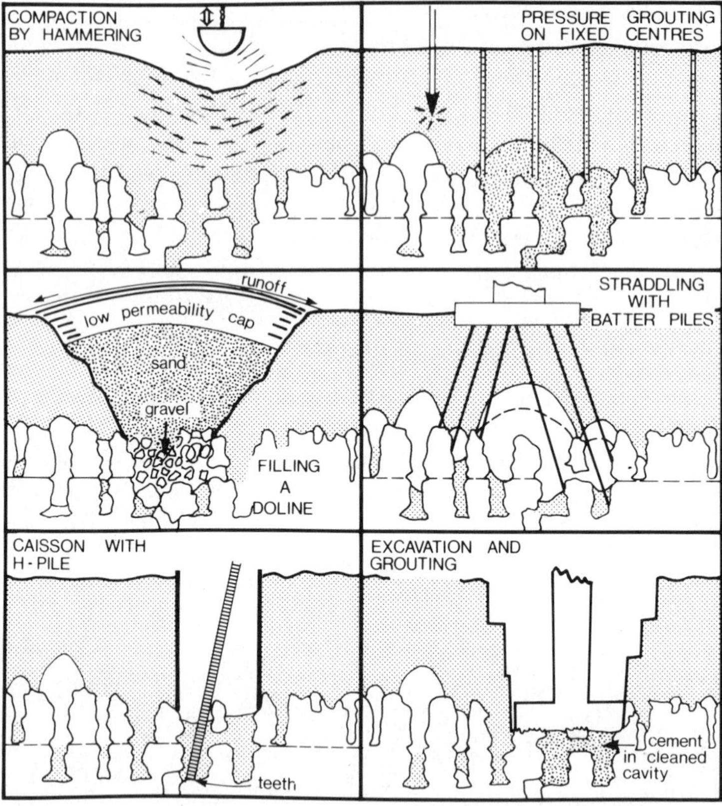

Figure 11.8 Illustrations of the principal types of foundation treatments in a soil-mantled karst. Based upon figures in Sowers (1984).

horizontally bedded, the rock mass must be made continuous beneath the foundation to a depth equivalent to 0.6 or 0.7 of the column spacing in the planned structure. Remedial work should go deeper where strata are inclined or folded, and on slopes.

Cavities entirely within bedrock can pose dangers if they are at very shallow depth or if the planned stuctural load is considerable. Otherwise, they are generally ignored. Waltham *et al.* (1986) provide an example of the problem of high loading. Severe difficulties were encountered in footings for four piers for a motorway viaduct in Belgium, increasing its overall cost by 15%. One pier had to be shifted 15 m onto stronger rock. A standard programme of exploratory drilling had missed cavities 3 m wide.

Catastrophic doline development induced by groundwater extraction, by construction or other activities

The development of closed depressions so rapidly that structures and other installations threatened by them cannot begin to be protected until the holes are fully enlarged and quiescent, (or nearly so) may be defined as *catastrophic* (Fig. 11.9). In the extreme, they form within a few seconds and lives may be lost. Development is attributable primarily to collapse of tension domes (arches) created by suffosion of mantling sediments into underlying karst cavities. In very sandy soils there may be simple funnelling suffosion without arch formation. Settling *en masse* within an encircling fault plane occurs where there is steady compaction instead of void and dome formation.

The depressions typically range in dimension from ~ 1.0 m diameter and 0.5 m deep up to 100–200 m diameter and 10–50 m deep. Depressions in the smaller half of the local range are numerically predominant in most afflicted areas. Apart from overuse or pollution of aquifers, this is perhaps the most widespread problem encountered in karst terrains. Beck (1984) presents many sample studies, chiefly from North America.

The principal cause of suffosion with arch formation is lowering of the water table. This is most hazardous where it is lowered from a position in the mantling sediments deep into the epikarst zone or beneath it (Foose & Humphreville 1979); in turn, this implies that the particular karst will be shallowly buried or relict e.g. on the floodplain in a polje or between isolated towers in a tower karst.

In Alabama there have been 1700 collapses around five wellfield cones of depression in recent decades. In the Ukraine 1000 small collapses occurred in 1 km^2 about a particular well. At coal mines in Hunan (China) there have been 8500 collapses in this century attributed to drawdown. Every nation with karst can produce similar statistics.

A second major cause is the focusing of runoff from manmade impermeable surfaces such as roofs, highways and parking lots, into particular sinkpoints in the epikarst. This is doubly dangerous where the water table has been lowered also, as is often the case where comprehensive construction is in progress. At the small scale, runoff from a domestic roof may induce collapse beneath the downspout e.g. in Florida, where remedial action for a modest dwelling cost $20 000 in 1982. At larger scale, collapse induced by highway runoff may threaten adjoining property and road embankments, and even cause collapses in the road surfaces. There have been 6–18 such collapses each month in Florida during recent decades; their actual timing correlates poorly with antecedent precipitation, which makes prediction of time and place very difficult.

Leakage of water from canals, buried water pipes and sewage lines is a special case of focusing runoff that can be particularly rapid in its effect

Figure 11.9 The Winter Park suffosion doline, Florida. This feature developed during a 72-h period in May 1981, consuming a dwelling, part of a road, automobiles, etc. It developed in 30 m of sands and clayey sands burying densely karstified limestone and is believed to have been induced when the water table was lowered 6 m within the sands. The feature was 106 m in diameter and 30 m deep (top photo–May 1981). The bottom photo (August 1984) shows the doline after remedial action, including restoration of the water table to its original elevation.

because of the large volumes of flow that may be involved before the leak is detected. One unfortunate example was the collapse of a two-year old sewage lagoon built on glacial drift in Minnesota; it dumped 70 000 m^3 of effluent into an underlying karst aquifer (Alexander & Book 1984).

Raising the water table can also create collapse or subsidence by destroying the cohesion of susceptible clay soils. However, this is comparatively rare in karst areas. The load or vibration from heavy equipment can induce small collapses locally, especially beneath it. In historic times plowhorse teams have dropped; in modern times many tractors, haulage trucks, drilling rigs and military tanks have fallen. Rock blasting from quarries or foundation cutting, etc. often causes collapses of small to intermediate scale.

The solution mining of salt and use of waterflooding to enhance recovery from oil and gas reservoirs is a final cause of subsidence and collapse, especially the former. Pressure injection of water into the tops of salt diapirs has proved to be particularly dangerous.

It is obviously desirable to detect the cavities before they have stoped to the surface. Thought and money must be devoted to careful exploration before designs are finalized and construction begins in a karst area. At the reconnaissance scale aerial photography will display already existing dolines. Thermal infrared imagery is particularly useful because it may reveal patches of vegetation that are showing drought stress due to arches advancing beneath them; it also indicates sites of focused soil drainage into the epikarst.

Surface geophysical techniques are widely used. Kirk & Werner (1981) review them. Traditional methods may be divided into three groups – low energy ('hammer') seismology, electrical resistivity and gravimetry. These yield data for points along a traverse. Electromagnetic conductivity (EM) and ground penetrating radar (GPR) are new methods that provide continuous readings along a traverse, with the result that cavities are less likely to be missed. All techniques are imperfect, however. Use of a combination of them may reduce the error; nevertheless correct identification of shallow but waterfilled cavities and of any features under deep cover remain problematic. We endorse Waltham *et al.* (1986) when they write

> Geophysical exploration of cavernous limestone is very difficult to interpret, and normally can act only as an aid to efficient planning of a more conclusive drilling programme.

Design of exploratory drilling patterns is discussed by Benson & La Fountain (1984) and Newton (1984). In general, the cheaper percussion ('airtrack') drilling is preferable to core drilling because it detects cavities as readily. Once holes are drilled, downhole techniques (section 6.4) may yield useful information. Drilling is essential beneath any expensive or otherwise important structure, such as the pier of a bridge. Benson & La Fountain (1984) question its cost effectiveness beneath lesser structures, based on the likelihood of missing cavities when drilling on a fixed pattern. They point out that 1000 3-cm holes per acre (0.4 ha) are necessary to have a 90% chance of intercepting a cavity that is 2.5 m in diameter or less.

As an example of exploration difficulties, at the site for a commercial tower building in New Jersey there were 3000 m of seismic traverse, a gravimetric survey and 15 drill holes. No cavities were detected and excavation for foundations began. Problems arose almost immediately. A further 80 bore holes and 36 test pits were dug, encountering partly soil-filled cavities in all but one case. Extensive redesign was necessary.

The types of remedial measures used where foundations are to be emplaced are shown in Figure 11.8. In other circumstances the best remedy is to restore the former natural conditions i.e. return the water table to its previous elevation and reduce or eliminate sites of focused runoff. This will not arrest all stoping collapses that were initiated by the interference. Filling of the dolines is the standard treatment once they are revealed, using whatever material is cheapest in the locality. Costs for filling or merely

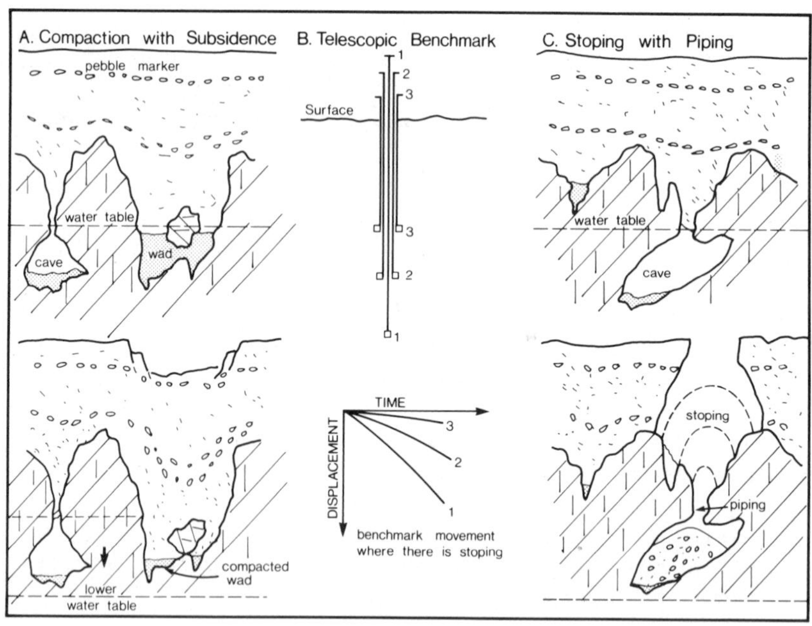

Figure 11.10 Catastrophic doline formation in the Far West Rand gold mining district, South Africa. A. Subsidence doline created by the compaction of dolomitic residuum ('wad') that occurs upon dewatering. B. The design of telescopic benchmarks. Where an arched cavity is stoping upwards (C), the deepest benchmark will move first and furthest, as in the Figure. Where compaction is occurring the shallow benchmark (No. 3) moves first and furthest; thus, subsidence and the more dangerous collapse can be discriminated. Adapted from Brink (1984).

stabilising the slopes of a large doline often will exceed 1 million dollars in the USA.

Perhaps the most celebrated examples of catastrophic doline collapse in this century are those of the Far West Rand goldfields in South Africa. Brink (1984) gives an entry to the extensive literature on the subject. Residuum and other unconsolidated sediments deeply bury a rugged corridor-and-pinnacle karst in dolomites that are underlain by gold-bearing quartzites. To assist in mining, the water table was lowered into the dolomites (Fig. 11.10). The first collapses came shortly after and were truly catastrophic. In the worst incident (December, 1962) a three-storey crusher plant with 29 occupants was lost in a few seconds at the West Dreifontein Mine. Geophysical exploration for new cavities has proved to be nearly useless (Brink, op. cit. p. 125). Arrays of simple telescopic benchmarks (Fig. 11.10) have been set out in hopes of detecting stoping arches before they breach the surface. Wagener & Day (1986) discuss construction techniques on these rocks.

Problems of dam construction on carbonate rocks

Dams and reservoirs for the generation of hydro-electricity, for water storage or for flood control are common on carbonate rocks. Many of them exceed 100 m in height and a few are greater than 200 m. Limestone or dolomite are chosen because, in most sedimentary terrains, they fulfil two of the three basic requirements for a dam site much better than other strata do. These requirements are (a) that at the site the valley is narrow and with steep sides, in order to minimize the quantity of construction material required, and (b) that the rock possesses sufficient mechanical strength to support the load of the dam and impounded water. Limestones and dolomites contain particularly narrow canyons precisely because tributary waters are routed underground and thus will not indent the valley flanks and, as shown in section 2.6, they are stronger than most shales and sandstones. But because they are karstified they may not meet the third basic requirement – that the rock in the foundations and abutments should be impermeable. When a carbonate site is selected, designers plan to overcome this problem by applying special treatments.

The evident dangers of dam construction are that by raising the water table, extremely steep (unnaturally steep) hydraulic gradients are created across the foundation and abutment rock with unnatural rapidity, and an unnaturally large supply of water is then provided that may follow those gradients. In any karst this must be a hazardous undertaking.

The Hales Bar Dam, Tennessee, is a notorious example of a simple and immediate response to raising the water table. There was leakage directly under the dam i.e. where hydraulic gradient was greatest. Building plans called for dam construction in a period of two years at a cost of 3 million

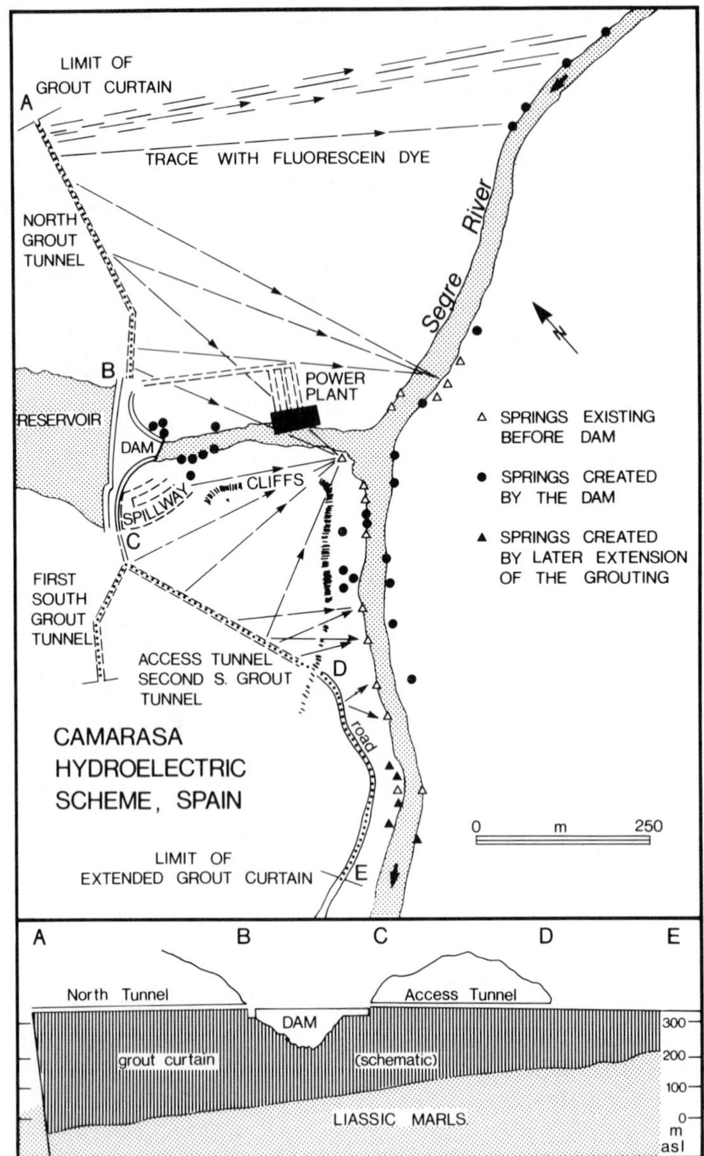

Figure 11.11 Camarasa hydroelectric dam, Noguera, Spain. The dam is built on thick–massive dolomites dipping at 17° in the upstream direction and with an aquiclude stratum of marls at shallow depth beneath them. The dam is a gravity arc concrete structure 92 m in height and 377 m in length at the crest. It and the abutments were grouted to the marls via North and South grout tunnels. Twenty five new springs were created when the reservoir filled, with an aggregate flow of ~ 12 m^3s^{-1}, an acceptable amount. From Thérond (1972).

dollars. Because of initial foundation problems in karst, this was extended to eight years and 11.5 million dollars. Two weeks after filling (November 1913) the first serious leaks appeared. Expensive remedial treatments continued for 30 years before leakage was reduced to an acceptable level (TVA 1949). Camarasa Dam, Spain (Fig. 11.11) is a good example of unexpected leakage via both the foundation and abutments.

Not only may modern fissures and conduits be flushed and enlarged at accelerated rates, but long-dormant relict karst and even filled and indurated paleokarst may be reactivated. Thérond (1972) cites an instance where raising the water level to +75 m induced only 1.6 m^3s^{-1} of leakage downstream; this increased to an unacceptable 8 m^3s^{-1} when the level was raised to +85 m. Evidently, sediment-filled conduits (undetected in the exploration) were flushed and reactivated. This emphasizes that conditions of hydraulic conductivity, etc. that are established by even the most detailed investigation during the site exploration stage, may change radically as the reservoir fills.

While leakage through dam foundations and abutments is most feared it is quite possible that there may be lateral leakage elsewhere in a reservoir. This is to say that problems with karst can arise where the dam itself is built on some other rock but there are karst rocks upstream of it. Montjaques Dam, Spain, was built to inundate a polje. It failed by leaking through tributary passages and the scheme was abandoned (Thérond 1972).

Karst problems with dams are summarized in some engineering design and construction textbooks. The Tennessee Valley Authority has accumulated vast experience of building them, chiefly in comparatively simple structural and lithologic situations. Their main report (TVA 1949) is still pertinent and Soderberg (1979) gives a more recent review of the work. Thérond (1972) and Eraso (in press) discuss European experience, which generally has been with more complex, mountainous sites.

Thérond (1972) identifies seven different major factors that contribute to the general problem. These are – type of lithology, type of geologic structure, extent of fracturing, nature and extent of karstification, physiography, hydrogeologic situation, and the type of dam to be built. For each factor, clearly, there are a number of significantly different conditions. In Thérond's estimation, these together yield a combination of 7680 distinct situations that could arise at dam sites on carbonate rocks! It follows that dam design exploration and construction must be specific to the particular site –

> 'There is perhaps no phase of engineering or construction which lends itself less readily to rule-of-thumb or handbook methods than does (dam) foundation' (TVA 1949, p. 93). 'When dealing with karstic foundations *all* geologic features must be evaluated no matter how small or insignificant they appear' and 'design as you construct' (Soderberg 1979, p. 425).

Table 11.1 Examples of dam grouting costs in the Spanish-speaking countries. From Eraso (in press).

	Grouting material (kg m^{-2})	Approximate cost (1980 US dollars m^{-2})	
Minimum	41	40	Mountain karst
Maximum	3500	500	
Average	366	105	
Minimum	30	25	Other faulted karst
Maximum	3000	300	
Average	343	81	
Minimum	11	12	Epikarst
Maximum	66	50	
Average	24	23	
Minimum	2	5	Karst beds alternate with others
Maximum	170	80	
Average	44	42	

If dam sites with limestone or dolomite must be used, the preferred location will be where there is a simple geologic structure, the least karst development and fissure frequency, and where there are shales or some other aquiclude strata at shallow depth so that a grout curtain can be extended to them economically (Fig. 11.11). At the site, all authorities agree that a large number of exploration boreholes (with core retrieval and downhole technology) is mandatory. Large cavities should be mined into, if possible. Natural fill should be removed from caves and replaced with plugs of grout. It is now standard practice to remove the epikarst completely from a centre line beneath the dam and its abutments. This may require trenching to 10–20 m or more.

Grout curtains are essentially dams built within the rock. Their design and density are discussed by the authors listed here, and many others. The normal practice is to place a main grout curtain beneath the dam, in the abutments and on the flanks (Fig. 11.11). A cut-off trench and second, denser curtain may be placed upstream in the foundation if there are grave problems there. In the main curtain a first line of airtrack grout holes will be placed on centres never more than 8–10 m apart and filled until there is back pressure. A second, offset line of holes is then placed and filled between them. Third and fourth lines may be used until the spacing reaches a desirable minimum that is normally not more than 2 m. Standard grouts are cement or bentonite (a clay that expands when wetted) and mixtures of the two, made up as slurries with differing proportions of water. Table 11.1 cites the experience of Eraso (in press) in a number of differing situations in the Spanish-speaking nations.

The Normandy Dam, Tennessee, provides a good example of exploration and grouting at a comparatively simple site. It is an earth fill dam 34 m in height, footed on horizontal limestones. In preliminary exploration, 4400 m of core holes were drilled. In a problem zone 80 m wide, 25 cm diameter holes were drilled on 50 cm centres for inspection by downhole TV camera. The epikarst was then removed along a cut-off trench 6–12 m in depth. Beneath it 100 cm grout holes were emplaced on 120 cm centres (i.e. overlapping), and 12 cm holes were high pressure grouted upstream and downstream of this line. The main grout curtain was then emplaced, on 3 m centres and to a depth of 25 m. All holes that accepted grout were reinforced by 1–3 further holes. The treatment was successful (Soderberg 1979).

All springs and piezometers must be monitored carefully as the reservoir fills behind a completed dam. Operators should be prepared to halt filling and drain the reservoir as soon as serious problems appear. In extreme cases the reservoir floor and sides may be sealed off e.g. by plastic sheeting.

Experience shows that remedial measures after a dam has been completed and tested are much more costly than dense grouting during construction.

Dam construction on gypsum and anhydrite

The surest principle here is to avoid construction on these rocks. The problems that may be encountered are considerable (James & Lupton 1978, James & Kirkpatrick 1980). These include the rapid enlargement of existing conduits and the creation of new ones, because hydraulic gradients are excessive (Fig. 3.19); the settling or collapse of foundations or abutments where gypsum is weakened by solution; heave of foundations where anhydrite is hydrated, and attack by sulphate-rich waters upon concrete in the dam itself (Eqn 3.60).

In the United States experience has been gained in simple geologic, low relief terrains in west Texas and New Mexico, where truly excessive hydraulic gradients and the problems of complex structure can be avoided. At one celebrated site, McMillan Dam, gypsum is present only in the abutments, further reducing the difficulties. No caves were detected when it was built in 1893; nevertheless, the reservoir drained dry via caves through the lefthand abutment within 12 years. Attempts to seal off the leaking area by a coffer dam failed because new caves developed upstream of it. Between 1893 and 1942 it is estimated that 50×10^6 m^3 of dissolution channels were created (James & Lupton 1978). McMillan Dam is now abandoned, as are nearby Avalon Dam and Hondo Dam. Dams can be built successfully in gypsum terrains where relief is low and geology simple (or where there are gypsum interbeds in carbonate strata), but comprehensive grouting is necessary and an impermeable covering over all gypsum outcrops is desirable (Pechorkin 1986). Periodic draining and regrouting will probably be needed also.

Dams in mountainous country appear to be especially hazardous. In California, St Francis Dam was built in 1928 on conglomerate containing clay and a high frequency of gypsum veinlets. Shortly after filling an abutment collapsed catastrophically; 400 persons were killed downstream. Collapse was attributed to dissolution of the gypsum which weakened the mechanical structure and permitted lubrication of the clay. This is an instance of failure in a rock that would not be classified as karstic at all, merely having a low proportion of gypsum ($<5\%$?) as fracture filling.

Tunnels and mines in karst rocks

Tunnels and mine galleries (adits or levels) will be cut through rocks in one of three hydrogeological conditions; (a) vadose; (b) phreatic but at shallow depth or where discharge is limited, so that the tunnel serves as a transient drain that permanently draws down the water table along its course (Fig. 5.8b); (c) phreatic, as a steady state drain (Fig. 5.8a) i.e. permanently waterfilled unless steps are taken to drain it. Long tunnels in mountainous country may start in the vadose zone at each end but pass into a transient zone, or even a steady state phreatic zone, in their central parts.

Vadose and transient zone tunnels are cut on a gentle incline to permit them to drain gravitationally. It was by this means that lead/zinc ore bodies within interfluves in the hilly country of Bohemia and Derbyshire were drained below the natural water tables in the 15th and 16th centuries in order to mine them. A modern example is the water supply tunnel of the city of Yalta in the Crimea. It passes through faulted limestone as a transient phreatic drain, and is 7 km in length with a fall of 50 m. Groundwater discharges were $\sim 1000\ m^3\ h^{-1}$ in the first year, declining to $\sim 350\ m^3\ h^{-1}$ over the next several years. The water enters on faults. Despite the general decline, discharge from some particular faults increased by a factor of 10 during the first 20 years of operation; evidently these routes are enlarging rapidly or capturing more of the epikarst drainage.

Where the tunnel or mine is a deep transient drain or is in the steady-state phreatic zone, gravitational drainage will not suffice, e.g. if the tunnel is below sea level. Three alternative strategies can then be adopted. The first is to pump from the tunnel itself, when necessary. The first steam engines were developed to apply this simple strategy below sea level in the coastal tin mines of Cornwall, UK. It continues to be the most popular method in small mines and some short, shallow tunnels. It is prone to failure if the pumps fail and to disaster (for the miners) if large water-filled cavities are encountered, causing catastrophic inrushes of water.

The second means is to grout the tunnel and then to pump any residual leakage as necessary. It is the essential method for transportation tunnels. Traditionally, tunnel surfaces were rendered impermeable by applying a sealant (e.g. concrete) as they became exposed. This does not deal with the catastrophic inrush problem. The first undersea tunnel was the Severn Railway Tunnel, cut in the 1860s in thick- to massively-bedded limestones beneath the Severn River estuary of Britain. It used the cut-and-seal method. At about 1 km from shore beneath a saltwater estuary a large *fresh water* spring was encountered. It flooded the tunnel, delaying completion for one year. It has been necessary to pump the site continuously ever since. The proposed English Channel Tunnel is to be cut in massive chalk just below the sea bed. We shall watch its progress with interest.

Modern practice is to drill a 360° array of grouting holes forward horizontally, then blast out and seal a section of tunnel inside this completed grout curtain. This largely deals with the hazard of catastrophic inrush i.e. a flooded cavity should be first encountered by a narrow bore drill hole that can be sealed off quickly. For example, the cooling water intake tunnel for Bruce B atomic power station, Ontario, was a 8 m diameter tunnel extending 600 m from shore beneath Lake Huron. It followed a corallian limestone formation just below the lake bed. Grouting forward proceeded in 20 m sections and the tunnel was cut in 8 m sections i.e. there was 60% overlap of successive grout curtains. A cavity was encountered that could not be grouted because it was too large. It was sealed off and the tunnel was then deflected around it without serious difficulty, but at substantial extra cost.

Grouting is not feasible in the extracting galleries of a mine. Here, a third and most elaborate strategy is to dewater the mine zone entirely i.e. maintain a cone of depression about it for as long as the mine is worked. A good modern example of the method is provided by the development of lead/zinc mines at Olkusz, Poland (Wilk in press). The ores are contained in filled dolines and cavities in a dolomite paleokarst at a depth of 200 to 300 m below a plain of Quaternary sediments that is in hydrologic contact with the bedrocks. Potentially, this was a very hazardous situation. An area of 500 km^2 was surveyed about the potential mine. It contained 70 natural springs and 600 wells. A further 1700 exploration boreholes were drilled. Piezometers were installed in 300 wells and bores for the conduct of pumping tests. From the latter it was estimated that 300×10^6 m^3 of groundwater would have to be pumped to establish the cone of depression for the mine. The cone was pumped out via vertical wells plus drainage adits with high capacity pumps that were cut beneath each extraction level before ore extraction began. By these means, maximum local inrushes of water were held to 1.5 m^3 s^{-1}, within the capacity of the pumps.

The first two strategies may have local effects upon the water table. Wells above the tunnel and springs at some distance from it may dry up. The effects of regional dewatering obviously will be more drastic. An example of coal and bauxite mining in Paleozoic limestones in Hungary is shown in Figure 11.12. The drawdown cone for the mine near Tapolca is 2500 km^2 in

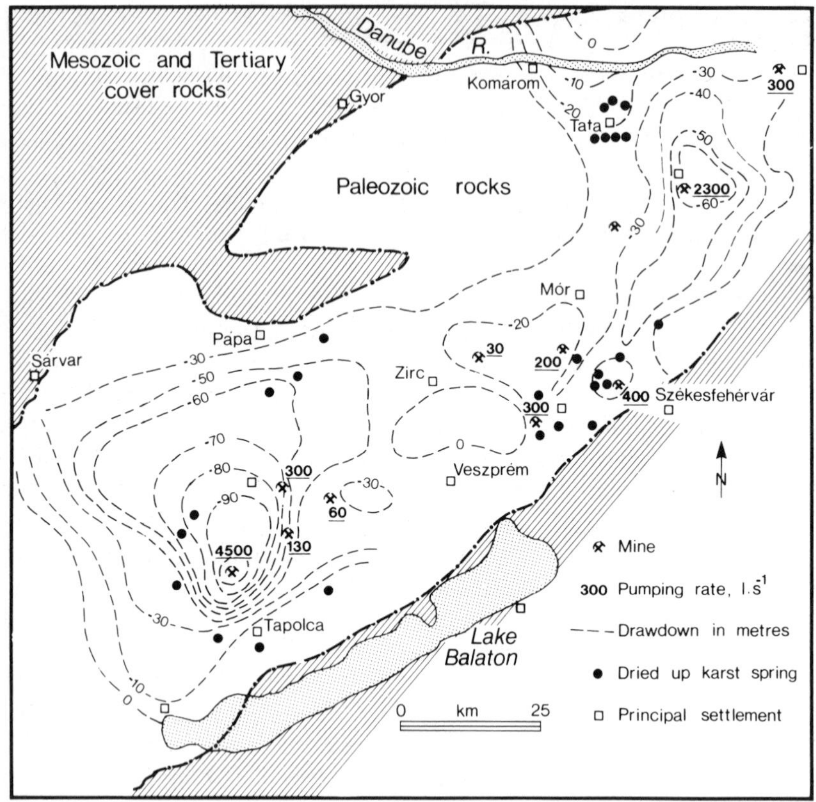

Figure 11.12 Drawdown of the karst groundwater table around bauxite mines in southwestern Hungary. Adapted from Bocker (in press).

extent. The water table has been lowered 113 m, requiring 20×10^6 m^3 of water pumped per metre of drawdown. Most of the pumped water is of potable quality; 28% of it is used as drinking water, compensating the wells and springs that have been dried up (Bocker in press).

The Olkusz mine (Wilk op. cit.) also dried up springs and wells. It was proposed to replace them with mine water. However, a paper mill 6 km north of the mine had disposed of highly contaminated waste water into the local Quaternary sand aquifer. With the drawdown these contaminants were drained into part of the mine water, ruining it as a source of supply. Many collapse dolines developed in the sands, one of them in the mill tailings pond which, as a result, discharged 30×10^3 m^3 of sludge into a sector of the mine. In a mature and complex karst terrain such as this, drastic interference with the water table can have unanticipated effects!

11.4 Urban hydrology of karst

Flooding, pollution and ground collapse are the main problems arising from the urbanization of karst terrains. It is well known that urbanization considerably increases the size of the mean annual flood in normal surface watersheds, but has progressively lesser influence on high magnitude, long recurrence interval events (Dunne & Leopold 1978). The effects are transmitted to karst when impervious rocks provide an allogenic input, and to a lesser extent are also seen in autogenic systems. Case studies demonstrating the hydrological impacts of urbanization on karst are provided by Betson (1977), Crawford (1981, 1984a), and White *et al.* (1984, 1986).

The city of Bowling Green is built upon the Sinkhole Plain of Kentucky. Crawford observes that almost all surface stormwater generated there escapes either to natural sinkholes or to drainage wells, over 400 of which have been drilled. Urban flooding occurs by three mechanisms: (a) by surface detention when storm flow input into sinkholes is greater than their discharge capacity, which may be reduced by clogging during building; (b) when the increased and rapidly concentrated runoff volumes typical of urban stormwater so surcharges underground tributary conduits that backflooding is produced upstream of constrictions right back to the surface; and (c) when the trunk conduit (the subterranean Barren River) is in flood due to rural runoff, as a consequence of which its backwater effect on vadose feeders causes them to backflood, so exacerbating the other tendencies. Subsidiary problems are also associated with these processes. Crawford suggests that increased runoff down drainage wells drilled into the bottoms of dolines causes increased subsurface erosion, which in turn leads to collapse of residual soils in about 10% of cases, so presenting an additional hazard for buildings. A similar problem is identified by White *et al.* (1984) in Pennsylvania. In addition, the quality of urban stormwater is poor with consequent deleterious impacts on the subterranean ecology, a problem that can be compounded by the dispersion through karst cavities of toxic and explosive fumes from chemical spills (Crawford 1984b). Flooding damage has been alleviated by restricting building in flood-prone areas and by requiring developers to construct floodwater detention basins. The concept of a sinkhole 'flood plain' has been adopted for flood insurance purposes, based on a three hour, 100 year return period rainfall event assuming no drainage from the bottom of the depressions, i.e. complete ponding. Quinlan (1986b) reviews legal aspects of sinkhole development and flooding in karst terrains from the perspective of American law.

In central Pennsylvania, White *et al.* (1984) conclude that human activities (a) increase hydraulic gradients through the lowering of the natural water table and (b) modify stormwater runoff patterns. However, only the latter has a major effect on the generation of dolines in the area. Urbanization strongly modifies the previously diffuse pattern of rainfall infiltration, focusing stormwater runoff from paved areas into natural sinkholes, swales or into the soil. This encourages the generation of suffosion dolines. Prevention of focusing through planning measures is much less expensive than engineering solutions but, where point discharge from paved areas is inevitable, the guiding principle should be to pipe stormwater directly to the subsurface bedrock without giving it the opportunity to excavate the soil.

11.5 Industrial exploitation of karst rocks and minerals

Limestone and dolomite

Limestone and dolomite are used for a wider range of purposes than any other rocks. To begin with the most costly, most stone sculpture and interior decorative stonework such as stairs uses the highest quality limestone i.e. with fewest imperfections. The preferred stone is true marble (section 2.3) because of its homogeneous crystalline composition and colour. However, any well textured limestone or dolomite that will take a good polish is marketed as 'marble' in many nations.

Limestone blocks are used for the construction of entire buildings, normally with cheaper grades of stone (more porous and friable) plus marble facings where desired. An early example was the use of a gleaming white limestone for facing the sandstone masses of pyramids at Giza, Egypt. More significant, perhaps, was the fact that limestone is the predominant local rock in Crete and Greece. Cretan palaces of the Minoan age and the Hellenic temples and public buildings of Greece used it almost exclusively (Fig. 11.13). The Romans followed with their buildings and statuary; at Rome itself a locally available lake travertine was especially favoured because it was soft and easy to cut. It is comparatively porous and friable and thus the classical buildings of Rome have not worn so well as those of Athens.

Classical use set the taste for building throughout much of Western history. In a majority of nations the principal palaces, churches, parliament buildings, etc. of the 12th to 19th centuries use limestone e.g. St Peter's, Rome and St Paul's, London. Entire cities such as Bath and Venice are built of it. Often, the stone was transported considerable distances. Eighteenth and 19th century buildings in Budapest and Vienna used 'Aurisina marble' from Trieste; see Cucchi & Gerdol (1985) for a thorough review of limestone masonry, with the Aurisina as an example.

Marble is prized in the Oriental cultures. It was used to build the Taj Mahal, in India, and whereas the Chinese and Japanese mainly built with wood, limestone and marble were used for courtyards and stairs in the

Figure 11.13 The Parthenon at Athens, the most celebrated building in Western architecture, is built of massive limestone with marble facings and sculpture, and stands on a pitted karren surface, now rounded down by the tread of countless visitors.

imperial palaces. Temple builders of the Western Hemisphere (Aztec, Inca, Maya, Toltec) used any local stone. In many instances this was limestone. The Mayan structures of the Yucatan Peninsula, for example, are of soft, very permeable Pleistocene limestones that have stood up surprisingly well to 800–1200 years of burial in secondary jungle.

At a more humble level limestones are widely used to build dry stone walls i.e. without cement. As field or garden walls these are features of rural landscapes in places as far apart as Galway Bay and Okinawa.

Nowadays, the use of whole stone for building is uncommon in Western countries. Concrete blocks with thin facings of limestone or dolomite have replaced it.

Coarsely ground to rubble or gravel sizes, carbonate rocks are widely used as flux for smelting iron and as aggregates for roadbeds or concrete. A quarry that is cutting good stone for facings or very pure limestone for cement may condemn less desirable intervening beds to this application.

Historically, in the European and Oriental cultures limestone was burned in kilns to produce 'quick lime' (CaO) as a calcium fertilizer. In long settled rural areas the landscapes are dotted with small extraction pits as a consequence; these are sometimes mapped as dolines! With the availability of powerful grinding machinery, pulverized limestone is now replacing burned lime.

High purity limestone is the principal ingredient of Portland cement. It is pulverized and fired at 1400–1650°C with proportions of iron compounds and silica for additional strength. Gypsum then is added for quick setting. Mixed with aggregate such as sand it forms concrete, which is now the world's principal building material. In Western countries modern concrete consumption averages between 0.1 and 0.5 tonnes per person per year.

Gypsum and anhydrite

Gypsum is used as a whole rock for statuary and interior decorative facings. It is too soft and soluble to be suitable for exterior use in most climates. It is most prized when it is pure white or pink, lustrous and macrocrystalline, termed 'alabaster'. As such, it was used in the baths of Minoan palaces, and can be found as small sculptures in tourist shops everywhere in the world today.

The principal use of gypsum is as plaster of Paris. With a proportion of silt or sand added, it is used for interior wall and ceiling plaster work, mouldings, etc. In North America it is most often made up as plaster board, to be cut to size during construction. As noted, pulverized gypsum is also an important ingredient of cement.

Oil and natural gas

Approximately 50% of the current production and known reserves of pool oil and natural gas are contained in carbonate rocks. North (1985) and Roehl & Choquette (1985) provide detailed reviews.

Carbonate reservoirs display a wider range of conditions than do reservoirs in the clastic rocks. The most productive oil wells on record (10^5 m^3 daily from 36 wells at Agha Jari in Iran) are from limestone; at the other extreme are tight, normally siliceous, limestones that will not yield at all unless artificially fractured and treated with acid.

There are three principal types of traps. In *limestones*, traps are chiefly paleokast features such as large depressions or marine–freshwater mixing zone caves, sealed by an unconformable cover of less permeable strata. There is increasing recognition of the importance of large depressions such as poljes and of reservoirs in what are probably sealed epikarst zones. In the world's most productive basin, the Persian Gulf, approximately 80% of carbonate reservoirs are limestone.

In *dolomites*, the predominant traps are patches, zones or particular beds of high porosity. These are normally created by secondary dolomitization during diagenesis, and have led to great emphasis on facies studies to determine the most likely sites of vuggy dolomite formation. One type locale is the upstanding reef, which will always be targeted during exploration for oil. In contrast to the Persian Gulf, most carbonate-hosted oil and gas in North America has been found to be in dolomites. Craig (1987) gives an excellent account of a 1 billion barrel oilfield in west Texas that was a coral island karst subsequently buried and dolomitized.

The third type of trap is the structural trap, which may be in limestone, dolomite, chalk or even marl. They are especially productive if naturally fractured. Carbonate caprocks above salt diapirs or anticlines belong in this category.

The world's first *drilled* oil well (as opposed to a dug well) was in southern Ontario in 1857. It encountered Silurian dolomites that had subsided and fractured where salt was dissolved beneath them. This trap was sealed tight by a cover of glacial clay; it can be described as an ancient paleokarst feature enhanced by Quaternary glacial action. It yielded a prolific flow of oil when drilled to a depth of a mere 60 m.

Carbonate-hosted ores

Approximately half of the world's present production of lead and zinc and the lion's share of its historic production comes from carbonate rocks. The minerals, galena and sphalerite, are associated with pyrite and marcasite, and with macrocrystalline ('sparry') dolomite plus minor fluorite and barite in some deposits. Ores typically assay a few per cent each of lead and zinc.

There is little silver. As noted in section 7.10, there appear to be three distinct modes of deposition: (a) from cool brines into karstic traps such as mixing zone caves or collapse sinkholes (e.g. Pine Point, Fig. 2.9); (b) from ascending thermal waters into earlier karst traps and (c) from thermal waters into cavities created simultaneously, a paragenetic mode as at Nanisivik (Fig. 7.31). Opinion is divided concerning the relative importance of the modes (Sangster 1987).

There are a great many deposits of iron ores in karst and paleokarst depressions. They occur chiefly as the minerals, limonite, siderite, hematite and goethite. They accumulated as weathering residua, most often being transported from adjoining non-karst rocks richer in iron, and normally are components of more complex, layered fillings. Most deposits are of low grade and small in volume, so that few are economic in the modern world. However, because they were widespread and generally easy to work, they were probably the most important sources of iron before the Industrial Revolution.

Bauxite is the ore of alumina, concentrated as a weathering residuum of the aluminosilicate minerals. Approximately 10% of global production is obtained from karst sites, the balance from laterites. The mineral is named for Les Baux, France, where mining in karst depressions began in the 1860s. Karstic bauxites are still mined widely in southern Europe but major production now is in Jamaica and other Caribbean islands, China, Vietnam and the Pacific islands. Its total exceeds 3 million tonnes per year.

Bauxite deposits are red, porous and earthy, interlayered with other continental sediments. Average composition is 35–50% Al_2O_3 plus lesser Fe_2O_3, trace minerals, and earth. The oldest known bauxites are from basal Cambrian strata and the youngest are Quaternary. There is an obvious geographical association with warm climates. Most deposits appear to have been trapped in closed depressions where depression deepening and bauxite accumulation proceeded simultaneously. Bardossy (1982) gives a comprehensive review.

A number of more exotic minerals such as antimony, copper, uranium and vanadium are also known in karst. Some are incorporated in depression fillings while others are probably thermal precipitates in caves and collapse breccias. None is of major economic importance today.

11.6 Recreational and scientific values of karstlands

If any geomorphologist were asked to nominate a short list of the great landscapes of the world, the celebrated tower karst of southern China would certainly rank high upon it. This landscape has been an enduring source of inspiration, entrancement and curiosity for travellers, artists and scientists of many dynasties from before Xu Xiake's time in the 17th century right up to the present. It is illustrated on the front cover of this book. We remain equally captivated by its beauty and mystery. It would be our first choice on a world heritage list of karst.

The great karsts of China also contain some of our most important sites for the study of human evolution. The discovery of 'Peking Man' or *Homo erectus pekinensis* in the late 1920s drew the attention of world science to excavations in cave deposits at Zhoukoudian, about 40 km southwest of Beijing. Modern paleomagnetic work on these cave sediments suggests the fossil has an age of at least 700 000 years (Ren *et al.* 1981). Indeed whether in karst of the 'Old World' or the 'New', we are constantly reminded of their immense importance as protected repositories of natural and cultural history. Caves are Nature's vaults, containing irreplaceable and datable records of biological, climatic and landscape history. Teeth and mandibles of the largest known primate, *Gigantophithicus blacki*, have also been found in Chinese caves, in Guangxi Province. When standing erect, this Orangutang-like creature may have been about 4 m tall. The world's tallest bird, *Dinornis maximus*, the now extinct giant Moa, was of comparable height, and is represented in many cave deposits in New Zealand, sometimes alonside the great extinct eagle *Harpagornis*, with a wingspan as large as the Condor. A recent discovery of fragments of Moa flesh preserved in the natural refrigerator of an alpine cave has just permitted investigation of its DNA structure – and so possibly opened the door to its genetic reconstruction. Paleontological discoveries in European caves since the 18th and 19th centuries (e.g. see *Reliquiae Diluvianae* by Buckland 1823) also revealed in their rhinoceros, hyaena, hippopotamus, etc. some of our first evidence for environmental change.

Ancient rock drawings of animals and hunting scenes provide other clues to the ecology of past times, as well as being superb examples of the artistic ability of our forebears. Most famous amongst locations where these can be seen are Altamira Cave near Santander in northern Spain, Lascaux Cave in the Dordogne valley in France, and Niaux Cave in the French Pyrenees (Fig. 11.14). Measured on any scale, these are internationally important sites of cultural heritage, but sadly they face conservation problems of great complexity, being inadvertently damaged by the very tourists who go to admire them (Andrieux 1983, Brunet *et al.* 1985, Villar *et al.* 1986). It was in these caves that the intricate and conflicting demands of tourist cave development and conservation were first appreciated. Less well known but still important sites with cave art occur in Africa, Australia and Papua New Guinea.

There are about 650 tourist caves in the world (Habe 1981), with those in the 'eastern hemisphere' alone being visited by over 15 million tourists a year (Maksimovich 1977). The pressure that this brings alters the atmosphere and energy balance of the subterranean world and so puts at risk the very features that are the source of attraction.

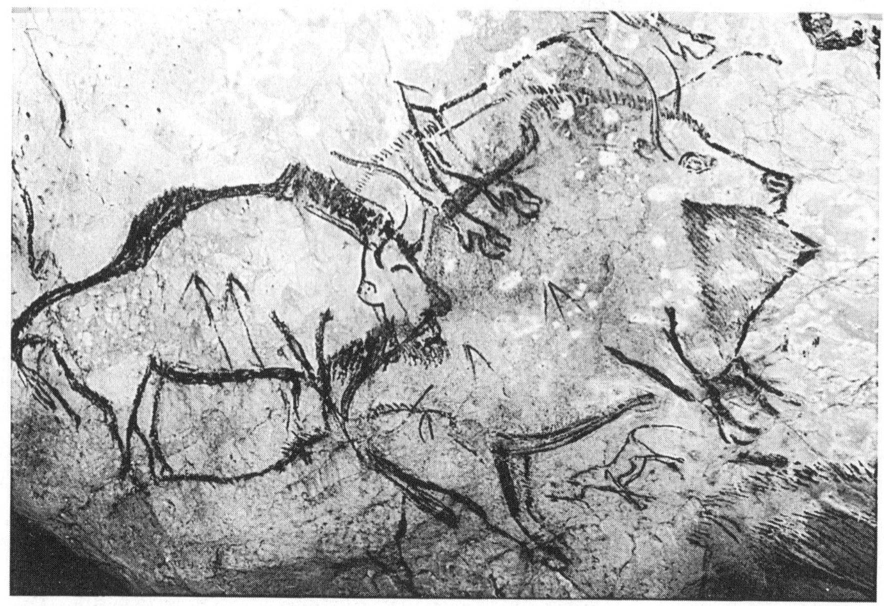

Figure 11.14 Painting of bison, Grotte de Niaux, Ariege, France. The work is estimated to be 12 000 years old.

The importance of conserving representative karst areas for science and recreation has been recognized in many countries by the designation of national parks and reserves. Of great value amongst those recently preserved is the Gunong Mulu National Park of Sarawak in northern Borneo. It contains magnificent humid tropical karst with its ecosystem intact – both below and above ground – a feature of immense significance at a time when tropical forests around the world are being destroyed. Rising precipitously above the forest are the spire-like pinnacle karren of Mt Api (Fig. 11.15), the best known example of this landform in the world. And piercing the mountain are huge allogenic river caves, including Lubang Nasib Bagus (Good Luck Cave) with its enormous void of Sarawak Chamber (Fig. 7.43).

Karst rocks are important in almost every facet of our lives, as we have shown. Limestones also contain the key to the sea level history of the earth for at least the last million years; the palaeotemperature record of the continents over a similar period; the habitats of troglobitic ecosystems that are barely investigated; and they support a morphology that is a microcosm of geomorphology as a whole. Carbonate rocks that cover perhaps 12% of the continents thus have a value much belied by their limited area.

The ravages of 20th century industrialization are consuming and polluting much of the world's karst, including the 'classical' terrain between

Figure 11.15 Spire-like pinnacle karren of Gunung Api (Mt Api) in the Gunong Mulu National Park, Sarawak. Photo by D. Dunkerley.

Yugoslavia and Italy. By focusing on only two of its facets, this book provides no more than a glimpse of the wider environmental significance of karst. Nevertheless, it is apparent to us through our wider professional contacts that the establishment of a world heritage series of karst sites is both justified and due. The International Association of Hydrogeologists, the International Speleological Union and the newly emerging international geomorphology association, working through UNESCO, are amongst the best placed to ensure that this is achieved.

References

Acaroglu, E. R. & W. H. Graf 1968. Sediment transport in conveyance systems 2: The modes of sediment transport and their related bedforms in conveyance systems. *Bull. Int. Assoc. Sci. Hydrol.* **13,** 123–35.

Adams, A. E., W. S. MacKenzie & C. Guilford 1984. *Atlas of sedimentary rocks under the microscope*. New York: Wiley.

Aitken, J. D. & D. G. Cook, 1969. Geology, Lac Belot, District of Mackenzie. *Geol. Surv. Can: Map 6*. 1969.

Akerman, J. H. 1983. Notes on chemical weathering, Kapp Linne, Spitzbergen. *Proc. 4th Internat. Conf. on Permafrost* 10–15. Washington, DC: Nat. Academy Press.

Aley, T. 1975. *A predictive hydrological model for evaluating the effects of land use on the quantity and quality of water from Ozark Springs*. Ozark Underground Laboratory, Missouri.

Alexander, E. C. & P. R. Book 1984. Altura Minnesota lagoon collapses. *Sinkholes: their geology, engineering and environmental impact*. B. F. Beck (ed.), 311–18. Boston: A. A. Balkema.

Allen, J. R. L. 1972. On the origin of cave flutes and scallops by the enlargement of inhomogeneities. *Rass. Speleo. Ital.* **24**, 3–19.

Allen, J. R. L. 1977. *Physical Processes of Sedimentation*. London: Allen & Unwin.

Ambert, P. 1981. Chronologie locale et synchronisme paleoclimatique. Actes du Colloque de L' A.G.F., '*Formations carbonates externes, tufs et travertins*', Ass. Francaise de Karstologie, Mem. 3, 201–6.

Andrieux, C. 1963. Etude crystallographique des pavements polygonaux des croutes polycristallines de calcite des grottes. *Bull. Soc. Franc. Miner. Crist.* **(86)**, 135–8.

Andrieux, C. 1965. Morphogenese des helictites monocristallines. *Bull. Soc. Franc. Mineral. Crist.* **88**, 163–71.

Andrieux, C. 1983. Etude des circulations d'air dans la Grotte de Niaux – consequences. *Karstologia* **1**, 19–24.

Arandjelovic, D. 1966. Geophysical methods used in solving some geological problems encountered in construction of the Trebisnjica water power plant (Yugoslavia). *Geophys. Prospecting*, **14(1)**, 80–97.

Arandjelovic, D., 1984. Application of geophysical methods to hydrogeological problems in Dinaric karst of Yugoslavia. In *Hydrogeology of the Dinaric karst*, B. F. Mijatovic (ed.), 143–59. Internat. Contribs. to Hydrogeology 4, Hannover: Heise.

Ashton, K. 1966. The analyses of flow data from karst drainage systems. Trans. Cave Research Group G.B., 7(2), 161–203.

Association Francaise de Karstologie, 1981. Formations Carbonates Externes, Tufs et Travertins. *Bull. Ass. Geogr. France Ass. Francaise de Karstologie*, Memoire 3.

Astier, J. L. 1984. Geophysical prospecting. In *Guide to the Hydrology of Carbonate Rocks*, P. E. LaMoreaux, B. M. Wilson & B. A. Memon (eds). *Studies and Reports in Hydrology No. 41*, 171–96. Paris: UNESCO.

Atkinson, T. C. 1977a. Carbon dioxide in the atmosphere of the unsaturated zone: an important control of groundwater hardness in limestones. *J. Hydrol.* **35**, 111–23.

Atkinson, T. C. 1977b. Diffuse flow and conduit flow in limestone terrain in the Mendip Hills, Somerset (Great Britain). *J. Hydrol.* **35**, 93–110.

Atkinson, T. C. 1983. Growth mechanisms of speleothems in Castleguard Cave, Columbia Icefields, Alberta, Canada. *Arctic and Alpine Res.*, **15 (4)**, 523–36.

Atkinson, T. C. 1985. Present and future directions in karst hydrogeology. *Ann. Soc. Geol. Belgique* **108**, 293–6.

Atkinson, T. C. & D. I. Smith 1976. The erosion of limestones. In *The science of speleology*, T. D. Ford & C. H. D. Cullingford (eds), 151–77. London: Academic Press.

Atkinson, T. C., T. J. Lawson, P. L. Smart, R. S. Harmon & J. W. Hess 1986. New data on speleothem deposition and paleoclimate in Britain over the last forty thousand years. *J. Quat. Sci.* **1(1)**, 67–72.

REFERENCES

Attewell, P. B. & I. W. Farmer 1976. *Principles of Engineering Geology*. London: Chapman and Hall.

Aubert, D. 1967. Estimation de la solution superficielle dans le Jura. *Bull. Soc. Vaudoise Sci. naturelles*. No. 324, **69(8)**, 365–76.

Aubert, D. 1969. Phenomenes et formes du karst jurassien. *Ecologiae geol. Helv*. **62(2)**, 325–99.

Ayers, J. F. & H. L. Vacher 1986. Hydrogeology of an atoll island: a conceptual model from detailed study of a Micronesian example. *Ground Water* **24(2)**, 185–98.

Babushkin, V. D., T. Bocker, B. V. Boirevsky, & V. S. Kovalevsky 1975. Regime of subterranean water flows in karst regions. In *Hydrogeology of karstic terrains*, A. Burger & L. Dubertret (eds), 69–78. International Union of Geological Sciences, Series B, 3.

Bachman, G. O. 1976. Cenozoic deposits of southeastern New Mexico and an outline of the history of evaporite dissolution. *J. Res., U.S. Geol. Survey* **4(2)**, 135–49.

Bachman, G. O. 1984. *Regional geology of the Ochoan evaporites, northern part of Delaware Basin*. New Mexico Bureau of Mines and Min. Res., Circ. 184.

Bachman, G. O. 1987. *Karst in evaporites in southeastern New Mexico*. Sandia National Labs., SAND 86-7078.

Back, W. & J. Zotl 1975. Application of geochemical principles, isotopic methodology, and artificial tracers to karst hydrology. In *Hydrogeology of karstic terrains*, A. Burger & L. Dubertret (eds), 105–21. International Union of Geological Sciences, Series B, 3.

Back, W., B. B. Hanshaw & J. N . van Driel 1984. Role of groundwater in shaping the Eastern Coastline of the Yucatan Peninsula, Mexico. In *Groundwater as a geomorphic agent*, R. G. LaFleur (ed.), 280–93. Boston: Allen & Unwin.

Backshall, D. G., J. Barnett, P. J. Davies, D. C. Duncan, N. Harvey, D. Hopley, P. J. Isdale, J. N. Jennings & R. Moss, 1979. Drowned dolines–the blue holes of the Pompey Reefs, Great Barrier Reef. BMR *J. Aust. Geol. Geophys*. **4**, pp. 99–109.

Bagnold, R. A. 1966. *An approach to the sediment transport problem from general physics*. U.S. Geol. Survey Prof. Paper 422-I.

Bakalowicz, M. 1973. Les grandes manifestations hydrologiques des karsts dans le monde. *Spelunca* **2**, 38–40.

Bakalowicz, M., 1976. Geochimie des eaux karstiques. Une methode d'etude de l'organisation des ecoulements souterrains. *Ann. Sci. Univ. Besancon* **25**, 49–58.

Bakalowicz, M. 1977. Etude du degre d'organisation des ecoulements souterrains dans les aquifères carbonates par une methode hydrogeochimique nouvelle. *C. R. Acad. Sc. Paris*, **284D**, 2463–6.

Bakalowicz, M. 1979. *Contribution de la geochimie des eaux a la connaissance de l'aquifere karstique et de la karstification*. These Doct. Sciences, Univ. Paris-6.

Bakalowicz, M. 1984. Water chemistry of some karst environments in Norway. *Norsk. geogr. Tiddsskr*. **38**, 209–14.

Bakalowicz, M. & A. Mangin 1980. L'aquifère karstique. Sa definition, ses characteristiques et son identification. *Mm. L. sr. Soc. geol. France* **11**, 71–9.

Bakalowicz, M. & C. Jusserand 1986. Etude de l'infiltration en milieu karstique par les methodes geochimiques et isotopiques. Cas de la Grotte de Niaux (Ariège, France) *Bull. Centre d'Hydrogéol., Univ. Neuchâtel*, **7**, 265–83.

Bakalowicz, M., B. Blavoux & A. Mangin 1974. Apports du tracage isotopique naturel à la connaissance du fonctionnement d'un système karstique-teneurs en oxygène-18 de trois systèmes des Pyrenees, France. *J. Hydrol*. **23**, 141–58.

Bakalowicz, M., A. Mangin, R. Rouch, C. Andrieux, D. D'Hulst, J. Daffis & A. Descouens 1985. *Caractere de l'environnement souterrain de la galerie d'entree de la grotte de Bedeilhac, Ariege*. Lab Souterrain du C.N.R.S., Moulis.

Bakalowicz, M. J., D. C. Ford, T. E. Miller, A. N. Palmer & M. V. Palmer 1987. Thermal Genesis of solution caves in the Black Hills, South Dakota. *Bull., Geol. Soc. Am*. **99**, 729–38.

Balazs, D. 1973. Relief types of tropical karst areas. In *IGU symposium on karst morphogenesis*, L. Jakucs (ed.), 16–32. Szeged: Attila Jozsef University.

Balch, E. S. 1900. *Glacières or Freezing Caverns*. Philadelphia: Lane & Scott.

Ball, J. W., E. A. Jenne & D. K. Nordstrom 1979. WATEQ 2 – a computerized chemical model for trace and major element speciation and mineral equilibrium of natural waters. In

Chemical modelling in aqueous systems. E. A. Jenne (ed.) 815–35. Washington: Am. Chem. Soc.

Bardossy, G. 1982. *Karst Bauxites. Bauxite deposits on carbonate rocks*. Budapest–Amsterdam: Akad. Kiado-Elsevier.

Bardossy, G. (in press). Bauxites in paleokarst. In *Paleokarst – a world review*, P. Bosak, D. C. Ford & J. Glazek (eds). Prague and Amsterdam: Academic Praha/Elsevier.

Barlow, C. A. & A. E. Ogden 1982. A statistical comparison of joint, straight cave segment, and photo-lineament orientations. *Bull., Nat. Speleo. Soc. Am.*, **44**, 107–10.

Barrett, P. J. 1963. The development of Kairimu Cave, Makaropa district, southwest Auckland. *N. Z. J. Geol. Geophys.* **6(2)**, 288–98.

Bastin, B. 1979. L'annalyse pollinique des stalagmites. *Ann. Soc. Geol. Belgique* **101,** 13–19.

Bastin, B. & M. Gewellt 1986. Analyse pollinique et datation ^{14}C de concretions stalagmitiques Holocenes. *Geog. physs. et Quat.* **XL(2)**, 185–96.

Batsche, H., F. Bauer, H. Behrens, K. Buchtela, H. J. Dombrowski, R. Geisler, M. A. Geyh, H. Hotzl, F. Hribar, W. Kass, J. Mairhofer, V. Maurin, H. Moser, F. Neumaier, J. Schmit, W. A. Schnitzer, A. Schreiner, H. Vogg & J. Zotl 1970. Kombinierte Karstwasser-untersuchungen in Gebiet der Donauversickerung (Baden-Wuttemberg) in den Jahren, 1966–1969, 1970. *Steir. Beitrage z. Hydrogeologie* **22**, 5–165.

Bauer, F. 1962. Nacheiszeitliche Karstformen in der osterreichischen Kalkalpen. *Proc. 2nd Internat. Congress Speleo. Bari*, 299–328.

Bear, J. 1972. *Dynamics of fluids in porous media*. New York: Elsevier.

Bebout, D., G. Davies, C. H. Moore, P. A. Scholle & N. C. Wardlaw 1979. *Geology of carbonate rock porosity*. Am Assoc. Petroleum Geologists, Continuing Education Course Note Series 11.

Beck B. F. (ed.), 1984. *Sinkholes: their geology, engineering and environmental impact*. Rotterdam: Balkema.

Beck, B. F. & W. L. Wilson (eds) 1987. *Karst hydrogeology: engineering and environmental applications*. Boston: A.A. Balkema.

Beede, J. W. 1911. The cycle of subterranean drainage as illustrated in the Bloomington Quadrangle (Indiana). *Proc. Indiana Acad. Sci.* **20**, 81–111.

Behrens, H., & M. Zupan 1976. Tracing with lithium chloride. In *Underground water tracing*, R. Gospodaric & P. Habic (eds), 187–92. Ljubljana: Inst. Karst Research Postojna.

Belloni, S., B. Martins & G. Orombelli 1972. Karst of Italy. In *Karst: important karst regions of the northern hemisphere*, M. Herak & V. T. Stringfield (eds), 85–128. Amsterdam: Elsevier.

Beniawski, Z. T. 1976. Rock mass classification in rock engineering. In *Exploration for rock engineering* Z. T. Beniawski (ed.), 97–106. Cape Town: A.A. Balkema.

Benson, R. C. & L. J. La Fountain 1984. Evaluation of subsidence or collapse potential due to subsurface cavities. In *Sinkholes: their geology, engineering and environmental impact*, B. F. Beck (ed.), 201–15. Boston: Balkema.

Bernasconi, R. 1976. The physico-chemical evolution of moonmilk. *Cave Geol.* **1**, 63–8.

Berner, R. A. 1971. *Principles of chemical sedimentology*. New York: McGraw-Hill.

Berner, R. A. 1978. Rate control of mineral dissolution under earth surface conditions. *Am. J. Sci.* **278**, 1235–52.

Berner, R. A. & J. W. Morse 1974. Dissolution kinetics of calcium carbonate in seawater. IV. Theory of calcite dissolution. *Am. J. Sci.* **274**, 108–34.

Betson, R. P. 1977. The hydrology of karst urban areas. In *Hydrologic problems in karst regions*, R. R. Dilamarter & S. C. Csallany (eds), 162–75. Bowling Green: Western Kentucky University.

Bini, A. 1978. *Appunti di geomorfologia ipogea: le forme parietali*, 19–46. 5th Conv. Reg. Speleo. Trentino-Alto Adige.

Bird, J. B. 1967. *The physiography of Arctic Canada* with special reference to the area south of Parry Channel. Baltimore: Johns Hopkins University Press.

Bischof, G. 1854. *Chemical and physical geology*. trans. Paul & Drummond, London.

Biswas, A. K. 1970. *History of hydrology*, Amsterdam, London: North-Holland.

Blackwell, B, H. P. Schwarcz & A. Debenath 1983. Absolute dating of hominids and paleolithic artefacts of the cave of La Chaise-de-Vouthon (Charente), France. *J. Archaeo. Sci.* **10**, 493–513.

REFERENCES

Blatt, H., G. V. Middleton & R. Murray 1980. *Origin of sedimentary rocks* 2nd edn. New Jersey: Prentice-Hall Inc.

Bloom, A. L. 1974. Geomorphology of reef complexes. In *Reefs in time and space*, L. F. Laporte (ed.) Soc. Econ. Palaeontol. Mineralogists, Spec. Pub. 18, 1–8, Tulsa, USA.

Blumberg, P. N. & R. L. Curl 1974. Experimental & theoretical studies of dissolution roughness. *J. Fluid Mech.* **75**, 735–42.

Bocker, T. 1969. Karstic water research in Hungary. *Internat. Assoc. Sci. Hydrol. Bull.* **14**, 4–12.

Bocker, T. 1973/4. Dynamics of subterranean karstic water flow. *Karszt-es Barlangkutatas* **8**, 107–45 (Budapest 1976).

Bocker, T. (in press). Karst hydrogeological problems of Hungarian bauxite and coal deposits occurring during their exploration, development and exploitation. In *Paleokarst – a world review*, P. Bosak, D. C. Ford & J. Glazek (eds). Amsterdam: Acad. Praha/Elsevier.

Bogli, A. 1956. Grundformen von Karsthohlenquerschnitten. *Z. Schweiz. Ges. Hohlenforsch.*, 56–62.

Bogli, A. 1960. Kalklosung und Karrenbildung. *Z. Geomorph, Supp/bd.* **2**, 4–21.

Bogli, A. 1961. Karrentische, ein Beitrag sur Karstmorphologie. *Z. Geomorph.* **5**, 185–93.

Bogli, A. 1964. Mischungskorrosion; ein Beitrag zum Verkarstungsproblem. *Erdkunde* **18(2)**, 83–92.

Bogli, A. 1970. *Le Holloch et son karst.* Ed. la Baconniere, Neuchatel.

Bogli, A. 1980. '*Karst hydrology and physical speleology*'. Berlin: Springer.

Bosak, P. 1981. The Lower Cretaceous paleokarst in the Moravian karst (Czechoslovakia). *Proc. 8th Int. Congr. Speleol.* **1**, 164–5.

Bosak, P., & I. Horacek 1981. The investigation of old karst phenomena of the Bohemian massif in Czechoslovakia: a preliminary regional evaluation. *Proc. 8th Int. Congr. Speleol.* **1**, 167–9.

Bosak, P., D. C. Ford & J. Glazek (in press). *Paleokarst – a world review.* Prague & Amsterdam: Academia Praha/Elsevier.

Bosak, P. (in press). Problems of origin and fossilisation of karst. In *Paleokarst – a world review*. P. Bosak, D. C. Ford & J. Glazek (eds). Prague and Amsterdam: Academia Praha/Elsevier.

Bourke, R. M. 1981. Preface to the report of the French speleological expeditions to Papua New Guinea. *Spelunca* **3**, 1.

Brady, B. M. G. & E. T. Brown 1985. *Rock Mechanics for underground mining*. London: Allen & Unwin.

Breisch, R. L. & F. L. Wefer 1981. The shape of 'gypsum bubbles'. *Proc. 8 Int. Cong. Speleo.* **2**, 757–9.

Bretz, J. H. 1942. Vadose and phreatic features of limestone caves. *J. Geol.* **50(6)** 675–811.

Brink, A. B. A. 1984. A brief review of the South African sinkhole problem. In *Sinkholes, their geology, engineering and environmental impact*. B. F. Beck (ed.). 123–7. Boston: Balkema.

Brod, L. G. 1964. Artesian origin of fissure caves in Missouri. *Bull., Nat. Speleo. Soc. Am.* **26(3)**, 83–112.

Brook, D. 1976. The karst and cave development of Filim Tel. *Trans, Brit. Cave Res. Assoc.* **3(3–4)**, 183–91.

Brook, G. A. 1981. An approach to modelling karst landscapes. *S. African Geog. J.* **63(1)**, 60–76.

Brook, G. A. 1983. Application of LANDSAT imagery to flood studies in the remote Nahanni karst, Northwest Territories Canada. *J. Hydrol.* **61**, 305–24.

Brook, G. A. & D. C. Ford 1978. The origin of labyrinth and tower karst and the climatic conditions necessary for their development. *Nature* **275**, 493–6.

Brook, G. A. & D. C. Ford 1980. Hydrology of the Nahanni karst, northern Canada, and the importance of extreme summer storms. *J. Hydrol.* **46**, 103–21.

Brook, G. A. & T. L. Allison 1983. Fracture mapping and ground subsidence susceptibility modelling in covered karst terrain: Dougherty County, Georgia. In *Environmental karst*, P. H. Dougherty (eds.), 91–108. Cincinnati: Geospeleo Pubs.

Brook, G. A., M. E. Folkoff & E. O. Box 1983. A world model of soil carbon dioxide. *Earth Surface Proc. & Landforms* **8**, 79–88.

Broughton, P. L. 1971. Origin and distribution of mineral species in limestone caves. *Earth Sci. J.* **5(1)**, 36–43.

Broughton, P. L. 1983a. Lattice deformation and curvature in stalactitic carbonate. *Int. J. Speleo.* **13 (1–4)**, 19–30.

Broughton, P. L., 1983b. Environmental implications of competitive growth fabrics in stalactitic carbonate. *Int. J. Speleo.* **13(1–4)**, 31–42.

Brown, M. C. 1972a. Karst hydrogeology and infrared imagery, an example. *Bull. Geol. Soc. Am.* **83(10)**, 3151–4.

Brown, M. C. 1972b. *Karst hydrology of the lower Maligne basin, Jasper, Alberta.* Cave Studies 13, Cave Res. Ass., Calif.

Brown, M. C. 1973. Mass balance and spectral analysis applied to karst hydrologic networks. *Water Resources Research* **9(3)**, 749–52.

Brown, R. J. E. 1970. *Permafrost in Canada.* Toronto: University of Toronto Press.

Brown, R. H., A. A. Konoplyantsev, J. Ineson & V. S. Kovalevsky (eds), 1972. *Groundwater Studies, Studies and Reports in Hydrology* **7**, Paris: UNESCO.

Brunet, J., P. Vidal & J. Vouve 1985. Conservation de l'art rupestre. *UNESCO etudes et documents sur le patrimoine culturel* **7**.

Buchtela, K. von, 1970. Aktivierungsanalyse in der Hydrogeologie. *Steirische Beitrage zur Hydrogeol.* **22**, 189–98.

Buckland, W. 1823. *Reliquae diluvianae.* Oxford.

Budel, J. 1982. *Climatic geomorphology* Princeton, N.Y.: Princeton University Press.

Buhmann, D. & W. Dreybrodt 1985a. The kinetics of calcite solution and precipitation in geologically relevant situations of karst areas. 1. open system. *Chem. Geol.* **48**, 189–211.

Buhmann, D. & W. Dreybrodt 1985b. The kinetics of calcite solution and precipitation in geologically relevant situations of karst areas. 2. closed system. *Chem. Geol.* **53**, 109–24.

Bull, P. A. 1975. Birdseye structures in caves. *Trans. Brit. Cave Res. Assoc.* **2(1)**, 35–40.

Bull, P. A. 1976. An electron microscope study of cave sediments from Agen Allwedd, Powys. *Trans. Brit. Cave Res. Assoc.* **3**, 7–14.

Bull, P. A. 1977a. Laminations or varves? Processes of fine grained sediment deposition in caves. *Proc., 7th Internat. Speleo. Cong.*, 86–7.

Bull, P. A. 1977b. Boulder chokes and doline relationships. *Proc. 7th Internat. Spel. Cong. (Sheffield)*, 93–6.

Bull, P. A. 1978a. Surge mark formation and morphology. *Sedimentology* **25**, 877–86.

Bull, P. A. 1978b. A study of stream gravels from a cave: Agen Allwedd, South Wales. *Zeit. Geomorph. N.F.* **22**, 275–96.

Bull, P. A. 1982. Some fine-grained sedimentation phenomena in caves. *Earth Surf. Proc. and Landforms* **6**, 11–22.

Bull, P. A. 1983. Chemical sedimentation in caves. In *Chemical sediments and geomorphology; precipitates and residue in the near-surface environment*, A. S. Goudie & K. Pye (eds), 301–20. London: Academic Press.

Bull, P. A. & M. Laverty 1982. Observations on phytokarst. *Z. Geomorph.* **26**, 437–57.

Burdon, D. J. & N. Papakis 1963. *Handbook of karst hydrogeology.* Athens: Institute for Geology and Subsurface Research/FAO.

Burdon, D. J. & C. Safadi, 1963. Ras-el-Ain: the great spring of Mesopotamia. *J. Hydrol.* **1**, 58–95.

Burdon, D. J. & C. Safadi 1964. The karst groundwaters of Syria. *J. Hydrol.* **2**, 324–47.

Burger, A. & L. Dubertret (eds) 1984. *Hydrogeology of karstic terrains: case histories. International Contributions to Hydrogeology* **1**, Hannover: Heise.

Burin, K., K. Spassov, D. Kolev, D. Apostolov & P. Deltchev 1976. Two experiments in tracing karst underground water with bromine, using neutron activation analysis in Bulgaria. *Papers 3 Internat. Symp. Underground Water Tracing*, 35–45. Ljubljana: Inst. Karst Res. Postojna,

Burri, E. 1986. Various aspects of the karstic phenomena in the urbanised area of Gissi and neighbouring areas (Southern Abruzzo, Italy). *Le Grotte d'Italia,* **4 (XIII)**, 143–61.

Busenberg, E. & L. N. Plummer 1982. The kinetics of dissolution of dolomite in CO_2–H_2O systems at 1.5° to 65°C and 0 to 1 atm PCO_2 *Am. J. Sci.* **282**, 45–78.

Butler, J. N. 1982. *Carbon dioxide equilibria and their applications.* Mass: Addison-Wesley.

Cabrol, P. 1978. *Contribution a l'étude du concretionnement carbonate des grottes du sud de la France; morphologie, génèse, diagénèse*. C.E.R.G.H., University of Montpellier, tome XII.

Cailleux, A. & J. Tricart 1963. *Initiation a l'etude des sables et des galets*. 3 vols. Centre Doc. Universitaire, Paris.

Carew, J. L., J. Mylroie, J. F. Wehmiller & R. S. Lively 1984. Estimates of late Pleistocene sea level high stands from San Salvador, Bahamas. In *Proc. 2nd Symp. Geology of the Bahamas*, J. W. Teeter (ed.), 153–75.

Casanova, J. 1981. Morphologie et biolithogénèse des barrages de travertins. '*Formations carbonates externes, tufs et travertines*', Ass. Francaise de Karstologie, Mem. 3, pp. 45–54.

Castany, G. 1982. *Principles et methodes de l'hydrologie*, Paris, Dunod.

Castany, G. 1984a. Hydrogeological features of carbonate rocks. In *Guide to the hydrology of carbonate rocks*, P. E. LaMoreaux, B. M. Wilson & B. A. Memon (eds), 47–67. Studies and Reports in Hydrology 41, Paris: UNESCO.

Castany, G. 1984b. Determination of aquifer characteristics. In *Guide to the hydrology of carbonate rocks*, P. E. LaMoreaux, B. M. Wilson & B. A. Memon (eds), 210–37. Studies and Reports in Hydrology No. 41, Paris: UNESCO.

Castellani, V. & W. Dragoni 1986. Evidence for karstic mechanisms involved in the evolution of Moroccan Hamadas. *Int. J. Speleo.* **15(1–4)**, 57–71.

Caumartin, V. 1964. Speleologie physique, biospeleologie et conservation des grottes: application au cas particulier de Lascaux. *Spelunca* **3**, 5–15.

Celico, P., R. Gonfiantini, M. Koizumi & F. Mangano 1984. Environmental isotope studies of limestone aquifers in central Italy. In *Isotope hydrology 1983*, 173–92. Vienna: IAEA.

Chafetz, H. S. & R. L. Folk 1984. Travertines: depositional morphology and the bacterially constructed constituents. *J. Sed. Pet.* **54(1)**, 289–316.

Chappell, J. 1983. A revised sea-level record for the last 300 000 years from Papua, New Guinea. *Search* **14**, 99–101.

Chappell, J. & N. J. Shackleton 1986. Oxygen isotopes and sea level. *Nature* **324**, 137–40.

Chen, Z. P., L. H. Song and M. M. Sweeting, 1986. The pinnacle karst of the Stone Forest, Lunan, Yunnan, China: an example of a subjacent karst. In *New Directions in Karst*, K. Paterson and M. M. Sweeting (eds), 597–607. Norwich, England: Geo Books.

Cherdyntsev, V. V. 1971. *Uranium 234*. Jerusalem: Israel Programme for Scientific Translations.

Choquette, P. W. & L. C. Pray 1970. Geological nomenclature and classification of porosity in sedimentary carbonates. *Am. Assoc. Petrol. Geol. Bull. 54*, 207–50.

Chorley, R. J. & B. A. Kennedy 1971. *Physical geography, a systems approach*. London: Prentice-Hall.

Christiansen, E. A. 1967. Collapse structures near Saskatoon, Saskatchewan, Canada. *Can. J. Earth Sci.* **4**, 757–67.

Christiansen, E. A. 1971. Geology of the Crater Lake collapse structure in southeastern Saskatchewan. *Can. J. Earth Sci.* **8(12)**, 1505–13.

Christopher, N. S. J. 1980. A preliminary flood pulse study of Russett Well, Derbyshire. *Trans. Brit. Cave Res. Assoc.* **7(1)**, 1–12.

Cigna, A. A. 1986. Some remarks on phase equilibria of evaporites and other karstifiable rocks. *Le Grotte d'Italia* **4(XII)**, 201–8.

Cigna, A. A. & R. Forti 1986. The speleogenetic role of airflow caused by convection. 1st contribution. *Int. J. Speleo.* **15**, 41–52.

Cinq-Mars, J. & B. Lauriol 1985. Le karst de Tsi-It-Toh-Choh: notes preliminaire sur quelques phenomenes karstiques du Yukon septentrional, Canada. *Comptes Rendus du Colloque International de Karstologie Applique*, Universite de Liege, 185–96.

Cita, M. B. & Ryan, W. B. F. (eds) 1981. Carbonate platforms of the passive-type continental margins: present and past. *Marine Geol.* **44(1/2)**.

Ciry, R. 1962. Le role du froid dans la speleogenese. *Spelunca Memoires* **2(4)**, 29–34.

Clark, P. J. & F. C. Evans 1954. Distance to nearest neighbour as a measure of spatial relationships in populations. *Ecology* **35**, 445–53.

Clarke, F. W. 1924. The data of geochemistry. *U.S. Geol. Surv. Bull.* **770**, 84 p.

Cocean, P. 1975. Sur la genese de la voute plane-horizontale de la grotte 'Pestera cu Apa din Valea Lesului'. *Travaux de L'Institute de Speleologie 'Emile Racovitza'* **XIV**, 189–96.

Corbel, J. 1956. A new method for the study of limestone regions. *Rev. Canad. Geogr.* **10**, 240–2.

Corbel, J. 1957. *Les karsts du nord-ouest de l'Europe*. Mems. Docs. Inst. Etudes Rhodaniennes Univ. Lyon **12**.

Corbel, J. 1959. Erosion en terrain calcaire. *Ann. Geog.* **68**, 97–120.

Courbon, P. & C. Chabert 1986. *Atlas des grandes cavites mondiales*. Union Internat. Speleo. and Fed. Francaise de Speleo.

Coward, J. M. H. 1975. *Paleohydrology and streamflow simulation of three karst basins in southeastern West Virginia*. PhD thesis, McMaster University, Canada.

Craig, D. H. 1987. Caves and other features of Permian karst in San Andres dolomites, Yates field reservoir, West Texas. In *Paleokarst*, N. P. James & P. W. Choquette (eds), 342–63. New York: Springer-Verlag.

Craig, H. 1961. Isotopic variations in meteoric waters. *Science* **133**, 1702–3.

Cramer, H. 1941. Die Systematik der Karstdolinen. *Neues Jb. Miner. Geol. Palaont.* **85**, 293–382.

Crawford, N. C. 1981. Karst flooding in urban areas: Bowling Green, Kentucky. *Proc. 8 Int. Congr. Speleol.* **2**, 763–5.

Crawford, N. C. 1984a. Sinkhole flooding associated with urban development upon karst terrain: Bowling Green, Kentucky. In *Sinkholes: their geology, engineering and environmental impact*, B. F. Beck (ed.), 283–92. Rotterdam: Balkema.

Crawford, N. C. 1984b. Toxic and explosive fumes rising from carbonate aquifers: a hazard for residents of sinkhole plains. In *Sinkholes: their geology, engineering and environmental impact*, B. F. Beck (ed.), 297–304. Rotterdam: Balkema.

Crowther, J. 1983. A comparison of the rock tablet and water hardness methods for determining chemical erosion rates on karst surfaces. *Z. Geomorph. NF.* **27(1)**, 55–64.

Cruden, D. M. 1985. Rock slope movements in the Canadian Cordillera. *Can. Geotech. J.* **22**, 528–40.

Cryer, R. 1986. Atmospheric solute inputs. In *Solute processes*, S. T. Trudgill (ed.), 15–84. New York: John Wiley.

Cser, F. & L. Maucha 1966. Contribution to the origin of 'excentric' concretions. *Karszt-es Barlang.* **6**, 83–100.

Cucchi, F. & S. Gerdol 1985. *I marmi del Carso triestino*, Trieste: Camera di Commercio.

Curl, R. L. 1965. Solution kinetics of calcite. *Proc. 4th Internat. Congr. Speleo.*, Ljubljana **3**, 61–6.

Curl, R. L. 1966. Scallops and flutes. *Trans. Cave Res. Gp., G.B.* **7(2)**, 121–60.

Curl, R. L. 1973a. Minimum diameter stalactites. *Bull. Nat. Speleo. Soc.* **34**, 129–36.

Curl, R. L. 1973b. Minimum diameter stalagmites. *Bull. Nat. Speleo. Soc.* **35(1)**, 1–9.

Curl, R. L. 1974. Deducing flow velocity in cave conduits from scallops. *Nat. Speleo. Soc. Am. Bull.* **36(2)**, 1–5.

Curl, R. L. 1986. Fractal dimensions and geometries of caves. *Math. Geol.* **18(8)**, 765–83.

Cvijic, J. 1893. Das Karstphaenomen. Versuch einer morphologischen Monographie. *Geog. Abhandl. Wien* **5(3)**, 218–329. (The section on dolines, pp. 225–76, is translated into English in Sweeting 1981).

Cvijic, J. 1901. Morphologische und glaciale Studien aus Bosnien, der Hercegovina und Montenegro: die Karst-Poljen. *Abhand. Geog. Ges. Wien* **3(2)**. 1–85.

Cvijic, J. 1918. Hydrographie souterraine et evolution morphologique du karst. *Rec. Trav. Inst. Geog. Alpine* **6(4)**, 375–426.

Cvijic, J. 1925. Types morphologiques des terrains calcaires. *C.R. Acad. Sci. (Paris)* **180**, 592, 757, 1038.

Damblon, F. 1974. Observations palynologiques dans la Grotte de Remouchamps. *Bull. Soc. Roy. Belge Anthrop. Prehist.* **85**, 131–55.

Danes, J. V. 1908. Geomorphologische Studien in Karstgebiete Jamaikas, 9. *Internat. Geog. Cong.* **2**, 178–82.

Danes, J. V. 1910. Die Karstphaenomene im Goenoeng Sewoe auf Java. *Tijdschr. K. ned. aardr. Genoot.* **27**, 247–60.

Danin, A. 1983. Weathering of limestone in Jerusalem by cyanobacteria. *Z. Geomorph.* **27(4)**, 413–21.

Danin, A. & J. Garty 1983. Distribution of cyanobacteria and lichens on hillsides in the Neger Highlands and their impact on biogenic weathering. *Z. Geomorph.* **27(4)**, 423–44.

Dansgaard, W. F. 1964. Stable isotopes in precipitation. *Tellus* **16**, 436–49.

Darcy, 1856. *Les fontaines publiques de le ville de Dijon*. Paris: Dalmont.

Darwin, C. 1842. *The structure and distribution of coral reefs*. London: Smith, Elder.

Davies, W. E. 1949. Features of cavern breakdown. *Bull. Nat. Speleo. Soc. Am.* **11**, 34–5.

Davies, W. E., 1951. Mechanics of cavern breakdown. *Bull. Nat. Speleo. Soc. Am.* **13**, 36–43.

Davies, W. E. 1960. Origin of caves in folded limestone. *Nat. Speleo. Soc. Bull.* **22**, 5–18.

Davies, W. E. & H. LeGrand 1972. Karst of the United States. In *Karst: important karst regions of the northern hemisphere*. M. Herak & V. T. Stringfield (eds), 467–505. New York: Elsevier.

Davis, D. G. 1980. Cave development in the Guadalupe Mountains: a critical review of recent hypotheses. *Bull. Nat. Speleo. Soc. Am.* **42(3)**, 42–8.

Davis, S. N. 1966. Initiation of groundwater flow in jointed limestone. *Nat. Speleo. Soc. Bull.* **28**, p. 111.

Davis, W. M. 1899. The geographical cycle, *Geog. Jour.* **14**, 481–504.

Davis, W. M. 1930. Origin of limestone caverns. *Geol. Soc. Amer. Bull.* **41**, 475–628.

Day, M. J. 1976. The morphology and hydrology of some Jamaican karst depressions. *Earth Surf. Process* **1**, 111–29.

Day, M., 1977a. Surface roughness in tropical karst terrain. *Proc. 7 Int. Spel. Cong. (Sheffield)*, 139–43.

Day, M. J., 1977b. Surface hydrology within polygonal karst depressions in northern Jamaica. *Proc. 7 Int. Spel. Cong. (Sheffield)*, 143–6.

Day, M. J. 1978. Morphology and distribution of residual limestone hills (mogotes) in the karst of northern Puerto Rico. *Bull. Geol. Soc. Am.* **89**, 426–32.

Day, M. J. 1979a. The hydrology of polygonal karst depressions in northern Jamaica. *Z. Geomorph., Suppl. 32*, 25–34.

Day, J. M. 1979b. Surface roughness as a discriminator of tropical karst styles. *Z. Geomorph. Suppl.-Bd.* **32**, 1–8.

Day, M. J. 1982. The influence of some material properties on the development of tropical karst terrain. *Trans. Brit. Cave Res. Assoc.* **9(1)**, 27–37.

Day, M. J. 1983. Doline morphology and development in Barbados. *Anns. Assoc. Amer. Geogr.* **73(2)**, 206–19.

Day, M. J. 1984a. Carbonate erosion rates in southwestern Wisconsin. *Physical Geog.* **5(2)**, 142–9.

Day, M. J. 1984b. Predicting the location of surface collapse within karst depressions: a Jamaican example. In *Sinkholes: their geology, engineering and environmental impact*, B. F. Beck (ed.), 147–51. Rotterdam: Balkema.

Day, M. J. & A. S. Goudie 1977. The Schmidt Hammer and field assessment of rock hardness. *Brit. Geomorph. Res. Gp., Tech. Bull.* **18**, 19–29.

Deal, D. E. 1962. *Geology of Jewel Cave National Monument, Custer County, South Dakota with special reference to cavern formation in the Black Hills*. M.S. thesis, University of Wyoming.

Deal, D. E. 1968. Origin and secondary mineralisation of caves in the Black Hills of South Dakota. *Proc. 4th Internat. Cong. Speleo., Yugoslavia* **3**, 67–70.

Debenham, N. C. & M. J. Aitken 1984. Thermoluminescence dating of stalagmitic calcite. *Archaeometry* **26(2)**, 155–70.

Dechant, M. 1967. Die Farbung von Lycopodiumsporen. *Steir. Beitr. z. Hydrogeologie* **18/19**.

Deike, G. H. & W. B. White 1969. Sinuosity in limestone solution conduits. *Am. J. Sci.* **267,** 230–41.

Delannoy, J. J. & R. Maire 1984. Les grandes cavités alpines. *Karstologia* **3**, 60–9.

DeMille, G., J. R. Shouldice & H. W. Nelson 1964. Collapse structures related to evaporites of the Prairie Formation, Saskatchewan. *Bull., Geol. Soc. Am.* **75**, 307–16.

Dincer, T. & B. R. Payne 1971. An environmental isotope study of the south-western karst region of Turkey. *J. Hydrology* **14**, 233–58.

Dincer, T., B. R. Payne, C. K. Yen & J. Zotl 1972. Das Tote Gebirge als – Entwasserungstypus der Karstmassive der nordostlichen Kalkhochalpen (Ergebnisse von Isotopenmessungen). *Steir. Beitr. z. Hydrogeol.* **24**, 71–109.

Douglas, I. 1968. Some hydrologic factors in the denudation of limestone terrains. *Z. Geomorph.* **12(3)**, 241–55.

Drake, J. J. 1980. The effect of soil activity on the chemistry of carbonate groundwater. *Water Resources Res.* **16(2)**, 381–6.

Drake, J. J. 1983. The effects of geomorphology and seasonality on the chemistry of carbonate groundwaters. *J. Hydrol.* **61(1/3)**, 223–6.

Drake, J. J. 1984. Theory and model for global carbonate solution by groundwater. In *Groundwater as a geomorphic agent*, R. G. LaFleur (ed.), 210–26. London: Allen & Unwin.

Drake, J. & D. C. Ford 1972. The analysis of growth patterns of two generation populations; the example of karst sinkholes. *Can. Geog.* **16,** 381–4.

Drake, J. J. & D. C. Ford 1973. The dissolved solids regime and hydrology of two mountain rivers. *Proc. 6th Int. Congr. Speleol. (Olomouc)* **4**, 53–6.

Drake, J. J. & D. C. Ford 1976. Solutional erosion in the Canadian Rockies. *Can. Geogr.* **xx(2)**, 158–70.

Drake, J. J. & R. S. Harmon 1973. Hydrochemical environments of carbonate terrains. *Water Resources Res.* **11**, 958–62.

Drake, J. J. & T. M. L. Wigley 1975. The effect of climate on the chemistry of carbonate groundwater. *Water Resources Res.* **11**, 958–62.

Dreiss, S. J. 1982. Linear kernels for karst aquifers. *Water Resources Res.* **18(4)**, 865–76.

Dreiss, S. J. 1984. Effects of lithology on solution development in carbonate aquifers. *J. Hydrol.* **70**, 295–308.

Drever, J. I. 1982. *The geochemistry of natural waters*, New Jersey: Prentice-Hall.

Drew, D. P. 1970. The significance of percolation water in limestone catchments. *Groundwater* **8(5)**, 8–11.

Drew, D. P. & D. I. Smith 1969. Techniques for the tracing of subterranean water. *British Geomorph. Research Group, Tech. Bull.* **2,** 36.

Drewry, D. 1986. *Glacial geologic processes*. London: Edward Arnold.

Dreybrodt, W. 1981. Kinetics of the dissolution of calcite and its applications to karstification. *Chem. Geol.* **31**, 245–69.

Dreybrodt, W. 1983. A possible mechanism for growth of calcite speleothems without participation of biogenic carbon dioxide. *Earth, Planet. Sci. Letters* **58**, 293–9.

Dreybrodt, W. 1988. *Processes in karst systems: Physics, Chemistry, and Geology*, Berlin: Springer, 288 p.

Drogue, C., A.-M. Laty & H. Paloc 1983. Les eaux souterraines des karsts mediterranéens. Exemple de la region pyreneo-provencale (France meridionale). *Bull. Bureau de Recherches Geologique et Minieres, Hydrogeologie-geologie de l'ingenieur* **4**, 293–311.

Droppa, A. 1966. The correlation of some horizontal caves with river terraces. *Studies in Speleology* **1(4)**, 186–92.

Drost, W., D. Klotz, A. Koch, H. Moser, F. Neumaier & W. Rauert 1968. Point dilution methods of investigating groundwater flow by means of radioisotopes. *Water Research Resources* **4(1),** 125–46.

Drost, W. & D. Klotz 1983. Aquifer characteristics. In *Guidebook on nuclear techniques in hydrology*, 223–567. Technical Reports Series No. 91, Vienna: IAEA.

Dublyansky, V. N. 1979. The gypsum caves of the Ukraine. *Cave Geol.* **1(6)**, 163–83.

Dublyansky, V. N. 1980. Hydrothermal karst in the alpine folded belt of southern parts of the U.S.S.R. *Kras i Speleologia* **3(12)**, 18–36.

Dublyansky, V. N & A. A. Lomaev 1980. *Karst caves of the Ukraine* (in Russian). Kiev: Naukova Dumka.

Dunham, R. J. 1962. Classification of carbonate rocks. *Am. Assoc. Petrol. Geol. Mem.* **1**, 108–21.

Dunkerley, D. L. 1979. The morphology and development of rillenkarren. *Zeitschrift fur Geomorphologie* **23(3)**, 332–48.

Dunkerley, D. L. 1983. Lithology and microtopography in the Chillagoe karst, Queensland, Australia. *Zeitschrift fur Geomorphologie* **27(2)**, 191–204.

Dunkley, J. R. & T. M. L. Wigley 1967. Caves of the Nullarbor. Sydney: Speleo. Res. Council.

Dunn, J. R. 1957. Stream tracing: mid-Appalachian region. *Nat. Speleo. Soc. Bull.* **2**, p. 7.

Dunne, T. R. & L. B. Leopold 1978. *Water in environmental planning*, San Francisco: Freeman.

Durozoy, G. & H. Paloc 1973. Le regime des eaux de la Fontaine de Vaucluse. *Min. du Dev. Industriel et Scientifique, Bureau de recherches geol. minières.*

Dzulynski, S. & M. Sass-Gutkiewicz (in press). Lead–zinc ores. In *Paleokarst – a world review*, P. Bosak, D. C. Ford & J. Glazek (eds). Amsterdam: Acad. Praha/Elsevier.

Eberentz, P. 1976. *Apports des méthodes isotopiques à la connaissance de l'aquifere karstique*. Thése 3e cycle, geol. dyn., Univ. Pierre-et-Marie Curie.

Ecock, K. E. 1984. *Karst hydrology of the White Ridges, Vancouver Island*. MSc thesis, McMaster University.

Egemeier, S. J. 1981. Cavern development by thermal waters. *Nat. Speleo. Soc. Bul.* **43**, 31–51.

Ek, C. 1961. Conduits souterrains en relation avec les terrasses fluviales. *Ann. Soc. Geol. Belg.* **84**, 313–40.

Ek, C. 1973. Analyses d'eaux des calcaires paleozoiques de la Belgique. *Prof. Paper 13, Service Geol. Belgique*.

Ek, C. & M. Gewelt 1985. Carbon dioxide in cave atmospheres. New results in Belgium and comparison with some other countries. *Earth Surf. Proc. Landforms* **10**, 173–87.

Embry, A. F. & J. E. Klovan 1971. A Late Devonian reef tract in northeastern Banks Island, Northwest Territories, *Can. Petrol. Geol. Bull.* **19**, 730–81.

Emiliani, C. 1966. Pleistocene paleotemperatures. *Science* **154**, 851–7.

Engh, L. 1980. *Karstomradet vid Lummelunds bruk, Gotland*. Lund University: Geog. Inst.

Eraso, A. 1975. Le role des facteurs physico-chimiques dans les processus de la karstification. *Ann. Speleol.* **30(4)**, 567–80.

Eraso, A. (in press). Paleokarst in civil engineering. In *Paleokarst: a world review*. P. Bosak, D. C. Ford & J. Glazek (eds). Amsterdam: Acad. Praha/Elsevier.

Erga, V. & S. G. Terjesen 1956. Kinetics of the heterogeneous reaction of calcium carbonate formation. *Acta Chem. Scand.* **10**, p. 872–4.

Erickson, E. 1983. Stable isotopes and tritium in precipitation. In *Guidebook on nuclear techniques in hydrology*, Technical Reports Series No. 91, 19–33. Vienna: IAEA.

Esteban, M. & C. F. Klappa 1983. Subaerial exposure. In P. A. Scholle, D. G. Bebout and C. H. Moore, (eds). *Carbonate Depositional Environments*. Am. Assoc. Petrol. Geologists, Me. **33,** pp. 1–54.

Even, H., I. Carmi, M. Magaritz & R. Gerson 1986. Timing the transport of water through the upper vadose zone in a karstic system above a cave in Israel. *Earth Surf. Proc. Landf.* **11**, 181–91.

Ewers, R. O. 1972. *A model for the development of subsurface drainage routes along bedding planes*. MSc thesis, University of Cincinnati.

Ewers, R. O. 1978. A model for the development of broadscale networks of groundwater flow in steeply dipping carbonate aquifers. *Trans. Brit. Cave Res. Assoc.* **5**, 121–5.

Ewers, R. O. 1982. *Cavern development in the dimensions of length and breadth*. PhD Thesis, McMaster University.

Ewing, A. 1885. Attempt to determine the amount and rate of chemical erosion taking place in the limestone valley of Center Co., Pennsylvania. *Ameri. J. Sci.* **3(29)**, 29–31.

Fairbanks, R. G. & R. K. Matthews 1978. The marine oxygen isotope record in Pleistocene coral, Barbados, West Indies. *Quat. Res.* (N.Y) **10**, 181–96.

Fang Lingchang 1984. Application of distances between nearest neighbours to the study of karst. *Carsologica Sinica* **3(1)**, 97–101.

Farnsworth, R. K., E. C. Barrett & M. S. Dhanju 1984. Application of remote sensing to hydrology including ground water. *Technical Documents in Hydrology* Unesco, Paris.

Faulkner, G. L. 1976. Flow analysis of karst systems with well developed underground circulation. In *Karst hydrology and water resources*, V. Yevjevich (ed.), vol. 1, 137–64. Fort Collins, Co: Water Resources Publications.

Fay, R. O. & D. L. Hart 1978. Geology and mineral resources (exclusive of petroleum) of Custer County, Oklahoma. *Oklahoma Geol. Surv., Bull.* **114**.

Feeney, C. M. 1977. *The karstification of drainage: a morphometric approach*. MA thesis, University of Auckland.

Fermor, J. 1972. The dry valleys of Barbados: a critical review of their pattern and origin. *Trans. Inst. Brit. Geogr.* **57**, 153–65.

Fish, J. E. 1977. *Karst hydrogeology and geomorphology of the Sierra de El Abra and the Valles-San Luis Potosi Region, Mexico*. PhD thesis, McMaster University.

Flenley, J. R. 1979. *The equatorial rain forest: a geological history*, London: Butterworths.

Folk, R. L. 1962. Spectral subdivision of limestone types. *Am. Assoc. Petrol. Geol., Mem. 1*, 62–84.

Folk, R. L. & R. Assereto 1976. Comparative fabrics of length-slow and length-fast calcite and calcitized aragonite in a Holocene speleothem, Carlsbad Caverns, New Mexico. *J. Sed. Pet.* **46(3)**, 486–96.

Folk, R. L., H. H. Roberts & C. M. Moore 1973. Black phytokarst from Hell, Cayman Islands, West Indies. *Geol. Soc. Americ. Bull.* **84**, 2351–60.

Fontes, J-Ch. 1980. Environmental isotopes in groundwater hydrology. In *Handbook of environmental isotope geochemistry*, P. Fritz & J. Fontes (eds), vol. 1, 75–140. Amsterdam: Elsevier.

Fontes, J.-Ch. 1983. Dating of groundwater. In *Guidebook on nuclear techniques in hydrology*, Technical Reports Series No. 91, 285–317. Vienna: IAEA.

Foose, R. M. & J. A. Humphreville 1979. Engineering geological approaches to foundations in the karst terrain of the Hershey Valley. *Bull. Assoc. Eng. Geologists*, **XVI(3)**, 355–81.

Ford, D. C. 1964. Origin of closed depressions in the central Mendip Hills. *20 Internat. Geog. Congr., London, Abstracts*, 105–6.

Ford, D. C. 1965a. The origin of limestone caverns: a model from the central Mendip Hills, England. *Nat. Speleo Soc. Amer. Bull.* **27**, 109–32.

Ford, D. C. 1965b. Stream potholes as indicators of erosion phases in caves. *Bull., Nat. Speleo Soc. Am.*, **27(1)**, 27–32.

Ford, D. C. 1968. Features of cavern development in central Mendip. *Trans., Cave Res. Gp. G.B.*, **10**, 11–25.

Ford, D. C. 1971a. Characteristics of limestone solution in the southern Rocky Mountains and Selkirk Mountains, Alberta and British Columbia. *Can. J. Earth Sci.* **8(6)**, 585–609.

Ford, D. C. 1971b. Geologic structure and a new explanation of limestone cavern genesis. *Cave Res. Gp., G.B., Trans.*, **13(2)**, 81–94.

Ford, D. C. 1973. Development of the canyons of the South Nahanni River, N.W.T. *Can J. Earth Sci.* **10(3)**, 366–78.

Ford, D. C. 1977. Genetic classification of solutional cave systems. *Proc. 7th Internat. Congr. Speleo. Sheffield*, 189–92.

Ford, D. C. 1979. A review of alpine karst in the southern Rocky Mountains of Canada. *National Speleological Society of America, Bulletin* **41**, 53–65.

Ford, D. C. 1980. Threshold and limit effects in karst geomorphology. In *Thresholds in Geomorphology*, D. L. Coates & J. D. Vitek (eds), 345–62. London: Allen & Unwin.

Ford, D. C. 1983. Effects of glaciations upon karst aquifers in Canada. *J. Hydrol.* **61**, 149–58.

Ford, D. C. 1984. Karst groundwater activity and landform genesis in modern permafrost regions of Canada. In *Groundwater as a geomorphic agent*, R. G. La Fleur (ed.), 340–50. London: Allen & Unwin.

Ford, D. C. 1985. Dynamics of the karst system; a review of some recent work in North America. *Ann. Soc. Geol. Belgique* **108**, 283–91.

Ford, D. C. 1986. Genesis of paleokarst and strata-bound zinc-lead sulfide deposits in a Proterozoic dolostone, northern Baffin Island, Canada: a discussion. *Econ. Geol.* **81(6)**, 1562–3.

Ford, D. C. 1987. Effects of Glaciations and Permafrost upon the Development of Karst in Canada. *Earth Surf. Processes and Landforms*, **12(5)**, 507–21.

Ford, D. C. & J. J. Drake 1982. Spatial and temporal variations in karst solution rates; the structure of variability. In *Space and time in geomorphology*, C. E. Thorn (ed), 147–70. London: Allen & Unwin.

Ford, D. C. & R. O. Ewers 1978. The development of limestone cave systems in the dimensions of length and breadth. *Can. J. Earth Sci.* **15**, 1783–98.

Ford, D. C. & J. A. Lundberg 1987. A review of dissolutional rills in limestone and other soluble rocks. *Catena, Suppl. 8*, 119–40.

Ford, D. C. & W. I. Stanton 1968. Geomorphology of the south-central Mendip Hills. *Geol. Assoc. Proc.* **79(4)**, 401–27.

Ford, D. C., H. P. Schwarcz, J. J. Drake, M. Gascoyne, R. S. Harmon, & A. G. Latham 1981. Estimates of the age of the existing relief within the southern Rocky Mountains of Canada. *Arctic & Alpine Res.* **13(1)**, 1–10.

Ford, T. D. 1977. *Limestones and caves of the Peak District*. Norwich: GeoBooks.

Ford, T. D. & R. J. King 1966. The Golconda Caverns, Brassington, Derbyshire. *Trans. Cave Res. Group. G.B.* **7**, 91–114.

Ford, T. D. & N. E. Worley 1977. Phreatic caves and sediments at Matlock, Derbyshire. *Proc. 7 Int. Congr. Speleol.*, 194–6.

Forti, P. & M. L. Gamberi 1983. Le pisoliti della Buca del Vasalo di Motrone e l'ipotesi del minimo e massimo diametro possibile. *Sottoterra*, **59**, 6 p.

Forti, P. & D. Postpischl 1984. Seismotectonic and paleoseismic analyses using karst sediments. *Marine Geol.* **55**, 145–61.

Forti, P. & D. Postpischl 1985. Relazioni tra terremoti e deviazioni degli assi di accrescimento delle stalagmiti. *Le Grotte d'Italia* **4(XII)**, 287–303.

Frank, R. M. 1965. *Petrologic study of sediments from selected central Texas caves*. MA thesis, University of Texas, Austin.

Franke, M. W. 1961. Formegesetze des Hohlensinters. *Symp. Int. Speleo. Varenna*, 185–209.

Franke, M. W. 1965. The theory behind stalagmite shapes. *Stud. Speleo.* **1**, 89–95.

Franke, M. W. & M. A. Geyh 1971. Radiokohlenstoff Analysen am Tropfsteinen. *Umschau in Wissen. und Tech.* **71(3)**, 91–2.

Franke, H. W., K. O. Munnich & J. C. Vogel 1958. Auflosung und Abscheidung von Kalk $-^{14}$C Datierung von Kalkabscheidungen. *Die Hohle* **9**, 1–5.

Freeze, R. A. & J. A. Cherry 1979. *Groundwater*. New Jersey: Prentice-Hall.

Friederich, H. & P. L. Smart 1981. Dye tracer studies of the unsaturated zone: recharge of the Carboniferous Limestone aquifer of the Mendip Hills, England. *Proc. 8th Int. Congr. Speleo., Kentucky, USA*, 283–6.

Friederich, H. & P. L. Smart 1982. The classification of autogenic percolation waters in karst aquifers: a study in G.B. Cave, Mendip Hills, England. *Proc. Univ. Bristol Spelaeo. Soc.* **16(2)**, 143–59.

Fritz, P. & J.-Ch. Fontes (eds) 1980. *Handbook of environmental isotope geochemistry, vol. 1. The terrestrial environment*. Amsterdam: Elsevier.

Fulweiler, R. E. & S. E. McDougal 1971. Bedded-ore structures, Jefferson City Mine, Jefferson City, Tennessee. *Econ. Geol.* **66**, 763–9.

Gale, S. J. 1984. The hydraulics of conduit flow in carbonate aquifers. *J. Hydrol.* **70**, 309–27.

Gams, I. 1962. Measurements of corrosion intensity in Slovenia and their geomorphological significance. *Geog. Vestnik* **34,** 3–20.

Gams, I. 1965. Uber die faktoren, die intensitat der sintersedimentation bestimmen. *Actes, 4th Int. Congr. Speleo. Ljubljana III*, pp. 107–15.

Gams, I. 1967. Faktorji in dinamika korozije na karbonatnih kameninah slovenskega dinarskega in alpskega krasa. *Geogr. vestnik* **38**, Ljubljana.

Gams, I. 1972. Effect of runoff on corrosion intensity in the northwest Dinaric karst. *Trans. Cave Res. Gp. GB* **14(2)**, 78–83.

Gams, I. 1973a. *Slovenska krasa terminologija (Slovene karst terminology)*, Zveza Geografskih Institucij Jugoslavije: Ljubljana.

Gams, I. 1973b. The terminology of the types of polje. *Slovenska Krasa Terminologija*, Zveza Geografskih Institucij Jugoslavije: Ljubljana, 60–7.

Gams, I. 1974. *Kras.*. Ljubljana, izdala Slovenska matica.

Gams, I. 1976. Hydrogeographic review of the Dinaric and alpine karst in Slovenia with special regard to corrosion. In *Problems of karst hydrology in Yugoslavia*, Mems. Serb. Geog. Soc. 13, 41–52.

Gams, I. 1978. The polje: the problem of its definition. *Z. Geomorph.* **22**, 170–81.

Gams, I. 1980. Poplave na Planinskem polju (Inundations in Planina polje). *Geog. zbornik* **20**, 4–30.

Gams, I. 1981. Comparative research of limestone solution by means of standard tablets. *Proc. 8th Int. Congr. Speleol. (Bowling Green, Kentucky)* **1**, 273–5.

Gandino, A. & A. M. Tonelli 1983. Recent remote sensing technique in freshwater marine springs monitoring: qualitative and quantitative approach. In *Methods and instrumentation for the investigation of groundwater systems*, 301–10. Proc. Internat. Symp., Noordwijkerhout, Netherlands: UNESCO/IAHS.

REFERENCES

Garres, R. M. & C. L. Christ 1965. *Solutions, minerals and equilibria*. New York: Harper & Row.

Gascoyne, M. 1977. Trace element geochemistry of speleothems. *Proc. 7th Int. Speleo. Cong.*, 205–7.

Gascoyne, M. 1979. *Pleistocene climates determined from stable isotope and geochronologic studies of speleothem*. PhD thesis, McMaster University.

Gascoyne, M. 1984. Twenty years of Uranium-Series dating of cave calcites. *Studies in Speleology* **v**, 15–30.

Gascoyne, M. & D. E. Nelson 1983. Growth mechanisms of recent speleothems from Castleguard Cave, Columbia Icefields, Alberta, Canada, inferred from a comparison of Uranium-series and Carbon-14 age data. *Arctic and Alpine Res.* **15(4)**, 537–42.

Gascoyne, M., G. J. Benjamin, H. P. Schwarcz & D. C. Ford 1979. Sea-level lowering during the Illinoian glaciation: Evidence from a Bahama 'blue hole'. *Science* **205**, 806–8.

Gascoyne, M., D. C. Ford & H. P. Schwarcz 1981. Late Pleistocene chronology and paleoclimate of Vancouver Island determined from cave deposits. *Can. J. Earth Sci. 18*, 1643–52.

Gascoyne, M., D. C. Ford & H. P. Schwarcz 1983. Rates of cave and landform development in the Yorkshire Dales from speleothem age data. *Earth Surf. Proc. Landforms* **8**, 557–68.

Gascoyne, M., H. P. Schwarcz & D. C. Ford 1978. Uranium series dating and stable isotope studies of speleothems. Part 1: Theory and techniques. *Trans., Brit. Cave Res. Assoc.* **5**, 91–112.

Gat, J. R. 1980. The isotopes of hydrogen and oxygen in precipitation. In *Handbook of environmental isotope geochemistry*. P. Fritz & J. Fontes (eds), vol. 1, 21–47. Amsterdam: Elsevier.

Gavrilovic, D. 1969. Kegelkarst-Elemente im Relief des Gebirges Beljanica, Jugoslavien. In *Problems of the karst denudation*. O. Stelcl (ed.), 159–66. Brno: Inst. Geog.

Gavrilovic, D. 1970. Intermittierende Quellen in Jugoslawien. *Die Erde* **101(4)**, 284–98.

Gerba, C. P., C. Wallis & J. L. Melnick 1975. Fate of wastewater bacteria and viruses in soil. *J. Irrig. and Drainage Div., Am. Soc. Civ. Eng.* **101**, 157–74.

Gerson, R. 1976. Karst and fluvial denudation of carbonate terrains under sub-humid Mediterranean and arid climates – principles, evaluation and rates (examples from Israel). In *Karst processes and relevant landforms*. I. Gams (ed.), 71–9. Int. Spel. Union, Comm. Karst Denudation Symp. Proc., Univ. Ljubljana.

Gerstenhaller, A. 1960. Der tropische Kegelkarst im Tabasco (Mexico). *Z. Geomorph.*, Suppl. 2, 22–48.

Gerstenhauer, A. 1968. Ein Karstmorphologischer Vergleich zwischen Florida und Yucatan. *Deutscher Geog., Bad Godesberg, Wiss. Abh.*, 332–44.

Gerstenhauer, A. & K.-H. Pfeffer 1966. Beitrage zur Frage der Losungsfreudigkeit von Kalkgestein. *Abh. z Karst-und Hohlenkunde* **2**, 1–46.

Geurts, M. A. 1976. Genese et stratigraphie des travertins au fond de vallee en Belgique. *Acta Geog. Lovaniensa* **16**, 87 p.

Gèze, B. 1964. Sur les profils normaux des entres de grottes et des galeries simples. *Spelunca Mem.* **4**, 26 pp.

Gèze, B. 1965. *La Speleologie Scientifique*. Paris: Ed. du Seuil.

Gèze, B. & J. Moinereau 1967. Sur l'evolution des sediments argileux dans les cavernes en fonction de la profondeur. *Spelunca Mem.* **5**, 58–62.

Gill E. D. & J. D. Lang 1983. Micro-erosion meter measurements of rock, wear on the Otway coast of southeast Australia. *Marine Geology*, **52,** 141–56.

Gilli, E. 1986. Les grandes cavites souterraines: etudes et applications. *Karstologia*, 3–11.

Gillieson, D. S. 1985. Geomorphic development of limestone caves in the Highlands of Papua New Guinea. *Z. Geomorph. N.F.* **29**, 51–70.

Gillieson, D. 1986. Cave sedimentation in the New Guinea highlands. *Earth Surface Proc. and Landforms* **11**, 533–43.

Gladysz, K. 1987. *Karren on the Quatsino Limestone, Vancouver Island*. BSc thesis, McMaster University.

Glazek, J. (in press). Tectonic conditions and paleokarst. In *Paleokarst – a world review*. P. Bosak, D. C. Ford & J. Glazek (eds), Amsterdam: Academia Praha/Elsevier.

Glazek, J., R. S. Harmon & K. Nowak 1980. Uranium-series dating of the hominid-bearing

travertine deposit at Bilzingsleben, G.D.R. and its stratigraphic significance. *Acta. Geol. Polonica* **30(1)**, 1–14.
Glazek, J., J. Rudnicki & A. Szynkiewicz 1977. Proglacial caves – a special genetic type of cave in glaciated areas. *Proc. 7th Internat. Speleo. Congr.*, Sheffield, 215–17.
Glennie, E. A. 1954. Artesian flow and cave formation. *Trans. Cave Res. Gp., G.B.* **3(1)**, 55–71.
Glew, J. R. & D. C. Ford 1980. A simulation study of the development of rillenkarren. *Earth Surface Processes* **5**, 25–36.
Glover, R. R. 1972. Optical brighteners – a new water tracing reagent. *Trans. Cave Res. Group G.B.* **14(2)**, 84–8.
Goede, A. & R. S. Harmon 1983. Radiometric dating of Tasmanian speleothems – evidence of cave evolution and climatic change. *J. Geol. Soc. Australia* **30**, 89–100.
Goede, A., D. C. Green & R. S. Harmon 1986. Late Pleistocene paleotemperature record from a Tasmanian speleothem. *Aust. J. Earth Sci.* **33**, 333–42.
Goldie, H. S. 1981. Morphometry of the limestone pavements of Farleton Knott, Cumbria, England. *Trans., Brit. Cave Res. Assoc.* **8**, 207–24.
Golwer, A. 1983. Underground purification capacity. In *Ground water in water resources planning*, Int. Assoc. Hydrological Sciences Pubn. 42 (Koblenz Symposium), 1063–72. Unesco.
Goodchild, J. G. 1890. Notes on some observed rates of weathering of limestone. *Geol. Mag.* **27**, 463–6.
Goodchild, M. F. & D. C. Ford 1971. Analysis of scallop patterns by simulation under controlled conditions. *J. Geol.* **79**, 52–62.
Goodman, R. E., D. G. Moye, A. von Schalkwyk and I. Javandel 1965. Ground water inflows during tunnel driving. *Eng. Geol.* **2**, 39–56.
Gorbunova, K. A. 1977a. Morphology of gypsum karst. *Proc. 7 Int. Spel. Cong. (Sheffield)*, 221–2.
Gorbunova, K. A. 1977b. Exogenetic gypsum tectonics. *Proc. 7 Int. Spel. Cong. (Sheffield)*, 222–3.
Gorbunova, K. A. 1979. *Morphology and hydrogeology of gypsum karst.* Univ. Perm. All-Union Karst and Speleology Institute (in Russian).
Gordon, D., P. L. Smart, J. N. Andrews, D. C. Ford, T. C. Atkinson, P. Rowe & N. S. J. Christopher 1989. The dating of United Kingdom interglacials and interstadials from speleothem growth frequency. *Quaternary Research.*
Gospodaric, R. & P. Habic 1976. *Underground water tracing*, Ljubljana: Inst. Karst Research, Postojna.
Gospodaric, R. & P. Habic 1978. Kraski pojavi Cerkniskega polja (Karst phenomena of Cerknisko polje). *Acta Carsologica* **8(1)**, 6–162.
Goudie, A. S. (ed.) 1981. *Geomorphological techniques*. London: Allen & Unwin.
Goudie, A. S. 1983. Calcrete. In *Chemical sediments and geomorphology*, A. S. Goudie & K. Pye (eds), 93–131. London: Academic Press.
Grabau, A. W. 1913. *Principles of stratigraphy*. New York: Seiler.
Graf, W. H. 1971. *Hydraulics of sediment transport*. McGraw-Hill: New York.
Green, H. S., C. B. Stringer, S. N. Collcutt, A. P. Currant, J. Huxtable, H. P. Schwarcz, N. Debenham, C. Embleton, P. A. Bull, T. I. Molleson, & R. E. Bevins 1981. Pontnewydd Cave in Wales – a new Middle Pleistocene hominid site. *Nature* **294(5843)**, 707–13.
Gregor, V. A. 1981. Karst and caves in the Turks and Caicos Island, B.W.I. *Proc., 8th Internat. Speleo. Congr., Kentucky*, 805–6.
Gregory, K. J. & D. E. Walling 1973. *Drainage basin form and process*. London: Arnold.
Grindley, G. W. 1971. Sheet S8 Takaka. Geol. Map of New Zealand 1:63, 360 DSIR, Wellington.
Grodzicki, J. 1985. Genesis of the Nullarbor Plain caves. *Z. Geomorph.* **29(1)**, 37–49.
Grun, R. 1985. *Beitrage zur ESR – Datierung*, Geol. Inst. Koln, 59, 157 p.
Grund, A. 1903. Die Karsthydrographie: Studien aus Westbosnien. *Geog. Abh. heraus. von A. Penck.* **7**, 103–200.
Grund, A. 1914. Der geographische Zyklus im Karst. *Ges. Erdkunde* **52**, 621–40. (Translated into English in Sweeting, 1981).
Guilcher, A. 1988. *Coral reef geomorphology*. Chichester: Wiley.

Gunn, J. 1978. *Karst hydrology and solution in the Waitomo district, New Zealand*. PhD Thesis, Auckland University.
Gunn, J. 1981a. Limestone solution rates and processes in the Waitomo district, New Zealand. *Earth Surf. Proc. Landforms* **6**, 427–45.
Gunn, J. 1981b. Prediction of limestone solution rates from rainfall and runoff data: some comments. *Earth Surf. Proc. Landforms* **6**, 595–7.
Gunn, J. 1981c. Hydrological processes in karst depressions. *Z. Geomorph.* **25(3)**, 313–31.
Gunn, J. 1982. Magnitude and frequency properties of dissolved solids transport. *Z. Geomorph.* **26(4)**, 505–11.
Gunn, J. 1983. Point recharge of limestone aquifers – a model from New Zealand karst. *J. Hydrol.* **61**, 19–29.
Gunn, J. 1986. Solute processes and karst landforms. In *Solute processes*, S. T. Trudgill (ed.), 363–437. Chichester: Wiley.
Gustavson, T. C., W. W. Simpkins, A. Alhades & A. Hoadley 1982. Evaporite dissolution and development of karst features in the Rolling Plains of Texas Panhandle. *Earth Surf. Proc. and Landforms* **7**, 545–63.
Gutierrez, M., M. J. Ibanez, J. L. Pena, J. Rodriguez & M. A. Soriano 1985. Quelques exemples de karst sur gypse dans la depresson de l'Ebre. *Karstologia* **6(2)**, 29–36.

Habe, F. 1981. Bericht der Kommission fur Karstschutz und Schauhohlen der UIS. *Proc. 8 Int. Speleol. Congr. (Bowling Green)* **2**, 442–3.
Hagen, G. 1839. Uber die Bewegung des Wassers in engen cylindrischen Rohren. *Poggendorff Annalen* **46**, 423–42.
Hallet, B. 1976. Deposits formed by subglacial precipitation of $CaCO_3$. *Bulletin, Geological Society of America* **87**, 1003–15.
Hallet, B. 1979a. Subglacial regelation water film. *J. Glac.* **23(89)**, 321–34.
Hallet, B. 1979b. A theoretical model of glacial abrasion. *J. Glac.* **23**, 39–50.
Halliwell, R. A. 1981. The geohydrology of the Ingleborough area, England. *Proc. 8 Int. Cong. Spel. (Bowling Green)* **1**, 126–8.
Hanshaw, B. B. & W. Back 1979. Major Geochemical Processes in the Evolution of Carbonate-Aquifer Systems. *J. Hydrol.* **43**, 278–312.
Harmon, R. S. 1975. *Late Pleistocene paleoclimates in North America as inferred from isotopic variations in speleothems*. PhD thesis, McMaster University.
Harmon, R. S., T. C. Atkinson & J. J. Atkinson 1983. The mineralogy of Castleguard Cave, Columbia Icefields, Alberta, Canada. *Arctic & Alpine Res.* **15(4)**, 503–16.
Harmon, R. S., D. C. Ford & H. P. Schwarcz 1977. Interglacial chronology of the Rocky and Mackenzie Mountains based on $^{230}Th/^{234}U$ dating of calcite speleothems. *Can. J. Earth Sci.* **14**, 2543–52.
Harmon, R. S., P. Thomspon, H. P. Schwarcz & D. C. Ford 1975. Uranium series dating of speleothems. *Bull. Nat. Speleo. Soc. Am.* **37**, 21–33.
Harmon, R. S., P. Thompson, H. P. Schwarcz & D. C. Ford 1978. Late Pleistocene paleoclimates of North America as inferred from stable isotope studies of speleothems. *Quat. Res.* **9**, 54–70.
Harmon, R. S., H. P. Schwarcz & D. C. Ford 1978. Late Pleistocene sea level history of Bermuda. *Quat. Res.* **9**, 205–18.
Harvey, E. J., J. H. Williams, & T. R. Dinkel 1977. Aplication of thermal imagery and aerial photography to hydrological studies of karst terrain in Missouri. *U.S. Geol. Survey Water-Resources Investigations 77–16*, 53 pp.
Haserodt, K. 1965. Untersuchungen zur Hohen-und Altersgliederung der Karstformen in den nordlichen Kalkalpen. *Munchner Geographische* **14** & **27**.
Hellden, U. 1973. Limestone solution intensity in a karst area in Lapland, Northern Sweden. *Geol. Ann.* **54A (3/4)**, 185–96.
Hendy, C. H. 1970. The use of C^{14} in the study of cave processes. In *Radiocarbon variations and absolute chronology*, I. U. Olson (ed.), 419–43. New York: Wiley.
Hendy, C. H. & A. T. Wilson 1968. Paleoclimatic data from speleothems. *Nature* **216**, 48–51.
Hennig, G. J. & R. Grun 1983. ESR dating in Quaternary geology. *Quat. Sci. Rev.* **2**, 157–238.
Hennig, G. J., U. Bangert, W. Herr & J. Freundlich 1980. Uranium series dating of calcite

formations in caves: recent results and a comparative study on age determinations via ^{230}Th/^{234}U, ^{14}C, TL and ESR. *Rev. d'Archaeometrie* **4**, 91–100.

Hennig, G. J., R. Grun & K. Brunnacher 1983. Speleothems, travertines and paleoclimates. *Quat. Res.* **29**, 1–29.

Herak, M., 1972. Karst of Yugoslavia. In Karst: *Important karst regions of the northern hemisphere*, M. Herak & V. T. Stringfield (eds), 25–83. Amsterdam: Elsevier.

Herak, M. 1976. The Yugoslav contribution to the knowledge of karst hydrology and geomorphology. In *Karst hydrology and water resources*, V. Yevjevich (ed.), vol. I, 1–30. Fort Collins Co.: Water Resources Publications.

Herak, M. & V. T. Stringfield (eds) 1972a. *Karst: important karst regions of the northern hemisphere*. Amsterdam: Elsevier.

Herak, M. & V. T. Stringfield 1972b. Historical review of hydrogeologic concepts. In *Karst: important karst regions of the northern hemisphere*, M. Herak & V. T. Stringfield (eds), 19–24. Amsterdam: Elsevier.

Herman, J. S. 1982. *The dissolution kinetics of calcite, dolomite and dolomitic rocks in the CO_2-water system*. PhD thesis, Pennsylvania State University.

Herman, J. S. & W. B. White 1984. Determination of carbonate hardness in karst waters from conductivity measurements. *Nat. Speleo. Soc. Am., Abstracts, Annual Conv., Cody, Wyo.*, p. 10.

Hewlett, J. D. & A. R. Hibbert 1967. Factors affecting the response of small watersheds to precipitation in humid areas. In *Forest hydrology*, W. E. Sopper & H. W. Lull (eds). 275–90. Oxford: Pergamon.

High, C. 1970. Cited in Spate *et al.* (1985).

High, C. & G. K. Hanna 1970. A method for the direct measurement of erosion of rock surfaces. *Brit. Geomorph. Res. Gp., Tech. Bull.* **5**, 24 pp.

Hill, C. A. (ed.) 1981. Saltpeter: a symposium. *Bull., Nat. Speleo. Soc.*, **43(4)**, 83–131.

Hill, C. A. 1982. Origin of black deposits in caves. *Bull., Nat. Speleo. Soc.* **44**, 15–19.

Hill, C. A. 1987. Geology of Carlsbad Caverns and other caves of the Guadalupe Mountains, New Mexico and Texas. *N.M. Bureau Mines & Miner. Resources* **Bull. 117**. 150 pp.

Hill, C. A. & P. Forti 1986. *Cave minerals of the world*. Huntsville, USA: Nat. Speleo. Soc.

Hillaire-Marcel, C., J. M. Soucy & A. Cailleux 1979. Analyse isotopique de concretions sous-glaciaires de l'inlandsis laurentidien et teneur en oxygene 18 de la glace. *Can. J. Earth Sci.* **16**, 1494–8.

Hillel, D. 1982. *Introduction to soil physics*. London: Academic Press.

Hladnik, J. & A. Kranjc 1977. Fluvio-glacial cave sediments – A contribution to speleochronology. *Proc. 7th Internat. Congr. Spel., Sheffield*, 240–3.

Hobbs, S. L. & P. L. Smart 1986. Characterization of carbonate aquifers: a conceptual base. *Proc. 9th Internat. Speleo. Congress, Barcelona* **1**, 43–6.

Holbye, U. 1983. Greft Marmoren. *Norske Grotteblad* **II**, 25–8.

Homann, W. 1969. Experimentelle Ergebnisse zum Wachstum rezenter Hohlenperlen. *5th Internat. Congr. Speleo., Stuttgart* **2**, 5/1–5/19.

Hopley, D. 1982. *The geomorphology of the Great Barrier Reef*. New York: Wiley.

Howard, A. D. 1963. The development of karst features. *National Speleological Society, Bulletin* **25**, 45–65.

Hubbert, M. K. 1940. The theory of groundwater motion. *J. Geol.* **48**, 785–944.

Hunt, R. E. 1984. *Geotechnical engineering investigation manual*. New York: McGraw-Hill.

Hutton, J. 1781. *A tour to the caves, in the environs of Ingleborough and Settle, in the West Riding of Yorkshire*. 2nd edn. London & Kendall: Richardson & Urquhart.

Hutton, J. 1795. *Theory of the Earth, with proofs and illustrations*, vol. II, Edinburgh.

Ikeya, M. 1985. Dating methods of Pleistocene deposits and their problems: IX electron spin resonance. *Geoscience Can. Reprint Ser.* **2**, 73–87.

Institute of Hydrogeology and Engineering Geology 1982. *Hydrogeologic atlas of the People's Republic of China*. New York: Academic Press.

International Atomic Energy Agency 1981. Stable isotope hydrology, *Technical Report Series No. 210*, Vienna: IAEA.

International Atomic Energy Agency 1983. Guidebook on nuclear techniques in hydrology. *Technical Reports Series No. 91*, Vienna: IAEA.

REFERENCES

International Atomic Energy Agency 1984. *Isotope hydrology 1983*. Vienna: IAEA.

International Union of Geological Sciences 1979. *Magnetostratigraphic polarity units*. *Geol.* **7**, 578–83.

Ireland, P. 1979. Geomorphological variations of 'case hardening' in Puerto Rico. *Z. Geomorph. Suppl.-Bd.* **32**, 9–20.

Issar, A. 1983. Emerging groundwater, a triggering factor in the formation of the makhteshim (erosion cirques) in the Negev and Sinai. *Israel J. Earth Sci.* **32**, 53–61.

Issar, A., J. L. Quijano, J. R. Gat & M. Castro 1984. The isotope hydrology of the groundwaters of central Mexico. *J. Hydrol.* **71**, 201–24.

Ivanovich, M. & R. S. Harmon 1982. *Uranium series disequilibrium: applications to environmental problems*. Oxford: Oxford University Press.

Ivanovich, M. & P. Ireland 1984. Measurements of uranium series disequilibrium in the case-hardened Aymamon limestone in Puerto Rico. *Z. Geomorph.* **28**, 305–19.

Jacobson, G. & P. J. Hill 1980. Hydrogeology of a raised coral atoll, Niue Island, South Pacific Ocean. *J. Australian Geol. and Geophys.* **5(4)**, 271–8.

Jakeman, A. J., M. A. Greenaway & J. N. Jennings 1984. Time-series models for the prediction of stream flow in a karst drainage system. *J. Hydrol. (N.Z.)* **23(1)**, 21–33.

Jakucs, L. 1959. Neue Methoden der Hohlenforschung in Ungarn und ihre Ergebnisse. *Die Hohle* **10(4),** 88–98.

Jakucs, L. 1977. *Morphogenetics of karst regions: variants of karst evolution*. Budapest: Akademiai Kiado.

James, A. N. & A. R. R. Lupton 1978. Gypsum and anhydrite in foundations of hydraulic structures. *Geotechnique* **28**, 249–72.

James, A. N. & I. M. Kirkpatrick, 1980. Design of foundations of dams containing soluble rocks and soils. *Q. J. Eng. Geol.* **13**, 189–98.

James, N. P. & Choquette, P. W. 1984. Diagenesis. 6. Limestone – the seafloor diagnetic environment. *Geoscience Can.* **10(4)**, 162–79.

James, N. P. & Choquette, P. W. (eds) 1988. *Paleokarst*. New York: Springer.

Jarvis, R. S. 1981. Specific geomorphometry. In *Geomorphological techniques*, A. Goudie (ed.), London: Allen & Unwin.

Jefferson, G. T. 1976. Cave faunas. In *The science of speleology*, T. D. Ford & C. H. D. Cullingford (eds), 359–421. London: Academic Press.

Jennings, J. N. 1968. Syngenetic karst in Australia. In *Contributions to the study of karst*, P. W. Williams & J. N. Jennings (eds), 41–110. Australian National University: Res. School of Pacific Studies, Pub. G5.

Jennings, J. N. 1971. *Karst*. Cambridge (Mass.) & London: MIT Press.

Jennings, J. N. 1972a. The Blue Waterholes, Cooleman Plain, N.S.W., and the problem of karst denudation rate determination. *Trans. Cave Res. Gp. G.B.* **14**, 109–17.

Jennings, J. N. 1972b. Observations at the Blue Waterholes, March 1965 – April 1969, and limestone solution on Cooleman Plain, N.S.W. *Helictite* **10(1–2)**, 1–46.

Jennings, J. N. 1972c. The character of tropical humid karst. *Z. Geomorph.* **16(3)**, 336–41.

Jennings, J. N. 1975. Doline morphometry as a morphogenetic tool; New Zealand examples. *N.Z. Geogr.* **31,** 6–28.

Jennings, J. N. 1976. A test of the importance of cliff-foot caves in tower karst development. *Z. Geomorph.,* **Suppl. 26**, 92–7.

Jennings, J. N. 1982a. Karst of northeastern Queensland reconsidered. *Chillagoe Caving Club, Tower Karst Occasional Paper No. 4* (ISSN0729–1183), 13–52.

Jennings, J. N. 1982b. Principles and problems in reconstructing karst history. *Helictite* **20(2),** 37–52.

Jennings, J. N. 1983a. The problem of cavern formation. In *Perspectives in Geomorphology*, H. S. Sharma (ed.), 223–53. New Delhi: Concept.

Jennings, J. N. 1983b. The disregarded karst of the arid and semiarid domain. *Karstologia* **1**, 61–73.

Jennings, J. N. 1985. *Karst geomorphology*. Oxford: Basil Blackwell.

Jennings, J. N. & M. J. Bik 1962. Karst morphology in Australian New Guinea. *Nature* **194**, 1036–8.

Jennings, J. N. & M. M. Sweeting 1963. The limestone ranges of the Fitzroy Basin, Western Australia. *Bonner. Geogr. Abh.* **32**.

Johnson, A. I., L. Carbognin & L. Ubertini (eds) 1986. *Land subsidence*. Int. Assoc. Hydrological Sciences Pub. 151. Wallingford: IAHS.

Julian, M., J. Martin & J. Nicod 1978. Les karsts Mediterraneens. *Mediterranee* **1 & 2**, 115–31.

Kaddu-Mulindwa, D., Z. Filip & G. Milde 1983. Survival of some pathogenic and potential pathogenic bacteria in groundwater. In *Ground water in water resources planning*, Int. Assoc. Hydrological Sciences Pubn 42, (Koblenz Symposium), 1137–45. Unesco.

Karanjac, J. & G. Gunay 1980. Dumanli Spring, Turkey – the largest karstic spring in the world? *J. Hydrol.* **45**, 219–31.

Karolyi, M. S. & D. C. Ford 1983. The Goose Arm karst, Newfoundland, Canada. *J. Hydrol.* **61**, 181–5.

Kass, W. 1967. Erfahrungen mit Uranin bei Farbversachen. *Steir. Beitr. z. Hydrogeologie* **18/19**, 123–340.

Kastning, E. H. 1983. Relict caves as evidence of landscape and aquifer evolution in a deeply dissected carbonate terrain: southwest Edwards Plateau, Texas, U.S.A. *J. Hydrol.* **61**, 89–112.

Katzer, E. 1909. *Karst und Karsthydrographie. Zur Kunde der Balkan halbinsel*, **8**, Sarajevo.

Kayser, K., 1934. Morphologische Studien in Westmontenegro II: Rumpftreppe Cetinje, Formenschatz der Karstabtragung. *Z. Ges. Erdk. Berlin*, 26–49, 81–102.

Kayser, K., 1955. Karstrandebene und Poljeboden. Zur Frage der Entstehung der Einebnungsflachen in Karst. *Erdkunde* **9**, 60–4.

Kemmerly, P. R. 1976. Definitive doline characteristics in the Clarksville quadrangle, Tennessee. *Bull. Geol. Soc. Am.* **87,** 42–6.

Kemmerly, P. R. 1982. Spatial analysis of a karst depression population: clues to genesis. *Geol. Soc. Amer. Bull.* **93,** 1078–86.

Kemmerly, P. R. & S. K. Towe 1978. Karst depressions in a time context. *Earth. Surf. Proc.* **35**, 355–62.

Kempe, A. L. W. & H. G. Thode 1968. The mechanism of the bacterial reduction of sulphate and sulphite from isotope fractionation studies. *Geochim et Cosmochim. Acta* **32**, 71–91.

Kempe, S. & C. Spaeth 1977. Excentrics: their capillaries and growth rates. *Proc., 7th Int. Speleo. Congr.*, 259–62.

Kempe, S., A. Brandt, M. Seeger & G. Vladi 1975. 'Facetten' and 'Laugdecken', the typical morphological elements of caves developed in standing water. *Ann. Speleo.* **30(4)**, 705–8.

Kendall, A. C. & P. L. Broughton 1978. Origin of fabrics in speleothems of columnar calcite crystals. *J. Sed. Pet. 48(2)*, 519–38.

Kharaka, Y. K. & I. Barnes 1973. SOLMNEQ: solution-mineral equilibrium computations. *U.S. Geol. Survey, Water Resources Paper 73–002,* 88 pp.

Kiraly, L. 1975. Rapport sur l'etat actuel des connaissances dans le domaine des caractères physiques des roches karstiques. In *Hydrogeology of karstic terrains*. A. Burger & L. Dubertret (eds), 53–67. Internat. Union Geol. Sci., Series B, 3.

Kiraly, L., B. Mathey & J.-P. Tripet 1971. Fissuration et orientation des cavites souterraines. *Bull Soc. Neuchatel de Sci. Nature.* **94**, 99–114.

Kirk, R. M. 1977. Rates and forms of erosion on intertidal platforms at Kaikoura Peninsula South Island, New Zealand. *N.Z. Geol. Geophys.* **20,** 571–613.

Kirk, K. G. & E. Werner 1981. *Handbook of geophysical cavity-locating techniques*. US Dept. of Transport Pub. FHWA–IP–81–3.

Kirkland, D. W. & R. Evans 1980. Origin of castiles on the Gypsum Plain of Texas and New Mexico. *New Mexico Geol. Soc., Guidebook* **v. 31**, 173–8.

Klimchouk, A. B. 1986. Genesis and development history of the large caves of the Ukraine. *Le Grotte d'Italia,* **4(XIII)**, 51–71.

Klimchouk, A. B. & V. N. Andrejchouk 1986. Geological and hydrogeological conditions of gypsum karst development in the western Ukraine. *Le Grotte d'Italia* **4(XII)**, 349–58.

Knisel, W. G. 1972. Response of karst aquifers to recharge. *Hydrol. Pap. Colo. State Univ., Fort Collins. Colo.* **60**, 48 p.

Kopper, J. S. & K. M. Creer 1973. Cova dets Alexandres, Majorca: paleomagnetic dating and archaeological interpretation of its sediments. *Caves and Karst* **15(2)**, 13–20.

Korotkov, A. N. (ed.) 1974. *Caves of the Pinego-Severodvinskaja karst* (in Russian). Geog. Soc. USSR, Leningrad.
Kovacs, J. 1983. Practical application of continuum approach to characterize the porosity of carbonate rocks. In *Methods and instrumentation for the investigation of groundwater systems*, 185–93. Proc. Internat. Symp., Noordwijkerhout, Netherlands: UNESCO/IAHS.
Kovacs, J. & P. Muller 1980. A Budai-hegyek hevizes tevekenysegenck kialakulasa es nyomai. *Karszt-s Barlang* **(ii)**, 93–8.
Kowalski, K. 1965. Cave studies in China today. *Studies in Speleol.* **1(2–3)**, 75–81.
Kozary, M. T., J. C. Dunlap & W. E. Humphrey 1968. Incidence of saline deposits in geologic time. *Geol. Soc. Am., Spec. Paper* **88**, 43–57.
Kral, Z. 1971. Studie vznika a barevnosti krapnikovych utvara. *Cesk, Kras* **23**. 7–15.
Kranjc, A. 1981. Sediments from Babja Jama near Most na Soci. *Acta Carsologica* **X/9**, 201–11.
Kranjc, A. 1982. Prod iz Kacne jame. *Nase jame* **23/4**, 17–23.
Kranjc, A. 1985. The lake of Cerknisko and its floods. *Geografski Zbornik* **25(2)**, 71–123.
Kranjc, A. & F. Lovrencak, 1981. Poplavni svet na Kocevskem polju (Floods in Kocevsko polje). *Geografski zbornik* **21**, 1–39.
Kruger, P. 1971. *Principles of activation analysis*. New York: Wiley-Interscience.
Kruse, P. B. 1980. *Karst investigations of Maligne Basin, Jasper National Park, Alberta*. MSc Thesis, University of Alberta, Canada.
Kuffner, D. 1986. Deckenkarren-ein Beitrag zur Spelaogenese. *Die Hohle* **3(37)**, 157–67.
Kunaver, J. 1975. On quantity, effects and measuring of the karst denudation in western Julian Alps – Kanin Mts. In *Karst Processes and Relevant Landforms*, I. Gams (ed.), 117–26. Univ. Ljubljana: Internat. Speleo. Union.
Kunaver, J. 1982. Geomorphology of the Kanin Mountains with special regard to the glaciokarst. *Geogr. Zbornik XXII*, 200–343.
Kunsky, J. 1950. *Kras a jeskyne*. Prague: Priro, Naklad Praz.
Kunsky, J. 1954. *Homes of primeval man*. Prague: Artia.

Lallemand, A. & G. Grison 1970. Contribution a la selection de traceurs radioactifs pour l'hydrologie. *Isotopes in hydrology*, Proc. Symp. 833–9. Vienna: IAEA.
LaMoreaux, P. E. & B. M. Wilson, 1984. Remote sensing. In *Guide to the hydrology of carbonate rocks*, P. E. LaMoreaux, B. M. Wilson & B. A. Memon (eds), 166–71. Studies and Reports in Hydrology No. 41, Paris: Unesco.
LaMoreaux, P. E., B. M. Wilson & B. A. Memon (eds) 1984. *Guide to the hydrology of carbonate rocks*. Studies and Reports in Hydrology No. 41, Paris: Unesco.
Lange, A. L. 1968. The changing geometry of cave structures. *Cave Notes* **10(1–3)**, 1–10, 13–19, 26–7, 29–32.
Langmuir, D 1971. The geochemistry of some carbonate groundwaters in central Pennsylvania. *Geochim et Cosmochim Acta* **35**, 1023–45.
Latham, A. G. 1983. *Paleomagnetism, rock magnetism and U–Th dating of speleothem deposits*. PhD thesis, McMaster University.
Latham, A. G., H. P. Schwarcz, D. C. Ford & W. G. Pearce 1979. Palaeomagnetism of stalagmite deposits. *Nature* **280 (5721)**, 383–5.
Latham, A. G., H. P. Schwarcz & D. C. Ford 1986. The paleomagnetism and U–Th dating of Mexican stalagmite, DAS2. *Earth and Planetary Sci. Letters* **79**, 195–207.
Lattman, L. H. & R. P. Parizek 1964. Relationship between fracture traces and the occurrence of groundwater in carbonate rocks. *J. Hydrol.* **2**, 73–91.
Laudermilk, J. D. & A. O. Woodford 1932. Concerning rillensteine. *American Journal of Science* **223**, 135–54.
Launay, M., M. Tripier, J. Guizerix, M. Firiot & J. Andre 1980. Pyranine used as a fluorescent tracer in hydrology: pH effects in determination of its concentration. *J. Hydrol.* **46,** 377–83.
Lauritzen, S.-E. 1981. Simulation of rock pendants – small scale experiments on plaster models. *Proc., 8th Internat. Congr. Speleo. Kentucky*, 407–9.
Lauritzen, S.-E. 1982. The paleocurrents and morphology of Pikhaggrottene, Svartisen, North Norway. *Norsk Geogr. Tidsskr.* **4**, 184–209.
Lauritzen, S.-E. 1984a. A symposium: arctic and alpine karst. *Norsk Geografisk Tidsskrift* **38**, 139–214.

Lauritzen, S.-E. 1984b. Evidence of subglacial karstification in Glomdal, Svartisen. *Norsk Geogr. Tids.* **38(3–4)**, 169–70.

Lauritzen, S.-E. 1986. Kvithola at Fauske, northern Norway: an example of ice-contact speleogenesis. *Norsk Geol. Tidds.* **66**, 153–61.

Lauritzen, S.-E., J. Abbott, R. Arnesen, G. Crossley, D. Grepperud, A. Ive & S. Johnson 1985. Morphology and hydraulics of an active phreatic conduit. *Cave Science* **12(4)**, 139–46.

Lauritzen, S.-E., D. C. Ford & H. P. Schwarz 1986. Humic substances in speleothem matrix. *Comm. 9th Int. Speleo. Congr. Barcelona*, 77–9.

Lauritzen, S.-E., A. Ive & B. Wilkinson 1983. Mean annual runoff & the scallop flow regime in a subarctic environment. *Trans., Brit Cave Res. Assoc.* **10(2)**, 97–102.

Laville, H. 1973. *Climatologie et chronologie du Paleolithique en Perigord: etude sedimentologique de deposits en grottes et sous abris*. DES Thesis, Univ. Bordeaux. 3 vols.

Lehmann, H. 1936. Morphologische studien auf Java. *Geog. Abhandl.* **III**, Stuttgart, 114 p.

Lehmann, H. (ed.) 1954. Das Karstphaenomen in den verschiedenen Klimazonen. *Erdkunde* **8**, 112–39.

Lehmann, H. W., H. K. Krommelbein & W. Lotschert 1956. Karstmorphologische, geologische und botanische Studien in der Sierra de los Organos auf Cuba. *Erdkunde* **10**, 185–204.

Leighton, M. W. & C. Pendexter 1962. Carbonate rock types. *Am. Ass. Petrol. Geol.* **Mem. 1**, 33–61.

Leroi-Gourhan, A. 1967. Pollens et datation de la grotte de la Vache (Ariege). *Bull. Soc. Prehist. Ariege* **22**, 115–27.

Leve, G. W. 1984. Relation of concealed faults to water quality & the formation of solution features in the Floridan aquifer, northeastern Florida, U.S.A. *J. Hydrol.* **61(1/3)**, 251–66.

Lewis, I. & P. Stace 1980. *Cave diving in Australia*. Adelaide: I. Lewis.

Lewis, D. C., Kriz, G. J. & R. H. Burgy 1966. Tracer dilution sampling technique to determine hydraulic conductivity of fractured rock. *Water Resources Res.* **2(3)**, 533–42.

Ley, R. G. 1979. The development of marine karren along the Bristol Channel coastline. *Zeitschrift fur Geomorphologie, Supplementband* **32**, 75–89.

Lloyd, J. W. (ed.) 1981a. *Case-studies in groundwater resources evaluation*. Oxford: Clarendon.

Lloyd, J. W. 1981b. Environmental isotopes in groundwater. In *Case-studies in groundwater resources evaluation*, J. W. Lloyd (ed.), 113–32. Oxford: Clarendon.

Lundberg, J. 1977. An analysis of the form of Rillenkarren from the tower karst of Chillagoe, North Queensland, Australia. *Proc. 7th Internat. Congr. Speleol. (Sheffield)*, 294–6.

Lyons, R. 1983. *A study of palaeomagnetism in New Zealand cave deposits*. Unpublished MSc thesis, University of Auckland.

Lyons, R. G., W. B. Wood & P. W. Williams 1985. Determination of alpha efficiency in speleothem calcite by nuclear accelerator techniques. In *ESR dating and dosimetry*, M. Ikeya & T. Miki (eds), 39–48. Tokyo: Ionics.

Macejka, M. 1976. The most abundant risings in Yugoslavia. In *Problems of karst hydrology in Yugoslavia*, D. Gavrilovic (ed.), 85–95. Belgrade: *Mems. Serbian Geog. Soc.* **13**.

Maclay, R. W. & T. A. Small 1983. Hydrostratigraphic subdivisions and fault barriers of the Edwards aquifer, south-central Texas, U.S.A. *J. Hydrol.* **61**, 127–46.

Magaldi, D. & U. Sauro 1982. Landforms and soil evolution in some karstic areas of the Lessini Mountains and Monte Baldo (Verona, Northern Italy). *Geogr. Fis. Dinam. Quat.* **5**, 82–101.

Mahaney, W. C. (ed.) 1984. *Quaternary dating methods*, New York: Elsevier.

Maillet, E. 1905. *Essais d'hydraulique souterraine et fluviale*. Paris: Hermann.

Mainguet, M. 1972. *Le modele des gres: problemes generaux*. Paris: Inst. Geog. Nat.

Maire, R. 1977. Les karst haut-alpins de Plate, du Haut-Giffre et de Suisse Occidentale. *Rev. Geog. Alpine* **65**, 403–23.

Maire, R. 1981a. Giant shafts and underground rivers of the Nakanai Mountains (New Britain). *Spelunca, Supp. to No. 3*, 8–9.

Maire, R. 1981b. Karst and hydrogeology synthesis. *Spelunca, Supp. to No. 3*, 23–30.

Maire, R. 1981c. Inventory and general features of PNG karsts, *Spelunca, Supp. to No. 3*, 7–8.

Maire, R. 1986. A propos des karsts sous-marins. *Karstologia* **7**, p. 55.

Maire, R. & Y. Quinif 1984. Un complexe sedimentaire karstique en milieu alpin: les depots de la galerie Aranzadi Atlantique. *C.R. Acad. Sc. Paris, 1298 serie* **2(5)**, 183–5.

Maksimovich, G. A. 1977. Man's utilization of caves throughout the ages. *Proc. 7th Internat. Speleo Congr. (Sheffield)*, p. 310.
Mandelbrot, B. B. 1983. *The fractal geometry of nature*. San Francisco: Freeman.
Mangin, A. 1969a. Nouvelle interpretation du mecanisme des sources intermittentes. *C.R. Acad. Sci., Paris* **269**, 2184–6.
Mangin, A. 1969b. Etude hydraulique du mecanisme d'intermittence de Fontestorbes (Blesta-Ariege). *Ann. Speleo.* **24(2)**, 253–98.
Mangin, A. 1973. Sur la dynamiques des transferts en aquifère karstique. *Proc. 6 Internat. Congr. Speleo., Olomouc* **6**, 157–62.
Mangin, A. 1975. *Contribution à l'étude hydrodynamique des aquifères karstiques*. DES thesis, Univ. Dijon, France (*Ann. Speleo. 1974* **29(3)**, 283–332; **29(4)**, 495–601; 1975, **30(1)**, 21–124).
Mangin, A. 1981a. Utilisation des analysis correloire et spectrale dans l'approche des systemes hydrologiques. *C.R. Acad. Sci. Paris* **293**, 401–4.
Mangin, A. 1981b. Apports des analyses correlatoire et spectrale croisees dans la connaissance des systemes hydrologiques. *C.R. Acad. Sc. Paris* **293(II)**, 1011–14.
Mangin, A. 1984a. Pour une meilleure connaissance des systemes hydrologiques à partir des analyses correlatoire et spectrale. *J. Hydrol.* **67**, 25–43.
Mangin, A. 1984b. Ecoulement en milieu karstique. *Anns. des Mines* (Mai-Juin), 1–8.
Margat, J. 1980. *Carte hydrogeologique de la France a 1 : 1 500 000*. Orleans: Bureau de Recherches Geologiques et Minieres.
Margrita, R., J. Guizerix, P. Corompt, B. Gaillard, P. Calmels, A. Mangin & M. Bakalowicz 1984. Reflexions sur la theorie des traceurs: applications en hydrologie isotopique. In *Isotope hydrology 1983* (Proc. Vienna Symp.), 653–78. Vienna: IAEA.
Marker, M. E. 1976. Cenotes: a class of enclosed karst hollows. *Z. Geomorph. Supplbd.* **26**, 104–23.
Marker, M. E. 1985. Factors controlling microsolutional karren on carbonate rocks of the Griqualand West Sequence. *Cave Science* **12(2)**, 61–5.
Marshall, P. & M. C. Brown 1974. Ice in Coulthard Cave, Alberta. *Can. J. Earth Sci.* **11(4)**, 510–18.
Martel, E. A. 1921. *Nouveau traité des eaux souterraines*. Paris: Editions Doin.
Martin, G. N. 1973. Characterization of simple exponential baseflow recessions. *J. Hydrol. (NZ)* **1291**, 57–62.
Martin, R. & A. Thomas 1974. An example of the use of bacteriophage as a groundwater tracer. *J. Hydrol.* **23**, 73–8.
Maurin, V. & J. Zotl. 1959. Die Untersuchung der Zusammenhange unterirdischer Wasser mit besonderer Berucksichtigung der Karstvarheltnisse. *Steir. Beitr. z Hydrologeologie (Graz)* **11**, 5–184.
Maurin, V. & J. Zotl 1967. Saltwater encroachment in the low altitude saltwater horizons of the islands of Kephallinia. In *Hydrology of fractured rocks, vol. 2*, Internat. Assoc. Sci. Hydrology. Paris: Unesco.
Mayr, A. 1953. Blutenpollen und pflanzliche Sporen als Mittel zur Untersuchung von Quellen und Karstwassern, *Anz. math.-natw. Kl. Osterr. Ak. Wiss., Wien*.
McConnell, H. & J. M. Horn 1972. Probabilities of surface karst. In *Spatial analysis in geomorphology*, R. J. Chorley (ed.), 111–33. London: Methuen.
McDonald, B. S. & J. S. Vincent 1972. Fluvial sedimentary structures formed experimentally in a pipe, and their implications for interpretation of subglacial sedimentary environments. *Geol. Surv. Can. Paper* **72–77**, 30 p.
McDonald, R. C. 1976a. Limestone morphology in south Sulawesi, Indonesia. *Zeit. Geomorph.* **Suppl. 26**, 79–91.
McDonald, R. C. 1976b. Hillslope base depressions in tower karst topography of Belize. *Zeit. Geomorph.* **Suppl. 26**, 98–103.
McDonald, R. C. 1979. Tower karst geomorphology in Belize. *Z. Geomorph.* **Suppl. 32**, 35–45.
McIlwreath, I. A. & N. P. James 1978. Facies Models – 13, Carbonate slopes. *Geoscience Can.* **5(4)**, 189–99.
McNutt, M. & H. W. Menard 1978. Lithospheric flexure and uplifted atolls. *J. Geophys. Res.* **83**, 1206–12.
Merrill, G. K. 1960. Additional notes on vertical shafts in limestone caves. *Bull. Nat. Speleo. Soc.* **22(2)**, 101–5.

Mesolella, K. J., R. K. Matthews, W. S. Broecker & D. L. Thurber 1969. The astronomical theory of climatic change: Barbados data. *J. Geol.* **77**, 250–74.

Meyer-Peter, E. & R. Muller 1948. Formulas for bedload transport. *Proc. 3rd Conf. Int. Assoc. Hydraulic Res., Stockholm*, 39–64.

Middleton, G. V. 1976. Hydraulic interpretation of sand size distributions. *J. Geol.* **84**, 405–26.

Mijatovic, B. F. 1984a. Problems of sea water intrusion into aquifers of the coastal Dinaric karst. In *Hydrogeology of the Dinaric karst*, B. F. Mijatovic (ed.), *International contributions to Hydrogeology* **4**, 115–42. Hannover: Heise.

Mijatovic, B. F. 1984b. Karst poljes in Dinarides. In *Hydrogeology of the Dinaric karst*, B. F. Mijatovic (ed.), *International contributions to hydrogeology* **4**, 87–109. Hannover: Heise.

Milanovic, P. T. 1976. Water regime in deep karst. Case study of the Ombla spring drainage area. In *Karst hydrology and water resources, vol. 1. Karst hydrology*, V. Yevjevich (ed.), 165–91. Colorado: Water Resources Publications.

Milanovic, P. T. 1981. *Karst hydrogeology*. Colorado: Water Resources Pubs.

Miller, T. E. 1982. *Hydrochemistry, hydrology and morphology of the Caves Branch karst, Belize*. PhD Thesis, McMaster University.

Milske, J. A., E. C. Alexander & R. S. Lively 1983. Clastic sediments in Mystery Cave, southeastern Minnesota. *Nat. Speleo. Soc. Bull.* **45**, 55–75.

Miotke, F. D. 1968. Karstmorphologische Studien in der glazialuberformten Hohenstufe der 'Picos de Europa', Nordspanien. *Jb. geogr. Gesell. Hannover Sonderheft* **4**, 161 p.

Miotke, F.-D. 1974. Carbon dioxide and the soil atmosphere. *Abh. Karst-u. Hohlenkunde, A9*, Munich, 52 pp.

Moeschler, P., I. Muller, U. Schotterer and U. Siegenthaler, 1982. Les organisms vivants, indicateur naturels dans l'hydro-dynamique du karst, confrontes aux donnees isotopiques, chimiques et bacteriologiques. *Beitrage zur Geologie der Schweiz, Hydrogeologie* **23**, p. 213.

Monroe, W. H. 1964. The origin and interior structure of mogotes. *20th Int. Geog. Cong., Abstracts*, 108.

Monroe, W. H. 1966. Formation of tropical karst topography by limestone solution and reprecipitation. *Caribbean J. Sci.* **6**, 1–7.

Monroe, W. H. 1968. The karst features of northern Puerto Rico. *Nat. Speleo. Soc. Bull.* **30**, 75–86.

Monroe, W. H. 1974. Dendritic dry valleys in the cone karst of Puerto Rico. *J. Res. U.S. Geol. Survey* **2(2)**, 159–63.

Moon, B. P. 1985. Controls on the form and development of rock slopes in fold terrane. In 'Hillslope processes', A. D. Abrahams (ed.). *Program & Abstacts, 16th Annual Binghamton Symposium*, 22 pp.

Moore, C. H. 1979. Porosity in carbonate rock sequences. In Geology of Carbonate Porosity. *Am. Assoc. Petrol. Geol., Continuing Education Course Note Series 11*.

Moore, D. L. & M. T. Stewart 1983. Geophysical signatures of fracture traces in a karst aquifer (Florida, U.S.A.). *J. Hydrol.* **61**, 325–35.

Moore, G. W. 1954. The origin of helictites. *Nat. Speleo. Soc., Occ. Paper* **1**, 16 p.

Moore, G. W. 1981. Manganese speleothems. *Proc., 8th Int. Congr. Speleo., Kentucky*, 642–4.

Moore, J. D. 1980. Groundwater application of remote sensing. *U.S. Geol. Survey Open File Report 82–240*, Eros Data Center.

Morner, N.-A. 1983. Sea levels. In *Mega-geomorphology*, R. Gardner & H. Scoging (eds), 73–91. Oxford: Clarendon.

Morrow, D. W. 1982. Diagenesis 1. Dolomite-Part 1: the chemistry of dolomitization and dolomite precipitation. Part 2. Dolomitisation models and ancient dolostones. *Geoscience Canada* **9**, 5–13; 95–105.

Moser, H., V. Rajner, D. Rank & W. Stichler 1976. Results of measurements of the content of deuterium, oxygen-18 and tritium in water samples from test area taken during 1972–1975. In *Underground water tracing*, R. Gospodaric & P. Habic (eds), 93–117. Inst. Karst Research Postojna.

Moser, M. & M. Geyer 1979. Seismospelaologic-Erdbebenzerstorungen in Hohlen am Beispiel des Gaislochs bei Oberfellendorf (Oberfranken, Bayern). *Die Hohle* **4**, 89–102.

Mozetic, M. 1976. Virological examinations. In *Underground water tracing*, R. Gospodaric & P. Habic (eds), 119–22. Inst. Karst Research, Postojna.

Mueller, M. 1987. *Takaka valley hydrogeology (preliminary assessment)*. Nelson Catchment Board and Regional Water Board.

Muller, P. & I. Sarvary 1977. Some aspects of developments in Hungarian speleology theories during the last ten years. *Karszt-es Barlang*, 53–9.

Murray, A. N. & W. W. Love 1929. Action of organic acids upon limestone. *Am. Ass. Petrol. Geol. Bull.* **13**, 1467–75.

Myers, A. J. 1962. A fossil sinkhole. *Oklahoma Geol. Notes* **22**, 13–15.

Mylroie, J. E. (ed.) 1988. *Field guide to the karst geology of San Salvador Island, Bahamas*. San Salvador Is., Bahamas: College Center of the Finger Lakes.

Nancollas, G. H. & M. N. Reddy 1971. The crystallization of calcium carbonate. – II. Calcite growth mechanism. *J. Colloid Interface Sci.* **37**, 824–30.

Neumann, A. C. 1968. Biological erosion of limestone coasts. In *Encyclopedia of geomorphology*, R. W. Fairbridge (ed.), 75–81. New York: Reinhold.

Newitt, D. M., J. F. Richardson, M. Abbott, & R. B. Turtle 1955. Hydraulic conveying of solids in horizontal pipes. *Trans. Inst. Chem. Engrs.* **33**, 93–110.

Newson, M. D. 1971. The role of abrasion in cave development. *Trans., Cave Res. Gp., G.B.* **13(2)**, 102–8.

Newton, J. G. 1984. Review of induced sinkhole development. In *Sinkholes: their geology, engineering and environmental impact*, B. F. Beck (ed.), 3–9. Boston: Balkema.

Nicod, J. 1972. *Pays et paysages du calcaire*. Paris: Presses Univ. de France.

Nicod, J. 1976. Karst des gypses et des evaporites associees. *Ann. de Geogr.* **471**, 513–54.

Noel, M. 1986. The palaeomagnetism and magnetic fabric of sediments from Peak Cavern, Derbyshire. *Geophys. J.R. astr. Soc.* **84**. 445–54.

Noel, M. & P. A. Bull 1982. The palaeomagnetism of sediments from Clearwater Cave, Mulu, Sarawak. *Cave Sci.* **9(2)**, 134–41.

Nordstrom, D. K. & 18 others 1979. A comparison of computerized chemical models for equilibrium calculations in aqueous systems. In *Chemical modelling in aqueous systems*, E. A. Jenne (ed.), 856–92. Washington: Am. Chem. Soc.

North, F. K. 1985. *Petroleum geology*. Boston: Allen & Unwin.

Nunn, P. 1986. Implications of migrating geoid anomalies for the interpretation of high-level fossil coral reefs. *Geol. Soc. Am. Bull.* **97**, 946–52.

Obert, L. & Duvall, W. I. 1967. *Rock mechanics and the design of structures in rock*. New York: John Wiley.

Olive, W. W. 1957. Solution subsidence troughs, Castile Formation of the gypsum plain, Texas and New Mexico. *Geol. Soc. Amer. Bull.* **68**, 351–8.

Ollier, C. D. 1975. Coral island geomorphology – the Trobriand Islands. *Z. Geomorph* **19(2)**, 163–90.

Olson, R. A. 1984. Genesis of paleokarst and strata-bound zinc-lead sulfide deposits in a Proterozic dolostone, northern Baffin Island, Canada. *Econ. Geol.* **79**, 1056–103.

O'Neil, J. R., L. H. Adami & S. Epstein 1975. Revised value for the ^{18}O fractionation factor between H_2O and CO_2 at 25°C. *U.S. Geol. Surv. J. Res.* **3**, 623–4.

Osborne, R. A. L. 1984. Lateral facies changes, unconformities and stratigraphic reversals: their significance for cave sediment stratigraphy. *Trans. Brit. Cave Res. Assoc.* **11(3)**, 175–84.

Osmaston, H. and M. M. Sweeting 1982. Ch. 5, Geomorphology (of the Gunung Mulu National Park). *Sarawak Museum Journal* **30** (51, new series), 75–93.

Ozis, U. & N. Keloglu 1976. Some features of mathematical analysis of karst runoff. In *Karst hydrology and water resources*, vol. 1, V. Yevjevich (ed.), 221–35. Colorado: Water Resources Pubs.

Ozoray, G. 1961. The mineral filling of the thermal spring caves of Budapest. *Rass. Speleo. Ital. Mem.* **3**, 152–70.

Palmer, A. N. 1975. The origin of maze caves. *Bull. Nat. Speleo. Soc. Am.* **37(3)**, 56–76.

Palmer, A. N. 1981. *A geological guide to Mammoth Cave National Park*. Teaneck, NJ; Zephyrus Press.

Palmer, A. N. 1984. Geomorphic interpretation of karst features. In *Groundwater as a geomorphic agent*, R. G. LaFleur (ed.), 173–209. London: Allen & Unwin.

Palmer, M. V. & A. N. Palmer 1975. Landform development in the Mitchell Plain of southern Indiana: origin of a partially karsted plain. *Z. Geomorph. N. F.* **19(1)**, 1–39.
Palmer, R. J. 1986. Hydrology and speleogenesis beneath Andros Island. *Cave Science* **13(1)**, 7–12.
Palmer, R. J. & L. M. Heath 1985. The effect of anchialine factors and fracture control on cave development below eastern Grand Bahama. *Cave Science* **12(3)**, 93–7.
Paloc, H. 1970. La Fontaine de Vaucluse et son bassin d'alimentation. *Anns. Soc. Horticulture et Histoire Naturelle d'Hérault* **110(3)**, 130–4.
Paloc, H. 1975. Cartographie des eaux souterraines en terrains calcaires. In '*Hydrogeology of karstic terrains*' A. Burger & L. Dubertret (eds), 137–48. Paris: Int. Assoc. Hydrogeologists.
Paloc, H. & J. Margat 1985. Report on hydrogeological maps of karstic terrains. In 'Hydrogeological mapping in Asia and the Pacific region', W. Grimelmann, K. D. Krampe & W. Struckmeier (eds), *International Contributions to Hydrogeology* **7**, 301–15. Hannover: Heise.
Panos, V. & O. Stelcl 1968. Physiographic and geologic control in development of Cuban mogotes. *Z. Geomorph.* **12(2)**, 117–73.
Parizek, R. P. 1976. On the nature and significance of fracture traces and lineaments in carbonate and other terranes. In *Karst hydrology*, vol. 1, V. Yevjevich (ed.), 47–108. Colorado: Water Resources Pubs.
Pasini, G. 1975. Sull'importanza speleogenetica dell' 'erosione antigravitativa'. *Le Grotte d'Italia* **4**, 297–326.
Paterson, K. 1979. Limestone springs in the Oxfordshire Scarplands: the significance of spatial and temporal variations in their chemistry. *Z. Geomorph. N.F. Suppl.-Bd.* **32**, 46–66.
Pechorkin, A. N. 1986. On gypsum and anhydrite distribution in zones near to the surface of sulphate massifs. *Le Grotte d'Italia* **4(XII)**, 397–406.
Pechorkin, A. N. & G. V. Bolotov 1983. *Geodynamics of relief in karstified massifs*. (in Russian). Univ. Perm.
Pechorkin, A. N., I. A. Pechorkin, & V. N. Kataev 1982. Experimental study of calcium sulphate solubility and hydration of anhydrite. *Geol. Applicata e Idrogeol.* **17(2)**, 243–53.
Pechorkin, I. A. 1986. Engineering geological investigations of gypsum karst. *Le Grotte d'Italia* **4(XII)**, 383–8.
Penck, A. 1900. Geomorphologische Studien aus der Hercegovina, *Z. Deut. Osterreich. Alpenver* **31**, 25–41.
Perlega, W. 1976. Der Nachweis von Fluoreszenzfarbstoffen mittels Aktivkohle. *Papers 3 Internat. Symp. Underground Water Tracing*, 195–201. Postojna: Inst. Karst Res.
Perna, G. & U. Sauro 1978. *Atlante delle microforme di dissoluzione carsica superficiale del Trentino e del Veneto*. Trento: Museo Tridentino.
Peterson, G. M. 1976. Pollen analysis and the origin of cave sediments in the Central Kentucky Karst. *Bull. Nat. Speleo. Soc. Am.* **38(3)**, 53–8.
Peterson, J. A. 1982. Limestone pedestals and denudation estimates from Mt. Jaya, Irian Jaya. *Aust. Geogr.* **15**, 170–3.
Petrovic, J. 1969. Pojava dubinske karstificaje u juznom delu starog massiva Kopaonika. *Zbornik za prir. nauke Matice srpske* **36**, 97–108.
Pfeiffer, D. 1963. Die geschichtliche Entwicklung der Anschauungen uber das karstgrundwasser. *Beihefte zum geologischen Jahrbuch* **57**, 111 pp. (Hannover).
Pfeffer, K.-H. 1973. Flachenbildung in den Kalkgebieten. In A. E. Semmel (ed.) Neue Ergebnisse der Karstforschung in den Tropen und in Mittelmeerraum, *Geog. Zeit. Beiheft (Erdkundl. Wissen H.* **32**), Wiesbaden, 111–32.
Pfeffer, K.-H. 1976. Probleme der Genese von Oberflachenformen auf Kalkgestein. *Z. Geomorph.*, **Suppl. 26**, 6–34.
Pham, K. 1985. The development of karst landscapes in Vietnam. *Acta Geologica polonica* **35(3/4)**, 305–19.
Picknett, R. G. 1972. The pH of calcite solutions with and without magnesium carbonate present and the implications concerning rejuvenated aggressiveness. *Trans. Cave Res. Group G.B.* **14(2)**, 141–50.
Picknett, R. G., L. G. Bray & R. D. Stenner 1976. The chemistry of cave waters. In *The science of speleology*, T. D. Ford & C. H. D. Cullingford (eds). London: Academic Press.
Pielou, E. C. 1969. *An introduction to mathematical ecology*. New York: Wiley.

Pinder, G. F. & J. F. Jones 1969. Determination of groundwater component of peak discharge from chemistry of total runoff. *Water Resources Res.* **5**, 438–45.
Pitty, A. F. 1966. *An approach to the study of karst water.* Univ. Hull, Occasional Papers in Geography 5.
Pitty, A. F. 1968a. The scale and significance of solutional loss from the limestone tract of the southern Pennines. *Proc. Geol. Ass. Lond.* **79(2)**, 153–77.
Pitty, A. F. 1968b. Calcium carbonate content of water in relation to flow-through time. *Nature* **217**, 939–40.
Pluhar, A. & D. C. Ford 1970. Dolomite karren of the Niagara Escarpment, Ontario, Canada. *Zeitschrift fur Geomorphologie* **14(4)**, 392–410.
Plummer, L. N. 1975. Mixing of seawater with calcium carbonate ground water: quantitative studies in the geological sciences. *Geol. Soc. Am. Mem.* **142**, 219–36.
Plummer, L. N. & E. Busenberg 1982. The solubilities of calcite, aragonite and vaterite in CO_2–H_2O solutions between 0 and 90°C, and an evaluation of the aqueous model for the system $CaCO_3$–CO_2–H_2O. *Geochim. Cosmochim. Acta* **46**, 1011–40.
Plummer, L. N. & T. M. L. Wigley 1976. The dissolution of calcite in CO_2 saturated solutions at 25C and 1 atmosphere total pressure. *Geochim. Cosmochim. Acta.* **40**, 191–202.
Plummer, L. N., D. C. Parkhurst & T. M. L. Wigley 1979. Critical review of the kinetics of calcite dissolution and precipitation. In *Chemical modelling in aqueous systems*, E. A. Jenne (ed.), 537–73. Washington: Am. Chem. Soc.
Plummer, L. N., T. M. L. Wigley & D. L. Parkhurst 1978. The kinetics of calcite dissolution in CO_2-water systems at 5 to 60°C and 0.0 to 1.0 atm CO_2. *Am. J. Sci.* **278**, 179–216.
Pohl, E. R. 1955. Vertical shafts in limestone caves. *Nat. Speleo. Soc. Am., Occasional Paper No. 2*, 24 p.
Poiseuille, J. M. L. 1846. Recherches experimentales sur le mouvement des liquides dans les tubes de tres petits diamètres. *Acad. Sci. Paris Mem. sav. étrang.* **9**, 433–545.
Popov, I. V., N. A. Gvozdetsky, A. G. Chikischev & B. I. Kudelin 1972. Karst of the U.S.S.R. In *Karst: important karst regions of the Northern Hemisphere*, M. Herak & V. Stringfield (eds), 355–416. Amsterdam: Elsevier.
Pouyllan, M. & M. Seurin 1985. Pseudo-karst dans les roches grés quartzitiques de la formation Roraima. *Karstologia* **5(1)**, 45–52.
Prestwich, J. 1854. Swallow holes on the Chalk hills near Canterbury. *Quat. J. Geol. Soc.* (London) C, 222–4.
Price, N. J. 1966. *Fault and joint development in brittle and semi-brittle rock.* Oxford: Pergamon.
Priesnitz, K. 1972. Formen, prozesse und faktoren der Verkarstung und Mineralum-bildung in Ausstrich salinarer Serien. *Gottinger Geogr. Abhand.* **60** 317–39.
Priesnitz, K. 1974. Losungsraten und ihre geomorphologische Relevanz. *Abh. Ak. d. Wiss. Gottingen. Math-Phys. Klasse 3, Folge* **29**, 68–84.
Pringle, J. M. 1973. *Morphometric analysis of surface depressions in the Mangapu karst.* MSc thesis, University of Auckland.
Pulido-Bosch, A. 1986. Le karst dans les gypses de Sorbas (Almeria): aspects morphologiques et hydrogeologiques. *Karstologia Mems.* **1**, 27–35.
Pulina, M. 1971. Observations on the chemical denudation of some karst areas of Europe and Asia. *Studia Geomorph. Carpatho–Balcanica* **5**, 79–92.
Pulinowa, M. Z. & M. Pulina 1972. Phénomenes cryogènes dans les grottes et gouffres des Tatras. *Biuletyn Peryglacjalny* **21**, 201–35.
Purdy, E. G. 1974. Reef configurations: cause and effect. In *Reefs in time and space*, L. F. Laporte (ed.), Soc. Econ. Palaeontol. Mineralogists, Spec. Pub. 18, 9–76. Tulsa, USA.

Quinif, Y. 1981. Thermoluminescence: a method for sedimentological studies in caves. *Proc., 8th Internat. Speleo. Congress* **1**, 308–13.
Quinif, Y. (in press). Paleokarst in Belgium. In *Paleokarst – a world review.* P. Bosak, D. C. Ford & G. Glazek (eds). Amsterdam: Academia Praha/Elsevier.
Quinlan, J. F. 1976. New fluorescent direct dye suitable for tracing groundwater and detection with cotton. *3rd Int. Symp. Underground Water Tracing*, vol. 1, 257–62, Ljubljana-Bled, Yugoslavia.

Quinlan, J. F. 1978. *Types of karst, with emphasis on cover beds in their classification and development*. PhD Thesis, University of Texas at Austin.
Quinlan, J. F. 1983. Groundwater pollution by sewage, creamery waste, and heavy metals in the Horse Cave area, Kentucky. In *Environmental karst*, P. H. Dougherty (ed.), 52. Cincinnati: Geo. Speleo. Pubs.
Quinlan, J. F. 1986a. Discussion of 'Ground Water Tracers', *Ground Water* **24(2)**, 253–9.
Quinlan, J. F. 1986b. Legal aspects of sinkhole development and flooding in karst terranes: 1. review and synthesis. *Environ. Geol. Water Sci.* **8(1/2)**, 41–61.
Quinlan, J. F. & R. O. Ewers 1981. Hydrogeology of the Mammoth Cave Region, Kentucky. In *Geol. Soc. Am. Cincinnati 1981 Field Trip Guidebooks*, vol. 3, T. G. Roberts (ed.), 457–506.
Quinlan, J. F. & R. O. Ewers 1985. Ground water flow in limestone terranes: strategy, rationale and procedure for reliable, efficient monitoring of ground water quality in karst areas. *National Symposium and Exposition on Aquifer Restoration and Ground Water Monitoring, Proceedings*, 197–234. Worthington, Ohio: National Water Well Association.
Quinlan, J. F. & R. O. Ewers 1986. Reliable monitoring in karst terranes: it can be done, but not by an EPA-approved method. *Ground Water Monitoring Review* **6(1)**, 4–6.
Quinlan, J. F. & J. A. Ray 1981. *Groundwater basins in the Mammoth Cave region, Kentucky*. Friends of the Karst, Occ. Pub. 1.
Quinlan, J. F. & P. L. Smart, 1976. Identification of dyes used in water-tracing: a suggestion to improve communication. *3rd Int. Symp. Underground Water Tracing* (Ljubljana–Bled) Yugoslavia, vol. 2, 263–7
Quinlan, J. F., R. O. Ewers, J. A. Ray, R. L. Powell & N. C. Krothe 1983. Groundwater hydrology and geomorphology of the Mammoth Cave region, Kentucky, and the Mitchell Plain, Indiana. In *Field trips in mid-western geology*, R. H. Shaver & J. A. Sunderman (eds), 1–85. Bloomington, Indiana: Geol. Soc. Am. and Indiana Geol. Survey.
Quinlan, J. F., A. R. Smith & K. S. Johnson 1986. Gypsum karst and salt karst of the United States of America. *Le Grotte d'Italia* **4(13)**, 73–92.

Racovitza, Gh. 1972. Sur la correlation entre l'evolution du climat et la dynamique des depots souterrains de glace de la grotte Scarisoara. *Trav. Inst. Spéléo, 'Emile Racovitza,'* **XI**, 373–92.
Ramljak, P., A. Filip, P. Milanovic & D. Arandjelovic 1976. Establishing karst underground connections and responses by using tracers. In *Karst hydrology and water resources*, V. Yevjevich (ed.), 237–57. Colorado: Water Res. Pubs.
Rathjens, C., 1954. Zur Frage der Karstrandebenen im Dinarischen Karst. *Erdkunde* **8,** 114–15.
Rauch, H. W. & W. B. White 1970. Lithologic controls on the development of solution porosity in carbonate aquifers. *Water Resources Res.* **6**, 1175–92.
Rauch, H. W. & W. B. White 1977. Dissolution kinetics of carbonate rocks: 1. Effects of lithology on dissolution rate. *Water Resources Res.* **13(2)**, 381–94.
Reams, M. W. 1968. *Cave sediments and the geomorphic history of the Ozarks*. PhD thesis, Washington University, Missouri.
Reddy, M. M. 1988. Acid rain damage to carbonate stone: a quantitative assessment based on the aqueous geochemistry of rainfall runoff from stone. *Earth Surf. Proc. Landforms* **13(4)**, 335–54.
Reeckmann, A. & G. M. Friedmann 1982. *Exploration for carbonate petroleum reservoirs*. New York: John Wiley.
Reilly, T. E. & A. S. Goodman 1985. Quantitative analysis of saltwater–freshwater relationships in groundwater systems – a historical perspective. *J. Hydrol.* **80**, 125–60.
Reitz, H. M. & D. S. Eskridge 1977. Construction methods which recognise the mechanics of sinkhole development. In *Hydrologic problems in karst regions*, R. R. Dilamarter & S. C. Csallany (eds), 432–8. Bowling Green: University of West Kentucky.
Rellensmann, O. 1957. Rock mechanics in regard to static loading caused by mining excavation. *Second Symp. on Rock Mech., Colo. Sch. of Mines*, p. 52.
Ren Meie, Liu Zhenzhong, Wang Feiyan & Yu Jinbiao 1982. *The morphological characteristics of karst in China*. Nanjing University, Geography Dept, 24 p.
Ren, M., Z. Liu, J. Jin, X. Deng, F. Wang, B. Peng, X. Wang & Z. Wang 1981. Evolution of

limestone caves in relation to the life of early man at Zhoukoudian, Beijing. *Scientia Sinica* **14(6)**, 843–51.

Renault, H. P. 1968. Contribution a l'etude des actions mechaniques et sedimentologiques dans la spéléogenèse. *Ann. Speleo.* **22**, 5–21, 209–67, **23**, 259–307, 529–96; **24**, 313–37.

Renault, P. 1970. *La formation des cavernes*. Presses univ. de France, Paris.

Renault, P. 1979. Mesures periodiques de la PCO_2 dans les grottes francaises du cours de ces dix dernieres annes. *Actes du Symp. Internat. sur l'erosion karstique*, Aix en Provence-Marseille-Nimes, 17–33.

Renault-Miskovsky, J. 1972. Contribution a la paleoclimatologie du Midi mediterraneen pendant la derniere glaciation et le post-glaciaire, d'apres l'etude palynologique de remplissage des grottes et abris-sous-roche. *Bull. Mas. Anthrop. Prehist. Monaco* **18**, 145–210.

Rhoades, R. & N. M. Sinacori 1941. Patterns of groundwater flow and solution. *J. Geol.* **49**, 785–94.

Rhodes, D., E. A. Lantos, J. A. Lantos, R. J. Webb & D. C. Owens 1984. Pine Point ore-bodies and their relationship to structure, dolomitization and karstification of the Middle Devonian barrier complex. *Econ. Geol.* **70**, 991–1055.

Richards, K. 1982. *Rivers: form and process in alluvial channels*. London: Methuen.

Richter, 1907. Beitrage zur Landeskunde Bosniens und der Herzegowina. *Wiss. Mitt. Bosnien. Herzegowina* **10**, 383–545.

Ristic, D. M. 1976. Water regime of flooded poljes. In *Karst hydrology and water resources*, V. Yevjevich (ed.), 301–18. Colorado: Water Res. Pubs.

Roberge, J. 1979. *Geomorphologie du karst de la Haute-Saumons, Ile d'Anticosti, Quebec*. MSc thesis, McMaster University.

Roberge, J. & D. Caron 1983. The occurrence of an unusual type of pisolite: the cubic cave pearls of Castleguard Cave, Columbia Icefields, Alberta, Canada. *Arctic and Alpine Res.* **15(4)**, 517–22.

Robinson, L. A. 1977. Erosive processes on shore platforms of north-east Yorkshire, England. *Marine Geol.* **23**, 339–61.

Robinson, V. D. & D. Oliver 1981. Geophysical logging of water wells. In *Case-studies in groundwater resources evaluation*. J. W. Lloyd (ed.), 45–64. London: Clarendon.

Roehl, P. O. & P. W. Choquette 1985. *Carbonate petroleum reservoirs*. New York: Springer.

Roglic, J. 1972. Historical review of morphological concepts. In *Karst: important karst regions of the northern hemisphere*, M. Herak & V. T. Stringfield (eds), 1–18. Amsterdam: Elsevier.

Roglic, J. 1974. Les caractères specifiques du karst Dinarique. *Centre Nat. Recherche Sci., Mems et Docs.* **15**, 269–78.

Roglic, J. 1976. Depth of water circulation and dimensions of cavities in the Dinaric karst. In *Problems of karst hydrology in Yugoslavia*, D. Gavrilovic (ed.), 29–40. Belgrade: Mems Serbian Geog. Soc. 13.

Roglic, J. 1981. Les barrages de tuf calcaire aux lacs de Plitvice. Actes du Coll. de L'A.G.F., '*Formations carbonates externes, tufs et travertines*', Assoc. Francaise de Karstologie, Mem. 3, 137–44.

Romero, J. C. 1970. The movement of bacteria and virus through porous media. *Ground Water* **8(2)**, 37–48.

Roques, H. 1962. Considerations theoriques sur la chimie des carbonates. *Ann. Speleo.* **17**, 1–41, 241–84, 463–7.

Roques, H. 1964. Contribution a l'etude statique et cinetique des systemes gaz carbonique–eau-carbonate. *Ann. Speleo.* **19**, 255–484.

Roques, H. 1969. Problemes de transfert de masse posé par l'evolution des eaux souterraines. *Ann. Speleo.* **24**, 455–94.

Rose, L. & P. Vincent 1986a. The kamenitzas of Gait Barrows National Nature Reserve, north Lancashire, England. In *New directions in karst*, K. Paterson & M. M. Sweeting (eds), 473–96. Norwich, England: Geo Books.

Rose, L. & P. Vincent 1986b. Some aspects of the morphometry of grikes: a mixture model approach. In *New directions in karst*, K. Paterson & M. M. Sweeting (eds), 497–514. Norwich, England: Geo Books.

Rouse, W. C. 1984. Flowslides. In *Slope instability*, D. Brunsden & D. B. Prior (eds), 491–521. New York: Wiley.

Rozanski, K. & T. Florkowski 1979. Krypton-85 dating of groundwater. In *Isotope hydrology 1978* (Proc. Symp. Neuherberg 1978, vol. 2, p. 949). Vienna: IAEA.

Ryder, P. F. 1975. Phreatic network caves in the Swaledale area, Yorkshire. *Trans. Brit. Cave Res. Assoc.* **2(4)**, 177–92.

Salvamoser, J. 1984. Krypton-85 for groundwater dating. In *Isotope hydrology 1983*, 831–2. Vienna: IAEA.

Salvigsen, O. & A. Elgersma 1985. Large-scale karst features and open taliks at Valdeborgsletta, outer Isfjorden, Svalbard. *Polar Res.* **3(2)**, 145–53.

Sangster, D. F. 1987. Breccia-hosted lead-zinc deposits in carbonate rocks. In *Paleokarst*, N. P. James & P. W. Choquette (eds), 102–16. New York: Springer.

Saunderson, H. C. 1977. The sliding bed facies in sands and gravels: a criterion for full-pipe (tunnel) flow. *Sedimentology* **24**, 623–38.

Sawicki, L. R. von 1909. Ein Beitrag zum geographischen Zyklus im Karst. *Geogr. Zeitschr.* **15**, 185–204, 259–81.

Schillat, B. 1965. Nachweis von Erdbeben in Hohlen. *Verband der Deutschen Hohlen und Karstforscher Mitt.* **11**, 100–7.

Schmid, E. 1958. *Hohlenforschung und Sedimentanalyse. Ein Beitrag sur Datierung des alpines Palaolithikums*, p. 13. Basel: Schriften des Inst. fur Urund Frungeschichte der Schweiz.

Schmidt, K.-H. 1979. Karstmorphodynamik und ihre hydrologische Steuerung. *Erdkunde* **33(3)**, 169–78.

Schmidt, V. A. 1982. Magnetostratigraphy of clastic sediments from caves within the Mammoth Cave National Park, Kentucky. *Science* **217**, 827.

Schmotzer, J. K., W. A. Jester & R. R. Parizek 1973. Groundwater tracing with post sampling activation analysis. *J. Hydrol.* **20**, 217–36.

Schoeller, H. 1962. *Les eaux souterraines*. Paris: Masson & Cie.

Scholle, P. A., D. G. Bebout & C. H. Moore (eds) 1983. *Carbonate depositional environments*. Am. Assoc. Petrol Geologists, Mem. 33.

Schroeder, J. 1979a. *Le developpement des grottes dans la region du Premier Canyon de la Riviere Nahanni Sud, T.N.O.* PhD Thesis, University of Ottawa.

Schroeder, J. 1979b. Developpement de cavites d'origine mécanique dans un karst froid (Nahanni, T.N.O., Canada). *Ann. Soc. Geol. Belgique* **102**, 59–67.

Schroeder, J. & Ford, D. C. 1983. Clastic sediments in Castleguard Cave, Columbia Icefields, Alberta, Canada. *Arctic and Alpine Res.* **15(4)**, 451–61.

Schroeder, J., M. Beaupre & M. Cloutier 1986. Ice-push caves in platform limestones of the Montreal area. *Can. J. Earth Sci.* **23**, 1842–51.

Schwarcz, H. P. 1980. Absolute age determinations of archaeological sites by uranium dating of travertines. *Archaeometry* **22(1)**, 3–24.

Schwarcz, H. P., R. S. Harmon, P. Thompson & D. C. Ford 1976. Stable isotope studies of fluid inclusions in speleothems and their paleoclimatic significance. *Geochim. et Cosmochim. Acta* **40**, 657–65.

Selby, M. J. 1980. A rock mass strength classification for geomorphic purposes: with tests from Antarctica and New Zealand. *Z. Geomorph.* **24**, 31–51.

Selby, M. J. 1982. *Hillslope materials and processes*. Oxford: Oxford University Press.

Serban, M. & M. Domas 1985. Sur le micro relief de corrosion de le Pestera Vintului (Monts Padurea Crainlui, Roumaine) et la morphogenese de la voûte plane dans les conduites forcées. *Theoretical and Applied Karstology* **2**, 97–121.

Shackleton, N. J. & N. D. Opdyke 1973. Oxygen isotope and paleomagnetic stratigraphy of equatorial Pacific core V28–238: oxygen isotope temperatures and ice volumes on a 10^5 and 10^6 year scale. *Quat. Res.* **3**, 39–55.

Shreve, R. L. 1966. Statistical law of stream numbers. *J. Geol.* **74**, 17–37.

Shuster, E. T. & W. B. White 1971. Seasonal fluctuations in the chemistry of limestone springs: a possible means of characterising carbonate aquifers. *J. Hydrol.* **14**, 93–128.

Siegenthaler, U., U. Schotterer & I. Muller 1984. Isotopic and chemical investigations of springs from different karst zones in the Swiss Jura. In *Isotope hydrology 1983*, 153–72. Vienna: IAEA.

Siffre, A. & M. Siffre 1961. Le faconnement des alluvions karstique. *Ann. Speleo.* **16**, 73–80.

Sklash, M. G., R. N. Farvolden & P. Fritz 1976. A conceptual model of watershed response to

rainfall, developed through the use of oxygen-18 as a natural tracer. *Can. J. Earth Sci.* **13**, 271–83.
Skrivanek, F. & J. Rubin 1973. *Caves in Czechoslovakia*. Prague: Academia.
Smart, C. C. 1983a. *Hydrology of a glacierised alpine karst*. PhD Thesis, McMaster University.
Smart, C. C. 1983b. The hydrology of the Castleguard Karst, Columbia Icefields, Alberta, Canada. *Arctic and Alpine Res.* **15(4)**, 471–86.
Smart, C. C. 1984a. Glacier hydrology and the potential for subglacial karstification. *Norsk Geogr. Tids.* **38(3–4)**, 157–61.
Smart, C. C. 1984b. Overflow sedimentation in an alpine cave system. *Norsk Geog. Tids.* **38(3–4)**, 171–6.
Smart, C. C. 1984c. The hydrology of the Inland Blue Holes, Andros Island. *Cave Sci.* **11(1)**, 23–9.
Smart, C. C. & M. C. Brown 1981. Some results and limitations in the application of hydraulic geometry to vadose stream passages. *Proc., 8th Internat. Cong. Speleo., Kentucky*, 724–5.
Smart, C. C. & D. C. Ford 1986. Structure and function of a conduit aquifer. *Canadian J. Earth Sci.* **23(7)**, 919–29.
Smart, P. L. 1976a. The use of optical brighteners for water tracing. *Trans Brit. Cave Res. Assoc.* **3(2)**, 62–76.
Smart, P. L. 1976b. Use of optical brightener/cellulose detector systems for water tracing. *Papers 3 Internat. Symp. Underground Water Tracing*, Inst. Karst Res. Postojna; 203–13.
Smart, P. L. 1982. A review of the toxicity of 12 fluorescent dyes used for water tracing. *Beitr. Z. Geol. der Schweiz – Hydrologie* **28**, 101–12.
Smart, P. L. 1984. A review of the toxicity of twelve fluorescent dyes used in water tracing. *Nat. Spel. Soc. Bull.* **46(2)**, 21–33.
Smart, P. L. 1986. Origin and development of glacio-karst closed depressions in the Picos de Europa, Spain. *Z. Geomorph, N.F.* **30(4)**, 423–43.
Smart, P. L. & M. C. Brown 1973. The use of activated carbon for the detection of the tracer dye Rhodamine WT. *Proc. 6 Internat. Speleo. Cong. Olomouc*, CSSR, **4**, 285–92.
Smart, P. L. & H. Freiderich 1982. An assessment of the methods and results of water-tracing experiments in the Gunung Mulu National Park, Sarawak. *Trans Brit. Cave Res. Assoc.* **9(2)**, 100–12.
Smart, P. L. & P. Hodge 1980. Determination of the character of the Longwood sinks to Cheddar resurgence conduit using an artificial pulse wave. *Trans. Brit. Cave Res. Assoc.* **7(4)**, 208–11.
Smart, P. L. & I. M. S. Laidlaw 1977. An evaluation of some fluorescent dyes for water tracing. *Water Resources Res.* **13**, 15–23.
Smart, P. L. & D. I. Smith 1976. Water tracing in tropical regions, the use of fluorometric techniques in Jamaica. *J. Hydrol.* **30**, 179–95.
Smart, P. L., T. C. Atkinson, I. M. S. Laidlaw, M. D. Newson, M. D. & S. T. Trudgill 1986. Comparison of the results of quantitative and non quantitative tracer tests for determination of karst conduit networks: an example from the Traligill basin, Scotland. *Earth Surface Processes & Landforms*, **11**, 249–61.
Smart, P., T. Waltham, M. Yang & Y. Zhang 1986. Karst geomorphology of western Guizhou, China. *Trans. British Cave Res. Assoc.* **13(3)**, 89–103.
Smith, D. I. 1965. Some aspects of limestone solution in the Bristol region. *Geog. J.* **131(1)**, 44–9.
Smith, D. I. 1972. The solution of limestone in an Arctic environment. *Inst. Brit. Geogr. Spec. Pub.* **4**, 187–200.
Smith, D. I. & T. C. Atkinson 1976. Process, landforms and climate in limestone regions. In *Geomorphology and climate*, E. Derbyshire (ed.), 369–409. London: Wiley.
Smith, D. I. & M. A. Greenaway 1983. Fluorometric dye techniques: their application to groundwater tracing and borehole studies. *Papers of the International Conference on Groundwater and Man.* Sydney December 1983, **1**, 311–20. Canberra: Australian Government Publishing Service.
Smith, D. I. & M. D. Newson 1974. The dynamics of solutional and mechanical erosion in limestone catchments on the Mendip Hills, Somerset. In *Fluvial processes in instrumented watersheds*, K. J. Gregory & D. E. Walling (eds), Inst. Brit Geogs Spec. Pub. 6, 155–67.
Smith, D. I., T. C. Atkinson & D. P. Drew 1976. The hydrology of limestone terrains. In *The*

science of speleology, T. D. Ford & C. H. D. Cullingford (eds), 179–212. London: Academic Press.

Smith, D. I., D. P. Drew & T. C. Atkinson 1972. Hypotheses of karst landform development in Jamaica. *Trans. Cave Res. Gp. GB* **14**, 159–73.

Sneed, E. D. & R. L. Folk 1958. Pebbles in the Lower Colorado River, Texas. A study of particle morphogenesis. *J. Geol.* **66**, 114–50.

Snow, D. T. 1968. Rock fracture spacings, openings, and porosities. *J. Soil Mech. Found. Div., Amer. Soc. Civil Engineers* **94**, 73–91.

Snow, D. T. 1969. Anisotropic permeability of fractured media. *Water Resources Res.* **5**, 1273–89.

Soderberg, A. D. 1979. Expect the Unexpected: Foundations for Dams in Karst. *Bull., Assoc. Engineeering Geologists* **16(3)**, 409–25.

Song, L. 1981. Some characteristics of karst hydrology in Guizhou plateau, China. *Proc. 8th Int. Congr. Speleol.* **1**, 139–42.

Song, L. 1986. Karst geomorphology and subterranean drainage in south Dushan, Guizhou Province, China. *Trans. British Cave Res. Assoc.* **13(2)**, 49–63.

Song, L., Y. Zhang, J. Fang & Z. Gu 1983. Karst development and the distribution of karst drainage systems in Dejiang, Guizhou Province, China. *J. Hydrol.* **61**, 3–17.

Sorriaux, P. 1982. *Contribution a l'étude de la sedimentation en milieu karstique: Le systeme de Niaüx-Lombrives-Sabart, Pyrenees Arigéoises*. Thesis, 3rd cycle, Univ. Paul Sabatier, Toulouse.

Souchez, R. A. & M. Lemmens 1985. Subglacial carbonate deposition: an isotope study of a present-day case. *Palaegeog., Palaeoclim., Paleoecol.* **51**, 357–64.

Sowers, G. F. 1984. Correction and protection in limestone terrane. In *Sinkholes: their geology, engineering and environmental impact*, B. F. Beck (ed.), 373–8. Boston: Balkema.

Spate, A. P., J. N. Jennings, D. I. Smith & M. A. Greenaway 1985. The micro-erosion meter: use and limitations. *Earth Surf. Proc. Landforms* **10**, 427–40.

Spencer, T. 1981. Micro-topographic change on calcarenites, Grand Canyon Island, West Indies. *Earth Surface Processes & Landf.*, **6**, 85–94.

Spencer, T., D. R. Stoddart & C. D. Woodroffe 1987. Island uplift and lithospheric flexure: observations and cautions from the South Pacific. *Zeit, f. Geomorph N.F., Suppl. Bd.* **63**, 87–102.

Spring, U. & K. Hutter 1981a. Conduit flow of a fluid through its solid phase and its application to intraglacial channel flow. *Int. J. Engng. Sci.* **20(2)**, 327–63.

Spring, U. & K. Hutter 1981b. Numerical studies of Jokulhaups. *Cold Regions Science and Technology* **4**, 227–44.

Spring, W. & E. Prost 1883. Etude sur les eaux de la Meuse, *Ann. Soc. geol. Belg.* **XI**, 123–220.

Stanton, W. I. & P. L. Smart 1981. Repeated dye traces of underground streams in the Mendip Hills, Somerset. *Proc. Univ. Bristol Speleo. Soc.* **16(1)**, 47–58.

Stewart, M. K. & C. J. Downes 1982. Isotope hydrology of Waikoropupu Springs, New Zealand. In *Isotope studies of hydrologic processes*, E. C. Perry & C. W. Montgomery (eds), 15–23. DeKalb: Northern Illinois University Press.

Stewart, M. & P. W. Williams 1981. Environmental isotopes in New Zealand hydrology 3: isotope hydrology of the Waikoropupu Springs and Takaka River, Northwest Nelson. *NZ J. Sci.* **24**, 323–37.

Stoddart, D. R., T. Spencer & T. P. Scoffin 1985. Reef growth and karst erosion on Mangaia, Cook Islands: a reinterpretation. *Z. Geomorph. Suppl.-Bd.* **57**, 121–40.

Strahler, A. N. 1957. Quantitative analysis of watershed geomorphology. *Am. Geophys. Union Trans.* **38**, 913–20.

Stringfield, V. T. & H. E. LeGrand 1971. Effects of karst features on circulation of water in carbonate rocks in coastal areas. *J. Hydrol.* **14**, 139–57.

Stumm, W. & J. J. Morgan 1980. *Aquatic chemistry: introduction emphasizing equilibria in natural waters*. New York: John Wiley.

Sunartadirdja, M. A. & H. Lehmann 1960. Der Tropische Karst von Maros und Nord-Bone im SW-Celebes (Sulawesi). *Zeit fur Geomorph.* **Suppl. 2**, 49–65.

Sundborg, A. 1956. The river Klarälven, a study of fluvial processes. *Geogr. Annaler* **38**, 127–316.

Sustersic, F. 1979. Some principles of cave profile simulation. *Actes Symp. Internat. sur l'erosion karst*, 125–31. Aix-en-Provence.
Sutcliffe, A. J. 1976. Cave paleontology. In *The science of speleology*, T. D. Ford & C. H. D. Cullingford (eds), 495–520. London: Academic Press.
Sweeting, M. M. 1950. Erosion cycles and limestone caverns in the Ingleborough district of Yorkshire. *Geog. J.* **115**, 63–78.
Sweeting, M. M. 1958. The karstlands of Jamaica. *Geog. J.* **124,** 184–99.
Sweeting, M. M. 1966. The weathering of limestones, with particular reference to the Carboniferous Limestones of northern England. In *Essays in geomorphology*, G. H. Dury (ed.), 177–210. London: Heinemann.
Sweeting, M. M. 1972. *Karst landforms*. London: Macmillan.
Sweeting, M. M. 1978. Some observations on New Zealand limestone areas. in *Landform evolution in Australia*, J. L. Davies & M. A. J. Williams (eds), 250–8. Canberra: A.N.U. Press.
Sweeting, M. M. 1979. Weathering and solution of the Melinau Limestones in the Gunung Mulu National Park, Sarawak, Malaysia. *Ann. Soc. Geol. Belg.* **102**, 53–7.
Sweeting, M. M. (ed.) 1981. *Karst geomorphology: benchmark papers in geology* **59**, Stroudsburg, Penn: Hutchinson-Ross.
Sweeting, M. M. & G. S. Sweeting 1969. Some aspects of the carboniferous limestone in relation to its landforms with particular reference to N.W. Yorkshire and County Clare. *Rech. Mediterr.* **7**, 201–8.
Swinnerton, A. C. 1932. Origin of limestone caverns. *Bull., Geol. Soc. Am.* **43**, 662–93.
Szczerban, E. & F. Urbani 1974. Carsos de Venezuela, Parte 4: Formas carsicas en areniscas precambricas del Territorio Federal Amazonas y Estado Bolivar. *Bol. Soc. Venezolana Espel.* **5(1)**, 27–54.
Szunyogh, G. 1984. Theoretical investigation of the origin of spherical caverns of thermal origin (Second Approach). *Karszt-es Barlang*, 19–24.

Talbot, C. J. & M. P. A. Jackson 1987. Salt tectonics. *Sci. Amer.* **257(2)**, 70–9.
Tate, T. 1879. The source of the R. Aire. *Proc. Yorks. Geol. Soc.* **VII**, 177–87.
Terjesen, S. C., O. Erga & A. Ve 1961. Phase boundary processes as rate determining steps in reactions between solids and liquids. *Chem. Eng. Sci.* **74**, 277–88.
Theis, C. V. 1935. The relation between the lowering of the piezometric surface and the rate and duration of discharge of a well using ground water storage. *Trans. Am. Geophys. Union* **2**, 519–24.
Thérond, R. 1972. *Recherche sur l'étancheite des lacs de barrage en pays karstique*. Paris: Eyrolles.
Thomas, T. M. 1974. The South Wales interstratal karst. *Trans. Brit. Cave Res. Ass.* **1**, pp. 131–52.
Thomson, G. M., D. N. Lumsden, R. L. Walker and J. A. Carter, 1975. Uranium series dating of stalagmites from Blanchard Springs Caverns, U.S.A. *Geochim. Cosmochim. Acta*, 39, pp. 1211–18.
Thorp, J. 1934. The asymmetry of the Pepino Hills of Puerto Rico in relation to the Trade Winds. *J. Geol.* **42**, 537–45.
Thorp, M. J. W. & G. A. Brook 1984. Application of double Fourier series analysis to ground subsidence susceptibility mapping in covered karst terrain. In *Sinkholes, their geology, engineering and environmental impact*. B. F. Beck (ed.), 197–200. Boston: Balkema.
Thrailkill, J. 1968. Chemical and hydrological factors in the excavation of limestone caves. *Bull., Geol. Soc. Am.* **79**, pp. 19–46.
Thrailkill, J. 1985. Flow in a limestone aquifer as determined from water tracing and water levels in wells. *J. Hydrol.* **78**, pp. 123–36.
Tintilozov, Z. K. 1983. *Akhali Atoni Cave System*, Metsniereba, Tbilisi, U.S.S.R., 150 p.
Todd, D. K. 1980. *Groundwater Hydrology*. Wiley, New York.
Torbarov, K. 1976. Estimation of permeability and effective porosity in karst on the basis of recession curve analysis. In V. Yevjevich (ed.) *Karst Hydrology and Water Resources*: vol. 1 Karst Hydrology 121–36. Colorado: Water Resources Publications.
Tricart, J. and A. Cailleux 1972. *Introduction to climatic geomorphology*. London: Longman.

Trimmel, H. 1968. *Hohlenkunde*. Vieweg, Braunschweig.
Troester, J. W., E. L. White & W. B. White 1984. A comparison of sinkhole depth frequency distributions in temperate and tropical karst regions. In *Sinkholes: their Geology, Engineering and Environmental Impact*, B. F. Beck (ed.), 65–73. Rotterdam: Balkema.
Trombe, F. 1952. *Traité de Speleologie*, Paris: Payot.
Trudgill, S. T. 1976. The marine erosion of limestones on Aldabra Atoll, Indian Ocean. *Zeits. f. Geomorph., Suppl.* **26**, 164–200.
Trudgill, S., 1985. *Limestone Geomorphology*. London: Longman.
Trudgill, S., High, C. J. & Hanna, F. K. 1981. Improvements to the micro-erosion meter. *Brit. Geom. Res. Gp. Tech. Bull.* **29**, 3–17.
Trupak, N. G. 1956. Cited in Milanović, P. T. 1981, p. 237.
Tsui, P. C. & D. M. Cruden 1984. Deformation associated with gypsum karst in the Salt River Escarpment, northeastern Alberta. *Can. J. Earth Science* **21**, 949–59.
Tsykin, R. A. (in press). Paleokarst in the Union of Soviet Socialist Republics. In *Paleokarst: a world review*, P. Bozak, D. C. Ford & J. Glazek (eds), Amsterdam: Academia Praha/Elsevier.
Turekian, K. K. & E. Nelson 1976. Uranium decay series datings of the travertines of Caune de l'Arago (France). *Union des Sci. Prehist. Proto Hist., IX Congr., Nice*, 172–9.
Turner, G. M. & R. G. Lyons 1987. A paleomagnetic secular variation record ca 120 000 yr BP from New Zealand cave sediments. *Geophys. J. Roy. Astr. Soc.* **87**, 1181–92.
TVA 1949. *Geology and foundation treatment*. Tennessee Valley Authority Projects, Tech. Report No. 22.

UNESCO 1970. *International legend for hydrogeological maps*. Paris: Unesco.
UNESCO/IAHS 1983. *Methods and instrumentation for the investigation of groundwater systems*. Proc. International Symposium, Noordwijkerhout, Netherlands.
United States Department of Energy 1982. *National waste terminal storage program: criteria for mined geologic disposal of nuclear waste repository performance and development criteria*, public draft. Washington.
United States Dept. of the Interior 1981. *Ground water manual*. New York: Wiley.
United States Environmental Protection Agency 1979. *Methods for chemical analysis of water and wastes*. Doc. EPA-600 4–79–020.
United States Environmental Protection Agency & United States Geological Survey 1988. *Application of dye-tracing techniques for determining solute-transport characteristics of ground water in karst terranes*. Atlanta: US EPA.
United States Geological Survey 1970. Study and Interpretation of the Chemical Characteristics of Natural Water. *Water Supply Paper 1473*.
Urai, J. L., C. J. Spiers, H. J. Zwart & G. S. Lister 1986. Weakening of rock salt by water during long-term creep. *Nature* **324**, 554–7.

Van Everdingen, R. O. 1981. Morphology, hydrology and hydrochemistry of karst in permafrost near Great Bear Lake, Northwest Territories. *National Hydrological Research Institute of Canada, Paper 11*.
Vennard, J. K. & Street, R. L. 1976. *Elementary fluid mechanics*. 5th Ed., S.I. version. New York: Wiley.
Verges, V. 1985. Solution and associated features of limestone fragments in calcareous soil (lithic calcixeroll) from southern France. *Geoderma* **36**, 109–22.
Verosub, K. L. 1977. Depositional and post-depositional processes in the magnetization of sediments. *Rev. Geophys. Space Phys.* **15**, 129–43.
Verstappen, H. th. 1964. Karst morphology of the Star Mountains (central New Guinea) and its relation to lithology and climate. *Z. Geomorph.* **8**, 40–9.
Vierhuff, H. 1983. Hydrogeological requirements for a site for a mined repository of nuclear waste. In *Ground water in water resources planning. International Association of Hydrological Sciences, Pub.* **142(2)**, 1209–12.
Viles, H. 1984. Biokarst: review and prospect. *Progress in Physical Geography* 8(4), 523–42.
Viles, H. A. 1987. Blue-green algae and terrestrial limestone weathering on Aldabra Atoll: an S.E.M. and light microscope study. *Earth Surface Processes and Landforms* **12**, 319–30.

Viies, H. A. & S. T. Trudgill 1984. Long term measurements of microerosion meter sites, Aldabra Atoll, Indian Ocean. *Earth Surf. Proc. Landforms* **9**, 89–94.
Villar, E., P. L. Fernandez, I. Gutierrez, L. S. Quindos & J. Soto 1986. Influence of visitors on carbon dioxide concentrations in Altamira Cave. *Cave Science* **13(1)**, 21–3.
Vincent, P. J. 1987. Spatial distribution of polygonal karst sinks. *Z. Geomorph. N.F.* **31(1)**, 65–72.

Wadge, G. & T. H. Dixon 1984. A geological interpretation of Seasat-SLAR imagery of Jamaica. *J. Geol.* **92**, 561–81.
Wagener, F. V. M. & P. W. Day 1986. Construction on dolomite in south Africa. *Environ. Geol. Water Sci.* **8(1/2)**, 83–9.
Waltham, A. C. 1970. Cave development in the limestone of the Ingleborough district. *Geog J.* **136**, 574–84.
Waltham, A. C. 1981. Origin and development of limestone caves. *Progress in Phys. Geog.* **5(2)**, 242–56.
Waltham, A. C. & D. B. Brook 1980. Geomorophological observations in the limestone caves of Gunung Mulu National Park, Sarawak. *Trans., Brit. Cave Res. Assoc.* **7(3)**, 123–40.
Waltham, A. C., P. L. Smart, H. Friederich & T. C. Atkinson 1985. Exploration of caves for rural water supplies in the Gunung Sewu Karst, Java. Comptes Rendus du Colloque International de Karstologie appliquée *Anns. Soc. Geol. Belg.* **108**, 27–31.
Waltham, A. C., G. Vandeven, & C. M. Ek, 1986. Site investigations on cavernous limestone for the Remouchamps Viaduct, Belgium. *Ground Engineering* **19(8)**, 16–18.
Wanless, H. R. 1979. Limestone response to stress: pressure solution and dolomitization. *J. Sed. Pet.* **49(2)**, 437–62.
Ward, R. C. 1975. *Principles of hydrology*, 2nd edn. London: McGraw Hill.
Warwick, G. T 1953. The origin of limestone caves. In *British Caving*, C. H. D. Cullingford (ed.), 41–61. London: Routledge & Kegan Paul.
Waterhouse, J. D. 1984. Investigation of pollution of the karstic aquifer of the Mount Gambier area in South Australia. In *Hydrogeology of karstic terrains*, A. Burger & L. Dubertret (eds). *International Contributions to Hydrogeology* **1(1)**, 202–5. Hannover: Heise.
Waterman, S. E. 1975. *Simulation of conduit network development on bedding planes*. MSc thesis, McMaster University.
Watson, R. A. & W. B. White 1985. The history of American theories of cave origin. *Geol. Soc. Am. Centennial Special Vol. 1*, 109–23.
Weisrock, A. 1981a. Morphogenèse des édifices tuffeux d'Imouzzer Ida Ou Tanane (Maroc). *Ass. Francaise de Karstologie,* **Mem. 3**, 157–70.
Weisrock, A. 1981b. Stratigraphie et petrographie des formations travertineuses. *Ass. Francaise de Karstologie,* **Mem. 3**, 187–90.
Weisrock A. 1985. Originalité karstique de l'Atlas atlantique marocain. *Karstologia* **5(1)**, 29–38.
Weyl, P. K. 1958. Solution kinetics of calcite. *J. Geol.* **66**, 163–76.
White, E. L. & W. B. White 1968. Dynamics of sediment transport in caves. *Bull., Nat. Speleo. Soc. Am.* **30(4)**, 115–29.
White, E. L. & W. B. White 1969. Processes of cavern breakdown. *Bull., Nat. Speleo. Soc. Am.* **31(4)**, 83–96.
White, E. L. & W. B. White 1979. Quantitative morphology of landforms in carbonate rock basins in the Appalachian Highlands. *Bull. Geol. Soc. Amer.* **90**, 385–96.
White, E. L., G. Aron & W. B. White 1984. The influence of urbanization on sinkhole development in central Pennsylvania. In *Sinkholes: their geology, engineering and environmental impact*, B. F. Beck (ed.), 275–81. Rotterdam: Balkema. Also published in *Environ. Geol. Water Sci.* **8(1/2)**, 91–7 (1986).
White, W. B. 1969. Conceptual models for carbonate aquifers. *Ground Water* **7(3)**, 15–21.
White, W. B. 1976. Cave minerals and speleothems. In *The science of speleology*, T. D. Ford & C. H. D. Cullingford (eds), 267–327. London: Academic Press.
White, W. B. 1977a. The role of solution kinetics in the development of karst aquifers. In *Karst hydrogeology*, J. S. Tolson & F. L. Doyle (eds). Internat. Assoc. Hydrogeol., Mem. 12, 503–17.

White, W. B. 1977b. Conceptual models for carbonate aquifers: revisited. In *Hydrologic problems in karst regions*, R. R. Dilamarter & S. C. Csallany (eds), 176–87. Bowling Green: Western Kentucky University.

White, W. B. 1981. Reflectance spectra and colour in speleothems. *Bull. Nat. Speleo. Soc.* **40**, 20–6.

White, W. B. 1984. Rate processes: chemical kinetics and karst landform development. In *Groundwater as a geomorphic agent*. R. G. LaFleur (ed.), 227–48. London: Allen & Unwin.

White, W. B. 1988. *Geomorphology and hydrology of carbonate terrains*. Oxford: Oxford University Press.

Wigley, T. M. L. 1971. Ion pairing and water quality measurements. *Can J. Earth Sci.* **8(4)**, 468–76.

Wigley, T. M. L. 1977. WATSPEC: a computer programme for determining the equilibrium speciation of aqueous solutions. *Brit. Geomorph. Res. Sp., Tech. Bull.* **20**.

Wigley, T. M. L. & M. C. Brown 1976. The physics of caves. In *The science of speleology*, T. D. Ford & C. H. D. Cullingford (eds), 329–58. London: Academic Press.

Wigley, T. M. L. & L. N. Plummer 1976. Mixing of carbonate waters. *Geochim. Cosmochim. Acta* **40**, 989–95.

Wilcock, J. D. 1968. Some developments in pulse-train analysis. *Trans. Cave Res. Gp. GB* **10(2)**, 73–98.

Wilford, G. E. 1966. 'Bell holes' in Sarawak caves. *Bull. Nat. Speleo. Soc. Am.* **28(4)**, 179–82.

Wilk, Z. (in press). Hydrogeological problems in the exploration, development and exploitation of the Cracow-Silesia Zn-Pb ore deposits. In *Paleokarst: a world review*, P. Bosak, D. C. Ford & J. Glazek (eds). Amsterdam: Academia Praha/Elsevier.

Williams, A. M. 1959. The formation and deposition of moonmilk. *Trans. Cave Res. Gp., G.B.* **5(2)**, 133–8.

Williams, H. R., Corkery, D. & Lorek, E. G. 1985. A study of joints and stress-release buckles in Paleozoic rocks of the Niagara Peninsula, southern Ontario. *Can. J. Earth Sci.* **22(2)**, 296–300.

Williams, P. W. 1963. An initial estimate of the speed of limestone solution in County Clare. *Irish Geog.* **4**, 432–41.

Williams, P. W. 1966a. Limestone pavements: with special reference to western Ireland. *Trans. Inst. Br. Geog.* **40**, 155–72.

Williams, P. W. 1966b. Morphometric analysis of temperate karst landforms. *Irish Speleology* **1**, 23–31.

Williams, P. W. 1968. An evaluation of the rate and distribution of limestone solution and deposition in the River Fergus basin, western Ireland. In *Contributions to the study of karst*, P. W. Williams & J. N. Jennings, 1–40. Australian National University, Res. School of Pacific Studies Pub. G5.

Williams, P. W. 1969. The geomorphic effects of groundwater. In *Water, earth and man*, R. J. Chorley (ed.), 269–84. London: Methuen.

Williams, P. W. 1970. Limestone morphology in Ireland. In *Irish geographical studies*, N. Stephens & R. E. Glasscock (eds), 105–24. Queens University, Belfast.

Williams, P. W. 1971. Illustrating morphometric analysis of karst with examples from New Guinea. *Z. Geomorph.* **15**, 40–61.

Williams, P. W. 1972a. Morphometric analysis of polygonal karst in New Guinea. *Bull. Geol. Soc. Am.* **83**, 761–96.

Williams, P. W. 1972b. The analysis of spatial characteristics of karst terrains. In *Spatial analysis in geomorphology*, R. J. Chorley (ed.), 136–63. London: Methuen.

Williams, P. W. 1977. Hydrology of the Waikoropupu Springs: a major tidal karst resurgence in northwest Nelson (New Zealand). *J. Hydrol.* **35**, 73–92.

Williams, P. W. 1978. Interpretations of Australasian karsts. In *Landform evolution in Australia*, J. L. Davies & M. A. J. Williams (eds), 259–86. Canberra: ANU Press.

Williams, P. W. 1982a. Karst landforms in New Zealand. In *Landforms of New Zealand*, J. Soons & M. J. Selby (eds), 105–25. Auckland: Longman Paul.

Williams, P. W. 1982b. Speleothem dates, Quaternary terraces and uplift rates in New Zealand. *Nature* **298**, 257–60.

Williams, P. W. 1983. The role of the subcutaneous zone in karst hydrology. *J. Hydrol.* **61**, 45–67.

Williams, P. W. 1985. Subcutaneous hydrology and the development of doline and cockpit karst. *Z. Geomorph.* **29(4)**, 463–82.
Williams, P. W. 1987a. Geomorphic inheritance and the development of tower karst. *Earth Surface Processes and Landforms* **12(5)**, 453–65.
Williams, P. W. 1987b. The importance of karst in New Zealand's National Parks. *N.Z. Geographer* **43(2)**, 84–94.
Williams, P. W. & R. K. Dowling 1979. Solution of marble in the karst of the Pikikiruna Range, northwest Nelson, New Zealand. *Earth Surf. Proc.* **4**, 15–36.
Williams, P. W., R. G. Lyons, X. Wang, L. Fang & H. Bao 1986. Interpretation of the paleomagnetism of cave sediments from a karst tower at Guilin. *Carsologica Sinica* **5(2)**, 113–26.
Williams, P. W., P. McDougall and J. M. Powell 1972. Aspects of the Quaternary geology of the Tari-Koroba area, Papua. *J. Geol. Soc. Australia* **18(4)**, 333–47.
Wilson, J. F. 1968. Fluorometric procedures for dye tracing. *Techniques of water resources investigations of the United States Geological Survey, Book 3*, Chapter A12.
Wilson, J. F., E. D. Cobb & F. A. Kilpatrick 1986. Fluorometric procedures for dye tracing. *Techniques of water resources investigation*, 03–A12, US Geol. Survey.
Wilson, J. L. 1974. Characteristics of carbonate platforms margins. *Am. Assoc. Petrol. Bull.* **58**, 810–24.
Wilson, J. L. 1975. *Carbonate facies in geologic history*. Heidelberg: Springer.
Wintle, A. G. & D. J. Huntley 1982. Thermoluminescence dating of sediments. *Quat. Sci. Rev.* **1**, 31–53.
Wissman, H. von, 1954. Der Karst der humiden heissen und sommerheissen Gebiete Ostasiens. *Erdkunde* **8**, 122–30.
Wittwen, R., H. Waser and B. Matthey 1971. Essai de fixation de la sulforhodamine B et de la sulforhodamine G extra sur charbon actif. *Act. 4 Congr. Suisse Speleol.*, Neuchâtel, 78–83.
Wolfe, T. E. 1973. *Sedimentation in karst drainage basins along the Allegheny Escarpment in southeastern West Virginia, U.S.A.* PhD thesis, McMaster University.
Wolman, M. G. & J. P. Miller 1960. Magnitude and frequency of forces in geomorphic processes. *J. Geol.* **68**, 54–74.
Woo, M.-k. & P. Marsh 1977. Effect of vegetation on limestone solution in a small high Arctic basin. *Can. J. Earth Sci.* **14(4)**, 571–81.
Wood, W. W. & G. G. Ehrlich 1978. Use of baker's yeast to trace microbial movement in ground water. *Ground Water* **16(6)**, 398–403.
Worthington. S. R. H. 1984. *The paleodrainage of an Appalachian Fluviokarst: Friars Hole, West Virginia*. MSc Thesis, McMaster, University.

Yaalon, D. H. & E. Ganor 1968. Chemical composition of dew and dry fall out in Jerusalem, Israel. *Nature* **217(5134)**, 1139–40.
Yan, Z. Y. L., S. L. M. Zhao & R. Z. D. Liu, 1984. Oxygen isotope composition, paleotemperature and ^{230}Th/^{234}U dating of speleothem from the Fourth Cave of Peking Man site. *Dev. in Geoscience, Acad. Sinica*, 177–83.
Yang, M. 1982. The geomorphological regularities of karst water occurrences in Guizhou Plateau. *Carsologica Sinica* **1(2)**, 81–92.
Yong, C. J., D. C. Ford, J. Gray & H. P. Schwarcz 1985. Stable isotope studies of cave seepage water. *Chem. Geol.* **58**, 97–105.
Yonge, C. J. 1982. *Stable isotope studies of water extracted from speleothems*. PhD thesis, McMaster University.
Yuan Daoxian 1981. *A brief introduction to China's research in karst*. Inst. Karst Geol., Guilin, Guangxi, China.
Yuan Daoxian 1983. *Problems of environmental protection of karst areas*. Inst. Karst Geol., Guilin, Guangxi, China.
Yuan Daoxian, 1985a. New observations on tower karst. *Proc. First Int. Conf. Geomorph., Manchester*, p. 676.
Yuan Daoxian, 1985b. On the heterogeneity of karst water. *Karst Water Resources* (Proc. Ankara-Antalya Symposium), I.A.H.S. Publ. 161, 281–92.
Yurtsever, Y. 1983. Models for tracer data analaysis. In '*Guidebook on nuclear techniques in hydrology*', Technical Report Series No. 91, 381–402. Vienna: IAEA.

REFERENCES

Zenger, D. H., J. B. Dunham & R. L. Etherington 1980. Concepts and Models of Dolomitisation. *Soc. Econ. Palaeontol. Mineral., Spec. Pub.* **28**.

Zhang, S.-Y. (in press). Paleokarst in China, In *Paleokarst – a world review*. P. Bosak, D. C. Ford & J. Glazek (eds). Amsterdam: Academia Praha/Elsevier.

Zhang, Z. 1980. Karst types in China: *Geo. Journal* **4.6**, 541–70.

Zhu Dehau 1982. Evolution of peak cluster depressions in the Guilin area and morphometric measurement. *Carsologica Sinica* **10(2)**, 127–34.

Zibret, Z. & Z. Simunic 1976. A rapid method for determining the water budget of enclosed and flooded karst plains. In *Karst hydrology and water resources*, V. Yevjevich (ed.), 319–39. Colorado: Water Res. Pubs.

Zotl, J. 1966. Ein fossiler semi-arider tropischer karst auf Ithaka. *Erdkunde* **20(3)**, 204–8.

Zotl, J. 1974. *Karsthydrogeologie*. Vienna: Springer.

Index

Figures in *Italic* Section references in **Bold**

Aach (German Federal Republic) 231, 239
abandonment suites 325
abrasion 306–7
abrasion platforms 498
abris (rock shelter) deposits 317
absolute dating 356–64
acid dissolution 46–7
acid rain 77–8
activity 48–51
activity coefficient 48, *49*
adsorption 94, 519
  artificial isotopes 230
adsorption layer 82
aeolianites 445
aerial photography 414–7, 527
aggressivity 50, 52, 164, 402, 441
alabaster 540
Aldabra Atoll (Indian Ocean) 396
algae 377, 396, 448
algal mats 16–17, 20
Aliou (France) 212–13
alkalinity 61, 283
allogenic (allochthonous) systems 98–9
allogenic inputs, associated landforms **9.7**
allogenic rivers 441, 458, 507
  effluent 424
  influent 426–7
  superimposed 470
allogenic waters 99–100, 114, 124
alluvial channels 302, 321
alluvium 433
alpine karst
  Canadian type 475–6
  Pyrenean type 475
alpine karst zonation 482
Alps 459, 475
alumina 350, 542
amidorhodamine 233
anastomoses 296, *297*
Andros Island (Bahamas) 504
anemolites 337
anhydrite 3, 34, 43, 348, 488
  dissolution 51–2, 93
  hydration of 28–9, 289, 464
ankerite 11
Antarctica 472, 495
anthodites 332
anticlines 464
Anticosti Island (Canada) *59*, *478*
antidune mode 321
antimony 542
apatite 27
aqueous sediment transport 318–21
aquiclude 127, 278, 532
aquifers 92, 127–9
  anisotropic 134–5, 151, 171
  artesian 222, 224
  boundary conditions 148–9, 173–4, 188
  characteristics 166–8
  classification 168–70, Table 5.8, Table 6.5
  coastal 177, 220
  conduit 177, 204, 483–4, 519–20;
    permeability 161–2, 251–61
  confined 127–8, *136*, 137, 174, *187*, 487
  controls on development **5.3**
  definition 127
  diffuse 204, 210
  epikarst 169, 208, 404, 440, 482
  exploration techniques **6.2**
  fissured 214
  heterogeneous 134
  homogenous 134
  isotropic 134–5, 138–9, 151
  karstified 214
  limits 188, 190
  organization of **6.7**
  perched 128–9, 162, 401–2
  pollution of **11.2**
  porosity of 199
  porous of 210, 214, 520–1
  rate of development **5.5**
  sandstone, and maze formation 280–1
  storage *136*, 137–8, 158–9, 188, 190
  transmissivity 137, 158, 177–8, 188, 190, 198
  unconfined 127, 133–4, *136*, 137, 174, *187*
  zonation 177–8
aquifuge 127–8
aquitards 127, 279
aragonite 10–11, 20, 30, 283
  dissolution 21
  precipitation 15, 95, 332–42
  saturation index 58
  solution of 57–8
aragonite cement 504
Aran Islands (Ireland) 396
Archangel (USSR) 464
arches *see* tension domes
arete-and-pinnacle karst 391
Argostoli (Greece) *504-5*

Arkansas (USA) 278
Arnhem Land (Australia) 29
artesian confinement 41
artesian maze caves 278–9
artesian storage 165
artesian wells 128
artificial isotopes for water tracing 229–30
Athabasca River (Canada) 107
atolls, karstification of 501–2
atolls, *see* coral atolls
Australia 445
  Arnhem Land 29
  Chillagoe 441
  Flinders Ranges 471–2
  Great Barrier Reef 411, 502
  Limestone Ranges 469–71
  Mount Gambier 520
  New South Wales 111
  Nullarbor Plain 246, 267, 314, 471–2
  Queensland 118, *468*
  Western Australia 161, *447*, *468*
Austria 355
  Hallstatt 509
  Vienna 2, 451
autogenic (autochthonous) systems 98
autogenic solution 116
Avalon Dam (USA) 533
Axel Heiberg Island (Canada) 496
Aymamon Limestone 445, 507

Bacillis subtilis 220
bacteria 64, 78, 377, 449, 519
Baget (France) 212–13
Bahamas 362, 500
  Andros Island 504
  San Salvador Island 500
Barbados 499
barite 283, 348, 541
Barren River (USA) 537
barrier reefs 500
baseflow recession 197–8, 211
baseflow reserves 211
baselevel control 437–8
Bashkir (USSR) 458–9
bassanite 348
bauxite 350, 509, 511, 542
Bay of Ha Long (Vietnam) 441
beach rock 502, *503*, 504
beams, mechanical strength 310–13
bedding planes 35–6, 150, 249–50, 267, 296, 496, 522
bedload 319–21, 327
bedload grain size 327
Belgium 368, 524
  Hainault 460
Belize 119, 123, 423, 444
  Caves Branch 428
  Searanxx *413*
  Sibun River *427*, 428
Beljanica Mountains (Yugoslavia) 509
bell holes 298–9
bending stress 311–12
bentonite 532
berms 326
Bermuda 362
bicarbonate waters 53–5
Big Horn Basin (USA) 289–90
bileca (Yugoslavia) 199
biochemical processes 375
bioherms 17
biokarst 118, 450–1
  see also phytokarst
biomicrite 18, 31
biostromes 17
bioturbation 329, 364
birdseye structure 329
birnessite 350
Black Sea 289, 512
blind valleys *425*, 426–7, 457, 459, 501–2
bloedite 348
blue holes 411, 502, 504
boehmite 350
bogaz 391
Bohemia (Czechoslavakia) 534
Boreal Optimum 355
borehole analysis **6.4**
  dye dilution tests 182, 238–9
  logging 191–3
  Lugeon tests 182–3
  pumping tests 182
  recharge tests 182–6
boreholes logging 191–3
Borneo 544
boundary layer detachment 380
boundary layers 81–2
  saturated, detachment of 305–6
boundary shear stress 318
Bowling Green (USA) 537
box valleys 391, 470–1
boxwork 286–7
breakdown 291, **7.12**
  asymmetric 312–13
  autogenic 321
  block 300–1, 313
  chip 313–14
  dam creation 325
  slab 313–14
breakdown barriers 326
breccia 21, 37, 283, 292, 509
  gypsum solution 496
  mega-breccia 465
breccia pipe collapse 495
breccia pipe residuals 465
breccia pipes 313, 405, 460, 487
brines 52
Brunhes (magnetic) epoch 364

Brunhes-Matuyama reversal 367
Budapest (Hungary) 284–5, 342
building foundations in karst areas 523–4
building stone 538–40
Bukovina (USSR) 281
buoyant force 409
buoyant support
  artificially reduced 411
  removal of *406*, 410
Burren Plateau (Ireland) 301
bypass passages 272–3

calcite 10–11, *13*, 20
  dissolution of 55–7
  dogtooth spar 340–1
  electrical conductivity 61
  high-Mg 30
  Iceland spar 283
  length-fast 331–2
  length-slow 332
  nailhead spar 283, 286, 341, 358
  precipitation 15, 80, 84, 94–5, 329–42; sub-glacial 479
  saturtion index 58
  secondary 37–8, 150
  solubility 63, 72
  solution of 57–8, 85–8
  subaqueous deposits 340–2
  see also sinter; speleothem; travertine
calcite crystal growth 95
calcite ice 94, 341
calcite rafts 283, 286, *339*, 341
calcite solubility *52*, 53
calcium carbonate concentration 45
calcium concentration 205
calcium ion concentrations 220
calcrete 21, 123, 444, 448, 464
caliche 21, 448
caliper logging 191, 193
Camarasa dam (Spain) *530*
Canada 461, 491
  Anticosti Island *59*, *478*
  Athabasca River 107
  Axel Heiberg Island (NWT) 496
  Bruce B atomic power station 535
  Columbia Icefield 487
  Ellesmere Island (NWT) 496
  Frank Slide (Alta.) 523
  Goose Arm (Newfoundland) 484, *485*
  Great Bear Lake (NWT) 494–5
  Hamilton (Ont.) *486*
  Howe Lake (Sask.) 487, *488*
  Lake Huron *395*, 396
  Mackenzie Mountains (NWT) 391–2, 478, 492, *493*
  Maligne River 163, 227–8, 483
  Medicine Lake 228, 483
  Nahanni region 362, 444, 471, 492, *493*
  Nanisivik 292–3, 307, 542
  Newfoundland 391
  North Saskatchewan River 107
  Nova Scotia 459
  Ontario 511, 541
  Pine Point (NWT) *27*, 292, 488, 511, 542
  Rocky Mountains 34, 65, 121, *481*
  Saskatchewan 460, 487
  Vancouver Island 265, 371, 386, *387*
  Vermilion Creek (NWT) *408*, 476
  Winnipeg *486*, 487
  Wood Buffalo National Park 465
Canadian Prairies 41
cantilevers 310–12
capillary flow 378
capillary suction 330
capillary tension 378
carbon dioxide
  concentration and availability 64–6
  dissolution; and degassing 80, 82–3; open and closed systems 66–70
  ground air CO_2 64–6, 101, 123
  and hypogene caves 282–7
  partial pressure 53–4, 64–6, 101, 123
  role of *54*, 55
  solubility of 53, Table 3.4
carbon isotopes 220–1, 283, 356, 369
carbon-14 dating 356
carbonate, terrestrial 20
carbonate coasts, erosion of 394–6
carbonate deposition 18, 342
  aeolian calcarenite 502
  beach rock 502, *503*, 504
  calcrete 21, 123, 444, 448, 464
  caliche 21, 448
  kunkur 448
  tufa 448–51
  vertical distribution 123–4
  see also speleothem
carbonate karst evolution (humid terrains)
  alternative models 454–8
  early concepts 451–4
carbonate minerals, conditions and effects affecting solubility **3.6**
carbonate mud *see* micrite
carbonate reservoirs 541
carbonate rocks
  as flux 540
  global distribution 6
carbonate rocks and minerals **2.4**
carbonate sand 15–16
carbonate solubility 72
carbonatite 26–7
carbonatite apatite 27
carbonic acid 7, 53, 55, 80–2, 87
Caribbean 33, 542
carnallite 28
case-hardening 123, 444–7, *502*

castiles 465
Caucasus Mountains (USSR) 289, 475
cave art 543
cave earths 323, 330, 350
cave interior deposits *316*, Table 8.1
cave passages, vadose 119, 160
cave pearls 338, *339*, 364
cave sediments, diagenesis of 329–30
cave streams, rates of incision 121–3
cave temperature effect 370–1
cavernous weathering 380
caves 6, 445, 543
  allogenic 544
  artesian, maze 325
  bathyphreatic 261–9
  breakdown 246, **7.12**
  classifications 342–8
  cliff-foot 435, *443*
  coastal 326, *501*
  cut-off 469
  definitions 242–3
  dendritic 285
  fissure 470
  flank margin *290*
  floodwater 318
  formed by gypsum replacement and solution 289–90
  fracture 500
  hydrothermal 282–7, 315; monogenetic 284–5
  hypergene *see* common caves
  hypogene 282–91
  longest and deepest Table 7.2
  maze 245, 273, 278–81, 283, 291–2, 469; multi-storey 281, 283–5
  minerals; non-carbonate 347–51; other **8.2**; principal Table 8.3
  mixing zone 290–2, 291
  multi-level (multi-storey) 114, 274, 285
  multi-loop *262*, 263
  multi-phase 245, 248, 256, 265–6, 269, 274–8
  multiple input 253–9
  non–integrated 242
  nothephreatic 253
  phreatic 71, 246, 261, *262*, 263–5, 267, 269, 271–3, 281, 318, 320, 325, 437, 469; drainage of 362; glacier-induced *476*; maze 285
  relict *316*, 492
  restricted input 259–61
  river 502
  single input 251–3
  solution 38
  solutional pit and fissure 281–2
  thermal 325, 342
  vadose 80, 124, 246, 267–9, 318, 325; drawdown 267–9; invasion 268; meandering channels 301–2
  water table 261, *262*, 263, *264*, 267, 269
  see also proto-caves
Actun Chek (Belize) *427*, 428
Agen Allwedd (UK) 326–7
Cueva del Agua (Spain) *276*, 277
Akhali Atoni (USSR) 289
Altamira (Spain) 543
Atlantida (USSR) *279*, 280–1, *288*
Bedeilhac (France) 66
Bullock Creek (New Zealand) 274
Carlsbad Cavern (USA) 76, 246–87, 289, 312, 315
Castleguard (Canada) *270*, 314, 320, 325–6, 327, *339*, 342, 346, 348, 354, 362, 484
Caune de l'Arago (France) 362
Cave Creek (USA) *260*
Cave Branch (Belize) 263, *264*, 265, *427*, 428
la Chaise-de-Vouthon (France) 362
Cockebiddy (Australia) 267
Coulthard (Canada) *352*, 353
Cserzegtomajikut (Hungary) 285
de la Deveze (France) 333
Dobsinska Ice Cave (Czechoslovakia) 353
Eisriesenwelt 354, 483
Endless Cave (USA) *288*
Eos Chasm (Greece) *268*
Friars Hole (USA) 245, *277*, 278, *425*, **8.7**
Fung Kui (China) *308*
Gaping Ghyll (UK) 31–7
Gaurdakskaya (USSR) 290
Good Luck (Lubang Nasib Bagus, Sarawak) 315, 544
Greft Stream (Norway) *40*
Grotte di Frasse (Italy) 289
Holloch (Switzerland) 36, 245, 256, 263, *264*, 267, 275
La Hoya de Zimapan (Mexico) *264*, 296
Hurricane River (USA) *302*
Jean Bernard (France) 268
Jewel Cave (USA) 245, *284*, 286, 341, 350, 358
Kacne (Yugoslavia) 327
Kvithola (Norway) 484, 487
Langtry (USA) *264*
Lascaux (France) 544
Lighthouse (Bahamas) 290
Lubul Ha (Belize) 428
Ludi (China) 336
Lummelund (Sweden) 279
McClungs' (USA) *299*
Mammoth-Flint Ridge-Roppel-Procter (USA) 36, 234, *236*, 258–9, 263, 275–6, 303, 312, 330, 364, *365*, 367
Mangapohue Cave (New Zealand) *105*
Mas d'Azil (France) 336
Metro (New Zealand) 121

Mullamullang (Australia) 471
Mulu (Sarawak) 246, 267, 315, 364
Na Spicaku (Czechslovakia) 308
Nanxu (China) *247*
Nare (Papua, New Guinea) 246
Nettlebed (New Zealand) 122–3
Niaux (France) 307, 323, 543
Ochtinska (Czechoslovakia) 308
Ogle Cave (USA) *335*
Optimists' (USSR) 37, 245, *279*, 281
Ozernaya (USSR) 281
Pierre St Martin (France) 301, *315*, Table 7.2
Poloska Jama (Yugoslavia) *276*, 277
Pontnewydd (UK) 362
Postojna (Yugoslavia) 345
Red Cave (USSR) *330*
St Anne de Tilff (Belgium) 274, *299*
Satorkopuszta (Hungary) *284*, 285
Scarisoara (Bulgaria) 355
Selminum Tem (Papua, New Guinea) 267, *276*, 277–8, *324*
Skocjanske (Yugoslavia) *272*
Sof Omar (Ethiopia) *273*, 274, 469
Sonora Caverns (USA) *276*, 277
Sotano del Arroyo (Mexico) *366*
el Sotano (Mexico) *315*
Spluga della Preta (Italy) *268*
Swildon's Hole (UK) 263, *264*, *275*, 276–7, 301
Tun Kul (Belize) *315*
Valerie (Canada) *351–2*, 354–5
Vaucluse (France) 363, *264*
Wind Cave (USA) 245, 286–7, *339*, 342
Zhoukoudian (China) 362
Zolushka (USSR) 281
cavitation 340
cavities, in bedrock 524
celestite 348
cementation 329, 504
cenote *410*, 411
Central Kentucky Karst 151, 234, 258–9, 456
Cerknisko polje (Yugoslavia) 218
Cetina River (Yugoslavia) 438
Chad, Lake Chad 20
chalk 21
chambers 315
Channel Tunnel 535
chemical denudation *see* solutional denudation
chemical energy, for solution 164–5
chemical weathering 329
chemograph analysis **6.6**
chert 23, 36
Chillagoe (Australia) 441
China (Peoples' Republic of) 168, 542
Guangxi 543
Guilin 442–4
Guizhou Plateau 438, 458, 507
Hunan 525
Quighai (Tibet) 458, 507
Sichuan basin 175
Stone Forest of Yunnan 391, *392*, 509
Zhoukoudian 543
chloride 61, 239
a natural tracer 220
chlorite 329
Chromobacterium violaceum 220
clastic rocks 373, **2.7**
clastic sediments **8.2**
clay 323–5, 364, 404
clay minerals 30, 329
alteration of 329
clay soils 527
claypan corridors 471
claypans 471–2
climatic change, effects of 506–7
climatic change and cave speleothems 368–9
clint-and-grike (pinnacle-and-cutter) *381*
clints 382, 389, 391
closed (sequential) system 66–68, 101
penetration with laminar flow 90–1
solution of calcite 79–80
coal seams 460
coastal karst **10.5**
cobbles 326–7
cockling 385
cockpit dolines 397
cockpit karst 119, 466, 502, 507–8, 510–11
contour pattern 414–15
coefficient of heterogeneity 201
coincident (open) systems 66–70, 101, 104, 151
bicarbonate dissolution-equilibriation sequence 82–3
dissolution and degassing of CO_2 80
collapse depressions 471
collapse domes 312
collapse *see* breakdown
colluvium 326
Columbia Icefield 487
common caves 246, **7.2**
common ion effect 72, 449
comparative dating 364–8
completely mixed reservoir model 222
compression 409
compressive strength 33, Table 2.5
concrete 540
condensation, seasonal 469
condensation corrosion 309
conduit drainage 492
conduit extension and enlargement 166
conduit flow 7, 144, 204, 222
conduit permeability 161–2, 251–61
see also protocaves
conduits

flushed and enlarged 531
propagation of 249, 250
solutional 68
cone of depression 402, 535
cone karst 509
conservation of karst 544–5
contamination 536
see also pollution
Cook Islands 457
copper 542
coral atolls 124, 500
coral islands 499–500
coral reefs 16–17, 126, 151–2, 456, 498, 500, 502
case-hardening 445, 447
karstic drainage 502
raised *498*
corals 16
Cornwall (UK) 534
corrasion 301, 441
corridor karst 492, 495, 510
see also giant grikelands
corridor-and-pinnacle karst 529
corrosion 104, 402, 407, 437–8, 441, 467
alternative methods of assessment 110–4
focused 399, 401–2, 409, 501
phreatic 400
subcutaneous 120–1
see also solution processes; solution denudation
corrosion bevels 307–8, 379
corrosion front, retreating 501
corrosion notches 292–3, 307–8
corrosion plains 433–7, *440*, 441, 452, 502, 507
County Clare (Ireland) 111, 117–18
County Leitrim (Ireland) 117
couplet laminae 326
crandallite 349
Crimea (USSR) 309, 489
water supply tunnel 534
critical spans 312
critobalite 351
Crveno Jezero (Yugoslavia) 405
crystallites 94, 329–31
crystals 84
Cuba 444
Sierra de los Organos 441
Cumberland River (USA) 261
cupolas *see* solution pockets
cutters 382
cyanobacteria 118
cyanophytes 377
cycle of erosion 451, 506
Czechoslovakia
Bohemia 534
Supikovice *508*
dam construction
on gypsum and anhydrite 533–4
problems on carbonate rocks 529–33
dam grouting 532–3
Danube River 231, 239, 284–5, 426
Darai Hills (Papua, New Guinea) *413*
darapskite 350
Darcy flux 142, 144
darcy units 132
Darcy–Weisbach equation 146
Darcy–Weisbach friction law 146–8
Darcy's law 129–30, 132, 134
applicability to karst **5.2**
dating
absolute methods 356–64
comparative methods 364–8
stable isotopic studies 368–71
de-watering 409, 509
of mine zones 535
regional 535
Dead Sea 20, 458
Death Valley (USA) 458
Debye-Huckel equations 48
decantation runnels/fluting *383*, 386, 388
dedolomitization 25
degassing, CO_2 445
dejecta 326
denudation 407, 409
chemical, concept of 6
Derbyshire (UK) 456, 534
desert karst **10.2**
dessication cracking 329
deuterium 220, 225
diagenesis 133
of cave sediments 329–30
of limestone 20–3; and dolomite 14–15
submarine 21
vadose 21, 444, 509
diamiction 321, 326
diffusion 92–3
diffusion control, potential significance of 93
Dinaric Karst (Yugoslavia) 178, 184, 429, Table 6.3, Table 9.5
Dinornis maximus 543
Direct Yellow 96 234–5
discharge 106, 158, 218–19
allogenic 424
dominant 306
point 285
specific 129, 137, 144–5, 177, 238
discharge responses 196
discontinuities, penetrative 522
dispersive models 222
dissociation 46, 51–2
dissolution 3, 21, 30, 477
acid 46–7
congruent and incongruent 42–3
reactions 81–5

dissolution channels 382–8
dissolution pockets *297*
dissolution processes, kinetics of
  calcite 85–8, 90–2
  dolomite 88–9, 92
  gypsum and anhydrite 92–4
dissolution reactions 3
dissolved solids transport 124–6
divides, phreatic and vadose 173–4
Dneister River 281
Dobsinska Ice Cave (Czechoslovakia) 353
doline densities 418, 420
doline ponds 404, 410
dolines 7, 34, *153*, 162, 278, 292, 426, 452, 454–5, 459, 466, 482, 495, 510, 512, **9.3**
  blockage of 494
  catastrophic development 525–9
  cenote *410*, 411, 492
  cockpit 397
  collapse 397, *398*, 399, *410*, 445, 456, 467, 476, 488, 501; origin of 405–11
  drawdown 400–2
  drowned *499*
  kotlic 477
  morphometric analysis of **9.6**
  point recharge 399–400, 426
  polygenetic 409
  relief attributes 415–16
  solution 152, 397, *398*, 402–5, 438, 502; origin and development of **9.4**; postglacial 477
  spatial location 415
  styles and patterns 413–14
  subjacent collapse 409
  subsidence *398*, 399
  suffosion *398*, 420, 438, 476–7, *502*, *526*; origin of 412
dolomicrite 25
dolomite 10–11, 14, *19*, 30, 43, 449, 529
  composition of 25
  dissolution of 56–7, 88–9
  dolostone 10
  electrical conductivity 61
  formation of 23–5
  hydrothermal origin 24
  protodolomite 14
  pseudodolomite 14
  saturation index 58
  solubility 72
  sparry 291–2, 541
dolomite layers 11, 14, 89
Dolomites (Italy) *387*
dolomitization 23–5, 283, 290, 292, 502
Doma Peaks (Papua, New Guinea) 510
domepits 269, 303
  flute *304*
  micro-domepits 298
dongas (dayas) 471–2
Dordogne (France) 456
doubled solvency effect 75, 77–8
downcutting rate 121–3, 361–2, 438
drainage
  deranged *485*, 489
  focused 492
  karstification of 421–2
draperies 333, 335
drawdown 140, 190, 271–2
drawdown cone 535–6
drawdown cone (cone of depression) 186, *187*, 190, 535–6
drawdown depression 455
drilling, exploratory 527–8
dripstone 345
dry stone walls 540
dry valleys 501
dual spectra 231
Duvanjsko Polje (Yugoslavia) 432
dye tracing *see* water tracing
dyes, as water tracers 230–9
dynamic volume 202–3

earth tides *164*, 165
earthquakes 371–2
eddy diffusion 81, 83
Edwards aquifer, Texas 191, *192*
effective porosity 198, 238, 268, 290
egg box relief 414
Egypt 433
  Giza 538
electrical conductivity 60–1, 106, 213, *214*
electrical logging 191
electromagnetic conductivity 527
electron spin resonance decay (ESR) dating 362–4
Elk Point Formation *461*, 462
Ellesmere Island (Canada) 496
Empire State Building *315*
entrance deposits 317
entrenchment 299–301
environmental isotopes 174, 220–6
  main applications 220–1
  radioactive isotopes 221–4
  stable isotopes 224–6
eolianties 33
  see also aeolianites
eosine 231, 233
epikarst 123, 346–7, 374, 455, 458
  removal of 532
  see also subcutaneous zone
epiphreatic zone 265, 440
epsomite 348
equilibrium calcite 369–71
equilibrium fractionation 369
equipotential lines 139, 142
erosion surfaces 506
erosione antigravitativa 273

see also paragenesis
Escherischia coli 519
eskers 321–2
estavelles *see* springs, reversing
etching, preferential 377
eustasy
  geoidal 496, 500
  glacio–eustasy 496
evaporite karst 467
evaporite rocks *5*, **2.4**
evaporite terrains **9.13**
evaporites 159, 405
  see also anhydrite; gypsum; rock salt
evapotranspiration 64, 66, 68, 218
evapoturbation 330, 348
exhumed karst 1, 507
exotic acids, inorganic 74–8

facets/facetting 308–9
facies
  of cave sediments 325–6
  of limestone **2.2**
faults 37–8, 150, 249, 522
  vertical 37–8
fauna, and comparative dating 367
felsenmeer 495
fengcong 443–4
fenglin 442
Fick's first law 83
filtration velocity 142, 238
finite state mixing cell model 222
fissure frequency 143, 263, 267, 278, 401, 437, 455
  coral reefs 456
  and dam location 532
  effective, increase with time 265–7
  and maze passages 273–4
  measures of 265
fissure intersection 253
fissure network 179
fissure porosity 179–80
fissures 90, 135, 144, 151, 272–3, 500
  flushed and enlarged 531
fissuring 400
flashfloods 274, 467
Flinders Ranges 471–2
flint *see* chert
flood flow recession curve 201, 211, 213
flood hydrographs 201, 203, 211
flood recession 201–3
flood water volume 203
flooding, urban 537–8
flora, and comparative dating 367–8
Florida (USA) 24, *410*, 411, 423, 440, *526*
flow envelopes (equipotential fields) 251, 255, 257
flow media, conceptual classification 168–70
flow nets 138–40
flow paths 496
flow rates 163
flow type 166–8
flowpaths
  development of 162–3, 165–6
  rate of development **5.5**
flowstone 332, 335–6, 345, 368, 371
flowthrough time 160, 227
fluid inclusions 369–70
fluid viscosity 52
fluorescent dyes 230–9
fluorite (fluorspar) 348, 541
fluorometric techniques 237–9
fluoroscein 230–1, 234, 238
flutes/fluting 298–9, 303, 384
  rillenkarren *383*
  theory of 305–6
fluviokarst 423
fold topography **2.9**
folding 267
Fontestorbes (France) 212–13
foreign ion effect 71–3
foundation treatments *524*
four-state model of speleogenesis 261–71
fractionation
  and hydrological cycle 224–5
  of oxygen 369
fracture caves 500
fracture porosity 180
fracture traces 38, 500
fractured spans 310
fracturing 285, 407
France
  Aliou 212–13
  Alps 111, 114
  Baget 212–13
  Dordogne 456, 543
  Fontestorbes 212–13
  Grandes Causses 391, 452–3
  Les Baux 511
  Pyrenees 212, 307, 543
  Vaucluse 263, *264*
Frank Slide (Canada) 523
fresh water lenses 500
fresh water/salt water interface/mixing zone 24, 73, 140–2, 177, 220, 290, 438, 456, 498, 500, *501*
friction factor 146–8
fringing reefs 500
  karstification of 501–2
frost pockets 314–15
frost shattering 314, 482, 495–6
fulvic acid 78, 95, 342, 344, 368
fungi 377

galena 541
General Systems Theory, cascading system 261

geological controls, developement of karst
  aquifers 148–51
Geomorphic Rock Mass Strength
  Classification 38–9
geophysical techniques (surface) 527
Georgia (USA) *304*, 511
geothermal gradients 282
German Degree of Hardness 44
German Federal Republic, Aach 231, 239
Ghyben-Herzberg principle 140–1, 177
giant grikelands 391–3, 444, 479
Gibbs free energy 47, 50
gibbsite 43, 350
Gigantophithicus blacki 543
Giza (Egypt) 538
glacial deposits 512
  eskers 321–2
  till (moraine) 292, 484
glacial entrenchment 487
glacial injection 483–4
glacial landforms
  cirques 480, *481*, 483
  kettle holes 477
  moraine-dammed valleys 478
  nunataks 475
glacial scour 476–7, 480, 482, *486*
glaciated karst (simple) 479–80
glaciated terrains
  karst development **10.3**
  surface karst morphology 476–82
glaciation, and limestone pavements 389
glacier flour 325
glacier flow 473
glacieres 489–90
glaciers
  dry–based 473
  jokulhaup 487
  N channels 475
  R channels 475
  salt 464
  wet-based 473, 479
glacio-eustatic fluctuations 411, 456
glaciokarst 473
glaciokarst effects 482–9
  deep injection 483–4; meltwater 487
  dissection 483
  erasure 482–3
  infilling 483
  preservation 484, 487
  sealing 484, 487
  shielding 484
glaciotectonic cavities *474*
goethite 542
Goose Arm Karst (Canada) 484, *485*
gorges 3, 424, 432–3
Gort Lowland (Ireland) 151, 433, *434–5*, 500
grain (particle) shape 318–21, 326–7
grain (particle) size 318–19, 321, 327 and
  texture 31–2
Grandes Causses (France) 391, 452–3
gravimetry 527
gravitational convection 298
Great Barrier Reef (Australia) 411, 502
Great Bear Lake (Canada) *485*, 494–5
Great Salt Lake (USA) 20
Great Slave Lake (Canada) 488
Greece 538
  Argostoli *504–5*
  Athens 538, *539*
  Lake Stymphalia 6
'grey box' nature of karst **6.1**
grikes 380, 382, 386, 389, 391, 444, 467, 470,
  477, 510
ground ice 465, 495
ground penetrating radar 527
ground-air CO_2 64–6, 101, 123
groundwater 3, 218
  movement in isotropic aquifers 138–40
  saturated 456
groundwater flow 7
groundwater flow paths 3
groundwater movement 238
groundwater reservoir development 195
grout curtain 532
grouting 533–5
growth layering 332, 336
Guadaloupe Mountains (USA) 287–9
Guangxi (China) 543
guano 349
Guatemala 423
Guilin (China) 442–4
Guizhou Plateau (China) 438, 458, 507
gypsum 3, 6, *13*, 20, 28, 33–4, 43, 51, 76–7,
  331, 348, 464, 467
  deposition and replacement *288*
  dissolution rates 92–3
  flowers 283, 348
  needles 348
  precipitation 314, 329–30
  solubility of 72
  uses of 540
  whiskers 283, 348
gypsum crusts 77
gypsum karst 27, 458–9
gypsum tumuli 464

Ha Long (Vietnam) 441, 504
Hagen-Poiseuille equation 145–6
Hainault (Belgium) 460
Hales Bar Dam (USA) 529, 531
half-flow period 198, 211
half-tubes, ceiling 63, 296, 298
halides 348
halite *see* rock salt
Hallstatt (Austria) 509
Hamilton (Canada) *486*

Hammada de Guir (Morocco) 469, *485*
hard water 44
hardness 204
Harpagornis 543
head
  fluctuations 152–3
  hydraulic 129
  pressure 129
headwater stripping *425*
heavy metal pollution 520
heavy metal abundances 329
helictites (eccentrics) 84, 336–7, 479
hematite 350, 542
Henry's Law 53, 63
heterogeneity, coefficient of 201
hexahydrite 348
hiatus 332, 336
hollandite 350
holokarst 452
Homo erectus pekinensis 543
Hondo Dam (USA) 533
Hortonian channels 386
Hortonian rills 384–5
Horton's laws 258
hot springs 283, 340
Howe Lake (Canada) 487, *488*
humic acid 78, 85, 95, 342, 344, 368
Hummingbird Trough 462
hums 482
Hunan (China) 525
Hungary 535–7
  Budapest 284–5, 342
  Transdanubian Mountains 181–2, 219, 282
huntite 342
Huon Peninsula (Papua, New Guinea) *498*, 499
hydration, anhydrite 51, 267, 464
hydration front 464
hydraulic conductivity 134–5, 140, 148, 178, 404, 424, 456
  estimation of by tracer dilution 238–9
  fissured network 179–80
  and permeability 130–2, 186
  vertical *153*, 402, 404
hydraulic gradient 129, 151–2, 252–3, 257, 272, 426, 426, 529
  increased 537
  low 494
hydraulic head 129, 504
  in glaciers 487
hydraulic potential 129, 139, 161, 437
hydrochloric acid 74–5
hydrogen isotopes 220, 221–2, 224–5, 228, Table 6.6
hydrogen sulphide 75–6
  and cave formation 287–90
  in caves 76
hydrogeological mapping 516–18
  hydrogeodynamic approach 518
  hydrogeolithological approach 516, 518
hydrograph analysis **6.5**
  poljes **6.8**
hydrograph recessions
  composite 198–203
  simple 196–8
hydrographic zones 452, Table 5.1
hydrologic budget 218
hydrological processes 374
hydrology, urban, karst **11.4**
hydrolysis, of alumino–silicates 329
hydromagnesite 337–8
hydrostatic pressure 63, 158
hydrothermal caves 315
hydroxyapatite 349

ice ages, late Tertiary-Quaternary 472–3
ice buildups 492
ice dam burst 487
ice (in caves) **8.5**
  distribution patterns 354–5
  extrusion 354
  firn 353
  glacier 353–4
  ground 354
  hoarfrost 353, 355
  intrusive 354
  seasonlity 355
  water ice 352–3, 355
ignimbrites 512
illite 329
imbrication 322
incasion *see* breakdown
inclusions 332, 369–70
incongruent dissolution 42–3, 282
Indonesia
  Sulawesi 435
  West Irian 117
induration 502, 504
industrialization, ravages of 544–5
infiltration 401, 412, 483, 502
  diffuse 500
  and pollution 519
inheritance 126, 495–6
  complex 489
injecta 484
interbedding, of clastic rocks **2.7**
internal energy 163–7
interparticle bonding 33
interstratal dissolution 407, 459–60, 461
interstratal karst/karstification 3, 6, 27, 43, 487, 509
ion balance error 62
ion exchange 519
ion substitution (poisoning) 333
ionic strength effect 72–3
ionic strength (I) 45, 48, 60–1

ions
  ion pairs 48, 57, 71–2, 307
  ion solution 220
Iran, Zagros Mountains 462, 464
Ireland 456
  Aran Islands 396
  Burren 301
  County Clare 111, 117–18
  County Leitrim 117
  Gort Lowland 151, 433, *434–5*, 500
  Lough Caherglassaum 500
  Lough Mask 396
iron compounds 75
iron ore 542
iron oxides 350
isotopes
  carbon 356
  uranium 356–62
Israel 433
Italy 459
  Bari *395*
  Dolomites *387*
  Prealps 509
  Rome 538
  Trieste 538
  Vaiont Dam 522

Jamaica 397, 451, 542
Japan, Okinawa *395*, 540
Java (Indonesia) 397, 451
Jefferson City Mine (USA) *284*, 291–2
joints 36–7, 149–50, 249, 253, 279, 496, 522
  enlargement of 509
Jordan River 6
Jura (Switzerland) 224

Kacne Jama (Yugoslavia) 326–7
kaolinite 43, 329
karra 2
karren 6, 31–2, 34, 90, 294, 375–96, 459, 467, 476, 480, 482, 510
  channel 495
  cutters 382
  flachkarren (clints) 380, 382, 386, 389, 391
  hohlkarren 386
  kamenitze 379
  karrenfeld 375, 389, 455
  kluftkarren (grikes) 382, 389, 444, 467, 470, 477, 510
  mäanderkarren 388
  microkarren 375, 377–8
  pitts 509
  rillenkarren 32, 94, 382–6, 462
  rillenstein 378
  rinnenkarren 386, *387*
  rundkarren 386, *387*
  solution pits/solution pans 33, 379, 394–5, 470
  solution scallops 32
  spitkarren 380, 383, 389
  tinajitas 379
  trittkarren (heelprints) 32, 379–80
  wallkarren 462, 470
  wandkarren 386, 388
  assemblages 389–93
  classification Table 9.1
  composite nature of 388–9
karren shaft 380, 386, 459
Karrentische 116–18
karst denudation rate 97
karst domes 464–5
karst margin plains 432–3
karst system, main features 2–3
karstic evolution 420–1
Karstrandebene *see* corrosion plains
kegelkarst 466
Kentucky (USA) 36, *236*, 258–9, 263, 269, 424, 456
kinematic wave 160, 227
kinetic fractionation 369
knickpoints 275, 301
Kocevsko polje (Yugoslavia) 218
Korana (Yugoslavia) 449
kotlic 477
kras 2
Krka River (Yugoslavia) 424, 438
krypton-85 224

labyrinth karst *see* giant grikelands
lacework 396
Lake Chad (Chad) 20
Lake Glomdal (Norway) *257*
Lake Huron (Canada) *395*, 396
Lake Stymphalia (Greece) 6
laminar flow 90, 132, 143–5, 161, 164
Landsat imagery 175
lapies *see* karren
Law of Mass Action 47, 50
lead 509, 541
lead/zinc deposits 291–3
lead/zinc mines 488, 535
lead/zinc ore bodies 534
leakage, from dams 531
leakage paths, vertical 401–2
Les Baux (France) 511
Li River (China) 443
lichens 377
limestone 10, 43, 544, **2.2**
  as building stone 538–40
  compositions and depositional facies **2.2**
  diagenesis and metamorphism 20–3
  petrological classification 17–20
  principle components Table 2.4
limestone and dolomite, industrial exploitation of 538–40
limestone pavement 382, 389–91

and permafrost 495
see also karren
limestone pedestals *see* Karrentische
Limestone Ranges (Australia) 469–71
limonite 75, 350, 542
linears/lineaments *see* fracture traces
lithification 502
lithology 194
and structure 374–5
lithostatic pressure 37, 128
littoral karren 393–6, 500
load casts 329
loess 459, 512
Lough Caherglassaun (Ireland) 500
Lough Mask (Ireland) 396
Lugeon tests 182–3
Lycopodium clavatum 240–1

Mackenzie Mountains (Canada) 391–2, 478, 492, *493*
McMillan Dam (USA) 533
magnesite 11
magnesium 79
magnetism, remanent 364, 367
magnetite 350, 364, 367
makatea 502
makhteshim 433
Malaysia, Selangor 435
Malham Tarn (Cove) (UK) 239, *390*, 433
Maligne River (Canada) 163, 227–8, 483
Mangaia Is. (Cook Islands) 457
manganese 350
manganese oxides 350
Manus Island (Papua, New Guinea) *503*
marble 25–6, 159
as building stone 538–40
marcasite 541
marine mixing cave genesis 292
marls 20
marmites *see* stream potholes
mass flux rating curves 107–8, 115
mass–balance, spring water 207
Matuyama (magnetic) epoch 364
maze passages 273–4
meandering channels, vadose caves 301–2
alluvial 302
ingrowing 301–2
intrenched 301
mechanical failure, cause of breakdown 314
Medicine Lake (Canada) 228, 483
melt waters 475
deep injection of 487–8
merokarst 452
meteoric water line 225
meteorites 305
Mexico, Yucatan Peninsula 73, 142, 291, 440, 500, 540
Mg:Ca ratio 23
Mg-calcite 89
micrite 15, 20, 33–4
Porcellaneous Band 31
micro-conduits 251–2
micro-erosion meter (MEM) 111–14, 118
micro-stylolites 23
microkarren 375, 377–8
microorganisms, naturally occurring 219–20
microrills 378
microscopic velocity 238
microspar 445
mirabilite 43, 348
Mississippi Valley-type deposits (MVT) 291–3
Missouri (USA) 31, 326
mixing corrosion 70–1, 253, 280, 298
mixture corrosion 500
Moa 543
mogote 441–2, 507–8, 510–11
case-hardening 444–5, *446*
molality 44
molar abundances 58
molarity 44
molecular diffusion 81–3, 85
molecular dissociation 92–3
Montjaques dam (Spain) 531
montmorillonite 329
moonmilk 340, 342
Morocco, Hammada de Guir 469
morphometric analysis, dolines **9.6**
mosses 448
mound springs 449
Mt Api (Sarawak) 391, 545
Mt Kaijende (Papua, New Guinea) 391
mud mountains 323
Mulu National Park (Sarawak) 315, 544

Nahanni karst (Canada) 362, 444, 471, 492, *493*
Nakanai Mountains (Papua, New Guinea) 409
namakiers 464
Nanisivik (Canada) 292–2, 307, 542
nearest neighbour analysis 418–20
Negev desert 433
neomorphism, aggrading 445
Nevesinjsko polje (Yugoslavia) 216
New Britain (Papua, New Guinea) 97, 246
New Guinea 41, 326
New Jersey (USA) *523*
New South Wales (Australia) 111
New Zealand 543
Riwaka River (basin) 100, 123
Takaka River 222, 228, 426, *514–15*
Waikoropupu 222, 504, Table 6.6
Waitomo 119, 301, *413*, 421–2, 428, 512
Newfoundland (Canada) 391
nitrate contamination 520–1
nitrate minerals 349–50

nitric acid 77
Niue Island 177, 499–500, 502
nival karst 489
non-fluorescent dyes 239
non-saturated zone 200–1
Normandy Dam (USA) 532
North Saskatchewan River (Canada) 107
Norway
  Glomdal *257*, 476
  Kvithola 476, 487
  Pikhauga *478*
notches 394–6
  coastal 500
  corrosion 292–3, 307–8
  intertidal *501–8*
  swamp 435, *443*
Nova Scotia (Canada) 459–8
nuclear contamination 521
nucleation, homogenous and heterogeneous 94–5
Nullarbor Plain (Australia) 246, 267, 314, 471–2
nunataks 475

oil and gas recovery 527
oil and natural gas 509, 541
Olkusz (Poland) 535, 536
Ombla (Yugoslavia) 199
oolites 33
opal 351
open systems *see* coincident (open) systems
optical brightners 234–5
ores in carbonate rocks 541–2
  deposition 542
ores in karst 291–3, 534
oulopholites 348
oxidation 75, 329
oxygen isotopes 220, 224, 283, 473
  in speleothems 369–71
  in spring water 205–6
palaeo-water lines 283
palaeo-water tables 307
palaeocaves 488
palaeodrainage 423
palaeoflow, direction 287
palaeokarst 1, 159, 177, 291–2, 462, 487, 507, *508*, 509–12
  dolomite 535
  major 510–12
  polyphase 511
  reactivated 531
  site of economic minerals 509
palaeomagnetism 364–7
palaeosols 445
palaeotemperature 544
palynology, Quaternary, and speleothems 368
Papua, New Guinea 404
  Darai Hills *413*
  Doma Peaks 510
  Huon Peninsula *498*, 499
  Manus Island *503*
  Mt Kaijende 391
  Nakanai Mountains 409
  New Britain 97, 246
paragenesis *272*, 281, 293
  see also erosione antigravitativa
partial pressure, CO_2 53–4, 64–6, 101, 123
particulate matter in karst systems 519
passages
  phreatic 294–6
  vadose 299–301
pathogenic organisms 519
pebbles 326–7
Pecos Basin (USA) 459
pediments 472
pediplains 434, 467, 470–1
Peking Man 543
pendants (Deckenkarren) 296, 298
peneplain 506
penetrative discontinuities 522
Pennsylvania (USA) 6, 96
percolation 123–4, 128, 150, 152, 407
  autogenic 120
percolation input 169
periclase 26
Perm (USSR) 458–9, 461
permafrost 128, 465, 479, 489–91
  active layer 490, 495
  continuous 490
  discontinuous 491
  discontinuous-continuous transition 494–5
  karst development 491–6
  sporadic 490–1
  taliks 490–1
  thermoactive layer 490–1, 495
permeability 130–2, 150, 158, 178–9, 346–7, 401
  secondary 135
  vertical 404
Persian Gulf 541
petroleum 76
pH 85
  measurement of 60
  normal rainwater 77
phosphate minerals 349
phreatic conduit 227
phreatic loops 437
phreatic passages 294–6
  draining of 361
  frozen 353
phreatic waters 114, 208
phreatic zone 75, 150, 159, 177, 205, 438
  importance of 211
  tunnels and mines 534–5
  see also saturated zone
phreatic zone storage 217

Phuket Island (Thailand) 504
phytokarst 118, *395*
  coastal 377
Picos de Europa (Spain) 118, 475
piezometric pressure 216–17
piezometric surface *see* water table
Pine Point (Canada) *27*, 292, 488, 511, 542
Pinero-Odvinsky (USSR) 114
Pingelap Atoll 177
pinnacle karren 544
pinnacle karst 391, 464
pinnacles *447*, 482
  fluted 509
pipe flow 129
  turbulent 146
pipefull flow 319–21
pipes 144
  in dune sand 445
piping 2, 412, 477
pipkrake 354
piracy, subterranean 246
pisolites 21
pisoliths 338
piston flow 160
  model 222–4
pitting 396
Planina polje (Yugoslavia) 218
plaster of Paris 540
plate tectonics, and karst 510
platea 391, 470–1
Plitvice Lakes (Yugoslavia) 340, 449
pocket valleys 500
Podolia (USSR) 77, 114
Poiseuille's law 129, 145
Poland, Olkusz 535, 537
polje hydrographs **6.8**
poljes 3, 7, 452, 457, 459, 478, 510, **9.8**
  baselevel 217, *429*, 431–2, 492
  border 426, *429*, 431
  enclosed 215
  open downstream 215
  open upstream 215
  structural *429*, 431–2
pollen and spores 367–8
pollution
  in aquifers **11.2**
  from chemical spills 537
polycyclic karst 506–7
polygenetic karst 506–7
polygonal karst 409, 413, 423, 428, 440–1
  final stage of evolution 440–1
  redeveloped *439*
ponors 487
  see also stream-sinks; swallow holes
Popovo polje (Yugoslavia) *215*, 218
porosity 32–3, 101, 199
  effective 133–4, 158, 198, 238, 268, 290
  enhancement of 133–4
  fabric-selective 33
  fissure 179–80
  fracture 180
  primary 32, 124, 133, 159, 161
  secondary 32, 124, 133, 142, 159
Portland cement 10, 540
potentiometric surface 128, 158, 174, 188
potholes 477
precipitation
  of carbonate 123
  and case-hardening 44
  chemical 329–30
  homogeneous 94
  hydrothermal caves 285
pressure head 129, 251
pressure injection tests *see* lugeon tests
pressure pulse 160, 227–8
  see also piston flow
pressure release spalling 313–14
prismatic ore bodies 292
proto-caves 81, 242, 400
  competitive development *254*, 255
  connection of 255–6
  primary and subsidiary tube 251–3
  principal tubes 257
provenance, of water 59, 174, 220
provenance studies, composition of particles 329
pseudokarst 1
Pseudomonas aeruginosa 519
psilomelane 350
Puerto Rico 441–2, 444, 507
pulse through time 227
pulse train analysis 174, 226–8
pulses
  artificial 228
  natural 226–8
  as tracers 226–8
pumping tests 182, 186–91
Pyrenees (France–Spain) 212, 307, *387*, 475
pyrite 36, 75, 307, 541

Qinghai (Tibet) (China) 458, 507
quadrat analysis 418, 420
quartz grains 327
quartzite 29, 43
Quaternary glaciers 389
quick lime 540

radioactive logging 193
radioisotopes 221–4, 356–64
radiometric dating, stalagmites and flowstone 345–6
Rand (South Africa) *528*, 529
rank/size correlation (caves) *245*
re-solution 332
recession coefficient 196, 198
recession constant 196–8

recession constant 196–8
recession curves 196–7
recharge 150, 167, 218–19
  a-periodic impedance 492, 494
  allogenic 151–2, *154*, 159, 161, 169, 204; lateral 502
  autogenic 151–2, 159–61, 204, 226, 455, 504
  concentrated 169
  diffuse 159–61, 169, *204*, 400, 456
  point 152, 159, 161, 222, 445, 454, 502
recharge pulse 205
recharge response 427–8
recharge tests 182–6
recreational value of karst 542–4
reefs *see* coral reefs
regional flow 7
rejuvenation 438, 458, 506
  of palaeokarst 488
relict karst 1, 507–9
  cones and towers 509
  reactivated 531
remote sensing 174–7
residual hills **9.10**
resistivity 527
resistivity surveys 175, *176*, 177
resurgences *see* springs
reversals 364, 367
Reynolds Number 143–5, *145*, 163, 305–6
Rhine River 6
rhodamine 231, 233–4
rhythmites 325–6
riegel 480
rilling 323–4
rimstone dams (gours) 340
rimstone pools 449
Riwaka Rivers (basin) (New Zealand) 100, 123
rock bursts 313
rock (cave) drawings 543
rock mills 301
rock purity 30
rock salt (halite) *13*, 43, 52, 467
rock strenght 404
rock-slide avalanches 522–3
Rome (Italy) 538
rooms *see* chambers
rotational failure 522
RUBE dating 358–9, 372
ruiniform karst 391
Rumania 274
runoff 96–7, 101–2, 217–18, 375, 444
  allogenic 400, 472
  focused 525, 528
Russia *see* Union of Soviet Socialist Republics

sabkhas 27
Saccharomyces cerevisiae 241
St Francis Dam (USA) 534
Salmonella typhimurium 519
salt 3, 6, 29, 34, 458
salt diapirs 462–4, 496, 541
  dangers in pressure injection 527
salt domes *463*
salt karst 27
salt slope 462
saltation 319
saltpetre 350
salts 239
San Juan River (USA) 301
sand 327, 364
  abrasive effects of 306–7
sands and sand structures, in caves 322–3
sandstone 29, 43
  maze formation by diffusion 280–1
Santander (Spain) 543
Sarawak
  Mt Api 391, 544
  Mulu National Park 315, 544
Sarisarinama Plateau, Venezuela 29
Saskatoon Low (Canada) 487–8
saturated zone 200–1
  see also phreatic zone
saturation index 50–2, 58–60
saturation ratio 50
sauter mean 306
scallops, fluted 388
scallops/scalloping 94, 148, 303, 305, 322
  dissolutional *305*
  formation theory 305–6
schichtreppenkarst *390*
schlotten 391, 459, 464, 476
  see also pinnacle karst
Schmidt hammer 33–4
scientific value of karst 544–5
sea level, variations in 496, 498
sea-estavelles 504
Searanxx (Belize) *413*
seawater 72–3
  precipitation in 27
seepage 412
seismospeleology 371–2
Selangor, (Malaysia) 435
selenite 348
sequential systems *see* closed systems
Serratia marcescens 220
settling 525
settling velocity 319
Severn Railway Tunnel 535
Severn River (UK) 535
shale 34
shear box testing 313
shelfstone 341
shillow 389
Sibun River (Belize) *427*, 428
Sichuan basin (China) 175
siderite 11, 75, 542

Sierra de los Organos (Cuba) 441
silica 30, 350–1
  amorphous 330; *see also* crystallites
silicate karst 510
silt 323–5, 327, 364
silver 509, 542
sinkholes 38, 535–6
  see also dolines
sinter 330
  see also speleothem; travertine; tufa
slab slides 522
sliding beds 319–22
slippage, differential 36, 41
Slovenia (Yugoslavia) 233, 323
slurries 318, 326, 532
smooth and rough topography 417
SMOW (standard mean ocean water) 224–5, 369
soda nitre 350
soft water 44
soil CO_2 54, 64–6, 69, 78, 101, 346
soils 64–5, 69
solifluction *494*
SOLMNEQ (program) 62
solubility, range of 43
solute concentration 106–7, 109
solute load 100, 105
solution 409
  and frost action 315
  vertical 434; distribution of 115–21
solution conduits, propagation of **7.2**
solution corridors 470
solution flutes 471–2
solution front 126
solution pipes 144
solution pockets 71, 283, 298
solution processes 46–60
solution rates 97
solution sag 412
solution subsidence troughs 460–1
solution denudation 3, 96–7, 467
  autogenic, allogenic and mixed systems 98–100
  calculation of 106–10
  carbonate 55–60
  global models 66–68, 101–3
  gross and net solution 100–1
  gypsum 51–2, 114
  halite 51–2
  link between chemical and environmental factors 101–3
rates 97–8, 101, 121–3, 126; measurement and calculation of **4.2**, **4.4**
  water availability 466–7
solvent penetration distance 252
solvent water flow 82–4
South Africa *see* Union of South Africa
Spain
  Almeria 114
  Camarasa dam *530*
  Montjaques dam 531
  Picos de Europa 118, 475
  Santander 544
spalling 77–8
sparite 18, 31, 33–4
spathite 333, 337
specific discharge 137, 144–5, 177, 238
specific permeability 183, 184, Table 6.3
specific storage 137, 158
specific stream power 163
specific yield 137–8
speleogenesis, hot arid regions 467, 469
speleothems 78, 100–1, 114, 121–4, 291, 330–1, 492, *501*
  calcite; distibution patterns 346–7; growth rate 344–6
  colour in 342, 344
  disordered 84
  drowned 411
  excentric 336–8; anemolites 337; globular (botryoida) 337–8; helictites 336–7; palettes (shields) 337–8
  facetting of 309
  growth in glaciated regions 362
  growth of 371
  principal forms *330*, *334–5*
  stable isotopic studies 368–71
  used for dating 362
  see also sinter; travertine
sphalerite 541
Spitzbergen 118, 495
spongework mazes 287, 290–1
spores and yeasts as water tracers 240–1
spring hydrograph analysis 513
spring hydrograph response 167
spring hydrographs *207*, **6.5**
  flood hydrographs 194–5, 201, 203
spring mounds 482
spring sapping 432–4
springs 3, 104, 215
  artesian 156, *158*, 449
  chemically complex 208, 210
  coastal, drowned 438
  conduit-fed 205
  confined 156, *157*, 158
  dammed 156, *157*
  diffuse flow 204
  episodic 7
  free draining 156, *157*
  hot springs 340
  intermittent 195
  mound springs 449
  periodic 158
  reversing (estavelles) 155, 215
  rhythmic 158
  seasonal 496

submarine (vruljas) 142, 174, 437, 504
tufa 340
vauclusian 158
world's largest Table 5.6
Aach (GFD) 231, 239
Argostoli (Greece) 504
Baget (France) *206*
Bileca (Yugoslavia) 199
Cheddar (UK) 199–200
Cnoy (Mexico) *209*
Fontestorbes (France) 212–13
Komlo (Hungary) *205*
Ombla (Yugoslavia) 199
Vaucluse (France) 263, *264*
Waikoropupu (New Zealand) 222, 504, Table 6.6
Wookey Hole (UK) 177
stable isotopes 224–6
stalactites 33, 333, 347
soda-straw 83, 333, *335*, 345–6, 371
stalagmites 81, 121, 335–6, 345–6, 359, 361, 368, 371
calcite geyser 283
Stari Trg Mine (Yugoslavia) 285
step and kink model (crystals) 84, 94
Stokes' law 319
Stone Forest of Yunnan (China) 391, *392*, 509
stopping 313, 407, 409, 460, 528
storage (aquifer), specific *136*, Table 5.4
storativity 137–8, 158, 188, 190
stored water 205–6
strata-bound ores 291–3
stream potholes 301
stream-sinks 7, *425*, 426
see also swallow holes
streamlines 139, *187*
streets 391
stress 407, 409, 464
strike passages 256
stromatolites *see* algal mats
strontium 11
stylolites 380
subcutaneous water 208
subcutaneous zone 120, 162, 206, 208, 255, 401, 404
see also epikarstic zone
subglacial karst 478–9
sublimation 305
submarine karst, deep 504
submergence 438, 440
subsidence 409
and roof failure 407
subsidence depressions, origin of 411–12
suffosion 412, 477, 525
Sulawesi (Indonesia) 435
sulphate 61
sulphate solutional denudation 114–15
sulphates 348
sulphation 77–8
sulphide deposits 307
massive 291–3
sulphur, native 76–7
sulphuric acid 75–8, 287
supersaturation 50, 124, 289
of calcite 449
silica 351
Supikovice (Czechoslovakia) *508*
surge marks 323–4
suspended load 320, 323–5
suspension, heterogeneous and homogeneous 320
swallets *see* swallow holes
swallow hole capacity 215–16
swallow holes 7, 152, 216–17
sea 504
swamp notches *308*, 435, *436*, *443*
Switzerland, Jura 224
sylvinite 28
sylvite 28, 43
syngenetic karst 445

tablets, weight loss by solution 110–11, 114
tabular ore bodies 292
taliks 128
taranakite 349
tectonic uplift 438
temperature 103
influence on flow 163–4
temperature changes 370
temperature logs 191
temperature variation 375
Tennessee (USA) 511, 531
tension 41
tension domes 310, *311*, 312
breakdown 313
collapse of 525
terra rossa 323
Texas (USA) 291, 533
Thailand, Phuket Island 504
thenardite 348
thermal convection 298
thermal infrared imagery 174, 527
thermal waters 63–4, 282
thermodynamic equilibrium 47, 71
thermokarst 1
thermoluminescence (TL) dating 362–4
through valleys 424–6
todorokite 350
tombstone terrain 445
topography, inherited 458
tortuosity 83–4
tower karst 266, 307, 441–2, *468*, 509–10, 542
fluted 118, 470–1
part submerged 504
rejuvenation of *439*
see also fenglin; mogote

trace element effects 79
Transdanubian Mountains (Hungary) 181–2, 219, 282
transmissivity 137, 158, 177–8, 188, 190, 198
traps 541
travertine 3, 20–1, 330, 362, 368, 448
as building stone 538
see also sinter; speleothem; tufa
triaxial compressive tests 311–12
Trieste (Italy) 538
tritium 220–2, 224, 228, Table 6.6
Trobriand Islands (Papua, New Guinea) 291, 502
troglobite ecosystems 544
troglobite flora and fauna 367
trogloxene fauna 367
trough subsidence 412
Tsi-It-Toh-Choh (Canada) 492
tufa 20, 95, 100–1, 448–51, 470
tufa dams 449
tufa springs 340
turbulent flow 92, 132, 144, 161
turloughs 477–8

Ukraine (USSR) 281, 525
undercutting 300–1
undersatuartion 50, 85, 103, 114
see also aggressivity
Union of South Africa 138, 411
Dreifontein 529
Rand *528*, 529
Union of Soviet Socialist Rupublics 491
Archangel 464
Bashkir 458–9
Bukovina 281
Caucasus Mountains 289, 475
Crimea 309, 489, 534
Perm 458–9, 461
Pinero–Odvinsky 114
Podolia 77, 114
Red Cave *330*
Siberia 492, 495
Ukraine 281, 525
Yalta 534
United Kingdom 452
Cornwall 534
Derbyshire 456, 534
Devil's Hole Mine 284
Malham Cove 239, *390*, 433
Mendip Hills 163, 177, 213, 265, 420, 494, 511
Peak District 494
Severn River 535
Yorkshire 31, 111, 117–18, 123, 267, 456
United States of America 456
Alabama 525
Arkansas 278
Avalon Dam (NM) 533
Big Horn Basin 289–90
Bowling Green (Ky) 537
Central Kentucky Karst 234, 258–9
Death Valley (Nev.) 458
Empire State Building (NY.) *315*
Florida 24, *410*, 411, 423, 440, *526*
Georgia *304*, 511
Guadaloupe Mountains (NM-Tx.) 287–9
Hales Bar Dam (Tenn.) 529, 531
Hondo Dam (NM) 533
Jefferson City mine (Tenn.) *284*, 291–2
Kentucky 30, *236*, 258–9, 263, 269, 424, 456
McMillan Dam (NM) 533
Minnesota 527
Missouri 31, 326
New Jersey *523*
New Mexico 280, *335*, 521, 533
Normandy Dam (Tenn.) 532
Pecos Basin (NM) 459
Pennsylvania 96
Rocky Mountains 475
St Francis Dam (Ca) 534
South Dakota 245, *284*, 286, 341, *350*, *508*
Tennessee 511, 531
Texas 291, 533
West Virginia 34, 123, 312, 322, 327, *425*, Table 8.2
Wisconsin 110
Yates (Tx.) *290*
Yellowstone Park (Wyo.) 340
unsaturated zone 128, 211
see also vadose zone
Uralian Troughs 462
uranine 230–1, 233
uranium 542
uranium isotopes 356–62
uranium series dating 356–62
urbanization, problems in karst regions **11.4**
Ursus spelaeus 367
USSR *see* Union of Soviet Socialist Republics
uvuals 452

vadose passages 227, 299–301
vadose shafts 303
vadose streams 124
vadose trenches 361
vadose waters 114
vadose zone 21, 69, 75, 123–4, 150, 162, 269, 303, 402, 456, 458
case-hardening 502
CO_2 in caves and fissures 66
perched 438
tunnels 534
see also epikarst; subcutaneous zone; unsaturated zone
Vaiont Dam (Italy) 522
valley penetration 428
vanadium 542

Vancouver Island (Canada) 265, 371, 386, *387*
varved clays 325
Vaucluse (France) 263, *264*
vauclusian springs *see* springs, artesian
Vava'u Island (Tonga) *503*
Venezuela, Sarisarinama Plateau 29
venturi mixing process 504
vermiculations 324
Vermilion Creek (Canada) *408*, 476
Vienna (Austria) 2, 451
Vietnam 542
  Ha Long 441, 504
viscosity 464
  kinematic 306
viscous flow, gypsum 464
voids 158–9
vorfluter 433
vruljas *see* springs, submarine
vugs 123–4
vulcanokarst 1–2

Waikoropupu (New Zealand) 222, 504, Table 6.6
Waitomo (New Zealand) 119, 301, *413*, 421–2, 428, 512
wall friction 146
WATEQ (programs) 62, 72
water availability 466–7
water balance equation 208
water balance estimation **6.9**
water balance technique 513
water budget equation 215
water leakage, effects of 525, 527
water quality, reliable monitoring of 520
water resource assessment **11.1**
water table 127–8, 138, 156, 217, 263, 437
  affected by construction work 535–6
  discontinuous 8–9
  lowering of 458
    cause of suffosion 525
  raising of 527
  regional 409, 507
  seasonal flustuations 409
  subcutaneous *153*
water temperature measurement 60
water tracing 174, 205, **6.10**
  artificial labels 219, 229–41
  fluorescent dyes 230–9
  fluorometric techniques 231
  natural labels 219–26
  pulses 219, 226–8
waters
  classifications of 43, Table 3.2
  connate 282
  juvenile 282
  thermal 282
WATSPEC (program) 62, 72
wellfield cone collapse 525
wells, ebbing and flowing *see* springs, rhythmic
West Irian (Indonesia) 117
West Virginia (USA) 34, 123, 312, 322, 327, *425*, Table 8.2
Winnipeg (Canada) *486*, 487
Wisconsin (USA) 110
wollastonite 26
Wood Buffalo National Park (Canada) 465

Yalta (USSR) 534
Yates (USA) *290*
Yellowstone Park (USA) 340
Yorkshire (UK) 31, 111, 117–18, 123, 267, 456
Yucatan Peninsula (Mexico) 73, 142, 291, 440, 500, 540
Yugoslavia 2, 389, 451
  Beljanica Mountains 509
  Bileca 199
  Cerknisko polje 218
  Crveno Jezero 405
  Dinaric Karst 178, 184, 429, Table 6.3, Table 9.5
  Duvanjsko Polje 432
  Kocevsko polje 218
  Korana 449
  Nevesinjsko polje *216*
  Ombla 199
  Planina polje 218
  Plitvice Lakes 340, 449
  Popovo polje *215*, 218
  Slovenia 111, 233, 323, 326–7, 345
  Trebisnjica 175, *176*

Zargros Mountains (Iran) 462, 464
zanjones 391
Zhoukoudian (China) 543
zinc 509, 541
zinc ores 291–3

TITLES OF RELATED INTEREST

Aeolian geomorphology
W. G. Nickling (ed.)

Catastrophic flooding
L. Mayer & D. Nash (eds)

Discovering landscape in England and Wales
A. S. Goudie & R. Gardner

Environmental change and tropical geomorphology
I. Douglas & T. Spencer (eds)

Experiments in physical sedimentology
J. R. L. Allen

The face of the Earth
G. H. Dury

Geomorphology: pure and applied
M. G. Hart

Geomorphology in arid regions
D. O. Doehring (ed.)

Geomorphology and engineering
D. R. Coates (ed.)

Geomorphological field manual
V. Gardiner & R. Dackombe

Geomorphological hazards in Los Angeles
R. U. Cooke

Geomorphological techniques
A. S. Goudie (ed.)

Glacial geomorphology
D. R. Coates (ed.)

Hillslope processes
A. D. Abrahams (eds)

The history of geomorphology
K. J. Tinkler (ed.)

Image interpretation in geology
S. Drury

Mathematics in geology
J. Ferguson

Models in geomorphology
M. J. Woldenberg (ed.)

Planetary landscapes
R. Greeley

A practical approach to sedimentology
R. C. Lindholm

Principles of physical sedimentology
J. R. L. Allen

Rock glaciers
J. Giardino *et al.* (eds)

Rocks and landforms
J. Gerrard

Sedimentology: process and product
M. R. Leeder

Tectonic geomorphology
M. Morisawa & J. T. Hack (eds)

Karst Geomorphology and Hydrology